Modern Soil Microbiology

Third Edition

Modern Soil Microbiology

Third Edition

Edited by
Jan Dirk van Elsas, Jack T. Trevors,
Alexandre Soares Rosado, and Paolo Nannipieri

CRC Press is an imprint of the
Taylor & Francis Group, an informa business

Cover images courtesy of:
Professor Emeritus Jan Dirk van Elsas, University of Groningen, Netherlands (upper picture); Professor Emeritus Dr. George Barron, University of Guelph, Canada (left and right lower picture inserts); and Professor Raquel Peixoto, Federal University of Rio de Janeiro, Brazil (middle lower picture insert).

CRC Press
Taylor & Francis Group
6000 Broken Sound Parkway NW, Suite 300
Boca Raton, FL 33487-2742

Printed on acid-free paper

International Standard Book Number-13: 978-1-4987-6353-0 (Hardback)

Library of Congress Cataloging-in-Publication Data

Names: Elsas, J. D. van (Jan D.), 1951- editor.
Title: Modern soil microbiology / editors: Jan Dirk van Elsas, Jack T.
Trevors, Alexandre Soares Rosado, Paolo Nannipieri.
Description: Third edition. | Boca Raton : Taylor & Francis, 2019.
Identifiers: LCCN 2018050977 | ISBN 9781498763530 (hardback : alk. paper) |
ISBN 9780429607929 (pdf) | ISBN 9780429602405 (epub) | ISBN 9780429596889
(mobi/kindle)
Subjects: LCSH: Soil microbiology. | Molecular microbiology.
Classification: LCC QR111 .M58 2019 | DDC 579/.1757—dc23
LC record available at https://lccn.loc.gov/2018050977

Visit the Taylor & Francis Web site at
http://www.taylorandfrancis.com

and the CRC Press Web site at
http://www.crcpress.com

Contents

Preface ix
Editors xi
Contributors xiii

SECTION I Fundamental Chapters

Chapter 1 The Soil Environment 3

Jan Dirk van Elsas

Chapter 2 The Seven Grand Questions on Soil Microbiology (Selman A. Waksman, Reexamined by Arthur D. McLaren) 21

Jan Dirk van Elsas and Paolo Nannipieri

Chapter 3 The Soil Microbiome—An Overview 37

Francisco Dini-Andreote and Jan Dirk van Elsas

Chapter 4 The Bacteria and Archaea in Soil 49

Jan Dirk van Elsas, Anton Hartmann, Michael Schloter, Jack T. Trevors, and Janet K. Jansson

Chapter 5 The Fungi in Soil 65

Roger D. Finlay and R. Greg Thorn

Chapter 6 The Viruses in Soil—Potential Roles, Activities, and Impacts 91

Akbar Adjie Pratama and Jan Dirk van Elsas

Chapter 7 Horizontal Gene Transfer and Microevolution in Soil 105

Kaare Magne Nielsen and Jan Dirk van Elsas

Chapter 8 The Protists in Soil—A Token of Untold Eukaryotic Diversity 125

Michael Bonkowski, Kenneth Dumack, and Anna Maria Fiore-Donno

Chapter 9 Microbial Interactions in Soil 141

Jan Dirk van Elsas, Akbar Adjie Pratama, Welington Luis de Araujo, and Jack T. Trevors

Chapter 10 Plant-Associated Bacteria and the Rhizosphere 163

Abdul Samad, Günter Brader, Nikolaus Pfaffenbichler, and Angela Sessitsch

Chapter 11 Microorganisms Cycling Soil Nutrients........ 179

Penny R. Hirsch

SECTION II Methods Chapters

Chapter 12 Methods to Determine Bacterial Abundance, Localization, and General Metabolic Activity in Soil 195

Lise Bonnichsen, Nanna Bygvraa Svenningsen, Mette Haubjerg Nicolaisen, and Ole Nybroe

Chapter 13 Soil Microbiome Data Analysis 215

Francisco Dini-Andreote

Chapter 14 Soil Metagenomics: Deciphering the Soil Microbial Gene Pool........ 227

Sara Sjöling, Jan Dirk van Elsas, Francisco Dini Andreote, and Jorge L. Mazza Rodrigues

Chapter 15 Analysis of Transcriptomes to Assess Expression and Activity Patterns of the Soil Microbiome........ 245

Stefanie Schulz, Anne Schöler, Michael Schloter, and Shilpi Sharma

Chapter 16 Metaproteomics of Soil Microbial Communities........ 257

Paolo Nannipieri, L. Giagnoni, and G. Renella

Chapter 17 Stable Isotope Probing—Detection of Active Microbes in Soil 269

Marie E. Kroeger and Klaus Nüsslein

Chapter 18 Isolation of Uncultured Bacteria 295

Ulisses Nunes da Rocha

Chapter 19 Statistical Analyses of Microbiological and Environmental Data........ 307

Alexander V. Semenov

SECTION III Applied Chapters

Chapter 20 Soil Microbial Communities and Global Change........ 331

Mark P. Waldrop and Courtney Creamer

Chapter 21 Soil Suppressiveness to Plant Diseases 343

Christian Steinberg, Véronique Edel-Hermann, Claude Alabouvette, and Philippe Lemanceau

Chapter 22 Plant Growth-Promoting Bacteria in Agricultural and Stressed Soils 361

Elisa Gamalero and Bernard R. Glick

Chapter 23 Biodegradation and Bioremediation of Organic Pollutants in Soil 381

Kam Tin Leung, Zi-Hua Jiang, Nouf Almzene, Kanavillil Nandakumar, Kurissery Sreekumari, and Jack T. Trevors

Chapter 24 The Impact of Metal Contamination on Soil Microbial Community Dynamics 403

David C. Gillan and Rob Van Houdt

Chapter 25 Management Strategies for Soil Used for Cultivation, Including Modulation of the Soil Microbiome 421

Alexandre Soares Rosado, Paolo Nannipieri, and Jan Dirk van Elsas

Glossary 431

Index 459

Preface

"Essentially, all life depends upon the soil. There can be no life without soil and no soil without life; they have evolved together." (Charles E. Kellogg, 1938).

Living soils are indispensable for human persistence as they are the environments where key biogeochemical processes take place. Such life-support processes, for example, the cycling of carbon, nitrogen, sulfur, and phosphorus, are primarily carried out by different members of the soil microbiomes. Moreover, soil microbiomes are useful given their potential capacities to produce compounds that are important for biotechnology, industry and medicine.

In light of its relevance for the functioning of Earth's ecosystem, soil microbiology has been termed one of the last "frontiers" in science.

Over the last decade, the study of soil microbiology has undergone significant changes, with respect to both the understanding of the diversity and functioning of soil microbial communities and the methods that are being used to dissect soil microbiomes into their components. In particular, the rapid development and use of DNA- or RNA-based molecular methods, giving rise to soil metagenomics and soil metatranscriptomics, respectively, have revolutionized our knowledge of the living soils across the globe. Moreover, soil proteomics has also undergone significant breakthroughs. Finally, the current capacities to generate big "molecular" soil data and analyze them using advanced algorithms and deep learning approaches have also enabled unprecedented advances in soil microbiology. In 2012, Paul and Nannipieri suggested that these important advancements have given rise to an emergent "second golden age of soil microbiology." The previous "golden age" of soil microbiology was named by Waksman in 1932 and reflected the important discoveries of diverse soil microorganisms being involved in key steps of the soil decomposition processes and of the nitrogen and sulfur cycles.

What are the current advances about? Several instances of outstanding breakthroughs come to mind. We have an improved knowledge of the acidobacteria and their diversity as major constituents of many terrestrial habitats. In addition, the role of archaea, previously believed to encompass mainly extremophiles, in soil systems, particularly in nitrogen transformations (ammonia oxidation), is becoming increasingly better understood. We also have an increased understanding of the significance and drivers of the astounding microbial diversity in soils and the relevance of this diversity for the mechanism of suppressing plant disease. The wealth of diverse niches in soil has given rise to a range of diversification processes, and we now appreciate the importance of the key horizontal gene transfer (HGT) processes in soil bacterial genome adaptations.

This book, *Modern Soil Microbiology, Third edition*, constitutes a thorough update of the highly successful second edition. It covers all relevant topics in soil microbiology, with the aim to provide a broad understanding of this fascinating interdisciplinary area that encompasses aspects of soil science, ecology, physiology, genetics, molecular biology, biotechnology, biochemistry, and biophysics. The book is divided into three sections that go from the fundamental, methods, to the applied side.

After depicting the soil and examining the seven grand questions posed by Selman A. Waksman, as reexamined by Arthur D. McLaren, Section I addresses the aspects of soil as the habitat (matrix) for microorganisms, describing the different organismal groups in it, including the bacteria, archaea, viruses, fungi, and protozoa. The section examines their diversities and adaptive responses, as well as their functioning in interactive and functional terms.

Section II describes the state of the art methods currently used in soil microbiology, examining the novel knowledge these have provided and will continue to do so. Thus, traditional soil microbiology methods, next to soil metagenomics, soil metatranscriptomics, and soil metaproteomics methods; stable-isotope probing, and novel methods of culturing are examined. Moreover, the use of appropriate statistical methods, next to machine learning approaches, in soil analyses is examined.

Section III focuses on a range of applied aspects of soil microbiology, including the effects of global warming, the nature of disease-suppressive soils, the use of biological control, biopesticides, bioremediation, and heavy metal stresses in soil. Finally, approaches to modulate soil microbiomes in production systems are discussed.

Modern Soil Microbiology will serve as a basic book for use in courses that aim to capture the current developments in this discipline. It will be useful as a basic work that provides research scientists with a quick entry into the specific topics and the most relevant literature.

The editors sincerely thank all authors and the publisher for their excellent cooperation and contributions to this text. Special thanks are due to Akbar Adjie Pratama for his contributions to several illustrations. As with all scientific and scholarly pursuits, we have expanded our knowledge based on the discoveries of past generations of dedicated scholars, such as Selman A. Waksman and Arthur D. McLaren. To them and other scholars, all of humanity is indebted. Our wish is that present and future generations will use the science knowledge in this text for the benefit of humanity. We welcome any comments from all who use this text.

Jan Dirk van Elsas
Jack T. Trevors
Alexandre Soares Rosado
Paolo Nannipieri

Editors

Jan Dirk van Elsas is an emeritus professor of microbial ecology at the University of Groningen, Groningen, the Netherlands. He obtained his MSc degree in chemical technology at the University of Delft, Delft, the Netherlands, and then earned his PhD degree in microbiology from the Institute of Microbiology Professor Paulo de Góes at the Federal University of Rio de Janeiro (IMPPG-UFRJ), Rio de Janeiro, Brazil. From Rio de Janeiro, he moved to Wageningen University and Research Centre, Wageningen, the Netherlands, where he occupied various positions, lastly as the leader of the Microbial Buffering group at the Plant Research International Institute. In 2003, he moved to the University of Groningen, Groningen, the Netherlands, where he was appointed to the chair professorship of Microbial Ecology. He is a specialist on bacterial survival and evolution by horizontal gene transfer in soil, the rhizosphere, and the mycosphere. He has published more than 300 peer-reviewed papers in this area and edited several books. He has been a leader in the exploration of horizontal gene processes and agents in soil activity hot spots, including the rhizosphere and mycosphere. He formally retired in 2017 and is currently involved in a range of science policy and management activities.

Jack T. Trevors is an emeritus professor of microbiology at the University of Guelph, Guelph, Ontario, Canada. He obtained his BSc (biology) and MSc (microbiology) degrees from Acadia University, Wolfville, Nova Scotia, Canada, and his PhD degree (microbiology) from the University of Waterloo, Waterloo, Ontario, Canada. His academic career was at the University of Guelph, Guelph, Ontario, Canada, from 1982 to 2012. His areas of expertise are applied and environmental microbiology, bioremediation, pathogens in the environment, gene transfer, bacterial survival and activities in the environment, gene expression, microbiological methods, basic cell biology and the origin and evolution of microorganisms. Professor Trevors has published over 450 publications, co-edited several books, has been a highly cited researcher, is a member of several science academies and is currently the editor-in-chief of the *Journal of Microbiological Methods and Water, Air and Soil Pollution.* He is also an editor with the journal *Antonie van Leeuwenhoek* and the book series editor for Environmental Pollution.

Alexandre Soares Rosado is currently a professor at the Federal University of Rio de Janeiro (UFRJ), Rio de Janeiro, Brazil, and is also a visiting professor at the Department of Land, Air, and Water Resources, University of California, Davis, California. He is the former director of the Institute of Microbiology Professor Paulo de Góes at UFRJ (IMPPG-UFRJ) and vice president of the Brazilian Society of Microbiology. He holds a BSc in biological sciences, an MSc in microbiology, and a PhD in microbiology from UFRJ. This included a sandwich period at Wageningen University and Research Centre, Wageningen, the Netherlands. Professor Rosado is an environmental microbiologist specializing in the molecular ecology of soil, extreme environments, and bioremediation.

Paolo Nannipieri is an emeritus professor at the University of Florence, Florence, Italy. He obtained the degree in biological sciences in December 1969, was a researcher at the Institute for Soil Chemistry, National Research Council, Pisa (1972–1986), and occupied the chair of Soil Chemistry at the Faculty of Agriculture at the University of Tuscia, Viterbo, Italy (1986–1990) and the chair of Agricultural Biochemistry at the Faculty of Agriculture at the University of Florence, Florence, Italy. In the latter Institute, he was head of the Department of Agrifood and Environmental Sciences and the Department of Agrifood Production and Environmental Sciences. He is the author and coauthor of about 250 publications (one in *Nature*), mostly in international scientific journals, and the editor of several books. He is the editor-in-chief of *Biology and Fertility of Soils*. He received the Lifetime Achievement Award "Terrestrial Enzymology" during the meeting "Enzymes in the Environment—Ecology, Activity & Applications," Bangor, Wales, July 26, 2016. He has been a highly-cited researcher in both 2015 (among 44 Italian researchers) and 2016, according to Thomson Reuters. Professor Nannipieri is a specialist of Soil Biochemistry.

Contributors

Claude Alabouvette
Agroécologie, AgroSup Dijon, CNRS, INRA
University of Bourgogne Franche-Comté
Dijon, France

Nouf Almzene
Department of Chemistry, Lakehead University (Thunder Bay Campus),
Ontario, Canada

Welington Luis de Araújo
Department of Microbiology
Institute of Biomedical Sciences
University of São Paulo
São Paulo, Brazil

Michael Bonkowski
Cluster of Excellence on Plant Sciences (CEPLAS)
Terrestrial Ecology Group, Institute of Zoology
University of Cologne
Köln, Germany

Lise Bonnichsen
Department of Plant and Environmental Sciences
Faculty of Science
University of Copenhagen
Frederiksberg, Denmark

Günter Brader
AIT Austrian Institute of Technology
Center for Health and Bioresources, Bioresources Unit
Tulln, Austria

Courtney Creamer
U.S. Geological Survey
Menlo Park, California

Francisco Dini-Andreote
Department of Microbial Ecology
Netherlands Institute of Ecology (NIOO-KNAW)
Wageningen, the Netherlands

Kenneth Dumack
Cluster of Excellence on Plant Sciences (CEPLAS)
Terrestrial Ecology Group, Institute of Zoology
University of Cologne
Köln, Germany

Véronique Edel-Hermann
Agroécologie, AgroSup Dijon, CNRS, INRA
University of Bourgogne Franche-Comté
Dijon, France

Jan Dirk van Elsas
Department of Microbial Ecology
GELIFES
University of Groningen
The Netherlands

Roger D. Finlay
Department of Forest Mycology & Pathology
Uppsala BioCenter, Swedish University of Agricultural Sciences
Uppsala, Sweden

Anna Maria Fiore-Donno
Cluster of Excellence on Plant Sciences (CEPLAS)
Terrestrial Ecology Group, Institute of Zoology
University of Cologne
Köln, Germany

Elisa Gamalero
Dipartimento di Scienze e Innovazione Tecnologica
Università del Piemonte Orientale
Alessandria, Italy

L. Giagnoni
Department of Agrifood Production and Environmental Sciences
University of Florence
Florence, Italy

David C. Gillan
Proteomics and Microbiology Lab, Bioscience Institute
Mons University
Mons, Belgium

Bernard R. Glick
Department of Biology
University of Waterloo
Waterloo, Ontario, Canada

Anton Hartmann
Helmholtz Zentrum München
Department of Environmental Sciences
Neuherberg/Munich, Germany

Penny R. Hirsch
Rothamsted Research
Harpenden, United Kingdom

Rob Van Houdt
Microbiology Unit, Interdisciplinary Biosciences
Belgian Nuclear Research Centre (SCK•CEN)
Mol, Belgium

Janet K. Jansson
Pacific Northwest National Laboratory
Washington

Zi-Hua Jiang
Lakehead University (Thunder Bay Campus)
Thunder Bay, Ontario, Canada

Marie E. Kroeger
Department of Microbiology
University of Massachusetts Amherst
Amherst, Massachusetts

Philippe Lemanceau
Agroécologie, AgroSup Dijon, CNRS, INRA
University of Bourgogne Franche-Comté
Dijon, France

Kam Tin Leung
Lakehead University (Thunder Bay Campus)
Thunder Bay, Ontario, Canada

Kanavillil Nandakumar
Lakehead University (Orillia Campus)
Orillia, Ontario, Canada

Paolo Nannipieri
Department of Agrifood Production and Environmental Sciences
University of Florence
Florence, Italy

Mette Haubjerg Nicolaisen
Department of Plant and Environmental Sciences
Faculty of Science
University of Copenhagen
Frederiksberg, Denmark

Kaare Magnus Nielsen
Department of Life Sciences and Health
Oslo Metropolitan University
Oslo, Norway

Klaus Nüsslein
Department of Microbiology
University of Massachusetts Amherst
Amherst, Massachusetts

Ole Nybroe
Department of Plant and Environmental Sciences
Faculty of Science
University of Copenhagen
Frederiksberg, Denmark

Nikolaus Pfaffenbichler
AIT Austrian Institute of Technology
Center for Health and Bioresources, Bioresources Unit
Tulln, Austria

Akbar Adjie Pratama
Department of Microbial Ecology, Microbial Ecology
Groningen Institute for Evolutionary Life Sciences, University of Groningen
Groningen, the Netherlands

G. Renella
Department of Agrifood Production and Environmental Sciences
University of Florence
Florence, Italy

Ulisses Nunes da Rocha
Helmholtz Centre for Environmental Research—UFZ
Leipzig, Germany

Jorge L. Mazza Rodrigues
Department of Land, Air and Water Resources
University of California, Davis
Davis, California

Alexandre Soares Rosado
Federal University of Rio de Janeiro
Rio de Janeiro, Brazil

Abdul Samad
AIT Austrian Institute of Technology
Center for Health and Bioresources, Bioresources Unit
Tulln, Austria

Michael Schloter
Helmholtz Zentrum München
Research Unit for Comparative Microbiome Analysis
Department of Environmental Sciences
Neuherberg/Munich, Germany

Anne Schöler
Helmholtz Zentrum München
Research Unit for Comparative Microbiome Analysis
Department of Environmental Sciences
Neuherberg, Germany

Stefanie Schulz
Helmholtz Zentrum München
Research Unit for Comparative Microbiome Analysis
Department of Environmental Sciences
Neuherberg, Germany

Alexander V. Semenov
Incotec Europe BV
Enkhuizen, the Netherlands

Angela Sessitsch
AIT Austrian Institute of Technology
Center for Health and Bioresources, Bioresources Unit
Tulln, Austria

Shilpi Sharma
Indian Institute of Technology
Department of Biochemical Engineering and Biotechnology
Delhi, New Delhi, India

Sara Sjöling
Environmental Science, School of Natural Sciences
Technology and Environmental Studies
Södertörn University
Huddinge, Sweden

Kurissery Sreekumari
Lakehead University (Orillia Campus)
Orillia, Ontario, Canada

Christian Steinberg
Agroécologie, AgroSup Dijon, CNRS, INRA
University of Bourgogne Franche-Comté
Dijon, France

Nanna Bygvraa Svenningsen
Department of Plant and Environmental Sciences
Faculty of Science
University of Copenhagen
Frederiksberg, Denmark

R. Greg Thorn
Department of Biology
University of Western Ontario
London, Ontario, Canada

Jack T. Trevors
School of Environmental Sciences
University of Guelph
Guelph, Ontario, Canada

Mark P. Waldrop
U.S. Geological Survey
Menlo Park, California

Section I

Fundamental Chapters

1 The Soil Environment*

Jan Dirk van Elsas
University of Groningen

CONTENTS

1.1 Introduction 3
1.2 Scales and Gradients 4
1.2.1 Introduction 4
1.2.2 Spatial and Temporal Scales in Soil 5
1.3 The Soil Physicochemical Environment 6
1.3.1 Water—The Essential Factor for Soil Life 6
1.3.1.1 Definitions 8
1.3.2 Soil as an Energy and Nutrient Source 9
1.3.3 Soil Temperature 11
1.3.4 Soil Light 14
1.3.5 Soil Atmosphere and Redox Potential 14
1.3.6 Soil pH 16
1.4 Concluding Remarks 18
References 18

1.1 INTRODUCTION

Soil is a structured environment that "holds" a wealth of organisms with diverse activities and functions. Among these organisms, microorganisms have a central place, as they play major roles in key soil processes. However, a key problem in any discussion about soil as a microbiological habitat is our conceptual visualization of soil. The most commonly used unit of reference is 1 g. This unit has been an optimum mass from which biodiversity indices indicative of habitat richness are gained. One gram of soil consists of inorganic and organic fractions. The inorganic particles of soil are classified into three major groups according to their size: sand, silt, and clay. The proportions of these in any soil determine the soil texture. The number of particles in 1 g of soil can range from 90 (pure coarse sand) to 90 billion (pure clay), and the proportions of sand, silt, and clay in the soil sample determine the soil textural class. Assuming spherical shapes, the surface area of soil particles can range from 11 to 8 million cm^2g^{-1}, illustrating the high degree of physical heterogeneity present in any given soil sample. Although the concept of an "average" soil is not entirely meaningful, we will consider a hypothetical soil aggregate (Figure 1.1) as the basic unit of the soil habitat. Many biogeochemical processes occur at scales more or less relevant to this unit, including processes such as gas diffusion and water movement, which create a mosaic of *microsites* and gradients (Paul and Clark 1996). Most, if not all, aggregates constitute potential habitats that allow the survival of one or more species of soil microorganisms. In our "hypothetical" habitat, a gram of soil and the arrangement of the soil particles (soil structure; see Table 1.1) are vital to the microbes present. About half the

* Modified from Standing and Killham (2007).

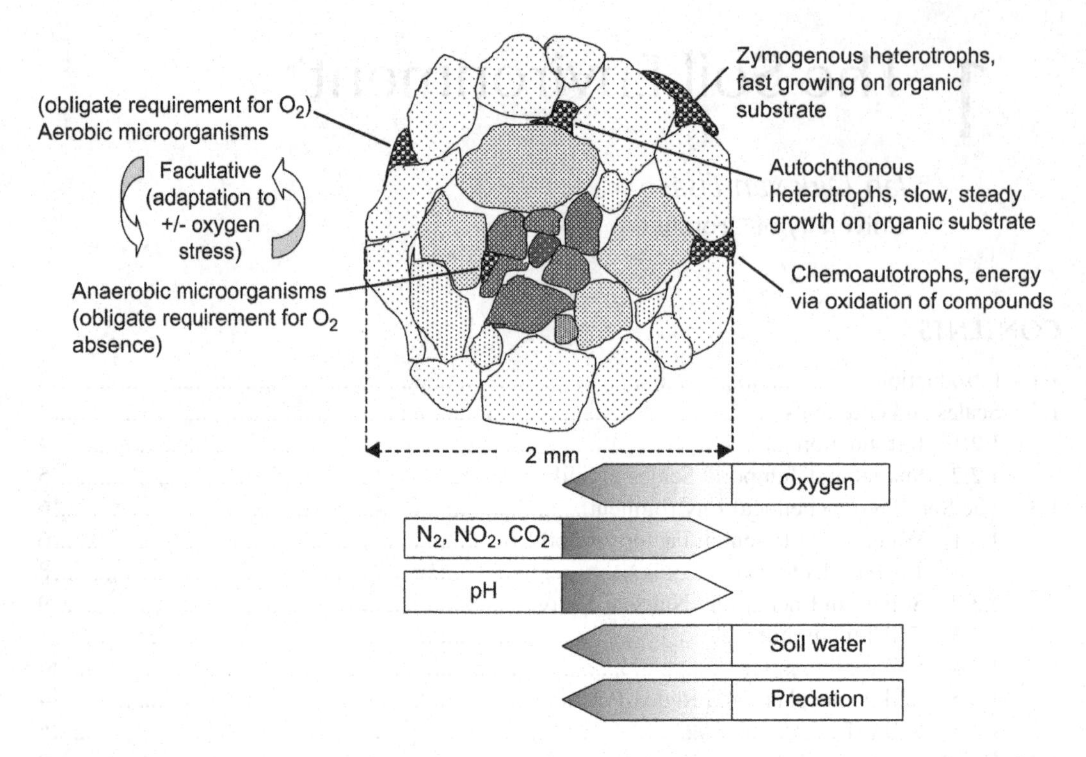

FIGURE 1.1 Schematic of a section through a 2 mm-diameter soil aggregate microsite yielding habitats for bacteria based on physiological requirements. The arrows show the main directions of diffusion for key processes. The increasing darkness of the soil particles indicates the increasing number of sites available for anaerobic bacterial function.

volume of an aggregate will be void spaces (pores) connected by tortuous pathways presenting a range of pore neck sizes. It is the relationship between the interconnecting pathways, channels, and pores in soil that provides the microhabitat space (*niche*, used here as the term to describe habitable space in soil) for the soil microbiota. The physical niches themselves will mainly be soil pore walls, although water in channels may contain, and transport, significant numbers of freely motile bacteria (Figure 1.2).

Soil water passes more rapidly through wider pores by means of gravity mass flow and diffusion (see below for a separate consideration of these topics) than through narrow pores. The physical niches in the wider pores are also mainly pore walls to which bacteria adhere.

1.2 SCALES AND GRADIENTS

1.2.1 Introduction

The microbial habitat in soil is, essentially, a porous medium that varies both spatially and temporally. The structural form of this porous medium is the product of many different processes. Biological, physicochemical, and mechanical processes act together to aggregate, compact, crack, and fragment the soil, resulting in a soil structure consisting of solids and pores. The soil pores form a tripartite group—*transmission pores* being the main conduits for water and nutrient flow, *storage pores* that are empty under the influence of matric potential, and *residual pores* (with pore neck diameters less than 0.3 μm) that remain water-filled. In most soils, the transmission pores form the majority of pore space in the soil microbial habitat.

TABLE 1.1
Approximate Dimensions (μm) of Soil Particles and Biota and Comparison of Water-Filled Pores and Water Films

	Type	Approximate Dimensions (μm)
Soil particles	Stones	>2,000
	Coarse sand	2,000–200
	Fine sand	200–50
	Silt	50–2
	Clay	2–0.2
Plant material	Roots	1,000
	Fine roots	1,000–50
	Root hairs	15–7
Microbes	Fungal hyphae	10–3
	Actinomycetes	1.5–1.0
	Bacteria	**0.5–1.0**
	Viruses	0.05–0.2
Some soil animals	Earthworms	5,000–2
	Mites	2,000–500
	Nematodes	2,000–500
	Protozoa	80–10
Water-filled pores	−10 kPa	<30
	−100 kPa	<3
	−1,000 kPa	0.3
Water films	−100 kPa	<0.0003
	−1,000 kPa	Few molecules thick

Bacteria are highlighted in bold for comparison with other components of the soil–microbe–plant system.

Historically, it has proven to be impossible to produce an overarching descriptor of soil habitats that fits all soil types (not to be confused with soil classification). It would be useful to have such a descriptor, as it may help us in understanding the observable conservation of microbial genes responsible for the processes carried out in soils, such as denitrification and carbon, nitrogen, and phosphorus mineralization.

Now entering a bit of theory: fractals (objects with fractional dimensions which possess self-similarity, composed of several parts, each of which is a small-scale copy of the whole) have been applied successfully to describe spatiotemporal, hierarchical, and complex systems. These fractal systems can be generated using scaling laws and iterative algorithms. To be classified as a fractal system, patterns must be self-similar over a range of scales (i.e., its properties are reproduced at a number of different spatial and temporal scales). As a result, no matter how intricate a particular pattern might be, its statistical properties allow its description, independent of scale (Bird and Perrier 2003, Perrier and Bird 2002). Within a given soil texture (except for very clayey soils), a description of soil as a system of fractals appears to be robust. However, high clay content soils do not display fractal particle size distributions (Millan et al. 2003).

1.2.2 Spatial and Temporal Scales in Soil

Spatial and temporal scales in soil can be best understood by way of an example. Here, we use a root with its surrounding soil to illustrate some key spatiotemporal aspects (Figure 1.3). The soil

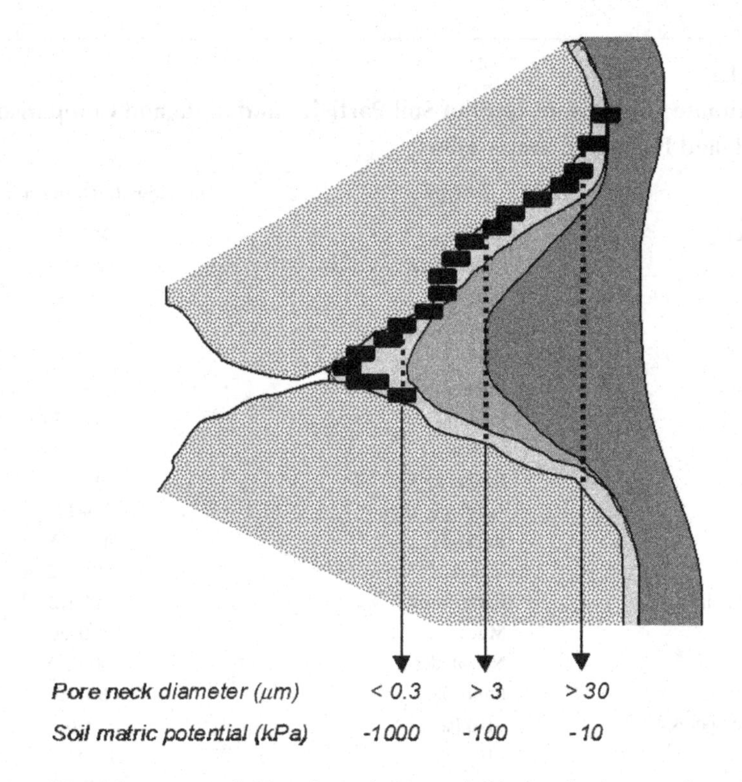

FIGURE 1.2 Possible niche space available to bacteria (shown in black, adhering to soil particle (gray hatching) surfaces showing the relationship between soil matric potential and pore size available.

habitat generally constitutes an aerobic, *oligotrophic* (nutrient-poor) environment, which is not conducive to high population densities and activities of microorganisms. In contrast, the *rhizosphere* (further discussed in Chapter 10) provides an environment in which elevated population sizes and activities of fast-growing (formerly called *zymogenous*) microorganisms are supported by plant-derived carbon substrates, in the form of exudates and cell lysates. This constitutes the soil–plant interface, a habitat that incites a wide spectrum of beneficial as well as detrimental associations with microorganisms.

Water movement and diffusion of molecules are key features of the soil and rhizosphere habitats in which microbial populations transmit information from one to another, for example, via *quorum sensing* (Whiteley et al. 2017). Bacterial interactions that are important in the rhizosphere (e.g., production of antibiotics and chitinases, biofilm formation, stationary phase, and motility) (Cha et al. 1998, Elasri et al. 2001, Pierson et al. 1998) are often driven by quorum sensing. See Chapters 9 and 10 for more detailed information about these interactions. Key issues arise here, such as the connectivity of the soil pores, which determines to a great extent the local concentrations of quorum sensing molecules. These concentrations ultimately determine the outcome of the quorum sensing-dependent processes.

1.3 THE SOIL PHYSICOCHEMICAL ENVIRONMENT

1.3.1 Water—The Essential Factor for Soil Life

"Understanding the movement of water in soil is understanding the most significant feature of the soil as a habitat for microbial life". Where water moves, so do ions and nutrients. Water carries dissolved gases and heat, and also bacteria and their predators. It protects microhabitats from

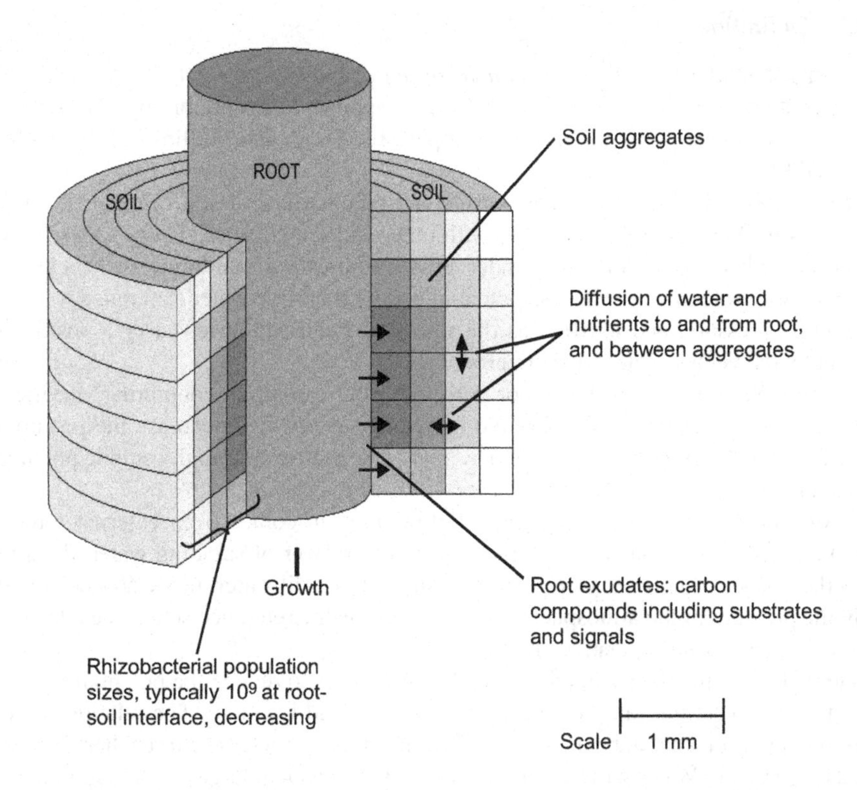

FIGURE 1.3 Schematic illustration of the soil environment around plant roots in terms of diffusion of materials (substrate carbon and nutrients) and information (signal molecules). Scale: 1 mm

desiccation and opens other potential habitats while closing others. The fundamental relationships between physics and chemistry that modulate soil water and biological activity are presented in the schematic picture in Figure 1.4. When considering any interaction of soil water and biological activity, it must be stressed that the four central boxes (representing physical and chemical laws) cannot be discussed in isolation: they all contribute to the interaction.

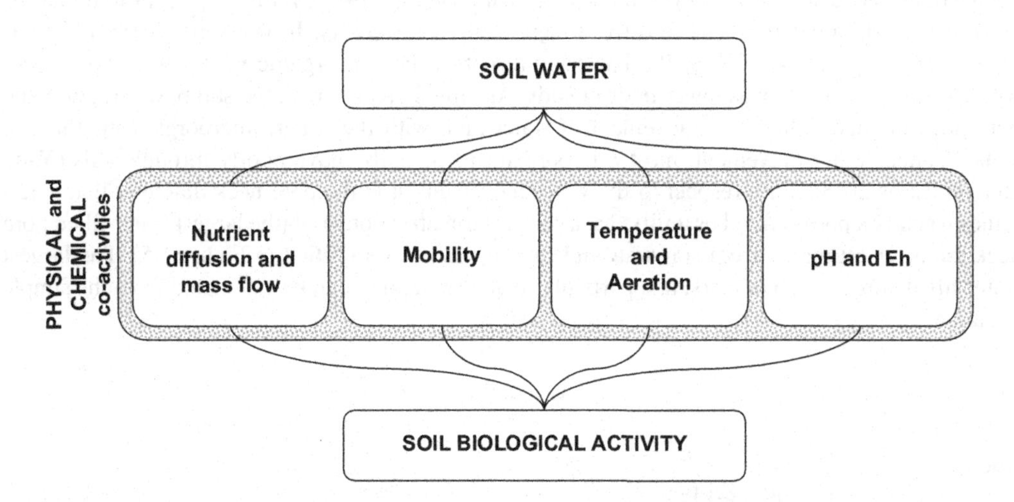

FIGURE 1.4 The range of water-influenced soil properties and processes which determine the microbial activity. Eh: redox potential. See Section 1.3.5.

1.3.1.1 Definitions

Soil water potential is the sum of the *matric, osmotic,* and *pressure* potentials; it is the key measure of the activity of water in the soil. These three component terms are briefly outlined in the following text. For a fuller consideration of these topics, see Smith and Mullins (2000) and Marshall and Holmes (1996).

Matric potential: Water molecules adsorb onto the surface of soil minerals through hydrogen bonding, as well as bonding cohesively with other water molecules. These adhesive and cohesive forces act together to hold soil water under tension against external forces such as gravity. Because of this, the soil water always has less potential energy than free water (reference water: at the same temperature, pressure, and location as the soil water) and can never carry a positive sign. Thus, matric potential is always negative or zero.

Osmotic potential: Soil water is not pure water but a solution containing varying amounts of osmotically active organic molecules and inorganic salts, which decrease the potential energy of soil water relative to a pure water reference. Thus, like matric potential, osmotic potential is always negative or zero.

Pressure potential: The pressure potential component comes from external forces (including gravity) exerted on soil water. In a flooded soil with a layer of standing water, the atmosphere as well as the surface water exerts a positive pressure on the soil water. In the absence of flooding, it is simply the pressure of the atmosphere alone, and this is the reference state. The additional pressure of ponded water creates a positive pressure.

Water is held dynamically in the soil by forces that act to reduce the potential energy relative to that of free water (at the same temperature, pressure, and location). The relationship of the water potential to microbial habitat can be visualized in a moisture release curve when drying soil is considered (Figure 1.5). Water will be drained gravitationally out of large pores first, followed by drainage from successively smaller pores. It should be emphasized here that it is the pore neck diameter that determines the rate of water movement, that is, a large water-filled pore with a small pore neck diameter will empty slower than a small pore with a large pore neck diameter. As soils rewet, it is not a simple reverse of drainage. Paradoxically, at a given suction force, a drying soil has higher water content than a wetting soil. This phenomenon is termed *hysteresis.* It is a function of the pore diameter generally being larger than the corresponding pore neck diameters; a full explanation is available in the work of Marshall and Holmes (1996).

Increases in soil moisture content can greatly affect bacterial population dynamics by providing connections between separate aggregates. Vargas and Hattori (1990), for example, showed that the level of bacterial predation was sensitive to aggregate connectivity. In wet soils (>60% of water-holding capacity), up to 90% of the bacteria present in the soil organic matter were consumed, whereas this percentage was lower in drier soils. As already mentioned, the soil pore structure and pore neck diameter allow the movement of water and, with the water, microorganisms through soils. Without water movement, most microorganisms actually move poorly through soils (Yang and van Elsas 2018). With respect to the water movement, it is the pore neck diameter that determines whether a pore is filled or unfilled at a particular matric potential (the largest water-filled pore neck diameter estimated from soil water release curves as exemplified in Figure 1.5). The largest water-filled pore neck diameters, at a particular matric potential, can be calculated from the simple equation:

$$\psi = 300/d$$

where

ψ is the matric potential (–kPa)

d is the neck diameter of the largest water-filled pore (μm)

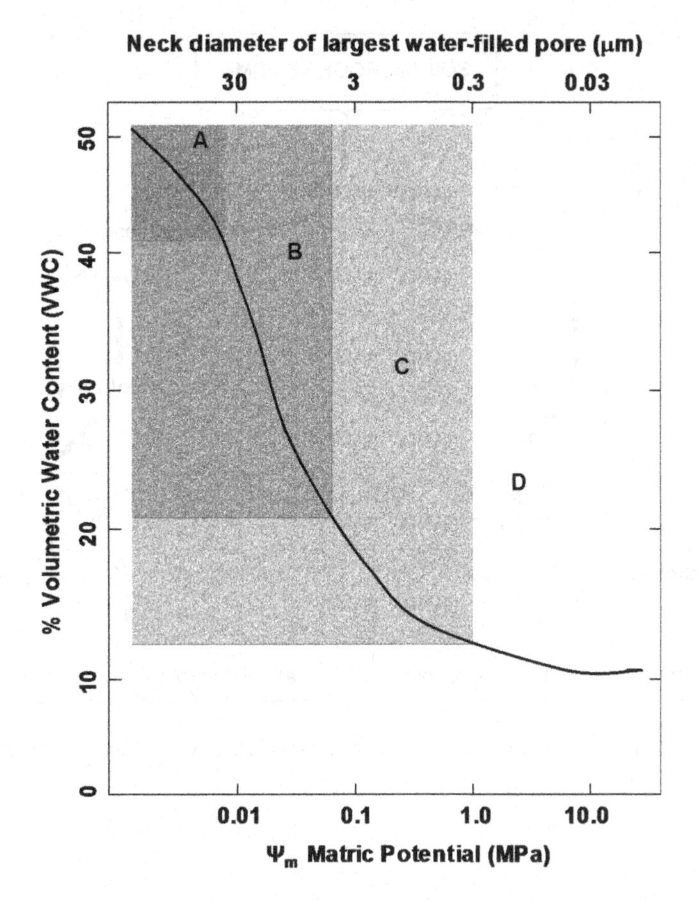

FIGURE 1.5 A hypothetical moisture release characteristic, demonstrating the relationship between water content, water potential, and neck diameter of the largest water-filled pore (Adapted from Gammack et al. 1991).

Thus, by applying a given matric potential to the soil, it is possible (with a finely nebulized inoculum in a precise volume with reference to the moisture release curve of the soil) to preferentially fill pore size classes (White et al. 1994). This elegant technique can be used to experimentally place bacteria in particular pore size classes of choice or to explore such topics as the consequences of grazing limitation (undergrazing of excluded pores and overgrazing of included pores) on community structure and function.

1.3.2 Soil as an Energy and Nutrient Source

Figure 1.6 shows how soil microorganisms are classified according to how they use their sources of energy and carbon. Light is obviously only present at the surface of the soil. The chemical energy for *chemoautotrophs* is supplied in the form of reduced inorganic chemicals, such as Fe^{2+}, and reduced sulfur or nitrogen compounds (e.g., SO_3^{2-}, S^{2-}, NH_4^+, and NO_2^-).

Organic matter provides the energy source for the *heterotrophic* microbial community of the soil. This organic matter ranges from the readily decomposable and more resistant ("recalcitrant") fractions of plant litter to the fractions of microbially processed organic matter that are either physically or chemically protected to some degree. Most importantly, the microbial biomass itself represents the "eye of the needle" through which all these fractions eventually pass (Figure 1.7). In the figure, degradation-resistant plant material is typified by *lignin* (a complex polymer of aromatic,

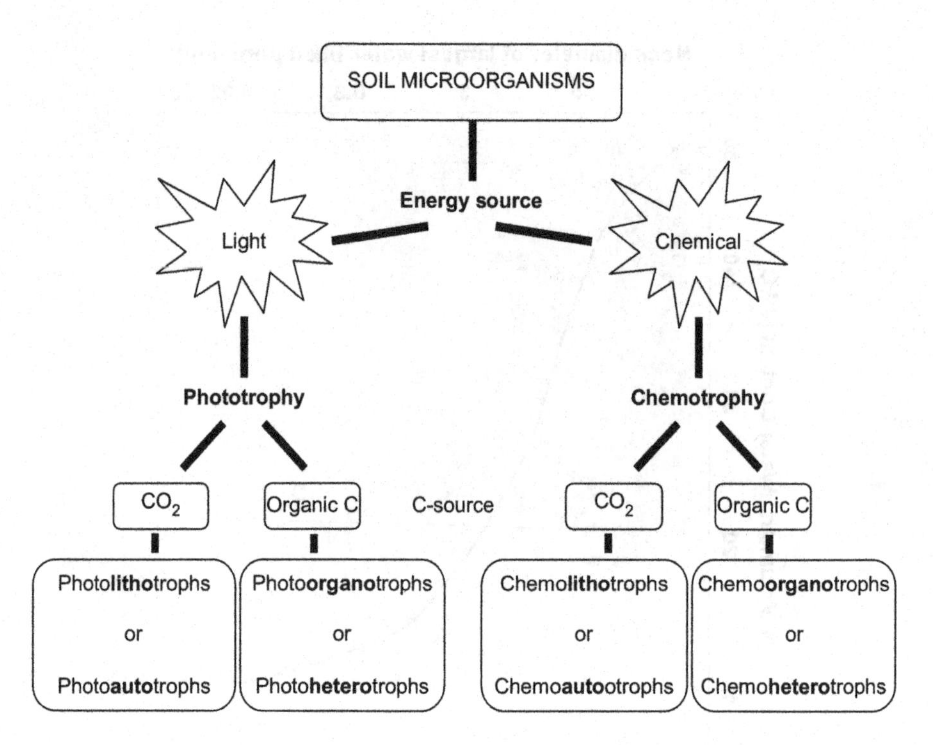

FIGURE 1.6 Classification of soil microorganisms in terms of use of carbon/energy sources.

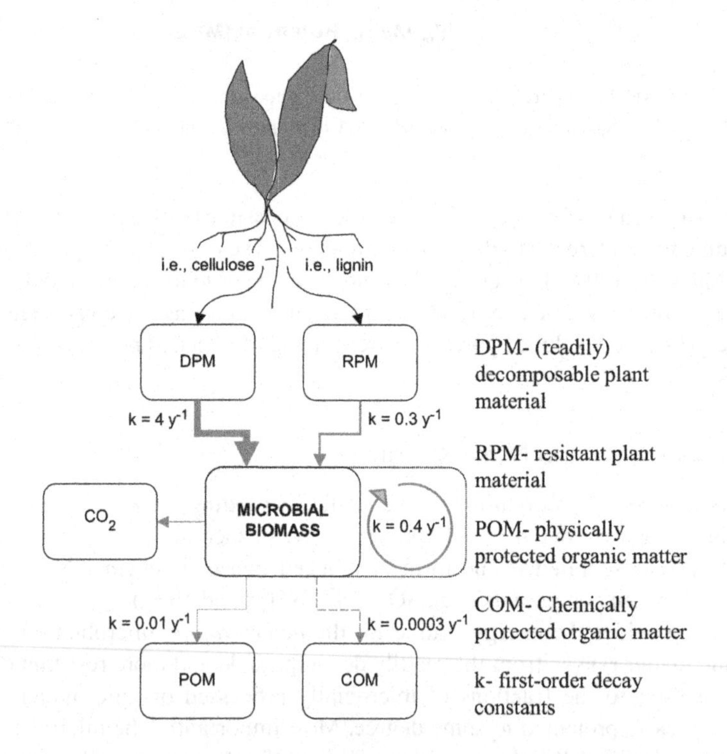

FIGURE 1.7 Main redox couples (terminal electron acceptor and redox product) and associated microbial processes operating in the soil system in relation to the redox potential at neutral pH.

phenyl propane, repeating units), whereas readily decomposable plant material is typified by *cellulose* (carbohydrate polymer with repeating glucose units). In addition to soil organic matter, the organic carbon that is directly released from plant roots represents a key energy source. This rhizosphere carbon flow, which comprises C in a variety of forms, can account for a considerable part (5%–25%) of the plant photosynthates (Lambers 1987); it is the main energy source for the heterotrophic microbes, particularly bacteria, that colonize the rhizosphere.

Not only does soil organic matter represent an energy source which is vital to the heterotrophic soil microorganisms, but it is also the main supplier of nutrients, as most of the N, P, and S reserves of the soil are tied up in organically bound forms. Some of the microbial transformations (and the associated enzymes) involved in the release of these bound nutrients are shown in the following text. These all produce small and simple compounds such as ammonia, phosphate, and sulfate, which can be directly used as nutrients by plants or microorganisms.

Deamination of amino acids (release of organically bound N)

COOH
|
H_2N—C—H —deaminase→ C=O + NH_3↑
|
CH
|
CH
|
COOH

glutamic acid → ketoglutamic acid +ammonia

Urease activity (release of organically bound N from urea)

$$CO(NH_2)_2 + H_2O \rightarrow 2NH_3 + CO_2$$

Phosphatase activity (release of organically bound P from phosphate monoesters)

$$\text{ROP(=O)(OH)OH} + H_20 \longrightarrow \text{ROH} + \text{HOP(=O)(OH)OH}$$

Sulfatase activity (release of organically bound S from sulfate monoesters)

$$ROSO_3^- + H_2O \rightarrow ROH + H^+ + SO_4^{2-}$$

1.3.3 Soil Temperature

Temperature in soil is a key determinant of both the distribution and the activity of soil microorganisms. In terms of activity, temperature directly affects microbial physiology, and it indirectly exerts its effects through changes in factors such as nutrient and substrate diffusion and water activity.

Soil temperature is the product of incident solar energy, modified by a series of factors. The first is reflectance (about one-third of the incident solar energy is reflected back by the soil–plant system, leaving two-thirds of the "net irradiation"), which is determined by soil color and vegetation type. The next factor is soil moisture. About 80% of the net radiation is used to evaporate water from the soil compared to 5% which is used for photosynthesis. So, only about 15% of the net radiation tends to warm the soil. The extent to which a unit of soil is warmed by a unit of net radiation is determined by the specific heat capacity of the soil and the soil moisture (the latter is important because of the specific heat capacity of water and the energy needed for the vaporization of water). In addition to these fundamental principles, soil temperature regime is affected by seasonal factors, as well as "edaphic" factors such as soil type/depth and the nature of the vegetation present.

Temperature is a strong selective force and is a key driver of physical, chemical, and physiological reactions. As such, it is of significant importance in soil, and soil microbial communities can be defined on the basis of temperature values alone (Figure 1.8). Within "average" *mesophilic* microbial communities, there is an approximate doubling of the rate of biochemical activity with every 10°C rise between 0°C and 30°C/35°C. This is referred to as the *Q10 relationship.* At the soil surface, the temperature will be primarily affected by the incident short-wavelength radiation (<2 μm), and it will fluctuate both diurnally and annually. These fluctuations have progressively less amplitude from the surface to deeper layers in the soil (Figure 1.9).

From the Q10 relationship, we can deduce that soil temperature is a critical regulator of microbial metabolism. Figure 1.9 illustrates how soil temperature can fluctuate with time and depth. How soil microorganisms respond to changes in temperature is not, of course, independent of the effects of temperature on the plants (and other organisms) with which they interact. For example, the quantity

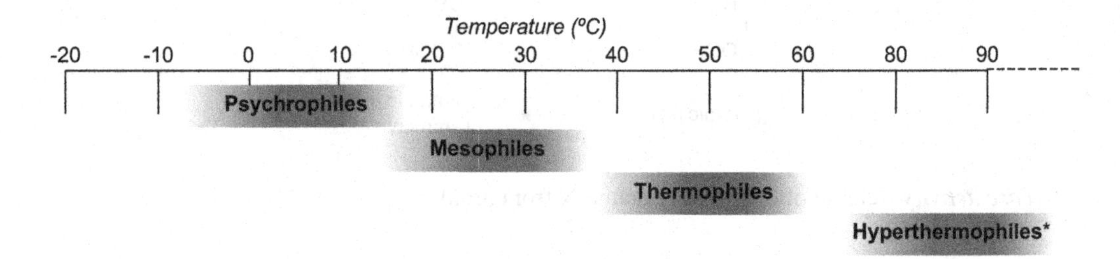

FIGURE 1.8 Schematic showing temperature preferences for soil bacteria.

* The majority of this group are archaea which are generally found in aqueous environments rather than soils.

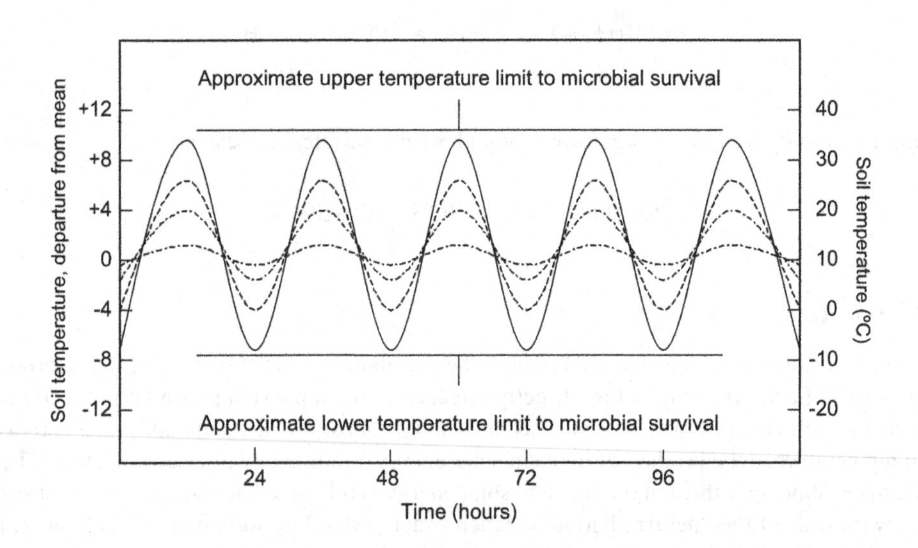

FIGURE 1.9 Trend of diurnal temperature fluctuation in soil as related to depth. The periodicity of fluctuation can be considered diurnally or seasonally.

and quality of rhizosphere C flow is critical to the diversity and activity of the rhizosphere microbial community and is strongly dependent on temperature (Meharg and Killham 1989), as is root growth and turnover.

Some soil microbial processes are particularly temperature sensitive. For instance, the accumulation of ammonium in temperate–climate soils in autumn and spring is due to the marked difference in low-temperature sensitivity of the processes of ammonification and nitrification, as outlined in the following scheme:

Ammonification (carried out by a very diverse group of soil microorganisms)		Nitrification (carried out by a very narrow range of soil microorganisms)
Organic matter	$\rightarrow NH_4^+ \rightarrow$	NO_3^-
		Selective low-temperature block

Table 1.2 shows the classification of soil microorganisms on the basis of temperature preference. Soil microorganisms are truly peculiar: some can live at temperatures that could not be imagined by higher life forms (see Box 1.1) (Killham 1994).

TABLE 1.2
Classification of Soil Microorganisms According to Temperature Preference

Environmental Class		Temperature Range (°C)	Optimum Growth (°C)
Psychrophile		–5~20	15
Mesophile		15~45	37
Thermophile	Moderate	40~70	60
	Extreme: (hyperthermophile)	65~95	85

BOX 1.1 EXTREME SOIL ENVIRONMENTS

That no region on Earth, with the possible exception of active volcanoes, is devoid of soil microbial life, is truly amazing in terms of the temperature and pH tolerances and adaptation of the microorganisms present. First, the thermophilic S-oxidizing bacteria and thermophilic algae that inhabit hot springs (often close to the boiling point of water) on the one hand, and the psychrophilic snow moulds that can decompose leaf litter below the snow cover at virtually zero degrees centigrade on the other hand, give us an indication of life across the spectrum of temperatures in our environments. This connects to pH tolerance as well. An extreme example of physiological pH preference occurs in some of the S- and Fe-oxidizing chemoautotrophic bacteria. *Thiobacillus ferrooxidans* and *T. thiooxidans*, for example, grow optimally at pH 2-3, while the soil bacterium *T. acidophilus* can be termed a true acidophile, being routinely cultured at pH 1.4. This is amazing, bearing in mind the logarithmic relationship between pH and hydrogen ion concetration. It means that *T. acidophilus* is growing at nearly a million times the acidity of many near-neutral agricultural soils!

1.3.4 Soil Light

Figure 1.10 highlights the interconversion of energy in the soil–plant system. The key driver for this interconversion is solar energy. Section 1.3.3 explained how solar energy determines soil temperature, which in turn influences soil microbial activity. Solar light also directly affects the microbial distribution and activity at or near the soil surface (where light can penetrate), providing energy (5% of the net solar radiation drives the photosynthetic activity of plants and microbes) and driving *photoautotrophic* soil microorganisms such as certain algae and cyanobacteria. It is generally only under the conditions of high light and moisture levels that such photoautotrophic soil microorganisms contribute significantly to soil microbial biomass and microbial energy flow through the soil. Furthermore, the contribution is often short-lived, as soils may dry out or become shaded. Under the conditions of high photoautotrophic activity, however, considerable amounts of polysaccharides can be produced in the soil, which play important roles in soil aggregate genesis and stabilization, thus enhancing the stability of the soil.

The most important effect of light, from a soil microbiological point of view, is that it stimulates plant seed germination, seedling establishment, and growth and is necessary for some algae and cyanobacteria to grow on surface soil soils and other surfaces. Plant roots penetrate and aerate the soil, redistribute the soil water and nutrients through hydraulic lift, and then provide the main source of carbon substrates for microbial growth. At low light intensity, root metabolic activity and respiration can be reduced by as much as 50% (Lambers 1987), impacting microbial activity, as carbon substrates will become limiting (Broughton and Gross 2000).

1.3.5 Soil Atmosphere and Redox Potential

The typical concentrations and diffusion coefficients of the major soil gases in the air and soil atmosphere are indicated in Table 1.3. Because soil water potential (matric and osmotic) is so critical to determining the extent and nature of soil microbial activity, as well as the diffusive supply of oxygen to the sites of *aerobic* activity (contrast the diffusion coefficient of oxygen in air and water), the characteristics of the soil atmosphere are strongly related to the soil water regime. In addition, temperature is also critical as a rate controller of microbial activity, assuming sufficient water is available.

In relation to the model soil aggregate shown in Figure 1.1, when the aggregate pore space is water-filled, the aggregate center will first become anaerobic when the demand for oxygen from

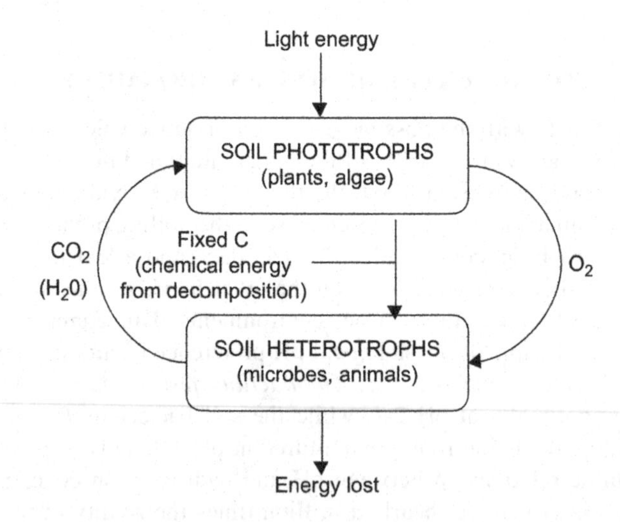

FIGURE 1.10 Interconversion of carbon and energy in the soil–plant system. Energy input is in the form of solar energy and output in the form of heat loss.

TABLE 1.3
Typical Composition of the Soil Atmosphere Compared to the Open Atmosphere

	N_2	O_2	CO_2
Typical concentration in air (%)	79	21	0.035
Typical concentration in soil atmosphere	79	20–21	0.1–1.0
Diffusion coefficient in air (cm^2s^{-1})	2.1	2.1	1.6
Diffusion coefficient in water (cm^2s^{-1})	1.6×10^{-4}	1.8×10^{-4}	1.8×10^{-4}
Solubility in water (cm^3L^{-1})	1.5	1.5	87.8

Diffusion coefficients and solubilities of the main gases are also included.

aerobic respiration cannot be met by the diffusive resupply from the surface of the aggregate. The central zone of anaerobiosis will grow outward until diffusive resupply is adequate to meet that demand (oxygen demand from roots and microbes can exceed 20 g $m^{-2}d^{-1}$ when activity is high; Russell 1973). In early work, Greenwood (1975) pioneered our understanding of this microsite onset of anaerobiosis in water-saturated aggregates, using a derivative of *Fick's law* (governing the rates of gas diffusion) to calculate the critical aggregate radius at which a water-saturated aggregate, with an active microbial community, generates an *anaerobic* center:

$$a^2 = 6CD/R$$

where
a is the critical aggregate radius
C is the distance in O_2 concentration between the aggregate surface and the center (mL O_2 mL^{-1})
D is the diffusion coefficient of O_2 in water ($cm^2\ s^{-1}$)
R is the respiration rate inside the aggregate (mL O_2 cm^{-3})

From Greenwood's equation, anaerobic conditions are expected to occur in water-saturated aggregates of 1 cm radius or greater for typical levels of soil respiration. From this, we can expect anaerobic microsites to develop reasonably often in soil, particularly when there is good aggregation and recent rainfall and/or poor drainage. In a completely water-saturated soil, more widespread anaerobic conditions will generally prevail if this system is biologically active. However, water-saturated soils are not always anaerobic. Flushed peat soils are often maintained in the aerobic state, with a continuous resupply of oxygen in the water that flushes through the soil.

Once oxygen is depleted by respirational demand from soil organisms and anaerobic conditions develop in soil, either in microsites or more widely, the availability of alternative electron acceptors determines the (anaerobic) processes that operate (Figure 1.10). These are entirely microbial processes, as plants and soil animals are obligate aerobes (i.e., they cannot function in the absence of oxygen). There is a sequence of reduction of terminal electron acceptors, and this is illustrated in Figure 1.11 along with the prevailing *redox potential*. The latter is simply a measure of the tendency of a substance to lose (oxidation) or gain (reduction) electrons and is described by the *Nernst equation* shown as follows:

$$E_h = E_o + 0.059/n \cdot \log(\mathrm{Ox/Red})$$

where
E_h is the measured (platinum electrode) redox potential (mV)
E_o is the standard potential of the system (mV)

n is the number of electrons in the system
Ox is the number of electrons lost
Red is the number of electrons gained

Understanding the redox chemistry of soils requires a consideration of available electron acceptors and an awareness of the general rule that the presence of an electron acceptor with a higher oxidation state will inhibit the operation of an acceptor with a lower oxidation state. So, sulfate reduction, for example, will be inhibited by the presence not only of oxygen but also of nitrate, manganese as Mn^{4+}, and iron as Fe^{3+} (Figure 1.11).

The spatial heterogeneity of soils in terms of structure, reducing power (primarily driven by available carbon supply such as from plant roots or organic residues of plants and animals), and availability of electron acceptors (oxygen tends to exhibit a gradient of concentrations from the surface to the edge of the rooting depth, and nitrate levels tend to reflect the localized zones of mineralization and nitrification) means that redox chemistry is also highly heterogeneous, demonstrating strong spatial as well as temporal variability.

1.3.6 Soil pH

The soil *pH* is the negative logarithm of the level of protons per unit volume:

$$\text{pH} = \log\left(1/\left[\text{H}^+\right]\right)$$

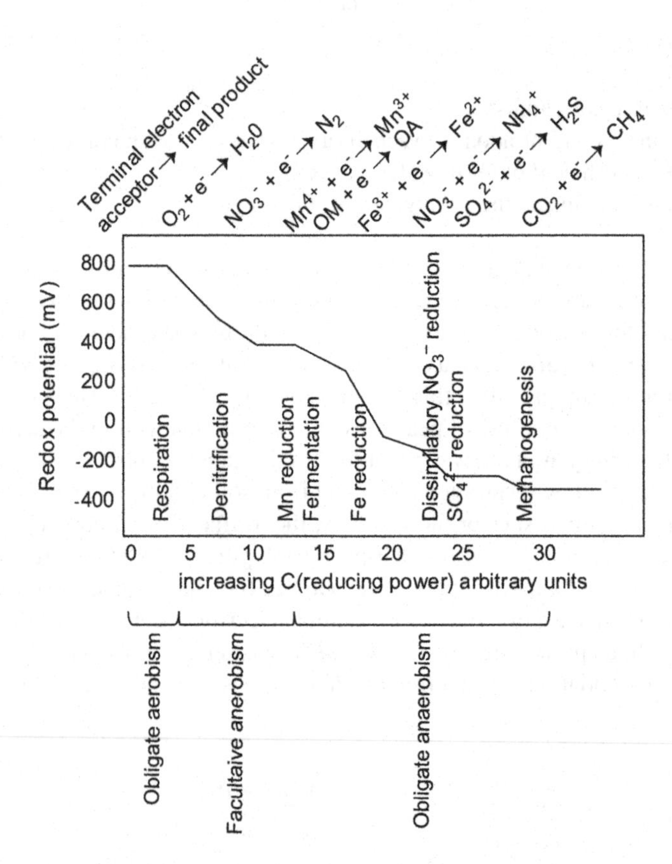

FIGURE 1.11 Main redox couples (terminal electron acceptor and redox product) and associated microbial processes operating in the soil system in relation to redox potential at neutral pH.

Soil pH is a major determinant of soil microbial distribution and activity (Rousk et al. 2010). The pH of a soil, or of a microsite in the soil, is the product of numerous factors and processes. First, it is determined by the parent material from which the soil is formed (acid soils tending to form from rocks such as granite and alkaline soils from chalks and limestones), as well as the degree to which mineral weathering has occurred since soil formation. Superimposed on this, a range of biological processes can modify the soil pH to varying degrees. Figure 1.12 illustrates these pH-modifying processes, occurring at a range of scales from the microsite (microbial habitat) to the entire bulk soil. Some of these processes are microbial, and others are animal based, whereas yet others are driven by the plant. These processes act to create a mosaic of variable pH conditions in the soil. Consequently, different microbial groups with physiological preferences for particular pH conditions (*acidophiles*, preferring low pH conditions; *alkaliphiles*, preferring high pH conditions), as indicated in Figure 1.12, are favored in soils differing in pH conditions.

Some microorganisms can also thrive in soils with low or high pH conditions (White et al. 1994) (see Box 1.1).

The distribution and activity of soil microbes as a result of pH is not simply determined by physiological pH preferences. Many soil microorganisms can tolerate particular pH conditions that are far from their optimum. For example, many of the fungi found in acid forest soils are not acidophiles (Killham 1994). When isolated from these environments, some fungi show a preference for near-neutral conditions, but they are highly competitive under considerable acidity (pH 3 is not uncommon in organic soils, for instance under conifers). A further complication in understanding pH-related microbial activity in soils is that most microbes are attached to soil

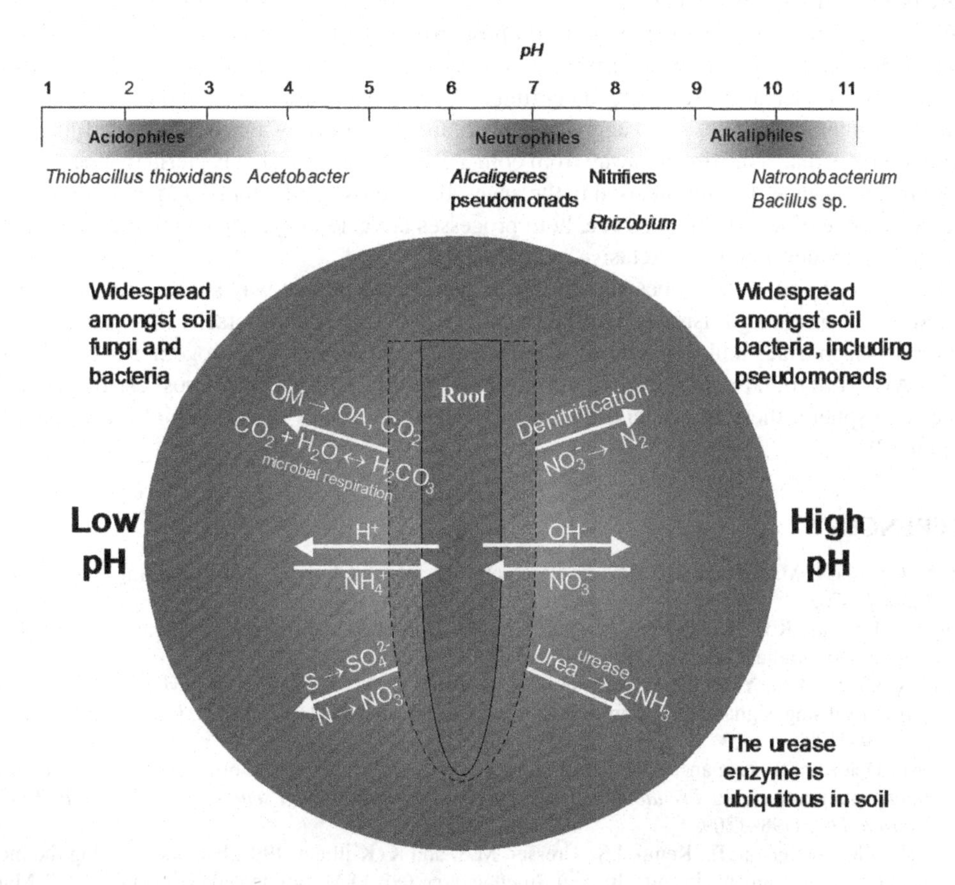

FIGURE 1.12 pH preference of soil microorganisms and the main pH-modifying processes which generate microhabitats with distinct pH conditions. Explanation: OM = organic matter; OA = organic acids.

surfaces and many grow in colonies and even biofilms, which can protect them from the pH prevailing in the soil solution. Growth in biofilms is further discussed in Chapters 4 and 9. We know, for example, that surface attachment can extend the pH range of chemoautotrophic nitrifiers in the soil, and we also know that protons can only diffuse slowly through extracellular polysaccharidic material that binds cells in microbial biofilms. Perhaps for these reasons, and because many processes in soil are carried out by microorganisms with different pH requirements and/or tolerance, the soil microbiologist should be wary about any considerations of pH exclusion of particular processes in soil. Again, using nitrification as an example, for a long time researchers believed that nitrification would not occur in soils with pH below 5. This assumption came from *in vitro* physiological evidence of the pH requirements of the relevant nitrifying bacteria, but it did not take into account the properties of surface-colonizing populations or of the heterotrophic nitrifiers, including certain fungi, which have wider pH ranges than their autotrophic counterparts (Killham 1994).

1.4 CONCLUDING REMARKS

This chapter examined how soil presents a dynamic and varied habitat, enabling the microorganisms present to interact with each other as well as with plants, animals, and the soil (in its solid, liquid, and gaseous phases) itself. Although, by necessity, key physical and chemical properties of soil have been considered separately, many of these properties are highly interactive (e.g., temperature and water), and they act in concert to determine the nature of the soil habitat and, consequently, the nature of the soil–organism interactions.

We further emphasize that variation in soil properties operates at an impressive range of scales. This is a vital issue, as variation down to the micrometer scale is relevant to organisms with micrometer dimensions, such as bacteria and many fungi. For example, variations in redox potential across the microporous structure of a soil aggregate ensure that facultatively anaerobic denitrifying bacteria function less than a millimeter away from obligately aerobic nitrifying bacteria that occur nearer the aggregate surface. The nitrate used in the anaerobic microsites of denitrification is coming from the aerobic zone of nitrification. Hence, both processes are able to operate simultaneously in close vicinity despite their mutually exclusive requirements.

The effect of soil on microbes should not be considered a "one-way street." To the contrary, the activity of microorganisms is continually shaping the soil habitat itself. For example, much of the polysaccharides and filamentous materials that hold soil particles together is of microbial origin. Also, microbial activity itself can strongly modify soil pH in microsites. In particular, in the rhizosphere, these are crucial features affecting the life of the microbial inhabitants (see Chapter 10).

REFERENCES

Bird, N.R.A. and E.M.A. Perrier. 2003. The pore-solid fractal model of soil density scaling. *Eur J Soil Sci* 54, 467–476.

Broughton, L.C. and K.L. Gross. 2000. Patterns of diversity in plant and soil microbial communities along a productivity gradient in a Michigan old-field. *Oecologia* 125, 420–427.

Cha, C., Gao, P., Chen, Y.-C., Shaw, P.D. and S.K. Farrand. 1998. Production of acyl-homoserine lactone quorum-sensing signals by Gram-negative plant-associated bacteria. *Molec Plant Microb Interact* 11, 1119–1129.

Elasri, M., Delorme, S., Lemanceau, P., et al. 2001. Acyl-homoserine lactone production is more common among plant-associated *Pseudomonas* spp. than among soil-borne *Pseudomonas* spp. *Appl Environ Microbiol* 67, 1198–1209.

Gammack, S.M., Patterson, E., Kemp, J.S., Cresser, M.S. and K. Killham. 1991. Factors affecting the movement of microorganisms in soils. In: *Soil Biochemistry*, (eds.) J.M. Bollag and G. Stotzky, vol. 7, Marcel Dekker, New York, pp. 263–305.

Greenwood, D.J. 1975. Soil physical conditions and crop production. *MAFF Bull* 29, 261–272.
Killham, K. 1994. *Soil Ecology*, Cambridge University Press, Cambridge.
Lambers, H. 1987. Growth, respiration, exudation in symbiotic associations: the fate of carbon translocated to the roots. In: *Root Development and Function*, (eds.) P.J. Gregory, J.V. Lake and D. Rose, Cambridge University Press, Cambridge, pp. 125–146.
Marshall, T.J. and J.W. Holmes 1996. *Soil Physics*, 2nd edition, Cambridge University Press, Cambridge, pp. 49–53.
Meharg, A.A. and K. Killham. 1989. Distribution of assimilated carbon within the plant and rhizosphere of *Lolium perenne*: influence of temperature. *Soil Biol Biochem* 21, 487–489.
Millan, H., Gonzalez-Posada, M., Aguilar, M., Dominguez, J. and L. Cespedes. 2003. On the fractal scaling of soil data. Particle-size distributions. *Geoderma* 117, 117–128.
Murr, L.E., Torma, A.E., Brierley, J.A. (eds.). 1978. *Metallurgical Applications of Bacterial Leaching and Related Microbiological Phenomena*, Academic Press, New York.
Paul, E.A. and F.E. Clark. 1996. *Soil Microbiology and Biochemistry*, Academic Press, New York.
Perrier, E.M.A. and N.R.A. Bird. 2002. Modelling soil fragmentation: the pore solid fractal approach. *Soil Tillage Res* 64, 91–99.
Pierson, E.A., Wood, D.W., Cannon, J.A., Blachere, F.M. and L.S. Pierson III. 1998. Interpopulation signaling via N-acyl-homoserine lactones among bacteria in the wheat rhizosphere. *Molec Plant Microb Interact* 11, 1078–1084.
Rousk, J., Baath, E., Brookes, P.C. et al. 2010. Soil bacterial and fungal communities across a pH gradient in an arable soil. *ISME J* 4, 1340–1351.
Russell, E.W. 1973. *Soil Conditions and Plant Growth*, 10th edition, Longman, London.
Smith, K. and C.E. Mullins (eds.). 2000. *Soil Analysis: Physical Methods*, 2nd edition, Marcel Dekker, London.
Standing, D. and K. Killham. 2007. The soil environment. In: *Modern Soil Microbiology II*, (eds.) J.D. van Elsas, J.K. Jansson and J.T. Trevors, CRC Press, New York, pp. 1–22.
Vargas, R. and T. Hattori. 1990. The distribution of protozoa among soil aggregates. *FEMS Microbiol Ecol* 74, 73–78.
White, D., Fitzpatrick, E.A. and K. Killham. 1994. Use of resin impregnated soil sections to assess spatial location of bacterial inocula introduced into soil. *Geoderma* 63, 245–254.
Whiteley, M., Diggle, S.P. and P. Greenberg. 2017. Progress in and promise of bacterial quorum sensing research. *Nature* 551, 313–320.
Yang, P. and J.D. van Elsas. 2018. Mechanisms and ecological implications of the movement of bacteria in soil. *Appl Soil Ecol* 129, 112–120.

2 The Seven Grand Questions on Soil Microbiology (Selman A. Waksman, Reexamined by Arthur D. McLaren)

Jan Dirk van Elsas
University of Groningen

Paolo Nannipieri
University of Firenze

CONTENTS

2.1 Introduction 21
2.2 Developments Examined by McLaren 22
2.3 What Organisms Are Active under Field Conditions and in What Ways? 24
2.3.1 Intracellular Activities in Soil Microbiomes 24
2.3.2 Activities of Extracellular Proteins in Soil 25
2.4 What Associative and Antagonistic Influences Exist among Soil Microflora and Fauna? 25
2.5 What Relationships Exist between SOM Transformations and Soil Fertility? 26
2.5.1 The Importance of SOM 26
2.5.2 The Concept of Soil Quality 27
2.6 What Is the Meaning and Significance of the Energy Balance in Soil, in Particular with Reference to C and N? 27
2.6.1 The Processes That Determine the Role of C in Soil 27
2.6.2 The Processes That Determine the Role of N in Soil 28
2.6.3 The Flow of Energy in Soil 28
2.7 How Do Cultivated Plants Influence Soil Transformations? 29
2.8 How Can One Modify Soil Populations and to What Ends? 29
2.9 What Interrelationships Exist between Physicochemical Conditions in Soil and Microbial Activities? 30
2.10 Conclusions and Perspectives 31
References 33

2.1 INTRODUCTION

Microbiology has flourished ever since the discovery and description of microorganisms as agents—invisible to the naked eye—that are ubiquitous and have key capacities of transforming matter and causing disease. The microbiology of soil has been a key area of study, as microbes drive most of the relevant biogeochemical processes that occur in soil, and immense diversities of microbial functions are present. World-famous basic microbiology courses, first given at Delft University of Technology (Beijerinck and disciples), and later taken to and developed in the United States by

Stanier, Phaff, and colleagues, have taken advantage of the immense microbial functional diversity that can be found in most soils. Thus, a huge diversity of microbial functions has been uncovered by the use of enrichments from soil, in which conditions were tuned to yield the targeted functional microbes. In this way, a plethora of nitrogen fixers, ammonia oxidizers, nitrate reducers, sulfate reducers, iron reducers, next to fermenters, and straightforward aerobic organic carbon degraders could be isolated from soil. Hence, soil has been depicted, for a long time, as an extremely rich source of functionally diverse microorganisms.

However, what has remained relatively unknown in the initial stages of the development of knowledge in soil microbiology was how the uncovered microbial processes actually "worked" *in situ* in the soil. Thus, although the detection of a functional activity following an enrichment may be taken as evidence for the contention that such a process is actually operational in soil, information about its scale, extent, and timing has been lacking. In his seminal book entitled *Principle of Soil Microbiology* (Williams & Wilkins, Baltimore, 1927), Selman A. Waksman described the seven grand outstanding questions about the living soil (reworded after McLaren 1977), which are as follows:

1. What organisms are active under field conditions and in what ways?
2. What associative and antagonistic influences exist among soil microflora and fauna?
3. What relationships exist between soil organic matter (SOM) transformations and soil fertility?
4. What is the meaning and significance of energy balance in soil, in particular with reference to C and N?
5. How do cultivated plants influence soil transformations?
6. How can one modify soil populations and to what ends?
7. What interrelationships exist between physicochemical conditions in soil and microbial activities?

As we will learn in this book, the living soil stands out as the most organism- and function-diverse habitat that exists on this planet. Moreover, the spatial and temporal intricacies encountered in most soils pose specific challenges to our understanding of the local processes that take place (see Chapter 1). Hence, when reconsidering the aforementioned seven grand questions that were posed almost 100 years ago, one may safely state that most of these still constitute the challenges in soil microbiology research of today. In an editorial in *Soil Biology and Biochemistry* written 50 years after Selman Waksman published his seven grand questions, Arthur Douglas McLaren (1977), the "father" of soil biochemistry, reexamined these questions. In this chapter, we discuss to what extent progress has been made with respect to each of the Waksman questions, as reexamined by McLaren (Figure 2.1).

2.2 DEVELOPMENTS EXAMINED BY McLAREN

McLaren (1977) briefly summarized the advances in the body of knowledge on the seven questions that emerged since the publication of the questions by Selman Waksman. These advances concerned the interactions between microbes and plants in soil, the distribution of microorganisms in the soil matrix, the ways to quantify nutrient transformations in soil (by using labeled compounds), and the possibility to set up quantitative mathematical models simulating nutrient dynamics in the soil–plant system. Moreover, the discovery of beneficial bacteria stimulating plant growth by producing phytohormones, the preferential growth of bacteria on some parts of the roots, the role of bacteria in the biological control of plant disease, and the role of mycorrhizae and other beneficial microorganisms in plant nutrition were reviewed. The first experiment on nutrient dynamics with labeled compounds was carried out by Norman and Werkman (1943), who showed that 20% of ^{15}N-enriched soybean residues is taken up by plant roots, whereas most of the remainder is

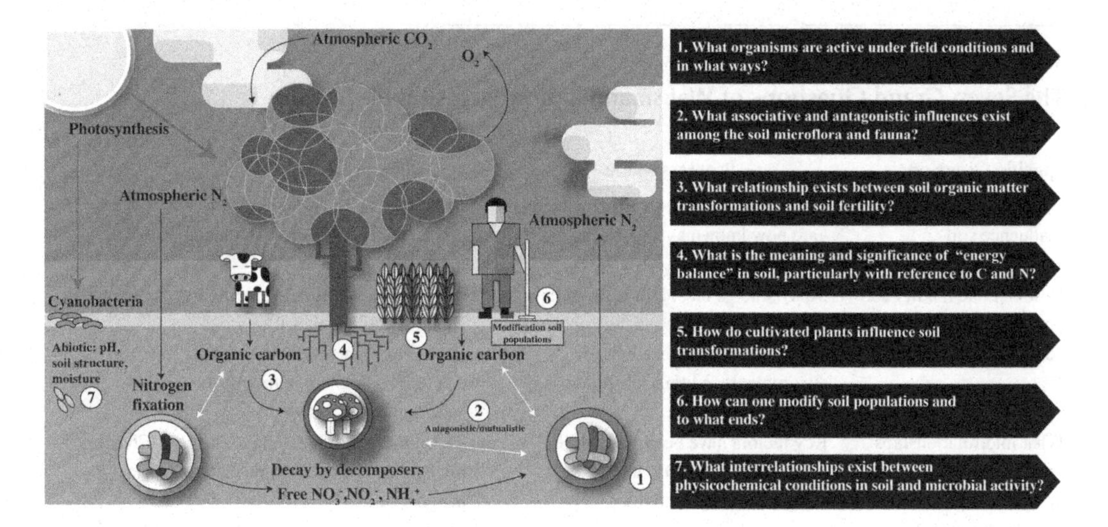

FIGURE 2.1 Depiction of the seven grand questions of Selman A. Waksman in soil (microbiome) functioning.

incorporated into the SOM. Later, Jenkinson and Powlson (1976) set up a method for determining the size of the soil microbiome, which enables to trace the dynamics of nutrients through soil microbial communities. This constituted an important advance in the quantification of nutrient behavior in the soil–plant system and in improving models simulating this dynamics (Table 2.1).

Major developments have taken place, with respect to all seven questions, in the decades that passed since these early achievements. Most of the developments have been driven by strong methodological advances in soil microbiology, as testified in various other chapters in this book (see, e.g., Chapters 12–19). Here, we briefly revisit each of the seven grand questions on our current understanding of the living soil.

TABLE 2.1
The Seven Grand Questions of Waksmann and Progress Made To Date

Specific Question	Progress Made	Future Prospects	References/ Remarks
What organisms are active under field conditions and in what ways?	Novel methods have slowly allowed overall and targeted analyses of activity	Soil activity is site- and condition-specific, and so targeted transcriptomics/proteomics approaches are required	Chapters 15 and 16
What associative and antagonistic influences exist among soil microflora and fauna?	Key rhizosphere inhabitants and their potential roles are understood. The complexity of the rhizosphere still poses a big hurdle	Targeted studies using selected novel methods, including meta-omics, are needed	Chapters 9, 21, and 22
What relationships exist between SOM transformations and soil fertility?	SOM is a key determinant of soil fertility. Soil fertility is now also denominated soil quality	Increasing the SOM content to counteract the release of CO_2 to the atmosphere	Garcia et al (2018); Chapter 20
What is the meaning and significance of energy balance in soil, in particular with reference to C and N?	Growth of heterotrophic microorganisms decreases when available C and/or N are low	To distinguish soil microenvironments according to their C/N availability	Schimel and Bennett (2004)

(Continued)

TABLE 2.1 (*Continued*)
The Seven Grand Questions of Waksmann and Progress Made To Date

Specific Question	Progress Made	Future Prospects	References/ Remarks
How do cultivated plants influence soil transformations?	Rhizospheres and mycorrhizospheres are now known as major "changers" of local activities	To simulate interactions between plants, microorganisms, and fauna in microcosms	Chapter 10
How can one modify soil populations and to what ends?	Many studies on soil inoculants have revealed the microbiostatic character of soil. Soil management, for example, by adding substrates, is a promising avenue	Future endeavors should attempt to create specific niches for (incoming) beneficials in soil, for example, by adding nutrients or specific microhabitats	Chapter 22
What interrelationships exist between physicochemical conditions in soil and microbial activities?	Key factors have been studied. Soil water, pH, and SOM are the key determinants of activities	Determination of these effects at the microscale level will allow better insights	Chapters 1 and 2

2.3 WHAT ORGANISMS ARE ACTIVE UNDER FIELD CONDITIONS AND IN WHAT WAYS?

2.3.1 Intracellular Activities in Soil Microbiomes

Questions about microbial activities in soil are broad and needs to be defined in the context of the research aims as well as the specific conditions. For instance, the question may pertain to the activity of specific ammonia oxidizers in soil under aerobic versus anaerobic conditions. Alternatively, it may be broader, addressing the aerobic degradation of lignocellulose material in the soil. Answers to each of these issues will require particular approaches and techniques, which are often molecularly based (see Chapters 12–17). However, there are still unresolved methodological problems that hamper progress in this area. Current advanced techniques to study soil microbiomes are based on analyses of soil DNA, RNA, and/or proteins. In addition, methods that allow to track and trace the fate of marker atoms (in molecules), for example, ^{13}C, ^{14}C, and ^{15}N, have enabled to reveal the efficiencies of key processes in soil, such as photosynthesis and translocation of photosynthate to belowground organisms (Drigo et al. 2009), organic carbon breakdown, nitrogen fixation, ammonia oxidation, and nitrate reduction. A key method here is stable isotope probing, in which the incorporation of stable label, for example, ^{13}C, into biomolecules (often DNA), is measured. See Chapter 17 for details. A combination of the aforementioned methods is recommended, as each offers a different perspective on (potential) function, with discrepancies being related to molecule stability (and thus timing of events). Also, the classical methods have shown puzzling results. For example, transmission electron microscopy of soil sections has shown that bacterial cells are mainly present in connection to plant residues in soil, often occurring in the interior of these debris (Foster et al. 1983). Supportive of the key presence of microbial cells within such refuges was the fact that chloroform fumigation primarily lyses cells of bacteria present between aggregates but not those of bacteria inside aggregates (Ladd et al. 1996). Unfortunately, the application of ultra-cytochemical activity tests did not detect active enzymes (and by inference active microbes) in soil due to the presence of cross-reacting electron-dense soil minerals and/or SOM (Ladd et al. 1996).

Research on the activity of microbes in soil also needs to consider the spatial and temporal aspects of soil, as these continuously influence the local cellular activities. It is possible that,

TABLE 2.2
Methodological Progress That Has Driven the Advancement in Our Knowledge about the Seven Grand Questions of Selman A. Waksman

Advanced Method	What Has It Brought Us?	Remarks/ References
Use of tracers/label for compounds	Broad view of fate of labeled compound, for example, C and N compounds	Nannipieri and Paul (2009)
Use of reporter genes/biosensors	Very sensitive *in situ* detection of specific microbial activities, including responses to soil conditions	Van Overbeek and van Elsas (1995)
Soil metagenomics	Overall vision of soil phylogenetic and functional diversity	Chapter 14
Soil metatranscriptomics	Specific vision of expressed genes at messenger RNA (mRNA) level	Chapter 15
Soil metaproteomics	Specifically expressed proteins	Chapter 16
Stable isotope probing	Enables monitoring an actual process in soil, for example, methane/methanol oxidation	Radajewski et al. (2002), Chapter 17

given the local nature of "conditions," a microbial population spread through the soil can show high activity levels at particular sites and/or times, while being virtually silent at others. Activity measurements, either considering output parameters such as released CO_2 or fixed N_2, or directly measuring expressed genes, will inevitably yield data that are averaged over a certain volume of soil. Such data do not precisely reflect the underlying plethora of local activities (Table 2.2).

2.3.2 Activities of Extracellular Proteins in Soil

A major characteristic of soil as a biological system is the capacity of surface-reactive soil particles to adsorb key biological molecules, such as proteins (and thus enzymes) and nucleic acids (Nannipieri et al. 2003). McLaren and Peterson (1967) reported that the adsorption of enzymes to the surface-reactive particles in soil protects them against proteolysis without activity losses. As discussed in Chapter 16, the extracellular protein- and peptide-N, stabilized by surface-reactive soil particles or entrapped in soil aggregates, may make up 30%–50% of the soil organic N. This is higher than the organic N (mostly protein-N) that is intracellularly localized (4% on average) in the soil matrix. The bibliography on stabilized extracellular proteins in soil is extensive (Nannipieri et al. 2012) and exceeds that related to extracellular DNA in soil (Pietramellara et al. 2009). Generally, a focus has been placed on extracellular stabilized hydrolases (such as urease, proteases, and phosphomonoesterases), and it was found that their stabilization depends on the adsorption onto surface-reactive particles (such as clays) or entrapment in the organo-mineral complexes. These enzymes can persist in soil as long as the aggregates or organo-mineral complexes entrapping them are not broken (Nannipieri et al. 2012).

2.4 WHAT ASSOCIATIVE AND ANTAGONISTIC INFLUENCES EXIST AMONG SOIL MICROFLORA AND FAUNA?

Since the incipience of the concept of microbial (associative and antagonistic) interactions in soil, great progress has been made in this area. This pertains to both the interactions between different members of soil microbiomes and those between these and components of the soil fauna and/or the roots of plants. In classical cultivation-based work, a large range of either antagonistic or synergistic interactions among soil microorganisms has been identified. Chapter 9 discusses some of the interactions that occur between the different soil microbes. It also places a focus on the way microbes occur in the soil (spatially explicit, often in biofilms adhering to surfaces). It is exactly here that the crux with respect to our understanding of the soil microbiome lies. Clearly, the heterogeneous and complex spatial structure that occurs in most soils has a strong bearing on the interactions, which are most often spatially explicit. Thus, it is precisely in this area that most of the progress has been and will continue to be made.

A particular case in point is found in the so-called hot spots for activity in soil, such as the rhizosphere, the mycosphere, and other such "spheres." In these habitats, we see a range of developments of our knowledge. In some cases, microbe–microbe interactions are understood to the finest molecular level, and such data have mostly come forward in studies performed in soil-mimicking systems. For instance, in a study performed on soil extract agar, Haq et al. (2017) found a five-gene cluster that is involved in energy generation (presumably at the expense of fungal-exuded glycerol and oxalate), to be highly upregulated in *Burkholderia* (recently reclassified to *Paraburkholderia*) *terrae* in interaction with the soil fungus *Lyophyllum* sp. strain Karsten. The gene cluster apparently endowed the organism with a "kick-start" machinery that enabled it to immediately respond to emerging fungal exudate. A large suite of other current developments with respect to the mechanisms that underlie interactions in soil also aims to unravel the molecular events that trigger particular associative or antagonistic responses (see Chapter 9). What emerges from the large bibliography on this topic is that, over evolutionary time, soil has acted as a complex and diverse matrix, in which a huge diversity of symbiotic, neutral, and/or antagonistic interactions has emerged, as driven by the Darwinian rules of evolution. However, we still are far from understanding how soil modulates such interactions, and so it is foreseen that, next to studies in soil-mimicking systems, real-world soil studies remain necessary in future research efforts.

2.5 WHAT RELATIONSHIPS EXIST BETWEEN SOM TRANSFORMATIONS AND SOIL FERTILITY?

2.5.1 The Importance of SOM

Already in early work, it became apparent that soil fertility was intrinsically linked to the organic matter in the soil, notwithstanding the existence of some exceptions to this rule. The exception illustrated by a curious finding described by McLaren (1977). They found an allophanic soil to have low fertility despite its high organic C content. A decrease in the mineralization processes (and thus of the release of nutrients that could be taken up by the plant) was found in this soil, and organic compounds turned out to be complexed with allophanes. In current research, the use of labeled organic compounds (see Section 2.3) allows us to determine the quantitative dynamics of nutrients in the soil–plant system. For example, the use of ^{15}N-enriched fertilizer can distinguish between the behavior of fertilizer N and that of soil N, and thus quantify what percentage of plant N uptake is satisfied by the added fertilizer compared to that of the N mineralized from soil organic N. The so-called priming effect, that is, the change in the mineralization rate of SOM due to the addition of an organic compound to soil, can be determined by using C-labeled compounds.

The role of SOM as the basis of soil fertility across many soils has thus been extensively documented. However, human activities over the past decades have significantly changed the regional and global balances of SOM, with effects on climate change. This has often been detrimental to soil fertility (Garcia et al. 2018). Moreover, there is a need to preserve or even enhance SOM in soils given the negative effects of SOM depletion on the climate. Therefore, both agricultural and forest management practices are increasingly aiming to store organic C in soil to counteract the release of CO_2 into the atmosphere (see also Chapter 25). In addition, interactions between SOM and complex human–natural systems require new research into regional and global SOM budgets.

2.5.2 The Concept of Soil Quality

In 1994, Doran and Parkin introduced the concept of soil quality, defined as "the capacity of a soil to function within ecosystem boundaries to sustain biological productivity, maintain environmental quality and promote plant and animal health." This definition is broader than that of soil fertility since it also contains the role of soil in plant and biological productivity and in attenuating the impact of environmental contaminants and pathogens. In addition, this definition addresses the importance of soil for plant, animal, and human health. It is important to determine changes in soil quality, since soils are affected by climatic changes as well as human activities, including intensive agricultural practices. Soil quality is related to the biological, chemical, and physical properties of soil, but unfortunately these parameters are interwoven and it is not possible to have an unequivocal evaluation of (changes in) soil quality. In a recent paper (Schloter et al. 2018), the need was emphasized to develop robust, reliable, and resilient molecular indicators for monitoring soil quality. Besides, biological, chemical, and physical indicators have been selected. However, the chemical and physical properties of soil are less sensitive than the biological ones, and this has a bearing on (subsequent) changes in SOM content, which is related to soil quality (Nannipieri et al. 2003). In addition, biological indicators of soil quality have, in the past, mainly focused on the visible part of the soil biota, neglecting the microbiome (Schloter et al. 2018). Hence, the need to develop indicators reflecting soil microbial activity has been stressed, as soil microbiomes play predominant roles in affecting soil functionality (Schloter et al. 2018, Stott et al. 2009). However, of the ten indicators that have been proposed for evaluating soil quality, only microbial biomass C and potential nitrification were based on microbiological parameters (Andrews et al. 2004). Therefore, Schloter et al. (2018) proposed to take into consideration genes encoding proteins that are involved in key microbial processes, such as several steps of the N, C, and P cycles, in soil. Several such genes were proposed for development into indicators and others can be derived from the extensive bibliography on molecularly based soil analyses.

2.6 WHAT IS THE MEANING AND SIGNIFICANCE OF THE ENERGY BALANCE IN SOIL, IN PARTICULAR WITH REFERENCE TO C AND N?

2.6.1 The Processes That Determine the Role of C in Soil

Organic C plays a key role in soil functionality, as it affects the biological, chemical, and physical properties of soil. The biological properties include soil biodiversity. Soil has a source/sink role in terms of carbon, so it can act as either a net carbon source or sink. As briefly discussed in chapter 1, carbon stably localized in the soil—in the SOM—is often present in highly complex macromolecules, whereas labile C encompasses smaller molecules that are more prone to utilization and conversion into biomass and CO_2. Current aims for soil (microbiome) management practices are to foster the organic C storage function of soil in order to counteract an increase in atmospheric CO_2 levels (Garcia et al. 2018). This implies that CO_2 fixation (autotrophic) processes are to be stimulated at the expense of C "burning" (heterotrophic) processes. In soil, the former processes are mainly carried out by plants, algae, and a range of autotrophic bacteria, including cyanobacteria.

The carbon-use efficiency by the soil microbiota has often been studied by a "holistic" approach, considering the soil microbiome as an entire pool (microbial biomass C) and the C respired as CO_2. Anderson and Domsch (1990) proposed the qCO_2 (the ratio between C–CO_2 and microbial biomass C) as an index to evaluate the amount of organic C that is incorporated into microbial biomass. High qCO_2 values were posed to be indicative of microbial stress, since under stress conditions, a higher amount of C is respired to repair metabolic damages due to the stress. However, the ratio also depends on factors other than stresses. For example, changes in the fungal/bacterial biomasses can affect the ratio, since fungi respire less C than bacteria per unit incorporated C (Wardle and Ghani 1995).

2.6.2 The Processes That Determine the Role of N in Soil

The second major element of importance to soil microbiomes is N. Soil N is subjected to cycling across the different chemical forms, as further outlined in Chapter 11. The N cycle in soil is characterized by many reactions and losses. Biological N fixation, nitrification, denitrification, N mineralization and immobilization/turnover, ammonia volatilization, N leaching, and N loss by runoff are processes that have been known even before the time of Waksman and later McLaren. In addition, even more cycling steps have been discovered later on. For example, N_2O can be produced not only through denitrification but also by nitrifier denitrification (Nannipieri and Paul 2009). Anaerobic ammonia oxidation (Anammox), the oxidation of ammonia to N_2 gas, with concomitant reduction of nitrite, can occur under anaerobic conditions in soil (Nannipieri and Paul 2009). The oxidation of ammonia to nitrite, the first step of the nitrification process, is carried out not only by ammonia-oxidizing bacteria but also by ammonia-oxidizing archaea. Also, a novel finding indicates the existence of a complete nitrification pathway in drinking water purification sand beds in one organism, *Nitrospira*, instead of in the classical two-species process carried out by *Nitrosomonas* (oxidation of ammonia to nitrite) and *Nitrosospira* (oxidation of nitrite to nitrate) (Daims et al. 2015). It is an open question to what extent the two-species consortium outcompetes, and thus prevails over, the single-organism system in soil settings. Finally, ammonia-oxidizing bacteria such as *Nitrosococcus* and *Nitrosospira* species also contain genes catalyzing the enzymes of the denitrification process (Norton et al. 2008). The N cycling processes are further detailed in Chapter 11.

Generally, if N is the limiting nutrient, the degradation of organic C will increase as it adds inorganic N to the soil. Similarly, if P and S are the limiting nutrients in soil, the degradation rate will increase if inorganic P and/or S compounds are added to soil, respectively. However, this is not always observed, since the quality of the soil organic C is also an important determinant. For example, the degradation of lignin decreases with high N availability, since ligninases that are responsible for lignin degradation are inhibited by inorganic N (Knorr et al. 2005). According to the ecological stoichiometry theory, elemental ratios in soil determine nutrient retention and biomass production (Sterner and Elser 2002). As a consequence, the elemental stoichiometry of soil microbial biomass (with a mean C:N:P ratio of 60:7:1) will determine the microbial C, N, and P demand in relation to the availability of these elements in soil (Sinsabaugh et al. 2009). The mean ratio of activities of four enzymes [β-1,4-glucosidase, β-1,4-*N*-acetylglucosaminidase, leucine aminopeptidase, and acid or alkaline phosphomonoesterase (the latter measured depending on soil pH)], involved in C, N, and P dynamics, was about 1:1:1 in all investigated soil and sediment samples (Sinsabaugh et al. 2009). However, whereas β-1,4-glucosidase and both phosphomonoesterase activities are proxies of organic C and organic P mineralization, respectively, there is no evidence that both β-1,4-*N*-acetylglucosaminidase and leucine aminopeptidase activities can serve as proxies of organic N mineralization (Nannipieri et al. 2018).

2.6.3 The Flow of Energy in Soil

Regarding the flow of energy in soil, Gray and Williams (1971) already suggested that most of the energy inputs into soil (from solar energy captured by plants, algae, and cyanobacteria) will serve to maintain microbiomes rather than allowing much microbial growth. This has resulted in the concept of extended microbial generation times in most soils. In other words, the (slow) building of new microbial biomass is offset by the gradual death of "old" parts of the biomass, and this results in a somewhat dynamic equilibrium. Both soil conditions and the rate of energy input would thus determine the biomass dynamics, resulting in the concept of a "carrying capacity," which is characteristic for each soil. The underlying average low growth rates in most soil microbiomes indicate that microbial transformations would be also slow, on average. However, these calculations have been based on microbial biomass as a whole, without considering that soils show hot spots (sites with enhanced microbial densities and activities) and "hot moments" that are characterized by the (sudden and ephemeral) presence of available substrates for microbial growth (Kuzyakov and Blagodatskaya 2017). The calculation of microbial growth rates in soil inside and outside of the hot spots (sites with enhanced microbial densities and activities) and/or hot moments (sudden and ephemeral) is still a challenging research aim, also considering that soil microorganisms can form resting structures under starvation conditions.

2.7 HOW DO CULTIVATED PLANTS INFLUENCE SOIL TRANSFORMATIONS?

It is well established that, in most soils, the soil under the influence of the roots (rhizosphere soil) has a larger microbial biomass with higher activity than the bulk soil. This so-called *rhizosphere effect* is due to the release of photosynthates and derived compounds produced by the plant, and to rhizodeposition of compounds from lysing root epithelial cells. Garbeva et al. (2004) examined the relative effects of soil type, plant type, and agricultural management on soil microbiomes. Soil type, as reflected in the soil properties, is often more important than plant species type in affecting the microbial diversity of rhizosphere soil, at least after a few years of monoculture. De Ridder-Duine et al. (2005) showed that *Carex arenaria*, a nonmycorrhizal plant species (chosen to avoid the confounding effect by different mycorrhizal colonization), cultivated in ten soils with different properties, did not affect the local bacterial diversities, as shown by polymerase chain reaction (PCR) followed by denaturing gradient gel electrophoresis. Instead, these depended on soil properties. Inceoğlu et al. (2010) revealed subtle differences in rhizosphere bacteriomes (assessed by DNA sequencing) exerted by the roots of different potato genotypes in the same soil. Hence, the effects are diverse and depend strongly on the system used as well as the methods of analysis. Moreover, these studies did not address the effects on microbial functions. With respect to these, a recent study showed that maize with a higher N-use efficiency (NUE) selected bacterial communities in the rhizosphere soil differing in β-glucosidase-encoding genes and with a higher β-glucosidase activity than maize with lower NUE (Pathan et al. 2015). Moreover, the former line also had higher protease activity and higher abundance of protease-encoding genes in its rhizosphere than the latter. It is important to underline that spatiotemporal effects are important in the rhizosphere, which can be studied under laboratory conditions, using rhizoboxes (Neuman and Romheld 2007). For example, root exudation is mainly confined to apical root zones and is not always constant but follows diurnal rhythms. A metatranscriptomics study showed the activities of many prokaryotic taxa were increased by root exudation from barley plants at the pre-dawn compared to the post-dawn stage (Baraniya et al. 2017). Therefore, it is plausible to hypothesize that, under field conditions, the rhizosphere effect comes in waves in dependency of sunlight, whereas it is lost when soil cropped to monoculture is plowed after harvest.

2.8 HOW CAN ONE MODIFY SOIL POPULATIONS AND TO WHAT ENDS?

The importance of soil type, reflected in soil properties, as a driver of the composition of soil microbiomes, was first shown, across a suite of soils, by Gelsomino et al. (1999) and later confirmed by several other studies. For instance, Delmont et al. (2014) recently studied the microbial diversity of two soils (one from a grassland of the long-term experimental site at the Rothamsted Experimental Station in Harpenden, England, and the other one from a forest in Vallombrosa, Firenze, Italy) treated to have either the same microbial community or that from the other soil. The study showed that the functional and taxonomic composition of the respective microbiomes was that of the soil receiving the inoculum and not that of the "donor" soil. Hence, soils may be rather "recalcitrant" to microbiome changes, with a presumed large effect of the soil matrix. Presumably, the type of soil matrix determines, to a large extent, the distribution of nutrients, water, and other "conditions" over the soil aggregates and pores. It is therefore the key driver of the distribution of habitable niche space for adapted microbial forms, resulting in the apparent recalcitrance of soil microbiomes to change.

However, for many purposes, such as the biological control of plant diseases, fertility enhancement, and/or bioremediation, it may be desirable to modify soil microbiomes, as already suggested by Waksman (1927). The aim would be to establish novel organisms that are beneficial to soil function, in terms of either fertility or plant health stimulation, or to promote the degradation of, for instance, polluting compounds. Such aims have been on the research agenda for several decades. These were also the main themes in the early research of Waksman (1927).

One of the key approaches taken to modify soil microbiomes has been the introduction of (micro) organisms with the desired function, whereas another approach has relied on the modulation of soil parameters, for instance, by the addition of "substrates" such as manure, biochar, and/or crop residues, with the aim to promote the occurrence as well as activity of naturally present beneficial organisms.

Although there are accounts of successful introductions, for instance, of rhizobia that support soil nitrogen acquisition from the atmosphere, major obstacles have been identified. These lie in the recalcitrance of soil to change following organism introductions. As discussed in the foregoing, the reason for this "recalcitrance" may to a large extent lie in the lack of an open niche for the incoming organisms (van Veen et al. 1997, Mallon et al. 2015), a phenomenon denoted as *soil microbiostasis*. Although we have greatly improved understanding of how plant-beneficial organisms may work (see Chapter 22), the success of introduction of many of such organisms is not guaranteed as a result of the lack of knowledge as to how to give them an "open niche" in the soil.

With respect to the second option of modulating the soil microbiome by soil amendments, the aforementioned review by Garbeva et al. (2004) is important, as it discusses the drivers of soil microbiome composition, which are as follows:

1. Plant driven
2. Soil type driven
3. Soil management driven

All the three groups of complex drivers are deemed to be important, and with respect to their effects on soil microbiomes, they were split up into several subfactors: types and levels of nutrients/organic matter (e.g., from root exudates, soil reshuffling by, for instance, plowing), pH level (e.g., as affected by soil processes), water regime (e.g., affected by plant water suction, agricultural regime, soil type), and other factors (e.g., biotic stresses exerted by predators). Although, as we have seen, soil microbiomes are often highly recalcitrant to change, they may be shifted, albeit slowly, when any of such soil parameters is changed/modulated. It appears the changes that can be effectuated are highly context-dependent (soil and climate type), and therefore, research is needed on a case-by-case basis, in order to develop the best practices for changing the activity, biomass, and composition of soil microbiomes in the planned direction.

2.9 WHAT INTERRELATIONSHIPS EXIST BETWEEN PHYSICOCHEMICAL CONDITIONS IN SOIL AND MICROBIAL ACTIVITIES?

It is currently known that, among other drivers, particular abiotic soil conditions drive the structure and activity of soil microbiomes. Here, it is important to discern the extent to which the different drivers operate, and to what extent they affect the different members of soil microbiomes. As a case in point, it has been amply shown, across a wide range of soils, that soil pH is an important driver of the structure of soil bacterial communities, with pH dropping to values below 5.0 being strongly selective (Rousk et al. 2010). See also Chapter 1 and Box 1.1. This finding is easily connected to our understanding of the behavior of bacteria under changing pH; a pH value below 5.0 is strongly prohibitive for most bacteria and allows a distinction of acid-sensitive from acidophilic types. On the other hand, soil fungi are less affected by changes in soil pH. For example, particular fungi, such as *Penicillium* and *Fusarium* spp., can grow at pH values of up to 13, whereas other fungi prefer slightly acidic conditions (around pH 5.2–5.5) (Thorn 2012; see Chapter 5). Fungal activity in the soil can thus be stimulated by the presence of (acidic) organic compounds of plant and animal origin. Their role in degrading cellulose, hemicellulose, lignin, and chitin is discussed further in Chapter 5.

As outlined in Chapter 1, the soil water regime is a strong regulator of soil microbial life. Soil water is obviously dependent on weather (rain versus drying) conditions, as well as soil drainage characteristics. Soils with poor drainage characteristics (e.g., clayey soils), when exposed to frequent heavy rainfall, exhibit prevailing conditions of water saturation, which incite anoxia. This stands in strong contrast to easily drained soils (e.g., sandy soils), which may turn out to be largely oxic, on average. However, the latter regime may imply that periods of drought also occur, which may be extensive. Under such dry conditions, the activities of local microbiomes will be virtually stalled. Thus, the consequences of a soil's prevailing water regime, resulting in water levels varying between 0% and 100% of the water-holding capacity, and thus oxia versus anoxia, are large. The respective microbiomes will need to shift their metabolisms from virtually stalled (period of drought) to aerobic respiration (oxia in most soil pores) to anaerobic respiration (using alternative terminal electron acceptors, see Chapter 1) or fermentation (anoxia in most pores).

Finally, agricultural management regime is an important determinant of soil microbiome dynamics, as local conditions in the soil can be heavily affected by measures such as plowing, tilling, crop rotation and fertilizer and animal waste application. The bibliography on the effects of different agricultural management regimes, causing different environmental effects, on soil microbial diversity is growing, but the respective findings should only be reviewed when data of different soils and agricultural managements are obtained. One key facet, discussed in many scientific fora, is the tendency of microbial diversity to become "eroded" with progressively enhanced intensity of (agricultural) land use. This phenomenon apparently has implications for important soil characteristics, for example, the suppressiveness to plant diseases (further discussed in Chapter 21).

2.10 CONCLUSIONS AND PERSPECTIVES

After about 100 years of research in soil microbiology, we can divide the scientific developments into several parts, i.e., the period from the early microbiology era (in which Winogradsky, Beijerinck and Waksman played pivotal roles), via the era typified by McLaren (soil biochemistry and processes) to the period up to today. Figure 2.2. provides a graphical depiction of the development of soil microbiology, highlighting some important findings. Considering this figure, it comes to our minds to ask : "where do we stand now with respect to understanding and fostering the functionality of our soils, and what are the decadal aims for the future?" The availability of a sufficient number of high-quality and highly fertile soils is at the basis of the sustenance of life on this planet. The key to survival of humans under the effects of a fast-growing population and the current climate change lies in our capabilities to preserve and/or improve our precious living soils in terms of their functioning and their quality. In particular, mitigation of the rising atmospheric CO_2 levels requires increased attention to the potential to enhance the inherent capacity of many soils to act as effective carbon sinks.

With respect to understanding the living soil and its functions, one may safely posit that enormous data sets have been generated that document the capacities of diverse soil organisms with respect to the transformation of key biochemical compounds in the soil, the interactions with other organisms, and the modulation of soil quality and fertility. However, we are still far from a complete understanding of the functioning of the living soil system. It is important to understand that this function should be considered in the context of the particular system studied (context dependency). In terms of the fate of key compounds, we can quantify the dynamics of at least some nutrients, by using a holistic approach and with the help of labeled compounds. For example, in the case of N, the

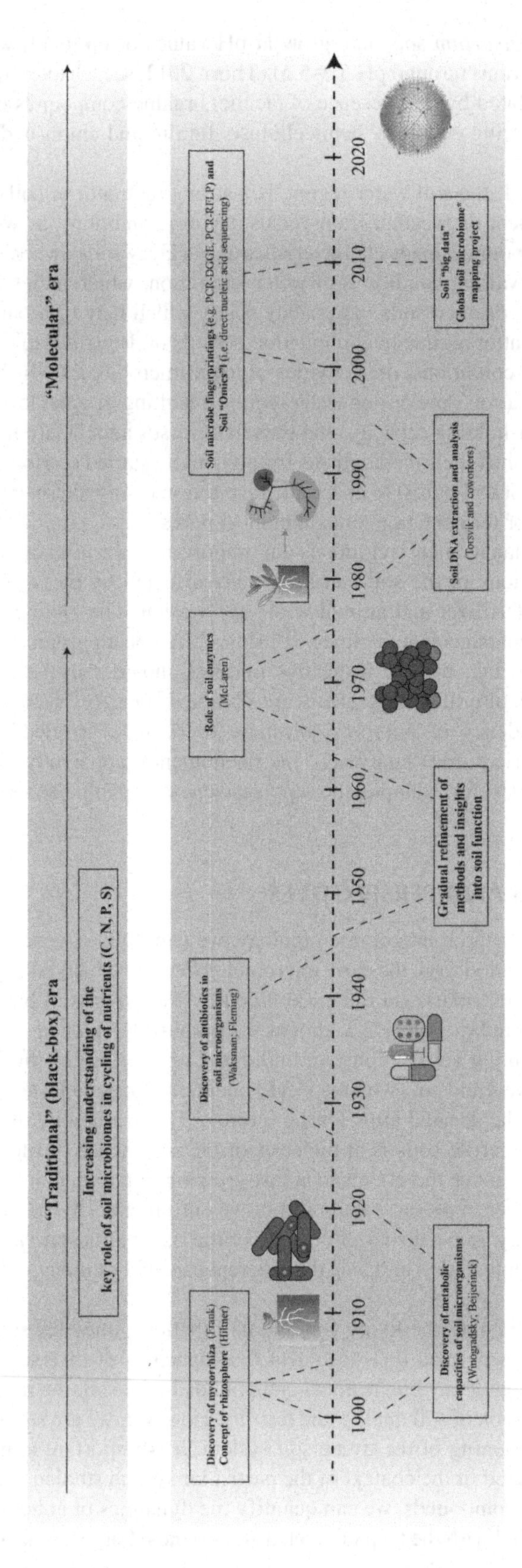

FIGURE 2.2 Time line depicting major developments and key scientists in soil microbiology over about 120 years. *Discussed further in Chapter 3.

main pools in soil are microbial biomass N, organic N, exchangeable and fixed (or non-exchangeable) ammonium-N, nitrate-N, and N taken up by plants. In one study, ammonium accounted for almost 25% of ^{15}N-enriched urea applied to a sorghum field, whereas the fertilizer taken up by the sorghum accounted for 21% and that immobilized into microbial biomass N was about 5% at harvest time (Nannipieri et al. 1999).

However, our analytical methods are still far from optimal and are not yet sufficiently fine-tuned for the analysis of a wide range of soil types. This, in combination with the enormous diversity of organisms in most soils, and the highly intricate network of interactions that can occur, in a spatiotemporally explicit manner, across these soils, prevents us from being all-encompassing in all aspects of soil microbiome analyses. As we learn in Chapters 1 and 3, even within a single soil the level of homogeneity may be low. Most soils in the world consist of a collection of "islands" for microbial life and function, where the ecological rules that determine soil function may all have to be locally determined.

How can we move forward in our analysis and exploration of soil functions? Given the immense diversity and variation across soils and their microbiomes, we advocate that future studies should (1) be performed on a case-by-case basis and (2) consider the local heterogeneity within the soil compartment that is being analyzed. In other words, each soil system needs to be considered as an intricate system in itself, taking into account its own intricacies and legacies. It may be too narrow to just compare experimental outcomes in one soil to those in other soils. Rather, a deeper understanding of each soil system must come from repeated studies of the same system, using variations of conditions and sampling scale, and including replicated time series of development. The proposal to define the Rothamsted soil—where over 100 years of historical study provide an excellent baseline—as a "model" soil for exploration by meta-omics techniques is a good example of such an approach (Vogel et al. 2009, Delmont et al. 2012). However, in the first analyses that were performed, only 34.5% of the metagenomics reads were assigned to function, with only <1% of the annotated sequences fitting already-sequenced genomes at 96% similarity. This indicates that most soil microbiomes are still understudied. Therefore, further improvements in technology and further imaginative research efforts are needed to bring sufficient progress with respect to our understanding of the microbiome and its diversity and activity in soil.

REFERENCES

Anderson, T.H. and K.H. Domsch. 1990. Application of eco-physiological quotients (qCO_2 and qD) on microbial biomass from soils of different cropping histories. *Soil Biol Biochem* 20, 107–114.

Andrews, S., Karlen, D.L. and C.A. Cambardella. 2004. The soil management assessment framework: a quantitative soil quality evaluation method. *Soil Sci Soc Am J* 68, 1945–1962.

Baraniya, D., Nannipieri, P., Kublik, S. et al. 2017. The impact of the diurnal cycle on the microbial transcriptome in the rhizosphere of barley. *Microb Ecol* 75, 830–833.

Daims, H., Lepedeva, E.V., Pjevac, P. et al. 2015. Complete nitrification in *Nitrospira* bacteria. *Nature* 528, 504–509.

De Ridder-Duine, A.S., Kowalchuk, G.A., Klein Gunnewick, P.J.A. et al. 2005. Rhizosphere bacterial community composition in natural stands of *Carex arenaria* (sand Sedge) is determined by bulk soil community composition. *Soil Biol Biochem* 37, 349–357.

Delmont, T.O., Prestat, E., Keegan, K.P. et al. 2012. Structure, fluctuation, and magnitude of a natural grassland and metagenome. *ISME J* 6, 1677–1687.

Delmont, T.O., Francioli, D., Jacquesson, S. et al. 2014. Microbial community development and unseen diversity recovery in inoculated sterile soil. *Biol Fertil Soils* 50, 1069–1076.

Doran, J.W. and T.B. Parkin. 1994. Defining and assessing soil quality. In: Doran, J.W., Coleman, D.C., Bezdicek, D.F., Stewart, B.A. (Eds). *Defining Soil Quality for a Sustainable Environment.* Soil Science Society of America and American Society of Agronomy, Inc., Madison, WI, pp. 3–21.

Drigo, B., van Veen J.A. and G.A. Kowalchuk. 2009. Specific rhizosphere bacterial and fungal groups respond differently to elevated atmospheric CO_2. *ISME J* 3, 1204–1217.

Foster, R.C., Rovira, A.D. and T.W. Cock. 1983. *Ultrastructure of the Root-Soil Interface*, American Phytopathological Society, St. Paul, MN.

Garbeva, P., van Veen, J.A. and J.D. van Elsas. 2004. Microbial diversity in soil—Selection of microbial populations by plant and soil type and implications for disease suppression. *Ann Rev Phytopathol* 42, 243–270.

Garcia, C., Nannipieri, P. and T. Hernandez. 2018. *The Future of Soil Carbon: Its Conservation and Formation*, Elsevier, Cambridge, MA.

Gelsomino, A., Keijzer-Wolters, A.C., Cacco G. et al. 1999. Assessment of bacterial community structure in soil by polymerase chain reaction and denaturing gradient gel electrophoresis. *J Microbiol Methods* 38, 1–15.

Gray, T.R.G. and S.T. Williams. 1971. Microbial productivity in soil. *Symp Soc Gen Microbiol* 21, 255–286.

Haq, I.U., Dini-Andreote, F. and J.D. van Elsas. 2017. Transcriptional responses of the bacterium *Burkholderia terrae* BS001 to the fungal host *Lyophyllum* sp. strain Karsten under soil-mimicking conditions. *Microb Ecol* 73, 236–252.

Inceoğlu O., Salles J.F., van Overbeek, L. et al. 2010. Effects of plant genotype and growth stage on the betaproteobacterial communities associated with different potato cultivars in two fields. *Appl Environ Microbiol* 76, 3675–3684.

Jenkinson, D.S. and D.S. Powlson. 1976. The effects of biocidal treatments on metabolism in soil. V. A method for measuring soil biomass. *Soil Biol Biochem* 8, 209–213.

Knorr, M., Frey, S.D. and P.S. Curtis. 2005. Nitrogen additions and litter decomposition: a meta-analysis. *Ecology* 86, 3252–3257.

Kuzyakov, Y. and E. Blagodatskaya. 2017. Microbial hotspots and hot moments in soil: concept & review. *Soil Biol Biochem* 83, 184–199.

Ladd, J.N., Foster, R., Nannipieri P. et al. 1996. Soil structure and biological activity. In: Stotzky, G and Bollag, J.-M. (Eds). *Soil Biochemistry*, vol. 9. Marcel Dekker, New York, pp. 23–78.

Mallon, C.A., van Elsas, J.D. and J.F. Salles. 2015. Microbial invasions: the process, patterns and mechanisms. *Trends Microbiol* 23, 719–729.

McLaren, A.D. 1977. The seven questions of Selman A. Waksman. *Soil Biol Biochem* 9, 375–376.

McLaren, A.D. and G.H. Peterson. 1967. Introduction to the biochemistry of terrestrial soils. In: McLaren, A.D., Peterson, G.H. (Eds). *Soil Biochemistry*, vol 1. Marcel Dekker, New York, pp. 1–8.

Nannipieri, P. and E.A. Paul. 2009. The chemical and functional characterization of soil N and its biotic components. *Soil Biol Biochem* 41, 2357–2369.

Nannipieri, P., Falchini, L., Landi, L. et al. 1999. Nitrogen uptake by crops, soil distribution and recovery of urea-N in a sorghum-wheat rotation in different soils under Mediterranean conditions. *Plant Soil* 208, 43–56.

Nannipieri, P., Ascher, J., Ceccherini, M.T. et al. 2003. Microbial diversity and soil functions. *Eur J Soil Sci* 54, 655–670.

Nannipieri, P., Giagnoni, L., Renella, G. et al. 2012. Soil enzymology: classical and molecular approaches. *Biol Fertil Soils* 48, 743–762.

Nannipieri, P., Trasar-Cepeda, C. and Dick, R.P. 2018. Soil enzyme activity: a brief history and biochemistry as a basis for appropriate interpretations and meta-analysis. *Biol Fertil Soils* 54, 11–19.

Neuman, G. and V. Romheld. 2007. The release of root exudates as affected by the plant physiological status. In: Pinton, R., Varanini, Z., Nannipieri, P. (Eds). *The Rhizosphere. Biochemistry and Organic Substances at the Soil-Plant Interface*. CRC Press, Boca Ratoon, FL, pp. 23–72.

Norman, A.G. and C.H. Werkman. 1943. The use of nitrogen isotope N^{15} in determining nitrogen recovery from plant materials decomposing in soil. *J Am Soc Agr* 35, 1023–1025.

Norton, J.M., Klotz, M.G., Stein, L.Y. et al. 2008. Complete genome sequence of *Nitrosospira multiformis*, an ammonia-oxidizing bacterium from the soil environment. *Appl Environ Microb* 74, 3559–3572.

Pathan, S.I., Ceccherini, M.T., Hansen, M.A. et al. 2015. Maize lines with different nitrogen use efficiency select bacterial communities with different β-glucosidase-encoding genes and β-glucosidase activity in the rhizosphere. *Soil Biol Biochem* 51, 995–1004.

Pietramellara, G., Ascher, J., Borgogni, F. et al. 2009. Extracellular DNA in soil and sediment: fate and ecological relevance. *Biol Fertil Soils* 45, 219–235.

Radajewski, S., Webster, G., Reay, D.S. et al. 2002. Identification of active methylotroph populations in an acidic forest soil by stable-isotope probing. *Microbiology* 148, 2331–2342.

Rousk, J., Baath, E., Brookes, P.C. et al. 2010. Soil bacterial and fungal communities across a pH gradient in an arable soil. *ISME J* 4, 1340–1351.

Schimel, J.P. and J. Bennett. 2004. Nitrogen mineralization: challenges of a changing paradigm. *Ecology* 85, 591–602.
Schloter, M., Nannipieri, P., Sorensen, S. and J.D. van Elsas. 2018. Microbial indicators for soil quality. *Biol Fertil Soils* 54, 1–10.
Sinsabaugh, R.L., Hill, B.H. and S. Follstad. 2009. Ecoenzymatic stoichiometry of microbial organic nutrient acquisition in soil and sediment. *Nature* 462, 795–799.
Sterner, R.W. and J.J. Elser. 2002. *Ecological Stoichiometry: The Biology of Elements from Molecules to the Biosphere*, Princeton University Press, Princeton, NJ.
Stott, D.E., Andrews, S.S., Liebig, M. et al. 2009. Evaluation of β-glucosidase activity as a soil quality indicator for the soil management assessment framework. *Soil Sci Soc Am J* 74, 107–119.
Thorn, R.G. 2012. Soil fungi. In: Huang, P.M., Li, Y. Summer, M.E. (Eds). *Handbook of Soil Sciences. Properties and Processes*, 2nd Edition, CRC Press, Boca Rato, FL, pp. 24-18–24-29.
Van Overbeek, L.S. and J.D. van Elsas. 1995. Root exudate-induced promoter activity in *Pseudomonas fluorescens* mutants in the wheat rhizosphere. *Appl Environ Microbiol* 61, 890–898.
Van Veen, J.A., van Overbeek, L.S. and J.D. van Elsas. 1997. Fate and activity of microorganisms following release into soil. *Microbiol Molec Biol Rev* 61, 121–135.
Vogel, T.M., Simonet, P., Jansson, J.K. et al. 2009. TerraGenome: a consortium for the sequencing of a soil metagenome. *Nat Rev Microbiol* 7, 252–253.
Wardle, D.A. and A. Ghani, 1995. A critique of the microbial metabolic quotient (qCO_2) as a bioindicator of disturbance and ecosystem development. *Soil Biol Biochem* 27, 1601–1610.
Waksman, S.A. 1927. *Principles of Soil Microbiology*, Williams and Wilkins, Baltimore, MD.

3 The Soil Microbiome—An Overview

Francisco Dini-Andreote
Netherlands Institute of Ecology (NIOO-KNAW)

Jan Dirk van Elsas
University of Groningen

CONTENTS

3.1 Introduction ... 37
3.2 Sampling of the Soil Microbiome ... 39
3.3 The Ubiquity of Soil Microbiomes and Their Structures ... 40
3.4 Determinants of Soil Microbiome Characteristics ... 41
3.5 Phylogeny and Definition of Operational Taxonomic Unit ... 43
3.5.1 The Marker Gene and Definition of Operational Taxonomic Unit and Sequence Variants ... 43
3.5.2 The Tree of Life ... 43
3.6 Does Soil Ecosystem Functioning Reflect Microbiome Structure? ... 44
3.7 Ecological Resistance and Resilience in Soil Microbiomes ... 45
3.8 Concluding Remarks ... 46
References ... 47

3.1 INTRODUCTION

Microbiomes are defined as the collection of all microbial inhabitants of a given system. This includes natural ecosystems (aquatic and terrestrial), and plant– and animal–host systems. For soils, most researchers agree to the tenet that soil microbiomes encompass four major groups of microorganisms: bacteria, archaea, fungi, and protozoa, as these are the main organisms for essential soil processes. Thus, the soil microbiome is composed of a *bacteriome*, an *archaeome*, a *fungome*, and a *protozoome*. In addition, one should note that viruses, in particular those infecting bacteria (bacteriophages), are also interactive parts of the system, as they may result in lysed, activated, or genetically altered host populations. Thus, the soil *virome* represents an additional, but often overlooked, part of the soil microbiome (see Chapter 6). Here, we will use the aforementioned (broad) definition of the soil microbiome and briefly address the constituent parts, that is, the bacterial, archaeal, fungal, and protist communities. The reason for addressing these organismal groups into one (micro)biome is that they all play essential roles in the key functions of soil. For example, in soil formation, nutrient cycling (see Chapter 11), disease suppression (see Chapter 21), and aggregate stability.

Despite the long history of soil microbiome research (formerly called soil microbiology research; see, e.g., Waksman (1927)), soil microbiologists still struggle with the complexity and variability of the microbiomes across soil types. This is evident when considering, for example, an apparently homogeneous agricultural field, a forest bed, or a soil from an extreme environment such as the Arctic. A hallmark article by Fierer (2017) summarizes the historical development of our knowledge

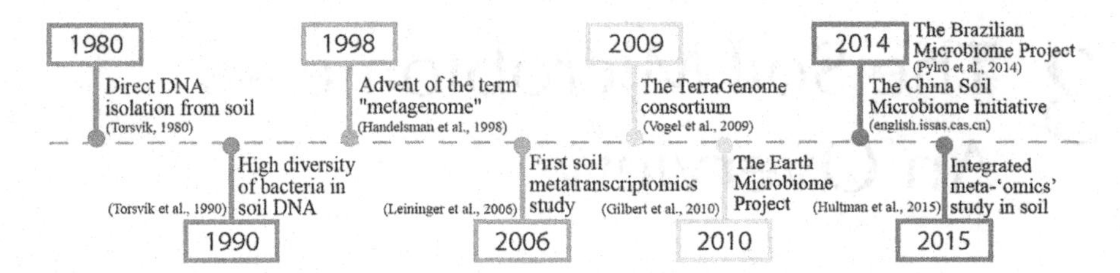

FIGURE 3.1 Schematic timeline depicting advances in "omics" methods and projects related to the investigation of soil microbiomes.

of soil microbiomes. Whereas, for a long time since the incipience of soil microbiology by pioneers such as Winogradsky, Beijerinck, Waksman and McLaren (see Chapter 2), soil microbiology has almost exclusively depended on cultivation- and microscopy-based approaches, the past three decades have witnessed an unprecedented development of novel techniques and large-scale projects that have revolutionized our understanding of soil microbiomes (see Figure 3.1 for a timeline). The great progress made with the advent of molecular techniques in recent years is remarkable. The tools that are now available have enabled researchers to leverage, at the molecular level, genomic information of the main soil taxa and predict their respective functional attributes. Considering these powerful developments, Schloter et al. (2018) recently provided arguments for the contention that soil microbiomes may now be dissected to the extent that robust function-based molecular indicators that describe soil quality can be established. In addition, such indicators should be applied at the levels where the intricacies of soil life are actually determined, that is, the (microscale) level of the organisms themselves (Box 3.1).

The basic question in this respect is how the aforementioned "omes" in soil can be properly examined. Of key importance is the fact that soils are inherently heterogeneous and diverse, meaning that it is unlikely that a "typical" soil microbiome exists. Basically, although it may appear that soil microbiome structures are fairly similar over a field or soil type, this similarity may dwindle away when either the sampling scale is diminished (e.g., milligrams instead of grams or kilograms of soil) or the taxonomic ranking is lowered (e.g., species or subspecies instead of family, order, or domain). As outlined in Chapter 1, the members of soil microbiomes live in diverse (micro)habitats that are scattered within and across soil types. In such complex microhabitats, the processes of microbial life (e.g., acquisition of nutrients, secretion of by-products, production and release of

BOX 3.1

- Soil constitutes one of the most biodiverse ecosystems on Earth. A handful of soil contains, on average, more than a thousand million cells encompassing bacteria, archaea, fungi, and protists—collectively called the soil microbiome.
- Despite the enormous microbial diversity in each soil, recent studies have revealed that soils are globally dominated by a well-defined core set of taxa.
- Surprisingly, the soil microbiome colonizes <1% of the total surface that is available in soil.
- The soil microbiome is responsible for driving the biogeochemical cycles that support life on Earth (e.g., those of carbon, nitrogen, phosphorus, and sulfur). Therefore, it is essential in a variety of ecosystem functions (e.g., stimulation of plant growth, influence on soil chemical and physical structures).
- Soil microorganisms can be dormant, with, on average, >95% of the total microbial biomass being composed of taxa that are inactive at a given point in time.

exopolysaccharides, ecological interactions—including synergism, antagonism and predation, viral lysis, and release of cellular components), take place. These processes are collectively affected by the physical and chemical conditions in the soil microhabitats. Thus, it is possible to envision, for each individual soil aggregate, a picture of an interconnected microbial community, in which a plethora of ecological interactions occurs in a dynamic manner. Such microscale processes, in turn, determine the local cycling of nutrients, the degradation of (xenobiotic) compounds, and/or the filtering of percolating water. However, these processes may be variable across the multitude of diverse soil microhabitats. Thus, the spatiotemporal variability across the soil microhabitats, that is, the microscale variation across space and time, poses fundamental challenges to the understanding of the dynamics of soil microbiome structure and activity.

In this chapter, we examine the current knowledge with respect to the overall structures of soil microbiomes, and the future developments that are required to deeper explore their composition, ecology, and functional attributes. Given the overriding effects of the microhabitat conditions on the local microbiomes, we will first place a focus on soil sampling, its scale, and its effect on our view of the soil microbiomes. Next, we provide an assessment of the taxonomic distribution of soil microbiome taxa and discuss the potential relationship between the phylogenetic structure of soil microbiomes and ecosystem functioning. Finally, we provide a view of the ecological properties of soil microbiomes with respect to their role in the resistance and resilience of soil to disturbances.

3.2 SAMPLING OF THE SOIL MICROBIOME

As outlined in Chapter 1, a 1 g amount of soil is often used as a "practical" sample size in soil microbiome studies. However, considering the dimensions of most members of soil microbiomes, which are around 1 to several micrometers, and their monitoring "range" (the volume of soil they can perceive), a 1.2 g sample (about 10,000 μm × 10,000 μm × 10,000 μm for a soil with a hypothetical density of 1.2 g cm^{-3}) still consists of a broad diversity of "within-range" microhabitats of about 20 μm × 20 μm × 20 μm. As such, it can be asserted that—in light of the heterogeneity of soil—the 1.2 g sample contains a multitude of distinct microenvironments that are just micrometer-scale distances apart. These microenvironments may differ significantly in their microbiome structure and abundance, rates of metabolic processes, and abiotic characteristics. For example, the concentration of oxygen in a small soil particle can vary from 20% at the outside to $<1\%$ at the inside of an individual soil aggregate (Sexstone et al. 1985). See also Figure 1.1. Thus, contrasting physiological groups, that is, aerobic versus (facultative) anaerobic taxa, are expected to thrive in the two microenvironments.

By making use of this practical sampling size in soil microbiome studies (i.e., up to about 1 g), it is widely and somewhat implicitly accepted that the different microhabitats are lumped together and hence analyzed jointly. In addition, in studies in which (so-called) representative soil samples are collected from a field, there are often post-processing steps that contribute to sample homogenization (e.g., sieving, aggregate separation and subsequent sample homogenization, and nucleic acid isolation). Thus, such sample processing approaches tend to lose resolution with respect to the actual microhabitats in which the individual microbiome components are present, that is, inter-aggregate and intra-aggregate environments. Despite the fact that information on the spatial location of individual cells relative to one another and their resources is lost, this approach is the most commonly and broadly used one, as can be found in the current literature.

Although not in common practice, there are some studies that highlight the importance of investigating and sampling the soil microbiome at the micro-aggregate level (e.g., Cosentino et al. 2006, Vos et al. 2013, Probandt et al. 2018). The study by Probandt et al. (2018) showed that bacterial communities differ significantly between sand grains and macroaggregates in soil. Also, Stefanic and Mandic-Mulec (2009) found that differently evolved *Bacillus subtilis* populations occur in closely spaced microsites in a soil. The process of soil aggregation *per se* may be considered to constitute an important driver of the (parallel) evolution of organisms across soil microbiomes. This phenomenon is further discussed in Chapters 6 and 7. It brings up the concept of soil as a collection of parallel "evolutionary incubators" (Rillig et al. 2017). The proper choice of soil sampling method holds the key to disentangling different facets of the ecology of soil microbiomes, for instance, with respect to taxon ecological interactions and eco-evolutionary dynamics. This includes aspects such as metabolic codependence and differential ecophysiological behavior.

The key question here is, therefore, "what constitutes an ideal soil sampling strategy for soil microbiome research?" The answer is that this will—to a large extent—depend on the scientific question (or hypothesis) that is being posed. For example, mapping soil microbiomes along an 8 km scale, as done in the work of Dini-Andreote et al. (2014), was useful to understand the distribution—over a large spatial scale—of soil microbiomes as well as their relation to ecosystem functioning. Moreover, this approach also served to investigate how different ecological mechanisms interrelate during microbial community assembly at an ecosystem scale (Dini-Andreote et al. 2015). However, it neglects the desirable observation of per-site (short-distance) microscale resolution of community structure. For instance, the selected sampling approach did not precisely map (quantify the relative abundance and location of) the free-living nitrogen fixers within the different soil aggregates, nor did it shed light on the effects of (natural or anthropogenic) disturbances that often occur at a given local scale.

Another aspect of the living soil that needs consideration is the fact that soil microbiomes are compositionally not stable, as they are continuously responding to changes in local conditions. In other words, soil microbiomes are in a dynamic flux (Shade et al. 2013). This dynamics operates at both the microscale and the macroscale level, and it varies in time. For example, whereas, at the microscale level, "endogenous" dynamics (i.e., ecological interactions among the taxa under shifting local environmental conditions) governs the relative abundance of community members and their patterns of gene expression, at a larger spatial scale, "exogenous" influences (e.g., variation in soil moisture, organic carbon, and/or pH), that are exerted by the overall environmental factors, prevail. The speed at which communities change varies in accordance with (1) the intrinsic nutrient dynamics at the local scale, which determines growth and metabolism (e.g., the organic matter and ammonium concentrations), and (2) extrinsic ecological factors, such as dispersal of organisms across soil space, which increase organismal variations at a local site. Similar to the proposed spatially explicit sampling of soil microbiomes, sampling strategies should also consider variations at the temporal scale. The decision as to what may constitute an ideal timeframe for soil sample collection will depend on the scientific question and should thus be guided by the conceptualization of the extrinsic and intrinsic factors that shape the local communities.

3.3 THE UBIQUITY OF SOIL MICROBIOMES AND THEIR STRUCTURES

When examining soils, one will consistently observe that various members of each of the distinct organismal types within microbiomes, that is, bacteria, archaea, fungi, and protists, are invariably present. Moreover, viruses are also found across all of the soils examined so far (Pratama et al. 2018; Chapter 6). In the simplest of terms, these findings of microbiome ubiquity indicate that relevant and sufficient numbers of niches for all of these microbiome groups exist in most soils, with the possible exception of extreme ones. Chapters 4–6 and 8 examine the drivers of diversity, activity, and growth of these organismal groups in soils in more detail. Interestingly, although most soils are teeming with microbial life, the fraction of the total surface area in a soil sample that is covered by soil microbes is consistently low, that is, <1% (Young and Crawford 2004). This, coincidentally, approximates the fraction of the surface area of Earth that is occupied by human beings. Hence, such rather sparse occupation of the surfaces in soil by microbial life seems to indicate that the *habitable space* allowing organismal survival is limited to specific sites in soil. A plausible explanation for this contention is that the conditions in large parts of the soil are often strongly "hostile" to microbiome members, and survival is limited to particular "habitable sites" at the microscale. In the hostile sites in soil, several biotic and abiotic stresses may occur. Examples are acidic and low water availability conditions, frequent disturbances (e.g., drying–rewetting and freezing–thawing cycles), high degrees of ecological competition (as evidenced by the widespread occurrence of antibiotic-producing and antibiotic-resistant mechanisms in bacteria), predation (e.g., by protozoa and/or nematodes), low O_2 tensions, and most often, a severely limited availability of organic carbon substrates. The latter, even when abundantly present, are unevenly distributed in most soils, across both space and time (Kuzyakov and Blagodatskaya 2015). Taking the analogy with humans on Earth a bit further, for microorganisms the soil habitable space is comparable to what particular habitable parts of Earth's solid surface mean to humans, with respect to the occupation of "fertile" or habitable spaces and the nonoccupation of infertile and/or inhabitable spots.

In light of these arguments, one may come to the conclusion that—due to spatiotemporal variations across and even within soils—there is no "typical" soil microbiome. That is, the relative abundances and diversities of bacteria, archaea, fungi, and protists in soils vary considerably across soil types and scales (Figure 3.2). However, a close examination of the main constituents of these groups provides an indication of the major commonly found taxonomic groups in soil, albeit grouped at high taxonomic rankings. In brief, Figure 3.2 depicts the molecularly determined abundance of the four groups across 66 unique soil samples (after Crowther et al. 2014). For all the organismal groups, we find typical rank–abundance curves across all soils, with just a few groups dominating and a larger number of groups present at increasingly lower densities. This distribution pattern, which is also commonly observed for other organismal groups, indicates the presence in soils of "dominant" niches for small subsets of the communities, followed by a suite of minor niches for larger subsets of the microbiome. Moreover, when the phylum level was used for distinction, broadly similar distribution curves were found across all soils, indicating that—roughly—most, if not all, soils generally contain representatives of all sampled groups defined at this level, in grossly similar relative abundances. Also, and in particular for the soil bacteriome structure at a lower taxonomic level, a recent study (Delgado-Baquerizo et al. 2018) found that a small fraction of the total soil bacterial diversity (about 2%, which encompasses about 500 distinct taxa) accounts for a large fraction (about 40%) of all taxa found in distinct soils across the globe. This indicates that, despite the enormous diversity of bacteria in soils, a large proportion of this diversity can be reasonably described by a short list of more or less ubiquitously occurring taxa. Additional aspects of the soil microenvironment will be examined in later chapters. However, it is known that, despite the similarities that may exist, soil microbiomes are diverse within and across the soil types. Moreover, these differences are known to be dynamically shaped by the interplay of biotic and abiotic factors that determine the status of soil microbiomes.

3.4 DETERMINANTS OF SOIL MICROBIOME CHARACTERISTICS

Across the distinct soil types, bacteria and fungi are generally the dominant microorganisms. The densities of both groups result in about 10^2- to 10^4-fold higher biomass values than those of the other organismal types within the soil microbiome (i.e., archaea, protists, and viruses) (Fierer 2017). A range of biotic and abiotic factors can influence the overall microbial biomass in soils, such as the abundance of microbial predators (e.g., predatory bacteria, protists, and nematodes), the amount and availability of soil organic carbon, soil moisture status, and a set of physicochemical constraints (e.g., pH, salinity). Conversely, variations in soil microbiome structure are similarly driven by key factors of particular importance, such as soil salinity, pH, organic carbon quality and quantity, and soil redox status.

The relative importance of these factors influencing the soil microbiome varies according to the soil type and the extent of variation of the sampled soils. For instance, when analyzing soil microbiomes across a range of particular constraining factors (e.g., salinity or pH), it is often assumed that differences in the microbiomes are attributed to each of the factors examined. However, these

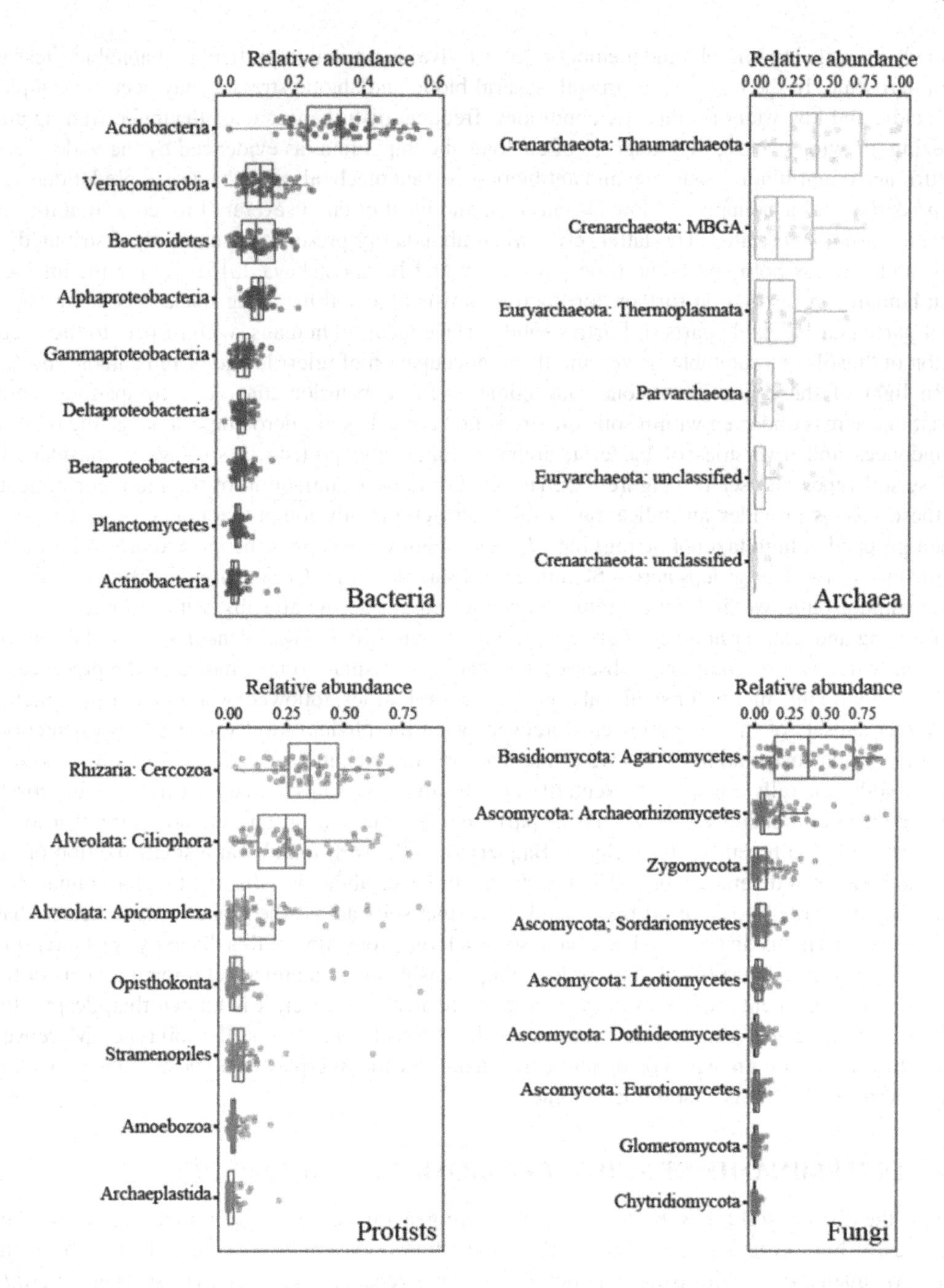

FIGURE 3.2 Structure of the soil microbiome. The figure depicts the bacteriome, archaeome, fungome, and protozoome in terms of the proportional abundance of each major group across 66 soil samples, as described by Crowther et al. (2014). (Original figure from Fierer (2017) used with permission)

factors may not be apparent when soil samples are collected across a narrow range of pH or salinity. In addition, at a finer scale, not all taxa within a given soil microbiome are expected to respond in a similar manner to the presence of a particular environmental constraint. In spite of the fact that generalities can be achieved by probing large-scale distributions of soil microbiomes and electing the best sets of environmental predictors, there is no single biotic or abiotic factor that consistently determines the structure of the soil microbiome across all soils.

3.5 PHYLOGENY AND DEFINITION OF OPERATIONAL TAXONOMIC UNIT

3.5.1 The Marker Gene and Definition of Operational Taxonomic Unit and Sequence Variants

Phylogenetic analysis aims at reconstructing potential evolutionary histories and relationships among individual organisms. The term "phylogeny" refers to both the processes by which organisms evolved by diverging from a common ancestor and the study of the evolutionary history of the organisms. The latter constitutes the basis for one of the most important organizing principles in biology. In investigations of the soil microbiome, properly defining a microbial species is an enormous challenge. This is due to the broad divergence observed even within species, resulting from the speciation process being heavily intertwined with horizontal gene transfer (HGT) processes (see Chapter 7). However, defining a species—or similar—unit is a key and desirable element in all soil microbiome assessments. In light of the intrinsic problems with microbial species definitions, soil microbiome researchers have adopted the concept of the "operational taxonomic unit" (OTU), which is based on the level of phylogenetic similarity across microbes. The use of OTUs for the bacterial, archaeal, and fungal fractions of the soil microbiome has a well-defined threshold, that is, the 97% nucleotide identity of the 16S or 18S ribosomal RNA (rRNA) gene sequence, also referred to as "small subunit (SSU) rRNA." Recently, researchers have proposed that this definition should be abandoned and replaced by single-nucleotide variants of these genes (Callahan et al. 2017), as it more precisely relates to organismal differences. This proposed change may become common practice in the near future in soil microbiome phylogenetic assessments.

There are several reasons for making use of rRNA gene sequences for phylogenetic inferences. First, rRNA genes are present in all living organisms as they are essential constituents of the ribosome, the organelle responsible for protein synthesis (translation). Second, the size of the gene (about 1,500–1,800 bp) is ideal for the analyses, as it is relatively small, thus facilitating amplification, whereas it still contains sufficient phylogenetic information and even encompasses both "variable" and "conserved" regions. Variable regions are areas of the molecule that are prone to an enhanced rate of mutation, hence being suitable for finer-scale distinctions between organisms. As a whole, the rRNA molecule constitutes an ideal "evolutionary chronometer" (Woese and Fox 1977). The conserved regions are often paired with each other in the functional fold of the rRNA molecule. As these are less prone to mutation, they can be used as anchoring sites for primers used in PCR-based amplifications of gene regions. Third, the rate of mutation, which results in differences along this gene, provides sufficient information to distinguish distinct microbial taxa, albeit that higher resolution is achieved for bacteria and archaea on the basis of the 16S rRNA gene than that for Eukarya based on the 18S rRNA gene.

3.5.2 The Tree of Life

In addition to the use of rRNA gene nucleotide sequences for microbiome assessments, these also constitute the basis for a valuable tool to determine the evolutionary relationships across organisms. Carl Woese and co-workers, in the 1980s, established a phylogenetic tree that relates all organisms by calculating differences in the nucleotide composition of the rRNA gene, and by pairwise alignments of several distinct and distantly related organisms. Still today, Woese's approach is widely used, with novel organismal phyla and clades being described and incorporated into the tree of life.

Although the relevance for soil microbiome studies is as yet unclear, a revolutionary update of the tree of life was recently proposed (Hug et al. 2016, Castelle and Banfield 2018). In brief, by using not only the SSU rRNA but also a concatenated set of gene sequences encoding 16 ribosomal proteins, and calculating their pairwise distinctness, Hug et al.'s phylogenetic reconstruction encompasses 92 bacterial and 26 archaeal phyla and all five of the eukaryotic supergroups (Opisthokonta, Excavata,

Archaeplastida, Chromalveolata, and Amoebozoa). The authors argued that this tree captures the expected organismal groupings at most of the taxonomic levels, while it is largely congruent with the tree calculated using traditional SSU rRNA gene sequence information.

This novel insight into evolutionary descent may have large implications for soil microbiome studies, as the limits of biodiversity in soil are still unknown. In particular, the diversity of the so-called rare biosphere is enigmatic. Thus, the well-known phylogenetic references, and the increasing resolution of curated databases, are crucial for improved molecular analyses of soil microbiomes. The improved insights in the diversity of, and evolutionary divergences in, such microbiomes may assist to infer their potential metabolic capacities, habitat preferences, ecological capabilities, and ecosystem contexts.

3.6 DOES SOIL ECOSYSTEM FUNCTIONING REFLECT MICROBIOME STRUCTURE?

The soil microbiome performs a multitude of functions that collectively modulate the physicochemical characteristics of the soil matrix. Thus, microorganisms play essential roles in the biogeochemical cycles of carbon, nitrogen, sulfur, phosphorus, and iron. They produce and consume atmospheric trace gases such as hydrogen, dinitrogen, methane, carbon dioxide, and nitrous oxide influence soil acidity and alter soil hydrophobicity and conductivity. As such, it is important not only to categorize the presence of each of these particular functions but also to understand the degree to which these distinct functions relate to the composition and structure of the microbiome.

It is, however, unknown to what extent a phylogenetic depiction of a soil microbiome can be taken as an indication of the functions of that microbiome. In an ideal scenario, the coupling of phylogenetic reconstruction—that is, determining phylogenetic relationships between organisms—to functional information would allow us to discern 1:1 relationships that allow an understanding of how environments drive evolutionary/adaptive processes. However, for one part, even within the same OTU, enormous functional divergences may exist (Preheim et al. 2013, Nguyen et al. 2016). Also, considerable proportions of genomic information may be similar between diverging organisms as a result of function-selective processes. As such, the functional capacities of such organisms may be—to a certain extent—similar, as well as their habitat preferences in soil. Hence, the use of "phylogenetic signal" for the inference of function and evolutionary relatedness may be unwarranted. By definition, a particular function may be inferred to "have a phylogenetic signal" (or to be phylogenetically conserved) if it is consistently present in a given clade in the phylogenetic tree. In the case of bacteria, given the widespread occurrence of HGT (see Chapter 7), some functions are known to have strong phylogenetic signals, whereas others may have none. For example, the ammonia monooxygenase structural gene *amoA* (involved in the first step of nitrification, ammonia oxidation; see Chapter 11) is present across a selected clade of the bacterial tree of life, and so is said to have a strong phylogenetic signal. This encompasses a few β- and γ-proteobacterial classes within the phylum Proteobacteria. Moreover, it is present in some representatives of the archaeal phylum Thaumarchaeota. In contrast, the copper-containing nitrite reductase structural *nirK* gene (involved in denitrification) has a weak phylogenetic signal, as it has a widespread distribution across the tree of life, being present across a range of distinct and phylogenetically widely divergent taxa (Graf et al. 2014).

Notwithstanding the occurrence of functions with "narrow" occurrence (i.e., having a phylogenetic signal), in most cases functions in soil microbiomes are encoded by genes present across a suite of distinct taxa, with diverse phylogenetic signals. This phenomenon has been referred to as "genetic" or "functional" redundancy. Genetic redundancy simply means that two or more organisms hold similar genetic information to perform a determined function. However, this term must be interpreted with caution, as it often does not imply "metabolic" redundancy (meaning that metabolic functions are fully interchangeable). A case in point is given by cellulose- or hemicellulose-degrading organisms that seemingly work in a metabolically similar fashion yet may secrete completely different sets of enzymes, each one functioning under (slightly) different conditions. Thus, cellular metabolisms vary across taxa and depend on the proximity and level of the nutritional source, as well as the cell's physiological status. Hence, metabolic functioning can vary, even across cells within the same species in a population context. For example, a population of *B. subtilis* in one (soil) habitat is known to be split up into a suite of subpopulations that are genetically similar yet perform a range of diverse functions (e.g., acting as protease secretors and as extracellular matrix builders, becoming competent for transformation, or committing suicide), all at the same time in the same habitat (Vlamakis et al. 2013).

In addition, the establishment of a link between the soil microbiome structure and soil functioning is complicated by the fact that a large proportion of the microorganisms in soil is often dormant. A recent study suggested that >95% of the total microbial biomass in soils is composed of taxa that are inactive at any given point in time (Blagodatskaya and Kuzyakov 2013). Also, soils contain DNA from cells with leaky or damaged cytoplasmic membranes, or from dead and damaged cells, which is collectively called "relic DNA". Such relic DNA in soil can be up to 40% of the total DNA yield, thus blurring our appreciation of the full extent of *living* (potentially active) microbial cells in soil (Carini et al. 2016).

3.7 ECOLOGICAL RESISTANCE AND RESILIENCE IN SOIL MICROBIOMES

Although multiple factors may blur the contention that the structure of soil microbiomes is linked with soil function, soil microbiome structure does match the *functional potential* behind the soil ecosystem functions. In this respect, differences in soil microbiome structures may reflect how these respond to ecological disturbances. Here, the fundamental question that can be posed is "what defines the stability of soil functioning?" Such functional stability may be a direct consequence of the status (structure and diversity) of the soil microbiome, in which the capacity to withstand functional changes in the face of disturbance is a key issue (*ecological resistance*). Alternatively, it may result from the ability of an impacted soil microbiome to return to its original status after being disturbed: *ecological resilience* (Allison and Martiny 2008).

Given the enormous diversity within most of the soil microbiomes, functions with high redundancies are likely to be homeostatic, that is, less prone to impacts from a perturbation event. In contrast, metabolically "exclusive" functions are more prone to be impacted and eventually lost. Each function within the soil microbiome is linked to the ecophysiological behavior of the key organisms hosting it, and so the survival of these key organisms underlies soil resilience. For example, particular members of the genus *Bacillus* (Firmicutes), a known spore-forming bacterial group, may rapidly recover after a disturbance, given that the respective spores persisted under the unfavorable conditions and germinate and grow following the relief of the stress. Moreover, it is known that the nature of disturbance events (e.g., heating, drought, wildfire) and their severity (e.g., long- versus short-term disturbances) influence how microbiomes respond to, and change under, local environmental adversities.

Apart from abiotic disturbances, soil microbiomes also experience, and respond differently to, biotic disturbances. Van Elsas et al. (2012) investigated how the degree of diversity in soil microbiomes reflects the ability of soil to resist an invasion, in this case by an apathogenic derivative of the severe pathogen *Escherichia coli* O157:H7. Several lines of evidence supported the conclusion that an inverse microbiome diversity–invasiveness relationship exists in the soil. In other words, the more diverse a soil microbiome is, the lower is the chance an invader can establish and survive. The mechanism underpinning this phenomenon was shown to be competition for the utilization of growth-limiting resources (i.e., sources of carbon), as a more diverse microbiome was found to more effectively utilize the available resources and thus occupy the available niches in soil. This displaced the invader comparatively more in high-diversity than in diversity-depleted soil microbiomes. The resistance concept extends beyond the topic of invasion, and the status of community structure and diversity possibly reflects the ability of soil microbiomes to withstand an additional vast array of both anthropogenic and natural disturbances.

Overall, there is no simple definition of the "ideal (beneficial) status of the soil microbiome" (i.e., supporting soil function and being highly resistant and/or resilient to disturbance). Community structure and diversity *per se* are measurable attributes of microbiomes that—in most cases—can be linked to resistance and/or resilience. The way these metrics are considered can vary as methodological differences (sequencing technology, database information, and taxonomic resolution) occur. Moreover, both metrics can only be interpreted within an ecological context. This implies that one cannot state that a more diverse microbiome invariably constitutes a "better" system. Also, the magnitude of the disturbances that a given soil may experience varies. Interestingly, it was found that soil microbiomes that had been exposed to a particular perturbation event will more effectively recover when confronted with a similar event (Jurburg et al. 2017).

3.8 CONCLUDING REMARKS

Soils are among the most biodiverse ecosystems on Earth. This diversity is linked to the inherent complexity and heterogeneity of the soil environment, providing a large diversity of microscale niches that are teeming with microbial life. Along with spatial heterogeneity, there is also a high degree of temporal variability across soils, giving rise to additional organismal diversity. It comes as no surprise that, as evidenced by the advanced methodologies that are currently available, the diversities across the four soil biomes discussed here, that is, the bacteriome, archaeome, fungome, and protozoome, are immense. The emergence of typical rank–abundance curves for all four biomes was intriguing, as it appears to indicate that, across a range of soils, ecological rules of community assembly exist, which dictate that each biome tends to hold similar organisms that represent a limited number of "major" niches next to a large number of "minor" ones. It will be a challenge in future research to come to understand the immense diversity and variability across soil microbiomes and the factors influencing these. Given the fact that soil function is—to a variable extent—linked to soil biome makeup, the translation of findings based on phylogenetic inference to actual function constitutes an additional challenge. As living entities, the organisms in soil microbiomes may or may not completely occupy a niche, in a system in which the niches are continuously changing. Unfortunately, our capabilities to pinpoint and understand these niches, and the niche occupants, at a fine level, may—even with the great progress made with molecular techniques—be still limited, which is comparable with efforts of physicists to come to grips with the uncertainty principle.

The challenge ahead lies in moving beyond the overall taxonomic and functional descriptions of soil microbiomes, towards a better understanding of:

i. The mechanisms that cause such variation in patterns.
ii. The consequences of these patterns for the particular ecosystem services and functions that microbiomes provide (e.g., enhancement of soil fertility, reduction of emissions of gases such as methane, carbon dioxide, and nitrous oxide).

These goals can be achieved by systematically categorizing the constituents of the soil microbiome on the basis of their ecological strategies, by narrowing down the analytical scale to the particular sections of the soil interface where interactions and metabolisms take place. Moreover, our insights can be fostered by developing and testing prospective experimental designs towards a better conceptual framework that enables us to identify and explain the patterns in soil microbiomes. Collectively, these efforts will improve our understanding of the "building blocks" of soil microbiomes and enable the emergence of better strategies to monitor, manipulate, and manage soil microbiomes in natural and agricultural settings.

REFERENCES

Allison, S.D. and J.B.H. Martiny. 2008. Resistance, resilience, and redundancy in microbial communities. *Proc Natl Acad Sci USA* 105, 11512–11519.

Blagodatskaya, E. and Y. Kuzyakov. 2013. Active microorganisms in soil: critical review of estimation criteria and approaches. *Soil Biol Biochem* 67, 192–211.

Callahan, B.J., McMurdie, P.J. and S.P. Holmes. 2017. Exact sequence variants should replace operational taxonomic units in marker-gene data analysis. *ISME J* 11, 2639–2643.

Carini, P., Marsden, P.J., Leff, J.W. et al. 2016. Relic DNA is abundant in soil and obscures estimates of soil microbial diversity. *Nat Microbiol* 2, e16242.

Castelle, C.J. and J.F. Banfield. 2018. Major new microbial groups expand diversity and alter our understanding of the tree of life. *Cell* 172, 1181–1197.

Cosentino, D., Chenu, C. and Y. le Bissonnais. 2006. Aggregate stability and microbial community dynamics under drying–wetting cycles in a silt loam soil. *Soil Biol Biochem* 38, 2053–2062.

Crowther, T.W., Maynard, D.S., Leff, J.W. et al. 2014. Predicting the responsiveness of soil biodiversity to deforestation: a cross-biome study. *Global Change Biol* 20, 2983–2994.

Delgado-Baquerizo, M., Oliverio, A.M., Brewer, T.E. et al. 2018. A global atlas of the dominant bacteria found in soil. *Science* 359, 320–325.

Dini-Andreote, F., Pereira e Silva, M.C., Triadó-Margarit, X. et al. 2014. Dynamics of bacterial community succession in a salt marsh chronosequence: evidences for temporal niche partitioning. *ISME J* 8, 1989–2001.

Dini-Andreote, F., Stegen, J.C., van Elsas, J.D. et al. 2015. Disentangling mechanisms that mediate the balance between stochastic and deterministic processes in microbial succession. *Proc Natl Acad Sci USA* 112, E1326–E1332.

Fierer, N. 2017. Embracing the unknown: disentangling the complexities of the soil microbiome. *Nat Rev Microbiol* 15, 579–590.

Gilbert, J.A., Meyer, F., Antonopoulos, D. et al. 2010. Meeting report: the terabase metagenomics workshop and the vision of an Earth microbiome project. *Stand Genomic Sci* 3, 243–248.

Graf, D.R., Jones, C.M. and S. Hallin. 2014. Intergenomic comparisons highlight modularity of the denitrification pathway and underpin the importance of community structure for N_2O emissions. *PLoS One* 9, e114118.

Handelsman, J., Rondon, M.R., Brady, S.F. et al. 1998. Molecular biological access to the chemistry of unknown soil microbes: a new frontier for natural products. *Chem Biol* 5, R245–R249.

Hug, L.A., Baker, B.J., Anantharaman, K. et al. 2016. A new view of the tree of life. *Nat Microbiol* 1, e16048.

Hultman, J., Waldrop, M.P., Mackelprang, R. et al. 2015. Multi-omics of permafrost, active layer and thermokarst bog soil microbiomes. *Nature* 521, 208–212.

Jurburg, S.D., Nunes, I., Brejnrod, A. et al. 2017. Legacy effects on the recovery of soil bacterial communities from extreme temperature perturbation. *Front Microbiol* 8, e1832.

Kuzyakov, Y. and E. Blagodatskaya. 2015. Microbial hotspots and hot moments in soil: concept & review. *Soil Biol Biochem* 83, 184–199.

Leininger, S., Urich, T., Schloter, M. et al. 2006. Archaea predominate among ammonia-oxidizing prokaryotes in soils. *Nature* 442, 806–809.

Nguyen, N.P., Warnow, T., Pop, M. et al. 2016. A perspective on 16S rRNA operational taxonomic unit clustering using sequence similarity. *NPJ Biofilms Microbiomes* 2, e16004.

Pratama, A.A. and J.D. van Elsas. 2018. The 'neglected' soil virome—Potential role and impact. *Trends Microbiol* 1533, 1–14.

Preheim, S.P., Perrotta, A.R., Martin-Platero, A.M. et al. 2013. Distribution-based clustering: using ecology to refine the operational taxonomic unit. *Appl Environ Microbiol* 79, 6593–6603.

Probandt, D., Eickhorst, T., Ellrott, A. et al. 2018. Microbial life on a sand grain: from bulk sediment to single grains. *ISME J* 12, 623–633.

Pylro, V.S., Roesch, L.F., Ortega, J.M. et al. 2014. Brazilian Microbiome Project: revealing the unexplored microbial diversity—Challenges and prospects. *Microb Ecol* 67, 237–241.

Rillig, M.C., Muller, L.A. and A. Lehmann. 2017. Soil aggregates as massively concurrent evolutionary incubators. *ISME J* 11, 1943–1948.

Schloter, M., Nannipieri, P., Sorensen, S.J. and J.D. van Elsas. 2018. Microbial indicators of soil quality. *Biol Fertil Soils* 54, 1–10.

Sexstone, A.J., Revsbech, N.P., Parkin, T.B. et al. 1985. Direct measurement of oxygen profiles and denitrification rates in soil aggregates. *Soil Sci Soc Am J* 49, 645–651.

Shade, A., Caporaso, J.G., Handelsman, J. et al. 2013. A meta-analysis of changes in bacterial and archaeal communities with time. *ISME J* 7, 1493–1506.

Stefanic, P. and I. Mandic-Mulec. 2009. Social interactions and distribution of *Bacillus subtilis* pherotypes at microscale. *J Bacteriol* 191, 1756–1764.

Torsvik, V.L. 1980. Isolation of bacterial DNA from soil. *Soil Biol Biochem* 12, 15–21.

Torsvik, V.L., Goksøyr, J. and F.L. Daae. 1990. High diversity in DNA of soil bacteria. *Appl Environ Microbiol* 56, 782–787.

Van Elsas, J.D., Chiurazzi, M., Mallon, C.A. et al. 2012. Microbial diversity determines the invasion of soil by a bacterial pathogen. *Proc Natl Acad Sci USA* 109, 1159–1164.

Vlamakis, H., Chai, Y., Beauregard, P. et al. 2013. Sticking together: building a biofilm the *Bacillus subtilis* way. *Nat Rev Microbiol* 11, 157–168.

Vogel, T.M., Simonet, P., Jansson, J.K. et al. 2009. TerraGenome: a consortium for the sequencing of a soil metagenome. *Nat Rev Microbiol* 7, 252.

Vos, M., Wolf, A.B., Jennings, S.J. et al. 2013. Micro-scale determinants of bacterial diversity in soil. *FEMS Microbiol Rev* 37, 936–954.

Waksman, S. 1927. *Principles of Soil Microbiology*, The Williams and Wilkins company, Baltimore, MD.

Woese, C.R. and G.E. Fox. 1977. Phylogenetic structure of the prokaryotic domain: the primary kingdoms. *Proc Natl Acad Sci USA* 74, 5088–5090.

Young, I. and J. Crawford. 2004. Interactions and self-organization in the soil-microbe complex. *Science* 304, 1634–1637.

4 The Bacteria and Archaea in Soil

Jan Dirk van Elsas
University of Groningen

Anton Hartmann and Michael Schloter
Helmholtz Zentrum München

Jack T. Trevors
University of Guelph

Janet K. Jansson
Pacific Northwest National Laboratory

CONTENTS

4.1 Introduction 49
4.2 Strategies for Life 50
4.2.1 Growth Strategies 50
4.2.2 Generalists versus Specialists 51
4.3 What Bacteria and Archaea Do We Find in Soil? 54
4.3.1 Data from Traditional and Advanced Cultivation-Based Approaches 54
4.3.2 Data from Cultivation-Independent Approaches 57
4.3.2.1 Bacterial Diversity 57
4.3.2.2 Archaeal Diversity 57
4.4 Coping with Life in Soil 58
4.4.1 Physiological and Genetic Adaptations to Soil Conditions 58
4.4.2 Physiological Response to Nutrient Limitation 59
4.4.3 Implications for Survival of Bacterial Inoculants in Soil 61
4.5 Concluding Remarks 62
References 63

4.1 INTRODUCTION

For a long time, it has been a challenge to distinguish Bacteria and Archaea, as viewed through the microscope, as both groups of microorganisms appear to have similar cell shapes and sizes. This is the reason why Archaea have classically been defined as Bacteria, and still today, the term often used is "archaebacteria." In 1977, hallmark work by the team of Carl Woese revealed Archaea and Bacteria to be very distinct groups based on differences in the sequences of the 16S ribosomal RNA (rRNA) genes (Woese and Fox 1977). Today, we know that these two groups differ not only in their 16S rRNA sequences but also in their cell membranes and several coenzymes. Based on these and other findings, the researchers proposed the existence of three separate domains of

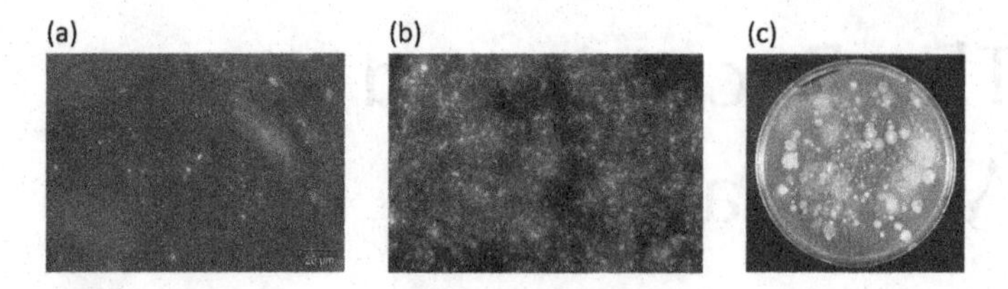

FIGURE 4.1 **(See color insert)** Bacterial cells in soil. (a) Microscopic impression of bacterial cells in soil. Soil extract was directly observed using FISH. Note the scattered occurrence of cells (yellow-green). (b) Cells extracted from soil stained with 4′-6-diamidino-2-phenylindole (DAPI) and examined under a fluorescence microscope. DAPI stains cells by interacting with their DNA. Note the great abundance and diversity of appearances of the cells. (c) Bacterial colonies from soil growing on a general growth agar. Note the diversity of the morphologies of the emerged colonies. (Courtesy of P. Hirsch, Rothamsted, Harpenden, UK.)

life: Eukarya, Bacteria, and Archaea (Woese et al. 1990). Since that time, Bacteria and Archaea are considered as two different domains that, together, constitute the prokaryotes. Their grouping together is mostly based on the finding that both groups share a large number of metabolic pathways. Several researchers have also pointed out that Archaea might have evolved from Gram-positive bacteria. The similarity of the cell wall structure of Gram-positive bacteria and Archaea, as well as a number of highly conserved proteins, forms the basis for this hypothesis (Gupta 2000). In contrast, gene expression mechanisms in Archaea are often more related to those of eukaryotes than to those in the Bacteria. Taking the different lines of evidence together, the term "prokaryote," which appears to indicate that a close relationship between Bacteria and Archaea exists, should be considered with care.*

As outlined in Chapter 1, soils are highly variable and dynamic in time and space. As a result of the enormous number of niches present in a given soil type, and the huge variety of different soil types that exist, soils have been considered as true "nurseries" for microbial life on Earth. In investigations on top soils, the number of bacteria per gram of soil ranges between 10^7 and 10^{10} (Artursson 2005). For Archaea, such numbers are 10- to 100-fold lower. Thus, cells of prokaryotes are abundant and diverse in most soils, as can be visualized using microscopy. Subsets of these can be cultured on agar plates, as further detailed in Section 4.3.1 and Chapter 18. An impression of the occurrence of bacteria in soil is given in Figure 4.1a, b and c.

4.2 STRATEGIES FOR LIFE

4.2.1 Growth Strategies

Both *autotrophic* and *heterotrophic* bacteria and archaea are found in soils. *Autotrophs* are able to produce carbohydrates, fats, and proteins from simple substances present in its surroundings, generally using energy produced from light (*phototrophs*) or from reduced chemical compounds, such as ammonia (used as electron donors—*chemotrophs*). See also Figure 1.6. For example, nitrifiers use ammonia or nitrite as electron donors. Also, methane oxidizers can be considered to be autotrophs as they use methane as electron donor and oxidize it to CO_2. Cyanobacteria are important phototrophs in the soil and are often considered as pioneering microbes colonizing raw materials or bare soil. However, as they require light for energy, they only occur on the surface of soils and are often outcompeted when plant development commences.

* As is common in microbiology, we will use the terms "Bacteria" and "Archaea" to indicate the formal phylogenetic/taxonomic groupings, and the terms "bacteria" and "archaea" to indicate these groups in a generic manner.

Heterotrophic bacteria and archaea take up organic carbon in order to be able to produce energy and synthesize compounds to maintain life. They have traditionally been divided into *autochthonous* and *zymogenous* types. Autochthonous microorganisms are common soil inhabitants that tend to remain at relatively constant levels in soil, in spite of the occurrence of fluctuations in local conditions, including organic matter levels. Many of these autochthonous bacteria are *oligotrophic*, which means that they can grow under conditions of scarcity of resources (i.e., under nutrient limitation) at low growth rates. This may explain their (often) raised abundance and greater diversities in the microbiomes of soil than in those of other habitats that are richer in nutrients (e.g., the gastrointestinal tract). Microorganisms classified as zymogenous will only grow when resources are abundant, and these are thus opportunistic in trophic terms. Zymogenous microorganisms are usually *copiotrophic*, which means that they require access to sufficient amounts of organic material for growth. They have been equated to the ecologically defined ***r*-strategists** (organisms that can grow fast when nutrient supply is abundant, and are fit in uncrowded situations). In contrast, the oligotrophs resemble **K-strategists** (organisms that can grow, albeit slowly, with a lower supply of nutrients, as well as in relatively crowded situations; see Table 4.1). Given the fact that soil microbiologists consistently find bacteria representing both *r*- and K-strategists in each soil that is examined, soils are thought to consist of a large variety of diverse niches that allow the coexistence of both physiological groups of microorganisms (Stenström et al. 2001). For instance, the surfaces of young roots are preferentially colonized by *r*-strategists because these use root exudates most efficiently and achieve fast growth rates. In microhabitats in which nutrients become scarce and/or microorganisms crowded, K-strategists may come up and prevail. Overall, based on the bacterial numbers and diversities in soil, soil appears to provide considerable hospitable niche space for a range of bacteria and archaea, despite the apparent limitations to their growth.

4.2.2 Generalists versus Specialists

Given the plethora of diverse niches that can be found in soil, it comes as no surprise that bacteria and archaea have adopted a range of different lifestyles that allow them to grow and survive under a range of conditions. We have examined the vagaries of the soil system in Chapter 1. In some cases, soil may offer strongly dominating (all-overriding) conditions to its microbial inhabitants. Thus, an extremely low or high soil pH may favor *acidophilic* or *alkaliphilic* bacteria or archaea (see Chapter 1), whereas completely anoxic conditions, such as those following soil flooding, may

TABLE 4.1
Bacterial Growth Strategies: *r*- versus K-Strategists

Characteristic	*r*-Strategist	K-Strategist
Growth rate (intrinsic)	High	Low
Population size near carrying capacity	No	Yes
Energy supply[a]	Bountiful	Scarce
Pioneer colonization ability	High	Low
Preferred carbon source (degradability)	Easy	Difficult
Typical colonization density	High	Low
(Nutrient) competitive ability	High	Low
Persistency of colonization	Low	High
Survival potential (e.g., spores, cysts)	Low	High
Preferred ecosystem type	Young, unstable	Mature, equilibrated[b]
Synonym (traditional term)	"Zymogenous"	"Autochthonous"

[a] That allows abundant growth.
[b] Mature/equilibrated: ecosystem at equilibrium, having an established, more or less stable community.

favor anaerobic bacteria and archaea, such as fermenters and methanogens. Moreover, the presence of specific plants like legumes may favor specific bacterial types, such as the rhizobia that form root nodules and carry out nitrogen fixation in symbiosis with legumes (see Chapter 11). In addition, the provision of accessible organic material such as (plant-, insect-, or fungal-derived) compounds rich in cellulose or chitin may incite the prevalence of specific bacterial types that are able to live off this material, for example, cellulose or chitin degraders. In most soils, there is not just one all-overriding condition, but the conditions at different sites are quite diverse, for example, with respect to nutrient or oxygen availability. Hence, the function and activities of soil-dwelling bacteria and archaea are largely driven by the soil type as well as the overriding abiotic and biotic conditions.

Many bacteria and archaea in soil can be characterized as soil "generalists"; that is, they are found (often in high numbers) in virtually every soil on Earth. Examples of such soil generalists are members of the genus *Bacillus* and related genera, actinobacteria, *Paraburkholderia*, *Pseudomonas*, and *Stenotrophomonas* species. The acidobacteria also encompass various soil generalists, with basically a K-type metabolic strategy, with high diversity, versatility, and adaptability to specific soil conditions (see Box 4.1 and Figure 4.2 for a specific example). In Table 4.2, an overview of key bacterial genera hypothesized to be soil generalists is presented. Such generalists can, however, adapt to the selective processes that are present in a particular microsite in the soil. If such local conditions are different per microsite, this adaptive process may lead to the emergence of different ecologically adapted types (*ecotypes*) within the species.

BOX 4.1 ACIDOBACTERIA ARE VERY ABUNDANT AND SUCCESSFUL SOIL BACTERIA

From the abundance of genes in soil metagenomics studies, as well as from an increasing amount of cultured and genome-sequenced isolates, it is known that the phylum Acidobacteria is widespread in soils. Across numerous 16S rRNA amplicon sequencing studies, 20%–60% of the sequences are often acidobacterial. Currently, the phylum has 26 highly diverse subdivisions. More than 50 different species have cultured representatives, spread over eight subdivisions. Most soil Acidobacteria are members of subdivisions 1, 3, 4, 6, 8, or 23. They occur in agricultural, grassland, forest, peatland, and tundra soils, both in temperate and in subtropical regions.

In subdivision 1, eight genera now have cultured species: *Acidobacterium*, *Terriglobus*, *Granulicella*, *Edaphobacter*, *Silvibacterium*, *Perracidiphilus*, *Terriglobus*, and *Candidatus Koribacter versatilis* Ellin345. Subdivision 3 has Acidobacteria bacterium, *Bryobacter aggregans*, and *Candidatus Solibacter usitatus* Ellin6076. Subdivision 4 has *Chloracidobacterium thermophilum* and *Pyrinomonas methylaliphatogenes*. Within subdivision 6, *Luteitalea pratensis* was isolated from temperate grassland soil and *Vicinamibacter silvestris* from semiarid subtropical savannah soil (Huber et al. 2016). Subdivision 8 has *Geothrix fermentans* and *Holophaga phoetida*, whereas subdivision 23 has *Thermoanaerobaculum aquaticum* (see concatenated tree from the work of Eichorst et al. (2018)).

Already in 1997, Acidobacteria had been identified in agricultural soil using FISH approaches. New 16S and 23S rRNA-directed oligonucleotide probes for different subdivisions of the Acidobacteria were developed by Meisinger et al. (2007). Several improved isolation techniques, as outlined in Chapter 18, brought better success (Huber et al., 2016). The application of phylum- or subdivision-specific 16S rRNA gene primers improved the identification of candidate colonies.

Comparative genomics of the soil-derived acidobacteria allowed insights into their potential life strategies in soil (Kielak et al., 2016; Eichorst et al., 2018). Multiple bacteriophage integrations were found, along with other mobile elements, influencing the structure and plasticity of these genomes. Low- and high-affinity respiratory oxygen reductases were found, reflecting the capacity to grow across oxygen gradients. The capacities to use a wide variety of carbohydrates and nitrogen sources via extracellular peptidases may assist in the adaptation to fluctuating nutritional conditions in the soil. Moreover, the potential to scavenge atmospheric H_2 was found in members of subdivisions 1, 3, and 4. The responsible enzyme, 1 h/5 hydrogenase, was first found in *Streptomyces avermitilis* (Constant et al. 2011), and it is also present in *Pyrinomonas methylaliphatogenes* (Greening et al. 2015). Scavenging of hydrogen is thought to be a mechanism that allows cells to remain energized in nutrient-starved soil ecosystems, contributing substantially to acidobacterial persistence.

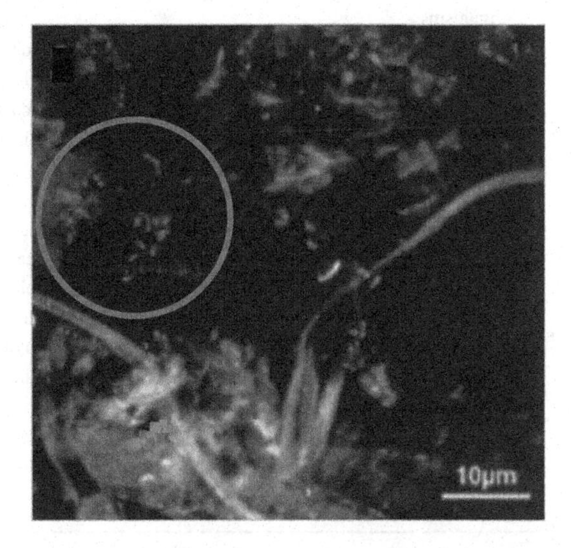

FIGURE 4.2 **(See color insert)** Detection of uncultured *Holophaga* sp. by fluorescent *in situ* hybridization (FISH) of soil. Blue (here in grey): *Holophaga* sp. cells stained using a specific ("IROG1") probe.

Perhaps a bit counterintuitive on the basis of the foregoing, some (plant-interactive) rhizobia, for example, *Bradyrhizobium* spp., can also be called generalists, as they have a non-plant-associated, saprophytic phase in their mode of life. In this phase, they have to survive in the soil and behave like common soil inhabitants. Several rhizobia can colonize a large range of plants efficiently—apart from their specific symbiotic partners. Other bacteria in the soil may show preference for either only specific soil types or soil conditions. These can be typified as *specialists*, although, at first glance, examples of these may be difficult to find. However, a paradigm of such specialists is given by organisms that thrive under the more extreme conditions in the soil. For instance, *Ralstonia metallidurans*, a heavy metal-resistant organism, was found to be dominant in heavy metal-polluted soils in Belgium, being rare in other soils. Hence, this organism may be called a specialist that is adapted to the heavy metal stress conditions. See Chapter 24 for further details.

TABLE 4.2
Bacterial and Archaeal Phyla Commonly Found in Soil and Their Lifestyles and Survival Strategies

Taxon	Lifestyle/Survival Strategy	Remarks
Bacillus and relatives	Spore formers; endospores are survival forms	Ubiquitous soil bacteria found in numerous non-extreme and extreme soils
Paenibacillus	Spore formers	Important inhabitants of rhizospheres and mycorrhizospheres
Actinobacteria	Some exhibit pleomorphic growth (*Arthrobacter*) or the ability to grow via hyphal extension (*Streptomyces*); some form spores and/or dormant, (nonspore) survival states	Ubiquitous organisms in soils Produce secondary metabolites, including antibiotics Many are good degraders of recalcitrant compounds
Paraburkholderia	Excellent survival of vegetative cells, "hardy" organisms	Often interactive with soil fungi and/or plants
Pseudomonas	Copiotrophs, fast growers upon substrate supply	Adapted to life in the rhizosphere
Serratia	Avid growth upon easily available substrates	Adapted to life in the rhizosphere
Rhizobium	Saprophytic, free-living, and symbiotic plant-interactive phases	Symbiotic nitrogen fixers that form nodules on legumes
Agrobacterium	Interactive with plants, taking profit of plant-released compounds	Include causal agents of crown gall disease in fruit and flower plants
Acidobacteria	Highly diverse phylum, with presumably numerous diverse lifestyles	Most evidence from direct molecular studies. Are increasingly being cultured
Verrucomicrobia	Unknown	Most evidence from direct molecular studies
Candidate division TM7	Unknown	Most evidence from direct molecular studies

4.3 WHAT BACTERIA AND ARCHAEA DO WE FIND IN SOIL?

To date, a broad range of diverse bacterial and archaeal types has been found to occur in soils. In light of the fiercely disputed species concept for prokaryotes, it is a continuing challenge to find a meaningful way to characterize these organisms. As outlined in Chapter 3, the modern phylogeny approaches developed over the past decades have allowed the rapid dissection of soil microbiomes in terms of their composition. Chapter 3 also addresses other strategies to identify and characterize the diversity of soil microorganisms. In the following section, we will examine traditional and modern approaches that have led to the current understanding of prokaryotes and their various roles in soil. We will also briefly review the drivers of soil microbiome composition, and examine some striking features of the lifestyles that prokaryotes have adopted to survive in soil.

4.3.1 Data from Traditional and Advanced Cultivation-Based Approaches

Over decades, the traditional cultivation-based studies using dilution plating on standard agar media or cultivation/enrichment in liquid media have taught us that a wide range of different bacteria, belonging to several different bacterial phyla, inhabit soil. However, we now know that only a minor fraction of the extant bacterial and, in particular, archaeal diversity has been unlocked so far, as is further outlined in Chapters 3, 12, and 18 (see also Figures 4.1a–c and 4.2). Another cultivation-based approach, that is, selective enrichment, also enables specific bacteria with the capacity to grow under defined conditions to become dominant. This strategy can, for instance, be used to obtain isolates that have the capacity to grow on specific carbon sources, including xenobiotic compounds (Figure 4.3). Chapter 23 further discusses the biodegradation of xenobiotic compounds in soil.

The conventional cultivation and enrichment approaches used to date have proven to be generally ineffective for the isolation of hitherto unknown microorganisms from soil, whose specific growth requirements are unknown. Hence, in addition to the direct molecular approaches, researchers have made recent improvements in cultivation methods (see Chapters 12 and 18). For instance, it has been demonstrated that minor changes in the methods, such as using polymers as carbon sources and applying longer incubation times and lower incubation temperatures, can result in the isolation of novel soil bacteria. Chapter 18 further examines current progress in isolations from soil. However, regardless of the progress achieved, it is foreseeable that there will still be microorganisms in soil that are recalcitrant to cultivation, as they—for instance—will only grow syntrophically, that is, in co-culture with other organisms (see also Chapter 9).

The bacteria isolated from soil by culturing so far have been shown to collectively possess an immense diverse metabolic capacity. The metabolic versatility of soil bacteria has led the pioneering microbiologists of the "Delft School of Microbiology," in particular Beyerinck and Baas-Becking, to propose the concept "Everything is everywhere, and the environment selects." The latter word might be replaced by "dictates," to emphasize the key relevance of environmental conditions for the outcome of the selection process. As an example of the catabolic diversity of microorganisms, a website has been established with lists of 150 pathways and more than 900 reactions, and these are primarily for xenobiotic chemical compounds (http://umbbd.ahc.umn.edu).

The roughly 100 years of soil microbiology have, thus, revealed a remarkable diversity of isolated organisms, of which specific groups are "common" to soils, whereas other groups are less frequently found. The common isolates have often been reported to belong to the class Proteobacteria (subclasses α-, β-, γ-, δ-, and ε-Proteobacteria) and the phyla Firmicutes, Actinobacteria, and Bacteroidetes (formerly named *Cytophaga–Flavobacterium–Bacteroides* group). Details about the phylogenetic relationships, ecology, and physiology of the prokaryotes can be found in *Bergey's Manual of Systematics of Archaea and Bacteria* (Whitley 2010; online publication 2015; https://onlinelibrary.wiley.com/doi/book/10.1002/9781118960608). This reference work has kept abreast with the roughly 100 new genera and more than 600 new species that are described per year (see also Chapter 3).

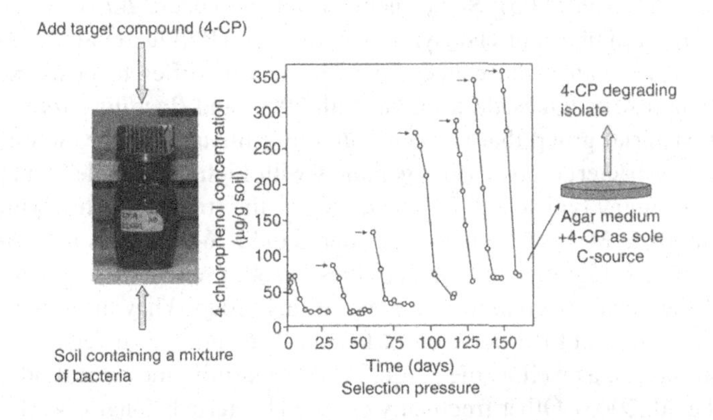

FIGURE 4.3 A strategy to isolate soil bacteria with specific biodegradative capacity-selective enrichment. Explanation: 4-CP = 4-chlorophenol, here used as the specific growth substrate (carbon and energy source) in the enrichment.

The Proteobacteria form a physiologically highly diverse group, encompassing organisms with different lifestyles. For instance, several α-Proteobacteria show saprophytic lifestyles, but others can interact with eukaryotic hosts. Functions assigned to the saprophytic α-Proteobacteria occurring in soil range from phototrophic CO_2 fixation (*Rhodobacter* spp.) to oxidation of methane and/or other C1 compounds (e.g., *Methylomonas*). The host-interactive α-Proteobacteria are *r*-strategists that can establish pathogenic as well as symbiotic relationships with plants and animals. With plants, these bacteria often colonize above- and belowground surfaces as well as endophytic sites. Well-known examples are the formation of nitrogen-fixing nodules in legumes by rhizobia and of crown gall diseases by *Rhizobium tumefaciens* (formerly *Agrobacterium tumefaciens*). The lifestyle of rhizobia is further detailed in Chapter 11. Also, bacteria that are classically not known as symbiotic or plant-pathogenic, like *Azospirillum* spp. and *Gluconacetobacter* spp., can form associations with plant roots, some of which occur endophytically in plant tissues but do not cause disease symptoms or are even beneficial. Other α-Proteobacteria are able to infect invertebrates (*Wolbachia* spp. in the springtail Collembola) or are even pathogens of humans (*Brucella abortus* or *Rickettsia* spp.). The β- and γ-Proteobacteria include many fast-growing *r*-strategists that can colonize plant roots efficiently and can be fairly abundant in various soils. Some of these bacteria are known as pathogens of plants and/or animals (e.g., *Erwinia carotovora* and *Pseudomonas aeruginosa*), but they also have the potential to control bacterial and fungal plant pathogens by producing antibiotics and other bioactive compounds (e.g., *P. fluorescens*, *Serratia plymuthica*—see Chapter 22 for further details). Typical β-Proteobacteria with particularly high affinity to plant roots are members of the genera *Paraburkholderia*, *Azoarcus*, *Herbaspirillum*, and *Variovorax*. Moreover, *Paraburkholderia* types are also known for their capacity to interact with fungi in soil. Some other β-Proteobacteria are also potential degraders of xenobiotic compounds or denitrifiers. Also, the nitrifying bacteria, belonging to the genera *Nitrosomonas* and *Nitrobacter*, fall into the group of β-Proteobacteria. Finally, the sulfate-reducing bacteria make up the greatest part of the δ-Proteobacteria.

Another frequently found group of soil bacteria can be found in the Firmicutes. These Gram-positive bacteria have genomes of a low G+C content (generally <50 mole%). The diverse genera *Bacillus*, *Paenibacillus*, and *Clostridium* fall into this group. In particular, members of *Bacillus* and *Paenibacillus* are frequently found by cultivation-based analyses of soils. The members of these genera are characterized by their capability to form endospores and survive for long periods of time in soil. They can grow rapidly on readily available carbon sources and can thus be considered as organisms that use the ecological *r*-strategy of niche occupation. Therefore, they can occur in rather high abundance in bulk soil, in the rhizosphere (Seldin et al. 1998), on the surface of plant residues, in the gut of invertebrates (Jensen et al. 2003), or in association with mycorrhizal fungi (Artursson 2005). Some members of the genera *Bacillus* and *Paenibacillus* are found as endophytes in plants or endosymbionts in fungi (Bertaux et al. 2003) without harming their hosts. However, others have become pathogens in different hosts, such as *Bacillus thuringiensis* in some insects (Quesada-Morada et al. 2004) and *Bacillus anthracis* in mammals. A third important bacterial group that is often found by cultivation is the Actinobacteria. These are also Gram-positive bacteria, possessing genomes with high G+C mole content, often >60%. Actinobacteria harbor many typical soil bacteria, such as the streptomycetes, which are very versatile in their metabolism. Most of them are considered to be K-strategists because of their rather low growth rates and their highly persistent activities in soil, even under low nutrient availability. Many members of the genus *Streptomyces* belong to this group. They are typical inhabitants of agricultural soils, as well as of the litter layer of forests. They produce diverse secondary metabolites, like many antibiotics, as well as the sesquiterpene geosmin, the typical odor compounds of earthy smell (Gust et al. 2003). Other frequently cultured bacteria belonging to the Actinobacteria are members of the genera *Rhodococcus*, *Arthrobacter*, and *Micrococcus*. Not surprisingly, many of these bacteria have high potential for the degradation of complex and recalcitrant natural and anthropogenic organic compounds. In addition, some Actinobacteria can form associations with plants. For instance, some *Frankia* spp. develop a nitrogen-fixing symbiosis with roots of trees, forming nodules (e.g., on Alder), whereas others can be plant pathogenic or are beneficially associated with the surface or the interior of plant roots.

4.3.2 Data from Cultivation-Independent Approaches

4.3.2.1 Bacterial Diversity

At present, numerous phyla in the domain Bacteria are recognized, on the basis of their 16S rRNA gene sequence similarity (Hug et al. 2016). See Chapter 3 for a description of the cultivation-independent approaches to the phylogeny of soil bacteria. The phyla are based on data obtained with isolates and on those obtained with direct molecular evidence. At the moment, a large fraction of the known phyla does not have cultured and described representatives. Such prokaryotic groups, which are characterized only by directly obtained sequences, are termed *candidate divisions* in order to indicate their unconfirmed status as a prokaryotic division. Many of the environmental phylogenetic groups (or *clades*), which are represented only by sequences in the database, are actually abundant and widely distributed in soil ecosystems, and may therefore be ecologically important and contribute significantly to key soil processes.

Molecular analyses of different soils across six continents (using 16S rRNA gene clone libraries) indicated that most soil bacteria fall into nine major phylogenetic groups (Delgado-Baquerizo et al. 2018): Proteobacteria, with the subclasses α-, β-, γ-, δ-, and ε-Proteobacteria, Actinobacteria, Acidobacteria, Planctomycetes, Chloroflexi, Verrucomicrobia, Bacteroidetes, Gemmatimonadetes, and Firmicutes. Representatives of the Planctomycetales, Verrucomicrobiales, and Acidobacteria had been uncovered a short time span before this major phylogeny-based "fishing" expedition (Buckley and Schmidt 2003). This listing is not exhaustive, as a variety of additional bacterial phyla can be found in the aforementioned candidate divisions, such as TM6, TM7, and OP11 (Hugenholtz et al. 2001). Even today, and in the phylogenetic groups that comprise many cultured species, new species are being described continuously (Fierer et al. 2013). This points to the fact that we are still far away from a complete knowledge of the diversity of the bacteriomes and archaeomes in soils.

Strikingly, members of the phylum Acidobacteria have been detected in almost every analysis of soil DNA (see Box 4.1). This phylum is therefore abundant and probably plays an ecologically important role in soil ecosystems. It has nearly as deep phylogenetic branchings as that of the Proteobacteria and is phylogenetically very diverse, which may indicate that it is also diverse in metabolic terms. In addition, the phylum Verrucomicrobia is also globally widespread and abundant in soil, although relatively few verrucomicrobia have been isolated (e.g., Bergmann et al. 2011). For instance, 16S rRNA gene sequences from cloned soil DNA show that, on average, about 7% (ranging from 0% to 21%) of the 16S rRNA sequences in soil clone libraries are affiliated with this phylum (Janssen 2006). Especially, members of the class "Spartobacterium" (subdivision 2) are predominant in soil. Also, members of subdivisions 3 and 4 of the Verrucomicrobia are often detected in soil. Among the isolated verrucomicrobia are members of the *Prosthecobacter* (subdivision 1) species and Ultramicrobacteria (subdivision 4), which often occur as "dwarf cells," being less than 0.1 μm^3 in volume.

4.3.2.2 Archaeal Diversity

Archaea in soil have been overlooked for a long period of time, as they are hard to isolate due to their slow growth rates. Moreover, they have poorly understood physiological traits, which prevent the development of suitable media and isolation strategies. Compared to the phylogeny of Bacteria, the tree of life for Archaea is still under discussion. Based on whole-genome sequencing, and in addition to the classically described Crenarchaeota, Korarchaeota, and Euryarchaeota (Woese et al. 1990), two new main phyla, namely Thaumarchaeota and Aigarchaeota (Nunoura et al. 2010), have been published. Using 16S rRNA-based construction of phylogenetic trees, even more novel phyla have been postulated to exist, including the Proteoarchaeota, Bathyarchaeota, Geoarchaeota, and Lokiarchaeota (Petijean et al. 2014; Koonin et al. 2015). For soils, mainly members of the Euryarchaeota and Thaumarchaeota are thought to play important roles, despite the fact that members of other phyla—like Crenarchaeota—have also been detected. The latter, for example, were found as part of the plant-associated microbiome (Taffner et al. 2018). Based on the molecular data, today it is obvious that many archaea play important roles in soils and contribute mainly to C and N turnover processes.

A functional trait that is unique to archaea (belonging to the phylum Euryarchaeota) is the production of methane under anoxic conditions. These *methanogens* have gained attention in the past decades, as the release of methane from soils (flooded, highly compacted, or permafrost affected) contributes to global warming. For a long time, it was thought that methanogens occur exclusively in such oxygen-depleted environments, as studies with the very few pure cultures that were obtained indicated that oxygen causes damage to the F420-hydrogenase complex, an essential enzyme in methanogenesis (Schönheit et al. 1981). However, molecular studies revealed the presence of methanogens also in soils that can be considered to be well aerated, like desert soils or soils from forests, glacier forefields, and/or subalpine meadows. The majority of the sequences related to methanogens obtained by the analysis of DNA from soils all over the world could be assigned to the genera *Methanosarcina* and *Methanocella*, which have been considered by many authors as part of the autochthonous microflora of soils (Aschenbach et al. 2013). For a minority of habitats, sequences of other methanogens were also retrieved, including the classes Methanobrevibacter, Methanobacteria, Methanomicrobia, and Methanococci (Praeg et al. 2014). Methanogens also colonize the digestive tract of soil invertebrates, where they convert the products of fermentation, like acetic acid, to hydrogen and methane (Sustr et al. 2014). Interestingly, often methanogens are also capable of fixing nitrogen, making such archaea important contributors to this process mainly in anoxic systems (Bae et al. 2018).

About one decade ago, the key archaeal involvement in N transformation, namely, the oxidation of ammonia to hydroxyl amine and nitrite—part of the nitrification process—was discovered (Leininger et al. 2006). Originally, the phylum Crenarchaeota was considered as the most important group in this aspect, constituting up to 5% of the total number of prokaryotic cells in the system. However, after the genome of *Cenarchaeum symbiosum* was sequenced and found to differ significantly from that of other members of the hyperthermophilic phylum Crenarchaeota, the phylum Thaumarchaeota was established, which contains, besides *C. symbiosum*, also the ammonia-oxidizing archaea (AOA) *Nitrosopumilus maritimus*, *Nitrososphaera viennensis*, and *Nitrososphaera gargensis* (Brochier-Armanet et al. 2008). Since the first description of the AOA, many studies addressed the question whether ammonia-oxidizing bacteria (AOB) and AOA are functionally redundant or have their own specific niche in soil (see also Chapter 11). Measurements of the abundance of AOB versus AOA in many cases revealed that AOA outcompete their bacterial counterparts. However, the N turnover rates of AOB are significantly higher than those of AOB, at least when comparing isolates. Several lines of evidence indicate that AOA prefer soils with lower pH values and reduced ammonia input, whereas AOB are mostly found in fertilized soils with neutral or slightly acid pH, indicating a niche separation between both types of ammonia oxidizers in soil (Stempfhuber et al. 2015).

4.4 COPING WITH LIFE IN SOIL

4.4.1 Physiological and Genetic Adaptations to Soil Conditions

What makes bacteria and archaea living in soil successful, in ecological and evolutionary terms? As a result of the great variability in conditions across a soil, it is difficult to give a straightforward and direct answer to this question. It is possible that the complex and heterogeneous nature of soil, resulting in a myriad of local selection pressures, is reflected in the tremendously diverse metabolic and adaptive strategies that one can find across soil bacteria and archaea. Probably more than any other type of organism, these organisms can quickly adapt to fluctuations in environmental conditions, in both physiological and genetic terms. The possibilities for (genetic) adaptation are discussed in Chapters 3 and 7. With respect to physiology, it has become known that soil prokaryotes possess a broad array of mechanisms that allow them to sense the chemistry of their immediate surroundings and provide the adequate physiological response to it. For instance, the soil bacterium *Bacillus subtilis* possesses a wide variety of environmental sensing/response systems (*two-component regulatory systems*; see Chapter 9) that allow it to actively integrate into its immediate environment. Thus, life in soil has evolved with bacteria fine tuning their expression systems—and genomes—over evolutionary time, allowing them to directly cope with the local conditions offered by the soil environment. Many bacteria thus fine-tune their gene expression using autoinduced diffusible metabolites that provide them with valuable information about habitat quality and population density (Hense and Schuster 2015). What has this selection for variability and alertness to a changing environment resulted in? Studies of whole bacterial genomes have revealed one key feature typical for soil bacteria: their genomes are generally large, especially compared to those of marine prokaryotes or of intracellular pathogens. The largest bacterial genomes known so far belong to soil bacteria; for example, the myxobacterium *Sorangium cellulosum* has a genome size of 12.2 Mbp, the fungal associate *Paraburkholderia terrae* one of about 11.5 Mbp, and the slow-growing plant symbiont *Bradyrhizobium japonicum* one of about 9.1 Mbp. Collectively, these and other genomes of soil bacteria are larger than those of any randomly chosen comparator group of clinical or marine bacteria. Strikingly, the genomes invariably contain a substantial number of open reading frames with as yet unknown functions. It is possible that many of these unknown putative functions are related to the bacterial struggle for survival under the fluctuating conditions encountered in the soil environment. Also, metagenomics-based analyses of soil microbial communities have provided clear indications of the presence of high diversities of potential functions—of which many unknown—in this habitat (Tringe et al. 2005). See also Chapter 14 for an extensive review of developments in soil metagenomics.

4.4.2 Physiological Response to Nutrient Limitation

One consequence of the nutrient (e.g., carbon, nitrogen, or phosphorus) starvation that bacteria and most likely archaea experience in soils is a reduction in their cell size. Generally, most bacteria (Gram-negative as well as Gram-positive ones) found in soil are smaller than those growing under nutrient-rich conditions in the laboratory. These small cells are formed as a result of either the occurrence of cell divisions without a concomitant increase in biomass or of the degradation of endogenous cell material, leading to condensed cytoplasms. See Figure 4.4. for an example. The cell size reduction is thought to be part of a highly programmed process, governed by regulatory networks present in bacterial cells, allowing for these to survive starvation. The cells will eventually enter a nongrowth phase if one or several essential nutrients are lacking. Under favorable conditions *in vitro*, some bacteria can grow at generation times of around 20 min. This means that one bacterial cell will result in 8 cells after 1 h and 64 after 2 h. If growth would continue for 24 h, we would get astronomic values of both cell numbers and biomass. It is known that such unlimited growth will never occur in soil, as the lack of one or more nutrients or other conditions will become strongly restrictive to growth.

There are different recognizable physiological responses to nutrient scarcity in soil. Cells may differentiate into spores (such as observed in species of the Gram-positive *Bacillus* and *Clostridium*) or shifts in their physiology may occur, as revealed by van Overbeek et al. (1995). The spore formers, for example, *Bacillus* species, form endospores during starvation as well as desiccation conditions in soil. This nongrowing cellular form allows for the genome of the cells to remain intact during harsh conditions. The endospore is highly resistant to mechanical and chemical stresses and thus allows the organism to persist for many years to maybe centuries in the soil. Once conditions

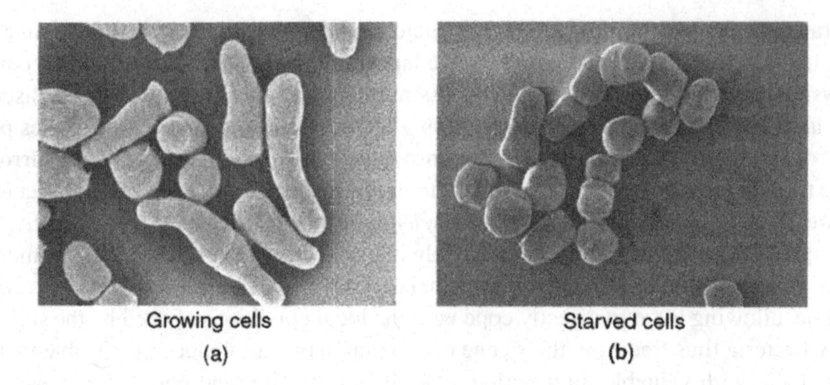

FIGURE 4.4 The change in morphology of cells of the soil bacterium *Arthrobacter chlorophenolicus* upon exposure to nutrient-limited conditions in the soil. (a) Cells with adequate nutrient supply occur mainly in the rod form. (b) Cells after nutrient deprivation form smaller rounded forms.

become favorable again, the endospore can germinate, converting back to a vegetative cell. In addition, some non-sporulating Gram-positive bacteria, for example, *Micrococcus* species, can accumulate reserve carbon polymers when nutrients are in excess, and subsequently utilize these polymers during nutrient deprivation. This allows for the cell to sustain a basal level of energy in order to remain viable.

In bacteria, the starvation response is part of a highly controlled regulatory mechanism that has mainly been studied in Gram-negative bacteria. In these bacteria, the process can be divided into three phases. The first phase, the *stringent control phase*, includes a decline in macromolecule synthesis and increased protein degradation. The second phase is characterized by the degradation of cellular reserve material, a shift in the fatty acid content of the cell membrane, and an onset of tolerance towards other stress conditions. The changes in fatty acid profiles in the cell membrane include, for instance, an increase in the *trans/cis* ratio of the fatty acids that make up the membrane and the ratio of saturated to unsaturated fatty acids. Starvation-induced stress tolerance can enable cells to become more tolerant to, for instance, oxidative stress, temperature changes, and osmotic challenges. Gram-negative bacteria can also become more hydrophobic and develop filaments, to increase adhesion and aggregation for protective purposes. The cell wall composition is also affected as a result of starvation, possibly to protect the cell from lysis. Under conditions of nutrient limitation, bacteria can also induce the expression of transport systems that are optimized for the uptake of nutrients at low concentrations. The third phase is characterized by a slow decline of respiratory activity as well as a further decline in the synthesis of RNA, protein, and peptidoglycan. Other changes can be found in the content of lipids and carbohydrates, and a stronger condensation of the DNA was also observed. In this starvation-induced state, the cells may still carry out essential functions, such as maintaining a functional protein synthesis machinery, which will allow the cells to recover rapidly upon the addition of nutrients. Interestingly, the ribosome content of starved bacteria is often higher than what would be needed for the immediate physiological requirements. Instead, it is thought that the high ribosome content will allow the cells to rapidly respond to a sudden input of nutrients.

Proteins specific for the response to starvation are sequentially synthesized during the different starvation-induced phases. Especially, the early proteins have been shown to be essential for starvation survival. Interestingly, a few of the starvation-induced proteins in *Escherichia coli* have been shown to be common with the proteins synthesized during heat shock and osmotic or oxidative stress. This may explain the higher tolerance to stress conditions, like those imposed by heat, oxidative stress, osmotic stresses, and low temperatures found for bacteria subjected to starvation.

We conclude from the foregoing that, under the nutrient hardship often found in soil, bacteria change their physiologies to become "generically" more resistant to stresses. They use a physiological mechanism that is dedicated to a generally slow utilization of resources. With respect to this, bacteria that are able to utilize endogenous molecules at a slower rate appear to be better survivors than those that are unable to do so.

Although different bacteria have, thus, developed different strategies to survive periods of nutrient deprivation, there are indications that a common type of regulation is operational during starvation survival. An interaction between starvation survival and quorum sensing—that is, cell-to-cell signaling between bacteria to monitor population density (see Chapter 9)—has been shown. The need to rapidly respond to fluctuating conditions has led to the evolution of sophisticated means of communication between cells, which prevents starvation by estimating (sensing) the population density and responding by density-dependent growth rate regulation (Hense and Schuster 2015).

4.4.3 Implications for Survival of Bacterial Inoculants in Soil

Some bacteria have properties that are of interest for environmental biotechnological applications, such as *plant growth-promoting bacteria* and *biological control agents* (see Chapter 22) and organisms involved in the bioremediation of *xenobiotic* compounds in soil (see Chapter 23). We have seen that limitation of available nutrients in soils can adversely affect the indigenous bacterial communities, and the effect on introduced bacteria can be even more extreme. In addition, the bacteria that are introduced into a habitat like soil, in addition to nutrient limitation, have to cope with other harsh conditions to which they are not adapted, making competition with the indigenous microorganisms for nutrients difficult (van Elsas et al. 2012). It has been observed that inoculant bacteria stand a better chance of survival in soil systems that have been disturbed, for instance, by the addition of resources from manure, or by (partial) sterilization. In such cases, it is plausible that the enhanced availability of resources temporarily alleviates the stress encountered in the system, which offers an enhanced chance for the inoculant cells to establish and persist. Similarly, the introduction of inoculants as seed coats offers higher chances for successful colonization of the emerging and growing roots in competition with the indigenous microflora.

It is important to detect bacterial inoculants after their introduction into soil, in order to determine their distribution, survival, and efficacy. Enumeration of the introduced bacteria has traditionally been conducted by counting of colony-forming units on agar media, and a decline in colony numbers is then usually correlated with a decline in the number of viable cells. However, immunofluorescence techniques and molecular biological tools, which have become available during the past two decades (see Chapters 12–17), have made it clear that bacterial cells may remain metabolically active, although they cannot be shown to divide on agar or in liquid medium. This state has often been referred to as the *viable but nonculturable* (*VBNC*) state (see Chapter 12), although this term has been challenged as viability has been traditionally defined as the ability to be cultivated. Also, VBNC cells may actually be culturable if conditions were optimized for their growth. Cells subjected to stressful conditions have also been suggested to enter a state of dormancy—or resting state—as part of the starvation response described above. Bacteria introduced into soil that enter a VBNC or dormancy state may be capable of long-term survival and may be resuscitated when conditions become favorable again. As outlined later (Chapter 12), molecular tools that are based on the detection of specific nucleic acid sequences cannot normally distinguish between different physiological states of the cells, such as viable, VBNC, dormant, dying, or dead. However, viability stains that detect intact versus injured membranes can be used in combination with microscopy or flow cytometry to distinguish dead from live cells.

In summary, inoculant bacteria, when exposed to soil, are likely to be affected by the stressful conditions of soil that will limit their establishment and survival. Disturbance of the soil ecosystem can often open niches and thus enhance inoculant establishment. It is key to the success of our applications that we fully understand the autecology of inoculants, for which we need specific techniques that detect inoculant fate as well as activity. Strain-specific quantification methods based on genome-borne molecular targets as well as specific labeling methods have recently been successfully applied with respect to these requirements (Stets et al. 2015).

4.5 CONCLUDING REMARKS

We have seen that, much as in macro-ecological communities, the communities of bacteria and archaea in soil are almost invariably composed of a range of diverse types with different ecophysiological characteristics. Analyses of the structures of such communities in soil by either cultivation or direct molecular techniques have shown that the bacterial phylogenetic and functional types found in similar soil types most often fall into largely the same broad categories. In other words, there appears to be a relative constancy or "stability" in the bacterial community structures in soil. A plausible explanation for the apparent coexistence of the underlying bacterial populations and the stability of bacterial community structures in soil is that the organisms that make up these communities have different lifestyles and survival strategies, which allow them to coexist as they occupy different niches, or even cooperate in a number of ways. The most striking example of coexistence is obviously the case when different bacterial types share the same microhabitat. This can occur if there is *syntrophy* or interdependency, for instance in nutritional terms, between the different microorganisms and if they are not ecologically incompatible (Chapter 9). However, our current-day molecular methods based on DNA extracted from 0.5 to 1 g of soil samples do not enable to consider the precise spatial interactions between microbes. Thus, visualization techniques play important roles in the assessment of interactions on the local scale, which is at the micrometer scale. Figure 4.1 shows a good example of this. Specific labeling techniques using molecular or immunological probes and high-resolution microscopic and image analysis techniques (Dazzo et al. 2015) allow insights into the small-scale microbial interactions in soil microhabitats, like in the rhizosphere. Since fluorescent probes have limitations in highly autofluorescent soils, so-called Gold-fluorescence *in situ* hybridization (FISH), which enables the detection of single microbial cells specifically by scanning electron microscopy, has been developed (Schmidt et al. 2012).

In contrast, coexistence of organisms that do not strictly share the microhabitat, since they occur some distance apart from each other in the soil, is a more common phenomenon. In such cases, there may still be mutual (chemical) interactions between the partners via diffusion through the soil water and gaseous phases, which may lead to mutualism or other interactive behavior. However, in other cases, the organisms that occur spatially separated in soil may occupy different niches as defined by their lifestyles. In the above cases, causally determined or *apparent* coexistence may be observed, and further study is required to work this out. Thus, there are different solutions to ecological success, and organisms that have developed these in their lifestyles can coexist in the same microhabitat or in semi-shared habitats. How can the organisms with different lifestyles coexist in light of the fact that they all depend on the limited resources present in soil? The answer lies in the fact that the different organisms in soil utilize the available resources in different ways, which places clear and distinct limits on their chances for growth and also differentiates their niches in space and time.

We have learned that in most soils, only a limited number of bacterial and archaeal types is dominant (present in relatively high numbers), whereas other species are more rare and can even be absent (or be present in numbers at or below the limits of current methods of detection). This is probably because the survival and growth of the different organisms are restricted to discrete sites in soil, in which the nutritional and other physicochemical conditions necessary for their establishment, growth, and survival are propitious. The presence and the number of these sites determine the *carrying capacity* of the soil for each of its inhabitants. And so, this carrying capacity can be different per inhabitant, depending on its physiological capacities and its required niche space.

From the foregoing, a picture emerges of soil as a very rich reservoir of bacteria and archaea, the sheer diversity of which we are presently only beginning to understand. These diversities can be maintained as a result of the presence of large numbers of niches in soil, which are determined by the local physical and nutritional conditions. It is a challenge for future work to more clearly establish the link between key habitat determinants and the lifestyles and survival strategies of soil bacteria. In this respect, the roughly 40% of open reading frames with unknown function that are commonly found in the genomes of many soil bacteria are very interesting, since these may provide the key to the enhanced understanding of the lifestyle of many members of soil microbiomes. From a genomic perspective, the microorganisms in a living and healthy soil collectively contain all genetic information that sustain our biosphere, and their further exploration will bring us closer to dissecting the "genetic gold" that is present just below our feet.

REFERENCES

Artursson, V. 2005. Bacterial-fungal interactions highlighted using microbiomics: potential applications for plant growth enhancement. Sveriges Landbrucknes Universitet, Faculty of Natural Resources and Agricultural Sciences, Doctoral Thesis no. 127, 2005.

Aschenbach, K., Conrad, R., Řeháková, K., Doležal, J., Janatková, K. and R. Angel. 2013. Methanogens at the top of the world: occurrence and potential activity of methanogens in newly deglaciated soils in high-altitude cold deserts in the Western Himalayas. *Front Microbiol* 4, 359.

Bae, H.S., Morrison, E., Chanton, J.P. and A. Ogram. 2018. Methanogens are major contributors to nitrogen fixation in soil of the Florida everglades. *Appl Environ Microbiol* 84, e02222-17.

Bergmann, G.T., Bates, S.T., Eilers, K.G., et al. 2011. The under-recognized dominance of *Verrucomicrobia* in soil habitats. *Soil Biol Biochem* 43, 1450–1455.

Bertaux, J., Schmid, M., Chemidlin Prevost-Boure, N., et al. 2003. *In situ* identification of intracellular bacteria related to *Paenibacillus* sp. in the mycelium of the ectomycorrhizal fungus *Laccaria bicolor* S238N. *Appl Environ Microbiol* 69, 4243–4248.

Brochier-Armanet, C., Boussau, B., Gribaldo, S. et al. 2008. Mesophilic crenarchaeota: proposal for a third archaeal phylum, the Thaumarchaeota. *Nat Rev Microbiol* 6, 245–252.

Buckley, D.H. and T.M. Schmidt. 2003. Diversity and dynamics of microbial communities in soils from agro-ecosystems. *Environ Microbiol* 5, 441–452.

Constant, P., Chowdhury, S.P., Hesse, L., Pratscher, J. and R. Conrad. 2011. Genome data mining and soil survey for the novel group 5 [NiFe]-hydrogenase to explore the diversity and ecological importance of presumptive high-affinity H_2-oxidizing bacteria. *Appl Environ Microbiol* 77, 6027–6035.

Dazzo, F.B., Yanni, Y., Jones, A. and A. Elsadany. 2015. CMEIAS bioimage informatics that define the landscape ecology in immature microbial biofilms developed on plant rhizoplane surfaces. *AIMS Bioeng* 2, 469–486.

Delgado-Baquerizo, M., Oliverio, A.M., Brewer, T.E., et al. 2018. A global atlas of the dominant bacteria found in soil. *Science* 359, 320–325.

Eichorst, S.A., Trojan, D., Roux, S., Herbold, C., Rattei, T. and D. Woebken. 2018. Genomic insights into the *Acidobacteria* reveal strategies for their success in terrestrial environments. *Environ Microbiol* 20, 1041–1063.

Fierer, N., Ladan, J., Clemente, J.C., et al. 2013. Reconstructing the microbial diversity and function of pre-agricultural tall grass prairie soils in the United States. *Science* 342, 621–624.

Greening, C., Carere, C.R., Rushton-Green, R., et al. 2015. Persistence of the dominant soil phylum *Acidobacteria* by trace gas scavenging. *Proc Natl Acad Sci USA* 112, 10497–10502.

Gupta, R.S. 2000. The natural evolutionary relationships among prokaryotes. *Crit Rev Microbiol* 26, 111–131.

Gust, B., Challis, G.L., Fowler, K., et al. 2003. PCR-targeted *Streptomyces* gene replacement identifies a protein domain needed for biosynthesis of the sesquiterpene soil odor geosmin. *Proc Natl Acad Sci USA* 100, 1541–1546.

Hense, B.A. and M. Schuster. 2015. Core principles of bacterial autoinducer systems. *Microbiol Molec Biol Rev* 79, 153–169.

Huber, K.L., Geppert, A.M., Wanner, G., Fösel, B.U., Wüst, P.K. and J. Overmann. 2016. The first representative of the globally widespread subdivision 6 Acidobacteria, *Vicinnamibacter silvestris* gen. nov., sp. nov., isolated from subtropical savannah soil. *Int J Syst Evol Microbiol* 66, 2971–2979.

Hug, L.A., Baker, B.J., Anantharaman, K., et al. 2016. A new view of the tree of life. *Nat Microbiol* 1, 16048.

Hugenholtz, P., Tyson, G., Webb, R., Wagner, A. and L. Blackall. 2001. Investigation of candidate division TM7, a recently recognized major lineage of the domain Bacteria with no known pure-culture representatives. *Appl Environ Microbiol* 67, 411–419.

Janssen, P.H. 2006. Identifying the dominant soil bacterial taxa in libraries of 16S rRNA and 16S rRNA genes. *Appl Environ Microbiol* 72, 1719–1728.

Jensen, G.B., Hansen, B.M., Eilenberg, J. and J. Mahillon. 2003. The hidden lifestyles of *Bacillus cereus* and relatives. *Environ Microbiol* 5, 631–640.

Kielak, A.M., Barreto, C.C., Kowalchuk, G.A., van Veen, J.A. and E.E. Kuramae. 2016. The ecology of *Acidobacteria*: moving beyond genes and genomes. *Front Microbiol* 7, e744.

Koonin, E.V. 2015. Archaeal ancestors of eukaryotes: not so elusive any more. *BMC Biol* 13, 84.

Leininger, S., Urich, T., Schloter, M., et al. 2006. Archaea predominate among ammonia-oxidizing prokaryotes in soils. *Nature* 442, 806–809.

Meisinger, D.B., Zimmermann, J., Ludwig, W., Schleifer, K.H., Wanner, G. and M. Schmid. 2007. *In situ* detection of novel *Acidobacteria* in microbial mats from a chemolithotrophically based cave ecosystem (Lower Kane Cave, WY, USA). *Environ Microbiol* 9, 1523–1534.

Nunoura, T., Takaki, Y., Kakuta, J., et al. 2010. Insights into the evolution of Archaea and eukaryotic protein modifier systems revealed by the genome of a novel archaeal group. *Nucleic Acids Res* 39, 3204–3223.

Petitjean, C., Deschamps, P., López-García, P. and D. Moreira. 2014. Rooting the Domain archaea by phylogenomic analysis supports the foundation of the new kingdom Proteoarchaeota. *Genome Biol Evol* 7, 191–204.

Praeg, N., Wagner, A.O. and P. Illmer. 2014. Effects of fertilisation, temperature and water content on microbial properties and methane production and methane oxidation in subalpine soils. *Eur J Soil Biol* 65, 96–106.

Quesada-Moraga, E., Garcia-Tovar, E., Valverde-Garcia, P. and C. Santiago-Alvarez. 2004. Isolation, geographical diversity and insecticidal activity of *Bacillus thuringiensis* from soils in Spain. *Microbiol Res* 159, 59–71.

Schmidt, H., Eickhorst, T. and M. Mußmann. 2012. Gold-FISH: a new approach for the *in situ* detection of single microbial cells combining fluorescence and scanning electron microscopy. *Syst Appl Microbiol* 35, 518–525.

Schönheit, P., Keweloh, H. and R.K. Thauer. 1981. Factor F420 degradation in *Methanobacterium thermoautotrophicum* during exposure to oxygen. *FEMS Microbiol Lett* 12, 347–349.

Seldin, L., Rosado, A.S., da Cruz, D.W., Nobrega, A., van Elsas, J.D. and E. Paiva. 1998. Comparison of *Paenibacillus azotofixans* strains isolated from rhizoplane, rhizosphere, and non-root-associated soil from maize planted in two different Brazilian soils. *Appl Environ Microbiol* 64, 3860–3868.

Stempfhuber, B., Engel, M., Fischer, D., et al. 2015. pH as a driver for ammonia-oxidizing Archaea in forest soils. *Microb Ecol* 69, 879–883.

Stenström, J., Svensson, K. and M. Johansson. 2001. Reversible transition between active and dormant microbial states in soil. *FEMS Microbiol Ecol* 36, 93–104.

Stets, M.I., Alqueres, S., de Souza, E.M., et al. 2015. Quantification of *Azospirillum brasilense* FP2 in wheat roots by strain-specific qPCR. *Appl Environ Microbiol* 81, 6700–6709.

Sustr, V., Chronáková, A., Semanová, S., Tajovský, K. and M. Simek. 2014. Methane production and methanogenic Archaea in the digestive tracts of millipedes (Diplopoda). *PLoS One* 9, e102659.

Taffner, J., Erlacher, A., Bragina, A., Berg, C., Moissl-Eichinger, C. and G. Berg. 2018. What is the role of Archaea in plants? New insights from the vegetation of alpine bogs, *mSphere* 3, e00122-18.

Tringe, S.G., von Mering, C., Kobayashi, A., et al. 2005. Comparative metagenomics of microbial communities. *Science* 308, 554–557.

van Elsas, J.D., Chiurazzi, M., Mallon, C.A., et al. 2012. Microbial diversity determines the invasion of soil by a bacterial pathogen. *Proc Natl Acad Sci USA* 109, 1159–1164.

van Overbeek, L.S., Eberl, L., Givskov, M., Molin, S. and J.D. van Elsas. 1995. Survival of, and induced stress resistance in, carbon-starved *Pseudomonas fluorescens* cells residing in soil. *Appl Environ Microbiol* 61, 4202–4208.

Whitley, W. (ed.). 2010. *Bergey's Manual of Systematic Bacteriology*, Vol. 2, Springer, USA.

Woese, C.R. and G. Fox. 1977. Phylogenetic structure of the prokaryotic domain: the primary kingdoms. *Proc Natl Acad Sci USA* 74, 5088–5090.

Woese, C.R., Kandler, O. and M.L. Wheelis. 1990. Towards a natural system of organisms: proposal for the domains Archaea, Bacteria, and Eucarya, *Proc Natl Acad Sci USA* 87, 4576–4579.

5 The Fungi in Soil

Roger D. Finlay
Swedish University of Agricultural Sciences

R. Greg Thorn
University of Western Ontario

CONTENTS

5.1 Introduction 66
5.2 Diversity of Fungi and Fungus-Like Organisms 66
  5.2.1 Chytrids: Chytridiomycota and Blastocladiomycota 68
  5.2.2 "Zygomycetes": Mucoromycotina and Zoopagomycotina 68
  5.2.3 Glomeromycota 68
  5.2.4 Ascomycota 69
  5.2.5 Basidiomycota 69
  5.2.6 Fungus-Like Organisms 69
5.3 Taxonomy and Evolution 70
5.4 Fungal Numbers, Biomass, Activity, and Identification 71
  5.4.1 Measurements of Biomass 71
  5.4.2 Measurements of Fungal Activity 71
  5.4.3 Identification of Fungi 72
5.5 Fungal Metabolites and Metabolism 72
5.6 Fungal Saprotrophs—Litter Decomposition 74
5.7 Fungal Parasites and Pathogens 76
5.8 Fungal Symbionts—Mycorrhiza, Lichens, and Endophytes 77
  5.8.1 Arbuscular Mycorrhiza (AM) 77
  5.8.2 Ericoid Mycorrhiza 78
  5.8.3 Ectomycorrhiza 78
  5.8.4 Orchid Mycorrhiza, Monotropid Mycorrhiza, and Mycoheterotrophs 79
  5.8.5 Other Types of Mycorrhiza 80
  5.8.6 Lichens, Endophytes, and Other Types of Symbioses 80
  5.8.7 Functional Effects of Mycorrhizal Fungi 81
5.9 Interactions of Fungi with Other Organisms 82
  5.9.1 Interactions with Bacteria 82
  5.9.2 Interactions with Other Fungi 83
  5.9.3 Interactions with Soil Microfauna and Mesofauna 83
5.10 Fungi in Biogeochemical Cycles and Nutrient Cycling 84
5.11 Applications of Soil Fungi 84
  5.11.1 Biological Control 86
  5.11.2 Bioremediation 86
  5.11.3 Biofertilization 87
5.12 Conclusions 87
References 88

5.1 INTRODUCTION

Fungi play a central role in many soil microbiological processes, influencing soil fertility, decomposition, cycling of minerals and organic matter, plant health, and nutrition. They also influence the structure and functioning of plant communities and soil ecosystems. Fungi are immensely diverse, both structurally and functionally, and have adopted different trophic strategies, occurring as *saprotrophs*, *symbionts*, and *pathogens*. Although this chapter is about "soil" fungi, the filamentous growth habit of many fungi, coupled with their different trophic strategies, implies that individual fungi can often simultaneously colonize different substrates, such as living or dead plant tissues, coarse woody debris, soil animals, and mineral substrates, in addition to the soil itself. Fungi, together with animals and plants, represent one of three major evolutionary branches of multicellular organisms, and their uniqueness is reflected in the fact that they have the status of a kingdom (the *Fungi*). The diversity of fungi is high, and although only about 120,000 species have been described so far, it is likely that this represents <5% of the true number of fungal species, which is estimated to be about 2.2–3.8 million (Hawksworth and Lücking 2017). Although fungi are structurally and functionally diverse, they share certain common features such as the fact that they are eukaryotic, contain a range of membrane-bound organelles such as mitochondria and vesicles, and possess membrane-bound nuclei containing several chromosomes. Fungi are heterotrophs, requiring external carbon sources, and typically display filamentous growth as a result of the hyphae that exhibit apical growth. They are thus able to colonize and penetrate solid substrates, forming a mycelium by repeated branching. Some fungi, such as single-celled yeasts, exhibit dimorphic growth and can reproduce in liquids by budding or fission, as well as colonizing other substrates by mycelial growth. Unlike other eukaryotes, fungi typically have haploid nuclei; however, the hyphae may have several nuclei in each compartment. Moreover, many budding yeasts are diploid. Fungi characteristically produce spores, and many can reproduce both sexually and asexually. They have cell walls composed of polymers of glucose, such as chitins and glucans, and secrete enzymes that degrade complex polymers at their hyphal tips, allowing them to take up smaller molecules. Fungi, being heterotrophic, require external carbon sources for energy and cellular synthesis. Three major groups—, saprotrophs, pathogens, and symbionts, —can be distinguished in accordance with the trophic strategies adopted to acquire the organic compounds. Apart from their major impact on natural terrestrial ecosystems, soil fungi have important, as yet largely unexploited, applications in the *biological control* of pathogens, the *bioremediation* of polluting compounds, and the *biofertilization* of soil. Soil fungi also produce a large range of secondary metabolites with potential for medical, biocontrol, or environmental applications. The secondary metabolites include antibiotics such as penicillins and cephalosporins, immunosuppressants, mycotoxins, and aflatoxins. Other compounds such as organic acids and siderophores may be involved in the release and sequestration of mineral nutrients as well as in antagonistic interactions with other organisms. Furthermore, fungi secrete a wide variety of enzymes used in either pathogenic interactions or the degradation of plant litter and woody substrates. Some of these enzymes find important applications in bioremediation of organic pollutants. Symbiotic mycorrhizal fungi produce mycelia growing from the roots of their host plants into the surrounding soil, connecting them to the heterogeneously distributed nutrients required for their growth, enabling the flow of energy-rich compounds required for nutrient mobilization while simultaneously providing conduits for the translocation of the mobilized products back to their hosts.

5.2 DIVERSITY OF FUNGI AND FUNGUS-LIKE ORGANISMS

In this section, the main groups of fungi and fungus-like organisms inhabiting soil are reviewed, together with their general structural and functional features (Table 5.1). These include the true fungi that can be divided into five groups: the chytrids, zygomycetes, Glomeromycota, Ascomycota, and Basidiomycota (Hibbett et al. 2007), as well as four main groups of fungus-like organisms:

TABLE 5.1
Features of Different Fungal Groups

Features	Chytrids: Blastocladiomycota and Chytridiomycota	Glomeromycota	"Zygomycetes": Mucoromycotina and Zoopagomycotina	Ascomycota	Basidiomycota	Oomycota
True fungi	Yes	Yes	Yes	Yes	Yes	No (Stramenopila)
Age/origin (million years (BP, before present)	1,156	980	1,049	908	791	
No. known species	1,000	250 (underestimate)	1,000	>70,000	32,000	500–800
Cell walls	Chitin/glucan	Chitin/glucan	Chitin/chitosan	Chitin/glucan	Chitin/glucan	Primarily glucans
Motile zoospores	Yes—one whiplash flagellum	No	No	No	No	Yes—biflagellate, with whiplash and tinsel flagella
Translocated carbohydrates	Polyols (sugar alcohols)	Polyols (sugar alcohols)	Polyols (sugar alcohols)	Polyols (sugar alcohols)	Polyols (sugar alcohols)	
Storage compounds	Glycogen, lipids, trehalose	Glycogen, lipids, trehalose	Glycogen, lipids, trehalose	Glycogen, lipids, trehalose	Glycogen, lipids, trehalose	Starch
Membrane sterols	Ergosterol	Ergosterol	Ergosterol	Ergosterol	Ergosterol	Fucosterol
Wall proteins	Proline present	Proline present	Proline present	Proline present	Proline present	Hydroxyproline present
Lysine synthesis	AAA pathway	AAA pathway	AAA pathway	AAA pathway	AAA pathway	DAP pathway

AAA, α-amino adipic acid; DAP, diaminopimelic acid.

the Oomycota, the cellular slime molds (Dictyosteliomycota and Acrasiomycota), the plasmodial slime molds (Myxomycota), and the plasmodiophorids (Plasmodiophoromycota).

5.2.1 Chytrids: Chytridiomycota and Blastocladiomycota

The chytrids include two lineages known from soil that are separated by DNA sequence data but united by the presence of zoospores with a single, posterior whiplash flagellum (James et al. 2006). Chytrids are considered to be the most ancient of the presently existing fungi. They occur as single, globose cells or simply branched chains, often attached to organic matter by root-like structures called rhizoids. These rhizoids are capable of secreting enzymes to aid in the degradation of the substrate and also assist in the uptake of nutrients. The chytrids are unique among the true fungi in that they are able to produce motile, flagellate zoospores, which typically have a single posterior, whiplash flagellum. About 1,000 species are currently known. Many chytrids are saprotrophs that can cause the decay of aquatic vegetation or degrade organic matter in soils. Two examples are *Allomyces macrogynus*, which occurs on organic debris in ponds and soil, and *Rhizophlyctis rosea*, a commonly occurring, strongly cellulolytic fungus in soil. Other species are parasites of algae or of insects. Some play important roles in soil, for instance, *Synchytrium endobioticum*, which causes galls on potato tubers, and *Catenaria anguillae*, which is a parasitoid of nematodes.

5.2.2 "Zygomycetes": Mucoromycotina and Zoopagomycotina

Similar to the chytrids, the group formerly referred to as Zygomycetes has been found by DNA-based phyogenetics to consist of several separate lineages, not united by a common ancestor. Two of these groups, the subphyla Mucoromycotina and Zoopagomycotina (or phyla Mucoromycota and Zoopagomycota; Spatafora et al. 2016), are important in soil. Zygomycetes are a heterogeneous group of fungi that are haploid, have broad hyphae that typically lack cross-walls (and are thus multinucleate), and produce thick-walled resting stages called *zygospores*. These zygospores can be formed by a sexual process involving the fusion of two hyphae with genes of opposite mating types. When the zygospores germinate, a sporangium is formed and haploid nuclei become partitioned to form sporangiospores that can be dispersed by wind or water. More commonly, sporangia and sporangiospores are formed asexually when conditions are favorable. The fungi in the zygomycetes are also characterized by their cell wall composition, which consists of chitin, chitosan, and/or polyglucuronic acid. The group is quite diverse, as it includes genera with different ecologies and morphologies, such as *Mucor*, *Rhizopus*, *Thermomucor*, and *Phycomyces*. Members of these genera can grow as saprotrophs in soil or on dung or decomposing fruits. Thermophilic species within this group (e.g., *Thermomucor*), growing at temperatures up to 60°C, are important in composts. Others, in the subphylum Zoopagomycotina, are important as parasitoids of invertebrate animals and protists.

5.2.3 Glomeromycota

The Glomeromycota, although of ancient origin, is a newly distinguished phylum that includes all the fungal species forming *arbuscular mycorrhiza* (AM), symbiotic associations with terrestrial plants (see Section 5.8.1), as well as the endocytobiotic fungus *Geosiphon pyrifomis*, forming a symbiosis by incorporating cyanobacteria in its cells (see Section 5.8.6) (Redecker et al. 2013). AM fungi are obligate symbionts and grow both inter- and intracellularly within plant roots. The fungal hyphae penetrate individual plant cell walls and branch repeatedly to form specialized structures called *arbuscules*. Some, but not all, species produce swollen *vesicles* that are thought to serve as storage structures containing abundant lipids. About 250 species of AMF have so far been described on the basis of the morphology of their spores, but molecular analyses suggest that this number may be a severe underestimate of the extant species diversity (Fitter 2005). The Glomeromycota are discussed in more detail, in particular in relation to mycorrhizal symbiosis (Section 5.8) and the taxonomy and evolution of the fungi (Section 5.3). Recent phylogenomic analyses suggest that the group be demoted to subphylum Glomeromycotina within the Mucoromycota (Spatafora et al. 2016).

5.2.4 Ascomycota

The Ascomycota constitute the largest and most diverse fungal phylum, containing more than 70,000 species (Schoch et al. 2009). Although we have a clear idea of the assignment to species of some groups, many of the phylogenetic relationships within the phylum still remain to be resolved, and in this endeavor, molecular methods will play a big role. The trophic strategies of many species are also as yet unknown, although the Ascomycota include well-known soil-colonizing fungi such as *Aspergillus*, *Fusarium*, and *Penicillium* species. In nature, numerous species form only asexual spores or conidia. However, when a sexual stage occurs, haploid nuclei from two compatible hyphae will fuse to form a diploid nucleus that undergoes meiosis, followed by mitotic division to produce eight haploid sexual ascospores contained within an ascus. Some species, such as the truffle-forming *Tuber* spp. and many Helotiales, form ectomycorrhizal associations with forest trees. On the other hand, many are economically important plant pathogens, such as the fungi associated with powdery mildews (e.g., *Blumeria graminis*) or vascular wilt diseases (*Fusarium* and *Verticillium* spp.). Many ascomycete fungi produce conidia but have sexual stages that are either rare, absent, or unknown. Earlier, these fungi were referred to as *Deuteromycota* or *Fungi imperfecti*. The sexual stage and asexual stage (if or when both are found) of a single species are now given a single name. Names used previously for asexual stages have been retained for particularly well-known fungi such as *Aspergillus* and *Penicillium*.

5.2.5 Basidiomycota

The Basidiomycota also form a large phylum, containing about 32,000 known species. The phylum is diverse and includes the familiar mushrooms and toadstools, but also yeasts, smuts, and rusts. The nuclei are typically haploid, but for most of the life cycle, the mycelium is segmented by septa into separate compartments, each containing two nuclei representing different mating compatibility groups. Such a mycelium is termed a *dikaryon*. As in the case of ascus formation in the Ascomycota, nuclear distribution in some Basidiomycota is maintained by backward-growing hyphae that merge with the cell behind, forming structures termed *clamp connections*. Basidia are a common feature of all basidiomycetes. When sexual reproduction occurs, compatible nuclei fuse within each basidium and meiosis occurs, leading to the production of sexual basidiospores on the exterior of the basidia. The Basidiomycota include a range of significant plant pathogens, such as the rust fungi (e.g., *Puccinia* spp.) and smut fungi (e.g., *Ustilago* spp.), but also symbiotic ectomycorrhizal fungi (*Amanita*, *Boletus*, *Cortinarius*, *Lactarius*, and *Russula* spp.). The fungi forming the familiar larger fruiting structures such as mushrooms, toadstools, brackets, and puffballs belong to the class Agaricomycetes (e.g., *Coprinus*, *Mycena*, *Marasmius*, *Trametes*, and *Lycoperdon*). They include a wide range of saprotrophs with the ability to degrade polymers such as cellulose, hemicellulose, or lignin in soil, leaf litter, and woody debris. Degradation of such polymers is a major feature of fungi, as outlined in Section 5.6.

5.2.6 Fungus-Like Organisms

Several other groups of fungus-like organisms, that is, the Oomycota, the cellular slime molds (Acrasiomycota and Dictyosteliomycota), the plasmodial slime molds (Myxomycota), and the plasmodiophorids (Plasmodiophoromycota), have traditionally been studied by mycologists, although they are not true fungi. The Oomycota include the economically important plant pathogens such as *Phytophthora* and *Pythium* spp. (see Section 5.7). Like fungi, these organisms form hyphae that exhibit apical growth and produce cell wall-degrading enzymes that allow them to penetrate plant cells. However, they possess diploid nuclei and have cell membranes and energy storage compounds that are more similar to those of plants and chromistan "algae" than to those of fungi. Their cell walls are composed of glucans and cellulose-like polymers rather than of chitin, and their cell membranes contain fucosterol rather than the characteristic fungal sterol ergosterol (Table 5.1). The cellular slime molds grow as unicellular, haploid amoebae that aggregate under conditions of nutrient limitation and differentiate to form stalked fruiting bodies that release spores. The dictyostelid slime mold *Dictyostelium discoideum* is the best-known example and it has become a model organism for studies of cellular communication and differentiation. The plasmodial slime molds (Myxomycota) (e.g., *Physarum polycephalum*) form *plasmodia*, multinucleate masses of protoplasm without cell walls, on moist, decaying wood where they feed on bacteria. When starved and exposed to light, the plasmodia transform into fruiting bodies that contain haploid spores. These fruiting bodies will germinate into amoeboid cells that also feed on bacteria and turn into flagellate swarmers in the presence of water or thick-walled cysts in the absence of nutrients. Mating and nuclear fusion between genetically different amoebae results in the initiation of a new diploid plasmodial phase. The best known plasmodiophorid is *Plasmodiophora brassicae*, which causes clubroot disease of crucifers, resulting in severe root deformation.

5.3 TAXONOMY AND EVOLUTION

Phylogenetic analyses of the small subunit rRNA gene sequence (18S rRNA gene) of the crown eukaryotes (animals, fungi, and plants) suggest that these groups diverged from each other roughly 1–2.5 billion years ago, undergoing rapid diversification and expansion, with chytrids estimated to have arisen 1.16 billion years ago and the youngest fungal group, the Basidiomycota, about 790 million years ago (Krings et al. 2017). For many years, the true fungi were divided into four phyla: the Chitridiomycota, Zygomycota, Ascomycota, and Basidiomycota. Although the Ascomycota and Basidiomycota are well described, the status of the Zygomycota was not considered to be sustainable on the basis of detailed molecular phylogenetic analysis. Thus, in 2001, a monophyletic clade was defined on the basis of small subunit rRNA gene sequence analysis, which was removed from the Zygomycota to form a fifth phylum, the Glomeromycota (Schüßler et al. 2001). The earliest fungi were possibly aquatic and they may have existed a billion years ago. The chytrids are considered to be the most ancient of the presently existing fungi, but the earliest available fungal fossils are from the Ordovician period (around 460 million years ago) and resemble the AM fungi of the Glomeromycota (Redecker et al. 2000). The fungi subsequently underwent a major and rapid diversification, giving rise to the Ascomycota and the Basidiomycota, which in turn diversified in parallel with the terrestrial land flora. These phyla are the best described fungal phyla and are both monophyletic with a common ancestor. The Ascomycota are currently divided into three subgroups: the Taphrinomycotina including the fission yeast *Schizosaccharomyces pombe*, the Saccharomycotina (true yeasts such as *Saccharomyces cerevisiae*), and the Pezizomycotina, many of which which produce ascocarps (Schoch et al. 2009). The Basidiomycota are divided into three major subgroups: the Pucciniomycotina (including the rust fungi), the Ustilaginomycotina (including the smut fungi), and the Agaricomycotina, which include the familiar mushrooms, toadstools, and puffballs. Most fungi today are closely associated with plants, as parasites, symbionts, or saprotrophs. Other trophic strategies may have developed within lineages of wood-degrading ancestors. For instance, the ectomycorrhizal fungi appear to have evolved concurrently with their plant hosts in the Pinaceae about 120 million years ago. Mycorrhizal symbioses have been alternatively viewed as stable derivatives of ancestral antagonistic interactions, or inherently unstable, reciprocal parasitisms, but recent phylogenetic analyses of free-living and mycorrhizal homobasidiomycetes (Tedersoo and Smith 2013) suggest that mycorrhizal symbionts have evolved repeatedly from saprotrophic precursors. Moreover, there may have been multiple reversals to a free-living condition, supporting the view that mycorrhiza are unstable, evolutionarily dynamic associations.

5.4 FUNGAL NUMBERS, BIOMASS, ACTIVITY, AND IDENTIFICATION

5.4.1 Measurements of Biomass

Estimates of fungal numbers in soil are complicated by their indeterminate mycelial growth habit. Accurate estimates are therefore only possible when the fungus is growing entirely as single cells or is present as spores. Usually, the length or biomass of the fungal mycelium is measured, using a variety of different techniques. Growth of hyphae onto agar films or into so-called ingrowth cores or mesh bags can be used to estimate fungal biomass, although the introduction of different artificial substrates into soil may produce artifacts. Direct microscopical methods involve the sampling of the soil, its homogenization and filtering, combined with staining with cotton blue for determining the total fungal length or with fluorescent vital stains such as fluorescein diacetate or Calcofluor white to detect vital hyphae. Estimates of the biomass (expressed in m g^{-1} soil) of fungal hyphae in both the organic and mineral soil horizons of boreal forest podzols are as high as 800 m of vital hyphae per gram of soil and 16 km of total hyphae per gram of soil. The mean annual production of ectomycorrhizal fungal mycelium in boreal forest soils has been estimated to amount to 400 kg ha^{-1} (Finlay and Clemmensen 2017). Most studies have not examined the mycelial distribution in different soil horizons. However, a clear vertical stratification of different ectomycorrhizal species has been found, and significant numbers of ectomycorrhizal roots occur in the deeper mineral soil horizons. In grassland ecosystems in temperate climate zones, the biomass of fungi may amount to 2–5 tonnes ha^{-1}, exceeding that of bacteria by a factor of 2. Although fungi are traditionally considered to be microorganisms, some have individual mycelia that are very large. One individual of *Armillaria bulbosa* growing in soil has been estimated to occupy at least 15 ha, to have a mass of almost 10,000 kg, and to have been genetically stable for over 1,500 years, making it one of the world's largest and oldest living organisms (Smith et al. 1992). In experimental laboratory systems, chitin can be measured in order to obtain an estimate of total (living + dead) fungal biomass. However, chitin is not present in oomycetes, and its presence in insect exoskeletons makes its use for the detection of fungi in field samples less suitable. *Ergosterol*, a sterol component of fungal membranes, has been used by a number of investigators as a fungal-specific marker; however, it may vary in concentration with age or between species (Weete et al. 2010).

5.4.2 Measurements of Fungal Activity

In experimental systems (micro- or mesocosms), in which the contribution of fungi can be measured separately from that of other organisms, estimates of fungal activities can include general measurements of parameters such as respiration, mass loss in decomposition processes, or release of ^{14}C-CO_2 from labeled substrates in studies of biodegradation or bioremediation. Fluorogenic 4-methyl-umbelliferyl-labeled substrates can be used to visualize and quantify the activity of chitinolytic enzymes *in situ*, and measurement of fluorescence following the degradation of fluorescein isothiocyanate-labeled proteins has also been used to quantify the proteolytic activity of ericoid and ectomycorrhizal fungi. Radioactive isotopes such as ^{14}C and ^{32}P, in combination with autoradiography, have been widely used to study the role of fungal mycelia in the translocation of carbon and nutrients. Field-scale measurements of carbon flow to plant roots and to mycorrhizal symbionts of forest trees have been taken by girdling the stems to restrict the flow of photoassimilate belowground and have shown a direct link between the flow of current assimilate and soil respiration. Similar measurements have been taken in grassland ecosystems (Johnson et al. 2002) using ^{13}C-labeled CO_2 and rotatable in-growth cores, suggesting that the movement of ^{13}C through AM hyphae is extremely rapid. Continuous imaging of amino acid transport in mycelia of *Phanerochaete velutina* has also been achieved using the non-metabolized ^{14}C-labeled amino acid analogue α-aminoisobutyric acid, a scintillation screen, and a photon camera. In addition, fluorescent or bioluminescent reporter genes fused to inducible promoters can be visualized using fluorescence microscopy or microscopes coupled to sensitive charge coupled device (CCD) cameras to provide quantitative estimates of different activities *in situ*. An additional method of interest is stable isotope probing (SIP) that employs DNA- or RNA-based analyses of the metabolically active fraction of the microbial community which incorporated ^{13}C or ^{15}N from a substrate labeled with either of these stable isotopes. SIP is further discussed in Chapter 17. This method was pioneered in studies of bacteria, but has since been applied to studies of fungal activity as well (e.g., Gkarmiri et al. 2017). Other nucleic acid-based methods such as real-time polymerase chain reaction (PCR), fluorescence *in situ* hybridization (FISH), large-scale gene expression studies based on cDNA microarrays, and metagenomics are described in more detail in Chapters 12–16.

5.4.3 Identification of Fungi

Morphological, physiological, and antibody-based methods for the detection and identification of fungi have now largely given way to DNA-based methods. Only about 17% of the about 120,000 species hitherto described have been successfully cultivated. Traditional identification methods based on morphology are limited by the fact that they often overlook asexual, cryptic, and obligately biotrophic species. Biochemical methods for environmental monitoring of fungi have been developed. These have been quite useful for the identification of culturable plant pathogens and commercially important strains. Substrate utilization assays—using 96-well microtitration plates (e.g., the commercially available BIOLOG plates)—were originally developed for the identification of single bacterial strains and can also be used to identify individual culturable fungi. However, they are not suitable for studies of fungal communities. Phospholipids are essential components of cell membranes, being rapidly degraded after cell death. Profiling of phospholipid fatty acids or fatty acid methyl esters has been used to monitor the overall changes in subsets of the microbial community, including the fungi. However, many fatty acids are common to large numbers of different fungal species. Much less information on the fatty acid composition of fungi than on that of—for instance—bacteria has thus been accumulated, so the approach is unlikely to be useful for detailed studies of fungal community structure. DNA-based methods for analyzing fungal communities have been in use since the early 1990s and have been progressively refined since then. These methods include PCR, cloning and sequencing methods, and a range of microbial community fingerprinting techniques (e.g., Taylor et al. 2017), but the development of high-throughput and single-molecule sequencing methods has enabled analyses of fungal community composition and function that were not previously possible (Jumpponen et al. 2010, Tedersoo et al. 2014). Now, hundreds or thousands of different fungi, many previously unknown, can be detected and differentiated from samples of soils or roots, based on 10^4–10^6 short (200–600 bp) DNA sequences per sample of a diagnostic or marker region, such as the internal transcribed spacer or a variable region in the small or large subunit rRNA gene. These methods are more fully described in Chapter 13. Their application to studies of fungi has revealed new information about the high diversity and spatial and temporal distribution of fungal communities (Clemmensen et al. 2013, Sterkenburg et al. 2018), as well as their interactions with other organisms such as bacteria (Deveau et al. 2018). For example, it has been shown that species other than known mycorrhizal fungi can associate with plant roots and be important in plant nutrition (Almario et al. 2017).

5.5 FUNGAL METABOLITES AND METABOLISM

As mentioned earlier, fungi have developed diverse trophic strategies, including saprotrophic, pathogenic, and symbiotic modes of growth, in order to obtain the organic compounds necessary for their growth and cellular synthesis. In addition to the primary metabolites required for normal growth, diverse secondary metabolites exist and can often be produced in abundance. Hyphal tips of fungi produce a wide spectrum of molecules, including organic acids, polysaccharides, antibiotics, siderophores, hormones, and enzymes. These molecules have a diverse array of functions and mediate cellular recognition, morphogenesis, antibiosis, pathogenesis, stress tolerance, substrate degradation, and nutrient mobilization, as well as symbiotic interactions, which are outlined in Table 5.2. Organic

TABLE 5.2
Fungal Metabolites and Products

Class of Compound	Specific Examples	Involved In
Organic acids	Citric acid	Tricarboxylic acid (TCA) cycle components—metabolic overflow products
	Gluconic acid	Metal chelation, mineral weathering
	Oxalic acid	Calcium binding/phosphorus acquisition
	Fumaric acid	Antibiosis/competition
		pH regulation
Sugars	Trehalose	Storage and translocation of carbon
Sugar alcohols (polyols)	Mannitol, arabitol	Maintenance of hyphal turgor/translocation
Polysaccharides	Chitin, chitosan	*N*-Acetylglucosamine polymers—cell wall components
Steroid hormones	Antheridiol, sirenin, oogoniol	Fungal and oomycete pheromones involved in mating
Antibiotics	Penicillins, cephalosporins, griseofulvin	Antibiosis/competition
Cyclic peptides	Cyclosporins	Antibiois/competition
Peptides/lipopeptides	Tremerogen	Sex pheromones
Pigments	Melanin	Protection of cells and propagules against UV radiation and microbial degradation
Proteins	Hydrophobins	Differentiation and hyphal interactions
Degradative enzymes	Cellulases, polygalacturonase, xylanse, tyrosinase, peroxidase, laccase, polyphenol oxidase	Degradation of structural components of litter with little N or P (cellulose, pectin, hemicellulose, monophenols, tannin, polyphenols, and lignin)
Degradative enzymes	Peptidases, proteinases, chitinase, acid phosphatase, phytase, phosphodiesterase ribonuclease, DNase	Hydrolysis of principal sources of organic N and P in litter (peptides, proteins, chitin, organic P monoesters, phytin, organic P diesters, nucleic acids)
Amino acid use	Transaminases	Transamination allows growth on a wide range of single amino acids
Siderophores	Hydroxamate	Iron chelation
Phosphate	Polyphosphate	P storage, reduced osmotic stress
N assimilation enzymes	GDH/GS GS/GOGAT (nitrate and nitrite reductases)	Most of the normal pathways of N assimilation found in other organisms.

acids (such as citric acid and oxalic acid) can be excreted as metabolic overflow products, with metals influencing biosynthesis. These acids are secreted by soil fungi and have effects on the solubilization and/or chelation of metals, the weathering of minerals, and the sequestration of phosphorus and base cations. Oxalic acid production in brown-rot fungi may lower the pH and aid in cellulose digestion but it may inhibit lignin degradation by white-rot fungi. Oxalic acid production by ectomycorrhizal fungi may also influence the tolerance of plants to metals such as Al and Cu (Sun et al. 1999). Fungi also produce hydroxamate siderophores, which are able to sequester iron that is subsequently used in cofactors of specific fungal enzymes. Fungal-specific carbohydrates include specific sugars, such as trehalose, and sugar alcohols and polyols, such as mannitol and arabitol. Polyols may also be important in the maintenance of turgor and the hydrostatic pressure gradients in fungal hyphae, which are necessary to sustain the translocation of carbohydrates and hyphal growth (Sun et al. 1999). Steroid hormones or peptides can aid in the communication between fungi during mating, and thus function as pheromones. Proteins such as *hydrophobins* confer hydrophobicity on fungal hyphae in aerial environments and may be important in surface interactions of hyphae. Many fungi produce pigments. One such specific pigment, *melanin*, may provide protection against ultraviolet (UV) radiation, whereas others may bind metals or provide protection against grazing, for example, by predatory protozoa. The production of antibiotics is also a well-known and exploited property of fungi. The release of these compounds may prevent competing organisms from utilizing digestion products in the microhabitat. However, the final stages of degradation may also be achieved by cell wall-bound, rather than secreted, enzymes, to ensure that easily utilizable monomers are less accessible to other organisms. Fungi are highly effective at sequestering phosphorus (P), a property of significance in the mycorrhizal–plant associations, where P can be stored as vacuolar polyphosphate to minimize osmotic stress before being transferred to the plant host. Nitrogen can be assimilated from simple sources such as glutamine using common pathways, either via the glutamate dehydrogenase/glutamine synthetase (GDH/GS) or the glutamine synthetase/glutamate synthase (GS/GOGAT) pathway. Many, but not all, fungi can utilize nitrate, converting it to ammonium with nitrate and nitrite reductases. However, in many natural soil ecosystems, organic polymers are quantitatively more important as sources of N. A range of degradative enzymes is produced in order to allow the decomposition of structural polymers, both as carbon sources and as sources of N and P. Fungal cell walls consist of polymers of *N*-acetylglucosamine containing N that can be internally recycled during mycelial differentiation or captured during competition or succession. Other structural polymers of plant litter, notably cellulose, hemicellulose, and lignin, can be attacked by different groups of fungi. These processes are described in more detail in the sections that follow.

5.6 FUNGAL SAPROTROPHS—LITTER DECOMPOSITION

Fungi play pivotal roles in the decomposition of organic polymers that occur in soil. In particular, they are important in breaking down and recycling plant cell walls, which are mainly composed of cellulose and hemicelluloses. Fungi also play a unique role in the degradation of plant-derived woody substrates containing *lignocellulose*, which is cellulose complexed with lignin. Litter decomposition usually proceeds through a series of well-characterized stages involving a succession of fungal communities with different degrees of enzymatic competence (Table 5.3). Microbes, weak parasites, and pathogens resident in plant material (or dung) are replaced by pioneer saprotrophic fungi that can utilize simple soluble substrates and storage compounds, but not the structural polymers. Pioneer saprotrophs such as members of Mucoromycotina and some Ascomycota are replaced by polymer-degrading fungi that are able to degrade the structural polymers. These fungi are usually able to "defend" their resources by using *antibiosis* (see also Chapter 9), are able to sequester nutrients, and have an extended growth phase. Cellulose is degraded by three different types of enzymes, collectively termed *cellulases*: endoglucanases break the cellulose chains at random points into successively smaller fragments, cellobiohydrolases cleave successive disaccharide units (cellobiose) from the ends of these fragments, and β-glucosidase cleaves the cellobiose to glucose. Chitin is degraded by different chitinolytic enzymes in an analagous way, which involves the activities of endochitinases, chitobiosidases, and *N*-acetylhexosaminidases. This may provide an important source of N during the secondary colonization of occupied woody substrates by wood-decomposing fungi. The polymer-degrading fungi are eventually replaced by fungal species that are able to degrade the more recalcitrant substrates, usually lignocellulose. Modification of this lignin is a necessary step in gaining access to this fraction of the cellulose since the lignin encrusts cell walls and prevents enzymes from accessing some of the more easily degradable cellulose and hemicelluloses. Fungi that are able to decay wood fall into three categories, which are as follows: *soft-rot fungi* grow in damp wood, in the lumen of woody cells, primarily decomposing cellulose-rich layers but leaving the lignin more or less intact; *brown-rot fungi* degrade most of the celluloses and hemicelluloses, leaving "brown" lignin, but despite this, their array of cellulase enzymes is depleted relative to white-rot fungi (Hori et al. 2013); and *white-rot fungi* such as *Phanerochaete chrysosporium*, *Trametes versicolor*, and *A. mellea*

TABLE 5.3
Features of Fungi as Saprotrophs

Fungal Group	Examples	Substrates	Enzymes
Endophytes, weak parasites and pathogens	Cladosporium *Alternaria*	Sugars Simple soluble substrates Storage compounds	
Pioneer saprotrophic fungi	*Pythium* spp.	Simple soluble substrates Storage compounds	
Polymer-degrading fungi	Fusarium *Trichoderma*	Cellulose, Hemicellulose Chitin	Chitinase Cellulase
Degraders of recalcitrant substrates	Soft-rot fungi *Chaetomium* *Ceratocystis*	(Damp wood) Cellulose Hemicellulose	Chitinase Cellulase
	Brown-rot fungi *Serpula lacrymans* *Schizophyllum commune*	Cellulose Hemicellulose Chitin	Chitinase Cellulase
	White-rot fungi *T. versicolor* *A. mellea* *Phanerochaete* spp.	Cellulose Hemicellulose Lignin	Glucose oxidase Laccases Manganese peroxidase Lignin peroxidase
Secondary opportunists		Dead fungi Insect exoskeletons (chitin)	Chitinases

Decomposition (arrow from top to bottom of table)

use cellulases to degrade cellulose but break down lignin by an essentially oxidative process. Laccase initiates the ring cleavage of phenolic compounds, and a series of oxidations is brought about by the activities of glucose oxidase, manganese peroxidase, and lignin peroxidase. Recent genomic studies indicate that white-rot and brown-rot mechanisms may be more of a continuum than a dichotomy as previously thought (Riley et al. 2014). Filamentous fungi growing in nutrient-poor substrates are nutrient-conservative, being able to redistribute their biomass to occupy fresh substrates, recycling nutrients within their own mycelium (Figure 5.1a–h). They also import N during the early stages of

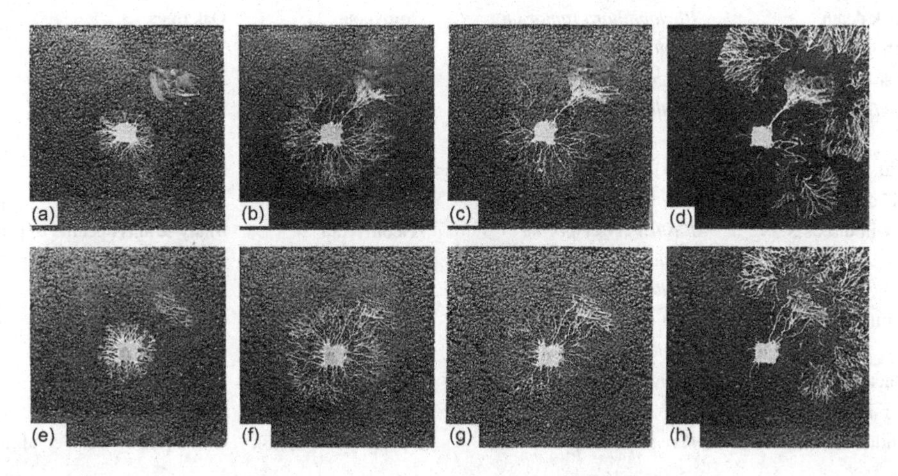

FIGURE 5.1 Resource capture and foraging patterns of the saprotrophic fungus *Hypholoma fasciculare*. The fungus (growing from a central woodblock) is able to reallocate its resources to colonize and degrade *Fagus sylvatica* leaves (a–d) or *Pinus sylvestris* needles (e–h).

decomposition and acquire nutrients through antagonistic interactions with other fungal mycelia. A significant fraction of the global organic carbon pool is located within woody plant tissues, and the fungal decomposition of these organic substrates makes an important contribution to biogeochemical cycling, which is discussed in more detail in Section 5.10 and Chapter 11.

5.7 FUNGAL PARASITES AND PATHOGENS

During evolution, the phylogenetic radiation of the plants has been paralleled by that of the fungi. The success of the strategy of obtaining carbon by acting as pathogens or parasites is evidenced by the fact that fungi, together with the fungus-like Oomycota, play key roles as plant pathogens and cause about 70% of all major plant diseases. Although disease-causing fungi do not all spread solely through soil, growth through soil or infection of roots may take place at certain stages of the life cycle. Fungi obtain carbon by acting as *necrotrophic* pathogens, which kill their hosts directly by invading them and producing toxins or digestive enzymes. Important classes of disease include the seed rots and seedling ("damping-off") diseases, which have a broad host range and are caused by fungi in a range of genera such as *Pythium*, *Rhizoctonia*, and *Fusarium*. *Phytophthora* (Oomycota) causes potato late blight, but also diseases that kill trees and shrubs. Other important diseases include "take-all" disease of cereals caused by *Gaeumannomyces graminis* and the vascular wilt diseases caused by *Fusarium* and *Verticillium* species. Basidiomycete species such as *A. mellea* and *Heterobasidion annosum* are also economically important pathogens of trees. Some of the main diseases and the fungi causing them are summarized in Table 5.4.

Another strategy of interaction with a host is to act as a *biotrophic pathogen*, that is, a nutritional parasite that does not kill its host. This trophic mode has evolved independently on several occasions over evolution. It includes the rust fungi, the powdery mildew fungi, and the downy

TABLE 5.4
Some Important Fungal Diseases of Plants

Disease	Fungus	Fungal Group	Host Plant(s)
Seed rots and seedling diseases "damping-off"	*Pythium*	Oomycota	Broad range of hosts
	Rhizoctonia solani	Basidiomycota	
	Fusarium spp.	Ascomycota	
Potato late blight	*Phytophthora infestans*	Oomycota	Potatoes, Solanaceae
Sudden oak death	*Phytophthora ramorum*	Oomycota	Oak trees
Phytophthora root rot	*Phytophthora cinnamomi*	Oomycota	Trees and shrubs
Heterobasidion root rot	*H. annosum*	Basidiomycota	Pine and spruce trees
Shoestring/Bootlace root rot	*Armillaria mellea*	Basidiomycota	Trees
Gray mold	*Botrytis cinerea*	Ascomycota	Broad host range
"Take-all" disease	*G. graminis*	Ascomycota	Cereals
Vascular wilt diseases	*Fusarium oxysporum*	Ascomycota	Many strains specific to particular crop species
	Verticillium dahliae	Ascomycota	
	V. albo-atrum		Wide host range
(Dutch Elm Disease)	*Ophiostoma ulmi*	Ascomycota	Elm trees
Smuts	*Ustilago* spp.	Basidiomycota	Maize, wheat
Rusts (black sten rust)	*Puccinia graminis*	Basidiomycota	Wheat
Powdery mildews	*B. graminis*	Ascomycota	Cereals
Downy mildews	*Perenospora* spp., *Bremia* spp.	Oomycota	Broad range (moist conditions)

Shaded portions indicate pathogens that primarily attack shoots, but which are included for the sake of completeness.

mildews, although these are usually classified as airborne pathogens and are only included for the sake of completeness. However, recent DNA-based surveys have shown that the moss-associated rust, *Eocronartium*, is common in soil.

5.8 FUNGAL SYMBIONTS—MYCORRHIZA, LICHENS, AND ENDOPHYTES

The third major trophic strategy fungal heterotrophs have adopted to obtain carbon is to form symbiotic associations with autotrophic host plants. The term *mycorrhiza* (plural *mycorrhizas*) was first used in 1885 by A.B. Frank to describe the modified root structures of certain forest trees and has since been extended to cover a range of symbiotic associations between fungi and plant roots. Mycorrhizal associations can be divided into seven basic types depending upon their morphological features and the fungal and plant species involved (Smith and Read 2008). These are outlined in Table 5.5 and described in more detail in the sections that follow.

5.8.1 Arbuscular Mycorrhiza (AM)

AM is the most ancient and widespread form of mycorrhizal symbiosis, and paleobotanical and molecular sequence data suggest that the first land plants already formed associations with fungi

TABLE 5.5
Features of Different Types of Fungal Symbioses

Type of Symbiosis	Intracellular Penetration	Sheath	Hartig Net	Mycobionts	Photobionts
AM	+	–	–	Glomeromycota	Bryophytes, Lycopodiophya, Polypodiophyta, Gymnosperms, Angiosperms
Ericoid mycorrhiza	+	–	–	Ascomycota Basidiomycota	Ericales
Ectomycorrhiza	–	+	+	Basidiomycota Ascomycota (Mucoromycotina)	Gymnosperms, Angiosperms
Orchid mycorrhiza	+	–	–	Basidiomycota	Orchidaceae
Monotropoid mycorrhiza	+	+	+	Basidiomycota	Monotropoideae
Arbutoid mycorrhiza	+	±	+	Basidiomycota	Ericales
Ectendomycorrhiza	+	±	+	Basidiomycota Ascomycota	Gymnosperms, Angiosperms
	Distinguishing features				
Lichens	Folicose, fructicose, squamulose, or crustose forms			Ascomycota Basidiomycota	Green algae and/or Cyanobacteria Green algae and/or Cyanobacteria
Geosiphon pyriformis	Unique symbiosis between a fungus and cyanobacterium, forms small bladders on soil surface			One fungus within the ancestral branches of the *Glomeromycota*	Cyanobacterium *Nostoc punctiforme*
Soil crusts and desert varnish	Mixed, loosely associated communities of filamentous fungi (many highly melanized) and filamentous cyanobacteria and green algae			Ascomycota	Cyanobacteria Green algae

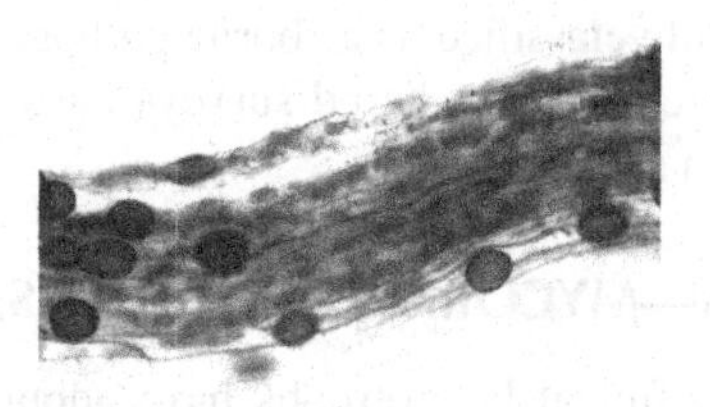

FIGURE 5.2 **(See color insert)** *Trifolium repens* root colonized by the AM fungus *Claroideoglomus* (formerly *Glomus*) *claroideum*—note vesicles and arbuscules.

from the Glomeromycota about 460 million years ago. This is estimated to be some 300–400 million years before the appearance of root nodule symbioses with nitrogen-fixing bacteria. However, recent studies have revealed a genetic overlap between the two types of symbiosis. A common set of plant genes is required for both, and it has been speculated that genes that had evolved in the context of the AM symbioses have been recruited much later during the evolution of root nodule symbiosis. AM symbioses can be formed with an estimated 250,000 plant species. Only about 250 fungal species have so far been distinguished that can form these symbioses, but molecular methods are now revealing a higher degree of diversity. Consequently, the earlier notion that there was only a very low level of specificity in these associations is giving way to the idea that there may be many more species-specific effects (Fitter 2005). The name "arbuscular" derives from the highly branched intracellular structures, the so-called arbuscules, which are thought to be sites of nutrient transfer between the fungus and the plant (Figure 5.2). These arbuscules maximize the surface area of contact between plant host and fungus, which is available for the exchange of nutrients. Intra- and intercellular hyphae grow within the root, without penetrating the host plasma membrane, and an extraradical mycelial phase extends into the soil, increasing the volume of soil that can be exploited for nutrient uptake. Swollen vesicles that contain compounds rich in lipids are also often formed (Figure 5.2)—explaining the older name *vesicular AM*. However, the vesicles are not present in about 20% of *Glomeromycota* species. Hence, the term "arbuscular mycorrhiza" is now more common, notwithstanding the fact that a diverse range of structures is formed in the roots of different plants.

5.8.2 Ericoid Mycorrhiza

Ericoid mycorrhizas are formed in three plant families: the Ericaceae, the Empetraceae, and the Epacridaceae, all belonging to the Ericales. These plants grow as more or less pure stands of dwarf shrubs in heaths and other nutrient-poor areas, such as the understory vegetation of boreal forests. Around 3,400 plant species form this type of association with various fungi from the Ascomycota, of which *Pezoloma* (earlier *Hymenoscyphus*) *ericae* is the only one studied in detail; Basidiomycota of the genus *Clavaria* have also been shown to form ericoid mycorrhiza. This fungus penetrates the cell walls of the plant in fine hair roots, forming coiled structures within each cell, without penetrating the host plasma membrane. The extraradical hyphae do not grow as extensively as in AM, and in many cases, it is likely that the main effect of the fungus is to mobilize N and P from organic matter, making them available to their plant hosts (Read and Perez-Moreno 2003).

5.8.3 Ectomycorrhiza

Ectomycorrhizas are formed between long-lived, woody perennial plants and fungi predominantly from the Basidiomycota and Ascomycota. As many as 10,000 fungal species may be involved, although the true number is not known. About 8,000 plant species are involved in these symbioses, which is only a small fraction of the total number of terrestrial plants. However, these are of great importance, since the families Pinaceae, Salicaceae, Betulaceae, and Fagaceae occupy a disproportionately large global area, forming vast tracts of forest in both the Northern and Southern hemispheres. The plant species involved are typically trees or shrubs from cool, temperate boreal, or montaine forests, but also include arctic–alpine dwarf shrub communities, Mediterranean/chaparral vegetation, and many species in the Dipterocarpaceae and leguminous Caesalpinoideae in tropical forests. The symbiosis is *ectotrophic* without fungal penetration of the host cells and is characterized by the presence of a more or less well-developed *mantle* or *sheath* around each of the short roots, as well as a network of intercellular hyphae penetrating between the epidermal and cortical cells, the so-called *Hartig net* (Figure 5.3). The Hartig net is the interface across which exchange of carbon and nutrients between the fungus and the host takes place. It is, much like the arbuscules in AM, an effective way of increasing the surface area of contact between the two partners in the symbiosis. Depending upon the fungal species involved, the mantle may be connected to a more or less well-developed extraradical mycelium (Figures 5.4 and 5.5), which may extend many centimeters from the mantle into the soil.

5.8.4 Orchid Mycorrhiza, Monotropid Mycorrhiza, and Mycoheterotrophs

In most types of mycorrhizal symbiosis, the fungus depends on its autotrophic host plant to supply carbon. However, plants in the Orchidaceae (orchids) are often achlorophyllous at first and have very small seeds with little or no reserve materials. Orchids form *orchid mycorrhiza* with a range of fungi, which were originally thought to be highly effective saprotrophs or parasites of other plants, which were thus capable of providing the orchids with the organic carbon compounds they needed

FIGURE 5.3 **(See color insert)** Longitudinal section of *Betula pendula* root showing colonization by the ectomycorrhizal fungus *Paxillus involutus* including the mantle and Hartig net.

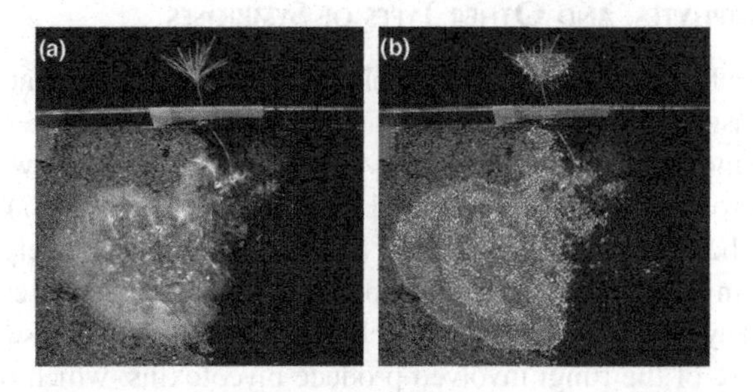

FIGURE 5.4 **(See color insert)** (a and b) Colonization of different substrates by extraradical mycelium of ectomycorrhizal fungus *Hebeloma crustuliniforme* showing active carbon allocation to E-horizon material after 2 months of hyphal growth.

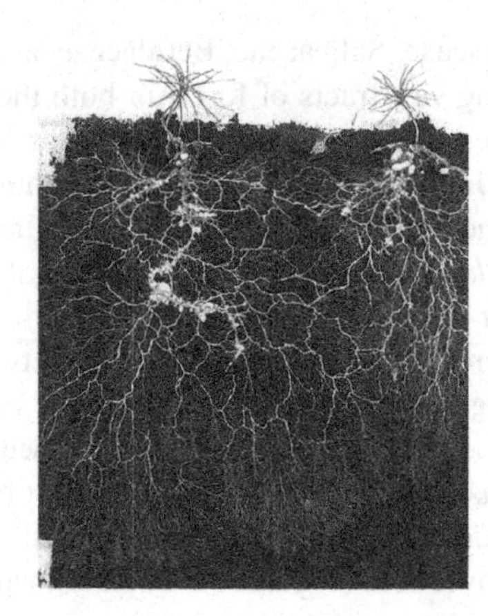

FIGURE 5.5 **(See color insert)** Extraradical mycelium of the ectomycorrhizal fungus *Suillus bovinus* colonizing *Pinus sylvestris* seedlings.

for germination and growth. Molecular analyses of the fungi that colonize orchids have since shown most of these fungi to be mycorrhizal fungi forming ectomycorrhizal associations with other plants (Leake 2005). These plants are "cheats" or *epiparasites*—parasites effectively deriving their carbon and nutrients via mycorrhizal associates of adjacent autotrophs. The "mycoheterotrophic" habit is common within the plant kingdom, with more than 30,000 species (10% of all plant species) obtaining their carbon in this way during the early stages of establishment from seeds. A further 400 plant species are fully achlorophyllous, including plants in the Monotropoideae that form *monotropoid* mycorrhiza. These are like ectomycorrhiza, usually with a well-developed mantle, but with a more superficial Hartig net with single hyphae growing into the epidermal cells, forming peg-like structures.

5.8.5 Other Types of Mycorrhiza

Two further classes of mycorrhiza have been distinguished. The *arbutoid mycorrhiza*, which are formed between fungi that are normally ectomycorrhizal and plants in the genera *Arbutus* and *Arctostaphylos* and the family Pyroloideae, show extensive intracellular fungal proliferation. The second group is called he *ectendomycorrhizas*, and these have characteristics of both ectomycorrhiza and endomycorrhiza. The conifer ectendomycorrhizas are often called "E-type" mycorrhizas, and the fungi are ascomycetes that form ectendomycorrhizas on *Pinus* and *Larix,* but normal ectomycorrhizas with other tree species.

5.8.6 Lichens, Endophytes, and Other Types of Symbioses

Fungi also can form symbiotic associations with photosynthetic partners that are either green algae or cyanobacteria. The association with algae is called a *lichen*. Most of the fungi-forming lichens are ascomycetes, but some basidiomycetes also have this capacity. They grow in/on diverse habitats, including semiarid environments, tundra soils, and rock surfaces (Gadd 2017). Lichens contribute to soil formation and have been proposed to represent a model system, which helps in the understanding of mineral weathering in the rhizosphere. Fungi occur widely as endophytes in the roots of many plants, without having any apparent effect on their hosts. Knowledge on these is still fairly sporadic, but it is known that some of the fungi involved produce mycotoxins, which may confer benefits on their host plants in terms of improved resistance to grazing or other types of environmental stress. Others are mycoparasites, defending plants from root-infecting pathogenic fungi, and some non-mycorrhizal root-associated fungi may also assist in plant nutrition (Almario et al. 2017).

5.8.7 Functional Effects of Mycorrhizal Fungi

Mycorrhizal fungi can have many different effects, which are of central importance to soil–plant interactions. Different types of mycorrhizal association work in different ways (Finlay 2008), and there is undoubtedly considerable variation between the different fungal associates. Some of the main effects are summarized in Figure 5.6.

AM fungi and ectomycorrhizal fungi often have well-developed extraradical mycelial phases, which are able to bridge nutrient depletion zones around the plant roots. They penetrate microsites in soil that are inaccessible to roots. The overall surface area of the root system that is available for the uptake of nutrients is thus greatly increased (at a minimal synthetic cost to the plant), having a quantitative effect on the uptake of mineral nutrients already in soil solution. In addition to this, the mycelia of some species may be able to solubilize nutrients or weather mineral particles through the production of organic acids, siderophores, and/or other chelating agents (Finlay et al. 2009). The mycelia are also able to access water that is trapped inside small soil pores. It has been suggested that water derived from hydraulic lift by roots may be allocated to hyphae that colonize dry substrates, which will maintain hyphal integrity and condition the fungal–soil interface to allow microbial interactions (Sun et al. 1999). Stress induced by toxic heavy metals, soil acidification, or increased levels of aluminum (Al) may also be reduced by metal chelation and the capture of base cations that would otherwise be removed by leaching. Many mycorrhizal fungi have a significant ability to mobilize N and P from organic polymers. They may even derive these and supply them to their host plants through competitive or antagonistic interactions with other organisms (Read and Perez-Moreno 2003, Finlay 2008). Other possible interactions include associative N fixation by mycorrhiza-associated bacteria or synergistic interactions with decomposers via allocation of C through the mycelium (Hodge et al. 2001). The central process driving most of these interactions is the allocation of carbon from the host plants (Finlay and Clemmensen 2017). This may also have effects on the stability of soil aggregates (Johansson et al. 2004) and the selective exploitation of soil heterogeneity. The effects of mycorrhizal symbionts at a micro-spatial scale may ultimately influence the functioning of whole plant communities and ecosystems. For instance, it has been established that the belowground diversity of mycorrhizal symbionts can have effects on both the productivity and the floristic diversity of grassland communities (van der Heijden et al. 1998).

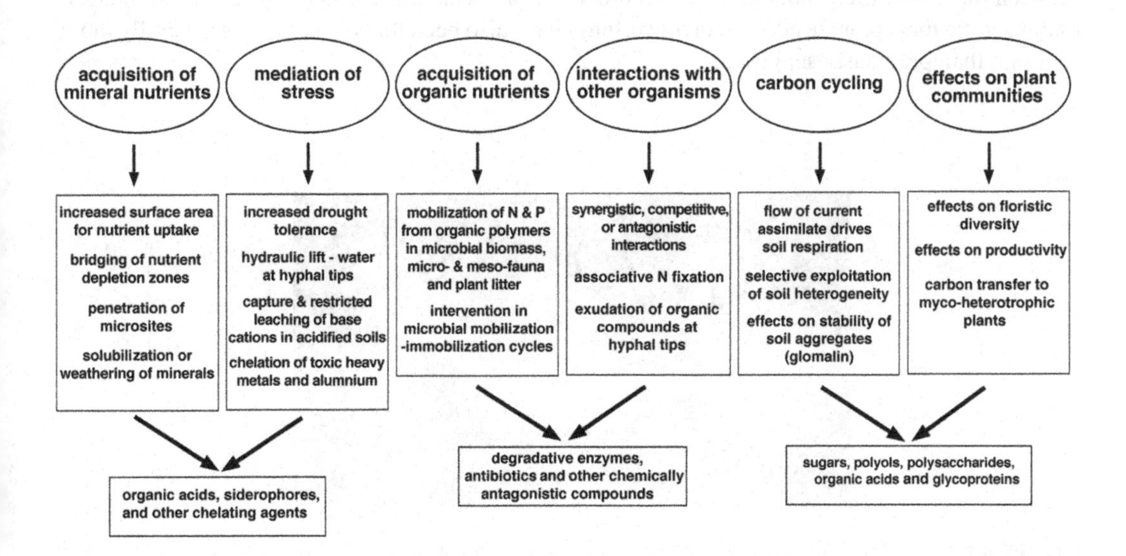

FIGURE 5.6 Effects of mycorrhizal symbiosis.

5.9 INTERACTIONS OF FUNGI WITH OTHER ORGANISMS

Interactions of fungi with other organisms are discussed in Chapter 9 in more detail, particularly with respect to the mechanisms involved, but some general aspects are mentioned here inasmuch as they specifically relate to the particular features of fungi. The filamentous growth form of many fungi means that they can simultaneously interact with different organisms, occupying different substrates. These mycelia also provide a very large surface area over which the interactions may take place.

5.9.1 Interactions with Bacteria

The preeminent ecological success of fungi in terrestrial decomposition systems, particularly bacterial niches associated with recalcitrant organic matter, has led not only to the loss of potential decomposition niches but also to the creation of new bacterial niches. Competition for simple substrates or scarce nutrients may lead to the development of *antagonistic* strategies, but in more recalcitrant substrates, both *competitive* and *mutualistic* strategies have evolved (de Boer et al. 2005). These interactions are discussed in detail in Chapter 9.

Interactions of mycorrhizal fungi with bacteria have a particular significance, since the mycorrhizal hyphae are connected to autotrophic plants that supply energy-rich carbon compounds (Smith and Read 2008). These sources of carbon and energy can be translocated to the fungal tissue and are then exuded as droplets (Figure 5.7) at the hyphal tips. It has been hypothesized that this process will condition the environment around the hyphal tips and facilitate interactions with other microbes at the soil–mycelial interface, even when soil water potential is low (Sun et al. 1999). Fungal *saprotrophs* also translocate carbon along a concentration gradient of sugars and polyols to the peripheral hyphae; for instance, the droplets exuded at the hyphal tips of *Serpula lacrymans* are well documented. Interactions of the mycorrhizal hyphae with bacteria in the mycorrhizosphere may have implications in a number of applied areas such as biocontrol, bioremediation, biofertilization, and sustainable agriculture, and are the subject of several reviews (Deveau et al. 2018, Johansson et al. 2004). Beyond these practical considerations, the recent discovery of apparently obligately endosymbiotic bacteria (with a reduced genome size) within hyphae of symbiotic mycorrhizal fungi (Bonfante and Desiro 2017) raises intriguing questions about their roles in fungal fitness and mycorrhizal symbiosis. Potential taxonomic relationships between these so-called *endobacteria* and other AM-associated bacteria are therefore of interest. Endobacteria that colonize ectomycorrhizal fungi have also been found and may be generally more common than previously supposed.

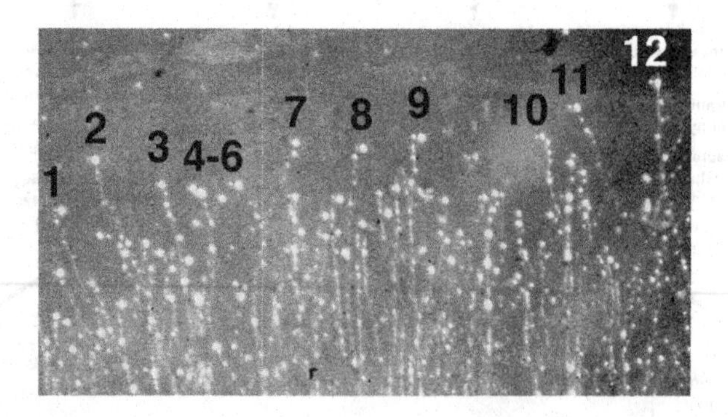

FIGURE 5.7 Droplets exuded from the tips of *Suillus bovinus* hyphae.

5.9.2 Interactions with Other Fungi

Fungal successions in relation to substrate quality are well documented in relation to decomposition (Deacon 2005). However, interactions between fungi with different primary trophic strategies have received less attention until recently. Antagonistic activity against saprotrophic fungi, coupled with nutrient transfer, has been shown in microcosm experiments (Figure 5.8). If this is widespread in ectomycorrhizal fungi occupying nutrient-poor environments, then nutrient cycling may be tighter than has been previously supposed. "Short-circuiting" of conventional decomposer pathways by mycorrhizal fungi with direct recycling of organic nutrients to plant hosts (Finlay 2008) would restrict nutrient immobilization in decomposer populations and lead to more efficient cycling of nutrients.

5.9.3 Interactions with Soil Microfauna and Mesofauna

Interactions of soil fungi with diverse soil animals may result in effects on the fungal mycelia, such as increased or decreased activity of mycorrhizal mycelia, depending upon the intensity of animal grazing. Fungi themselves may influence animal populations, stimulating them by acting as food, or depressing animal numbers in situations where the fungi themselves exploit the soil fauna as sources of nutrients. One well-documented example of this is the nematophagous fungi that can trap, parasitize, and digest nematodes as sources of food. It has been shown that the ectomycorrhizal fungus *Laccaria bicolor* can kill the soil microarthropod *Folsomia candida*, take up nitrogen from the killed animals, and transfer the derived N to its *Pinus strobus* host plants. Other experiments (Read and Perez-Moreno 2003) support this idea and have specifically demonstrated the ability of the ectomycorrhizal fungus *Paxillus involutus* to recycle nutrients from sources such as dead nematodes and pollen. Field experiments performed in upland grassland ecosystems, using stable isotopes, have also shown that grazing by soil microarthropods can disrupt the carbon flow through fungal networks.

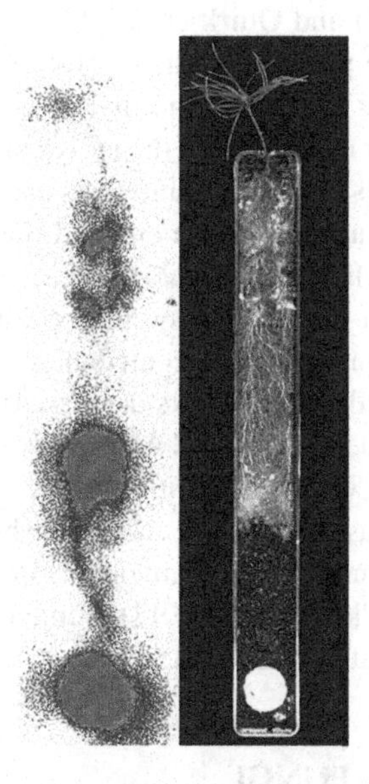

FIGURE 5.8 **(See color insert)** Interactions between ectomycorrhizal fungus *S. bovinus* and the saprotrophic fungus *H. fasciculare*, showing the capture of ^{32}P from the saprotroph by the mycorrhizal fungus.

5.10 FUNGI IN BIOGEOCHEMICAL CYCLES AND NUTRIENT CYCLING

Fungi play central roles in biogeochemical cycling because they link the allocation of carbon and sequestration of nutrients from organic and inorganic substrates (Read and Perez-Moreno 2003, Quirk et al. 2012). It is fair to say that the important roles that fungi play in element cycling in the biosphere have frequently been neglected within most microbiological and geochemical research. Thus, the saprotrophic fungi play key roles in the degradation of recalcitrant woody substrates that contain mainly lignin and lignocellulose. Both saprotrophic and ectomycorrhizal fungi exude large amounts of organic acids, which exert a variety of effects, including the weathering of minerals, chelation of metals, and the stimulation of soil respiration via their use as a microbial substrate. Microcosm experiments have shown the importance of the hyphal translocation of carbon for mineral solubilization (Jacobs et al. 2002). Moreover, decomposer communities in peatlands may shift in response to climate change variables, with potentially major consequences for the liberation of long-sequestered carbon from these widespread habitats (Asemaninejad et al. 2017). In addition, field experiments have demonstrated that soil respiration appears to be tightly linked to the allocation of current assimilates to roots and microbial symbionts (Högberg et al. 2001). Symbiotic fungi connect the primary producers of terrestrial ecosystems, that is, the plants, to the heterogeneously distributed nutrients that are required for their growth. This enables the flow of energy-rich compounds required for nutrient mobilization, while simultaneously providing conduits for the translocation of mobilized products back to their hosts. Recent studies by Sterkenburg et al. (2018) suggest that ectomycorrhizal fungi may either hamper or stimulate decomposition by different mechanisms, depending upon the stage of decomposition and location in the soil profile, and further studies of how the interactions of different functional guilds of fungi affect carbon cycling are needed (Averill et al. 2014, Fernandez and Kennedy 2016). Lindahl and Tunlid (2015) have argued that ectomycorrhizal fungi benefit from organic matter decomposition primarily through increased nitrogen mobilization rather than through the release of metabolic C. Recent findings support the view that some ectomycorrhizal fungi have the capacity to oxidize organic matter, either by "brown-rot" Fenton chemistry or by "white-rot" peroxidases (Shah et al. 2016, Sterkenburg et al. 2018).

Intervention of fungi in nutrient cycles involving mineral nutrients has been discussed by several authors, including Gadd (2010, 2017) and Quirk et al. (2012). The success of plants and fungi as biogeochemical engineers is in large part due to their ability to form symbiotic mergers. The role of fungi as symbionts of phototrophs in lichens and mycorrhizas is well known, and the ubiquity and significance of lichens as pioneer organisms in the early stages of mineral soil formation is also well understood. Successive increases in size of plant hosts and the extent of substrate colonization by their fungal symbionts (Quirk et al. 2012) have enabled them to have larger effects as biogeochemical engineers, affecting the cycling of nutrients and C at an ecosystem and global level. It has been claimed (Quirk et al. 2012) that forested ecosystems colonized by ectomycorrhizal fungi are major engines of biological weathering that have contributed to the historical drawdown of global CO_2 levels. Distribution of pulverized silicate rocks across terrestrial landscapes, ultimately leading to enhanced weathering and sequestration of C into ocean floor carbonates, has been proposed as a possible technology to remove CO_2 from the atmosphere (Taylor et al. 2016). It is known that ectomycorrhizal fungi may both release CO_2 and sequester carbon into stable organic and inorganic substrates, but the detailed mechanisms, the fungi involved, and the processes controlling them are still poorly understood (Finlay and Clemmensen 2017). Some of the mechanisms involved in these biogeochemical processes are illustrated in Box 5.1.

5.11 APPLICATIONS OF SOIL FUNGI

Soil fungi cause enormous economic losses through damage to food crops, timber, and manufactured goods, but they also have many beneficial activities and produce compounds that have great

BOX 5.1

The flow of plant-derived carbon through fungal hyphae to organic and inorganic substrates drives biogeochemical processes such as decomposition and weathering and influences patterns of C release and sequestration into stable organic and inorganic forms.

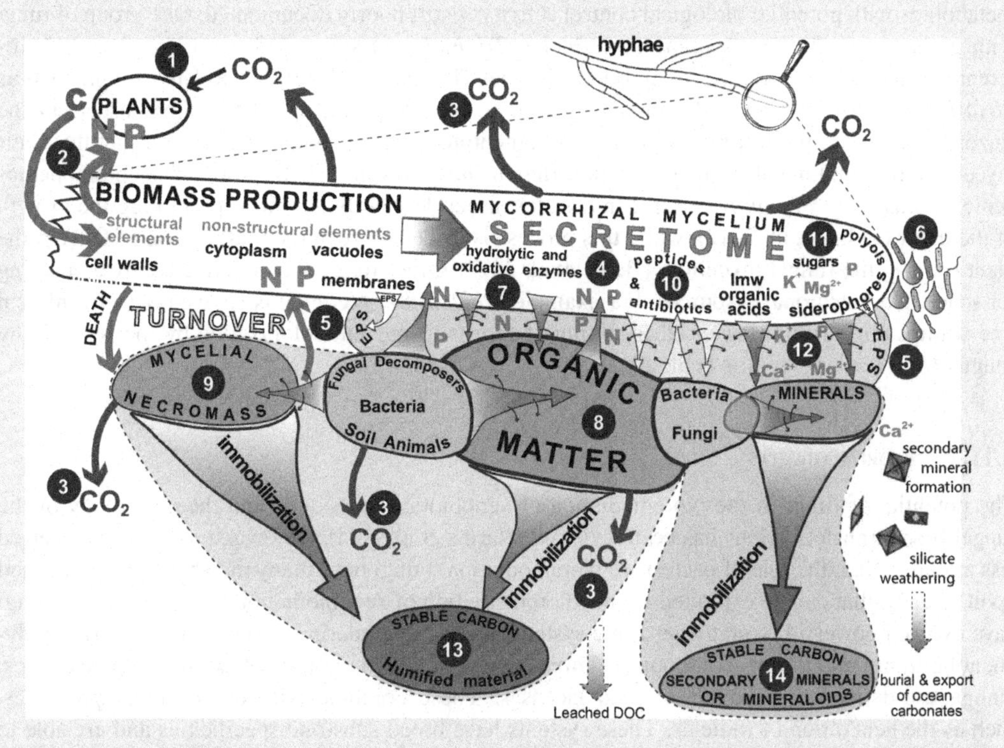

Carbon is assimilated by plants (**1**) and transferred directly to symbiotic mycorrhizal hyphae that transfer nutrients mobilised by the hyphae back to their hosts (**2**). Products of mycelial respiration are released to the atmosphere (**3**). The fungal secretome (**4**) consists of different labile compounds that can be translocated to different organic or inorganic substrates. These compounds may be released into an extracellular polysaccharide matrix (**5**) or as droplets that condition the hyphosphere, facilitating interactions with bacteria (**6**). Hydrolytic and oxidative enzymes (**7**) mobilise N and P from plant-derived organic substrates (**8**) or microbial necromass (**9**). Peptides and antibiotics play important roles in signalling and influencing microbiome structure (**10**), sugars and polyols maintain osmotic gradients and hyphal turgor (**11**) and low molecular weight organic acids and siderophores influence the mobilisation of P and base cations from minerals (**12**). Long term sequestration and stabilization of carbon can take place in recalcitrant organic substrates (**13**) and secondary minerals and mineraloids (**14**).

economic value. The latter include pharmaceuticals such as antibiotics, hormones, and immunosuppressants, as well as plant growth regulators, proteins, and enzymes with a range of industrial and biotechnological uses. The properties of soil fungi may thus be exploited for practical use in biocontrol of pathogens, plant growth stimulation, bioremediation, and as biofertilizers. These topics are dealt with in more detail in Chapters 21 to 23. A few general features of soil fungi, in relation to the properties that are of use in these application areas, are discussed in the sections that follow.

5.11.1 Biological Control

The role of fungi as *biological control agents* (BCAs) has been comprehensively reviewed (Butt et al. 2001, Jensen et al. 2017). Increasing numbers of fungal products are currently registered as BCAs and sold commercially. Different modes of action are involved, including competition for space and/or nutrients, predation, *hyperparasitism*, and antibiosis. The full range of fungal metabolites with potential biological control activity is still poorly documented. One group of fungi with great potential, the asexually reproducing *Trichoderma* spp., produces more than 100 different metabolites with known antibiotic properties (Harman et al. 2004). Related fungi, such as *Metarhizium,* possibly act as beneficial plant associates by reducing root-feeding insects in the rhizosphere. Assigning trophic strategies to filamentous fungi is complicated by the fact that their mycelia may simultaneously interact with different substrates in different ways. The entomopathogenic fungus *Metarhizium robertsii* can transfer insect-derived N to plants, promoting the growth of the latter (Behie et al. 2012), while this process is driven by reciprocal allocation of C from the plant roots to the fungal mycelium (Behie et al. 2017). Other fungi may be effective by competing for space and nutritional resources. For example, *Phlebiopsis gigantea* is sprayed onto freshly cut tree stumps during thinning or clear-cutting of forests to prevent colonization by the root-rotting fungus *H. annosum* (Menkis et al. 2012).

5.11.2 Bioremediation

The potential of fungi in the bioremediation of xenobiotic compounds and the physiology of the fungal-based transformations has been reviewed (Harms et al. 2011). However, this area has received less attention than the role of bacteria in bioremediation. Fungi have many metabolic traits, touched upon already, that can be exploited for the bioremediation of xenobiotic pollutants. White-rot fungi have evolved powerful oxidative enzyme systems for the degradation of naturally-occurring polyphenolic lignin and humus compounds. Three classes of enzyme are involved: lignin peroxidases, manganese-dependent peroxidases, and laccases. These combine with other fungal processes, such as the generation of radicals. These systems have broad substrate specificities and are able to degrade a wide range of highly recalcitrant organopollutants, namely, polycyclic aromatic hydrocarbons, chlorophenols, nitrotoluenes, polychlorinated biphenyls, and azo dyes, some of which are not typically degraded by bacteria under aerobic conditions. White-rot fungi require the addition of a primary growth substrate to provide carbon, and many cheap natural substrates can be used. Other advantages of white-rot fungi are that the key lignin-degrading enzyme systems will be expressed under conditions of nutrient limitation, which are prevalent in many soils, and that fungal hyphae are capable of extending into substrates, reaching pollutants that bacteria cannot reach. Fungi may also be used for the heterotrophic leaching of metals, and they are efficient producers of organic acids that supply both protons and metal-complexing anions. Mobilization may also be achieved by siderophores or through biomethylation. Fungi may take part in so-called *biosorption*. Biosorption is the removal or recovery of free metal ions from a solution by a prokaryotic and/or eukaryotic biosorbent. The biosorbent, unlike monofunctional ion-exchange resins, contains a variety of functional sites including carboxyl, imidazole, sulfhydryl, amino, phosphate, sulfate, thioether, phenol, carbonyl, amide, and hydroxyl moieties. Bioaccumulation includes all processes responsible for the uptake of bioavailable metal ions by living cells and thus includes biosorptive mechanisms, together with intracellular accumulation and bioprecipitation mechanisms. Chitin and chitosan are effective biosorbents of radionuclides, whereas fungal phenolic polymers and melanins possess many potential metal-binding sites. Fungi also produce a range of metal-binding peptides, proteins, and extracellular polysaccharides. There is conflicting evidence as to whether the interactions between mycorrhizal fungi and bacteria have positive or negative effects on the remediation of organic pollutants. Studies on the biodegradation of *m*-toluate have suggested that *m*-toluate degradation by fluorescent pseudomonads is synergistically improved under conditions when the bacteria interact with ectomycorrhizal fungi colonizing Scots pine seedlings. However, other studies have shown that the degradation of the polycyclic aromatic hydrocarbon fluorene is retarded in the ectomycorrhizosphere of Scots pine, which suggests that interactions between ectomycorrhizal fungi and bacteria may not always be synergistic.

5.11.3 Biofertilization

Biofertilization can be achieved by natural or enhanced nutrient-mobilizing activities of microorganisms. Although the term "biofertilization" is most frequently applied in relation to mineral substrates, the release of nutrients from organic residues can also be viewed as a kind of biofertilization. During the last decade, there has been renewed interest in the mineral-solubilizing activities of fungi. Fungi are important components of rock-inhabiting microbial communities. They are thus major biodeterioration agents of stone, wood, plaster, cement, and other building materials (Gadd 2010, 2017). However, their roles in geochemical processes have often been neglected, and their potential role in weathering interactions is still the subject of some debate (Finlay et al. 2009, Finlay and Clemmensen 2017). It is fairly well accepted that lichens participate in weathering interactions by producing organic acids that are manufactured with carbon derived from a *photobiont* (the photosynthetically active partner in the symbiosis), but delivered to mineral surfaces by a *mycobiont* (the fungal symbiotic partner). It is not unlikely that parallel processes occur in ectomycorrhizal symbioses, the only difference being that the size of the photosynthetic apparatus to provide the energy for nutrient mobilization and the size of the plant sink for mobilized nutrients are both much bigger. Mobilization of nutrients from crop residues or plant remains has been studied in AM systems (Hodge et al. 2001, Hodge and Fitter 2010), in which the presence of AM fungi appears to accelerate the release of nitrogen and its uptake by the host plants. Further research is necessary to distinguish between the direct capacity of the AM fungi to mobilize organic substrates and their possibly indirect effects on decomposition. The latter might be caused by the stimulation of decomposers and subsequent uptake of their decomposition products by mycorrhizal hyphae.

5.12 CONCLUSIONS

We can safely say that fungi play key roles in many soil biological processes. They are more or less ubiquitous in soils and very diverse, exhibiting a variety of growth forms and of trophic strategies. The filamentous nature of fungi complicates the methodology required to investigate their roles in soil microbiological processes, since many of the invasive sampling techniques used with bacteria disturb or destroy the functioning of the mycelium. A central, important feature of fungi forming a mycelium is that they are able to grow in heterogeneous environments with patchily distributed resources, simultaneously translocating carbon and nutrients in different directions. This makes such fungi ideally suited for growth in soil, since they can easily redistribute resources within their own mycelia. Since fungi are heterotrophs, they have developed three primary trophic strategies to obtain carbon compounds, achieving a preeminent degree of success as saprotrophs, pathogens, and symbionts in soil. As *saprotrophs*, the fungi play pivotal roles in the decomposition of plant residues, including cellulose and hemicelluloses from plant cell walls, as well as more recalcitrant components of woody plant litter such as lignin and lignocellulose. The degree of enzymatic competence differs between the different fungi, leading to *metabiosis*, the successive utilization and conversion of one substrate to another by organisms occurring through synergies or in successional sequences. Weak parasites and pioneer species are replaced by species that are able to degrade polymers with progressively greater degrees of recalcitrance. Some fungi may also attack, parasitize, or kill living organisms such as insects, nematodes, or other fungi to obtain carbon or nutrients. As *plant pathogens*, the fungi occupy a central role. More than 70% of all major crop diseases are caused by fungi or fungus-like oomycetes such as *Phytopthora* spp. Plant-pathogenic fungi differ in their nutritional relationships with their hosts, some being *necrotrophic*, killing the plant with toxins or degradative enzymes and others—such as the rusts and mildew fungi—entering into a biotrophic relationship and removing nutrients from the living host tissues. The third strategy that fungi have adopted to obtain carbon is *symbiosis*. Fossil and molecular evidence suggests that, in particular, fungi in the phylum Glomeromycota have developed a symbiotic habit early during evolution, forming symbiotic AM associations with early land plants some 460 million years ago. Mycorrhizal symbionts of plants have an array of different effects. Thus, different types of mycorrhizal symbiosis have evolved as adaptations to different suites of edaphic parameters, resulting in the characteristic vegetation types that dominate the different terrestrial biomes. The supply of photoassimilates from host plants to scavenging mycelial systems enables the fungal colonization of a range of organic and inorganic substrates. This provides energy for nutrient-mobilizing activities and conduits for the transport of mobilized nutrients back to the plants that act as sinks.

Irrespective of which of these three trophic strategies is involved, the coupled acquisition of plant-derived carbon and the mobilization of nutrients by soil fungi play important roles in soil biogeochemical cycles. These fungal processes have important consequences for plant health and productivity, soil fertility, and the structure and functioning of plant communities and ecosystems, and the global balance between ecosystem carbon sequestration in soil organic matter and secondary minerals and its release as carbon dioxide.

REFERENCES

Almario, J., Jeena, G. Wunder, J. et al. 2017. Root-associated fungal microbiota of nonmycorrhizal *Arabis alpina* and its contribution to plant phosphorus nutrition. *Proc Natl Acad Sci USA* 114, E9403–E9412.

Asemaninejad, A., Thorn, R.G. and Z. Lindo. 2017. Experimental climate change modifies degradative succession in boreal peatland fungal communities. *Microb Ecol* 73, 521–531.

Averill, C., Turner, B.L. and A.C. Finzi. 2014. Mycorrhiza-mediated competition between plants and decomposers drives soil carbon storage. *Nature* 505, 543–545.

Behie, S.W., Moreira, C.C., Sementchoukova, I., Barelli, L., Zelisko, P.M. and M.J. Bidochka. 2017. Carbon translocation from a plant to an insect-pathogenic endophytic fungus. *Nat Commun* 8, 14245.

Behie, S.W., P.M. Zelisko and M.J. Bidochka. 2012. Endophytic insect-parasitic fungi translocate nitrogen directly from insects to plants. *Science* 336, 1576–1577.

Bonfante, P. and A. Desiro. 2017. Who lives in a fungus? The diversity, origins and functions of fungal endobacteria living in Mucoromycota. *ISME J* 11, 1727–1735.

Butt, T.M., Jackson, C.W. and N. Magan. 2001. *Fungi as Biocontrol Agents: Progress, Problems and Potential.* Wallingford: CABI Publishing.

Clemmensen, K.E., Bahr, A., Ovaskainen, O. et al. 2013. Roots and associated fungi drive long-term carbon sequestration in boreal forest. *Science* 339, 1615–1618.

Deacon, J. 2005. *Fungal Biology.* 4th edition, Oxford: Blackwell Publishing.

de Boer, W., Folman, L.B., Summerbell, R.C. and L. Boddy. 2005. Living in a fungal world: impact of fungi on soil bacterial niche development. *FEMS Microbiol Rev* 29, 795–811.

Deveau, A., Bonito, G., Uehling, J. et al. 2018. Bacterial – fungal interactions: ecology, mechanisms and challenges. *FEMS Microbiol Rev*, fuy008. doi:10.1093/femsre/fuy008.

Fernandez, C.W. and P.G. Kennedy. 2016. Revisiting the 'Gadgil effect': do interguild fungal interactions control carbon cycling in forest soils? *New Phytol* 209, 1382–1394.

Finlay, R.D. 2008. Ecological aspects of mycorrhizal symbiosis: with special emphasis on the functional diversity of interactions involving the extraradical mycelium. *J Exp Bot* 59, 1115–1126.

Finlay, R.D. and K.E. Clemmensen. 2017. Immobilization of carbon in mycorrhizal mycelial biomass and secretions. In: *Mycorrhizal Mediation of Soil: Fertility, Structure, and Carbon Storage*, eds. N.C. Johnson, C. Gehring and J. Jansa, Elsevier: Amsterdam, pp. 413–440.

Finlay, R.D., Wallander, H., Smits, M. et al. 2009. The role of fungi in biogenic weathering in boreal forest soils. *Fungal Biol Rev* 23, 101–106.

Fitter, A.H. 2005. Darkness visible: reflections on underground ecology. *J Ecol* 93, 231–243.

Gadd, G.M. 2010. Metals, minerals and microbes: geomicrobiology and bioremediation. *Microbiology* 156, 609–643.

Gadd, G.M. 2017. Fungi, rocks, and minerals. *Elements* 13, 171–176.

Gkarmiri, K., Mahmood, S., Ekblad, A., Alstrom, S., Hogberg, N. and R. Finlay. 2017. Identifying the active microbiome associated with roots and rhizosphere soil of oilseed rape. *Appl Environ Microbiol* 83, e01938-17

Harman, G.E., Howell, C.R., Viterbo, A., Chet, I. and M. Lorito. 2004. *Trichoderma* species – Opportunistic, avirulent plant symbionts. *Nat Rev Microbiol* 2, 43–56.

Harms, H., Schlosser, D. and L.Y. Wick. 2011. Untapped potential: exploiting fungi in bioremediation of hazardous chemicals. *Nat Rev Microbiol* 9, 177–192.

Hawksworth, D.L. and R. Lücking. 2017. Fungal diversity revisited: 2. 2 to 3. 8 million species. Microbiology Spectrum 5. UNSP FUNK-0052-2016: doi:10.1128/microbiolspec. FUNK-0052-2016.

Hibbett, D.S., Binder, M., Bischoff, J.F. et al. 2007. A higher-level phylogenetic classification of the fungi. *Mycol Res* 111, 509–547.

Hodge, A., Campbell, C.D. and A.H. Fitter. 2001. An arbuscular mycorrhizal fungus accelerates decomposition and acquires nitrogen directly from organic material. *Nature* 413, 297–299.

Hodge, A. and A.H. Fitter. 2010. Substantial nitrogen acquisition by arbuscular mycorrhizal fungi from organic material has implications for N cycling. *Proc Natl Acad Sci USA* 107, 13754–13759.

Högberg, P., Nordgren, A., Buchmann, N. et al. 2001. Large-scale forest girdling shows that current photosynthesis drives soil respiration. *Nature* 411, 789–792.

Hori, C., Gaskell, J., Igarashi, K. et al. 2013. Genomewide analysis of polysaccharides degrading enzymes in 11 white- and brown-rot Polyporales provides insight into mechanisms of wood decay. *Mycologia* 105, 1412–1427.

Jacobs, H., Boswell, G.P., Ritz, K., Davidson, F.A. and G.M. Gadd. 2002. Solubilization of calcium phosphate as a consequence of carbon translocation by *Rhizoctonia solani*. *FEMS Microbiol Ecol* 40, 65–71.

James, T.Y., Kauff, F., Schoch, C.L. et al. 2006. Reconstructing the early evolution of Fungi using a six-gene phylogeny. *Nature* 443, 818–822.

Jensen, D.F., Karlsson, M. and B.D. Lindahl. 2017. Fungal-fungal interactions: from natural ecosystems to managed plant production, with emphasis on biological control of plant diseases. In: *The Fungal Community: Its Organization and Role in the Ecosystem*. 4th edition, eds. J. Dighton and J.F. White, Boca Raton: CRC Press, pp. 549–562.

Johansson, J.F., Paul, L.R. and R.D. Finlay. 2004. Microbial interactions in the mycorrhizosphere and their significance for sustainable agriculture. *FEMS Microbiol Ecol* 48, 1–13.

Johnson, D., Leake, J.R., Ostle, N., Ineson, P. and D.J. Read. 2002. In situ (CO_2)-C-13 pulse-labelling of upland grassland demonstrates a rapid pathway of carbon flux from arbuscular mycorrhizal mycelia to the soil. *New Phytol* 153, 327–334.

Jumpponen, A., Jones, K.L. and J. Blair. 2010. Vertical distribution of fungal communities in tallgrass prairie soil. *Mycologia* 102, 1027–1041.

Krings, M., Taylor, T.N. and C.J. Harper. 2017. Early fungi: evidence from the fossil record. In: *The Fungal Community: Its Organization and Role in the Ecosystem*. 4th edition, eds. J. Dighton and J.F. White, Boca Raton: CRC Press, pp. 37–51.

Leake, J.R. 2005. Plants parasitic on fungi: unearthing the fungi in myco-heterotrophs and debunking the 'saprotrophic' plant myth. *Mycologist* 19, 113–122.

Lindahl, B.D. and A. Tunlid. 2015. Ectomycorrhizal fungi – potential organic matter decomposers, yet not saprotrophs. *New Phytol* 205, 1443–1447.

Menkis, A., Burokiene, D., Gaitnieks, T. et al. 2012. Occurrence and impact of the root-rot biocontrol agent *Phlebiopsis gigantea* on soil fungal communities in *Picea abies* forests of northern Europe. *FEMS Microbiol Ecol* 81, 438–445.

Quirk, J., Beerling, D.J., Banwart, S.A., Kakonyi, G., Romero-Gonzalez, M.E. and J.R. Leake. 2012. Evolution of trees and mycorrhizal fungi intensifies silicate mineral weathering. *Biol Lett* 8, 1006–1011.

Read, D.J. and J. Perez-Moreno. 2003. Mycorrhizas and nutrient cycling in ecosystems – a journey towards relevance? *New Phytol* 157, 475–492.

Redecker, D., Kodner, R. and L.E. Graham. 2000. Glomalean fungi from the Ordovician. *Science* 289, 1920–1921.

Redecker, D., Schussler, A., Stockinger, H., Sturmer, S.L., Morton, J.B. and C. Walker. 2013. An evidence-based consensus for the classification of arbuscular mycorrhizal fungi (Glomeromycota). *Mycorrhiza* 23, 515–531.

Riley, R., Salamov, A.A., Brown, D.W. et al. 2014. Extensive sampling of basidiomycete genomes demonstrates inadequacy of the white-rot/brown-rot paradigm for wood decay fungi. *Proc Natl Acad Sci USA* 111, 9923–9928.

Schoch, C.L., Sung, G.H., Lopez-Giraldez, F. et al. 2009. The Ascomycota Tree of Life: a phylum-wide phylogeny clarifies the origin and evolution of fundamental reproductive and ecological traits. *Syst Biol* 58, 224–239.

Schüßler, A., Schwarzott, D. and C. Walker. 2001. A new fungal phylum, the Glomeromycota: phylogeny and evolution. *Mycol Res* 105, 1413–1421.

Shah, F., Nicolas, C., Bentzer, J. et al. 2016. Ectomycorrhizal fungi decompose soil organic matter using oxidative mechanisms adapted from saprotrophic ancestors. *New Phytol* 209, 1705–1719.

Smith, M.L., Bruhn, J.N. and J.B. Anderson. 1992. The fungus *Armillaria bulbosa* is among the largest and oldest living organisms. *Nature* 356, 428–431.

Smith, S.E. and D.J. Read. 2008. *Mycorrhizal Symbiosis.* 3rd edition, San Diego: Academic Press.

Spatafora, J.W., Chang, Y., Benny, G.L. et al. 2016. A phylum-level phylogenetic classification of zygomycete fungi based on genome-scale data. *Mycologia* 108, 1028–1046.

Sterkenburg, E., Clemmensen, K.E., Ekblad, A., Finlay, R.D. and B.D. Lindahl. 2018. Contrasting effects of ectomycorrhizal fungi on early and late stage decomposition in a boreal forest. *ISME J* 12, 2187–2197.

Sun, Y.P., Unestam, T., Lucas, S.D., Johanson, K.J., Kenne, L. and R. Finlay. 1999. Exudation-reabsorption in a mycorrhizal fungus, the dynamic interface for interaction with soil and soil microorganisms. *Mycorrhiza* 9, 137–144.

Taylor, J.D., Helgason, T. and M. Öpik. 2017. Molecular community ecology of arbuscular mycorrhizal fungi. In: *The Fungal Community: Its Organization and Role in the Ecosystem.* 4th edition, eds. J. Dighton and J.F. White, Boca Raton: CRC Press, pp. 3–25.

Taylor, L.L., Quirk, J., Thorley, R.M.S. et al. 2016. Enhanced weathering strategies for stabilizing climate and averting ocean acidification. *Nat Clim Change* 6, 402–406.

Tedersoo, L., Bahram, M., Polme, S. et al. 2014. Global diversity and geography of soil fungi. *Science* 1256688.

Tedersoo, L. and M.E. Smith. 2013. Lineages of ectomycorrhizal fungi revisited: foraging strategies and novel lineages revealed by sequences from belowground. *Fungal Biol Rev* 27, 83–99.

van der Heijden, M.G.A., Klironomos, J.N., Ursic, M. et al. 1998. Mycorrhizal fungal diversity determines plant biodiversity, ecosystem variability and productivity. *Nature* 396, 69–72.

Weete, J.D., Abril, M. and M. Blackwell. 2010. Phylogenetic distribution of fungal sterols. *PLoS One* 5, e10899.

6 The Viruses in Soil—Potential Roles, Activities, and Impacts

Akbar Adjie Pratama and Jan Dirk van Elsas
University of Groningen

CONTENTS

6.1 Introduction—Soil Viromes 91
6.2 Methods to Detect Viruses in Soil with an Emphasis on Phages 92
6.3 The Soil Habitat—Shaping the Virome via Organismal Hosts 92
  6.3.1 The Effect of Soil Habitat Structure on Virus–Host Interactions 92
  6.3.2 Soil Viromes—Mixed Communities of Agents Infecting Diverse Hosts 93
6.4 Abundance and Diversity of Phages in Soils 95
  6.4.1 Abundance 95
  6.4.2 Diversity 97
6.5 Ecological and Evolutionary Impact of Phages in Soil 100
  6.5.1 Phages as Controllers of Host Population Densities 100
  6.5.2 Phages as Agents of Horizontal Gene Transfer, Accelerating Bacterial Evolution 101
  6.5.3 Phages as Agents That Modulate Biofilms 101
6.6 Future Perspectives 101
  6.6.1 Three Approaches to Soil Virome Studies 101
  6.6.2 Spatiotemporal Aspects of Virus–Host Interactions 102
References 103

6.1 INTRODUCTION—SOIL VIROMES

In Chapters 1–5, we learned that soil plays a pivotal role in the functioning of the global biome, providing a multitude of different ecosystem services, activities, and interactions. This includes processes of the global biogeochemical cycles (C and N), as outlined in Chapter 11. Soil biomes encompass a number of diverse organisms (including bacteria, archaea, fungi, protozoa, nematodes, earthworms, arthropods, and plants). The interactions of viruses with organisms from all of these groups play essential roles in the activities of the latter, and so in the aforementioned soil functions. Although we do have increasingly detailed knowledge on microbiome abundance and diversities across a range of diverse soils, little is known with respect to the role of viruses.

Given the prevalence of bacteria in most soil microbiomes, it comes as no surprise that viruses infecting bacterial hosts, that is, bacteriophages (*phages*), have received the most interest. However, viruses of all other members of the soil biome also occur in soils, affecting the lifestyles of their respective hosts. We will briefly address the collective viruses (viromes) of nonbacterial hosts, after which we will examine the diversity and putative role of phages. What do we currently understand about these, in terms of their biology and their roles in soil microbiomes?

Since their discovery about one century ago, independently by Frederick Twort (1915) and Felix d'Herelle (1917), phages have been paramount in our understanding of the fundaments of biology. The hallmark work on the roles of phages, performed in the 1950s by Delbrück and co-workers, has advanced the development of molecular genetics (Salmond and Fineran 2015). A key finding was the contention that phages often are strong drivers of the evolution of their hosts. Studies of a suite of bacterial genomes, for example, those of *Escherichia coli* (Touchon et al. 2009) and *Mycobacterium* spp. (Pedulla et al. 2003), have yielded ample evidence for this contention, as multifold so-called phage regions were discovered in these genomes. However, what we cannot "read" from such genomes is how the phages "act" at the level of the members of soil microbiomes. Classically, phages have been implied in population control as well as horizontal gene transfer (HGT) through transduction (see also Chapter 7). Moreover, recent evidence has pointed to their roles as enhancers of cellular fitness in, for example, biofilms (Secor et al. 2015).

In this chapter, we examine the current understanding of virus and phage abundance, as well as the diversities, evolution, and putative roles of these agents in soil ecosystems. A brief examination of the methods used to detect viruses in soil is followed by the assessment of soil as a habitat for viral-affected life and of soil viromes, with emphasis on phages. We address the current knowledge as well as challenges in this area and suggest potential strategies that allow to foster our understanding of virus–host interactions in soil.

6.2 METHODS TO DETECT VIRUSES IN SOIL WITH AN EMPHASIS ON PHAGES

Whereas current advances in sequencing technologies have broadened the possibilities of soil virome studies, the early studies were exclusively based on enumeration methods, that is, via double-agar-layer (DAL) plating with susceptible host bacteria (Adams 1959) and phage observations/counts using direct transmission electron microscopy (TEM) and/or epifluorescence microscopy (EFM) (Ashelford et al. 2003, Williamson et al. 2005, Swanson et al. 2009). The (cultivation-based) DAL method has allowed researchers to detect and study those phages that are able to productively infect the host organism used (as it detects plaques produced by a lytic cycle following the infection of host cells), thus missing phages that perform poorly on the host under the conditions applied. TEM detects viruses via extractions from soil followed by purification, grid placement, and staining steps. EFM detects viruses based on the propensity of the fluorescent dye SYBR Green/Gold to strongly bind to double-stranded DNA (dsDNA) and RNA, giving high sensitivity of detection. Unfortunately, the efficiency of these methods, in particular EFM, is hampered by frequently occurring interactions of viral particles with soil compounds (Ashelford et al. 2003). Moreover, DAL and TEM based methods are also limited in their scope, as only cultivation (DAL) and morphological criteria (TEM) lie at their basis. For instance, the TEM-observable parameters (e.g., phage head and tail sizes and shapes) can be either explanatory or totally meaningless with respect to distinctions within and between viral families.

It is clear that one should complement the DAL, TEM, and EFM soil virome analysis methods with currently available advanced DNA sequencing methods (see Chapters 3 and 14). In this endeavor, multiple samples of (1) viromes, that is, collective viral particles separated from soil biomes, and (2) [micro]biomes need to be analyzed, in order to examine the extragenomic as well as intragenomic occurrences of viral sequences.

6.3 THE SOIL HABITAT—SHAPING THE VIROME VIA ORGANISMAL HOSTS

6.3.1 The Effect of Soil Habitat Structure on Virus–Host Interactions

Within soil aggregates, the habitats for microbes are intrinsically heterogeneous, as already outlined in Chapters 1 and 3. The spatial structure of soil thus generates multiple microhabitats, establishing diverse niches for members of the microbiome. Particular microhabitats in soil exhibit enhanced nutritional status, yielding microbial activity hot spots, often inside soil aggregates (Figure 6.1d). During its genesis, a soil aggregate may harbor a specific part of the soil microbiome, ranging from thousands to millions of individual cells (Rillig et al. 2017). The often encountered spatial isolation

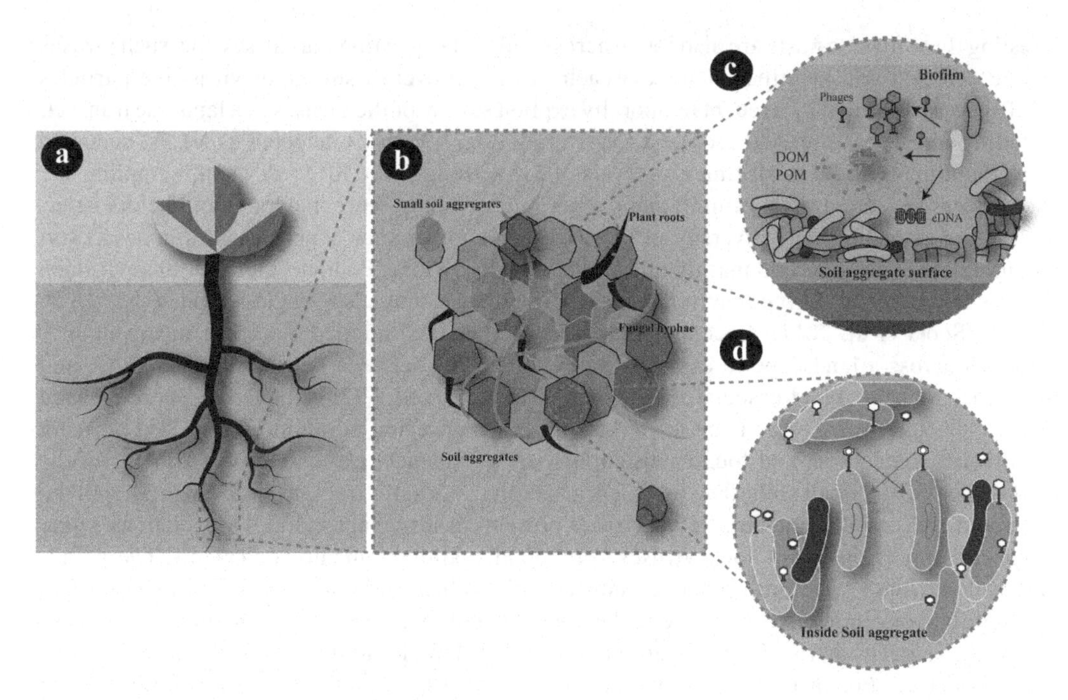

FIGURE 6.1 **(See color insert)** Role of viruses in soil microbiomes. (a) The heterogeneous nature of soil incites the presence of a great number of diverse organisms, as multiple microhabitats, including soil hot spots, are created. (b) Soil aggregates consist of microaggregates, soil organic matter, biotic binding agents (plant roots, fungal hyphae), water, and microbes. Thus, niches exist for bacteria, archaea, protozoa, nematodes, earthworms, and arthropods. (c) On surfaces, biofilms can emerge; in these, local lysis caused by phage activity introduces channels, in which low-molecular-weight compounds and extracellular DNA (eDNA) emerge. Both contribute to the dynamic development of the biofilm. (d) Soil aggregates in hot spots promote spatial isolation for bacteria, allowing separate evolution. Within a soil aggregate, pressure due to nutrient depletion may incite the excision of phages from host genomes. As a consequence, soil bacterial evolution is spurred. (Modified from Pratama and van Elsas (2017a,b)).

and lack of connectivity of the (sub) microbiomes inside the aggregates incite parallel microbial eco-evolutionary trajectories, resulting in conceptual metapopulations, even of the same species, that are distributed across isolated soil islands. This facet of life in the soil tends to enhance the diversity of soil microbiomes, which is reflected in the respective soil viromes (see Figure 6.1 for a depiction of such viral–host interactions across soil microhabitat islands). Incidentally, soil aggregates may disintegrate, enhancing the connectivity within the microbiome. In such circumstances, microbiomes may mix, both physically and genetically, for example, enabling horizontal gene transfers (HGT; Chapter 7). It is as yet unknown how and to what extent this dynamic scenario relates to viral activities. Potentially, the "activity," reflected in excision/lysis/insertion events, of phages is different under conditions of mixing versus life inside an aggregate, and hence, the fate of the respective host is differentially affected. However, the formation, persistence, and subsequent disintegration of soil aggregates undoubtedly affect the pools of free or integrated phages and hence soil microbiome diversity.

6.3.2 Soil Viromes—Mixed Communities of Agents Infecting Diverse Hosts

Given the fact that the soil ecosystem contains diverse organisms—including bacteria, fungi, archaea, protozoa, nematodes, earthworms, and arthropods, it comes as no surprise that viruses infesting these diverse hosts are also found across soils. The question thus arises how such viruses can be detected and identified. One approach lies in an overall survey of virus-like particles (VLPs), as exemplified by a recent temporally explicit survey of the viruses in a land-use transect, including forest, pasture, and cropland (Narr et al. 2017). Viruses detected as VLPs, occurred throughout the transect at all times, at levels of 2.4×10^7 to $5.9 \times 10^9 g^{-1}$ dry soil. Sampling site and, for less vegetated sites, rainfall and temperature regimes, were the dominant factors determining the abundance and diversity of the VLPs. In another survey performed in a desert soil (Namibia), Scola et al. found mainly tailed phages (Myoviridae, Podoviridae, and Siphoviridae), next to Iridoviridae, Phycodnaviridae, and Mimiviridae, but also single-stranded (ss) DNA viruses (Scola et al. 2017). Recently, Bernardo et al. (2018) applied a virome metagenomics approach across a landscape to examine the relationships between (agricultural) land use and the distribution of (plant-associated) viruses. They analyzed 1,725 samples of georeferenced plants collected over 2 years from $4.5 \times 4.5 km^2$ crates covering agricultural land and adjacent uncultivated vegetation, and found substantial viral prevalence, expressed as percentage of the total (25.8–35.7%), in all soils. The prevalence and diversities of viral families were higher in the cultivated areas, with some families of viruses presenting strong agricultural associations. Up to 94 species of previously unknown viruses were found, mainly of uncultivated plants. These findings indicate that agricultural practices substantially influence the distribution of plant viruses and highlight our immense ignorance of the diversity and roles of viruses in nature. These types of surveys indicate that soil viruses are prevalent, yet they do not tell us much more than this simple message. For instance, most of the putative viral hosts are still unknown. One would expect a majority of soil viruses to relate to the main cellular soil inhabitants (the ones that are numerically dominant), that is, the bacteria and fungi.

Considering viruses infecting the latter group (*mycoviruses*), a study found these to be widespread in soils, although not as free viral particles (Ghabrial et al. 2015). Thus far, viruses that interact with all major taxa of fungi have been found. In many cases, the infections these viruses cause are symptomless. Also, they do appear to have known natural vectors, and are mainly transmitted horizontally (via hyphal anastomosis) as well as vertically (through spores). These mycoviruses change the fungal host phenotype to a state called *hypovirulence*, without causing any clear deleterious (morphologically visible) symptoms in it. However, in certain cases, viral infections will affect the host's ability to cause disease in plants. Other changes in the fungal host include the loss of female fertility, the reduction of asexual sporulation, colony morphology changes, and a decreased accumulation of fungal metabolites such as oxalic acid (Ghabrial et al. 2015). As a case in point, four well-known mycosphere-induced hypovirulent viruses, the *Cryphonectria* hypoviruses 1–4 (CHV-1 through CHV-4), infect the chestnut blight fungus *Cryphonectria parasitica* and thus exert a control over the spread of chestnut blight disease by reducing parasitic growth and sporulation capacity of the fungus (Ghabrial et al. 2015).

Other soil virome studies are still in their infancy. A recent study based on RNA reverse transcription and subsequent sequencing of viruses of 70 arthropod species (Insecta, Arachnida, Chilopoda, and Malacostraca), including those of soil, showed a remarkable diversity across these viruses, as evidenced by viral genome organization criteria. This was confirmed by morphological data obtained via TEM observations. Overall, the study discovered 112 novel viruses, highlighting the role of these as vectors for the arthropod hosts (Li et al. 2015). Expectedly, in any given soil habitat, a plethora of viruses will occur as a reflection of the current or past presence of their hosts, and of their host's lifestyles. However, overall views of the diversities and putative roles of these viruses are still lacking. Here, we posit that, to enable an integral understanding of the soil ecosystem, it is of utmost importance to place more emphasis on whole-soil virome analyses. Such analyses are best performed taking into account issues of spatiotemporally explicit sampling, as discussed in Chapters 1 and 3. Given the fact that soil viromes are predicted to be dominated by bacteriophages, we will in the further text place a focus on these with respect to viral fate and effect (Table 6.1).

TABLE 6.1
Main Features of Viruses in Soils

Hosts	Viruses—Main Features	Typical Viruses Found	References
Bacteria	Viral type denominated (bacterio)phage. Genomes diverse: dsDNA, ssDNA, dsRNA, and ssRNA (+) (linear and circular). Diverse morphotypes, most with icosahedral capsid, with or without envelope. Filamentous and spherical to pleomorphic forms also found. Phages are divided into lytic versus lysogenic ones.	Caudovirales (Myoviridae, Podoviridae and Siphoviridae), Inoviridae, Plasmaviridae, and Leviviridae	Salmond and Fineran (2015)
Archaea	Viruses (phages) widespread, with hosts occurring across extreme and non-extreme environments. Genomes: dsDNA or ssDNA. Morphotypes divided into (1) archaea specific, without structural or genetic counterparts in other prokaryote or eukaryote domains and (2) "cosmopolitan", with structure and genetic makeup similar to viruses in other domains	Archaea specific: Ampullaviridae, Bicaudaviridae, Spiraviridae, and Fuselloviridae Cosmopolitan: Myoviridae, Siphoviridae, Podoviridae, Sphaerolipoviridae, and Turriviridae	Prangishvili et al. (2017)
Fungi	Viruses lack extracellular phase; transmitted via anastomosis or vertically. Often symptomless, may change host phenotype; hypovirulent, used in biocontrol. Diverse viral genomes: linear—dsRNA, ssRNA(+), and ssRNA(−), and circular—ssDNA	Totiviridae, Chrysoviridae, Partitiviridae, Megabirnaviridae, Quadriviridae, Reoviridae, Alphaflexiviridae, Endornaviridae, Barnaviridae, Gammaflexiviridae, Hypoviridae, Narnaviridae, and Mycomononegaviridae	Ghabrial et al. (2015)
Arthropods	Viruses infecting plants and mammals transported. Mostly RNA viruses: ssRNA(+), ssRNA(−)	Flavivirus, Arboviruses, Togaviruses, Bunyaviruses, Mesoniviruses, Reoviruses, Rhabdoviruses, Negiviruses, and Birnviruses	Li et al. (2015)
Nematodes	Some plant viruses transported. Orsay virus: first known virus that infects *Caenorhabditis elegans* or other nematodes. It has icosahedral virion and ssRNA(+) genome	Orsay virus infecting *Caenorhabditis elegans*	Guo et al. (2014)
Protozoa	Mostly RNA viruses. Examples: cryspoviruses: dsRNA, icosahedral capsids	Cryspovirus infecting apicomplexan, *Trichomonas vaginalis* virus, and *Giardia lamblia* viruses	Wang and Wang (1991)
Plants	Diverse viruses. Genomes: ssDNA, dsRNA, ssRNA(+), ssRNA(−), ssRNA (RT), and dsDNA (RT) viruses. Morphotypes: non-enveloped, round, twinned (geminate) incomplete icosahedral, double-capsid, filamentous, rigid helical rods, bacilliform. Also enveloped, allantoid, spherical, and highly flexuous forms	Tobacco mosaic virus, tomato spotted wilt virus, tomato yellow leaf curl begomovirus, potato virus Y (potyvirus), cauliflower mosaic virus, and potato virus X (potexvirus)	Rybicki (2015)

See Box 6.1 for the designation of viral nucleic acid type.

6.4 ABUNDANCE AND DIVERSITY OF PHAGES IN SOILS

6.4.1 Abundance

A modest bibliography is currently available on bacteriophages and their roles in soil (Pratama and van Elsas 2017a,b). Early work (based on the DAL and TEM approaches) performed in the 1970s addressed the potential role of phages, in particular of the soil bacterium *Bacillus*, in soil settings (Reanney and Marsh 1973). Later, studies on *Bacillus* phages in soil pinpointed the role of such phages as (1) controllers of *Bacillus* populations in soil and (2) agents of antibiotic resistance plasmid transfer (van Elsas and Penido 1982). Other work on viral/phage diversity in soil, based on TEM observations, coined the term *VLP* to identify the unit of detection. In these studies in rhizosphere and bulk soils, VLPs, constituting mostly predicted phages, showing tailed, spherical, rod-shaped, filamentous, or bacilliform morphs, were found (Swanson et al. 2009) (see Figure 6.2 for a depiction of diverse *Bacillus* phages detected from soil). Quantification of the VLP populations in six different soil types (agricultural, coastal, and piedmont types) in Delaware (Williamson et al. 2005) and the Namib desert, South Africa (Adriaenssens et al. 2015) revealed abundances of, on average, about 10^7–10^9 VLPs g^{-1} of soil. VLP abundances in the range of 10^8–$10^9\,g^{-1}$ have also been observed across desert, prairie, rain forest, Antarctic (Zablocki et al. 2014), and paddy soils (Kim et al. 2008). Moreover, studies on rhizospheres and rhizosheaths of *Beta vulgaris* var. Amethyst, *Poa pratensis* L., *Epilobium tetragonum* L. Griseb., *Senecio jacobaea* L., and *Cardamine flexuosa*, and respective bulk soils, yielded VLP counts in the range 10^6–$10^9\,g^{-1}$ dry soil (Ashelford et al. 2003, Swanson et al. 2009). Remarkably, the latter study found no significant differences in VLP abundance between rhizospheres and surrounding bulk soils ($1.18 \pm 0.014 \times 10^9\,g^{-1}$ dry soil for rhizosphere, $1.09 \pm 0.014 \times 10^9\,g^{-1}$ dry weight for rhizosheath, and $1.17 \pm 0.085 \times 10^9\,g^{-1}$ dry bulk soil) (Swanson et al. 2009). Possibly, the *piggy-back-the-winner* (*PtW*) model may have played a role in these phage–host populations (Knowles et al. 2016). According to this model, the high bacterial densities that may occur in activity hot spots, such as the rhizosphere, may favor phage integrations into host genomes over lytic events, yielding lysogens. These are insusceptible to subsequent infections with the same phage, as the resident phage offers superinfection immunity to its host. Overall, the PtW phenomenon may thus limit the abundance of free phage in crowded spots in environments like the rhizosphere.

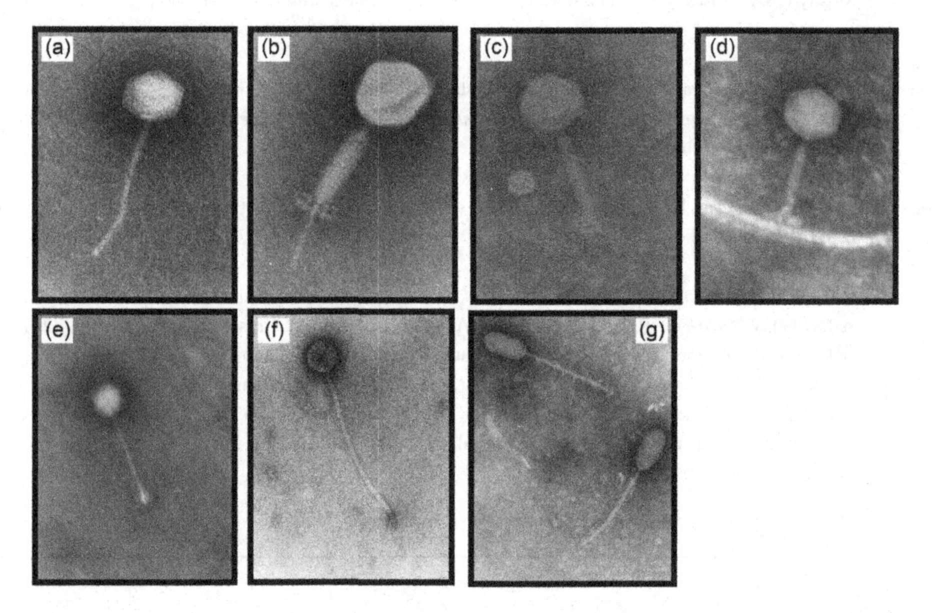

FIGURE 6.2 (See color insert) Diversity of soil phages associated with *Bacillus subtilis* and *Bacillus megaterium* (observed by TEM—staining with 2% uranyl oxalate). The phages were isolated from Brazilian soils using *B. subtilis* and *B. megaterium* isolated from the same soils as the hosts. (a) *B. subtilis* F6 phage FS2, magnification 258,000×; (b) *B. subtilis* F6 phage FS7, magnification 226,000×; (c) *B. subtilis* F6 phage IS1, magnification 170,000×; (d) *B. subtilis* F6 phage GS1, magnification 179,000×; (e) *B. megaterium* F4 phage MJ3, magnification 240,000×; (f) *B. megaterium* F4 phage MJ1, magnification 194,000×; (g) *B. megaterium* F4 phage MJ4, magnification 149,000×.

6.4.2 Diversity

How diverse are soil viromes? This question was initially addressed using the traditional microscopy-based approaches (see Box 6.1, last paragraph). TEM-based analyses of VLP morphology assessed viral diversity in six Delaware soils. Using as a proxy capsid size to assess soil viral diversity, analysis by Simpson's index (1/D) was applied to 450 observable VLPs. The results indicated that most soils were dominated by phages belonging to the Caudovirales (tailed phages) (Williamson et al. 2005). A separate TEM-based study of viromes from rhizosphere versus bulk soil indicated the presence of varied morphological types (tailed, spherical, rod-shaped, filamentous, or bacilliform), with no significant difference between bulk and rhizosphere soil samples (Swanson et al. 2009). The observed morphotype diversities might be related to the host community structures in the sampled soils. Lastly, the methods used to extract the viruses from soil particles may have affected and possibly "biased" the observed soil virus diversities.

BOX 6.1 HOW TO CLASSIFY VIRUSES?

In 2016, the International Committee on Taxonomy of Viruses (ICTV) has listed eight orders (Bunyavirales, Caudovirales, Herpesvirales, Ligamenvirales, Mononegavirales, Nidovirales, Picornavirales, Tymovirales, next to "Unassigned"), 125 families, and 4,403 species of viruses. The classification methods proposed by the ICTV subcommittee on bacterial and archaeal viruses are based on both biological properties (e.g., host range, habitat, and effect on host pathogenicity) and phage genome sequence information. Given the expanding metagenome data sets, Simmonds et al. (2017) recently recommended the incorporation of viral data from metagenomics-based pipelines into current virus taxonomy. Robust phylogenetic analyses based on the data are thought to be sufficient for most taxonomic assignments. Main criteria include the presence of genes that indicate phylogenetic relatedness, overall genome organization, gene complements and replication strategy, and the presence or absence of distinct motifs (Simmonds et al. 2017). Sequence data have enabled reconstructions of the evolutionary relatedness of viruses. They may also indicate their ecological significance and predict viral host ranges.

Using nucleic acid type, seven categories of viruses have been discerned: double-stranded (ds)DNA and single-stranded (ss)DNA, dsRNA, positive- and negative-sense ssRNA (ssRNA(+); ssRNA(−)), retro-transcribing (RT) ssRNA (RT-ssRNA), and RT-dsRNA viruses. On the basis of morphology, viral particles are discerned as a variety of types. Phages mostly fall in the Caudovirales (tailed viruses), in which three families are discerned: Myoviridae, Siphoviridae, and Podoviridae. These families all have "head and tail" morphologies. The icosahedral heads contain the genomes (usually dsDNA), and the tails differ in length and ability to contract. Phages in the Podoviridae have short and those in the Siphoviridae have long noncontractile tails, whereas Myoviridae have long contractile ones.

The classical approach to viral taxonomy based on morphology (capsid sizes and shapes), as well as nucleic acid type (ssRNA, ssDNA, dsRNA, and dsDNA), has progressively changed with the advances made in molecular viromics (Simmonds et al. 2017) (see Box 6.1). It is now recommended to classify viruses on the basis of genome-to-genome analyses of individual viral isolates, using signature genes such as the T4-like portal gene, the major capsid gene, and DNA polymerase gene *polA* (Adriaenssens and Cowan 2014). Classifications based on viral proteome sequences (Rohwer and Edwards 2002) and phage network clustering (Lima-Mendez et al. 2008) have also been proposed; these approaches constitute viable alternatives to the classification of viruses in cases where no "universal marker" genes are present (Table 6.2).

TABLE 6.2
Overview of Studies on Soil Viromes with an Emphasis on Bacteriophages

Soil Type/Habitat	Viral Abundance	Viral Diversity	Methods Used	Remarks	References
Soil devoid of vegetation and exposed to the sun, New Zealand	1.2×10^8–3.4×10^{10} pfu mL^{-1} suspension	Tailed phages	DAL on *Bacillus cereus*, *B. coagulans*, *B. licheniformis*, *B. megaterium*, *B. pumilus*, *B. sphaericus*, *B. subtilis*, *B. stearothermophilus*	This study highlights that *Bacillus* phages are agents that accelerate evolution	Reanney and Teh (1973)
Nile Valley, Egypt	$>10^6$ pfu mL^{-1} suspension	Tailed phages	DAL on *Azotobacter chroococcum* C32	This study assessed *Azotobacter* phages in soils	Hegazi and Jensen (1973)
Illinois and tropical soil, Brazil	10^9–10^{10} pfu mL^{-1} suspension	Tailed phage MJ-1 (Siphoviridae) and phages MP9–MP50 (Myoviridae and Siphoviridae)	DAL on *B. megaterium* F4 and *B. megaterium* QM BI55I (ATCC I2872), respectively	The papers highlight the potential role of *B. megaterium* phages in soil	van Elsas and Penido (1982); Vary and Halsey (1980)
Rhizosphere (*Beta vulgaris* var. Amethyst, *Poa pratensis* L, *Epilobium tetragonum* L. Griseb., *Senecio jacobaea* L., and *Cardamine flexuos*) and the respective bulk soils, the United Kingdom	1.5×10^7–1.5×10^8 VLPs g^{-1} dry soil	Tailed phages	Direct counting TEM, and DAL plaque counting on *Serratia quinivorans* CP6 and *Pseudomonas aeruginosa* PU21	Developed methods for counting the total numbers of bacteriophages in soil	Ashelford et al. (2003)
Agricultural soil, USA	4.0×10^7–2.15×10^8 VLPs g^{-1} dry soil	Tailed phages	TEM and EFM count (no significant difference from both methods)	This study showed that land use is a significant factor influencing viral abundance and diversity in soils	Williamson et al. (2005)
Rhizosphere (*Triticum aestivum*), bulk soil, the United Kingdom	1.1×10^9–$1.2 \times 10^9 g^{-1}$	Tailed phages (Caudovirales), spherical, bacilliform, filamentous, and rod-shaped viruses	Direct counting (TEM)	Reported that the majority of virus particles observed are autochthonous soil viruses	Swanson et al. (2009)
Hyperarid desert soil, Antarctic	2.3×10^8–6.4×10^8 VLPs g^{-1} dry soil	Dominant Caudovirales; signature gene: Adenoviridae, Bicaudaviridae, Hytrosaviridae, Retroviridae, and Rudiviridae	Virome metagenomics study	This study highlighted the factors that contribute to viral abundance and diversity in hot and cold deserts	Zablocki et al. (2014)

(*Continued*)

TABLE 6.2 (*Continued*)
Overview of Studies on Soil Viromes with an Emphasis on Bacteriophages

Soil Type/Habitat	Viral Abundance	Viral Diversity	Methods Used	Remarks	References
Namib desert soil, Africa	22-Mb virome reads	Tailed phages Iridoviridae, Phycodnaviridae, ssDNA viruses	Virome metagenomics study	This study evaluated the role of aridity and xeric gradient on the microbial community structure and function in Namib desert soil	Scola et al. (2017)
Soil metagenomics	10,009 viral contigs identified (IMG/VR)	18,470 viral groups	Mining viral signal from bacterial metagenomes	This study showed viral distribution across diverse ecosystems; strong habitat specificity for the vast majority of viruses, next to some cosmopolitan viruses	Paez-Espino et al. (2016)
Forest soil, pasture land, cropland	10^8–10^9 VLP g^{-1} dry soil	Not mentioned	Microscopic enumeration, virus community fingerprinting approach	This study showed that sampling site was a key factor for shaping the abundance and community structure of soil viruses	Narr et al. (2017)
Mycosphere	Up to 10^8 copies of major capsid gene mL^{-1}	Tailed phages (ϕ437)	Mitomycin C induction	Soil (inducible) prophage from *P. terrae* BS437	Pratama and van Elsas (2017a,b)

DAL, double agar layer; pfu, plaque-forming unit; TEM, transmission electron microscopy; EFM, epifluorescence microscopy; mitomycin C, prophage induction agent; IMG/VR, integrated microbial genome/virus (https://img.jgi.doe.gov/cgi-bin/vr/main.cgi); VLP, virus-like particle; ssDNA, single-stranded DNA.

The morphology-based analyses of viral diversities in South Western Africa desert soils showed the dominance of diverse tailed viruses (67%) belonging to the Myoviridae, Podoviridae, and Siphoviridae. Moreover, phages belonging to the Caudovirales (7%), next to *Geobacillus* phage E2 (6%), *Bacillus* phage phBC6A51 (4%), and a phage resembling a phage from a deep-sea thermophile (D6E; 1%) were also found. Finally, an ssDNA VLP belonging to the Microviridae (4%) was found (Adriaenssens et al. 2015). The study in Antarctic soil (Zablocki et al. 2014) found a high diversity of viruses, in particular tailed phages, next to ssDNA phages belonging to the Inoviridae, Microviridae, Circoviridae, Geminiviridae, and Nanoviridae. Finally, novel circular ssDNA viruses were discovered from the paddy soil (Kim et al. 2008). However, the viral hosts were not determined in the latter study, and the replication proteins of these ssDNA viruses were mostly unrelated to those of viruses in the database. This finding confirms our current lack of knowledge of the status of ssDNA soil viruses. Across diverse soils, diverse viromes are found, and it is tempting to assume that such viromes are potentially habitat specific (reflecting their host organisms). Most likely, phage-like particles dominate in most soil viromes, and so their diversities reflect the (high) species-level bacterial diversities.

6.5 ECOLOGICAL AND EVOLUTIONARY IMPACT OF PHAGES IN SOIL

6.5.1 Phages as Controllers of Host Population Densities

The main ecological effect of *lytic* phages is the control of host population sizes. In periods of host cell activity, lysis by phage attacks will reduce population sizes, concomitantly providing nutrients, in the form of *dissolved* and *particulate organic matter* (DOM and POM) to the organisms in the system. The role of lytic phages as (1) population size controllers and (2) providers of DOM/POM has been worked out well in marine settings, and it is possible that similar mechanisms are operational in soil. The population control process is governed by the rules of classical predator–prey relationships. A model, known as the *kill-the-winner* (KtW) model, describes the process. Bacterial cell populations that thrive under particular conditions in the soil system are, due to a mass action effect, the ones that are most prone to killing by infection by phages, and as a result, their population density will decrease. At the lowered population densities, phage–host cell encounters become less frequent, and so the intensity of lytic events will drop and cell lysis will eventually cease. Thus, a more or less stable, reduced host cell population size may arise. However, a mutated phage-resistant host cell population may emerge that, under the prevailing conditions, rebounds to higher population densities. In the extant phage population, mutants may have emerged that can infect the novel host cell population, and so the intensity of lytic events will, again, increase. Overall, this process incites a dynamically fluctuating host cell population, in which the diversity—with respect to interactions with phages, has shifted. KtW processes are operational at the highs in population density. They are thought to play key roles in highly dynamic soil sites where microbial activity is triggered, such as the rhizosphere and the mycosphere. A key corollary of the process is the enhancement of other microbial activities (due to effects of carbon released from lysed cells), resulting in nutrient cycling and growth of (parts of) local microbiomes.

In recent years, phages have received increasing interest as potential biocontrol agents of plant pathogens. However, phage lytic action may often be low and phage host ranges restricted, and so a careful selection and development of phage-based biocontrol agents (possibly based on mixtures of phages for different hosts or even for different mutant forms of the same host) is required. The advantages of the approach are as follows: (1) unlike chemical agents, phages occur naturally, and hence, applications are environmentally friendly, and (2) phages that only target pathogenic bacteria may selectively kill these, releasing nutrients and thus advancing the growth of potentially beneficial soil bacteria (Buttimer et al. 2017). As a case in point, two phages, LIMEstone1 and LIMEstone2, that lyse the potato rot pathogen *Dickeya solani*—when applied to potato tubers—did reduce the rot incidence *in situ* (Adriaenssens et al. 2012) (see Chapter 22 for further information).

6.5.2 Phages as Agents of Horizontal Gene Transfer, Accelerating Bacterial Evolution

Whereas, in traditional work, phages have been depicted as genetic "parasites" that limit the population sizes of their hosts, this picture has recently changed. Phages are currently increasingly being regarded as agents of HGT by transduction, and true reservoirs of fitness-enhancing traits (see also Chapter 7 for a discussion on transduction in soil). *Lysogenic* phages (forming *prophages*) may offer fitness advantages to their host following a process called *lysogenic conversion*, that is, the insertion of a phage genome copy, including a potential fitness-enhancing gene, into the organism's genome. Thus, the metabolic pathways of the host (for instance, introducing so-called *auxiliary metabolic genes*) may be influenced, or its ecological fitness, pathogenicity, or biofilm formation capacities enhanced (Secor et al. 2015). For example, the introduction of particular virulence genes may provide a host bacterium with the capacity to broaden its environmental niche by virtue of an enhanced capacity to infest host organisms. In return, the phage that establishes within a thus-modified host bacterium optimizes its own chances of persistence, as it has endowed the host with a fitness-enhancing trait.

With respect to the impact of phages on the evolutionary history of soil bacteria, a seminal study (Pal et al. 2007), in which the authors grew *Pseudomonas fluorescens* (strain SBW25) in soil in the presence or absence of lytic phages, revealed an accelerated evolution of the bacterium in the presence but not in the absence of phages. However, a follow-up microcosm study did not fully confirm this finding (Gomez and Buckling 2013). Both studies considered lytic phage as the driver of bacterial genome evolution, but failed to include temperate (lysogenic) phages. In addition, the heterogeneous nature of the soil was not considered (soil aggregates/hot spots). Host–phage coevolutionary events presumably also drove the fitness of the soil bacterium *Pseudomonas putida* KT2440 in another study, as in the KT2440 genome, 25 of the 105 genomic islands were found to relate to prophages (Quesada et al. 2012). The presence of one such prophage, denoted Pspu28, resulted in a competitive fitness advantage for the host in the rhizosphere, but the underlying mechanism remained unclear. There is, thus, increasing evidence for the key role phages play as modulators of the lifestyles of their hosts. As argued in the foregoing, in soil settings, their role has to be placed in the context of the compartmentalized "evolutionary incubator" model (Rillig et al. 2017). Major findings with respect to the exact roles of phages as enhancers of host fitness in soil settings will be forthcoming in future work, if aspects of this model are taken into account.

6.5.3 Phages as Agents That Modulate Biofilms

Phages may play major roles as promoters of the structure and "health" of bacterial biofilms. In soil settings, such biofilms are often found on surfaces, constituting key structures of the soil microbiome. For instance, the fungal networks in soil often contain biofilms of bacteria such as *Paraburkholderia* spp. In these biofilms, local phage-induced host cell lysis (related to unfavorable conditions caused by crowding, i.e., limited nutrient and oxygen availability, and accumulation of reactive oxygen species) may provide extracellular DNA (eDNA) that strengthens the biofilm structure. Moreover, the local phage-induced cell lysis events create channels in the biofilm, where nutrients become available. The new channels may also help to remove waste products, while transporting the new nutrients to the unlysed neighboring cells. Overall, this phenomenon of phage-assisted biofilm strengthening is thought to assist in fostering the biofilm structure and health (Secor et al. 2015).

6.6 FUTURE PERSPECTIVES

6.6.1 Three Approaches to Soil Virome Studies

Unfortunately, an integrative perspective of how viruses and phages shape soil (micro)biomes, distributed across the soil evolutionary "islands," is as yet not available. As a next step in the development of our knowledge, studies that intricately describe the dynamics and roles of the main soil virome components that occur at the microscale are required. Major challenges are posed by (1) the required adaptation and fine tuning of the available analytical methods in order to allow in-depth studies in different soils, (2) the paucity of current knowledge with respect to the triggers of phage activities in the soil, (3) the need to develop adequate spatiotemporally explicit sampling schemes, and (4) the translation of data on specific and local phage activities to effects on global microbiome functioning.

Three different approaches to soil virome studies are recommended. In the first one, physical separation of viral particles from soil is required, yielding a soil virome that is subsequently lysed and analyzed via genome sequencing and TEM. Unfortunately, the efforts made so far to optimize the methods have not yet resulted in robust procedures for all soils (Trubl et al. 2016). Moreover, the choice whether or not to amplify the extracted virome, and prepare a library, also affects the outcome of the studies (Solonenko et al. 2013). In the second approach, whole metagenomes generated directly from a soil microbiome are analyzed for prophage-specific sequences. However, this method does not take into account the "whole viral community," and so a proxy of a "partial" (integrated) viral community is produced. As an example, a recent study (Paez-Espino et al. 2016) analyzed 5 Terabase of metagenomics data, resulting in the discovery of vast amounts of novel viral groups, next to information that links members of viromes—across ecosystems—to their presumed hosts. On the positive side, bioinformatics pipelines for analyses of viromes, such as VirSorter (Roux et al. 2015), VirFinder (Ren et al. 2017), and integrated microbial genome/virus (IMG/VR) (Paez-Espino et al. 2017), have been developed. However, in spite of the ease of use of these pipelines, extensive manual curation of the data is still required.

In the third approach, the nature of the interactions between phages and their host cells in soil microhabitats needs to be addressed. The local nutritional status of the microhabitat may be the key determinant of the local phage–host processes, and so this factor needs to be taken as a lead parameter. Such studies on the intricacies of phage–host interactions, as related to soil microhabitat, pose great challenges to researchers. They can use either cultivation-based or cultivation-independent approaches, or both. The former approach requires prior knowledge on the phage and host and their relationship, that is, the relationship being lytic, lysogenic, and/or—in any way—fitness modulating. In contrast, cultivation-independent approaches allow direct assessments of phage (or other mobile genetic element [MGE]) prevalence. In a combined approach, the specificity of the former method may support the data of the second. Recent combined method-based studies in forest hot spots, in particular the mycosphere (Zhang et al. 2014), showed other MGEs, that is, IncP-1 and PromA plasmids, to be prevalent in the mycosphere. Moreover, the genome of the mycosphere-inhabitant *Paraburkholderia terrae* BS001 was found to contain a range of canonical phage genes (Haq et al. 2014), whereas another *P. terrae* strain, BS437, was found to excise a prophage, ϕ437, either "spontaneously" or following induction by mitomycin C (Pratama and van Elsas 2017a,b). An intriguing question is what type of soil trigger may spur the release of ϕ437 from its host, and what type of ecological effect this phage may provide. There is still a paucity of information as to the extent and ecological significance of the accelerated bacterial evolution expected to occur in such soil activity hot spots among the plethora of soil aggregates.

6.6.2 Spatiotemporal Aspects of Virus–Host Interactions

Given the fact that soil viruses–phages play fundamental roles in the ecologies of their hosts, key soil processes may be affected, at the microscale level. Hence, there are strong arguments that plea for the inclusion of spatiotemporal aspects of phage–host dynamics in future studies. The argument for this inclusion is the lack of connectivity of microhabitats (fragmented locations) that occurs in many soils, which results in the concept of compartmentalization and island-confined ecological and evolutionary phage–host trajectories. The extent to which such trajectories are shared across the islands, that is, the degree of mixing, is a potential focus in the research. The tailored technologies for virome detection and observation need to be optimized in order to address such issues.

As special island-like habitats, soil hot spots like the mycosphere and the rhizosphere need to be considered. In particular sites in the rhizosphere where nutrients abound, phage activity may have a strong effect on the ecology of the host organism, which can have a positive or negative effect on plant health. Thus, a better understanding of the dynamics of the eco-evolutionary events with respect to soilborne pathogens in terms of how it impacts plant health will be obtained. Moreover, the scope of exploration of the effects of viruses on hosts and processes in soil ecosystems should also include a temporal aspect, thus addressing the dynamics of the viromes, their interactions with hosts, and the effects on local microbiome members. Finally, most of the studies referred to in the foregoing have focused on dsDNA viruses, to the detriment of ssDNA, (ds/ss)RNAs, and Archaeal viruses. Although more difficult to study, the latter viruses require increased research focus.

REFERENCES

Adams, M.H. 1959. *Bacteriophages*, London: Interscience Publishers.

Adriaenssens, E.M. and D.A. Cowan. 2014. Using signature genes as tools to assess environmental viral ecology and diversity. *Appl Environ Microbiol* 80, 4470–4480.

Adriaenssens, E.M., van Vaerenbergh, J., Vandenheuvel, D. et al. 2012. T4-related bacteriophage LIMEstone isolates for the control of soft rot on potato caused by *'Dickeya solani.' PLoS One* 7, 33227.

Adriaenssens, E.M., van Zyl, L., de Maayer, P. et al. 2015. Metagenomic analysis of the viral community in Namib desert hypoliths. *Environ Microbiol* 17, 480–495.

Ashelford, K.E., Day, M.J. and J.C. Fry. 2003. Elevated abundance of bacteriophage infecting bacteria in soil. *Appl Environ Microbiol* 69, 285–289.

Bernardo, P.C.M., Charles-Dominique, T., Barakat, M. et al. 2018. Geometagenomics illuminates the impact of agriculture on the distribution and prevalence of plant viruses at the ecosystem scale. *ISME J* 12, 173–184.

Buttimer, C., McAuliffe, O., Ross, R.P. et al. 2017. Bacteriophages and bacterial plant diseases. *Front Microbiol* 8, 1–15.

d'Herelle, F. 1917. Sur un microbe invisible antagoniste des bacilles dysentériques. *Comptes Rendus de l'Académie des Sciences de Paris* 165, 373–375.

Ghabrial, S.A., Castón, J.R., Jiang, D. et al. 2015. 50-Plus years of fungal viruses. *Virology* 479–480, 356–368.

Gomez, P. and A. Buckling. 2013. Coevolution with phages does not influence the evolution of bacterial mutation rates in soil. *ISME J* 7, 2242–2244.

Guo, Y.R., Hryc, C.F., Jakana, J. et al. 2014. Crystal structure of a nematode-infecting virus. *Proc Natl Acad Sci USA* 111, 12781–12786.

Haq, I.U., Graupner, K., Nazir, R. et al. 2014. The genome of the fungal-interactive soil bacterium *Burkholderia terrae* BS001 - A plethora of outstanding interactive capabilities unveiled. *Genome Biol Evol* 6, 1652–1668.

Hegazi, N.A. and V. Jensen. 1973. Studies of *Azotobacter* bacteriophages in Egyptian soils. *Soil Biol Biochem* 5, 231–243.

Kim, K.H., Chang, H.W., Nam, Y.D. et al. 2008. Amplification of uncultured single-stranded DNA viruses from rice paddy soil. *Appl Environ Microbiol* 74, 5975–5985.

Knowles, B., Silveira, C.B., Bailey, B.A. et al. 2016. Lytic to temperate switching of viral communities. *Nature* 539, 123–123.

Li, C.X., Shi, M., Tian, J.H. et al. 2015. Unprecedented genomic diversity of RNA viruses in arthropods reveals the ancestry of negative-sense RNA viruses. *eLife* 4, 1–26.

Lima-Mendez, G., van Helden, J., Toussaint, A. and R. Leplae. 2008. Reticulate representation of evolutionary and functional relationships between phage genomes. *Mol Biol Evol* 25, 762–777.

Narr, A., Nawaz, A., Wick, L.Y., Harms, H. and A. Chatzinotas. 2017. Soil viral communities vary temporally and along a land use transect as revealed by virus-like particle counting and a modified community fingerprinting approach (fRAPD). *Front Microbiol* 8, 1975.

Paez-Espino, D., Eloe-Fadrosh, E.A., Georgios, A.P. et al. 2016. Uncovering Earth's virome. *Nature* 536, 425–430.

Paez-Espino, D., Min, I., Chen, A., Krishna P. et al. 2017. IMG/VR: A database of cultured and uncultured DNA viruses and retroviruses. *Nucl Acids Res* 45(D1), D457–D465.

Pal, C., Maciá, M.D., Oliver, A., Schachar, I. and A. Buckling. 2007. Coevolution with viruses drives the evolution of bacterial mutation rates. *Nature* 450, 1079–1081.

Pedulla, M.L., Ford, M.E., Houtz, J.M. et al. 2003. Origins of highly mosaic mycobacteriophage genomes. *Cell* 113, 171–182.

Prangishvili, D., Bamford, D.H., Forterre, P., Iranzo, J., Koonin, E.V. and M. Krupovic. 2017. The enigmatic archaeal virosphere. *Nat Rev Microbiol* 15, 724–739.

Pratama, A.A. and J.D. van Elsas. 2017a. A novel inducible prophage from the mycosphere inhabitant *Paraburkholderia terrae* BS437. *Sci Rep* 7, 9156.

Pratama, A.A. and J.D. van Elsas. 2017b. The 'neglected' soil virome – potential role and impact. *Trends Microbiol* 1533, 1–14.

Quesada, J.M., Soriano, M.I. and M. Espinosa-Urgel. 2012. Stability of a *Pseudomonas putida* KT2440 bacteriophage-carried genomic island and its impact on rhizosphere fitness. *Appl Environ Microbiol* 78, 6963–6974.

Reanney, D.C. and S.C.N. Marsh. 1973. The ecology of viruses attacking *Bacillus stearothermophilus* in soil. *Soil Biol Biochem* 5, 399–408.

Reanney, D.C. and C.K. Teh. 1973. The phages attacking *Bacillus* in soil. *Soil Biol Biochem* 8, 305–311.

Ren, J., Ahlgren, N.A., Lu, Y.Y., Fuhrman, J.A. and F. Sun. 2017. VirFinder: A novel k-mer based tool for identifying viral sequences from assembled metagenomic data. *Microbiome* 5, 69.

Rillig, M.C., Muller, L.A.H. and A. Lehmann. 2017. Soil aggregates as massively concurrent evolutionary incubators. *ISME J* 11, 1943–1948.

Rohwer, F. and R. Edwards. 2002. The phage proteomic tree : A genome-based taxonomy for phage. *J Bacteriol* 16, 4529–4535.

Roux, S., Enault, F., Hurwitz, B.L. and M.B. Sullivan. 2015. VirSorter: Mining viral signal from microbial genomic data. *PeerJ* 3, e985.

Rybicki, E.P. 2015. A top ten list for economically important plant viruses. *Archiv Virol* 160, 17–20.

Salmond, G.P.C. and P.C. Fineran. 2015. A century of the phage: Past, present and future. *Nat Rev Microbiol* 13, 777–786.

Scola, V., Ramond, J.B., Frossard, A. et al. 2017. Namib desert soil microbial community diversity, assembly, and function along a natural xeric gradient. *Microb Ecol* 75, 193–203.

Secor, P.R., Sweere, J.M., Lia, A.M. et al. 2015. Filamentous bacteriophage promote biofilm assembly and function. *Cell Host Microb* 18, 549–559.

Simmonds, P., Adams, M.J., Benkő M. et al. 2017. Consensus statement: Virus taxonomy in the age of metagenomics. *Nat Rev Microbiol* 15, 161–168.

Solonenko, S.A., Ignacio-Espinoza, J., Alberti, A. et al. 2013. Sequencing platform and library preparation choices impact viral metagenomes. *BMC Genomics* 14, 320.

Swanson, M.M., Fraser, G., Daniell, T.J., Torrance, L., Gregory, P.J. and M. Taliansky. 2009. Viruses in soils: Morphological diversity and abundance in the rhizosphere. *Ann Appl Biol* 155, 51–60.

Touchon, M., Hoede, C., Tenaillon, O. et al. 2009. Organised genome dynamics in the *Escherichia coli* species results in highly diverse adaptive paths. *PLoS Genet* 5, e1000344.

Trubl, G., Solonenko, N., Chittick, L., Solonenko, S.A., Rich, V.I. and M.B. Sullivan. 2016. Optimization of viral resuspension methods for carbon-rich soils along a permafrost thaw gradient. *PeerJ* 4, e1999.

Twort, F.W. 1915. An investigation on the nature of ultra-microscopic viruses. *The Lancet* 186, 1241–1243.

van Elsas, J.D. and E.G.C. Penido. 1982. Characterization of a new *Bacillus megaterium* bacteriophage, MJ-1, from tropical soil. *Antonie Van Leeuwenhoek* 48, 365–371.

Vary, P.S. and W.F. Halsey. 1980. Host-range and partial characterization of several new bacteriophages for *Bacillus megaterium* QM b1551. *J Gen Virol* 51, 137–146.

Wang, A.L. and C.C. Wang. 1991. Viruses of the protozoa. *Ann Rev Microbiol* 45, 251–263.

Williamson, K.E., Radosevich, M. and K.E. Wommack. 2005. Abundance and diversity of virus in six Delaware soils. *Appl Environ Microbiol* 71, 3119–3125.

Zablocki, O., van Zyl, L., Adriaenssens, E.M. et al. 2014. High-level diversity of tailed phages, eukaryote-associated viruses, and virophage-like elements in the metaviromes of Antarctic soils. *Appl Environ Microbiol* 80, 6888–6897.

Zhang, M., Visser, S., Pereira e Silva, M.C. and J.D. van Elsas. 2014. IncP-1 and PromA group plasmids are major providers of horizontal gene transfer capacities across bacteria in the mycosphere of different soil fungi. *Microb Ecol* 69, 169–179.

7 Horizontal Gene Transfer and Microevolution in Soil

Kaare Magne Nielsen
Oslo Metropolitan University

Jan Dirk van Elsas
University of Groningen

CONTENTS

7.1 Introduction 105
7.2 Mechanisms That Generate Genetic Diversity in Soil Bacterial Communities 106
7.2.1 Introduction of Genetic Changes through Random Mutation 106
7.2.2 Introduction of Genetic Changes through Horizontal Transfer of Chromosomal DNA 108
7.2.3 Introduction of Genetic Changes through Horizontal Transfer of Mobile Genetic Elements 109
7.2.4 Mosaic Structure of Many MGEs 110
7.3 Mechanisms of HGT in Soil 110
7.3.1 Natural Transformation in Soil 110
7.3.2 Conjugation in Soil 111
7.3.2.1 Mechanisms and MGEs Involved 111
7.3.2.2 Historical Overview of Studies on Conjugation in Soil 112
7.3.2.3 Conjugation Occurs Preferentially in Soil Hot Spots and in Biofilms 113
7.3.3 Transduction in Soil 114
7.3.3.1 Mechanism of Transduction 114
7.3.3.2 Transduction in Soil—Experimental Evidence 114
7.4 Methods Used to Study HGT in Soil 114
7.4.1 Retrospective Identification of Horizontally Transferred DNA Sequences in Bacterial Genes or Genomes 115
7.4.2 Experimental Studies of HGT into Defined Bacteria 116
7.5 Population-Scale Considerations of HGT Events 116
7.6 Studies on Selection in Soil 119
7.7 Concluding Remarks 119
Acknowledgments 120
References 120

7.1 INTRODUCTION

High-throughput (genomics and metagenomics) methods for DNA sequencing have now cataloged the composition of thousands of bacterial genomes, including those of many soil bacteria. Chapter 14 further discusses the use of metagenomics for soil studies. The list of genomes that are currently available can be examined at https://en.wikipedia.org/wiki/List_of_sequenced_bacterial_genomes and http://bacmap.wishartlab.com (accessed Feb. 6, 2019). Comparative analysis of these genomes has revealed that horizontal gene transfer (HGT) processes are major contributors to bacterial genome composition, diversity, and dynamics (Vos and Didelot 2009, Land et al. 2015, Soucy et al. 2015). Moreover, it has become apparent that bacterial populations and genomes are not likely to evolve in a uniform manner, with various populations and parts of a genome having their own evolutionary history and trajectory (Ray et al. 2009, Wilmes et al., 2009). Bacterial genomes are conceptually split into so-called *core* (present in all members of a particular species) and *accessory* (present in only one or some members of the species) parts. The core part of the genome is thought to largely follow a vertical inheritance path, whereas the accessory part is more prone to change as a result of HGT processes. The core genome is more influenced by concomitant mutation and stabilizing selection through purging of deleterious genetic variations (Popa et al. 2011, Oliveira et al. 2017).

In particular sites in soil, such as the rhizosphere and mycosphere, HGT is well known to promote the exchange of genetic material (van Elsas et al. 2003, Zhang et al. 2014, 2015). Whereas this generic contention has been demonstrated in a range of experimental studies in soil, other studies have identified HGT events retrospectively, on the basis of sequence analyses of bacterial genomes. Such HGT events are tentatively explained within the prevailing paradigms of gene flow, genetic drift, and natural selection. There is usually a time lag between the actual occurrence of the HGT event in a microbial community and the time point when it can be detected. This time interval is necessary for rare transfer events to multiply in the larger bacterial population, thereby facilitating their detection. The nature of such time lags is rarely understood, as we commonly ignore the drivers/conditions of the organisms involved in the period of time between the actual gene transfer event and the moment of detection. This incomplete understanding of the effects of time and other factors that determine the contemporary compositions of genomes in bacterial populations is problematic, as it hampers our ability to *a priori* identify the DNA compositions and alterations that benefit the long-term survival of the respective bacterial population in soil. Moreover, we currently have a limited knowledge of the genetic diversity extant in soil microbiomes; the biological functions thereof; the spatial and temporal variability of natural selection; and the stochastic component of gene acquisition, bacterial survival, and dispersal. This restricts our ability to accurately predict when, where, how, and at which rates genes flow and establish in natural bacterial communities. In spite of our still scattered knowledge, we will here examine the significance of HGT as an adaptive mechanism that can genetically shape soil bacterial communities, and also attempt to recommend strategies for future research.

7.2 MECHANISMS THAT GENERATE GENETIC DIVERSITY IN SOIL BACTERIAL COMMUNITIES

Bacterial evolution can be defined as "descent with modification," which implies the occurrence of cellular divisions in which not all progeny is clonal with the parental cells. This leads to the emergence of new genetic variants or traits in a bacterial population that appear as a result of random genetic drift and natural selection. Hence, the short-term adaptive and the long-term evolutionary potential in soil bacterial communities are limited both by processes that generate new variation within the cell (mutations and HGTs) and those that lead to changes in the frequencies of specific genotypes within populations (selection, genetic drift, and migration), as illustrated in Figure 7.1.

7.2.1 Introduction of Genetic Changes through Random Mutation

Mutational processes include point mutations, deletions, insertions, inversions, transpositions, and other intracellular mechanisms that cause changes in the genetic material of an organism. Mutations in actively growing bacteria commonly occur at the rates of ~5.4×10^{-10} base pair changes per generation (Drake et al. 1998). This observation is consistent over a range of species and genome sizes.

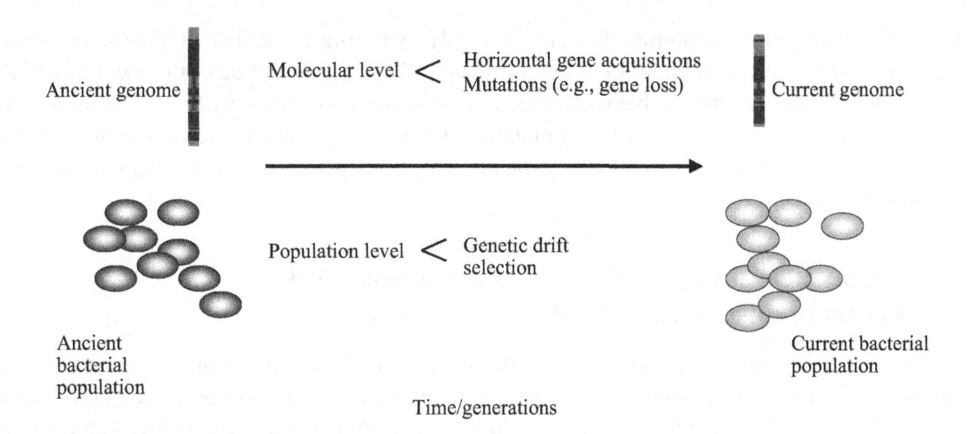

FIGURE 7.1 The combined effect of molecular-, cellular-, and population-scale factors determines the fate of mutations and HGT events.

The mutation rate produces about one mutant cell per 300 cells per generation. However, changes in the physiological state of a bacterium, for example, via nutrient starvation or other stress conditions, may transiently raise the mutation frequency at least 1,000-fold. Such raises may also be caused by mutation(s) in a small number of genes that regulate DNA damage repair and recombination (Matic 2017). Thus, heritable so-called *mutator phenotypes* often arise from spontaneous acquisition of deleterious mutations in genes important to the DNA maintenance and repair machinery of the bacterial cell. Some of these mutations, for instance, mutations leading to the inactivation of the methyl-directed DNA mismatch repair system, can also result in recombination events with DNA substrates from other species with of up to 20%–25% sequence divergence (Townsend et al. 2003). Although the mutator phenotypes are generally considered to be deleterious due to their overall increased mutation rate leading to many unfavorable changes, such phenotypes have been found at significant levels (1% to >4%) in natural bacterial communities, for instance, in those of *Escherichia coli* and *Pseudomonas aeruginosa*. Such cells with higher-than-usual mutation rates may be important in populations that live under critical conditions, as they can gain beneficial mutations more quickly than organisms with mutation rates closer to the population mean (Matic 2017). Nevertheless, numerous studies have shown that consistently high mutation frequencies are generally undesirable for a bacterial population, as the overwhelming majority of the mutations that arise will be deleterious and thereby reduce the fitness of the bacterial population over time. Although we are starting to get an improved understanding of mutator phenotypes from *in vitro* studies, their occurrence, dynamics, and distribution in natural soil microbiomes remain to be fully understood (Gómez and Buckling 2013).

We can approach this issue from a theoretical perspective. Probabilistic calculations can be used to understand the occurrence of particular mutations. For example, the likelihood of three point mutations (*A*,*B*,*C*) occurring within a bacterium and generation is $(5.4 \times 10^{-10})^3 = 1.6 \times 10^{-28}$, following the multiplication rule of probability:

$$p(A,B,C) = p(A) \times p(B) \times p(C).$$

Thus, point mutational processes occurring routinely within a genetically homogeneous population are, within a relevant ecological timescale, unlikely to produce multiple specific amino acid changes in a given protein (e.g., an enzyme), if such changes are urgently—and timely—required for bacterial adaptation and/or survival.

Here, HGT can facilitate multiple base pair changes in one event. The ratio of rates by which nucleotides become substituted by recombination versus mutation (*r*/*m*) varies between bacterial species and habitats, from genomes that are primarily changing through mutation to those that change mainly through recombination (Vos and Didelot 2009). For instance, the *r*/*m* ratio of the soil bacterium *Bacillus cereus* has been estimated to range between 0.2 and 1.6, suggesting roughly similar contributions of recombination and mutation. In contrast, the estimated *r*/*m* ratio of *Vibrio parahaemolyticus* is between 27 and 48, indicating a primary role of recombination (Vos and Didelot 2009).

7.2.2 Introduction of Genetic Changes through Horizontal Transfer of Chromosomal DNA

Horizontal transfer of chromosomal DNA enables bacteria to access the genetic diversity present in other members of its own population or in separately evolving organisms (Soucy et al. 2015). Bacteria can acquire chromosomal DNA through *conjugation*, *transduction*, and *transformation*. Conjugation is the directed transfer of DNA from cell to cell across cell membranes, transduction is the transfer of DNA mediated by bacteriophages that move from cell to cell, whereas transformation is the directed uptake of DNA fragments by cells. These classical mechanisms of gene transfer across bacteria have been reviewed in Thomas and Nielsen (2005) and Soucy et al. (2015). Recently, also various forms of vesicle-mediated gene transfer have been described (Domingues and Nielsen 2017). Regardless of the exact mechanism, the uptake of DNA fragments originating from distinct organisms with particular evolutionary histories can—in a single event—result in the acquisition of multiple genes (in case a novel DNA fragment is acquired) or nucleotide changes (in case of homologous recombination between similar DNA stretches). As with mutagenic processes, HGT can introduce beneficial changes (which is what is preserved in the genomes of the offspring) or deleterious ones (which are normally lost from the evolutionary record) (Nielsen et al. 2014). However, because HGT often introduces DNA that has been previously selected for a function that has contributed to some ancestor's success, it may more frequently result in neutral or beneficial changes in recipient cells than random point mutations. Moreover, recombination in chromosomal DNA often relies on integration through homologous recombination, a process that is critically dependent on DNA similarity. Thus, recombination is guided to occur at "preferred" genomic sites depending on the nucleotide composition of the acquired DNA (Ray et al. 2009).

Recombination rates between chromosomal DNA molecules have been found to drop log-linearly with increasing sequence divergence between the donor and the recipient bacterium (see Ray et al. 2009). The dependency on sequence similarity for *homologous recombination* to occur between species is well established. In contrast, *illegitimate recombination* (recombination at another, nonhomologous, site) is generally not considered to occur at significant frequencies in bacteria. Two additional processes have also been reported. In the process of "homology-facilitated illegitimate recombination" (de Vries et al. 2002), stretches of high DNA similarity of up to about 200-bp long initiate strand exchange and recombination events that continue into adjacent regions of lower similarity. This process has been shown to result in the additive integration (or substitution) of >1,000 bp-long heterologous fragments in the soil bacterium *Acinetobacter baylyi* (de Vries et al. 2002). Finally, an integration mechanism for short DNA fragments (<200 bp) that does not rely on the mechanism of homologous recombination has recently been described. This mechanism has been named "short patch double illegitimate recombination," and it allows the recombination with DNA fragments as short as 20 bp. Such recombination with short DNA fragments is based on microhomologies at the insertion site and the outcomes often resemble mutational processes, as the result of the recombination may be only a few base changes (Harms et al. 2016).

It is possible that these various recombination processes play roles in the evolution and adaptation of, for instance, bacilli (e.g., *Bacillus subtilis*) in soil. This organism occurs in microcolonies or biofilms in soil, in which it can become competent for transformation, being able to acquire fragments of DNA from its immediate surroundings. In these surroundings, most DNA may come from sibling cells, but DNA with other origins may also be available.

7.2.3 Introduction of Genetic Changes through Horizontal Transfer of Mobile Genetic Elements

Mobile genetic elements (MGEs) such as plasmids, transposons, integrons, genomic islands, and bacteriophages are collectively referred to as the drivers or vectors of the "horizontal gene pool," that is, the pool of genes that can move more or less "flexibly" between the genomes of diverse bacterial hosts (Thomas and Nielsen 2005, Wozniak and Waldor 2010, Domingues et al. 2015). Depending on their characteristics, these MGEs can move within and/or between genomes. Variability lies in their unit sizes, transfer frequencies, and host ranges. These features are discussed later in more detail. MGEs that encode all the genes necessary for their own transfer (next to replication) are referred to as conjugative or self-transmissible. They include plasmids and integrative conjugative elements (ICEs) (Wozniak and Waldor 2010). In soil, such MGEs transfer between bacterial cells via conjugation (van Elsas et al 2003, Heuer and Smalla 2012, Zhang et al. 2014). In addition, also transduction and transformation can - in some instances - transfer MGEs (Domingues et al. 2012).

ICE is a joint description of elements that excise by site-specific recombination into a circular form, self-transfer by conjugation, and subsequently integrate into the genome of the new host (Guglielmini et al. 2011). Genetic elements belonging to the ICE class include conjugative transposons, integrative plasmids, and genomic islands. Conjugative MGEs can also aid the transfer of other—nonconjugative (but mobilizable)—MGEs, since their transfer functions can act in *trans*, for example, on nonconjugative plasmids, or in *cis*, for example, on composite transposons integrated into larger ICEs (Frost et al. 2010, Smillie et al. 2010). Conjugative MGEs are, in general, less constrained in terms of their intercellular transfer capacity and persistence than foreign chromosomal DNA that is acquired by natural transformation or transduction. This is because they are self-organizational and encode their own transfer, replication, and/or integration machinery. Thus, the intercellular transfer of MGEs (such as plasmids) by conjugation is mainly dependent on their ability to establish in host cells (determining their host range), and not on prior sequence similarity. Nevertheless, the breadth of the host range of the integration-dependent ICEs will depend on the available mechanism for integration into the recipient cell's genome. The broad-host-range (BHR) potential of conjugative MGEs is exemplified by conjugative transposon Tn*916*, which is able to stably integrate in a wide range of bacterial species.

The host range of replication-proficient conjugative MGEs (i.e., plasmids) is determined by several factors. From the perspective of a plasmid, its ability to replicate and escape from the action of defense systems, for example, CRISPR-Cas9, DNA restriction/modification, or other systems (Doron et al. 2018), determines its immediate fate in a new host. BHR plasmids often have a reduced number of restriction cleavage sites, and their origins of replication are functional in a variety of bacterial genera. BHR plasmids, e.g., those of the IncP-1 class like pHB44 (Zhang et al. 2015) and those of the PromA group like pIPO2 (van Elsas et al 2003), confer genetic fluidity across a wide range of bacteria in soil. BHR plasmids may also contain genetic "addiction" systems that allow them to act as true genetic parasites of their hosts, as such systems will kill bacterial cells losing the plasmid. Various forms of such toxin-based plasmid addiction systems have been described that contribute to plasmid persistence in bacterial populations over time (Page and Peti 2016).

A major ecological factor that will limit the rapid conjugative transfer of MGEs through a soil system is the spatial isolation of bacterial cell populations, such as frequently encountered in bulk soil. See also Chapter 6. Hence, the greatest impact of conjugative processes will occur in nutritionally rich hot spots with high bacterial cell density and activity in soil, such as the rhizosphere (van Elsas et al. 2003) and the mycosphere (Zhang et al. 2014).

7.2.4 Mosaic Structure of Many MGEs

The structure of most MGEs can be considered to be composite (mosaic) elements combining several functional modules. Each one of such functional modules encodes a different aspect of an MGE's biological potential. Thus, some modules allow MGEs to transfer and be stably maintained by either integration into the host genome or replication. Insertion elements are the simplest MGEs, as they consist of only one functional module, that is, a recombinase that is responsible for their integration. On the other hand, the more complex ICEs consist of backbone modules that are essential for intercellular transfer and stability, in addition to a range of other modules often encoding catabolic properties. Larger MGEs have often acquired a wide variety of so-called accessory genes, which encode properties that provide the host bacterium with immediate selective advantages in changing environments, such as resistances to heavy metals and antimicrobial agents, virulence factors, or the ability to exploit any of a range of novel energy or carbon sources (Sen et al. 2011). Theoretically, the availability of diverse accessory genes (conferring the potential to have a fitness gain under particular conditions) within a larger bacterial population, combined with the transferability of such genes among cells, yields tremendous adaptive power to bacterial populations and communities in soil situations.

7.3 MECHANISMS OF HGT IN SOIL

7.3.1 Natural Transformation in Soil

Natural transformation is the physiologically regulated uptake of extracellular (plasmid or chromosomal) DNA into transformation-competent bacteria, that is, bacteria that have reached the state of natural competence. Whereas some bacterial species require the presence of specific short sequences (so-called uptake sequences) in the DNA, other species are capable of sequence-independent uptake of DNA and subsequent transport into the cytoplasm (Johnston et al. 2014). Many studies have identified the conditions that promote the development of natural competence in monocultures of soil bacteria grown *in vitro* in the laboratory. These studies have generally focused on defined model strains such as *Bacillus subtilis*, *Pseudomonas fluorescens*, and *Acinetobacter baylyi*, whereas hardly any data are available on the distribution of competence for natural transformation *in situ* among the diversity of bacteria present in various soils. Methodologically, it is a challenge to detect rare bacterial transformants among the high numbers of bacterial cells that are present in natural soils (Nielsen and Townsend 2004, Townsend et al. 2012).

Studies on *in vitro* transformation have been performed in soil microcosms, that is, small containers filled with soil placed under defined environmental conditions (Figure 7.2) or in infected plants in microcosms. Such microcosm studies have demonstrated that soil bacteria from several unrelated genera, including *Acinetobacter*, *Pseudomonas*, and *Bacillus*, can take up DNA by natural transformation when grown under soil conditions (e.g., Nielsen et al. 2000). Typically, transformation frequencies range from barely detectable, that is, 10^{-9} up to 10^{-1} transformants over the total number of recipient cells in a system. Many experimental studies in soil have relied on the addition of soil minerals (clays), nutrients, and high concentrations of DNA and recipient bacteria to promote transfers and thereby allow the detection of otherwise infrequent gene transfers. The degree of availability of extracellular DNA to bacterial cells in soil remains unknown and may be a limiting factor in the experimental designs used (Nielsen et al. 2007), although DNA fragments can persist on surfaces in the soil environment for long periods of time (Overballe-Petersen et al. 2013). One of the reasons for the variable availability of DNA to local bacteria is that DNA molecules have direct interactions with heterogeneous soil components. In contrast, DNA inside membrane vesicles released naturally from bacterial cells will be better protected inside the lumen of the vesicles (Soler et al., 2015), and the roles of these in HGT through transformation-like DNA uptake mechanisms in soil should be further explored. No data have hitherto been published with respect to the relative importance and occurrence of *in situ* events of natural transformation in undisturbed soils. Arguably, this is due to the experimental limitations and not to the lack of occurrence of such transfers.

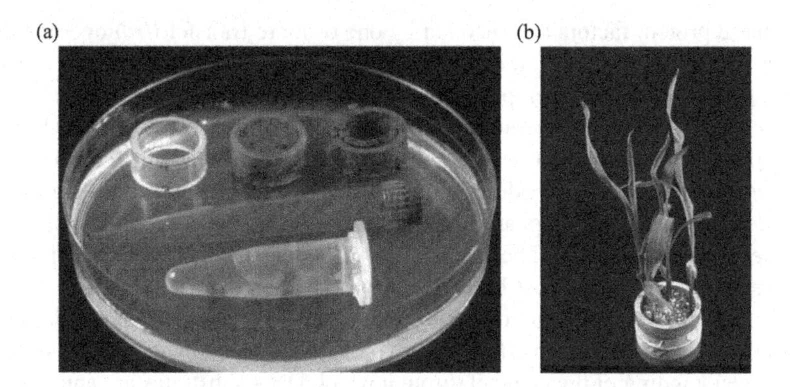

FIGURE 7.2 **(See color insert)** Two examples of soil microcosms used for the study of natural transformation. (a) Small soil microcosms consisting of polypropylene cylinders filled with soil. The number of bacterial transformants arising after the addition of bacteria and DNA to the soil is determined after sampling the central core of the soil using the wide end of a pipette tip and transferring the soil portion to an Eppendorf tube for suspension, dilution, and plating on transformant-selective agar plates. (b) Soil microcosm consisting of two plastic cylinders divided with a fine nylon mesh to allow bacterial activity and HGT to be investigated at the root–soil interface (rhizosphere). Various plant species can be germinated and grown in the upper cylinder, and the resulting soil–root interface on the lower cylinder sampled and transferred to a dilution tube for suspension, dilution, and plating on transformant-selective agar plates.

7.3.2 Conjugation in Soil

7.3.2.1 Mechanisms and MGEs Involved

Conjugation is a cell-to-cell contact-dependent mechanism of DNA transfer, which has been demonstrated to occur in most known bacterial genera. The known bacterial conjugation systems represent a subfamily of the so-called *type IV secretion systems* (T4SS), which share the capacity to transfer different macromolecules—such as DNA and proteins—from the cytoplasm across the cell envelope (Thomas and Nielsen 2005). Conjugation is a multistep process that is dependent on the presence of transfer (*tra*) genes on the mobile element, which can be divided into two components: one component of this system promotes DNA transfer and replication (*Dtr*), and another ensures effective mating pair formation (*Mpf*). The mating pair is formed through the attachment of a pilus of the donor to the surface of the recipient cell. Then, a pore is formed in the adjoining cell membranes. During the establishment of effective contact between the donor and recipient cell, a single-strand cut is made in the origin of transfer (*oriT*) of the MGE by a relaxase. A plasmid-encoded helicase then generates single-stranded plasmid DNA that is subsequently transferred to the recipient. The relaxase remains attached to the 5′-end of the transferred single-stranded plasmid DNA and promotes recircularization of the single-stranded plasmid in the recipient. Finally, after one round of replication, both donor and recipient cells end up with a double-stranded circular copy of the plasmid.

The best described mechanism of conjugative plasmid DNA transfer is provided by plasmid transfer in Gram-negative (G^-) bacteria. The mechanisms in Gram-positive (G^+) bacteria are not exactly known, but recent evidence points to an unexpected degree of similarity of the (T4SS-like) Gram-positive system to the classical Gram-negative one. However, clear differences between the G^+ and G^- transfer machineries concerning the putative membrane channel are apparent. These might be explained by the distinct structures of the G^+ and G^- cell envelopes. Thus, G^+ systems appear to lack three T4SS core complex components (VirB7, VirB9, and VirB10) (Goesman et al. 2016). However, much like in G^- systems, G^+ bacterial conjugation is driven by conjugative plasmids that contain minimally an origin of transfer (oriT), DNA-processing factors (a relaxase and accessory proteins), as well as proteins that constitute the trans-envelope transport channel, i.e., the Mpf proteins. All these protein factors are encoded by one or more transfer (*tra*) operons that together form the DNA transport machinery, the Gram-positive type IV secretion system. However, multicellular Gram-positive bacteria (exemplified by streptomycetes) appear to have developed another mechanism for conjugative plasmid transfer. The system is reminiscent of the machinery involved in bacterial cell division/sporulation, which transports double-stranded DNA from donor to recipient cells. *Streptomyces* plasmids encode a unique conjugative DNA transfer system as only a single plasmid-encoded protein, TraB, appears to be required to translocate a double-stranded circular DNA molecule into the recipient. The mechanism of conjugative transfer of ICEs also remains unresolved, but it is known that most ICEs transfer as single-stranded DNA and the *tra* genes of many ICEs are similar to those found in conjugative plasmids of both G^+ and G^- bacteria. Despite the current and past intense research efforts aimed to dissect the conjugative DNA systems, there is still a lack of understanding of the channel through which DNA substrates are apparently delivered across the donor and recipient cell envelopes.

7.3.2.2 Historical Overview of Studies on Conjugation in Soil

Experimental—as well as retrospective—studies have provided ample evidence for the importance of conjugative gene transfer in soil for local genetic diversity and adaptation (van Elsas et al. 2003). In experimental transfer studies, selectable plasmids have often been used. As a case in point, in early work a small tetracycline resistance plasmid, denoted pFT30, was shown to readily transfer between different *Bacillus* species from soil. In another sample study, *E. coli* strains that harbored a streptothricin resistance plasmid (encoding short and rigid pili) could transfer these to *E. coli* recipients in nutrient-amended sterile and non-sterile soils in microcosms (Pukall et al. 1996). Interestingly, several indigenous soil bacteria belonging to *Rhizobium* and *Flavobacterium* had also acquired the plasmid in non-sterile soil microcosms. The importance of conditions influencing plasmid transfers between bacilli in soil was highlighted by van Elsas et al. (1987). In later work, the same research team also showed that transfer of the BHR IncP1-type plasmid RP4 among pseudomonads was strongly stimulated in the rhizosphere of wheat (van Elsas et al. 1988). The authors pinpointed the raised nutritional status of the rhizosphere, as well as the provision of surfaces, as the drivers of this stimulus. BHR plasmids of the IncP1, next to PromA, type, were recently found to also be key agents of HGT in the mycosphere of ectomycorrhizal fungi at the forest floor (Zhang et al. 2015). A key soil PromA-type plasmid, pIP02, was originally isolated by triparental mating from an unknown host bacterium from the wheat rhizosphere (van Elsas et al. 2003). In a field test, it demonstrated self-transfer as well as mobilization of the nonconjugative IncQ plasmid pIE723, from the *P. fluorescens* donor to diverse indigenous proteobacteria in the wheat rhizosphere (van Elsas et al. 1998). Plasmid pIPO2 could also mobilize chromosomal genes and retromobilize (i.e., capture from an exogenous cell) IncQ plasmids. The fact that such plasmids can mobilize other, often unrelated, nonconjugative plasmids—as well as chromosomal genes—is an important observation, as it suggests that almost any fragment of DNA can potentially be transferred between cells in soil (van Elsas et al. 2003). In other work in soil, Goris et al. (2002) examined the organisms that acquired and expressed the catabolic BHR plasmids pJP4 and/or pEMT1, which encodes the degradation of 2,4-dichlorophenoxyacetic acid (2,4-D), in soil microcosms inoculated with plasmid donor populations. A suite of diverse proteobacterial recipients was identified, and they fell in the genera *Ralstonia* and *Burkholderia* (now renamed *Paraburkholderia*; *Paraburkholderia hospita* and *P. terricola*). Moreover, indigenous *Stenotrophomonas maltophilia* could also serve as the plasmid recipient when easily available carbon sources were added. This study was seminal in that it confirmed the importance of local conditions in the soil, as already identified by van Elsas et al (1987), in particular the effects of nutrients as well as selective pressure on plasmid transfers to diverse organisms. Finally, a big step forward in our understanding of the breadth of plasmid transfers in soil was achieved by the work of Klümper et al. (2015), who showed that BHR plasmids of the IncP and PromA types in suitable hosts can invade an unexpectedly diverse fraction of a soil bacterial community. While such plasmids are known to transfer to diverse hosts in pure culture, their ability to transfer to members of complex soil bacteriomes had not been comprehensively studied before. Three proteobacterial donor organisms were used to investigate the plasmid transfer ranges in a soil bacterial community. Transfers to members of 11 different bacterial phyla were found, including diverse G^+ taxa within the Firmicutes and Actinobacteria. This key finding suggests that heterogramic plasmid transfer (transfer across the Gram barrier) of IncP-1- and IncPromA-type plasmids may be a frequent phenomenon in soil. Interestingly, a so-called core super-permissive bacterial fraction was identified that had acquired different plasmids from diverse donor strains. The authors speculated that such a core bacterial group might serve as a collective genetic hub in the flow of the BHR IncP- and IncPromA-type plasmids across members of soil bacteriomes. This finding reinforces the evolutionary significance of these plasmids, acting as accelerators of bacterial evolution in soil.

As exemplified by the study by Goris et al. (2002), the acquisition of an MGE may result in the entrance of a whole set of new genes that may allow the occupation of a completely novel niche (in this case, the degradation of 2,4-D). The importance of such stepwise evolution for bacteria residing in soil is underscored by the fact that conjugative MGEs frequently carry genes that are highly advantageous for survival in specific environments, such as those for antimicrobial and heavy metal resistance, virulence, the ability to catabolize new carbon sources, the utilization of new nitrogen sources, DNA restriction systems, and decreased sensitivity to ultraviolet light. In accordance with the complexity of the genetic information offered, the MGEs can range in size from about 18 to more than 500 kb. In addition to the direct selective advantage that conjugative MGEs may offer when acquired, they also play key roles in the phenomenon of genome plasticity. The "metaphenomics" concept describes the outcome of the genetic diversity present in a microbial metagenome combined with the resources available to the microbiome in a given environment, thus representing the materialized genetic potential (Jansson and Hofmockel 2018).

7.3.2.3 Conjugation Occurs Preferentially in Soil Hot Spots and in Biofilms

In Section 7.3.2.2, we learned that conjugative transfer of plasmids or other MGEs may occur readily in soil, provided that local conditions allow sufficient cell-to-cell contact and bacterial activity. Such conditions are often found in soil hot spots for activity, such as the rhizosphere (Chapters 1 and 10; van Elsas et al. 1988), the mycosphere (Zhang et al. 2014), and in manure deposits (Heuer et al. 2011). In such soil hot spots, bacteria accumulate in biofilms or other cell agglomerates. As discussed in Chapter 1, biofilms are the predominant mode of life for bacteria in nature. The formation of a stable mature biofilm comes about as a result of social interactions that have evolved through adaptations enhancing group fitness. Bacteria living in biofilms are protected from harmful impacts from the environment, adapting readily to environmental changes. HGT processes such as conjugation are relevant in the context of bacterial occurrence in complex microbial communities in biofilms. Based on advanced reporter gene technology and high-resolution microscopic tools, plasmid transfer in complex bacterial communities can be quantified. Thus, high-frequency plasmid transfer was found to occur in biofilms *in vitro* (Hausner and Wuertz 1999), and this finding suggests occurrence in hot spots in soil where biofilms occur (van Elsas et al. 2003, Zhang et al. 2014). Nevertheless, spatial constraints within biofilms may hinder the spread of plasmids. Thus, the transfer of an IncP-1-type plasmid in an *E. coli* biofilm had spatial and nutritional constraints and occurred predominantly in the aerobic zone (Krol et al. 2011). Preexistence of a plasmid in the initial phases of biofilm formation might be key to plasmid transfer, which occurs if the biofilm-priming traits are encoded by the plasmid itself.

From recent studies, it has become clear that conjugative plasmids and biofilm structure (and function) are intertwined through complex interactions, ranging from the genetic to the community level. Conjugative plasmids may even promote the formation of biofilms or at least increase their formation. The MGEs that are involved in transfers are found to often harbor genes that promote temporally selective advantages to their hosts. We are, however, far from a complete understanding of the genetic, spatial, and temporal diversity of MGEs that occur in soil bacterial communities in biofilms. Moreover, there are pending questions with respect to the effects of the local (microhabitat) biotic and abiotic factors that determine the conjugative DNA transfer rates in soil and biofilms. In some cases, temporal effects of HGT are seen (Heuer et al. 2011). Finally, we need to improve our understanding of the immediate fitness effects that the MGEs will impose on their hosts *in situ*, as these will determine their temporal persistence and hence ecological and evolutionary success.

7.3.3 Transduction in Soil

7.3.3.1 Mechanism of Transduction

Transduction, that is, HGT between bacteria mediated by bacteriophages, is a process in which either random bacterial DNA from a bacterial host that was previously infected by a bacteriophage (generalized transduction), or host sequences flanking a previously integrated phage that got excised (specialized or restricted transduction) is packed into newly assembled bacteriophage particles and transferred to recipient cells (Penadés et al. 2015). Thus, bacterial DNA packaged in phage capsids can be injected into the cells of new hosts. However, there is a "compatibility issue" here, in that—to achieve successful transduction—suitable phage receptors must be present at the surface of the recipient cells, and the genetic processing machinery (leading to recombination/integration) should "recognize" the incoming DNA.

7.3.3.2 Transduction in Soil—Experimental Evidence

Transduction has been demonstrated to occur in natural habitats such as aquatic environments and on the surfaces of plant leaves (van Elsas et al. 2003). A suite of studies in soil (see Chapter 6) has indicated the wide occurrence of phages of diverse bacterial hosts (under which pseudomonads, bacilli, and streptomycetes) across soils, with an emphasis on soil hot spots such as the rhizosphere. Studies also showed the propensity of both phages and hosts to engage in transduction events. In generic terms, successful transductions in soil settings were suggested to depend on the local conditions. Such conditions should be conducive to phage–donor cell contact, cellular activity resulting in cell lysis and subsequent uptake of phage particles by suitable recipient cells. Our understanding of the soil factors that promote conduciveness to transduction indicates that the conditions in soil hot spots, where biofilms form, are often most conducive. A review of the key relevance of bacteriophages for HGT and other processes in soil was recently published (Pratama and van Elsas, 2018).

Transduction has not yet been unequivocally shown to occur *in situ* in indigenous soil bacteria, which is likely due to the technical difficulties of detecting the rare transduction events in soil. The transduction potential in soil may be underestimated. The average densities of phage particles in soil (see Chapter 6) are often on the order of 10^9 per g soil, so there is potential for successful transfers provided active hosts are present. The potential for transduction in soil is further underscored by observations that suggest that phages can persist in soil for long periods of time in the absence of bacterial hosts. In this respect, the production of large quantities of so-called *gene transfer agents* (GTAs; formerly called defective phages) by the soil bacterium *Bacillus subtilis* is worth mentioning. These soil bacilli, upon specific stress, can subvert their cellular metabolism to exclusively produce such GTAs intracellularly. The GTAs, which are released upon cell lysis, are phage particles (consisting of capsids and tails) that carry randomly packaged host genomic DNA. These GTAs are, in turn, able to infect newly appearing host cells that emerge. They thus appear to serve as unique survival mechanisms for the genes that make up the host genomes. This peculiar mechanism provides fuel to the "selfish gene" and Darwinian host fitness maximization concepts being operational in soil settings.

7.4 METHODS USED TO STUDY HGT IN SOIL

In the following sections, the information on HGT among bacteria in soil that can be obtained from comparative genome analyses (bioinformatics) and experimental approaches is briefly discussed with a focus on the methodological advances.

7.4.1 Retrospective Identification of Horizontally Transferred DNA Sequences in Bacterial Genes or Genomes

The development of high-throughput DNA sequencing methodology and data processing software has allowed entire bacterial genomes to be compared in respect of sequence homology, synteny, conservation of operon structure, and many other features. Sequencing of multiple genomes of the same species yields insight into the species *pangenome* (overall diversity found in the known members of the species) and the extent of presence of accessory genes (carried by some members of the species). Such analyses yield valuable information for distinguishing between compositional changes in bacterial genomes that are caused by intrachromosomal rearrangements and changes, for instance, via mutations, versus those caused by interchromosomal acquisitions (HGT). Table 7.1 summarizes the different analyses made at narrow and broad genomic levels. Large-scale genome analyses provide new opportunities to identify the current composition of genomes of soil bacteria and the relative impact of horizontally transferred DNA in such populations. In some cases, the sources and sizes of the transferred DNA fragments can be determined with relative certainty. However, since comparative DNA analysis examines the evolutionary outcome of the combined effects of cellular processes (e.g., for natural transformation: exposure to DNA, bacterial DNA uptake, integration of DNA into the genome, and mismatch repair) and population-scale processes (natural selection, random genetic drift, dispersal), inferences of the details of the exact processes involved in soil can rarely be made. Essential here is the factor time. That is, for more recent HGT events, the cellular processes may be of highest importance, whereas for older HGT events, the fate of the cellular processes will be determined by subsequent population processes alone (Pettersen et al. 2005, Townsend et al. 2012). For instance, HGT events may occur frequently in a bacterial population in soil without leaving detectable evidence when using comparative sequence analysis. In such scenarios, bacterial phenotypes that arise from frequent, but deleterious, HGT events will go extinct after negative (purifying) selection. The likelihood that rare bacterial phenotypes would rise in numbers by random genetic drift in large bacterial populations is exceedingly low (Pettersen et al. 2005). Thus, it is not possible to obtain accurate information from the current bioinformatics literature on the extent to which bacterial cytoplasms and genomes are—transiently—exposed to foreign DNA and on the effects of the biological factors governing HGT events in soil. On the other hand, and in cases of successful HGT (yielding positively selected clones), field and other studies (Rizzi et al. 2008) have proven useful to provide rough estimates of the frequencies of HGT.

TABLE 7.1
Genetic Variation Used in DNA Sequence Comparisons to Infer HGT Events in Bacterial Genomes

Level of Variation	Variation Patterns Compared	Indications of HGT from Observations
Nucleotide	SNPs	DNA regions with identical SNP patterns in polymorphic genes across a large collection of bacterial strains (linkage disequilibrium, *D*)
	GC% and codon usage	DNA regions with deviating patterns of base composition/codon usage
Gene	Composition of coding sequences	Closest database hits (sequence similarity) to genes in unrelated species
	Phylogeny	Phylogenetic gene trees with unexpected branch placements
Genome	Composition and synteny	Conflicting DNA distribution and composition patterns among related bacterial species and strains

7.4.2 Experimental Studies of HGT into Defined Bacteria

A classical way of experimentally approaching HGT in soil is by assessing the HGT frequencies into defined recipient bacteria *in vitro* or in soil microcosms (van Elsas et al. 2000, 2003). The main advantage of such experiments is that selected parameters (considered to be of importance for the HGT process) can be characterized and the HGT frequency quantified. This is often achieved by monitoring the transfer of selectable marker genes, which are detectable on laboratory media. The studies yield insights into the factors, for example, soil temperature, nutrients, salt concentrations, and pH, that affect HGT processes in the selected bacteria. Nevertheless, only a limited quantitative understanding of the incidence and population dynamics of HGT events in natural bacterial communities in soil can be obtained, since often artificially high concentrations of DNA or of bacterial donors and recipients have to be used. Moreover, the experiments may not encompass the variety of environmental factors that affect HGT events and their dynamics under natural conditions. Hence, to understand the real impact of HGT in natural microbial communities in soil, the full complexity of the natural soil, in addition to ecologically relevant time, should be taken into account, for example, in field studies.

7.5 POPULATION-SCALE CONSIDERATIONS OF HGT EVENTS

The former sections mainly focused on the likelihood and identification of horizontal transfer of genes to individual bacteria in soil. However, as illustrated in Figures 7.3 and 7.4, the long-term evolutionary outcome of an HGT event is defined not by the transfer frequencies alone, but, more importantly, by the population dynamics of the novel bacterial phenotype, as determined by either random genetic drift or changes in host fitness providing the ground for natural selection (Nielsen et al. 2014). Although selection is usually understood to act upon organisms, the discussion is

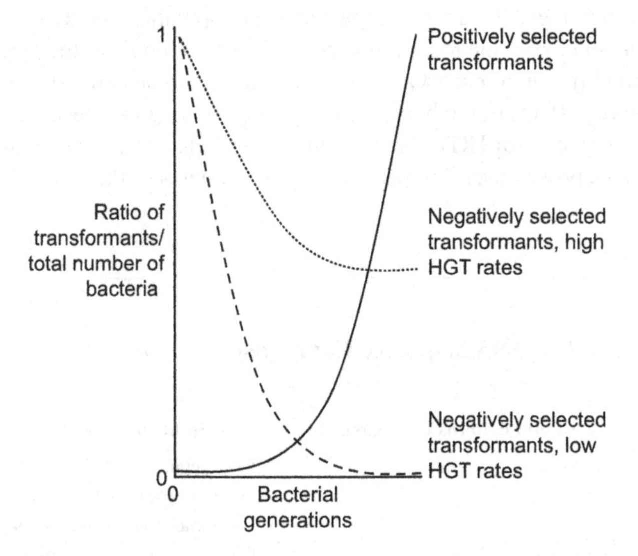

FIGURE 7.3 A schematic figure illustrating the outcome of strong directional selection of bacteria carrying horizontally acquired genes. As seen, the long-term persistence and prevalence of bacteria carrying horizontally acquired genes is determined mainly by selection and not the number of initial HGT events. The figure illustrates that negatively selected transformants initially present in high proportions will over time disappear from the population, whereas transformants with strong positive selection will raise to fixation over time. If the horizontal transfer rate is sufficiently high, even negatively selected bacterial transformants will persist in the population, since the high numbers of repeated HGT events will replace the transformants continually lost from the population.

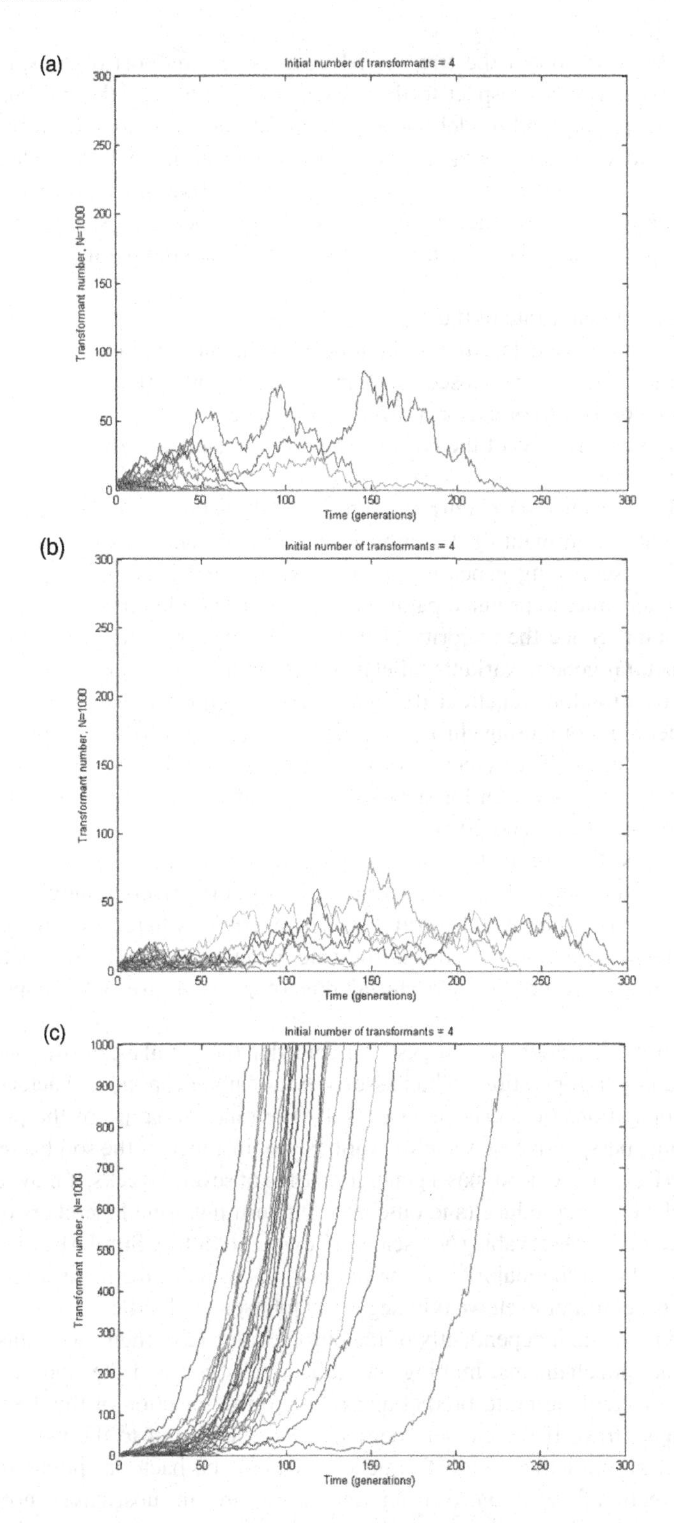

FIGURE 7.4 Computer modeling illustrating the stochasticity underlying the fate of four initial transformants in a population of 1,000 bacteria subject to no selection (a), weak positive selection (b), or strong positive selection (c) over 300 bacterial generations. Each line represents one of the 50 independently starting transformant populations (A. K. Pettersen, unpublished).

focused here on the selection of the transferred genes since the host organism is unnamed and is assumed to differ only with respect to the presence or absence of the horizontally transferred genetic material. To predict and model the long-term fate of a horizontally transferred trait, it is necessary to understand the key characteristics of the bacterial population, such as size, structure, and dispersal patterns, and generation time and annual cycles (Nielsen et al. 2014). The probability and time to fixation (i.e., the presence of the gene in all members of a specific population) of rare HGT events in a bacterial population can be related to the following parameters:

r is the horizontal transfer rate of the gene.
N is the population size (viable cells) of the total bacterial population.
n is the population size of the exposed recipient bacterial population.
s is the selection coefficient of bacteria carrying the gene.
t is the number of generations of the bacterial cells (Nielsen et al. 2014).

The number of HGT events (r) occurring within the larger bacterial population over time (t) can be an important determinant of the impact of HGTs on bacterial genomes. For instance, the recombination of housekeeping genes between closely related bacterial strains may occur with high frequencies over time, to produce patterns of single-nucleotide polymorphisms (SNPs) in the bacterial chromosome. Since the majority of these SNPs are likely to be neutral, they exemplify the potential for neutral genetic variants (alleles) to rise in numbers in populations due to frequent HGTs and subsequent random genetic drift. In contrast, as sequence divergence increases between species, the number of recombining chromosomal DNA molecules will decrease, and a much lower level of transfer is expected. The detection of such rare, sporadic HGT events will strongly depend on the chances of positive selection for survival, establishment, and amplification in the bacterial population (Nielsen and Townsend 2004).

Selection has, thus, the ability to rapidly change the distribution of various phenotypes in a bacterial population and ranges from negative (purifying), effectively removing new horizontally introduced genetic variability, via neutral (or near-neutral), where random genetic drift will determine the fate of the horizontally introduced genetic material, to positive (directional) selection, where the newly produced bacterial phenotype will rise to high proportions in the bacterial populations.

Figure 7.4 illustrates the stochastic processes involved in the initial establishment of rare bacterial transformants in a larger population. When assessing the fate of rare novel bacterial phenotypes in larger bacterial populations, it is crucial to consider the random nature of the processes involved and the relevant time perspective for the observations. For instance, if the soil bacterial species, into which the rare HGT event occurred, has a generation time of several weeks, it may take 10–100 years for the resulting clones to reproduce (and outcompete non-transformed members of the population) to numbers that would be observable (Nielsen and Townsend 2004). Similarly, once DNA has been acquired and is present in the majority of the populations, it will take many years before it is lost again if selection is neutral or even weakly negative (Johnsen et al. 2011).

MGEs that can replicate independently of the chromosome have their own transmission dynamics and maintenance mechanisms, making the unit of selection and the population dynamics of these elements more challenging to predict and relate to the selection of the host bacterium itself (Ghaly and Gillings 2018). If the element confers a beneficial trait to the host bacterium, it can, due to positive selection, achieve high frequencies within the bacterial population. However, in the absence of directional selection, transfer rates alone are, in most cases, not high enough to sustain the maintenance of MGEs over evolutionary time. It has been suggested that plasmids are maintained in a bacterial population through series of selective sweeps, depending on the trait they encode and confer to the host (Bergstrom et al. 2000). This model assumes that, without carrying beneficial genes, plasmids impose a biological/ecological cost to their hosts due to extra energy expenditure, and transfer rates alone will not be high enough to overcome this metabolic cost.

Hence, the plasmid will be inevitably lost on a theoretical basis. An increasing body of evidence suggests that newly acquired plasmids actually do impose a biological cost to their new hosts. However, these costs may be strongly reduced or eliminated, and in some cases reversed to a beneficial association, through subsequent compensatory mutations. Such mutations reduce the burden of the extra cost to the host in a manner which is not always fully understood. In addition, plasmid addiction systems (post-segregational killing systems) can maintain plasmids in populations over time (Diaz-Orejas et al. 2017).

7.6 STUDIES ON SELECTION IN SOIL

In most cases with rare HGT events, selection is considered to be the key factor that determines the fate of the novel DNA combinations. The immediate effect of selection is most easily seen in cases in which the HGT events confer strong selective advantages to the bacterial host. The classical example of an observable effect of the selection of newly acquired traits is the rapid amplification and dissemination of antimicrobial resistance genes in clinical settings and in animal husbandry. In soil, evidence for the selection of HGT events was found in the aforementioned study by Goris et al. (2002) on 2,4-D degradation: bacterial adaptation to xenobiotics in the environment.

Xenobiotics are defined as "man-made molecules foreign to life." Chapter 23 further discusses the varying mechanisms of biodegradation and bioremediation processes in soil. Xenobiotics can exert a strong selective pressure in soil, which spurs the selection and development of bacterial strains with the ability to degrade these. In a generic sense, the catabolic genes responsible for xenobiotic degradation are mainly plasmid-borne, and to some extent transposon-borne (see Chapter 23).

Bioaugmentation involves the release of an organism capable of utilizing pollutants as nutrients in order to accelerate the removal of these. The approach of Goris et al. (2002) aimed to release a host strain with a conjugative MGE harboring the relevant catabolic genes. The spread of these elements into competitive indigenous organisms counteracts the problems—which are commonly encountered—of inferior *in situ* competition of introduced strains. As another example, the conjugative plasmid RP4::Tn*4371*—after introduction into soil in the host *Enterobacter agglomerans* (unable to express the genes encoding degradation of biphenyl and 4-chlorobiphenyl)—was transferred to several members of the indigenous soil microflora. Raised rates of biphenyl breakdown were found in soil upon the addition of this donor strain (De Rore et al. 1994)

Additional evidence for the selection of transferred plasmid-mediated traits in soils was reported in a study on the transfer of plasmids in soil microcosms (Smets et al. 2003). Soil samples were taken from about 2.5 m below the ground surface at a site historically contaminated with heavy metals. IncQ plasmids with heavy metal resistance genes were introduced into *E. coli* as the delivery strain. The soil microcosms were exposed to increasing concentrations of $CdCl_2$. One such plasmid, pMOL222, induced an increase of heavy metal-resistant organisms in the soil, suggesting it had been mobilized by the indigenous community and, subsequently, transconjugants had been positively selected. This selective effect was highest at the highest doses of $CdCl_2$. Interestingly, addition of the conjugative plasmid RP4, without heavy metal resistance determinants, increased the level of heavy metal-resistant organisms, probably due to increased mobilization of the IncQ plasmids and possibly indigenous heavy metal resistance MGEs as well. These studies and others (van Elsas et al. 2003) demonstrate the potential for rapid selection for MGE-associated traits in soils.

7.7 CONCLUDING REMARKS

HGT between bacteria may occur under conditions in which the reproduction of host cells does not take place—in contrast to vertical gene transfer occurring at every bacterial cell division. Long periods of limited reproduction are thought to be the rule rather than the exception for most of the bacteria living in soil. Several independent mechanisms including transformation, conjugation, and transduction can facilitate HGT in different bacterial species. The potential selective advantage of such events provided to the host bacterium is affected by the complex interplay of the soil factors that govern the life of the bacterium in the niche taken (see Chapter 1). In spite of the accumulation of compelling experimental evidence that supports the notion of transformation, conjugation, and transduction as major mechanisms for bacterial HGT in soil, we lack a thorough understanding of the real, microhabitat-specific, impact of these processes in relation to the forces of selection that are operational in the same niche space.

Regardless of the impact of ecological factors on bacterial microevolution, recent whole-genome sequencing and comparative genome analysis do provide strong and compelling evidence for the importance of HGT in the general evolution of bacteria, including a range of typical soil bacteria. However, such evidence most often takes into account events that have taken place over longer periods of evolutionary time. Thus, in observing genome sequences shaped by selection and genetic drift, the analyses can obviously provide only limited information on the underlying mechanisms, frequencies, and factors that have governed HGT events from various sources in natural bacterial populations over time. Hence, field and laboratory studies will remain essential to elucidate the biotic and abiotic factors promoting or limiting HGT in bacterial communities in their natural habitats in soil. We emphasize, however, that the effects of HGT events are not defined by the singly or multiply occurring DNA transfer events and frequencies, but by the drivers of the population dynamics of the rare novel bacterial phenotypes in the larger bacterial community. Stochastic survival, dissemination processes, and natural selection will determine the outcome of the HGT events, where rare novel bacterial phenotypes that acquire beneficial genetic changes can establish in the population and community at large, at the expense of the less-successful phenotypes. These successful phenotypes will, as they increase in prevalence in the larger bacterial population, soon face competition from even newer phenotypes emerging as a result of ongoing mutations, HGT events, and migration—defining the ongoing process of bacterial microevolution.

ACKNOWLEDGMENTS

We thank Pål J. Johnsen and Jack Heinemann for comments on an earlier version of this manuscript.

REFERENCES

Bergstrom, C.T., Lipsitch, M. and Levin, B.R. 2000. Natural selection, infectious transfer and the existence conditions for bacterial plasmids. *Genetics* 155, 1505–1519.

de Rore, H., Demolder, K., de Wilde, K., Top, E., Houwen, F. and W. Verstraete. 1994. Transfer of the catabolic plasmid RP4::Tn*4371* to indigenous soil bacteria and its effect on respiration and biphenyl breakdown. *FEMS Microb Ecol* 15, 71–77.

de Vries, J. and W. Wackernagel. 2002. Integration of foreign DNA during natural transformation of *Acinetobacter* sp. by homology-facilitated illegitimate recombination. *Proc Natl Acad Sci USA* 99, 2094–2099.

Denamur, E. and I. Matic. 2006. Evolution of mutation rates in bacteria. *Mol Microbiol* 60, 820–827.

Diaz-Orejas R, Espinosa, M. and C. Chieng Yeo. 2017. The importance of the expendable: Toxin–Antitoxin genes in plasmids and chromosomes. *Front Microbiol* 8, 1479.

Domingues, S. and K.M. Nielsen. 2017. Membrane vesicles and horizontal gene transfer in prokaryotes. *Curr Opin Microbiol* 38, 16–21.

Domingues, S., Harms, K., Fricke, F.W., Johnsen, P.J., Da Silva, G. and K.M. Nielsen. 2012. Natural transformation facilitates transfer of transposons, integrons and gene cassettes between bacterial species. *PLoS Pathog* 8, e1002837.

Domingues, S., Da Silva, G. and K.M. Nielsen. 2015. Distribution patterns of common gene cassette arrays in integrons reveal an unsurpassed capacity for global transfer among species and environments. *Microbiology* 161, 1313–1337.

Doron, S., Melamed, S., Ofir, G., et al. 2018. Systematic discovery of antiphage defense systems in the microbial pangenome. *Science* 359, eaar4120.

Drake, J.W., Charlesworth, B., Charlesworth, D. and J.F. Crow. 1998. Rates of spontaneous mutation. *Genetics* 148, 1667–1686.

Frost, L.S. and G. Koraimann. 2010. Regulation of bacterial conjugation: balancing opportunity with adversity. *Future Microbiol* 5, 1057–1071.

Ghaly, T.M. and M.R. Gillings. 2018. Mobile DNA as ecologically and evolutionary independent units of life. *Trends Microbiol* 26, 904–912.

Goessweiner-Mohr, N., Arends, K., Keller, W. and E. Grohmann. 2014. Conjugation in Gram-positive bacteria. *Microbiol Spectr* 2, PLAS-0004-2013.

Gómez, P. and A. Bucking. 2013. Coevolution with phages does not influence the evolution of bacterial mutation rates in soil. *ISME J* 7, 2242–2244.

Goris, J., Dejonghe, W., Falsen, E., et al. 2002. Diversity of transconjugants that acquired plasmid pJP4 or pEMT1 after inoculation of a donor strain in the A- and B-horizon of an agricultural soil and description of *Burkholderia hospita* sp. nov. and *Burkholderia terricola* sp. nov. *Syst Appl Microbiol* 25, 340–352.

Grohmann, E., Muth, G. and M. Espinosa. 2003. Conjugative plasmid transfer in Gram-positive bacteria. *Microbiol Mol Biol Rev* 67, 277–301.

Guglielmini, J., Quintais, L., Garcillan-Barcia, M.P., de la Cruz, F. and E.P. Rocha. 2011. The repertoire of ICE in prokaryotes underscores the unity, diversity, and ubiquity of conjugation. *PLoS Genet* 7, e1002222.

Harms, K., Lunnan, A., Hülter, N., et al. 2016. Substitutions of short heterologous DNA segments of intra- or extragenomic origins produce clustered genomic polymorphisms. *Proc Natl Acad Sci USA* 113, 15066–15071.

Hausner, M. and S. Wuertz. 1999. High rates of conjugation in bacterial biofilms as determined by quantitative in situ analysis. *Appl Environ Microbiol* 65, 3710–3713.

Heuer, H. and K. Smalla. 2012. Plasmids foster diversification and adaptation of bacterial populations in soil. *FEMS Microbiol Rev* 36, 1083–1104.

Heuer, H., Schmitt, H. and K. Smalla. 2011. Antibiotic resistance gene spread due to manure application on agricultural fields. *Curr Opin Microbiol* 14, 236–243.

Jansson, J.K. and K.S. Hofmockel. 2018. The soil microbiome – from metagenomics to metaphenomics. *Trends Microbiol* 43, 162–168.

Johnsen, P.J., Townsend, J.P., Bøhn, T., Simonsen, G.S., Sundsfjord, A. and K.M. Nielsen. 2011. Retrospective evidence for a biological cost of vancomycin resistance in the absence of glycopeptide selective pressures. *J Antimicrob Chemother* 66, 608–610.

Johnston, C., Martin, B., Fichant, G., Polard, P. and J.P. Claverys. 2014. Bacterial transformation: distribution, shared mechanisms and divergent control. *Nat Rev Microbiol* 12, 181–196.

Klümper, U., Riber, L., Dechesne, A., et al. 2015. Broad host range plasmids can invade an unexpectedly diverse fraction of a soil bacterial community. *ISME J* 9, 934–945.

Krol, J.E., Nguyen, H.D., Rogers, L.M., Beyenal, H., Krone, S.M. and E.M. Top. 2011. Increased transfer of a multidrug resistance plasmid in *Escherichia coli* biofilms at the air-liquid interface. *Appl Environ Microbiol* 77, 5079–5088.

Land, M., Hauser, L., Se-Ran Jun, S.E., et al. 2015. Insights from 20 years of bacterial genome sequencing. *Funct Integr Genomics* 15, 141–161.

Matic, I. 2017. Molecular mechanisms involved in the regulation of mutation rates in bacteria. *Period Biol* 118, 363–372.

Nielsen, K.M. and J. Townsend. 2004. Monitoring and modeling horizontal gene transfer. *Nat Biotechnol* 22, 1110–1114.

Nielsen, K.M., van Elsas, J.D. and K. Smalla. 2000. Transformation of *Acinetobacter* sp. BD413(pFG4Δ*nptII*) with transgenic plant DNA in soil microcosms and effects of kanamycin on selection of transformants. *Appl Environ Microbiol* 66, 1237–1242.

Nielsen, K.M., Johnsen, P., Bensasson, D. and D. Daffonchio. 2007. Release and persistence of extracellular DNA in the open environment. *Environ Biosaf Res* 6, 37–53.

Nielsen, K.M., Bøhn, T. and J.P. Townsend. 2014. Detecting rare gene transfer events in bacterial populations. *Front Microbiol* 4, 415.

Oliveira, P.H., Touchon, M., Cury, J. and E.P.C. Rocha. 2017. The chromosomal organization of horizontal gene transfer in bacteria. *Nat Commun* 8, 841.

Overballe-Petersen, S., Harms, K., Orlando, L., et al. 2013. Natural transformation by degraded DNA allows for bacterial genetic exchange across geological time. *Proc Natl Acad Sci USA* 110, 19860–19865.

Page, R. and W. Peti. 2016. Toxin-antitoxin systems in bacterial growth arrest and persistence. *Nat Chem Biol* 12, 208–214.

Penadés, J.R., Chen, J., Quiles-Puchalt, N., Carpena, N. and R.P. Novick. 2015. Bacteriophage-mediated spread of bacterial virulence genes. *Curr Opin Microbiol* 23, 171–178.

Pettersen, A.K., Primicero, R., Bøhn, T., and Nielsen, K.M. 2005. Modeling suggests frequency estimates are not informative for predicting the long-term effect of horizontal gene transfer in bacteria. *Environ. Biosafety Res.* 4, 222–233.

Polz, M., Alm, E. and W. Hanage. 2013. Horizontal gene transfer and the evolution of bacterial and archeal population structure. *Trends Genet* 29, 170–175.

Popa, O, Hazkani-Covo, E., Landan, G., Martin, W. and T. Dagan. 2011. Directed networks reveal genomic barriers and DNA repair bypasses to lateral gene transfer among prokaryotes. *Genome Res* 21, 599–609.

Pratama, A.A. and J.D. van Elsas. 2018. The "neglected" soil virome – potential role and impact. *Trends Microbiol* 1533, 1–14.

Pukall, R., H. Tschape, and K. Smalla. 1996. Monitoring the spread of broad host and narrow host range plasmids in soil microcosms. *FEMS Microb Ecol* 20:53–56.

Ray, J.L., Harms, K., Wikmark, O.G., Starikova, I., Johnsen, P.J. and K.M. Nielsen. 2009. Sexual isolation in *Acinetobacter baylyi* is locus-specific and various 10 000 fold over the genome. *Genetics* 182, 1165–1181.

Rizzi, A., Pontiroli, A., Brusetti, L., et al. 2008. Strategy for *in situ* detection of natural transformation-based horizontal gene transfer events. *Appl Environ Microbiol* 74, 1250–1254.

Sen, D., van der Auwera, G.A., Rogers, L.M., Thomas, C.M., Brown, C.J. and E.M. Top. 2011. Broad-host-range plasmids from agricultural soils have IncP-1 backbones with diverse accessory genes. *Appl Environ Microbiol* 77, 7975–7983.

Smets, B.F., Morrow, J.B. and C.A. Pinedo. 2003. Plasmid introduction in metal-stressed, subsurface-derived microcosms: plasmid fate and community response. *Appl Environ Microbiol* 69, 4087–4097.

Smillie, C., Garcillán-Barcia, M.P., Francia, M.V., Rocha, E.P. and F. de la Cruz. 2010. Mobility of plasmids. *Microbiol Mol Biol Rev* 74, 434–452.

Soler, N., Krupovic, M., Marguet, E. and P. Forterre. 2015. Membrane vesicles in natural environments: a major challenge in viral ecology. *ISME J* 9, 793–796.

Soucy, S.M., Huang, J. and J.P. Gogarten. 2015. Horizontal gene transfer: building the web of life. *Nat Rev Genet* 16, 472–482.

Thomas, C.M. and K.M. Nielsen. 2005. Mechanisms and barriers to horizontal gene transfer between bacteria. *Nat Rev Microbiol* 3, 711–721.

Townsend, J.P., Nielsen, K.M., Fisher, D. and D.L. Hartl. 2003. Horizontal acquisition of divergent chromosomal DNA in bacteria: effects of mutator phenotypes. *Genetics* 164, 13–21.

Townsend, J.P., Bøhn, T., Nielsen, K.M. 2012. Probability of detecting horizontal gene transfer in bacterial populations. *Front. Microbiol.* 3, art. 27.

van Elsas, J.D., McSpadden Gardener, B.B., Wolters, A.C. and E. Smit. 1998. Isolation, characterization, and transfer of cryptic gene-mobilizing plasmids in the wheat rhizosphere. *Appl Environ Microbiol* 64, 880–889.

Van Elsas, J.D., J. Fry, P. Hirsch, and S. Molin. 2000. Ecology of plasmid transfer and spread. In: *The Horizontal Gene Pool. Bacterial Plasmids And Gene Spread*, (C. Thomas ed.) p. 175. Harwood academic publishers, Amsterdam.

Van Elsas, J.D., Govaert, J.M. and J.A. van Veen. 1987. Transfer of plasmid pFT30 between bacilli in soil as influenced by bacterial population dynamics and soil conditions. *Soil Biol Biochem* 19, 639–647.

Van Elsas, J.D., Trevors, J.T. and M.E. Starodub. 1988. Bacterial conjugation between pseudomonads in the rhizosphere of wheat. *FEMS Microbiol Ecol* 53, 299–306.

van Elsas, J.D., Turner, S. and M.J. Bailey. 2003. Horizontal gene transfer in the phytosphere. *New Phytol* 157, 525–537.

Vos, M. and A. Didelot. 2009. Comparison of homologous recombination rates in bacteria and archaea. *ISME J* 3, 199–208.

Wilmes, P., Simmons, S.L., Denef, V.J. and J.F. Banfield. 2009. The dynamic genetic repertoire of microbial communities. *FEMS Microbiol Rev* 33, 109–132.

Wozniak, R.A.F. and M.K. Waldor. 2010. Integrative and conjugative elements: mosaic mobile genetic elements enabling dynamic lateral gene flow. *Nat Rev Microbiol* 8, 552–563.

Zhang, M., Pereira e Silva, M.C., Chaib de Mares, M. and J.D. van Elsas. 2014. The mycosphere constitutes an arena for horizontal gene transfer with strong evolutionary implications for bacterial-fungal interactions. *FEMS Microbiol Ecol* 89, 516–526.

Zhang, M., Visser, S., Pereira e Silva, M.C. and J.D. van Elsas. 2015. IncP-1 and PromA group plasmids are major providers of horizontal gene transfer capacities across bacteria in the mycosphere of different soil fungi. *Microb Ecol* 69, 169–179.

8 The Protists in Soil—A Token of Untold Eukaryotic Diversity

Michael Bonkowski, Kenneth Dumack, and Anna Maria Fiore-Donno
University of Cologne

CONTENTS

8.1 Introduction 125
8.2 Short Description of the Main Eukaryotic Groups 126
8.3 Classification of Protists 129
8.4 Surveys of Protists in Soils 129
8.4.1 Molecular Approaches 129
8.4.2 Cultivation-Based Approaches and Protistan Abundance in Soil 131
8.5 Sexuality in Protists 131
8.6 Protistan Feeding Modes 133
8.7 Protists in the Soil Food Web 135
8.8 Protists and Plant Growth 137
8.8.1 Roles of Protists in Plant Nutrient Uptake 137
8.8.2 The Protistan Microbiome of Plants 138
8.9 Concluding Remarks and Outlook 138
Acknowledgments 138
References 139

8.1 INTRODUCTION

Protists have been defined as unicellular eukaryotes, in contrast to multicellular ones, such as plants, fungi, and animals. This classification is highly artificial since it focused on the visible organisms, lumping together completely unrelated microscopic eukaryotes. Phylogenetic inferences based on hundreds of genes and taxa ("phylogenomics") have recently helped in defining the main eukaryotic groups and their respective evolutionary paths.

Much debated is the position of the root of the tree of eukaryotic life, essentially because it points to the last common ancestor—the first eukaryote—and to the origin and evolution of the eukaryotic cell. A great obstacle in phylogenetic inferences is the large genetic drift between prokaryotes and eukaryotes. The breakthrough was achieved by analyzing either eukaryotic genes acquired by ancient lateral transfers from eubacteria or genes with a mitochondrial function, using the corresponding bacterial genes to root the tree. Two recent analyses have converged to a main eukaryotic dichotomy between the newly named clades Opimoda (acronym that stands for OPIsthokonta and aMOebozoa) and Diphoda (DIscoba and diaPHOretickes) (Derelle et al. 2015). The root indicates that the first lineage to diverge from the remaining eukaryotes would have been biflagellated and that the eukaryotic diversity can be summarized into only four supergroups (Derelle et al. 2015) (Figure 8.1): Amoebozoa, Obazoa, Corticata, and Euglenozoa. Still, some of the mutual relationships between the main lineages are much debated. In particular, the tree is challenged by the unstable position of some taxa, in particular *Malawimonas* (morphologically an excavate but rarely branching in this group, making the Excavata polyphyletic) and the recently discovered *Collodictyon*.

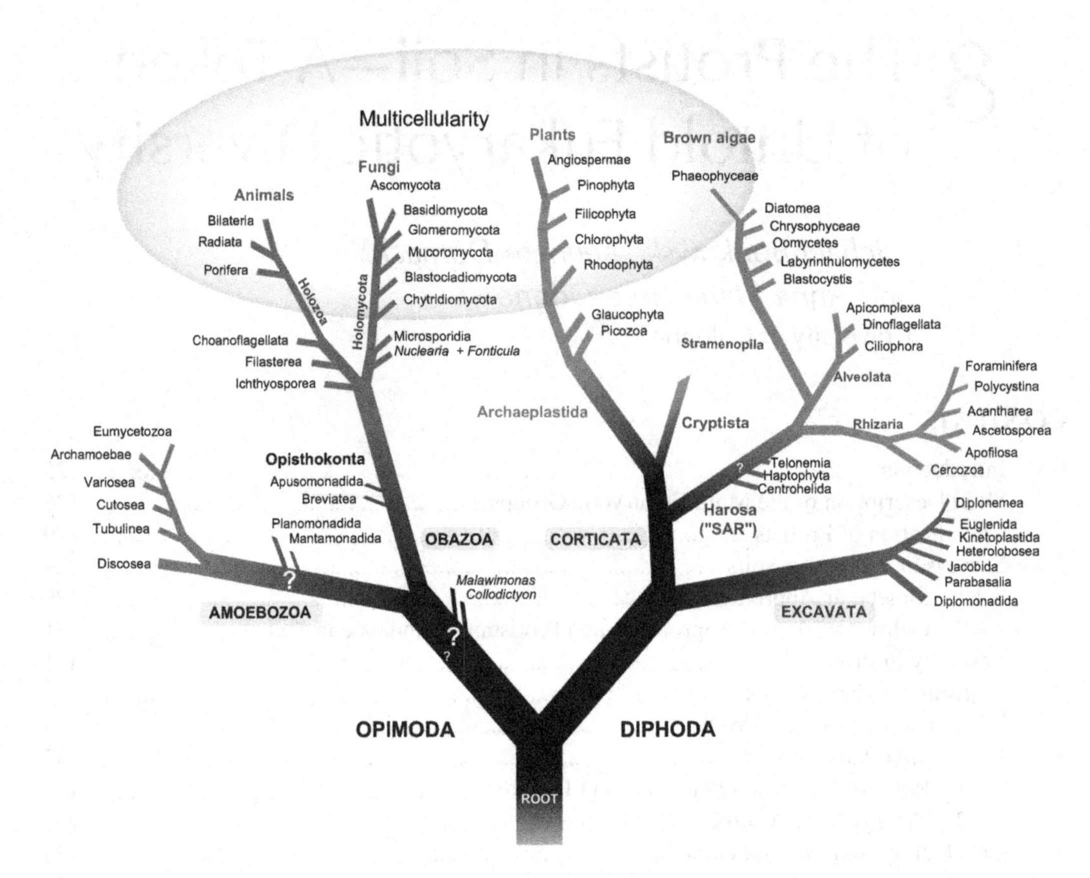

FIGURE 8.1 Schematic tree of the phylogenetic relationships between the four main eukaryotic groups (shaded in gray, names capitalized), divided into Opimoda and Diphoda (Derelle et al. 2015). Topologies in each group are drawn from Derelle et al. (2015) and other recent publications. The multicellular taxa are highlighted (animals, plants, fungi, and brown algae), showing that the greatest part of the eukaryotic diversity is represented by unicellular organisms (protists). Note that there are protistan taxa at the base of each of the multicellular branches, making the distinction between protists and "higher taxa" obsolete. Question marks indicate taxa of uncertain position. (Drawing by AMFD.)

8.2 SHORT DESCRIPTION OF THE MAIN EUKARYOTIC GROUPS

OPIMODA (Derelle et al. 2015)

Amoebozoa (Cavalier-Smith 1998)

Amoebozoa is a key group for eukaryote evolutionary history, which is of immense ecological importance. Amoebozoa comprise a wide variety of amoebae and flagellates, many of which are key inhabitants of soil. They measure from few microns to several meters across, are free-living or parasites, and are often found to be the dominating group of protists in many ecosystems. Amoebozoa may count more than 2,400 species (Pawlowski et al. 2012), and their phylogeny is still not consensual. See, for example, Cavalier-Smith et al. (2016) and Kang et al. (2017). Morphologically, the broadly defined Discosea (Smirnov et al. 2011) comprise common soil amoebae, like the fan-shaped vannellids and the "spiky" *Acanthamoeba* (Figure 8.2). Tubulinea, a highly supported clade in most trees, includes the very diverse and ecologically important testate amoebae Arcellinida. These two groups plus the newly described Cutosea are purely amoeboid, never have flagella, and mostly have blunt and broad pseudopodia. The three following clades, named Conosa, comprise amoeboid, amoeboflagellate, and flagellate lineages, next to amoebae with often more pointed or even

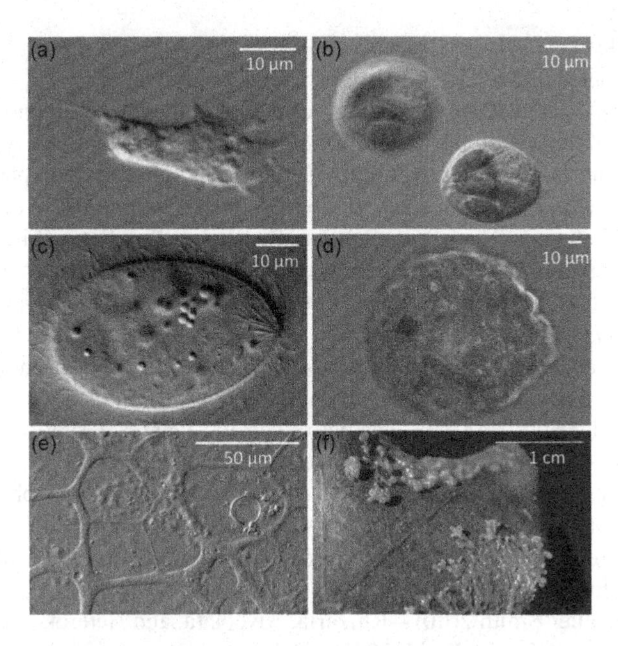

FIGURE 8.2 **(See color insert)** Examples of protistan morphological diversity. (a) *Acanthamoeba*, a very common soil amoeba with "spiky" pseudopods. (b) *Cercomonas*, a common amoeboflagellate, with ingested green algae (*Characium* sp.) cells in food vacuoles. (c) *Parenchelys terricola*, a predatory ciliate with extrusomes at the apical end. (d) *Amphizonella* (Amoebozoa), a shell-bearing amoeba with purple pigments and various ingested eukaryotes. (e) A multinucleate variosean amoeba (Amoebozoa) forming a network. (f) A multinucleate giant amoeba of Myxomycetes of the order Physarales (Amoebozoa) on a dead leaf, starting to form fruiting bodies. (Photographs by the authors.)

mildly branched pseudopodia (Smirnov et al. 2011). Conosa is composed of Archamoebae (mostly adapted to anoxic environments, including the human parasite *Entamoeba histolytica*), Variosea, and the group including the fruiting body-forming Eumycetozoa (mainly the mostly macroscopic Myxomycetes and their relatives). It is probably in Conosa, and in particular in Variosea and Myxomycetes, that the greatest unexplored soil amoebozoan diversity resides. Sequences broadly assigned to unknown varioseans are often found in molecular environmental surveys, whereas Myxomycetes, whose worldwide distribution is based mostly on the occurrence of fruiting bodies, possess such highly divergent ribosomal sequences that they are excluded from both primer-based environmental surveys and reference databases (Fiore-Donno et al. 2016). Metatranscriptomics-based soil environmental surveys will probably reveal a much wider diversity of amoebozoans, in conjunction with single-cell transcriptomics, to complete the reference databases.

Obazoa (Brown 2013)—Fungi and animals and their protistan relatives

The grouping of fungi and animals was firmly established about 30 years ago, challenging the past classification of fungi as plants. The name "Opisthokonta" means "with a posterior flagellum"—a unique characteristic among flagellates. With the progress of phylogenetics and the discovery of new species, the Opisthokonta group expanded to include a variety of protists. One group, the choanoflagellates, are most closely related to animals and the single-cell equivalent of multicellular sponges. Groups that diverged earlier are the parasitic Ichthyosporea and the thread-like pseudopodia-bearing Filasterea (animals + Choanoflagellata + Ichtyosporea + Filasteria = Holozoa). In parallel, Holomycota (Liu et al. 2009) refers to fungi plus the parasitic Microsporidia and the amoeboid Nuclearia; with the adjunction of Fonticula, this group is called Nucletmycea (Brown et al. 2009). At the base of the Opisthokonta, a few groups finally found their place, of which the position had been changing with every new phylogeny: the "proboscis"-bearing flagellates Apusomonodida and the biflagellate Breviatea—collectively called Obazoa.

DIPHODA (Derelle et al. 2015)—Excavata and Corticata

Excavata (Cavalier-Smith 2002)—cells with a feeding groove

The excavates share clear ultrastructural characteristics, although the monophyly of this group is still not confirmed (uncertain position of *Malawimonas*). The cell body typically has a longitudinal groove, where flagella collect food by beating. The phylogenetic position of this group close to the root of the tree makes it a key group to infer the characteristics of the first eukaryote, along with the Amoebozoa. The group contains free-living flagellates, like the photosynthetic euglenids and the common *Bodo*, feeding on bacteria. Also, it encompasses the amoeboflagellate Heterolobosea, which are mostly free-living, but contain some dangerous opportunistic parasites in the genus *Naegleria*, called the "brain-eater." Finally, obligate parasites are included, such as *Trypanosoma cruzi* and *T. brucei*, the agents of Chagas disease and sleeping sickness, respectively. Heterolobosean amoebae are frequently observed in soil samples.

Corticata (Cavalier-Smith 2003)—"one with a cortex," Plantae, Cryptista, and Harosa

Plantae (Adl 2005)—the land plants and their related algae

The algal relatives of plants are the green and red algae and the glaucophytes. Molecular evidence supports the establishment of an endosymbiotic relationship between a heterotrophic protist and a cyanobacterium, at the origin of the Archaeplastida (= Plantae *sensu lato*), the only primary chloroplastic endosymbiosis in the eukaryotes.

Harosa (Cavalier-Smith 2010)—Rhizaria, Alveolata, and Heterokonta

Rhizaria (Cavalier-Smith 2002)—Cercozoa, Foraminifera, and Radiolaria

Cercozoa (Cavalier-Smith 1998) is an immense group of organisms that have tremendous ecological importance. The group often represents, together with the Amoebozoa, the dominant protistan group in terrestrial habitats. The feeding habits and, therefore, the ecological roles of such an immense phylum are multiple. They include heterotrophs, parasites, or autotrophs. Morphologies include amoebae (mostly testate but also naked) usually with thread-like pseudopodia, a huge group of flagellates (e.g., in soils the very common Glissomonadida), and amoeboflagellates (cercomonads), among others (Figure 8.2). Cercozoa also contain important parasites and vectors of viruses in crop plants (e.g., *Plasmodiophora*, *Spongospora*, and *Polymyxa*). Foraminifera (Zborzewsky 1834) represent one of the most ecologically important groups of marine heterotrophic amoebae that are mostly benthic. Their fossilized delicately ornate shells are key indices used in geology to date sedimentary rocks. The shell (when present) consists, in most cases, of calcium carbonate and is perforated (foraminifera = hole bearers), with slender pseudopodia forming an extended net. Bacteria attached to the pseudopodia are slowly driven towards the center of the cell. Foraminifera also occur in terrestrial ecosystems, but they are, unfortunately, rarely observed there due to their fragility. Radiolaria (Müller 1858) is another group of (mostly marine, planktonic) amoebae, with intricate ornamented shells made of silica (Polycystina) or strontium sulfate (Acantharea). They are also diagnostic in dating siliceous rocks. As for foraminiferans, their pseudopodia widely extend through the perforations of the shell, but without forming a net.

Alveolata (Cavalier-Smith 1991) possess a unique characteristic of the cell membrane, as it is internally covered with vesicles that form a continuous layer. They include three diverse groups: ciliates, dinoflagellates, and apicomplexans. The ciliates are one of the better-studied groups of (soil) protists, being often quite conspicuous and well recognizable, with their cilia-covered body (Figure 8.2) and their fast-swimming back and forth bumping into soil grains. They are heterotrophic and can be found in all kinds of environments, although they are rarely dominating. Typical dinoflagellates are covered with armor-like plates and found in all aquatic habitats. They are mostly *phagotrophic*, but can also be myxotrophic or parasitic. Notorious are the dinoflagellate endosymbionts of corals in the genus *Symbiodinium*. Apicomplexa are obligatory parasites, with a structure at the apex of the cell called apicoplast that is used to perforate the host cells. They display complex life cycles with motile, nonflagellated stages. However, they are mostly known in their infectious form, causing devastating illnesses like malaria (*Plasmodium falciparum*). Recent molecular surveys suggest that they could be surprisingly abundant in tropical forest soils (Mahé et al. 2017).

Heterokonta (Cavalier-Smith 1986), also called stramenopiles, are mostly algae, ranging from the giant multicellular kelp to the unicellular diatoms, which are a primary component of plankton. The name "heterokont" refers to the two unequal flagella, one of which is covered with lateral bristles (mastigonemes). Very common in soil are the *Spumella*-like flagellates. Most important from an economic perspective are the oomycetes, whose terrestrial members contain some of the most common and devastating plant pathogens (e.g., *Pythium*, *Phytophthora*, *Albugo*), next to saprotrophs and animal parasites.

8.3 CLASSIFICATION OF PROTISTS

The desirable aim of reaching a consensual classification of protists is hampered by early distinctions between photosynthetic protists (unicellular algae), heterotrophic "protozoa," and protists that have previously been considered as fungi (Oomycota, amoebozoan slime molds). Two nomenclatural codes are currently still in use: the zoological for "protozoa" and the botanical for algae, Oomycota, and slime molds.

Two initiatives, the International Society of Protistologists (Adl et al. 2012) and the Integrated Taxonomic Information System (ITIS), have attempted to unify the classification (Ruggiero et al. 2015). These approaches differed in at least two aspects. While the latter kept the traditional hierarchical classification and description of taxa, Adl et al. (2012) did not consider hierarchic rank endings (e.g., -ida for orders and -aeae for families). In the nomenclatural code, changes in rank will be followed by changes in suffices, followed by the author's name, to ensure traceability. The classification of Adl et al. (2012) favored name stability during changes in classification, at the expense of traceability. The classical enumeration of nested taxa was replaced by a "node-based phylogenic description," which reads "the least inclusive clade containing … " followed by a list of the "popular" representatives (called the "specifiers")—certainly an arbitrary choice. In addition, Adl et al. (2012) did not accept names for polyphyletic or paraphyletic groups. This is a somewhat radical approach, since many taxa may be paraphyletic in some phylogenies, and will stay so until proven monophyletic. Protistologists thus have the choice to use one or the other of these classifications if not one of the many others, according to their evolutionary hypotheses, specific needs, and personal preferences!

8.4 SURVEYS OF PROTISTS IN SOILS

8.4.1 Molecular Approaches

The diversity of protists in the environment seems to exceed that of multicellular organisms (de Vargas et al. 2015). Such a statement may be surprising; how do we reach this conclusion and why only now? For almost two centuries, progress in protistology was synonymous with progress in microscopy. The study of soil protists largely depended on cultivation-based techniques, with their known limitations and biases. The advent of molecular environmental sampling opened a new era for soil protists, but their study still lags far behind their marine and freshwater counterparts. The latter is exemplified by the comprehensive worldwide inventory realized by the Tara expedition in the oceans (de Vargas et al. 2015). To our knowledge, only 20 molecular environmental sampling surveys of soil protists (excluding studies focusing on specific taxa) have been undertaken since the first pioneering study in 1999 up to the beginning of 2017 (Figure 8.3). The number of reads per study was limited when Sanger sequencing was used (a few hundreds to a few thousands). As from 2013, it was replaced with the now defunct Roche 454 sequencing (1–2 million reads per run) and in

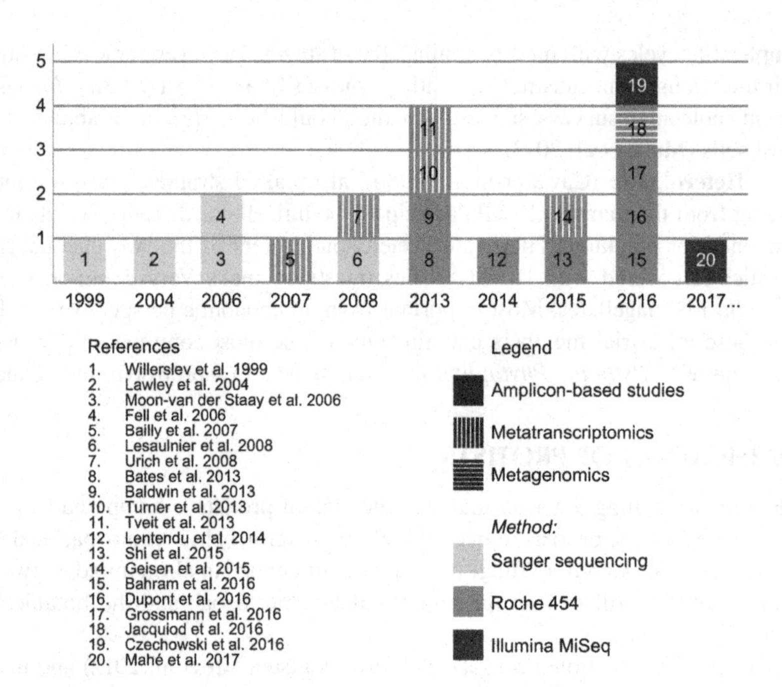

FIGURE 8.3 **(See color insert)** Summary of the molecular environmental studies targeting protists, per year, and indicating the method.

the last years by Illumina sequencing, with more than 10 million reads per run, finally reaching the depth needed to conduct large-scale biodiversity surveys.

Amplicon-based studies comprise the majority of the studies conducted so far. The markers most frequently used for protists are sections of the ribosomal protein genes (18S, variable regions V4 and V9), the internal transcribed spacers (ITS1 and ITS2), next to mitochondrial and chloroplastic markers. However, these studies suffer from a severe primer bias. The immense diversity and the polyphyly of «protists» make it obvious that no protistan universal primers exist. Most general primers will amplify a majority of fungi (Bates et al. 2013, Lentendu et al. 2014, Baldwin et al. 2013, Shi et al. 2015, Bahram et al. 2016, Dupont et al. 2016, Czechowski et al. 2016)—a waste of sequencing effort when targeting protists. In addition, the primers in use appear to be biased towards the most represented taxa in the databases (del Campo et al. 2014), for example, the ciliates (Alveolata). Their dominance in the surveyed soils (Bates et al. 2013, Baldwin et al. 2013, Shi et al. 2015) appears now as artefactual, since RNA-centered transcriptomics approaches (without a primer-based amplification step) depicted the phyla Amoebozoa and Cercozoa as the major terrestrial protistan groups (Urich et al. 2008, Geisen et al. 2015, Grossmann et al. 2016, Turner et al. 2013). Alveolata may dominate only in peatlands (Tveit et al. 2013, Geisen et al. 2015). Amoebozoa was underrepresented in all studies conducted using various "universal eukaryotic primers" (Bates et al. 2013, Baldwin et al. 2013, Fell et al. 2006, Bahram et al. 2016). In-depth, large-scale environmental surveys have revealed that parasites (Apicomplexa) may dominate the soil protistan landscape (Dupont et al. 2016, Mahé et al. 2017); that primary producers (Chloroplastida, most Chrysophyceae, and Dinoflagellata) are present in soils, especially in arid areas (Bates et al. 2013, Fell et al. 2006); and that groups that had been considered to be mainly or exclusively aquatic (e.g., foraminiferans and choanoflagellates) can also be present in soils (Bates et al. 2013, Geisen et al. 2015). In Arctic peat soils, protists were found to be the most abundant eukaryotes, dominating over fungi (Tveit et al. 2013) (Figure 8.3).

The incompleteness and the biases of the reference databases (del Campo et al. 2014) still hamper molecular environmental surveys. Identification of soil protists is more difficult than that of their marine counterparts, principally because the reference molecular databases are biased towards marine species (Mahé et al. 2017). Only one metagenomics-based study of soil protists has been conducted so far (Jacquiod et al. 2016), probably because there are too few genomes of free-living eukaryotes available, which reduces the strength of taxonomic assignment.

The few large-scale surveys of soil protists that have been performed to date do not allow drawing a general picture, although it is certain now that protists do follow biogeographical patterns (Bates et al. 2013). Protistan communities in soil seem to be strongly structured by climatic conditions that regulate annual soil moisture availability (Bates et al. 2013), and their diversity increased with water availability (Shi et al. 2015). Protistan diversity decreased with altitude in the Antarctic region (Czechowski et al. 2016). At a smaller scale, pH could be an important driver of protistan communities (Dupont et al. 2016) (Figure 8.3), although it may affect more parasites compared with the bacterivores or autotrophs.

8.4.2 Cultivation-Based Approaches and Protistan Abundance in Soil

Very different methods have been developed to quantify protists in soil (Geisen and Bonkowski 2017), and each method was found to bring its own specific biases. The two most common approaches are the most probable number (MPN) method and the more recent liquid aliquot method (LAM). MPN is based on serial dilutions of a soil suspension until no protists are observed anymore. MPN determines a maximum threshold of protistan abundance in the studied sample. The microscopical observation of protists that is required in MPN assessments is hampered by the high content of soil particles. Accordingly, protists are lumped into three rough categories: amoebae, flagellates, and ciliates. In contrast, the LAM is based on a high *initial* dilution of soil. This eliminates soil particles, predators, and competitors of protists from the beginning. Protists can replicate without predation pressure, and cultures of single species can be more reliably determined by light microscopy. Most importantly, clonal cultures can be established from LAM, which is a prerequisite to link molecular and morphological taxonomic work. Cultures also support subsequent laboratory experiments to gain further insight on protistan functions in soil.

Estimates of densities of different protistan groups in soil strongly depend on the method used. Testate amoebae can be directly extracted by wet sieving of soil and living cells can be stained. Therefore, estimates of their numbers in soil are the most reliable. According to Darbyshire (1994), testate amoebae can range between about 1 and 100×10^5 individuals per gram. In a study comparing different European conifer forests, the biomass of testate amoebae on average reached ~4 kg ha^{-1}, which was in the range of two- to almost fivefold greater than the combined biomass of all soil animals (Schröter et al. 2003).

There has been a long-lasting debate about the "best" method to estimate the numbers of active protists in soil. However, activity periods of protists can be highly variable over short timescales, as protists can encyst and excyst in less than one hour. Cultivation-based methods inherently have a strong bias. Since protists have specific growth requirements, cultivation-based methods inevitably underestimate their numbers. Much more effort has to be spent on developing reliable quantitative polymerase chain reaction (qPCR) methods for the direct quantification of soil protists in the future.

8.5 SEXUALITY IN PROTISTS

There is a general agreement that sex appeared with the origin of the eukaryotes. While it is the predominant mode of reproduction in higher eukaryotes, many protists were previously considered to be originally asexual, as exemplified by amoebae. Recent work demonstrates that sexuality is present in every major lineage of the eukaryotic tree (see Speijer et al. 2015 and citations therein).

Sexual reproduction should ensure the persistence of species by increasing their genetic variability (through meiosis) and thus the fitness and adaptability of the progeny in changing environments

(Best Man hypothesis). The paradox of maintenance of sex arises from the fact that it has many immediate costs, like the costs of conducting meiosis, in encountering a compatible mating type, generating offspring that may have lost a favorable combination of genes. In contrast, asexual species, thanks to a faster reproduction rate, should be successful colonizers, but they are also prone to extinction by the accumulation of deleterious mutations (Müller's ratchet) and parasites (Red Queen hypothesis). The general mode of reproduction of most protistan groups is best described by *facultative* sexual reproduction: clonally propagating cell lines with infrequent rounds of sex (see Speijer et al. 2015 and citations therein). Soil protists, with their ability to switch to resting stages (cysts or spores), thus combine all advantages of sexual and asexual reproduction—they are fast colonizers of ephemeral favorable niches as well as being resistant to adverse conditions. Moreover, they display the genetic plasticity required for long-term adaptations to changing environments.

Sexual reproduction has been observed only in ciliates (Ciliophora, Alveolata), diatoms (Diatomea, Stramenopila), green algae (Chlorophyta, Plants), and Foraminifera (Rhizaria, Harosa), and in the Amoebozoa, Myxomycetes, and Dictyostelia. In the Excavata, it is only recently that sexuality has been reported in the sporulating genus *Acrasis* (Heterolobosea) and in the Kinetoplastida.

Why is it so difficult to observe sexuality in protists? Because the sexual cycle is probably infrequent, being initiated by intrinsic or extrinsic factors that are not reproducible in culture. Moreover, it may be hampered by the lack of suitable mating types (most cultures are clonal). It may be possible that we fail to recognize sexual reproduction, especially in protists with complex life cycles that involve an alternation not only of ploidy but also of uninucleate and multinucleate cells, active and resting stages, with distinct morphotypes. This is well documented for parasites (in the Apicomplexa and Kinetoplastida), foraminiferans (gamonts and agamonts), and among Amoebozoa (in the Dictyostelia and Myxomycetes). The variation in morphotypes in the latter group is astounding: haploid cells (actually gametes) can be biflagellated or amoeboid, and the diploid zygote is multinucleate (called plasmodium) and can reach conspicuous dimensions (several centimeters to meters), giving rise to often macroscopic colorful fruiting bodies bearing spores. In cases where sex may happen in culture, the fusion of two (or more) cells may be confused with plain mitosis without a prolonged observation under the microscope. Some often observed types of behavior might hint at sexual reproduction: aggregation of multiple cells forming "rosettes" (Figure 8.4). As in ciliates, genetic exchange could occur between adjacent cells by transferring nuclei, without cell fusion. In the Variosea (Amoebozoa), the occurrence of single-spore fruiting amoebae that were previously described as Protostelia also hints towards a sexual cycle. A simple way to test this assumption would be the use of DNA staining of successive life stages to reveal a change of ploidy by an increase in fluorescence intensity. Another system involves the transfection of parental cells with a gene encoding green or red fluorescent proteins. In the progeny, if sexuality occurred, there will be co-expression of both proteins, resulting in a yellow signal. Molecular methods have been

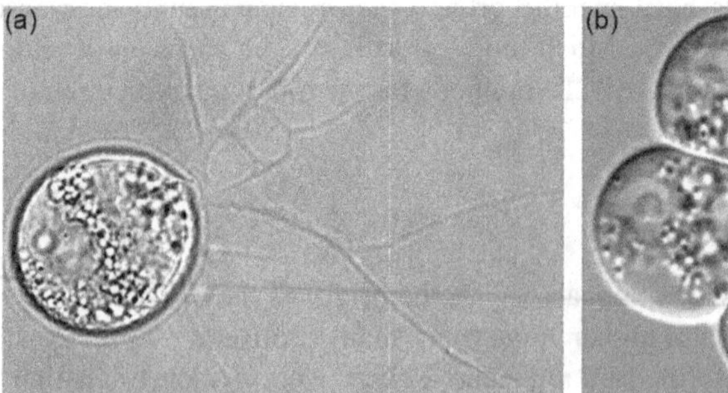

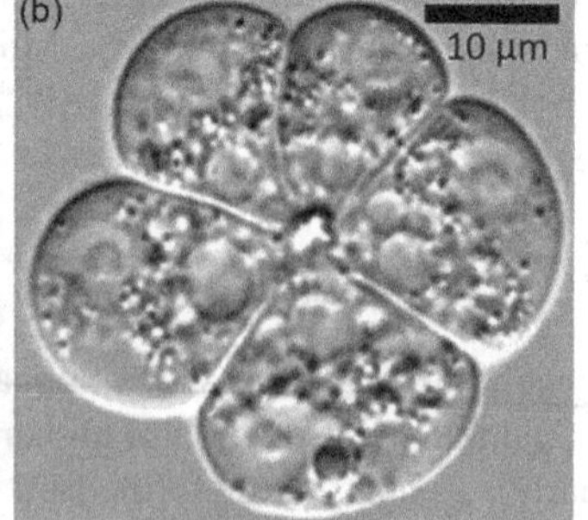

FIGURE 8.4 *Fisculla terrestris* (Cercozoa) is common testate amoeba in soils. (a) A single cell with extended filopodia and (b) rosette-like aggregation of cells hinting towards sexual reproduction. (Photographs by KD.)

largely applied in screenings for genes that are essential for meiosis and fusion of gametes or nuclei that have allowed, for example, to detect sexuality in the choanoflagellates (see Speijer et al. 2015 and citations therein).

Why and how sex evolved in the early eukaryotes is still an open question and a topic of hot debate. Investigating it will also help to define species boundaries (are morphospecies composed of a mixture of asexual and sexually reproducing strains?), next to the fundamentals of the evolution of sex (how do asexual strains arise from sexual ones? how long do they survive?). The new molecular and fluorescent methods, and the combination of both, will probably allow exciting discoveries in times to come.

8.6 PROTISTAN FEEDING MODES

The common mode of food intake is *phagocytosis* (Box 8.1), which may take place all over the cell body, as in many amoebae, or at one specialized feeding site, as in most ciliates. A food vacuole is formed during ingestion and maintained during migration through the cell body until the food is digested by lysosomal enzymes. Undigested food is expelled from the food vacuole, sometimes at a fixed location. A number of protists can directly ingest soluble molecules by *pinocytosis*. Thus, taxa like *Acanthamoeba* can be easily maintained in axenic culture on media containing only simple organic molecules. Some protists directly absorb nutrients through the cell membrane without making use of food vacuoles (*osmotrophy*).

Protists distinguish different food qualities and particle sizes, and the selectivity differs according to their feeding types. Large amoebae and filter feeders show lower degrees of selectivity than small flagellates. The latter, which ingest bacteria one by one, have long handling times and therefore need to be more selective with respect to food quality. Certain flagellates have even been shown to "spit out" an unpalatable prey (Boenigk and Arndt 2002). Even the giant amoeba *Physarum polycephalum* has been shown to select between different proportions of nutrients in accordance with

BOX 8.1 PROTISTAN FEEDING MODES AND STRATEGIES

PROTISTAN FEEDING MODES

Phagocytosis: Ingestion of particles in a food vacuole.
Pinocytosis: Ingestion of nutrients in solution in a food vacuole.
Osmotrophy: Absorbing nutrients through the cell membrane, osmotrophs are commonly parasites, endophytes, or saprotrophs.

PROTISTAN FEEDING STRATEGIES

Interception feeding: Feeding on suspended bacteria, either passively (**filter feeding**) or actively by the moving predator (**raptorial** or **direct interception feeding**).
Browsing: Feeding by moving over biofilms of attached bacteria; depending on the locomotion of the predator, free-swimming **transient** or **gliding** browsers have been classified.
Engulfing: Enclosing large amounts of prey (bacteria), or large prey items in a food vacuole, but also gradual digestion of prey items too long to fit into a food vacuole (e.g., filaments).
Gulping: Enclosing whole protists through an oral opening; usually after paralyzing the prey by **extrusomes** (membrane-bound structures that may be triggered to discharge their content, often toxic).
Sucking: Describes a turgor-driven uptake of cell content after piercing preyed cells.
Trapping: Feeding strategy where agile prey is immediately immobilized upon contact with sit-and-wait predators (often aided by extrusomes) before being ingested.

its optimal diet (Dussutour et al. 2010). Protistan grazing rates on bacteria show high variations and strongly depend on the concentration, type, and quality of bacteria, as well as whether the food is suspended or attached in biofilms (see Parry 2004).

Interception feeding on suspended bacteria, where each prey item is dealt with separately, is found among small protists (flagellates). Interception feeding on bacteria can be passive (filter feeding), or it can be actively supported by the movement of the predator (raptorial or direct interception feeding). Kraken (Cercozoa, Rhizaria) with its extended, slender filopodia is an example of a surface-attached interception-feeding sit-and-wait predator. Other typical examples of sessile filter feeders in soil are the "*Spumella*-like flagellates" (Chrysophyceae, Stramenopila), which rotate around an attached stalk, or the characteristic peritrich ciliates like *Vorticella* (occurring especially in litter). Ciliates also contain a number of freely swimming transient filter feeders, like *Bursaria*, which swims like a huge vacuum cleaner through water films in bacteria-rich litter layers of soils.

Browsing over surfaces is a common feeding strategy in soil. *Transient browsers* actively detach prey from biofilms and are found among free-swimming ciliates (e.g., colpodids) or oxytrichs that "walk" over surfaces and detach bacteria with bundled cilia on their ventral sides. By moving over and out of the surfaces, these predators can spatially avoid the often high concentrations of bacterial antibiotics in biofilms. *Gliding browsers* feed on surface-attached bacteria, as exemplified by the (amoebo)flagellates glissomonads and cercomonads (Cercozoa, Rhizaria).

Engulfing is a typical feeding strategy of amoebae, where the attached prey is enclosed in a food vacuole formed by pseudopodia. Engulfing is not restricted to bacteria, but may encompass yeast cells, fungal spores, or algae. The prey does not always have to be fully enclosed in a food vacuole. Large prey can be captured by small protists from one end and subsequently lysed in the food vacuole, as shown for the testate amoebae *Difflugia* (Amoebozoa) that feed on nematodes, or ciliates that are specialized in feeding on cyanobacterial filaments, like *Pseudomicrothorax* (Ciliophora).

Gulping is a strategy that is often found in slow-mobile hunters among the predatory ciliates. These predators immediately paralyze their prey (often other ciliates) using *extrusomes* (membrane-bound structures that may be triggered by environmental stimuli to discharge their content; see Box 8.2) before engulfing it as a whole by a greatly expandable oral area. Extrusomes vary in complexity from simple adhesives to complex anchoring devices in amoebae and flagellates, and to harpoon-like trichocysts for the immobilization of large prey. Common ciliate genera with this feeding mode are *Encheliodon*, *Enchelys*, and *Spathidium*.

Sucking in prey content is driven by the turgor of the pierced prey cell. Some predators attach themselves to large prey, like nematodes (*Cryptodifflugia*, Amoebozoa), fungal hyphae (*Grossglockneridae*, Ciliophora), and algae (Vampyrellidae, Cercozoa). After piercing the cuticle or cell wall of the prey, the protistan predators appear to suck in the prey content; however, it is probably a passive process.

BOX 8.2 TROPHIC LEVELS IN SOIL PROTISTS

Autotrophs: Primary producers, ability to fix carbon by photosynthesis.
Saprotrophs: Consumers of dead organic matter.
Mixotrophs: Can switch from autotrophy to being consumers.
Consumers: Consume living organisms, in categories as below:

Bacterivores: Obligate consumers of bacteria.
Fungivores: Obligate consumers of fungi.
Algivores: Obligate consumers of algae.
Eukaryvores: Consumers of whole other protists, fungi, and algae.
Omnivores: Consumers of eukaryotes and bacteria.
Parasites: Consumers living at the expense of a host.

Trapping is a hunting strategy used by sit-and-wait predators on large mobile prey. It is best exemplified by activities of actinophryids (Stramenopila). After rewetting air-dried forest litter samples, passively floating "sun-like" spherical cells with long radiating pseudopods will gradually appear after 2 weeks of protistan succession. Any mobile prey coming into contact with their sticky extrusome-rich pseudopods is immediately immobilized.

Osmotrophy is the least studied mechanism of food intake in protists and maybe more widespread than thought. It is attested in saprotrophic and parasitic Oomycota (Stramenopila) and animal parasitic Apicomplexa (Alveolata). It is sporadically found in Myxomycetes (Amoebozoa) and perhaps in many other protists.

8.7 PROTISTS IN THE SOIL FOOD WEB

Sequencing of environmental DNA has dramatically changed the view on protistan functional roles in soil. As an artifact of culturing techniques, soil protists have previously been regarded mainly as bacterivores. This reductionist view has been severely questioned in the past years, as protists are present at all trophic levels, for example, as autotrophs, mixotrophs, saprotrophs, bacterivores, fungivores, algivores, predators, omnivores, and parasites (Geisen and Bonkowski 2017). Here, we briefly discuss each of these roles.

Bacterivory is probably the most important functional role of protists in soil systems (Trap et al. 2016). Bacterivorous protists play dominant roles in soil nitrogen cycling, especially during the early stages of litter decomposition and in the rhizosphere of plants (Bonkowski and Clarholm 2012). Bacterivores further regulate the composition of bacterial communities (Trap et al. 2016). Common examples of obligate bacterivores in soil systems are most of the glissomonads (Cercozoa) and most colpodid ciliates (Alveolata). Less is known about fungivores that feed on fungal spores, hyphae, or both. These might be much more widespread than previously thought (Geisen et al. 2016), challenging the paradigm of the functional prevalence of fungivorous microarthropods in soil. Strictly fungivorous taxa, such as *Grossglockneridae* (Ciliophora) or *Platyreta* (Cercozoa), appear to be rare. Consumption of yeast cells and fungal spores, however, is frequent, but often these protists must be considered as omnivores, as they feed on more trophic levels (e.g., bacteria, fungi, algae). Common examples are *Rhogostoma* (Cercozoa) and some species of the Myxomycetes (Amoebozoa). Some omnivores like *Acanthamoeba* (Amoebozoa) completely suppress the spread of filamentous fungi when added to freshly sterilized soil (Figure 8.5).

Protists can even be primary decomposers: by quantifying the C flux into a decomposer soil food web by isotope labeling, Kramer et al. (2016) identified species of Peronosporomycetes (Oomycetes, Stramenopiles) as saprotrophs, competing with bacteria and fungi for plant-derived carbon sources. Autotrophic protists (unicellular algae), although less abundant in soil than in aquatic systems, are a significant component of biological soil crusts, which cover most bare soil surfaces worldwide, especially in drylands. Their predators, algivores like *Chilodonella* (Ciliophora), Viridiraptoridae, or some Vampyrellidae (both in Cercozoa), are common in litter and in biological crusts (Seppey et al. 2017). Mixotrophs like some Spumella-like flagellates that can switch from phototrophy to bacterivory have also been found in soils.

Many *eukaryvorous* predators that are specialized in the hunting of other protists are found in the Ciliophora (e.g., *Spathidium, Dileptus*). Only recently, apicomplexan (Alveolata) *parasites* of soil animals were evidenced as major components of the soil biome, probably shaping the evolution of Metazoa (Mahé et al. 2017). In addition to soil fauna, the Cryptosporidia (Alveolata) can infect other soil protists. Finally, the protistan plant endophytes (Box 8.3) (Oomycetes—Stramenopila, Phytomyxea—Cercozoa) form a continuum of severe parasitic plant pests to taxa that do not cause any disease symptoms (Neuhauser et al. 2014). However, fundamental information about their distribution in natural plant communities is entirely lacking.

The omnipresence of protists in, and their influence on, all trophic levels has striking consequences for food web relationships in soil. The rhizosphere and phyllosphere (Box 8.3), with their

FIGURE 8.5 Soil protists in their natural habitat; soil particles are shown in dark grey. (1) Green algae (Chlorophyta) and diatoms (Stramenopila) cover soil surfaces. (2) Testate amoeba feeding on algae. (3) An amoeba with long filopods binding soil particles (e.g., Kraken and *Limnofila*, Cercozoa). (4) Spumella-like flagellates (Stramenopila) attached to soil particles. (5) A colpodid ciliate browsing in the interstices. (6) A "walking" oxytrich ciliate. (7) A gliding browser (cercomonad, Cercozoa). (8) Fungal hyphae and yeasts (grey filaments and cell aggregates). (9) Fungivorous amoeba (Vampyrellida, Cercozoa) feeding on hyphae. (10) Large omnivorous amoeba preying on various eukaryotes and bacteria. (Drawing by Kenneth Dumack.)

BOX 8.3 PROTISTS AND PLANTS

The *microbial loop* assumes a steady release of ammonia from consumed bacteria by protists in the rhizosphere and is driven by the constant liberation of exudates at root tips. The exudates provide bacteria in the rhizosphere with energy and enable microbial mining of nitrogen from SOM (see Bonkowski and Clarholm, 2012). Due to stoichiometric reasons, only part of the consumed nitrogen in bacterial biomass can be utilized by protists. About one-third of the nitrogen will be released as ammonia and can be directly taken up by plants or mycorrhiza (see Trap et al. 2016). The microbial loop is of major functional importance, because it shifts the competition for nutrients between rhizosphere bacteria and plants in favor of the latter (see Chapter 6 and Geisen et al. 2018 for possible roles of phages and predatory bacteria in the microbial loop).

Protistan abundance in soil strongly increases towards the rhizosphere (a radial zone around roots influenced by root metabolites) and the rhizoplane (root surface), but protists also colonize the phyllosphere (leaf surfaces); those being able to overcome the plant defense systems colonize the endosphere (the inner parts of plants), the latter living as endophytes, and endophytes causing plant diseases are known as plant pathogens.

cortege of associated bacterial and fungal taxa, will probably be the focus of many forthcoming studies on protists. To date, it was found that higher numbers of bacterivorous protists, nematodes, and fungi occur in the rhizosphere of oat and pea compared to that of wheat; all rhizospheres were enriched in bacterivorous protists (Turner et al. 2013). Virtually unexplored are the trophic networks of the bacteria and fungi that co-occur with their protistan counterparts. The first approaches to study these indicate deterministic changes in bacterial community composition and function in response to protistan predation (Flues et al. 2017). Importantly, the whole concept of separate bacterial and fungal energy channels in soil may be in need of revision (Kramer et al. 2016). The trophic links between the protistan functional groups described earlier are somewhat blurred by reciprocal feeding relationships. There are examples of bacteria (e.g., *Chromobacterium violaceum*) and fungi preying on protists (Michel et al. 2014), protists preying on nematodes, nematodes preying on protists (Geisen and Bonkowski 2017), and (other) protists (e.g., Oomycetes) switching from saprophagy to plant parasitism. Chapter 9 discusses some of these interactions. In light of evolution, these reciprocal trophic links may prevent the dominance of any microbial group, partly thwarting earlier concepts of directed energy flows that have been at the basis of soil food webs.

8.8 PROTISTS AND PLANT GROWTH

8.8.1 Roles of Protists in Plant Nutrient Uptake

The release of N as ammonia from bacteria consumed by soil protists (i.e., the microbial loop) (Box 8.3) is long known as a central mechanism of plant N acquisition (Bonkowski and Clarholm 2012) (Figure 8.6). A detailed review on the roles of protists in plant nutrition is found in Geisen et al. (2018). More recently, the microbial loop was shown to be central for the acquisition of N from soil organic matter (SOM) for plant uptake by symbiotic arbuscular mycorrhizal fungi (AMF) (Koller et al. 2013), because they lack enzymes for N mineralization. These results gained support by independent investigations by Bukovská et al. (2016), who quantified plant uptake of N via AMF from different organic sources in natural field soil. According to their model (Figure 8.6), ammonia is first released from SOM by protistan grazing on bacteria and the soil microbial loop. AMF then preferentially use the ammonia released by the protists, but a significant part of it is first oxidized to nitrite/nitrate before being assimilated by AMF and/or roots (Bukovská et al. 2018); this, again, is fostered by protists (Griffiths 1989). In both scenarios, the soil microbial loop has a central role.

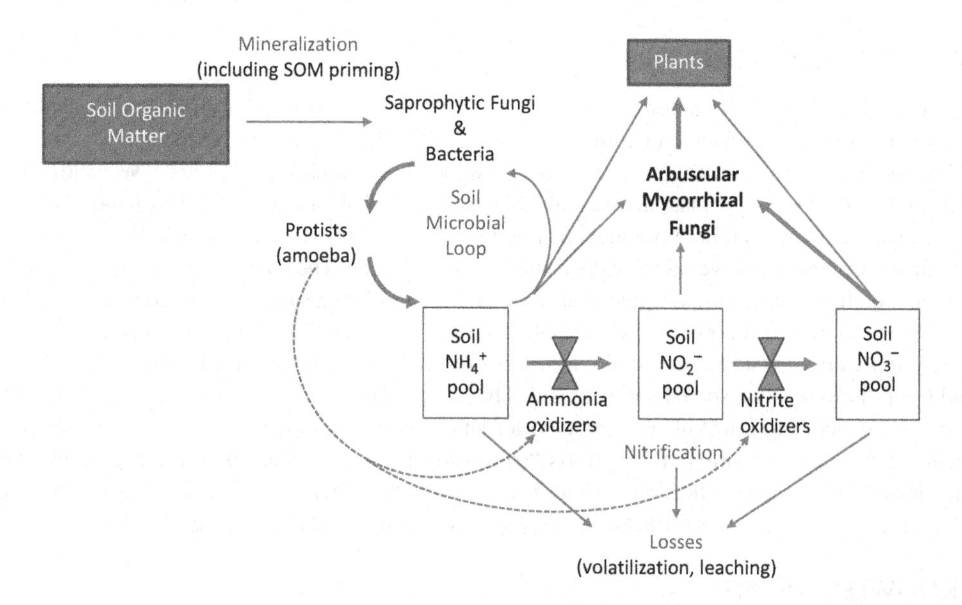

FIGURE 8.6 Schematic representation of the main pathways of soil nitrogen (re)cycling (arrows) from SOM to plants (modified from Bukovská et al. 2018), highlighting the main players (shown in black) and the main processes (shown in grey). The thickness of the lines indicates the assumed importance of the individual pathways for the flux of nitrogen from organic matter to plants. Dashed lines indicate frequently observed but mechanistically unresolved positive feedbacks between protists and nitrifying bacteria (Griffiths 1989).

8.8.2 The Protistan Microbiome of Plants

Recent studies have shown that plants selectively recruit microbes from the soil to establish a complex, yet stable and quite predictable community of bacterial microbiota in and on their roots (van der Heijden and Schlaeppi 2015). Chapter 10 further discusses the role of the rhizosphere (Box 8.3) microbiome. This "plant microbiome" has been considered as "the plants' second genome," as it provides a large number of traits that are crucial for the survival of plants in their natural environment (Friesen et al. 2011). Current models suggest that root exudates have a major role in shaping rhizosphere bacterial community composition. In line with this, the abundance of bacterivorous protists shows a steep increase towards the plant rhizosphere, where predation by protists is known to stimulate bacterial turnover and rapidly alters bacterial community composition (Bonkowski and Clarholm 2012). The grazing-induced changes in bacterial community composition are surprisingly predictable. Both the grazing preferences of protists and microbiome structure shifts due to growth–defense trade-offs in bacteria seem responsible. As shown above, bacterivores differ in their feeding modes, and therefore, higher species richness may lead to increased exploitation of bacterial prey. However, certain bacterial taxa respond to predation by specific phenotypic (i.e., formation of filaments, alteration of shape and size) and functional changes (production of antibiotics, upregulation of type VI secretion systems). In other words, well-defended bacterial taxa will strongly benefit from protistan predation, while their competitors are consumed (see Geisen et al. 2018 for a recent review). Thus, the assembly of bacterial rhizosphere communities may be shaped by both bottom-up processes—via resource supply from rhizosphere deposits—and top-down processes (through protistan grazing). A recent study gives another twist to this mechanism. Sapp et al. (2017) have shown that communities of bacterivorous cercozoan protists assembled in accordance with the bacterial communities on roots and leaves of *Arabidopsis thaliana*, a result that has also been found in a number of crop species (unpublished data). These results are consistent with the view that the bacterial community selects a specific community of grazers adapted to grow and survive on it. Taken together, these coupled feedback mechanisms may provide a first clue for a mechanistic understanding of the assembly of rhizosphere microbiomes. Future work has to concentrate on finding recurrent patterns in the protistan-induced shifts of bacterial community assembly, and must link these to feed back on protistan community assembly.

8.9 CONCLUDING REMARKS AND OUTLOOK

Soil protistology is entering a new era, as we have now efficient molecular approaches to study the diversity and function of protistan communities directly in terrestrial environments.

However, we are far from having answered even basic questions about soil protists. We neither know which protistan groups dominate across soils, nor the roles of abiotic (e.g., soil type) and biotic (e.g., plant, fungal, and bacterial communities) factors in shaping their communities. We also do not know the main drivers of their biogeographic distribution in each biome. The huge functional diversity of protistan taxa outlined earlier implies questions that can be posed regarding their functional redundancy. Thus far, we do not even know at which spatial or temporal scales studies have to be conducted—do we find differences in protistan communities at scales of millimeters, meters, or kilometers? See Chapters 1 and 6 for discussions on spatial scales in soil. The complex interactions of the soil food web, with or without the structuring effect of plant roots, offer an exciting field of almost untapped investigations. Protistologists need to continue their efforts "(i) to sequence representatives of the neglected groups available at public culture collections, (ii) to increase culturing efforts, and (iii) to embrace single cell genomics to access organisms refractory to propagation in culture" (del Campo et al. 2014).

ACKNOWLEDGMENTS

This work is a contribution to the Cluster of Excellence on Plant Sciences (CEPLAS), and was supported by the Grant BO 1907/18-1 in the frame of the priority program SPP 1374 *Biodiversity Exploratories* and by the Grant BO 1907/19-1 in the frame of the priority program SPP 1991 *TaxonOmics* of the German Research Foundation (DFG).

REFERENCES

Adl, S.M., Simpson, A.G.B., Lane, C.E. et al. 2012. The revised classification of Eukaryotes. *J Eukaryotic Microbiol* 59, 429–514.

Bahram, M., Kohout, P., Anslan, S. et al. 2016. Stochastic distribution of small soil eukaryotes resulting from high dispersal and drift in a local environment. *ISME J* 10, 885–896.

Bailly, J., Fraissinet-Tachet, L., Verner, M.C. et al. 2007. Soil eukaryotic functional diversity, a metatranscriptomics approach. *ISME J* 1, 632–642.

Baldwin, D.S., Colloff, M.J., Rees, G.N. et al. 2013. Impacts of inundation and drought on eukaryote biodiversity in semi-arid floodplain soils. *Molec Ecol* 22, 1746–1758.

Bates, S.T., Clemente, J.C., Flores, G.E. et al. 2013. Global biogeography of highly diverse protistan communities in soil. *ISME J* 7, 652–659.

Boenigk, J. and H. Arndt. 2002. Bacterivory by heterotrophic flagellates: community structure and feeding strategies. *Antonie Van Leeuwenhoek* 81, 465–480.

Bonkowski, M. and M. Clarholm. 2012. Stimulation of plant growth through interactions of bacteria and protozoa: testing the auxiliary microbial loop hypothesis. *Acta Protozool* 51, 237–247.

Brown, M.W., Spiegel, F.W. and J.D. Silberman. 2009. Phylogeny of the "forgotten" cellular slime mold, *Fonticula alba*, reveals a key evolutionary branch within Opisthokonta. *Mol Biol Evol* 26, 2699–2709.

Bukovská, P., Bonkowski, M., Konvalinková, T. et al. 2018. Utilization of organic nitrogen by arbuscular mycorrhizal fungi—is there a specific role for protists and ammonia oxidizers? *Mycorrhiza*. doi:10.1007/s00572-018-0825-0.

Bukovská, P., Gryndler, M., Gryndlerová, H., Püschel, D., and J. Jansa. 2016. Organic nitrogen-driven stimulation of arbuscular mycorrhizal fungal hyphae correlates with abundance of ammonia oxidizers. *Front Microbiol* 7, 711.

Cavalier-Smith, T., Chao, E.E. and R. Lewis. 2016. 187-gene phylogeny of protozoan phylum Amoebozoa reveals a new class (Cutosea) of deep-branching, ultrastructurally unique, enveloped marine Lobosa and clarifies amoeba evolution. *Mol Phylogenet Evol* 99, 275–296.

Czechowski, P., Clarke, L.J., Breen, J., Cooper, A. and M.I. Stevens. 2016. Antarctic eukaryotic soil diversity of the Prince Charles Mountains revealed by high-throughput sequencing. *Soil Biol Biochem* 95, 112–121.

Darbyshire, J. 1994. *Soil Protozoa*. CAB International, Wallingford, 209.

de Vargas, C., Audic, S., Henry, N. et al. 2015. Eukaryotic plankton diversity in the sunlit ocean. *Science* 348(6237). doi:10.1126/science.1261605.

del Campo, J., Sieracki, M.E., Molestina, R. et al. 2014. The others: our biased perspective of eukaryotic genomes. *Trends Ecol Evol* 29, 252–259.

Derelle, R., Torruella, G., Klimeš, V. et al. 2015. Bacterial proteins pinpoint a single eukaryotic root. *Proc Natl Acad Sci USA* 112, E693–E699.

Dupont, A., Griffiths, R.I., Bell, T. and D. Bass. 2016. Differences in soil micro-eukaryotic communities over soil pH gradients are strongly driven by parasites and saprotrophs. *Environ Microbiol* 18, 2010–2024.

Dussutour, A., Latty, T., Beekman, M. and S.J. Simpson. 2010. Amoeboid organism solves complex nutritional challenges. *Proc Natl Acad Sci USA* 107, 4607–4611.

Fell, J.W., Scorzetti, G., Connell, L. and S. Craig. 2006. Biodiversity of micro-eukaryotes in Antarctic Dry Valley soils with <5% soil moisture. *Soil Biol Biochem* 38, 3107–3119.

Fiore-Donno, A.M., Weinert, J., Wubet, T. and M. Bonkowski. 2016. Metacommunity analysis of amoeboid protists in grassland soils. *Sci Rep* 6, 19068.

Flues, S., Bass, D. and M. Bonkowski. 2017. Grazing of leaf-associated cercomonads (Protists: Rhizaria: Cercozoa) structures bacterial community composition and function. *Environ Microbiol* 19, 3297–3309.

Friesen, M.L., Porter, S.S., Stark, S.C., Von Wettberg, E.J., Sachs, J.L. and E. Martinez-Romero. 2011. Microbially mediated plant functional traits. *Ann Rev Ecol Evol Syst* 42, 23–46.

Geisen, S. and M. Bonkowski. 2017. Methodological advances to study the diversity of soil protists and their functioning in soil food webs. *Appl Soil Ecol*. doi:10.1016/j.apsoil.2017.05.021.

Geisen, S., Koller, R., Hünninghaus, M. et al. 2016. The soil food web revisited: diverse and widespread mycophagous soil protists. *Soil Biol Biochem* 94, 10–18.

Geisen, S., Mitchell, E.A.D., Adl, S. et al. 2018. Soil protists: a fertile frontier in soil biology research. *FEMS Microbiol Rev*. doi:10.1093/femsre/fuy006.

Geisen, S., Tveit, A.T., Clark, I.M. et al. 2015. Metatranscriptomic census of active protists in soils. *ISME J* 9, 2178–2190.

Griffiths, B.S. 1989. Enhanced nitrification in the presence of bacteriophagous protozoa. *Soil Biol Biochem* 21, 1045–1051.

Grossmann, L., Jensen, M., Heider, D. et al. 2016. Protistan community analysis: key findings of a large-scale molecular sampling. *ISME J.* doi:10.1038/ismej.2016.10.

Jacquiod, S., Stenbæk, J., Santos, S.S. et al. 2016. Metagenomes provide valuable comparative information on soil microeukaryotes. *Res Microbiol* 167, 436–450.

Kang, S., Tice, A.K., Spiegel, F.W. et al. 2017. Between a pod and a hard test: the deep evolution of amoebae. *Mol Biol Evol.* doi:10.1093/molbev/msx162.

Koller, R., Rodriguez, A., Robin, C., Scheu, S. and M. Bonkowski. 2013. Protozoa enhance foraging efficiency of arbuscular mycorrhizal fungi for mineral nitrogen from organic matter in soil to the benefit of host plants. *New Phytol* 199, 203–211.

Kramer, S., Dibbern, D., Moll, J. et al. 2016. Resource partitioning between bacteria, fungi, and protists in the detritusphere of an agricultural soil. *Front Microbiol* 7, 1524.

Lawley, B., Ripley, S., Bridge, P. and Convey P. 2004. Molecular analysis of geographical patterns of eukaryotic diversity in Antarctic soils. *Appl Environ Microbiol* 70, 5963–5972.

Lentendu, G., Wubet, T., Chatzinotas, A. et al. 2014. Effects of long-term differential fertilization on eukaryotic microbial communities in an arable soil: a multiple barcoding approach. *Mol Ecol* 23, 3341–3355.

Lesaulnier, C., Papamichail, D., McCorkle, S. et al. 2008. Elevated atmospheric CO_2 affects soil microbial diversity associated with trembling aspen. *Environ Microbiol* 10, 926–941.

Liu, Y., Steenkamp, E.T., Brinkmann, H. et al. 2009. Phylogenomics analyses predict sistergroup relationship of nucleariids and Fungi and paraphyly of zygomycetes with significant support. *BMC Evol Biol* 9, 272.

Mahé, F., de Vargas, C., Bass, D. et al. 2017. Parasites dominate hyperdiverse soil protist communities in neotropical rainforests. *Nat Ecol Evol* 1, 0091.

Michel, R., Walochnik, J. and P. Scheid. 2014. Article for the "Free-living amoebae Special Issue": isolation and characterisation of various amoebophagous fungi and evaluation of their prey spectrum. *Exp Parasitol* 145, S131–S136.

Moon-van der Staay, S.Y., Tseneva, V.A., van der Staay, G.W.M. et al. 2006. Eukaryotic diversity in historical soil samples. *FEMS Microbiol Ecol* 57, 420–428.

Neuhauser, S., Kirchmair, M., Bulman, S. and D. Bass. 2014. Cross-kingdom host shifts of phytomyxid parasites. *BMC Evol Biol* 14, 33.

Parry, J.D. 2004. Protozoan grazing of freshwater biofilms. *Adv Appl Microbiol* 54, 167–196.

Pawlowski, J., Adl, S.M., Audic, S. et al. 2012. CBOL protist working group: barcoding eukaryotic richness beyond the animal, plant and fungal kingdoms. *PloS Biol* 10, e1001419.

Ruggiero, M.A., Gordon, D.P., Orrell, T.M. et al. 2015. A higher level classification of all living organisms. *PLoS One* 10, e0119248.

Sapp, M., Ploch, S., Fiore-Donno, A.M., Bonkowski, M. and L. Rose. 2017. Protists are an integral part of the *Arabidopsis thaliana* microbiome. *Environ Microbiol* 20, 30–43.

Schröter, D., Wolters, V. and P.C. De Ruiter. 2003. C and N mineralisation in the decomposer food webs of a European forest transect. *Oikos* 102, 294–308.

Seppey, C.V.W., Singer, D., Dumack, K. et al. 2017. Distribution patterns of soil microbial eukaryotes suggests widespread algivory by phagotrophic protists as an alternative pathway for nutrient cycling. *Soil Biol Biochem* 112, 68–76.

Shi, Y., Xiang, X., Shen, C. et al. 2015. Vegetation-associated impacts on Arctic tundra bacterial and microeukaryotic communities. *Appl Environ Microbiol* 81, 492–501.

Smirnov, A., Chao, E.E., Nassonova, E. and T. Cavalier-Smith. 2011. A revised classification of naked lobose amoebae (Amoebozoa: Lobosa). *Protist* 162, 545–570.

Speijer, D., Lukeš, J. and M. Eliáš. 2015. Sex is a ubiquitous, ancient, and inherent attribute of eukaryotic life. *Proc Natl Acad Sci USA* 112, 8827–8834.

Trap, J., Bonkowski, M., Plassard, C., Villenave, C. and E. Blanchart. 2016. Ecological importance of soil bacterivores for ecosystem functions. *Plant Soil* 398, 1–24.

Turner, T.R., Ramakrishnan, K., Walshaw, J. et al. 2013. Comparative metatranscriptomics reveals kingdom level changes in the rhizosphere microbiome of plants. *ISME J* 7, 2248–2258.

Tveit, A., Schwacke, R., Svenning, M.M. and T. Urich. 2013. Organic carbon transformations in high-Arctic peat soils: key functions and microorganisms. *ISME J* 7, 299–311.

Urich, T., Lanzen, A., Qi, J., Huson, D.H., Schleper, C. and S.C. Schuster. 2008. Simultaneous assessment of soil microbial community structure and function through analysis of the meta-transcriptome. *PLoS One* 3.

Van der Heijden, M.G. and K. Schlaeppi. 2015. Root surface as a frontier for plant microbiome research. *Proc Natl Acad Sci USA* 112, 2299–2300.

Willerslev, E., Hansen, A.J., Christensen, B., Steffensen, J.P. and P. Arctander. 1999. Diversity of holocene life forms in fossil glacier ice. *Proc Natl Acad Sci USA* 96, 8017–8021.

9 Microbial Interactions in Soil

Jan Dirk van Elsas and Akbar Adjie Pratama
University of Groningen

Welington Luis de Araujo
University of São Paulo

Jack T. Trevors
University of Guelph

CONTENTS

9.1 Introduction ... 142
 9.1.1 Microbial Interactions and the Spatial Structure of Soil ... 142
 9.1.2 Microbial Function Is Modulated by Environmental Sensing ... 142
 9.1.3 Interaction Studies Performed at Different Scales ... 142
9.2 Microbial Assemblages and Their Relationship to Function ... 143
 9.2.1 Microbial Assemblages, Microcolonies, and Biofilms ... 143
 9.2.2 Function-Driven Microbial Communities ... 144
 9.2.3 Gene Expression and Effects on Biofilm Structure ... 145
9.3 Molecular Mechanisms Underlying Microbial Interactions ... 145
 9.3.1 Molecular Sensing and Signaling ... 145
 9.3.2 Quorum Sensing (QS)—The Paradigm of Signaling between Microorganisms in Nature ... 146
9.4 Ecology of Microbial Interactions in Soil ... 148
 9.4.1 Concepts and Terms to Describe Microbial Interactive Systems ... 148
 9.4.2 Interactions Classified with Respect to Ecological Effects ... 149
  9.4.2.1 Mutualistic Interactions ... 150
  9.4.2.2 Antagonistic Interactions ... 151
9.5 Interactions between Bacteria and Fungi in Soil ... 152
 9.5.1 How Bacteria Interact with Fungi in Soil ... 152
 9.5.2 How Fungi "Cope" with Bacteria and Other Organisms in Soil ... 154
 9.5.3 Mutualistic Interactions and Niche Construction ... 155
 9.5.4 Molecular Recognition Mechanisms in Interactions between Bacteria and Eukaryotes in Soil ... 155
9.6 Predatory Interactions ... 156
 9.6.1 Protozoa as Predators of Bacteria ... 156
 9.6.2 Predatory *Bdellovibrio* ... 157
 9.6.3 Myxobacteria—Slime Molds ... 158
9.7 Concluding Remarks ... 158
References ... 159

9.1 INTRODUCTION

9.1.1 Microbial Interactions and the Spatial Structure of Soil

As discussed in Chapter 1, soil constitutes a heterogeneous habitat that is dominated by solid components, with often fluctuating gas and water phases. In simple terms, soil can be thought of as an environment representing a myriad of habitable microsites, each one of which might be different from the next one. In this habitat, the principles of *island biogeography* can be applied to most microorganisms. Island biogeography is a concept from macro-ecology, which is defined as a framework of parallel systems, within which local eco-evolutionary events reign, and between which events of migration are possible. Thus, in soil, each microhabitat may be like an island that is geographically isolated from other islands, containing its own characteristic microbial community that is specifically adapted to functioning under local conditions and interacting with local organisms. The functional connectivity of the island to other islands may depend on water flow connections and/or the presence of, for example, a rhizosphere or a mycosphere. Connectivity may be restricted most of the time (e.g., under field conditions). As a result of such spatial constraints, interactions between members of the soil microbiota may be more likely to occur **within** an island than **between** different islands. Notable exceptions to the general rule of the island biogeography of soils occur at sites where hypha—or chain-forming organisms, such as *Bacillus mycoides*, actinomycetes, and fungi; mobile organisms such as protozoans, nematodes, and earthworms; and macroorganisms such as plants dwell. At such sites, interconnectivity between microsites, and therefore breaking of the island concept, is possible (Yang and van Elsas 2018).

9.1.2 Microbial Function Is Modulated by Environmental Sensing

Microorganisms in soil perform a diverse range of functions under a range of environmental conditions. From one perspective, an aliquot of soil, of 1 g or smaller, can be viewed as a microbial biochemical gene library (Trevors 2010). The millions of microorganisms contained in this 1 g portion are the drivers of the biogeochemical transformations that take place in the soil site. A considerable interplay exists between the information present in the genomes of microorganisms and the (often-changing) local environmental conditions. *Niche construction* is a phenomenon that describes the role of the organisms, through their metabolism, interactions, and choices (changing their own niche and modifying both biotic and abiotic forces of natural selection; Odling-Smee et al. 1996). Bacteria in soil shape their niche by producing biofilms, detoxifying or releasing, for example, protons, enzymes, and metabolites into their surroundings, thus modulating the composition of local microbiomes, including at host organisms (McNally and Brown 2015).

The microorganisms dwelling in soil possess mechanisms that enable them to monitor, and respond to, the local conditions (e.g., the level of a range of compounds, temperature, pH, and water). For bacteria, the cellular systems involved mostly fall into the class of two-component sensory/response systems (further discussed in Section 9.3.1). The local natural selection—occurring dynamically—will be perceived by the organism and determine its fate or fitness, reflected in cellular growth, reproduction, survival, dormancy, spore formation, gene transfer interactions (see Chapter 7), and/or death.

9.1.3 Interaction Studies Performed at Different Scales

Given the fact that the immediate surroundings of cells in soil microhabitats may consist of other (microbial or macro) species, interactions with other cells are central to the functioning of these microorganisms. Hence, the study of the interrelationships among soil microorganisms, as well as those between microorganisms and organisms such as protozoa and plants, is central to understanding the activities of microbes in soils. Cell-to-cell connectivity is central to the rate of interactions. Most of our current knowledge of microbial interactions has been obtained from *in vitro* laboratory studies. However, research using soil microcosms, greenhouses, and the field (Kvas et al. 2017) has increasingly been performed. The study of soil microbial interactions is obviously different from that of single microbial species in the laboratory. In the latter, the complexity and heterogeneity of soil are often absent and spatial and temporal variabilities are reduced. It is foreseen that, in the next decade, research in this area will increasingly use "real-world" systems, in which the functioning of microorganisms is addressed within the complexity of the soil ecosystem.

In soil, stoichiometric decomposition and/or nutrient mining involves different microbial actors that target different soil organic matter (SOM) pools and are affected by environmental conditions, including those resulting from climate change (Razanamalala et al. 2018). The interactions occur at much smaller scale than many studies have used. For instance, Bach et al. (2018) showed that soil microaggregates support highly diverse microbial communities that are related to many functions in the soil matrix. Other studies have described microbial communities within water-stable aggregates (Davinic et al. 2012), both microaggregates (<2 mm; Davinic et al. 2012) and macroaggregates (>2 mm; Rillig et al. 2015). For the detailed information about microbial interactions in the soil, and relations to soil microaggregates and spatial structure, the reader is referred to reviews by de Boer et al. (2005) and Jansson and Hofmockel (2018).

In this chapter, the interactions of microorganisms with their surrounding biota in soil will be discussed with an emphasis on prokaryotes. Different types of interactions will be addressed, including the molecular background of each interaction where possible. Interactions among bacteria, as well as between bacteria and higher organisms such as fungi, protozoa, *Bdellovibrio* (Otto et al. 2016), and plant roots, will be central. As outlined in Chapter 6, bacteriophages are also important, as they represent nonliving structures that interact with their hosts—the bacteria—in a parasitic or genetic fashion. Thus, they can behave as, respectively, "controllers" of bacterial population densities and/or drivers of evolutionary (adaptive) processes. For further information, the reader is referred to Chapters 6 and 7. Here, we will examine examples of the aforementioned interactions as paradigms for the varied ways soil microorganisms interact with their physically proximal neighbor organisms.

9.2 MICROBIAL ASSEMBLAGES AND THEIR RELATIONSHIP TO FUNCTION

9.2.1 Microbial Assemblages, Microcolonies, and Biofilms

Microorganisms in most natural systems—including soil—do not normally occur as single cells in suspension, as is often the case in liquid laboratory cultures. Rather, they occur in assemblages (conglomerates) of cells, which can encompass either single (*homogeneous* assemblages) or multiple (*heterogeneous* assemblages) species. The cells in these assemblages can be tightly assembled and may constitute interactive structures that have some of the characteristics of a multicellular organism. Such assemblages may even involve, or be associated with, higher organisms, such as fungi, protozoa, or plants. The functioning and interactions of microbial communities in soil are thus affected by the occurrence of cells in assemblages. Assemblages in their most evolved form are called *biofilms*, whereas the term *microcolonies* indicates primitive biofilms containing low cell numbers (Lennon and Lehmkuhl 2016). The formation of a biofilm (or microcolony) involves several steps, including reversible adsorption, adherence, growth and cell division, and secretion of protective extracellular polysaccharidic material). These steps are outlined in Figure 9.1. Depending on the local conditions (mainly the presence of colonizable surfaces, of water and of nutrients), both types of assemblages, that is, larger biofilms next to smaller microcolonies, can be present in soil. These assemblages may occur either within or across island habitats, and they may include either single or multiple bacterial species, fungi, protozoa, plant roots, and other organisms. The extracellular materials of polysaccharidic nature (exopolymers) thus form matrices surrounding the multi-organism assemblages and binding these to colonizable surfaces (Sinha 2013). In soil, the biofilms/microcolonies are mostly attached to solid surfaces, such as specific (nutrient-rich) soil particles,

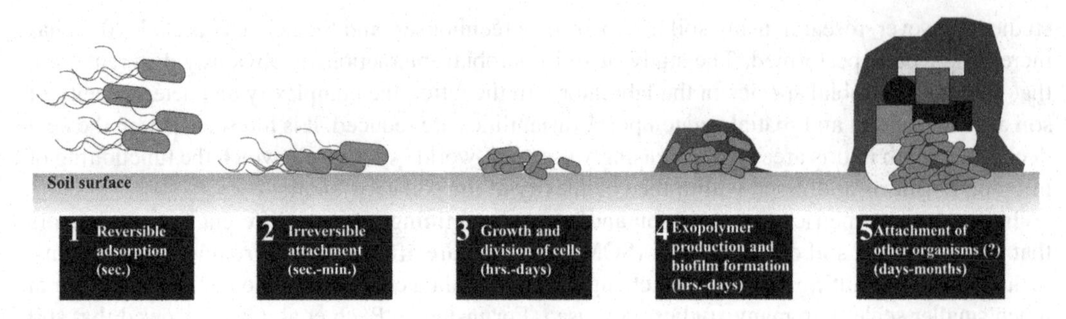

FIGURE 9.1 Scheme of formation of a biofilm in soil. (Figure prepared by Akbar Adjie Pratama, University of Groningen.)

SOM domains, plant roots, and decaying organisms. The communities comprising the biofilm may have both a specific community structure (species composition) and a "physical/biological" architecture, referring to the relationships between the microbial cells, exopolymers, and the microenvironments. There is evidence that biofilm formation is a trait that is selected for by water tension stress, as water conditions within a biofilm are more propitious to microbial life than those outside of the biofilm (Lennon and Lehmkuhl 2016). Hence, biofilm formation can be regarded as a form of niche construction. The study by Lennon and Lehmkuhl further provides evidence for the view that biofilm production is an important functional trait that can contribute to the distribution, abundance, and functioning of microorganisms in soil.

9.2.2 Function-Driven Microbial Communities

Microbial function is the key determinant of the dynamics within a microbial community that is forming a microbial assemblage. Some microbial processes are carried out by just one particular group of organisms, and therefore result in a largely homogeneous (one-organism) assemblage or biofilm. However, other processes cannot be performed by just one microbial group. Such processes are thus dependent on a (function-driven) microbial assemblage that consists of more than one species (heterogeneous biofilm). For example, the process of nitrification (conversion of ammonia to nitrate) generally needs the combined action of ammonia- and nitrite-oxidizing bacteria (under which *Nitrosomonas* and *Nitrobacter* spp., respectively; see also Chapter 11). Both organisms derive energy from the conversion, with the nitrite oxidizers removing the nitrite produced by the ammonia oxidizers. Ideally, both bacterial species occur in the same physical location in soil, in sufficient cell numbers, and are metabolically active. Alternatively, reaction intermediates and final products are transported (in soil water) to the location where the other species is present. Similarly, the digestion of industrial waste and municipal effluent under anoxic conditions in soil or aqueous systems is most efficient when metabolically cooperating microbial aggregates form biofilms, often in sludge granules or flocs. This may require the action of a cometabolic or commensal microbial consortium (Xie et al. 2018). Some bacteria can be metabolically completely dependent on other bacteria, a phenomenon called *syntrophy*. For instance, the obligately syntrophic bacterium *Pelotomaculum schinkii* can grow in anoxic soil habitats where consortia consisting of bacteria and archaea collaborate to degrade and ferment organic matter, generating CO_2 and CH_4. *P. schinkii* can only grow on propionate, transforming it to acetate and H_2. This reaction only works efficiently if other organisms, notably methanogenic archaea, transform these compounds to methane. Interdependency of bacterial species also exists in aerobic metabolic processes in which different species interact, such as in the degradation of xenobiotic compounds. Thus, whole microbial consortia instead of single strains are commonly required for an efficient remediation of soil with respect to toxic chemicals such as benzyl alcohol and toluene aromatic compounds (see Chapter 23).

9.2.3 Gene Expression and Effects on Biofilm Structure

The physiological attributes of microbes in soil exert strong influences on the development of local (homogeneous or heterogeneous) biofilms. During biofilm development, differential gene expression occurs in the microbial cells that grow in the biofilm (Trevors 2010, Haq et al. 2017). Complex regulatory networks will lead to the concerted and temporally controlled expression of suites of genes in cells of the biofilm. The gene expression can be different per microsite occupied by specific cells within the biofilm, for instance, between cells at the biofilm surface—which is rich in oxygen—as compared to cells in lower, often oxygen-depleted, layers. Moreover, disposal of products of cellular metabolism is commonly different per biofilm layer. A mature biofilm consists of various layers in depth that are linked by channels, through which the transport of (waste or nutritive) compounds occurs. Interestingly, recent evidence points to the fact that bacteriophages are agents that make holes/channels in biofilms of their hosts, thus enhancing biofilm health that may lead to functional augmentation (Simmons et al. 2018). Per layer or microsite of the biofilm, the local conditions are different, which will instantaneously drive differential gene expression. For instance, in *B. subtilis* biofilms, genes for the regulation of flagellar-driven motility, osmolarity, oxygen limitation, and high cell density are among those that are differentially expressed (Sinha 2013).

With respect to mixed (heterogeneous) microbial biofilms, much is still unknown about their dynamics and specific interactions at the molecular level, but it is clear that cooperative and competitive cell-to-cell interactions are influenced by the spatial arrangements of microbial genotypes within the biofilm community (Nadell et al. 2016).

Such cell-to-cell interactions, via signaling, play important roles in differentiation processes of individual cells and coordinate group behavior. In the signaling, secreted small molecules that diffuse through water films or through gaseous space (volatiles) are important, as will be discussed later in this chapter. The regulation of a range of bacterial functions is related to cell-to-cell communication, where higher cell concentrations often provide an ecological advantage in the natural environment and hence trigger gene expression shifts. Ostensibly, a suite of different diffusible signal molecules has a direct role in activating gene expression in cells of the biofilm, modulating the ability of cell populations to respond to the external environment. This suggests the occurrence of high phenotypic plasticity in the cells, in response to the coexistence with other species and in dealing with differential nutrient availabilities (Hansen et al. 2017).

9.3 MOLECULAR MECHANISMS UNDERLYING MICROBIAL INTERACTIONS

9.3.1 Molecular Sensing and Signaling

As discussed, microorganisms in soil can respond to their surroundings by sensing local conditions of, for example, water content, pH, temperature, and also the level of particular chemical compounds including nutrients. Most microorganisms have developed sophisticated genetic systems, under which two-component sensor/response systems that adequately monitor their immediate environment and rapidly adapt cellular physiology to it. With respect to the chemical sensing, a suite of compounds, under which signaling, nutritive, and toxic compounds can be perceived, taken up and, if relevant, metabolized. Microorganisms can also excrete compounds that allow them to capture molecules with nutritional or other value. One example is the production of so-called siderophores (iron-chelating compounds of different nature) that allow microbial cells to actively sequester iron from their surroundings (see Chapters 10 and 22 for further details). In light of this evolved fine tuning of (mostly) metabolic capacity, it is not surprising that microorganisms in soil can also release and sense molecules as signals to or from other cells. These mechanisms allow cross-communication between the cells in the soil ecosystem and extend from communication between cells of the same population to that with cells of other organisms, within the same community. There are numerous chemical compounds, as discussed later, that can be employed for bacterial cell-to-cell communication. A particular case in point is given by soil myxobacteria, which show group (swarming and eating) behavior in soil. The two processes are regulated by a suite of diverse molecules (Muñoz-Dorado et al. 2016). Moreover, the regulation of cell physiology is complex in many soil bacteria, resulting in the fact that most bacteria in soil are continuously excreting as well as receiving multiple messenger molecules. The quorum sensing (QS) system has turned into the paradigm of bacterial cell-to-cell signaling, and hence, Section 9.3.2 discusses this key mechanism that drives gene expression in microorganisms, in more detail.

9.3.2 Quorum Sensing (QS)—The Paradigm of Signaling between Microorganisms in Nature

QS was first discovered in the 1980s, and it has been the topic of intensive research ever since. It has been reviewed recently in several key reviews (see Papenfort and Bassler 2016). QS is the regulation of gene expression in response to fluctuations in cell densities. It involves cell-to-cell signaling within a microbial assemblage. By their collective production of signaling compounds referred to in the foregoing, the cells within the assemblage will raise the concentration of the signaling compound above a threshold level, upon which a specific cellular physiological response is induced. The most common signaling compounds belong to the class of the *N*-acyl-L-homoserine lactones (AHLs), but compounds such as long-chain fatty acids (e.g., palmitic acid methyl ester) or small peptides and quinolones are also used. The response to QS within the microbial assemblage can be regarded as a community response. The signaling molecules (also referred to as autoinducers) are secreted by the cells and accumulate in the external environment as the cells grow. As shown in Figure 9.2, the detection of accumulated signaling molecules (in this case, AHLs are depicted) by a specific receptor, that is, the AHL receptor protein, activates the expression of a specific ecologically relevant genetic system. In complex regulatory systems, a whole signaling cascade may be induced. The complex regulatory networks that are associated with QS suggest that these systems, in creating positive feedback loops, may be utilized for rapid signal amplification of exogenous signals. QS relies on two proteins, an AHL synthase and an AHL receptor protein, which functions as a transcriptional regulator (see Figure 9.2).

The ability of bacteria to produce the diffusible signaling molecules has been amply demonstrated, not only for members of the α-, β-, and γ-subclasses of the Proteobacteria but also for Firmicutes and Actinobacteria. Thus, QS has been observed to influence various phenotypes in Gram-negative and Gram-positive bacteria (Papenfort and Bassler 2016). In Gram-positive bacteria, processed peptide signaling molecules are secreted, usually by a dedicated ATP-binding cassette exporter protein. This signaling substructure is a phosphorylation/dephosphorylation two-component cascade of varying complexity and regulatory factors. In brief, the secreted peptide autoinducers increase in concentration as a function of cell population density. Sensor kinases detect the peptide signals and interact with these, initiating a series of phosphorylation events that culminate in the phosphorylation of an associated response regulator protein. The phosphorylation of the response regulator activates it, allowing it to bind DNA and alter the transcription of the QS-controlled target gene(s).

Bacteria use QS communication to regulate a diverse array of physiological activities within microbial communities. Phenotypes associated with AHL production include bioluminescence, production of virulence factors, conjugational transfer of tumor-inducing (Ti) plasmids in *Agrobacterium tumefaciens*, antibiotic production, response to starvation (*Rhizobium leguminosarum, Vibrio* spp.), motility, and biofilm formation (Whiteley et al. 2017). Thus, these processes may include symbiosis, virulence, competence for transformation, conjugation, antibiotic production, motility, sporulation, and biofilm formation. All of these processes are relevant activities of the respective organisms in soil and the larger biosphere.

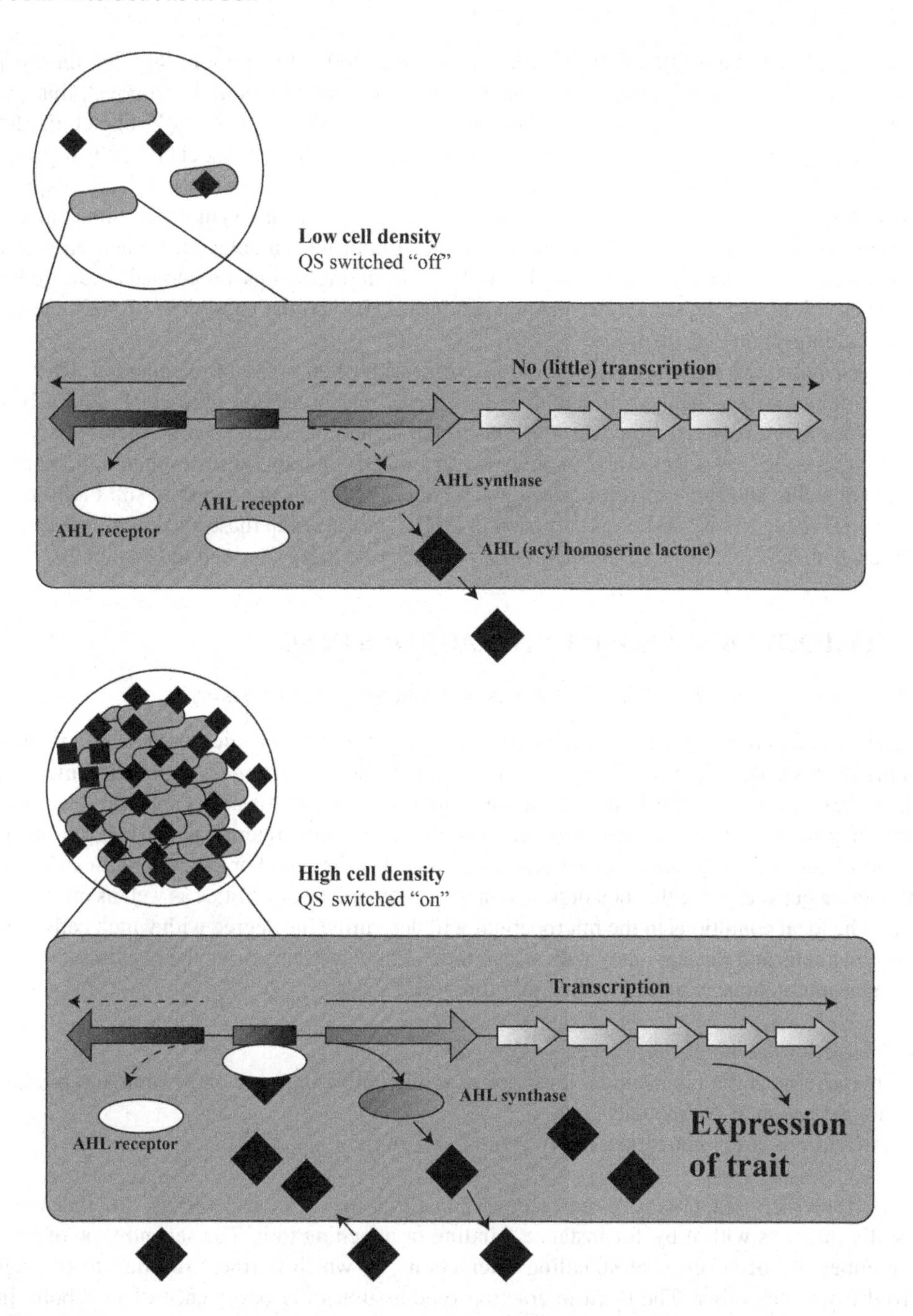

FIGURE 9.2 Principles of QS, an important sensing and regulatory mechanism in soil bacteria. See Whitely et al. (2017) for further details. (Figure prepared by Akbar Adjie Pratama, University of Groningen.)

QS may be intrinsically involved in the survival strategy of soil bacteria. The nature of the chemical signals, the signaling relay mechanisms, and target genes controlled by QS systems differ, but the ability to communicate with one another allows the cells in bacterial populations to coordinate gene expression and, consequently, population behavior. However, little information exists about the role of *in situ* AHL-mediated gene regulation in soil. There are clear ecological aspects that allow bacteria to benefit from the presence of this density-dependent regulatory system. Competing bacterial populations in a particular niche could attempt to thwart each other by targeting and inactivating one another's QS circuits (Miller and Bassler 2001). In addition, *Bacillus subtilis* produces an enzyme, AiiA (a lactonase, homologous to zinc-binding metallohydrolases), that inactivates the *Erwinia carotovora* AHL autoinducer (Dong et al. 2000). Similarly, Zhang et al. (2016a) reported the QS-regulated behavior of *Serratia marcescens*, which is associated with wilting disease in *Pseudostellaria heterophylla*, to be mediated by root exudates. Interestingly, the expression of the aiiA gene in a co-occurring *B. thuringiensis* reduced the disease symptoms. Sun et al. (2016) have identified QS-regulated synthesis of metabolites implicated in antagonistic activities against plant pathogens, such as phenazines and pyoverdines, in fluorescent pseudomonads. Several bacterial species colonizing the same microhabitat produce AHLs, so the possibility of AHL cross talk in soil is an important consideration.

AHL-mediated QS signaling is a community trait, with often >65% of members of diverse taxa in a community being engaged in signaling and quenching processes (Tan et al. 2015). In fact, microbiomes may evolve in the context of QS-driven signaling. Thus, specific members with QS signaling properties are selected and members with anti-QS signaling traits excluded, fostering the presence of stable and functional microbial assemblages. QS is a key process in soil biofilms, with cell densities rising to $>10^{10}$ mL^{-1}, as diffusion of AHLs is reduced in these and so critical levels are easily reached. The effects of QS molecules on plants are further examined in Chapter 10.

9.4 ECOLOGY OF MICROBIAL INTERACTIONS IN SOIL

9.4.1 Concepts and Terms to Describe Microbial Interactive Systems

In the island biogeography concept of soil, the *in situ* interaction of microorganisms in soil with other organisms is dependent on the physical proximity between cells, often at the micrometer-scale distance. Physically close organisms share the same microhabitat ("island") and can thus interact, whereas physically distant ones are on different islands and cannot interact other than following the transport of cells or compounds (the latter, for instance, via the liquid or gaseous phases). The level of interaction between the cells then depends on their interactive capabilities as well as on local conditions. The local conditions in the microhabitat will determine the degree with which cells "sense" neighboring cells and the adequacy with which their genetic systems drive the response.

Cellular interactions can be classified into three categories:

1. Interactions at the physical level
2. Interactions at the biochemical level, for example, following gene expression and production/secretion of compounds
3. Interactions at the nutritional level

The first type of interaction between cells assumes that an organism "senses" another one and physically interacts with it by, for instance, binding or attaching to it. The second type of interaction assumes the occurrence of signaling interactions, in which partners respond to the signals received from each other. The third interaction type assumes the occurrence of metabolic interactions between the partners, like those discussed in Section 9.2.2. Obviously, the three types of interaction can occur simultaneously, between the same organisms, which can be simple (e.g., two organisms involved) or complex (e.g., more than three organisms involved).

An alternative way of categorizing microbial interactions in soil is based on the interacting species that take place in the interactions. Using this as a basis, all microbial interactions in soil can be broadly grouped into the following categories: microorganism–microorganism, microorganism–plant, and microorganism–other organism interactions (see Table 9.1 for some examples). More specifically, bacteria (or archaea) can interact with other bacteria (bacterial/archaeal–bacterial interactions), fungi (bacterial/archaeal–fungal), protozoa (bacterial/archaeal–protozoal), and plants (bacterial/archaeal–plant). A similar categorization can be made for fungi.

TABLE 9.1
Some Examples of Key Microbial Interactions in Soil and Plants

Organismal Type	Interaction with	Type	Examples/Remarks
Bacteria	Bacteria	Syntrophy	*Propionibacterium/P.maculans*: obligate syntrophy by *Propionibacterium*
		Antagonism	Bacteriocin-mediated antagonisms
		Predation	*Bdellovibrio bacteriovorus* predating on soil bacteria
	Fungi	Amensalism	Diverse bacteria associated with soil fungi
		Antagonism	*Pseudomonas fluorescens* producing DAPG against Ggt[a]
		Predation	*Collimonas fungivorans* consuming hyphal contents
		Mutualism	*Paraburkholderia* sp. induce sporulation in *Rhizopus*, increasing dissemination
	Plants	Symbiosis	Rhizobia nodulating legumes
		Parasitism	Agrobacteria forming tumors on plants
		Pathogenic	*Ralstonia solanacearum* causing wilting of tomato
	Bacteriophages	Parasitism	Bacteriophages can cause lytic events in bacterial populations in soil (see also Chapter 6).
	Other organisms	Mutualism	Actinobacteria in soil ants antagonizing pathogenic fungi
Fungi	Bacteria	Symbiosis	AMF harbor endosymbiotic *Glomeribacter* (see also Chapter 5).
		Predation	Fungi (e.g., *Pleurotus ostreatus*) use bacteria as nutrient sources
	Fungi	Antagonism	*Trichoderma* spp. antagonize other fungi
		Competition	*Phlebiopsis gigantea* competes with *Heterobasidion annosum*
		Succession	Fungi deriving nutrients from the tissues of other fungi during decomposition processes
	Plants	Pathogenic	Fungi causing plant diseases, for example, Fusarium spp.
		Endophytic	*Neotyphodium* spp. colonizing grasses
		Symbiosis	Fungal partners in mycorrhiza and lichens use photosynthates
	Viruses	Reduced virulence	*Cryphonectria parasitica* causing chestnut blight reduced in virulence (see Chapter 6).
	Other Organisms		
	Nematodes	Predation	Nematophagous and ectomycorrhizal fungi can use nematodes as nutrient sources
	Collembola	Predation	Ectomycorrhizal fungus (e.g., *Laccaria bicolor*) kills *Collembola* and assimilates N derived from it

[a] Ggt: *Gaeomannomyces graminis* var *tritici*, the causal agent of take-all disease in wheat. DAPG: 2,4-diacetyl phloroglucinol.

In the following sections, the interactions between microorganisms of different kinds will be examined. Microorganism–plant interactions are described in Chapters 10 and 22, and will only be discussed here with respect to some of the basic principles involved.

9.4.2 Interactions Classified with Respect to Ecological Effects

Table 9.1 lists the types of interactions between partner organisms with respect to the ecological benefits gained by either partner. For the simplest of communities (two partners, denoted A and B), these can be delineated as follows:

Neutralism: No interactions (e.g., due to distance, no overlap, or complementarity of processes)

Commensalism: A takes profit, whereas B gains no disadvantage (products, pH)

Mutualism: A and B have mutual benefit (e.g., symbiosis, cross-feeding)
Parasitism: A takes profit of B (A>B)
Predation: A consumes B (A>B)
Amensalism: A is limited by B (A<B; e.g., by a toxic product or pH lowering)
Competition: A and B compete for a limiting factor

Thus, interactions can be either *mutualistic* (or cooperative, leading to a *positive* effect on partners of the interaction), neutral or *antagonistic* (in which a *negative* effect on at least one partner of the interaction can be seen). The effects can be observed by assessing the population dynamics of the partners of the interactome, in the presence as compared to the absence of the interactive partner. In some cases, the outcome of the interaction can be variable, as it depends on the ecological conditions that prevail. In almost all interactive microbial systems, the interaction is chemical in nature; that is, chemical compounds produced by either of the partners of the microbial system are the drivers of the interaction. Such compounds may be either *signaling* or *antibiotic* compounds (antibiotics), the latter in negative (amensalistic) interactions. Currently, a multitude of interactions between organisms in soil environments is known. Hereafter, some examples of mutualistic and antagonistic interactions will be discussed.

9.4.2.1 Mutualistic Interactions

A key example of mutualism is metabolic cooperation. In metabolic cooperation between microorganisms (syntrophy), two or more microorganisms each contribute different elements of a common metabolic pathway, resulting in the net synthesis or degradation of specific compounds, to the benefit of all participants in this metabolic network. Nitrification is one example of such cooperation between bacteria, as discussed in the foregoing. Another example is given by the degradation of the herbicide atrazine by the syntrophic soil bacteria *Clavibacter* and *Pseudomonas*, via products of overlapping genes (Douglas 2004). In these cases, the cooperation is enhanced by the close physical proximity of the microorganisms in the microhabitat, as the distances (and time required) over which compounds have to be transported are minimized. Physical proximity between partners is obtained by two routes: (1) formation of a microbial assemblage in which cell-to-cell contact is promoted and (2) intracellular symbiosis, in which one microbe is enclosed within cells of the other one.

Christensen et al. (2002) showed another example of metabolic cooperation between bacteria, in this case *Pseudomonas putida* (a benzoyl alcohol degrader) and *Acinetobacter* sp. (a more efficient degrader). Both organisms derive their carbon and energy from the substrate. Remarkably, the *P. putida* strain was found to use the intermediate compound benzoate efficiently, whereas *Acinetobacter* was inefficient in this step. The organisms were shown to form a structured biofilm, in which a layer of *Acinetobacter* cells was covered by *P. putida* cells, at a fivefold lower density. Presumable, the *P. putida* cells captured the intermediate compound released by *Acinetobacter*, further processing it. Apparently, this interaction provided an ecological solution to the needs of both partners in the metabolic conversions. The biofilm structure turned out to be essential, as consortia without it performed at much lower rate.

Cooperation in soil also takes place between bacteria and fungi. As discussed in Chapter 5, arbuscular mycorrhizal fungi (AMF) transfer plant photosynthate (C) to the soil where phosphate-solubilizing bacteria (PSB) occur. Phosphate availability was found to drive the outcome of this interaction (Zhang et al. 2016b). Under low phosphate levels, PSB competed with AMF, whereas at elevated levels, the former bacteria enhanced AMF hyphal growth. In turn, PSB activity was stimulated by the AMF. Thus, this mutualistic interaction was phosphate level dependent, only becoming evident above particular levels.

Another study on bacterium–fungal interactions was performed by Jiang et al. (2018), who described an auxotrophic interaction between the plant-beneficial endophytic fungus *Serendipita indica* and the soil bacterium *Bacillus subtilis*. The fungus is auxotrophic for thiamine, a key cofactor in central carbon metabolism. The authors showed that *S. indica* was not able to grow in thiamine-free media, but did grow when co-cultured with *B. subtilis*, which apparently furnished thiamine. Here, the success of this interaction was also dependent on the spatial and temporal organization of the two-partner system. Hence, the ability of this fungus to colonize the rhizoplane may be dependent on this physical association with an organism that supplies the missing nutrient thiamine.

9.4.2.2 Antagonistic Interactions

Antagonistic interactions can derive from trophic/competitive (competition for the same nutrient sources) versus antibiotic/antagonistic mechanisms (creating a competitor-suppressive microenvironment). With respect to the latter, microbially produced compounds are often the mediators. The antagonistic interactions can easily be visualized on dual-culture plates, such as often used when biocontrol agents against pathogenic fungi are searched for (Figure 9.3; see also Chapter 22). The compounds that underlay these negative interactions are often diffusible substances that are released from the bacterial cells. Antibiotics of different types and bacteriocins are two classes of compounds that are often involved.

Antibiotics: Antibiotics are biochemical compounds produced by microorganisms that negatively affect (micro)organisms, being either -cidal (killing) or -static (growth-inhibiting). A range of different chemical classes of antibiotics exists, and the reader is referred to the relevant literature for specific information. With respect to their mode of action, antibiotics are grossly divided into four classes, which are as follows:

1. Interference with cell wall synthesis (e.g., penicillins)
2. Affecting cell membrane integrity (e.g., polymyxins)
3. Affecting nucleic acid metabolism (e.g., rifampicin)
4. Interference with protein synthesis (e.g., streptomycin, tetracyclines)

Although antibiotics are naturally produced by a variety of prokaryotic and eukaryotic organisms in soil, and affect another range of organisms, their precise role in soil ecological processes is still insufficiently understood. It is reasonable to assume that antibiotics modulate the competition for space between the producers and the other microorganisms in the local soil habitat. Clearly, the capacity to produce antibiotics, often as secondary metabolites, is an important ecological trait that occurs in many soil microorganisms. Soil may even constitute the key environment on Earth that selects antibiotic producers, as its heterogeneous (structured) nature promotes competitive interactions.

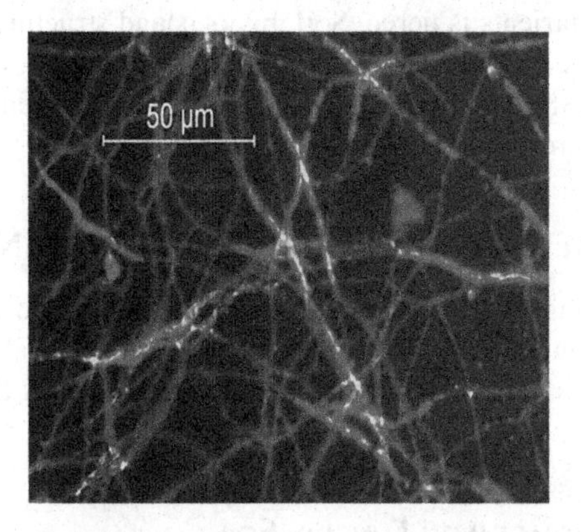

FIGURE 9.3 **(See color insert)** Biofilm formation by the soil bacterium *Paraburkholderia terrae* on the soil fungus *Trichoderma asperellum*. Yellow-green (here whitish): fungal tissue. Red (here grayish): bacteria adhering to fungal tissue. (Courtesy of Dr. D. Tazetdinova, University of Kazan, Russia.)

Mathematical models have indicated that the capacity to produce antibiotics confers a selective advantage to the bacteria dwelling in a structured habitat, even those with reduced growth rate, when low numbers of fast-growing competitors invade. The hypothesis that antibiotic production is enhanced in structured environments is supported by the models, which show persistence of the producers in such habitats, even when the antibiotic production is energetically costly. However, the tenet that antibiotic production in soil is costly could not be experimentally proven in recent work (Garbeva et al. 2011).

Key examples of antibiosis can be found in the biological control of plant pathogens. With respect to bacterial–bacterial interactions, Cronin et al. (1997) found that the biocontrol agent *Pseudomonas fluorescens* F113, by the production of the antibiotic 2,4-diacetyl phloroglucinol (DAPG), controlled the potato soft rot pathogen *Erwinia carotovora* ssp. *atroseptica*. Concerning bacterial–fungal interactions, Khalid et al. (2018) showed that the germination of the plant-pathogenic fungus *Aspergillus flavus* is shifted by lipopeptides (called ralstonins) produced by the plant-pathogenic bacterium *Ralstonia solanacearum*. The ralstonins were found to downregulate the expression of a cryptic biosynthetic gene cluster associated with the production of imizoquin, an alkaloid that serves as a protector against oxidative stress, which is essential for *A. flavus* spore germination. In addition, the imizoquin was found to suppress the growth of *R. solanacearum*. Thus, a reciprocal small-molecule-mediated antagonism affected the pathogenicity and survival of both interactive species.

Moreover, some organisms may use the antibiotic production of bacteria to their own advantage. Remarkably, some specific ants use bacterial antibiotic producers (streptomycetes) in their "farming" of fungi, to ward off key pathogens (Seipke et al. 2011). Similarly, evidence is emerging for the fact that specific beetles also harbor streptomycetes with this capacity to their advantage (Human et al. 2017, van Arnam et al. 2018). Hence, evolution has harnessed the employment of the antibiotic production capacity of soil microbes to serve specific eco-evolutionary fitness aims, spurring the survival of a suite of eukaryotic organisms.

Bacteriocins: Bacteriocins are proteinaceous compounds produced by bacteria, which are toxic to tightly related bacteria (Nishie et al. 2012). Thus, bacteriocins mediate, in a negative way, the interactions between closely knit organisms. The best-studied example is the production by *Escherichia coli* of so-called colicins, of which more than 25 different types have been described. Colicins are produced when competition for nutrients is high, that is, when per-cell nutrient levels are low (also called conditions of "crowding"). Bacteriocins such as colicins are released by cell lysis and transported into susceptible cells, where they degrade DNA, inhibit protein synthesis, or destroy membrane integrity, thus killing the cell. They do not lyse their own cells, as these produce an immunity protein that binds to, and inactivates, the toxin.

This sophisticated mechanism that allows a subtle interaction with closely related organisms is thought to play an important role in the interactions that take place in a microhabitat in soil, when competition for space and nutrients is fierce. Soil, by its island structure, promotes the coexistence of different types of bacteriocin producers (Pagie and Hogeweg 1998), which parallels the effect on antibiotic producers. Hence, the spatial structure of the soil environment has major implications for interactions between the microorganisms that inhabit it.

9.5 INTERACTIONS BETWEEN BACTERIA AND FUNGI IN SOIL

Bacteria and fungi in soil may show diverse behavior, ranging from explorative, antagonistic to mutually beneficial. In the latter case, niche construction may even take place. We examine these possibilities and also address the molecular recognition mechanisms that may be involved in the sections that follow.

9.5.1 How Bacteria Interact with Fungi in Soil

Using microscopic techniques, bacteria have often been observed to occur on fungal tissues (hyphae, mycorrhizal roots, spores, and interiors of fruiting bodies). A range of species from diverse bacterial taxa, including *Variovorax*, *Pseudomonas*, *Dyella*, *Paenibacillus*, and *Paraburkholderia* (formerly *Burkholderia*), has been found as occupants of fungal surfaces (Li et al. 2016). Using direct molecular techniques, even fungal-associated Archaea have been detected. An example of a fungus prone to being colonized by the soil bacterium *Paraburkholderia terrae* (producing a thin biofilm) is shown in Figure 9.3. The surfaces offered by fungi to bacteria in their surroundings represent specialized niches, in which nutrients and colonizable surfaces become available. The mycosphere-associated bacteria are thought to live off fungal "exudates," which contain compounds such as trehalose, oxalic acid, glycerol, acetic acid, formic acid, and mannitol. For instance, growth on oxalic acid or oxalate (oxalotrophy) is widespread among soil bacteria, and it has been recently linked to the bacterial associates of fungi (Deveau et al. 2018).

However, there are differences across different soil fungi, in that not all fungi appear to be hospitable to soil bacteria. Guennoc et al. (2017) recently found that whereas the hyphae of *Laccaria bicolor* were colonized by many soil bacteria, those of other fungi (belonging to the Ascomycetes) were not. Bacterial biofilm formation was thus modulated by the fungus and its tree host and was found to rely on filaments made up of extracellular DNA.

Bacteria confronted with fungal hyphae in soil can show different responses:

1. They may compete with, and antagonize, the fungal partner, for example, producing antifungal compounds such as HCN, siderophores, antibiotics, lytic enzymes, and/or a suite of volatile compounds (see Figure 9.4 for an example).
2. They may take a direct profit of the presence of the fungus. This may boil down to a nutritional interaction, as the small organic molecules produced by the hyphae (e.g., products from lignocellulose degradation) may be avidly consumed. Some bacteria may even live off consumption of the fungal tissue. For instance, chitinolytic *Collimonas* species isolated from acid dune sand (de Boer et al. 2005) can live off hyphal tips. Myxobacteria, a specialized group of "predatory" bacteria, may also consume soil fungi (as well as bacteria) by collective eating (Muñoz-Dourado et al. 2016). As such, these bacteria show group behavior and are called micropredators.
3. They may modulate fungal behavior in such a way that the fungus will "open" a novel niche for them. This mechanism was recently suggested by Wang et al. (2014), who showed that particular bacteria mobilize nematode-killing fungi to promote the killing effect.

Conversely, in particular cases, fungi in soil have developed systems that counteract bacteria. For instance, they may produce antibacterial compounds, acidify the environment (for instance,

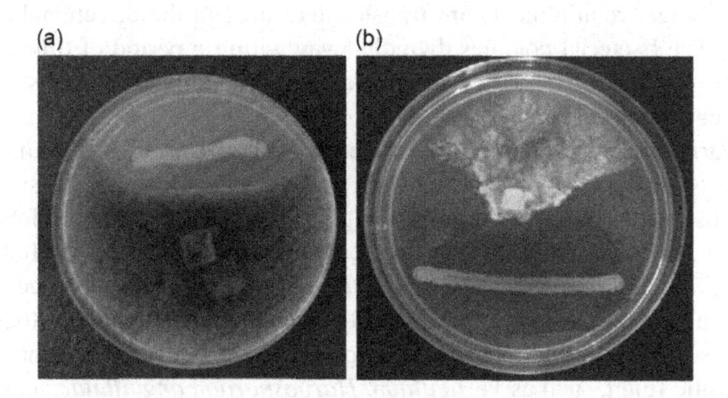

FIGURE 9.4 **(See color insert)** Antagonistic interaction between a soil bacterium and soil fungi revealed on dual-culture plates. Inhibition of (a) *Ceratocystis paradoxa* and (b) *Phytophthora parasitica* by the bacterium *Paraburkholderia seminalis*. (Courtesy of Priscila J.R.O. Gonçalves, University of São Paulo, São Paulo, Brazil.)

by releasing oxalic acid), or produce highly hydrophobic cell wall structures, making bacterial attachment difficult.

Particular bacteria in soil, for example, *Pseudomonas*, *Sphingomonas*, and *Paraburkholderia* species, may possess the metabolic capabilities that result in an amensalistic or even mutualistic interaction with soil fungi (such bacteria are called *mycosphere competent*). The interaction is thought to drive their fitness. A case-in-point is given by *Paraburkholderia terrae*, which has been found to be extremely fine-tuned to interacting with diverse soil fungi. In terms of ecological behavior, several *P. terrae* strains had the capacity to migrate towards, as well as along with, growing fungal hyphae. In the process, several mechanisms were used. Thus, flagellar motility (resulting in chemotaxis), the presence of a type III secretion system (T3SS) and of type-4 pili, in addition to the capture and utilization of fungal-released compounds such as glycerol, acetate, formate, and oxalate, were likely the bacterial capacities underlying the mycosphere competence (Haq et al. 2018). Hence, *P. terrae* may have evolved in soil in interaction with fungal partners and adjusted its genome accordingly (Haq et al. 2014). A hallmark review (de Boer et al. 2005) discusses how soil fungi provide niches in (soil) ecosystems in which bacteria can thrive, and how such fungi potentially select for bacteria.

On the basis of the foregoing evidence, we posit that many fungi in soil provide microhabitats for specific bacteria. The selective effect these microhabitats exert is likely to be both nutritional and spatial. That is, the fungus may provide both specific nutrients and hospitable (colonization) sites to the bacteria. In early work, it was shown that, for instance, pseudomonads specifically take profit of the conditions offered by ectomycorrhizal hyphae, which provide hospitable microniches. A key association is the further endomycotic presence of bacteria. In this association, bacterial cells occur exclusively within the fungal tissue, potentially evolving towards an organelle (Biancotto and Bonfante 2002).

9.5.2 How Fungi "Cope" with Bacteria and Other Organisms in Soil

Particular fungi have evolved metabolic capabilities to take advantage of the bacteria with which they share the habitat. For instance, the soil fungi *Pleurotus ostreatus*, *Lepista nuda*, *Agaricus brunnescens*, and *Coprinus quadrifidus* attack soil bacteria (e.g., *Pseudomonas* and *Agrobacterium*), using these as nutrient sources. The fungi sense the bacterial colonies (by a chemoattractive mechanism) in their vicinity and initiate specialized directional hyphae that grow towards these. The fungi then secrete compounds in advance of the directional hyphae that lyse the bacterial cells. After penetrating the bacterial colony, a coralloid mass of assimilative hyphae is formed, and there is no further proliferation of hyphae in the vicinity of it. Nutrients in excess of the immediate fungal requirements are translocated through the directional hyphae for utilization elsewhere. The bacterial colonies dwindle away within a period of time as short as 24 h. This process gives the fungus a considerable advantage in obtaining nutrients, especially under nutrient-limiting conditions in soil.

Agaricus bisporus, the cultivated mushroom, can mineralize dead *B. subtilis* cells, using the cellular components as sole sources of N and C for growth. This illustrates the significant role that some bacterial cells may have in providing nutrients for particular soil fungi. Moreover, numerous fungal species can attack nematodes or their eggs. Such nematode-destroying fungi, for example, *Stylopage hadra*, *Dactylella gephyropaga*, and *Arthrobotrys oligospora*, produce extensive hyphal systems. At intervals along the hyphae, trapping devices that catch and hold the live nematodes are produced. These are then penetrated by the hyphae and the body contents metabolized or consumed. Endoparasitic fungi, such as *Verticillium*, *Harposporium anguillulae*, and *Meria* sp., have no extensive hyphal development outside of the body of the host. Instead, conidiophores and conidia attach to the nematodes, germinate, penetrate the cuticle, and form an infection hypha in the body cavity. Over several days, the body of the nematode is filled with fungal hyphae, some of which break out. A commonly studied soil nematode that is destroyed by fungi is *Panagrellus redivivus*.

9.5.3 Mutualistic Interactions and Niche Construction

Next to strategies that allow to compete with fungal counterparts, bacteria may also have developed into organisms that are mutualistic with fungal hosts. Such interactions may increase the fitness of both species in the environment. This coexistence may shape the specific microenvironment in an event-denominated niche construction. In fact, this phenomenon may range from elaborate and adaptive extracellular mechanisms that yield a better environment, enhancing nutrient supply and shelter to the inevitable environmental degradation, which may be associated with microbial succession, due to the consumption of limiting resources (McNally and Brown 2015).

The rice seedling blight fungus *Rhizopus microsporus* harbors the bacterium *Paraburkholderia rhizoxinica* in its hyphae, which stimulates vegetative sporulation (Partida-Martinez et al. 2007). The bacterium was found to produce a phytotoxin as well as a virulence factor (Partida-Martinez and Hertweck 2005), indicating that it changes fungal physiology, which allows both partners to occupy the specific ecological niche in rice plants. For the formation of this endosymbiosis, the T3SS of the bacterial endosymbiont was required, which allowed to release chitinolytic enzymes (chitinase, chitosanase) and chitin-binding proteins, allowing the active invasion of the fungus by the bacteria (Moebius et al. 2014). Similarly, Nazir et al. (2014) found that *Paraburkholderia terrae* BS001 formed a tight association with the fungus *Lyophyllum* sp. strain Karsten. In this association, it blocked primordium setting (thus prolonging existence of the fungus as environment-exploring hyphae) and provided a layer around the fungal hyphae that protected these against the deleterious effects of toxic compounds such as cycloheximide in soil.

9.5.4 Molecular Recognition Mechanisms in Interactions between Bacteria and Eukaryotes in Soil

Deveau et al. (2018) recently reviewed the interactions between bacteria and fungi in a range of environmental settings. A suite of diverse mechanisms has evolved in partners of the associations. What is basic in all these interactions is that small molecules, of different nature, play diverse roles. Such molecules are secreted by the bacteria and perceived by the partner organism, upon which a physiological response follows. Thus, adhesive or signaling molecules are expressed on bacterial surfaces that promote interactions with the counterpart organism or with soluble macromolecules. So-called microbial-associated molecular patterns (MAMPs) or pathogen-associated molecular patterns are molecules that allow to increase fitness during interaction with eukaryote cells, regulating the interactions with different host species. These MAMPs are highly conserved molecules or structures present in microorganisms that are recognized via direct interaction with a host receptor. In the host cells, modifications of molecules are induced by the activity of the microbial effector (defined as microbe-induced molecular patterns) (Mackey and McFall 2006).

MAMPs are recognized by so-called pattern recognition receptors, a family of transmembrane receptors equipped with an extracellular leucine-rich repeat (LRR). They elicit an innate immune response via a mitogen-activated protein kinase and elicitation of phytohormone signaling in a Ca^{2+}-dependent manner (Newton and Dixit 2012, Vadassery et al. 2009). Bacterial flagellin protein constitutes a typical MAMP; it has a 22-amino acid peptide (fgl22) that is recognized by the FLS2 receptor in plant cells (Chinchilla et al. 2006). Peptidoglycan constitutes another MAMP that interacts with NOD-like receptors, thereby triggering, in both plant and animals, innate immune responses as well as physiological shifts (Philpott et al. 2014, Wheeler et al. 2014).

In soil, fungal hyphae—when confronted with bacteria—likely respond accordingly, allowing the interaction or triggering a defense response. For this, fungi developed a mechanism for the recognition of these bacterial species, regulating defense and metabolite secretion. As yet, little is known about bacterial MAMPs and their interaction with soil fungi, which could trigger not only the immunity response but also metabolism related to competitiveness in the environment. In recent work, Ipcho et al. (2016) evaluated the transcriptome of *Fusarium graminearum* in the presence of peptidoglycan, flagellin, and lipo-oligosaccharides (LOSs; lipopolysaccharide without the O-antigen), each having a different molecular structure. Although such MAMPs did not induce detectable effects on fungal morphology, mitochondrial activity rapidly increased in their presence. In addition, as a response to the MAMPs, the fungus upregulated genes related to efflux pumps, iron metabolism, and siderophore uptake, allowing cell detoxification and strong iron uptake from the soil, respectively. This, in turn, could reduce the capability of the bacteria to colonize the environment and compete with the fungus. Furthermore, an overlap of gene expression profiles related to both fungal growth in planta and fungal growth in the presence of bacteria was observed, suggesting that the mechanisms for fungal pathogenicity on plants may also be used for fungal defense against bacteria.

9.6 PREDATORY INTERACTIONS

Predation is a widespread phenomenon among soil microorganisms. In fact, the "cheapest" substrates for soil microbes are present in other—edible—organisms. Some microorganisms have developed the capability to degrade other organisms, for which **engulfment** and **lysis** are major mechanisms. We will here briefly discuss three examples of predatory relationships that have a role in soil: predation by protozoa, *Bdellovibrio*, and myxobacteria.

9.6.1 Protozoa as Predators of Bacteria

Protozoa and their roles in soil are discussed in Chapter 8. We will here only address their roles in the interaction with selected bacteria in soil. Protozoa, divided into amoebae, flagellates, and ciliates, are unicellular eukaryotes, which move through soil in order to feed by predation. Bacterial cells are found in small colonies that cover a small fraction (0.1%) of the available soil surface area. Protozoa graze only provided water tension is adequate. As we have seen in Chapter 8, free-living (phagotrophic) protozoa exhibit three main different feeding habits: filter, raptorial, and diffusion feeding. Filter feeding involves the filtering of suspended particles such as bacteria. The filter-feeding ciliate *Paramecium caudatum* can ingest 400 bacterial cells per second. Raptorial feeding is the individual ingestion of suspended bacterial cells and cells surface-bound to clays and organic matter by small flagellates and amoebae. Diffusion feeding by some amoeboid protozoa, such as foraminifera and small heliozoans, involves the extension of the pseudopodia to engulf prey, and then the formation of a food vacuole.

The interrelationship between bacteria and their predatory protozoa can be exemplified by studies on *Rhizobium*–protozoa interactions in soil. Rhizobial cell numbers declined as a result of predation by protozoa, but rhizobia were not totally eliminated—while an increase in protozoan density was observed. In sterile soil inoculated with both bacteria and protozoa, protozoa also will not eliminate all the bacterial cells. The inability of the protozoa to remove all bacterial prey is a result of protozoan numbers reaching a population density where the energy expended in finding prey equals that gained from feeding. However, protozoa do not always respond to bacterial cells added to soil. For instance, protozoan numbers do not increase upon the addition of *Arthrobacter globiformis* and/or *Bacillus thuringiensis* cells to soil. *B. mycoides* spores added to soil caused an increase in numbers of selected protozoans. It is not known whether this response is associated with germination of the *Bacillus* spores. Even though *E. coli* is generally not considered to be a common soil microorganism, protozoans increased in response to its addition in a manner similar to that when *B. mycoides* was added.

A long-standing question revolves around the food selectivity exhibited by soil protozoa. Free-living phagotrophic protozoa may select their food by size, not distinguishing between clay particles and bacteria, and ingest both. Filter-feeding ciliates can ingest latex beads (diameters 0.09–5.7 μm) as readily as bacteria and size selection may be a function of the size of the protozoan mouth. If selective feeding behavior by indigenous protozoans is common, it may be possible to tinker with bacterial community structure in soil by manipulation of predation, or vice versa. Bacterial cells in contact with clay particles were found to offer protection from predation by protozoa by increasing the number of protective soil microhabitats available to bacteria. Also, protozoans may ingest clay particles indiscriminately, thereby sparing bacterial cells. This may reduce their food intake and decrease their reproductive potential.

The presence of protozoa in soil increases turnover and mineralization of carbon and nitrogen. Given the central role of protozoa as nutrient cycling enhancers, bacteria–protozoa interactions in soil are of central importance to soil nutrient cycles.

9.6.2 Predatory *Bdellovibrio*

Bdellovibrio constitutes a group of predatory bacteria in soil. Their life cycle comprises two developmental phases, as depicted in Figure 9.5. During the first, free-living phase, cells are small (0.2–0.5 µm in width by 0.5–1.4 µm in length) and motile due to the presence of a single polar flagellum. During the second phase, designated the attack phase, *Bdellovibrio* cells encounter a suitable prey (e.g., another Gram-negative) cell and attach to it ("head-on," i.e., at the nonflagellated pole). They then penetrate the outer membrane and cell wall of the prey, losing their own flagella (Figure 9.5). Once inside the periplasm of the prey cell, *Bdellovibrio* alters the prey cell physiology in such a way that it cannot be attacked by other attack-phase cells. The prey cell becomes a spherical structure referred to as the bdelloplast. The prey cell cytoplasm is a rich source of nutrients. In this phase, denoted the intraperiplasmic growth phase, *Bdellovibrio* elongates into a coiled aseptate filament. As nutrients are depleted, the filament divides into single cells that then develop

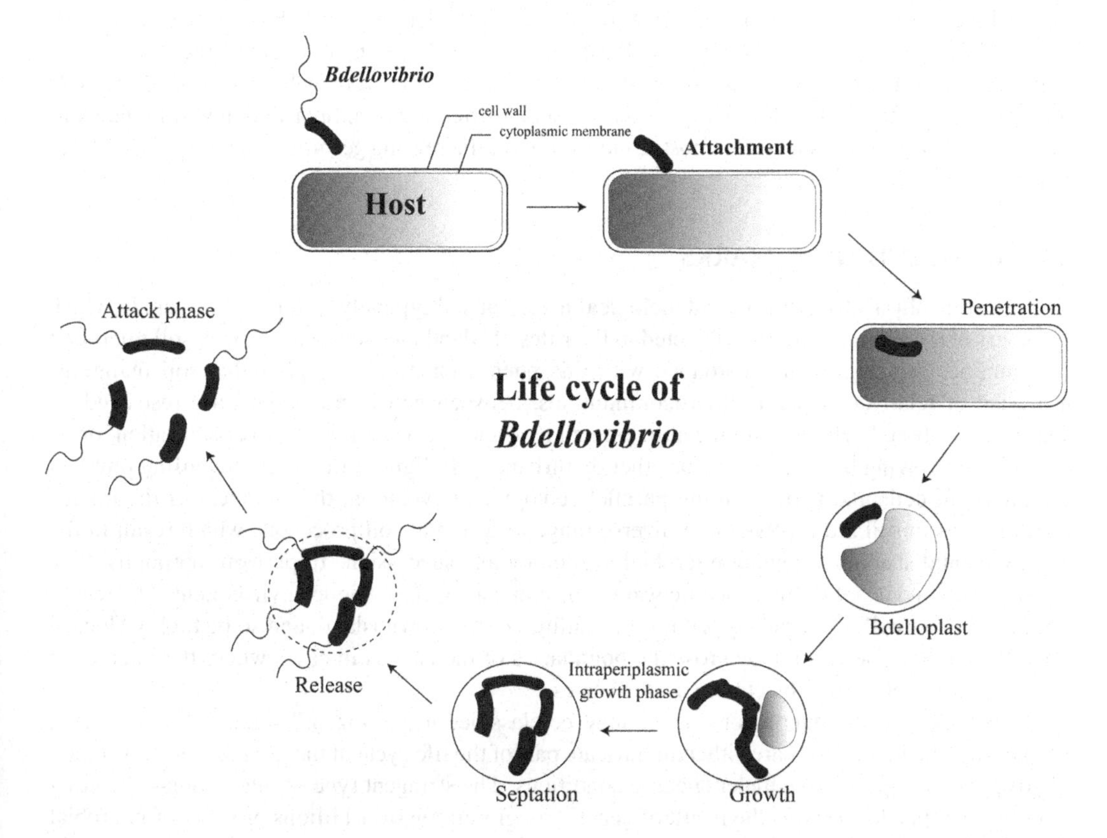

FIGURE 9.5 Life cycle of *Bdellovibrio bacteriovorus* during prey capture. (Figure prepared by Akbar Adjie Pratama, University of Groningen.)

new flagella required for a next attack phase. In this phase (the release phase), the bdelloplast is lysed and the cycle can be repeated. The entire cycle can occur in 4 h at 30°C with *E. coli* as the prey. The exact cycle duration in soil is not known. The ability to attack and utilize other cells for nutrition offers a considerable advantage to *Bdellovibrio* cells in the often nutrient-limited soil environments.

9.6.3 Myxobacteria—Slime Molds

Myxococcus xanthus is the key organism used as a model for the myxobacteria. It is a so-called *social* bacterium that moves and feeds cooperatively in predatory groups. The molecular ecology of its behavior has been recently reviewed by Muñoz-Dorado et al. (2016). In search of a prey, *M. xanthus* vegetative cells move over surfaces in a coordinated manner. The dynamic multicellular groups are referred to as swarms. Within the swarms, there is intense interaction between the cells, where two separate locomotion systems function. So-called adventurous motility drives the movement of individual cells. It is associated with the secretion of slime, forming trails at the leading edge of the swarms. The cellular traffic along these trails may contribute to the social behavior via so-called stigmergic regulation. In contrast, most of the cells travel in groups by using social motility, which is cell contact dependent and requires large cell numbers. Exopolysaccharides and type IV pili at alternate cell poles drive the social motility. When the cell swarms encounter prey, the population lyses and takes up nutrients from nearby cells. This feeding behavior results in the pool of hydrolytic enzymes and other secondary metabolites secreted by the entire group being shared by the community to optimize feeding. This multicellular behavior is especially observed in nutrient-scarce conditions. The *M. xanthus* swarms can organize the gliding movements of numerous rod-shaped cells, synchronizing rippling waves of oscillating cells. They form macroscopic fruiting bodies, with subpopulations of cells showing a division of labor: a small fraction of cells either develops into resistant myxospores or remains as peripheral rods, whereas most of the cells die, providing nutrients to allow aggregation and spore differentiation. Sporulation within the fruiting bodies may promote survival in hostile environments, increasing germination (and growth) rates when conditions turn favorable.

9.7 CONCLUDING REMARKS

The dynamic physical, chemical, and biological nature of soil typically results in a system in which members of the microbiome are subjected to the rules of island biogeography. Hence, soil microbial communities are shaped in accordance with this phenomenon. This implies that soil maintains a myriad of relatively separate microcommunities, between which interactions are restricted by spatial separation. Such separation can be broken by events of water flow or (over)saturation, or by events of burrowing by earthworms or other disturbances. In light of the often-prevailing physical separation of cells and their resulting parallel ecology and evolution, the members of the microcommunities in soil are involved in a diverse range of different soil processes, which result in the environmental shaping, allowing microbial adaptation and succession. To perform optimally, they interact with and adapt to (in a specific way) their biotic as well as abiotic environment. The resulting interactions will occur mostly within the confines of their own island, and so be highly "local." Such interactions, however, may cross the boundaries of the islands in cases where the interactive biota is large, motile, or subjected to perturbances.

With respect to the interactions, these may be classified as "stringent" versus "relaxed." This implies that the interactions are either an intricate part of the life cycle of the microorganisms or facultative and only play a role under specific conditions. The stringent type of interaction—resulting in obligate dependencies—is the result of genetic programming over billions of years of microbial evolution in the context of other interacting species. In contrast, the relaxed one may not be under strong selective force, as it may only confer a fitness advantage when an ecological opportunity arises. Furthermore, the extent to which microbial interactions in soil occur is subjected to the often-changing overall soil environmental conditions, which can override all locally determined effects. For instance, severe drought in a soil system—even in the presence of sufficient amounts of nutrients—has as an immediate consequence the halting of microbial activities and so also the interactions between microbial species. Hence, the expression of the microbial interactive systems will depend on both the overall and the local soil conditions.

On the basis of the foregoing, one may thus posit that it is fairly common for microorganisms that are locally connected in soil to interact with other local microorganisms, either of their own kind (within the population) or of those of other taxa. Genetic systems that enable microorganisms to obtain a fitness (survival) advantage from such interactions have evolved. The QS system that bacteria use for cell-to-cell communication within the population is currently known as the paradigm of such signaling systems, and there is increasing insight that this system is also used between populations of different kinds. The system induces a range of physiological responses in microorganisms that benefit survival, including the production of antibiotics.

In the Darwinian struggle for survival with the outcome being organismal growth and division, microbial interactions may be antagonistic or mutualistic/cooperative. The outcome of particular interactions in soil may even be variable in dependence of the local conditions. Antibiotics and bacteriocins have been hypothesized to constitute agents that allow microorganisms to antagonize other microorganisms. These agents mediate the antagonisms with broad- or narrow-spectrum activities, respectively. It is possible that these antagonistic interactions primarily play a role at the microhabitat level. On the other hand, mutualism/metabolic cooperation/syntrophy are promoted in a spatially structured environment like soil, and in close proximity to cells.

REFERENCES

Bach, E.M., Williams, R.J., Hargreaves, S.K., Yang, F. and K.S. Hofmockel. 2018. Greatest soil microbial diversity found in micro-habitats. *Soil Biol Biochem* 118, 217–226.

Biancotto, V. and P. Bonfante. 2002. Arbuscular mycorrhizal fungi: a specialised niche for rhizospheric and endocellular bacteria. *Antonie Van Leeuwenhoek* 81, 365–371.

Chinchilla, D., Bauer, Z., Regenass, M., Boller, T. and G. Felix. 2006. The *Arabidopsis* receptor Kinase FLS2 binds flg22 and determines the specificity of Flagellin perception. *Plant Cell* 18, 465–476.

Christensen, B.B., Haagensen, J.A.J., Heydorn, A. and S. Molin. 2002. Metabolic commensalisms and competition in a two-species microbial consortium. *Appl Environ Microbiol* 68, 2495–2502.

Cronin, D., Moënne-Loccoz, Y., Fenton, A., Dunne, C., Dowling, D.N. and F. O'Gara. 1997. Ecological interaction of a biocontrol *Pseudomonas fluorescens* strain producing 2,4-diacetylphloroglucinol with the soft rot potato pathogen *Erwinia carotovora* subsp. *atroseptica*. *FEMS Microbiol Ecol* 23, 95–106.

Davinic, M., Fultz, L.M., Acosta-Martinez, V., et al. 2012. Pyrosequencing and mid-infrared spectroscopy reveal distinct aggregate stratification of soil bacterial communities and organic matter composition. *Soil Biol Biochem* 46, 63–72.

de Boer, W., Folman, L.B., Summerbell, L.C. and L. Boddy. 2005. Living in a fungal world: impact of fungi on soil bacterial niche development. *FEMS Microbiol Rev* 29, 795–811.

Deveau, A., Bonito, G., Uehling, J., et al. 2018. Bacterial – fungal Interactions: ecology, mechanisms and challenges. *FEMS Microbiol Rev* 42, 335–352.

Dong, Y.-H., Xu, J.-L., Li, X.-Z. and L.-H. Zhang. 2000. AiiA, an enzyme that inactivates the acylhomoserine lactone quorum-sensing signal and attenuates the virulence of *Erwinia carotovora*. *Proc Natl Acad Sci USA* 97, 3526–3531.

Douglas, A.E. 2004. Strategies in antagonistic and cooperative interactions. In: Miller, R., Day, M. (eds.), *Microbial Evolution*, ASM Press, Washington, DC, pp. 275–289.

Garbeva, P., Tyc, O., Mitja, N., et al. 2011. No apparent costs for facultative antibiotic production by the soil bacterium *Pseudomonas fluorescens* Pf0-1. *Plos One* 6, e27266.

Guennoc, C.M., Rose, C., Labbe, J. and A. Deveau. 2017. Bacterial biofilm formation on soil fungi: a widespread ability under controls. *bioRxiv* 2017, 130740.

Hansen, L.B.S., Ren, D., Burmølle, M. and S.J. Sørensen. 2017. Distinct gene expression profile of *Xanthomonas retroflexus* engaged in synergistic multispecies biofilm formation. *ISME J* 11, 300–303.

Haq, I.U. Graupner, K., Nazir, R. and J.D. van Elsas. 2014. The genome of the fungal-interactive soil bacterium *Burkholderia terrae* BS001 – A plethora of outstanding interactive capabilities unveiled. *Genome Biol Evol* 6, 1652–1668.

Haq, I.U., Dini-Andreote, F. and van Elsas, J.D. 2017. Transcriptional responses of the bacterium *Burkholderia terrae* BS001 to the fungal host *Lyophyllum* sp. strain Karsten under soil-mimicking conditions. *Microb Ecol* 73, 236–252.

Haq, I.U., Zwahlen, R.D., Yang, P. and J.D. van Elsas. 2018. The response of *Paraburkholderia terrae* strains to two soil fungi and the potential role of oxalate. *Front Microbiol* 9, 989.

Human, Z.R., Slippers, B., de Beer, Z.W., Wingfield, M.J. and S.N. Venter. 2017. Antifungal actinomycetes associated with the pine bark beetle, *Orthotomicus erosus* In South Africa. *S Afr J Sci* 113, 1–7.

Ipcho, S., Sundelin, T., Erbs, G., Kistler, H.C. Newman, M.-A. and S. Olsson. 2016. Fungal Innate immunity induced by bacterial microbe-associated molecular patterns (MAMPs). *G3 (Bethesda)* 6, 1585–1595.

Jansson, J.K. and Hofmockel, K.S. 2018. The soil microbiome—from metagenomics to metaphenomics. *Curr Opin Microbiol* 43, 162–168.

Jiang, X., Zerfaß, C., Feng, S., et al. 2018. Impact of spatial organization on a novel auxotrophic interaction among soil microbes. *ISME J* 12, 1443–1456.

Khalid, S., Baccile, J.A., Spraker, J.E., et al. 2018. NRPS-derived isoquinolines and lipopeptides mediate antagonism between plant pathogenic fungi and bacteria. *ACS Chem Biol* 13, 171–179.

Kvas, S., Rahn, J., Engel, K., et al. 2017. Development of a microbial test suite and data integration assessing microbial health of contaminated soil. *J Microbiol Methods* 143, 66–77.

Leiman, P.G., Basler, M., Ramagopal, U.A., et al. 2009. Type VI secretion apparatus and phage tail-associated protein complexes share a common evolutionary origin. *Proc Natl Acad Sci USA* 106, 4154–4159.

Lennon, J.T. and B.K. Lehmkuhl. 2016. A trait-based approach to bacterial biofilms in soil. *Environ Microbiol* 18(8), 2732–2742.

Li, L., Yang, M., Luo, J., et al. 2016. Nematode-trapping fungi and fungus-associated bacteria interactions: the role of bacterial diketopiperazines and biofilms on *Arthrobotrys oligospora* surface in hyphal morphogenesis. *Environ Microbiol* 18, 3827–3839.

Mackey, D. and A.J. McFall. 2006. MAMPs and MIMPs: proposed classifications for inducers of innate immunity. *Mol Microbiol* 61, 1365–1371.

McNally, L. and S.P. Brown. 2015. Building the microbiome in health and disease: niche construction and social conflict in bacteria. *Phil Trans R Soc B* 370, 20140298.

Miller, M.B. and B.I. Bassler. 2001. Quorum sensing in bacteria. *Ann Rev Microbiol* 55, 165–199.

Moebius, N., Üzüm, Z., Dijksterhuis, J., Lackner, G. and C. Hertweck, 2014. Active invasion of bacteria into living fungal cells. *eLife* 3, e03007.

Muñoz-Dorado, J., Marcos-Torres, F.J., García-Bravo, E., Moraleda-Muñoz, A. and J. Pérez. 2016. Myxobacteria: moving, killing, feeding, and surviving together. *Front Microbiol* 26, 781.

Nadell, C.D., Drescher, K., Foster, K.R. 2016. Spatial structure, cooperation and competition in biofilms. *Nat Rev Microbiol* 14, 589–600.

Nazir, R., Tazetdinova, D.R. and J.D. van Elsas. 2014. *Burkholderia terrae* BS001 migrates proficiently with diverse fungal hosts through soil and provides protection from antifungal agents. *Front Microbiol* 5, 598.

Newton, K. and Dixit, V.M. (2012) Signaling in innate immunity and inflammation. *Cold Spring Harb Perspect Biol* 4, a006049.

Nishie, M., Nagao, J. and K. Sonomoto. 2012. Antibacterial peptides "bacteriocins": an overview of their diverse characteristics and applications. *Biocontrol Sci* 17, 1–16.

Odling-Smee, F.J., Laland, K.N. and M.W. Feldman. 1996. Niche construction. *The American Naturalist* 147, 641–648.

Otto, S., Bruni, E.P., Harms, H., et al. 2016. Catch me if you can: dispersal and foraging of *Bdellovibrio bacteriovorus* 109J along mycelia. *ISME J* 11, 386–393.

Pagie, L. and P. Hogeweg. 1998. Colicin diversity: a result of eco-evolutionary dynamics. *J Theor Biol* 196, 251–261.

Papenfort, K. and B.L. Bassler. 2016. Quorum sensing signal–response systems in Gram-negative bacteria. *Nature Rev Microbiol* 14, 576–588.

Partida-Martinez, L.P. and C. Hertweck. 2005. Pathogenic fungus harbours endosymbiotic bacteria for toxin production. *Nature* 437, 884–888.

Partida-Martinez, L.P., Monajembashi, S., Greulich, K.-O. and C. Hertweck. 2007. Endosymbiont-dependent host reproduction maintains bacterial-fungal mutualism. *Curr Biol* 17, 773–777.

Philpott, D.J., Sorbara, M.T., Robertson, S.J., Croitoru, K. and S.E. Girardin. 2014. NOD proteins: regulators of inflammation in health and disease. *Nat Rev Immunol* 14, 9–23.

Razanamalala, K., Razafimbelo, T., Maron, P.-A., et al. 2018. Soil microbial diversity drives the priming effect along climate gradients: a case study in Madagascar. *ISME J* 12, 451–462.

Rillig, M.C., Aguilar-Trigueros, C.A., Bergmann, J., Verbruggen, E., Veresoglou, S.D. and A. Lehmann. 2015. Plant root and mycorrhizal fungal traits for understanding soil aggregation. *New Phytol* 205, 1385–1388.

Seipke, R.F., Barke, J., Brearley, C., Hill, L.,Yu, D.W., Goss, R.J.M. and M.I. Hutchings. 2011. A single streptomyces symbiont makes multiple antifungals to support the fungus farming ant *Acromyrmex octospinosus*. *Plos One* 6, e22028.

Simmons, M., Drescher, K., Nadell, C.D. and V. Bucci. 2018. Phage mobility is a core determinant of phage–bacteria coexistence in biofilms. *ISME J* 12, 532–543.

Sinha, N.N. 2013. Visualizing patterns of gene expression in growing *Bacillus subtilis* biofilms. *Doctoral dissertation*, Harvard University, Boston, USA.

Sun, S., Zhou, L., Jin, K., Jiang, H. and Y.-W. He. 2016. Quorum sensing systems differentially regulate the production of phenazine-1-carboxylic acid in the rhizobacterium *Pseudomonas aeruginosa* PA1201. *Sci Rep* 6, 30352.

Tan, C.H., Koh, K.S., Xie, C. et al. 2015. Community quorum sensing signalling and quenching: microbial granular biofilm assembly, *NPJ Biofilms Microbiomes* 1, 15006.

Trevors, J.T. 2010. One gram of soil: a microbial biochemical gene library. *Antonie van Leeuwenhoek* 97, 99–106.

Vadassery, J., Ranf, S., Drzewiecki, C., et al. 2009. A cell wall extract from the endophytic fungus *Piriformospora indica* promotes growth of Arabidopsis seedlings and induces intracellular calcium elevation in roots. *Plant J* 59, 193–206.

van Arnam, E.B., Currie, C.R. and J. Clardy. 2018. Defense contracts: molecular protection in insect-microbe symbioses. *Chem Soc Rev* 47, 1638–1651.

Wang, X., Li, G.-H., Zou, C.-G., et al. 2014. Bacteria can mobilize nematode-trapping fungi to kill nematodes. *Nat Commun* 5, 5776.

Wheeler, R., Chevalier, G., Eberl, G. and I.G. Boneca. 2014. The biology of bacterial peptidoglycans and their impact on host immunity and physiology. *Cell Microbiol* 16, 1014–1023.

Whiteley, M., Diggle, S.P. and E.P. Greenberg. 2017. Progress in and promise of bacterial quorum sensing research. *Nature* 551, 313–320.

Xie, X., Liu, N., Ping, J. Zhang, Q., Zheng, X. and J. Liu. 2018. Illumina MiSeq sequencing reveals microbial community in HA process for dyeing wastewater treatment fed with different co-substrates. *Chemosphere* 201, 578–585.

Yang, P. and J.D. van Elsas. 2018. Mechanisms and ecological implications of the movement of bacteria in soil. *Appl Soil Ecol* 129, 112–120.

Zhang, L., Guo, Z., Gao, H., et al. 2016a. Interaction of *Pseudostellaria heterophylla* with quorum sensing and quorum quenching bacteria mediated by root exudates in a consecutive monoculture system. *J Microbiol Biotechnol* 26, 2159–2170.

Zhang, L., Xu. M., Liu, Y., Zhang, F., Hodge, A. and G. Feng. 2016b. Carbon and phosphorus exchange may enable cooperation between an arbuscular mycorrhizal fungus and a phosphate-solubilizing bacterium. *New Phytol* 210, 1022–1032.

10 Plant-Associated Bacteria and the Rhizosphere

Abdul Samad, Günter Brader, Nikolaus Pfaffenbichler, and Angela Sessitsch
AIT Austrian Institute of Technology

CONTENTS

10.1 Introduction 163
10.2 Niche Differentiation of Bacterial Communities at the Soil–Root Interface 164
 10.2.1 Non-vegetated Soil 164
 10.2.2 The Rhizosphere 166
  10.2.2.1 The Rhizosheath 166
  10.2.2.2 The Rhizoplane 167
  10.2.2.3 Root Border Cells and Mucilage 168
 10.2.3 Root Endosphere 168
10.3 Techniques to Separate Root-Associated Bacterial Communities 169
10.4 Host Factors Shaping Root Microbiota 169
 10.4.1 Root-Mediated Spatial Patterns of Bacterial Communities 170
  10.4.1.1 Root-Mediated Radial Patterns of Bacterial Communities 170
  10.4.1.2 Root-Mediated Lateral Patterns of Bacterial Communities 170
 10.4.2 Root Exudates 171
 10.4.3 The Plant Immune System 171
10.5 Bacterial Traits Needed for Root Colonization 172
 10.5.1 Rhizosphere and Rhizoplane Competence 172
 10.5.2 Root Endophyte Competence 172
10.6 Root Core Microbiome 173
10.7 Interactions between Plant-Associated Bacteria and Their Hosts 174
 10.7.1 Beneficial Bacteria 174
 10.7.2 Pathogenic Bacteria 174
 10.7.3 Commensal Bacteria 175
10.8 Conclusions 175
References 175

10.1 INTRODUCTION

Plants interact with diverse soil microorganisms providing niches for growth, and in particular, plant roots represent hot spots for multitrophic interactions. The system comprising the plant, its associated microbiome, and other organisms is termed the plant *holobiont*. In this system, all partners, primarily the plant and its associated microbiota, interact and contribute to the overall holobiont fitness and stability (Vandenkoornhuyse et al. 2015). The microbial partners provide important traits to the host plant, such as traits for nutrient acquisition, stress tolerance, and disease resistance. At the same time, the host plant provides organic carbon to its associated microbiome.

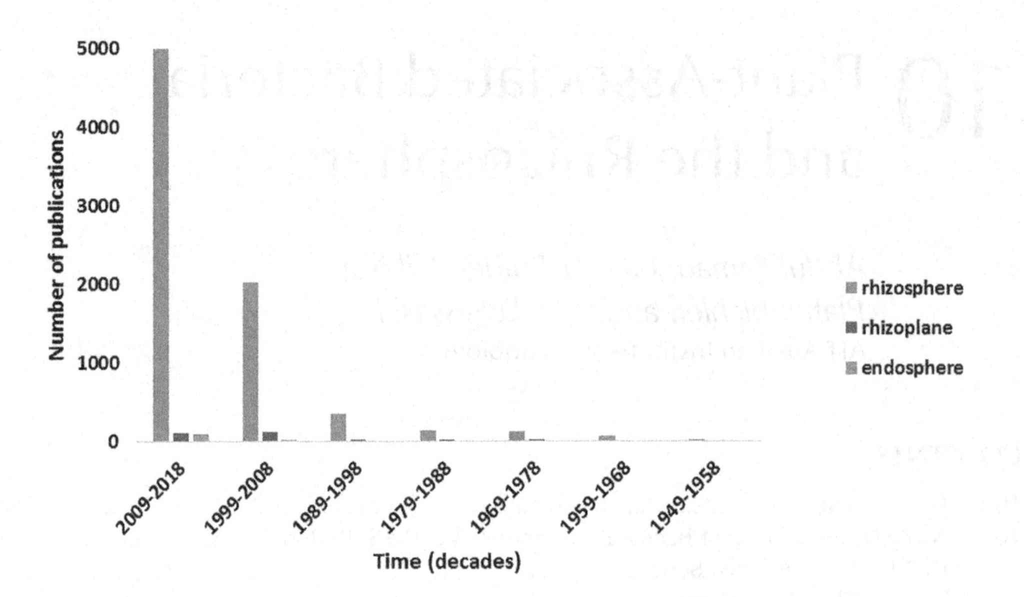

FIGURE 10.1 Number of publications found on PubMed from 1949 to February 28, 2018, featuring the keywords rhizosphere, rhizoplane, and endosphere.

Plants—as sessile organisms—develop different strategies that allow to adapt to changing environmental conditions. This includes the recruitment of beneficial microorganisms.

The soil under the influence of plant roots, that is, the rhizosphere, is considered to constitute a hot spot for plant–soil–microbe interactions. The proper surface of the root is denoted as the rhizoplane, whereas the root interior is called the endosphere. Over the last few years, research into the interactions and processes in the rhizosphere has increased significantly. To illustrate this, 7,708 published manuscripts were found on PubMed featuring the keyword "rhizosphere"; among these, about 65% were published in the last decade (Figure 10.1). The studies on model species, cultivated crops, and wild plants have improved our understanding of microbial ecology and function in the rhizosphere of different ecosystems, notably agroecosystems and natural ecosystems (Finkel et al. 2017, Bulgarelli et al. 2012, Samad et al. 2017).

However, our knowledge about the factors that shape microbial communities at the root–soil interface, their interactions, and the consequences for functioning still remains limited. A better understanding of these factors will help to manage the root microbiota in order to achieve a more sustainable crop production. In this chapter, we present the recently obtained knowledge on the belowground microbial compartments of the plant microbiome, elucidating the differences between the rhizosphere, the rhizoplane, and the endosphere.

10.2 NICHE DIFFERENTIATION OF BACTERIAL COMMUNITIES AT THE SOIL–ROOT INTERFACE

10.2.1 Non-vegetated Soil

An exceptional high number of microbial cells, including bacteria, archaea, fungi, and protists, live in soil. For example, about 10^8 bacterial cells are commonly estimated in 1 g of non-vegetated (bulk) soil (Raynaud and Nunan 2014). These soil microorganisms represent the reservoir of many important functions, including microorganisms available to colonize and interact with plants. Six dominant soil-inhabiting bacterial phyla found by the molecular methods in a wide range of soils are Acidobacteria, Verrucomicrobia, Bacteroidetes, Proteobacteria, Planctomycetes, and Actinobacteria, as shown by Fierer (2017). Moreover, soil pH, organic carbon (quantity and quality), moisture availability, availability of nitrogen and phosphorus, texture, and structure were found to drive these communities (Fierer 2017).

Given the fact that the soil provides the "start inoculum" of plant microbiomes, soil microbiome composition is important. Bacteria from this inoculum will colonize the rhizosphere, the rhizoplane, and the endosphere and so affect the plant. For example, a recent study on grapevine-associated microbiota showed that the bacterial communities associated with aerial plant parts (leaves, flowers, and grapes) share a higher proportion of operational taxonomic units (OTUs) with soil bacteriomes than with one another, suggesting a soil origin for the root and the aboveground communities (Zarraonaindia et al. 2015). On the other hand, there is growing evidence that also vertical transmission routes exist (through seeds, pollen, and vegetative reproduction organs) (Frank et al. 2017).

The type and degree of interactions between microorganisms and plants depend on the intimacy of their association. The root-associated layers rhizosphere, rhizoplane, and root endosphere are considered to be gradually "enriched" with specific microorganisms (Figure 10.2), as a specific microenvironment is provided to inoculum originating from the bulk soil.

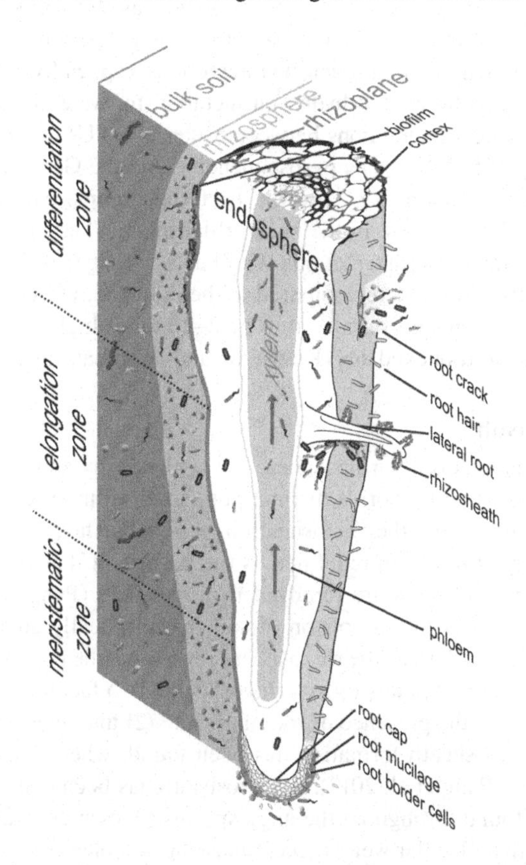

FIGURE 10.2 (See color insert) Schematic illustration of bacterial colonization and distribution in different root microhabitats. Gradual shifts in microbiota composition occur in root-associated habitats from the bulk soil to the root endosphere. Root cracks and sites of lateral roots are among the hot spots of bacterial colonization. Only microorganisms with special traits (e.g., cell wall-degrading enzymes) can enter root tissues. Arrows indicate the translocation of microorganisms, here bacteria, inside the vascular tissue. Root border cells originating from the root cap may be involved in defense against pathogens. The mucilage (shown in light green) secreted from the root and microorganisms leads to the formation of the rhizosheath, which enhances nutrient uptake and water-holding capacity of the root. The bacteria with the ability to form a biofilm and attach to the surface are good colonizers of the rhizoplane.

10.2.2 The Rhizosphere

The term *rhizosphere* was introduced in 1904 by the German scientist Lorenz Hiltner. It is defined as "the narrow layer of soil that is under the direct influence of plant roots" (Hiltner 1904). Hiltner first reported that the plant's nutrition is substantially affected by the composition of its rhizosphere microbial community. This soil plant–interface is considered as a hot spot for microorganisms, which can comprise 10^{10}–10^{12} cells and up to 10^{12} functional genes per gram rhizosphere soil (Prosser 2015). The rhizosphere is strongly influenced by the metabolism of the plant through the secretion of a variety of compounds in the root exudates (such as organic acids, amino acids, fatty acids, phenolics, plant growth regulators, nucleotides, sugars, organic amines, sterols, and vitamins) and the release of oxygen, which mediate a selective pressure for colonizing microorganisms. Overall, due to the root exudates and the structuring of the soil through the roots as well as the availability of decaying root cells, the rhizosphere represents a habitat that is highly dynamic and different from non-vegetated soil. These differences are also evident in the microbial communities of both compartments at the structural and functional levels (Ofek-Lalzar et al. 2014, Colin et al. 2017). For instance, Yan et al. (2017) investigated the role of bacterial diversity in the recruitment of microorganisms by the rhizosphere using 16S rRNA gene amplicon sequencing as well as shotgun metagenomics, and found different microbial communities with different structures and functions between rhizosphere and bulk soil. Key functions known for plant–microbe interactions, especially related to transporters, the Embden–Meyerhof–Parnas pathway, and hydrogen metabolism, were enriched in the rhizosphere. Moreover, they noticed stronger interactions between bacterial OTUs in the rhizosphere than in the bulk soil. Similarly, using 16S rRNA gene pyrosequencing and the GeoChip approach, Colin et al. (2017) reported a selective enrichment of specific bacterial groups, particularly Burkholderiales, Caulobacterales, and Rhodobacterales, in the beech rhizosphere. They found genes related to sulfur cycling, metal homeostasis, and interactions with bacteriophages (CRISPR-Cas systems) to be significantly enriched in the rhizosphere compared to the bulk soil. Overall, in the rhizosphere, the fraction of active microorganisms is considered to be 2–20 times higher in the rhizosphere than in the bulk soil primarily due to root exudates (Kuzyakov and Blagodatskaya 2015).

10.2.2.1 The Rhizosheath

The rhizosheath—a sub-habitat of the rhizosphere—is a physically and chemically enriched compartment in the rhizosphere that harbors a distinct population of microorganisms. In 1887, Georg Volkens defined, for the first time, the rhizosheath as a peculiar layer, composed of agglutinated particles of sand. Recently, it has been redefined as the portion of the soil that is adhering to root hairs and mucilage when plant roots are removed from the field (Pang et al. 2017). It is, thus, a defined compartment that refers to the portion of soil that physically attaches to the root system with the help of the mucilage, whereas the rhizosphere represents the portion of soil under the influence of roots (Table 10.1). Not all plants form a rhizosheath. Two factors are assumed to be critical for rhizosheath formation: (1) the presence of root hairs and (2) mucilage production. A correlation between root hairs and rhizosheath formation has been found, whereas the root hair length does not play an important role (Pang et al. 2017). A rhizosheath has been commonly reported to occur in grasses, and it can be found throughout the angiosperms (Brown et al. 2017). The mucilage is a gelatinous material of high molecular weight, encompassing complex polysaccharides that originate from the root (particularly from root caps and border cells) and its associated microbes. Microbial mucilage may vary in chemical composition as compared to root mucilage due to its higher protein contents and more hydrophilic nature. The mucilage, which arises from root and microorganisms, helps to bind soil particles to root hairs and, thus, substantially contributes to the acquisition of water and nutrients. The size of the rhizosheath has been linked to the acquisition of nutrients and drought tolerance in some grass species (Brown et al. 2017).

The presence of root-associated microorganisms is thought to play an important role in rhizosheath formation and nutrient acquisition through the production of exudates. Rhizosheaths have

TABLE 10.1
List of Different Root Microhabitats with Their Components and Their Generally Accepted Definitions

Rhizocompartment/Component	Definition
Bulk soil	The soil outside the rhizosphere
Rhizosphere	Soil influenced by roots
Rhizoplane	Root epidermis, mucigel, and adhering soil
Rhizosheath	Soil adhering to root hairs and mucilage
Root border cells	Cells detached from the root tips that disperse immediately into suspension after their contact with water
Root hair	Long tubular-shaped outgrowth from root epidermal cells
Cortex	The tissue of a root confined externally by the epidermis and internally by the endodermis
Endodermis	A single layer of cells enclosing the vascular tissue of the root
Mucilage	A high-molecular-weight mixture of various polysaccharides secreted from the root cap and border cells
Root endosphere	Internal root tissue
Rhizodeposits	Root-derived carbon sources released to soil from growing plants

been shown to host arbuscular mycorrhizal fungi, as well as a wide range of bacteria, including members of the genera *Agrobacterium*, *Enterobacter*, *Burkholderia*, *Bacillus*, and *Paenibacillus*. Unno et al. (2005) isolated about 300 bacterial strains from the rhizosheath and rhizoplane of *Lupinus albus* (L.) with the ability to utilize phytate (Na-inositol hexa-phosphate; Na-IHP) as C and P source. Most of the isolates were identified as *Burkholderia*. Numerous rhizosheath isolates were found to utilize phytate as a C source, whereas numerous rhizoplane isolates could mobilize P. Bacterial isolates, particularly those belonging to the genera *Azospirillum*, *Bacillus*, *Arthrobacter*, *Pseudomonas*, and *Enterobacter*, showing plant growth-promoting traits, were isolated from the rhizosheath of wheat. Wheat plants inoculated with *Bacillus* sp. T-34 and *Azospirillum* sp. WS-1 increased the root growth as well as the dry weight of rhizosheath soil as compared to uninoculated control plants (Tahir et al. 2015). Rhizobiales were found to be more abundant in pearl millet inbred lines with higher amount of root-adhering soil dry mass per root tissue dry mass, whereas Bacillales were more prevalent in lines with a low ratio (Ndour et al. 2017). Further understanding of the role of rhizosheaths in plant stress tolerance, as well as on how bacteria mediate rhizosheath formation, will foster future sustainable crop production, especially under conditions of drought.

10.2.2.2 The Rhizoplane

The term *rhizoplane* was coined by Francis E. Clark in 1949; it was defined as the "external root surfaces together with any closely-attached soil particles" (Clark 1949). This root compartment is thought to be colonized by microorganisms, which strongly adhere to the outer layers of the root. It also refers to a combination of the epidermis and rhizosheath, and it is the zone of the root where nutrient and water uptake and root exudation take place (York et al. 2016). The rhizoplane is believed to play a critical "gating" role in regulating microbial entry into the root tissue. Compared to the rhizosphere, this compartment has gained much less attention (Figure 10.1). Due to the sloughing off of dead cells from the growing root surface, the rhizoplane environment is constantly changing. Recent analysis showed that the rhizoplane hosts distinct microbiomes as compared to the rhizosphere and the endosphere (Edwards et al. 2015). This investigation, additionally, showed a gradual decrease in species richness from the bulk soil to the endosphere, in which the rhizoplane can be considered as a transitional zone for microbial life outside and inside the root. Niche- and host-associated functional genes required for root colonization have been observed in the rhizoplane of different crops (wheat and cucumber) (Ofek et al. 2014). Schmidt and Eickhorst (2014) applied catalyzed reporter deposition–fluorescence *in situ* hybridization (CARD-FISH) in a spatiotemporally explicit study of microbes colonizing the rhizoplane of rice. They identified the sites of preferential microbial colonization, such as root tips, root elongation zones, and lateral root cracks, as entry points into the rice roots. Moreover, they found a significantly higher population of methanotrophic bacteria at flowering stage in the rhizoplane. This suggests a stronger effect of root exudates in the rhizoplane as compared to rhizosphere soil. Bulgarelli et al. (2012) further support the hypothesis of recruitment of root endophytes through the rhizoplane by detecting the dominant root endophytic taxa on the surface of roots by using a combination of CARD-FISH and 16S rRNA gene-based microbiome analyses.

10.2.2.3 Root Border Cells and Mucilage

During root growth, border or border-like cells detach from root tips into the rhizosphere, which is considered to be crucial for plant–microbe interactions. Border cells are defined as cells that detach from the root tip as single cells or as small aggregates of cells, whereas border-like cells are detached as sheets of cells that remain attached to each other. Border cell production and detachment can increase in response to pathogen attack, whereas other factors including metals, secondary metabolites, carbon dioxide, and soil type also influence the production and release of border or border-like cells. Border cells have activities that are different from those of root tips, as evidenced by metabolomic and transcriptomic analyses (Watson et al. 2015). In addition to the border cells, root caps are also known to release mucilage, which consists of proteins and extracellular DNA, with antimicrobial properties against certain bacteria and fungi. Mucilage can also influence rhizosphere microbiota by providing distinct carbon sources and stabilizing soil particles. However, there is not much known about the genetic and physiological differences between border and border-like cells and their effects on root–microbe interactions.

10.2.3 Root Endosphere

Internal root tissues are referred to as the root endosphere, and microorganisms that are able to colonize internal root tissues represent microbial endophytes. These endophytes play an important role in plant fitness, growth, and development. The term *endophyte* (Greek word *endo* meaning "within" and *phyton* meaning "plant") was first introduced by de Bary in 1886; since then, many definitions have been suggested for endophytes, most of them defining endophytes as microorganisms not causing any pathogenicity to their hosts. A recent definition has been proposed by Hardoim et al. (2015), who define endophytes as those microorganisms that live at least part of their lifetime inside plant tissues without concluding on any functional traits. This was considered essential, as function (including pathogenicity) is dependent on a number of parameters such as plant and microbial genotype, as well as the environment and interactions within the holobiont (Brader et al. 2017).

Root endophytes are usually recovered from the surface-sterilized roots and can be visualized inside the root tissues using microscopic techniques (Reinhold-Hurek and Hurek 2011). The entry of endophytes into root tissues often occurs through root cracks or emergence points of lateral roots (Compant et al. 2005). The root endosphere usually hosts a more specific microbial community than the outer root compartments (rhizosphere and rhizoplane). Only a subset of the rhizoplane microflora colonizes the endosphere, as only few microorganisms are able to manipulate plant defense, evade host recognition, and produce enzymes facilitating entry into the root. The composition of the root endophytic microflora is usually significantly different from that of the rhizosphere- or rhizoplane-associated community (Reinhold-Hurek et al. 2015, Beckers et al. 2017, Samad et al. 2017). For example, Xiao et al. (2017) reported substantial differences in the composition of the bacterial community in four different root compartments of legumes (soybean and alfalfa): the nodule endosphere, the root endosphere, the rhizosphere, and the root zone. They concluded that a hierarchical filtration of the microbiota had occurred in these root compartments.

In contrast to the rhizosphere, where soil factors primarily determine microbial community structure, the root endosphere microbiota has been shown to be strongly influenced by host-related factors such as host genotype, developmental stage, immune system, health, and fitness (Kandel et al. 2017). A recent study showed that 18 grass species recruited specific root microbiota in a manner that showed a clear correlation with host phylogeny. However, this correlation was affected when drought stress was applied (Naylor et al. 2017). Unfortunately, the root zone environmental factors that influence colonization of the root interior by bacteria are still largely unexplored.

10.3 TECHNIQUES TO SEPARATE ROOT-ASSOCIATED BACTERIAL COMMUNITIES

As there is no physical boundary between the rhizosphere, the rhizoplane, and the bulk soil, there are different definitions and methods in use regarding the sampling of these compartments. Contrasting rhizosphere effects (strong to weak) have been observed in different studies. These were due to the difficulties associated with rhizosphere sampling methods, which do not precisely separate the rhizo compartments. Microbiomes of the root-associated compartments (rhizosphere, rhizoplane, and endosphere) are commonly separated using the following main steps: (1) the loosely attached bulk soil is removed by manually shaking off the root (leaving about 1 mm layer of soil around the root); (2) rhizosphere soil is detached from the root by shaking, brushing, or washing; (3) then, the rhizoplane microbes are separated from the surface of the root using sonication (30 s at 50–60 Hz); (4) to study the endophytic community, root surface cells are removed by repeated sonication (twice sonication after removing rhizosphere soil) or by chemical treatment (using ethanol and NaOCl) (Richter-Heitmann et al. 2016, Edwards et al. 2015). However, sonication techniques may not always be effective in fully removing microorganisms from the root surface. Therefore, for the analysis of endophytic microorganisms, a careful application of chemicals is considered to be more effective, as it removes the rhizoplane-attached cells (Richter-Heitmann et al. 2016). CARD-FISH or SYBR Green-based fluorescence microscopy may be applied to assess the efficacy of the separation techniques. An example of the visualization of a root-associated population of specific beneficial bacteria is shown in Figure 10.3.

10.4 HOST FACTORS SHAPING ROOT MICROBIOTA

Root exudates released by the host plant are considered to be critical for shaping the root-associated bacterial community. The profile of root exudates can vary resulting from the factors such as host genotype, root physiology, root morphology, plant immune system, and biotic and abiotic stresses.

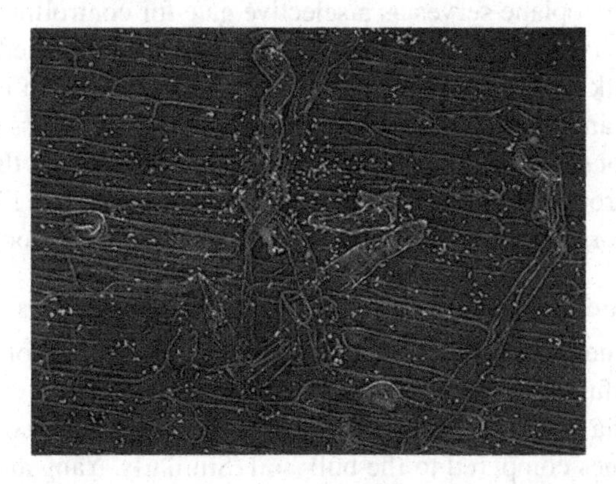

FIGURE 10.3 (See color insert) Confocal microscopy image of a maize root surface gray (maize root) and whitish (beneficial bacterium). (Photograph credit: Stéphane Compant.)

Plant genotype was identified to be a major driver of the root-associated microbiota, especially in similar soils under similar environmental conditions. For instance, Samad et al. (2017) found that substantially different bacterial communities were associated with roots of weeds and grapevine growing side by side under natural conditions. Edward et al. (2015) reported that rice root-associated microbiomes were affected by the plant genotype and soil type under controlled conditions, whereas in natural conditions, cultivation practices (organic versus conventional) and the geographical location represented the main factors of microbial community assemblage. Plants can recruit specific microbial taxa, presumably with specific functions, at certain development stages. For example, Pfeiffer et al. (2017) found bacterial taxa (OTUs) in potato that were enriched at specific developmental stages (emergence, flowering, and senescence). Likewise, Chaparro et al. (2014) identified 81 unique transcripts in *Arabidopsis* that were differentially expressed at different vegetation stages. Moreover, the root-associated microbiomes can be influenced by plant nutritional status, root biomass, and plant diversity (Yang and Crowley 2000).

10.4.1 Root-Mediated Spatial Patterns of Bacterial Communities

Roots growing through soil induce physicochemical changes in the surrounding soil, thus providing a different environment as compared to non-vegetated soil. The root microbiota can be affected by its spatial position to roots in two ways: (1) the radial position of microbes in relation to the root and (2) the lateral orientation of microbes in relation to the root.

10.4.1.1 Root-Mediated Radial Patterns of Bacterial Communities

Plant roots influence the surrounding microbiome by releasing a complex mixture of organic compounds, by modulating oxygen and pH gradients, and by depleting nutrients. Root exudates and mucilage-derived nutrients attract microbial communities in the rhizosphere and rhizoplane environments. A highly specific bacterial community with special traits for root entry and internal establishment colonize the root endosphere. This root-driven microbial selection causes niche differentiation among different root-associated microhabitat inhabitants (Beckers et al. 2017).

Reinhold-Hurek et al. (2015) proposed a three-step enrichment model to describe the root microbiota assembly. According to this model, bulk soil serves as the microbial reservoir and soil physicochemical characteristics and processes mainly shape the soil microbial community. Microbial diversity gradually decreases and the degree of specialization increases when going from the bulk soil to the root endosphere, in which the exchange of nutrients at the soil–root interface plays a key role. Edwards et al. (2015) further supported this three-step model by performing a time series experiment on the spatial resolution of the rhizo compartments of rice. They found substantial differences in the microbiota of the three root compartments, rhizosphere, rhizoplane, and endosphere, and concluded that the rhizoplane serves as a selective gate for controlling bacterial entry into the root tissue. Other authors (van der Heijden and Schlaeppi 2015) speculated that, during "filtration" of microbes from the bulk soil to the root endosphere, gradual shifts in microbiome composition occur due to enrichment and depletion processes. The enrichment process starts in the rhizosphere at a distance from the root surface, whereas depletion seems to initiate at the rhizoplane and occurs more strongly inside the root compartment. The plant immune system is likely to play a major role in the exclusion of microorganisms, starting from the rhizoplane to the root endosphere.

10.4.1.2 Root-Mediated Lateral Patterns of Bacterial Communities

In addition to the aforementioned radial position, the lateral position of microbial communities along the root can also influence their structure. For example, DeAngelis et al. (2009) found a lower richness of Actinobacteria and Bacteroidetes in the rhizosphere soil near the root tip and higher richness in older root zones compared to the bulk soil. Similarly, Yang and Crowley (2000) found distinct bacterial communities in different root zones (primary and secondary root tips, older roots, and sites of lateral root emergence). In another study, differences in microbial community structures were observed between tips and bases of lateral roots, as well as between different types of lateral roots of *Brachypodium distachyon* (Kawasaki et al. 2016). This microbial differentiation along the root has been correlated with spatially distinct exudation patterns of root metabolites or nutrient uptake (Sasse et al. 2018). The full extent of the shaping of root-associated microbiomes by spatial patterns of root exudation and growth remains unclear and may be highly context-dependent (plant type, plant age, and local conditions).

10.4.2 Root Exudates

It is known that compounds in root exudates shape microbial communities in the rhizosphere. Most plants secrete about 5%–30% of their photosynthates into the soil surrounding the roots, and the released compounds can either stimulate or inhibit rhizosphere microbiota. Root exudates thus may contain multiple different molecules, the majority classified as soluble sugars, organic acids, amino acids, fatty acids, proteins, and sterols. Their amount and composition vary depending on many factors, including plant genotype, size, age, and photosynthetic activity, as well as soil conditions (Dennis et al. 2010). For example, 19 *Arabidopsis* accessions showed distinct root exudate profiles, and interestingly, variability in exudate profiles was found to be metabolite class dependent. Glucosinolates showed the highest variation followed by flavonoids and phenylpropanoids (Mönchgesang et al. 2016). The soil nutrient status may also influence the profile of root exudates. For example, under phosphate-deficient conditions, coumarin and oligolignol exudation increased in *Arabidopsis* (Ziegler et al. 2016). Similarly, Zn deficiency and/or salinity stress increased the exudation of a phytosiderophore in wheat roots (Daneshbakhsh et al. 2013). Root exudates can also mediate nutrient cycling in the soil; for instance, root exudates showing either negative or positive effects on soil nitrification have been reported to occur in different crops, including wheat, rice, and sorghum. This ultimately affects the N-use efficiency of these crops (Coskun et al. 2017).

The amount of compounds exuded by roots correlates with the rhizosphere microbial biomass and root diversity (Eisenhauer et al. 2017). Differential exudation is presumably a major factor by which plants modulate their interaction with microorganisms. It has often been reported that compounds in root exudates can act as signaling molecules that initiate the communication between plants and associated bacteria (Sasse et al. 2018).

10.4.3 The Plant Immune System

Plants possess a range of defense mechanisms that control colonization by microorganisms. Plant defenses rely mainly on two types of innate immune systems: (1) microbe-associated molecular pattern (MAMP)-triggered immunity (MTI) and (2) effector-triggered immunity (ETI). As discussed in Chapter 9, MAMPs are conserved bacterial features such as lipopolysaccharides (LPSs) and flagella. Plants recognize microorganisms through pattern-recognition receptors (PRRs) that bind MAMPs and thereby activate MTI to control microbial colonization (Boller and Felix 2009). Recognition of MAMPs leads to the elicitation of systemic acquired resistance (SAR)/induced systemic resistance (ISR) pathways in the plant, and thereby enhances plant resistance against the organisms carrying the MAMPs, often pathogens. Such microorganisms thus induce the salicylic acid (SA)-dependent SAR pathway, whereas others (e.g., rhizobacteria) trigger the jasmonic acid (JA) and ethylene-dependent ISR signaling pathways. Pathogens sometimes can evade MTI using virulence effectors. However, some plants can detect virulence effectors via nucleotide-binding leucine-rich-repeat receptors, which result in the activation of ETI.

Plant roots are surrounded by diverse soil bacterial communities but it is thought they "recruit" many microbial taxa in a specific manner. However, how plants sense and recruit beneficial bacteria in the rhizosphere and within the roots largely remains as yet unknown. Lebeis et al. (2015) proposed that phytohormones such as SA play critical roles in the recruitment of root microbiota. They compared the root microbiota of an *Arabidopsis thaliana*-mutant line that lacks the ability to produce phytohormones (SA, JA, and ethylene) to those associated with the wild-type plant. The mutant line recruited specific bacterial families that differed from those recruited by the wild type. Moreover, they found less diverse endophytic bacterial communities, with altered composition, as compared to the rhizosphere. In another study, the effect of JA on the *Arabidopsis* rhizosphere microbiome was investigated. Mutants (*myc2* and *med25*) with disrupted jasmonate pathways showed different exudation profiles as well as distinct microbiomes compared to the wild type (Carvalhais et al. 2015). The plant defense response can differ between species and even within a species. For example, some *Arabidopsis* accessions supported the growth of *Pseudomonas fluorescens* strains, whereas others inhibited the same strains in their roots (Haney et al. 2015).

10.5 BACTERIAL TRAITS NEEDED FOR ROOT COLONIZATION

On the basis of plant-driven microbial colonization, plant-associated bacteria can be divided into free-living soil bacteria, rhizosphere colonizers, rhizoplane colonizers, and endophytes. The nature of interaction needs specialization of the bacteria themselves, which enable them to survive in the diverse environmental conditions and niches or to convert to different lifestyles (facultative endosymbionts or opportunistic pathogens).

10.5.1 Rhizosphere and Rhizoplane Competence

Plant roots, through the secretion of exudates, attract and recruit microorganisms, some of which supply nutrients and help plants to survive under stressful conditions. Soil bacteria require a set of features for successful rhizosphere colonization that mainly includes (1) microbial ability to catabolize diverse exudate compounds; (2) root sensing using receptors for root exudate compounds; (3) chemotaxis and motility towards the root, and (4) ability to compete with other microbes for root colonization (substrate competition, bacterial growth rate, and antibiotic secretion). Moreover, for efficient rhizoplane colonization, the capacity to form biofilms and attach to surfaces is important. Other bacterial traits known to confer rhizosphere and rhizoplane competence are capacities to produce/perform quorum sensing, flagella, fimbriae, outer membrane protein, type IV pili, agglutinin, site-specific recombinase, amino acid and vitamin B1, and siderophores (Compant et al. 2010).

Ghirardi et al. (2012) studied the colonization of the tomato rhizosphere by 23 fluorescent *Pseudomonas* strains (isolated from different environments). They identified two main traits that determine rhizosphere competitiveness: (1) the ability to utilize various organic compounds as electron donors and acceptors (nitrogen oxides and oxygen) and (2) the ability to produce antibiotic compounds (e.g., phenazines and diphloroglucinol). Also, Lopes et al. (2016) investigated the bacterial traits and processes of rhizosphere colonization in sugarcane and identified polygalacturonase activity as an important trait for rhizosphere colonization, next to pointing to the role of horizontal gene transfer. Using genome-wide mapping of the root colonizer *Pseudomonas simiae*, Cole et al. (2017) identified more than 100 genes encoding traits that are important for competitive colonization of *A. thaliana* roots. Motility, sugar metabolism, and defense-related genes were found to be the most important for root colonization. Some bacterial strains showed plant species-dependent rhizosphere competence, whereas others could colonize the rhizosphere independently of host species and genotype (Schreiter et al. 2018).

10.5.2 Root Endophyte Competence

Several rhizobacteria have special traits that enable them to enter plant roots as well as to establish endophytically. Many of these endophytes have shown positive effects on plant growth (Santoyo et al. 2016). Bacteria enter into plant roots through wounds, lenticels, nodules, root cracks, and stomata (Compant et al. 2010, Santoyo et al. 2016). Root colonization by bacteria most likely starts with the adhesion of cells to the root surface, in which process exopolysaccharides (EPSs) synthesized by bacteria play an important role. For instance, during rice root colonization by *Gluconacetobacter diazotrophicus* Pal5 (a competent endophyte), EPS was identified as an essential factor in biofilm formation and plant colonization (Meneses et al. 2011). Moreover, EPS was recently shown to protect Pal5 cells from oxidative damage; the hydrogen peroxide (H_2O_2) hypersensitivity of EPS-defective mutants decreased with the addition of purified EPS (Meneses et al. 2017).

Metagenomic analysis of a rice root endophytic community revealed important characteristics that are required for plant surface attachment and colonization by endophytes. These predicted features included flagella, cell wall-degrading enzymes, protein secretion systems, iron acquisition and storage, and quorum sensing (Sessitsch et al. 2012). Balsanelli et al. (2010) found that the endophyte *Herbaspirillum seropedicae* required LPS for surface attachment as well as internal colonization of maize tissues. Additionally, the O-antigen portion in the LPS was identified as essential for successful *H. seropedicae* colonization of maize roots (Balsanelli et al. 2013). Using a *pilT* mutant of *Azoarcus* sp. (strain BH72), Böhm et al. (2007) found that twitching motility is required for invasion and endophytic colonization, but not for rhizoplane colonization. Similarly, a strain BH72 mutant lacking flagella was impaired in endophytic colonization (Buschart et al. 2012).

10.6 ROOT CORE MICROBIOME

The root microbiota represents a subset of the soil microbial reservoir and is shaped by host genetics and physiology. A core microbiome may exist, consisting of microorganisms that are associated with a given plant species or genotype grown in different soils. For example, Lundberg et al. (2012) identified the core microbiome of *Arabidopsis* sampled from distinct soils and regions; they found a more diverse core microbiome in the rhizosphere than in the root. Recently, Pfeiffer et al. (2017) found a core potato (*Solanum tuberosum*) microbiome that comprised 40 OTUs, in which *Bradyrhizobium*, *Sphingobium*, and *Microvirga* dominated. Interestingly, core OTUs appeared to be tightly associated with potato independent of soil characteristics, climatic conditions, and plant developmental stage. In grapevine, Zarraonaindia et al. (2015) found core root-associated OTUs belonging to Micrococcaceae, Hyphomicrobiaceae, and Pseudomonadaceae, irrespective of the region and climate in which the crop was grown. Similarly, Edwards et al. (2015) identified a set of 32 OTUs representing the core microbiome of rice roots, sampled from multiple field sites; these microbiomes were characterized by dominating δ-proteobacteria, α-proteobacteria, and Actinobacteria.

Distinct plant species generally host highly distinct microbiomes. However, a number of taxa is shared by different plant species. Recently, Yeoh et al. (2017) identified shared taxa across roots of multiple plant phyla, including angiosperms, gymnosperms, lycopods, and ferns. The core root OTUs represented about 33.2% of the root microbiomes, particularly including the genera *Burkholderia*, *Rhizobium*, *Bradyrhizobium*, and *Azospirillum*. In another study, grapevine and four different weed species shared several taxa in the rhizosphere (145 OTUs) as well as in roots (9 OTUs). The dominant shared OTUs were assigned to *Arthrobacter*, *Flavobacterium*, and *Variovorax*. Strikingly, some bacterial families, notably Pseudomonadaceae and Hyphomicrobiaceae, were invariably found (in eight different studies) in sugarcane, *Arabidopsis*, grapevine, and some wild plants independent of sampling method and growing area (Samad et al. 2017). However, in several studies, core microbiota has been identified at rather high taxonomical levels (order, family, and genus), and so there might still be distinct taxa at the lower taxonomic levels (species or subspecies level).

In addition to the aforementioned taxonomy-based core microbiomes (characterized by taxonomic markers, e.g., the 16S rRNA gene), a new concept that identifies a *functional core microbiome* (based on functional traits) has been proposed (Lemanceau et al. 2017). Functional core microbiomes are defined on the basis of consistency at the functional level, comprising sets of functional genes or traits irrespective of the taxonomy of the microorganisms carrying these. Burke et al. (2011) investigated the structure and function of bacteriomes associated with the green alga *Ulva australis*. They found high similarities in different samples at the functional level (about 70%), but comparatively low similarities at the taxonomic level (15%). Metagenomic analysis of the microbiomes of two different plant species grown in the same soil revealed a core set of functional genes that were linked with root colonization in both species (Ofek-Lalzar et al. 2014).

10.7 INTERACTIONS BETWEEN PLANT-ASSOCIATED BACTERIA AND THEIR HOSTS

Plant–microbe interactions are complex, and the outcome can be positive (e.g., beneficial mutualists, decomposers, pathogen antagonists, or phytohormone producers), negative (pathogens), or neutral (commensal).

10.7.1 Beneficial Bacteria

Positive interactions consist of symbiotic and associative interactions of plants with beneficial microbes, like nitrogen-fixing and/or plant growth-promoting bacteria (PGPB; see Chapter 22). Beneficial microbes promote plant growth either directly by providing nutrients and protecting plants from various abiotic stresses (drought, salinity) or indirectly through protecting the plant from biotic stresses (diseases and infections). Essential nutrients provided by PGPB typically include nitrogen, phosphorus, and iron. PGPB may modulate endogenous hormone levels in association with a plant by synthesizing phytohormones, which usually include auxin, cytokinin, and gibberellin. Additionally, some PGPB can produce the enzyme 1-aminocyclopropane-1-carboxylate (ACC) deaminase, which decreases the level of ethylene in the plant by cleaving ACC, the immediate precursor of ethylene in all higher plants. Several bacteria, through the production of ACC deaminase, have been reported to enhance plant growth, including *Pseudomonas* spp., *Arthrobacter* spp., and *Bacillus* spp. (Gaiero et al. 2013).

Indirect promotion of plant growth occurs when PGPB inhibit the growth and activity of phytopathogens. This usually occurs through the production of antibiotics, lytic enzymes, pathogen-inhibiting volatile compounds, and siderophores. PGPB can also influence host growth by decreasing plant ethylene levels and inducing plant defense reactions (induced systemic resistance).

10.7.2 Pathogenic Bacteria

Bacteria that cause plant damage through infection or production of phytotoxic compounds are called pathogenic bacteria. Compared to bacteria, fungi and nematodes are considered as major plant pathogens. However, a few devastating bacterial pathogens are known. Many of them are assumed to be soilborne, and they penetrate into plant tissue through a wound or natural opening. A well-known example is *Ralstonia solanacearum*, the causal agent of bacterial wilt in tomato. Another example is given by *Agrobacterium tumefaciens*, which causes crown gall in dicotyledonous plants (Mansfield et al. 2012). Some *Streptomyces* spp., such as *Streptomyces acidiscabies*, *S. turgidiscabies*, *S. ipomoeae*, and *S. scabies*, are phytopathogenic and can cause scab-like lesions on tomato and potato. *P. syringae* is comprised of more than 40 different pathovars that can infect various plant species and cause symptoms such as wilt, leaf spots, blights, and cankers. Other known plant-pathogenic bacteria include *Xanthomonas oryzae*, *X. campestris*, *X. axonopodis*, *Erwinia amylovora*, *Xylella fastidiosa*, *Dickeya dadantii*, *D. solani*, and *Pectobacterium carotovorum* (Mansfield et al. 2012). These pathogens manipulate plant host physiology by secreting phytotoxins, proteins, and phytohormones and thus cause disease. Pathogenic streptomycetes share root colonization mechanisms with endophytic streptomycetes such as *Microbispora* and *Streptomyces*. However, unlike endophytic species, pathogenic streptomycetes live inside the plant host for short periods of time and cause observable disease symptoms by secreting phytotoxins, proteins, and phytohormones (Seipke et al. 2012).

10.7.3 Commensal Bacteria

Many bacteria have no direct or obvious effect on plants or plant pathogens. They are referred to as commensal bacteria. However, they may have effects on other bacteria through complex interactions and may have indirect effects on plants or plant pathogens. The rhizosphere is considered to constitute a combat zone where such commensal, next to beneficial, bacteria can influence the outcome of pathogen infections by competing with the latter for space and nutrients. Moreover, certain beneficial microorganisms—such as fluorescent pseudomonads—can adversely affect the growth of soil pathogens by secreting secondary antimicrobial compounds (e.g., phenazines and hydrogen cyanide), lytic enzymes, and/or effectors. On the other hand, pathogenic bacteria harbor a diverse array of defense mechanisms (using efflux transporters) to combat such antagonism.

10.8 CONCLUSIONS

The rhizosphere is a key global environment of microbial activity, with 2–20 times higher cell densities than are encountered in the surrounding bulk soil. Recent studies using advanced molecular technologies suggest that within the rhizosphere, distinct microhabitats host substantially different microbiomes at the structural as well as functional level. Several factors, including soil, environment, host plant genetics, and physiology, shape the root-associated microbiota. Given this complexity of drivers, the composition of the root-associated microbiome needs to be considered in the context of these (context dependence). However, there may be consistency at the taxonomical or functional level. Evidence that a given plant genotype recruits a core microbiome at both these levels, even in different soil and environmental conditions, is increasing. Future studies should focus on improving our understanding of the mechanisms involved in the cross talk between plant roots and their associated microbiota. Furthermore, the key genes in the host plant that control root colonization by microorganisms should be identified. A detailed understanding of such plant–microbe interactions in the rhizosphere will provide new avenues to the application of microorganisms for the improvement of plant growth and health.

REFERENCES

Balsanelli, E., Serrato, R.V., de Baura, V.A., et al. 2010. *Herbaspirillum seropedicae rfbb* and *rfbc* genes are required for maize colonization. *Environ Microbiol* 12, 2233–2244.

Balsanelli, E., Tuleski, T.R., de Baura, V.A., et al. 2013. Maize root lectins mediate the interaction with *Herbaspirillum seropedicae* via N-acetyl glucosamine residues of lipopolysaccharides. *PloS One* 8, e77001.

Beckers, B., Op de Beeck, N., Weyens, N., Boerjan, W. and J. Vangronsveld. 2017. Structural variability and niche differentiation in the rhizosphere and endosphere bacterial microbiome of field-grown poplar trees. *Microbiome* 5, 25.

Böhm, M., Hurek, T. and B. Reinhold-Hurek. 2007. Twitching motility is essential for endophytic rice colonization by the N_2-fixing endophyte *Azoarcus* sp. strain BH72. *Mol Plant Microbe Interact* 20, 526–533.

Boller, T. and G. Felix. 2009. A renaissance of elicitors: perception of microbe-associated molecular patterns and danger signals by pattern-recognition receptors. *Ann Rev Plant Biol* 60, 379–406.

Brader, G., Compant, S., Vescio, K., et al. 2017. Ecology and genomic insights into plant-pathogenic and plant-nonpathogenic endophytes. *Ann Rev Phytopathol* 55, 61–83.

Brown, L.K., George, T.S., Neugebauer, K. and P.J. White. 2017. The rhizosheath – a potential trait for future agricultural sustainability occurs in orders throughout the angiosperms. *Plant Soil* 418, 115–128.

Bulgarelli, D., Rott, M., Schlaeppi, K., et al. 2012. Revealing structure and assembly cues for *Arabidopsis* root-inhabiting bacterial microbiota. *Nature* 488, 91–95.

Burke, C., Steinberg, P., Rusch, D., Kjelleberg, S. and T. Thomas. 2011. Bacterial community assembly based on functional genes rather than species. *Proc Natl Acad Sci USA* 108, 14288–11493.

Buschart, A., Sachs, S., Chen, X., Herglotz, J., Krause, A. and B. Reinhold-Hurek. 2012. Flagella mediate endophytic competence rather than act as MAMPS in rice-*Azoarcus* sp. strain BH72 interactions. *Mol Plant Microbe Interact* 25, 191–199.

Carvalhais, L.C., Dennis, P.G., Badri, D.V., Kidd, B.N., Vivanco, J.M. and P.M. Schenk. 2015. Linking jasmonic acid signaling, root exudates, and rhizosphere microbiomes. *Mol Plant Microbe Interact* 28, 1049–1058.

Chaparro, J.M., Badri, D.V. and J.M. Vivanco. 2014. Rhizosphere microbiome assemblage is affected by plant development. *ISME J* 8, 790–803.

Clark, F.E. 1949. Soil microorganisms and plant roots. In: *Advances in Agronomy*, Vol. 1, A.G. Norman (ed.). New York: Academic Press, pp. 241–288.

Cole, B.J., Feltcher, M.E., Waters, R.J., et al. 2017. Genome-wide identification of bacterial plant colonization genes. *PLoS Biol* 15, e2002860.

Colin, Y., Nicolitch, O., van Nostrand, J.D., Zhou, J.Z., Turpault, M.-P. and S. Uroz. 2017. Taxonomic and functional shifts in the beech rhizosphere microbiome across a natural soil toposequence. *Sci Rep* 7, 9604.

Compant, S., Clément, C. and A. Sessitsch. 2010. Plant growth-promoting bacteria in the rhizo- and endosphere of plants: their role, colonization, mechanisms involved and prospects for utilization. *Soil Biol Biochem* 42, 669–678.

Compant, S., Reiter, B., Sessitsch, A., Nowak, J., Clément, C. and E. Ait Barka. 2005. Endophytic colonization of *Vitis vinifera* L. by plant growth-promoting bacterium *Burkholderia* sp. strain PsJN. *Appl Environ Microbiol* 71, 1685–1693.

Coskun, D., Britto, D.T., Shi, W. and H.J. Kronzucker. 2017. How plant root exudates shape the nitrogen cycle. *Trends Plant Sci* 22, 661–673.

Daneshbakhsh, B., Khoshgoftarmanesh, A.H., Shariatmadari, H. and I. Cakmak. 2013. Phytosiderophore release by wheat genotypes differing in zinc deficiency tolerance grown with Zn-free nutrient solution as affected by salinity. *J Plant Physiol* 170, 41–46.

DeAngelis, K.M., Brodie, E.L., DeSantis, T.Z., Andersen, G.L., Lindow, S.E. and M.K. Firestone. 2009. Selective progressive response of soil microbial community to wild oat roots. *ISME J* 3, 168–178.

Dennis, P.G., Miller, A.J. and P.R. Hirsch. 2010. Are root exudates more important than other sources of rhizodeposits in structuring rhizosphere bacterial communities? *FEMS Microbiol Ecol* 72, 313–327.

Edwards, J., Johnson, C., Santos-Medellín, C. et al. 2015. Structure, variation, and assembly of the root-associated microbiomes of rice. *Proc Natl Acad Sci USA* 112, E911–E920.

Eisenhauer, N., Lanoue, A., Strecker, T. et al. 2017. Root biomass and exudates link plant diversity with soil bacterial and fungal biomass. *Sci Rep* 7, 44641.

Fierer, N. 2017. Embracing the unknown: disentangling the complexities of the soil microbiome. *Nat Rev Microbiol* 15, 579–590.

Finkel, O.M., Castrillo, G., Herrera Paredes, S., Salas González, I. and J.L. Dangl. 2017. Understanding and exploiting plant beneficial microbes. *Curr Opin Plant Biol* 38, 155–163.

Frank, A.C., Saldierna Guzmán, J.P. and J.E. Shay. 2017. Transmission of bacterial endophytes. *Microorganisms* 5. doi:10.3390/microorganisms5040070.

Gaiero, J.R., McCall, C.A., Thompson, K.A. et al. 2013. Inside the root microbiome: bacterial root endophytes and plant growth promotion. *Am J Bot* 100, 1738–1750.

Ghirardi, S., Dessaint, F., Mazurier, S. et al. 2012. Identification of traits shared by rhizosphere-competent strains of fluorescent pseudomonads. *Microb Ecol* 64, 725–737.

Haney, C.H., Samuel, B.S., Bush, J. and F.M. Ausubel. 2015. Associations with rhizosphere bacteria can confer an adaptive advantage to plants. *Nat Plants* 1. doi:10.1038/nplants.2015.51.

Hardoim, P.R., van Overbeek, L.S., Berg, G. et al. 2015. The hidden world within plants: ecological and evolutionary considerations for defining functioning of microbial endophytes. *Microbiol Mol Biol Rev* 79, 293–320.

Hiltner, L. 1904. Über neuere Erfahrungen und Probleme auf dem Gebiete der Bodenbakteriologie und unter besonderer Berücksichtigung von Grund und Brache. *Arbeiten der Deutschen Landwirtschaftlichen Gesellschaft* 98, 59–78.

Kandel, S.L., Joubert, P.M. and S.L. Doty. 2017. Bacterial endophyte colonization and distribution within plants. *Microorganisms* 5. doi:10.3390/microorganisms5040077.

Kawasaki, A., Donn, S., Ryan, P.R. et al. 2016. Microbiome and exudates of the root and rhizosphere of *Brachypodium distachyon*, a model for wheat. *PloS One* 11, e0164533.

Kuzyakov, Y. and E. Blagodatskaya. 2015. Microbial hotspots and hot moments in soil: concept & review. *Soil Biol Biochem* 83, 184–199.

Lebeis, S.L., Paredes, S.H., Lundberg, D.S., et al. 2015. Salicylic acid modulates colonization of the root microbiome by specific bacterial taxa. *Science* 349, 860–864.

Lemanceau, P., Blouin, M., Muller, D. and Y. Moënne-Loccoz. 2017. Let the core microbiota be functional. *Trends Plant Sci* 22, 583–595.

Lopes, L.D., Pereira e Silva, M.C. and F.D. Andreote. 2016. Bacterial abilities and adaptation toward the rhizosphere colonization. *Front Microbiol* 7. doi:10.3389/fmicb.2016.01341.

Lundberg, D.S., Lebeis, S.L., Paredes, S.H., et al. 2012. Defining the core *Arabidopsis thaliana* root microbiome. *Nature* 488, 86–90.

Mansfield, J., Genin, S., Magori, S., et al. 2012. Top 10 plant pathogenic bacteria in molecular plant pathology. *Mol Plant Pathol* 13, 614–629.

Meneses, C., Gonçalves, T., Alquéres, S. et al. 2017. *Gluconacetobacter diazotrophicus* exopolysaccharide protects bacterial cells against oxidative stress in vitro and during rice plant colonization. *Plant Soil* 416, 133–147.

Meneses, C.H.S.G., Rouws, L.F.M., Simoes-Araujo, J.L., Vidal, M.S. and J.I. Baldani. 2011. Exopolysaccharide production is required for biofilm formation and plant colonization by the nitrogen-fixing endophyte *Gluconacetobacter diazotrophicus*. *Mol Plant Microbe Interact* 24, 1448–1458.

Mönchgesang, S., Strehmel, N., Trutschel, D., Westphal, L., Neumann, S. and D. Scheel. 2016. Plant-to-plant variability in root metabolite profiles of 19 *Arabidopsis thaliana* accessions is substance-class-dependent. *Int J Mol Sci* 17.

Naylor, D., DeGraaf, S., Purdom, E. and D. Coleman-Derr. 2017. Drought and host selection influence bacterial community dynamics in the grass root microbiome. *ISME J* 11, 2691.

Ndour, P.M.S., Gueye, M., Barakat, M. et al. 2017. Pearl millet genetic traits shape rhizobacterial diversity and modulate rhizosphere aggregation. *Front Plant Sci* 8, 1288.

Ofek, M., Voronov-Goldman, M., Hadar, Y. and D. Minz. 2014. Host signature effect on plant root-associated microbiomes revealed through analyses of resident vs. active communities. *Environ Microbiol* 16, 2157–2167.

Ofek-Lalzar, M., Sela, N., Goldman-Voronov, M., Green, S.J., Hadar, Y. and D. Minz. 2014. Niche and host-associated functional signatures of the root surface microbiome. *Nat Commun* 5, 4950.

Pang, J., Ryan, M.H., Siddique, K.H.M. and R.J. Simpson. 2017. Unwrapping the rhizosheath. *Plant Soil* 418, 129–139.

Pfeiffer, S., Mitter, B., Oswald, A. et al. 2017. Rhizosphere microbiomes of potato cultivated in the high Andes show stable and dynamic core microbiomes with different responses to plant development. *FEMS Microbiol Ecol* 93. doi:10.1093/femsec/fiw242.

Prosser, J.I. 2015. Dispersing misconceptions and identifying opportunities for the use of 'omics' in soil microbial ecology. *Nat Rev Microbiol* 13, 439–446.

Raynaud, X. and N. Nunan. 2014. Spatial ecology of bacteria at the microscale in soil. *PloS One* 9, e87217.

Reinhold-Hurek, B., Bünger, W., Burbano, C.S., Sabale, M. and T. Hurek. 2015. Roots shaping their microbiome: global hotspots for microbial activity. *Ann Rev Phytopathol* 53, 403–424.

Reinhold-Hurek, B. and T. Hurek. 2011. Living inside plants: bacterial endophytes. *Curr Opin Plant Biol* 14, 435–443.

Richter-Heitmann, T., Eickhorst, T., Knauth, S., Friedrich, M.W. and H. Schmidt. 2016. Evaluation of strategies to separate root-associated microbial communities: a crucial choice in rhizobiome research. *Front Microbiol* 7, 773.

Samad, A., Trognitz, F., Compant, S., Antonielli, L. and A. Sessitsch. 2017. Shared and host-specific microbiome diversity and functioning of grapevine and accompanying weed plants. *Environ Microbiol* 19, 1407–1424.

Santoyo, G., Moreno-Hagelsieb, G., Orozco-Mosqueda, M.C. and B.R. Glick. 2016. Plant growth-promoting bacterial endophytes. *Microbiol Res* 183, 92–99.

Sasse, J., Martinoia, E. and T. Northen. 2018. Feed your friends: do plant exudates shape the root microbiome? *Trends Plant Sci* 23, 25–41.

Schmidt, H. and T. Eickhorst. 2014. Detection and quantification of native microbial populations on soil-grown rice roots by catalyzed reporter deposition-fluorescence in situ hybridization. *FEMS Microbiol Ecol* 87, 390–402.

Schreiter, S., Babin, D., Smalla, K. and R. Grosch. 2018. Rhizosphere competence and biocontrol effect of *Pseudomonas* sp. RU47 independent from plant species and soil type at the field scale. *Front Microbiol* 9, 97.

Seipke, R.F., Kaltenpoth, M. and M.I. Hutchings. 2012. *Streptomyces* as symbionts: an emerging and widespread theme? *FEMS Microbiol Rev* 36, 862–876.

Sessitsch, A., Hardoim, P., Doring, J. et al. 2012. Functional characterization of an endophyte community colonizing rice roots as revealed by metagenomic analysis. *Mol Plant Micr Interact* 25, 28–36.

Tahir, M., Mirza, M.S., Hameed, S., Dimitrov, M.R. and H. Smidt. 2015. Cultivation-based and molecular assessment of bacterial diversity in the rhizosheath of wheat under different crop rotations. *PloS One* 10, e0130030.

Unno, Y., Okubo, K., Wasaki, J., Shinano, T. and M. Osaki. 2005. Plant growth promotion abilities and microscale bacterial dynamics in the rhizosphere of lupin analysed by phytate utilization ability. *Environ Microbiol* 7, 396–404.

van der Heijden, M.G.A. and K. Schlaeppi. 2015. Root surface as a frontier for plant microbiome research. *Proc Natl Acad Sci USA* 112, 2299–2300.

Vandenkoornhuyse, P., Quaiser, A., Duhamel, M., Le Van, A. and A. Dufresne. 2015. The importance of the microbiome of the plant holobiont. *New Phytol* 206, 1196–1206.

Watson, B.S., Bedair, M.F., Urbanczyk-Wochniak, E., et al. 2015. Integrated metabolomics and transcriptomics reveal enhanced specialized metabolism in *Medicago truncatula* root border cells. *Plant Physiol* 167, 1699–1716.

Xiao, X., Chen, W., Zong, L. et al. 2017. Two cultivated legume plants reveal the enrichment process of the microbiome in the rhizocompartments. *Mol Ecol* 26, 1641–1651.

Yan, Y., Kuramae, E.E., de Hollander, M., Klinkhamer, P.G.L. and J.A. van Veen. 2017. Functional traits dominate the diversity-related selection of bacterial communities in the rhizosphere. *ISME J* 11, 56–66.

Yang, C.H. and D.E. Crowley. 2000. Rhizosphere microbial community structure in relation to root location and plant iron nutritional status. *Appl Environ Microbiol* 66, 345–351.

Yeoh, Y.K., Dennis, P.G., Paungfoo-Lonhienne, C. et al. 2017. Evolutionary conservation of a core root microbiome across plant phyla along a tropical soil chronosequence. *Nat Commun* 8, 215.

York, L.M., Carminati, A., Mooney, S.J., Ritz, K. and M.J. Bennett. 2016. The holistic rhizosphere: integrating zones, processes, and semantics in the soil influenced by roots. *J Exp Bot* 67, 3629–3643.

Zarraonaindia, I., Owens, S.M., Weisenhorn, P., et al. 2015. The soil microbiome influences grapevine-associated microbiota. *mBio* 6. doi:10.1128/mBio.02527-14.

Ziegler, J., Schmidt, S., Chutia, R., et al. 2016. Non-targeted profiling of semi-polar metabolites in *Arabidopsis* root exudates uncovers a role for coumarin secretion and lignification during the local response to phosphate limitation. *J Exp Bot* 67, 1421–1432.

11 Microorganisms Cycling Soil Nutrients

Penny R. Hirsch
Rothamsted Research

CONTENTS

11.1 Introduction 179
11.2 The Soil Carbon Cycle 180
11.2.1 Organic Carbon Cycling 181
11.2.2 CO_2 Fixation 181
11.2.3 The Methane Cycle 181
11.3 The Nitrogen Cycle 182
11.3.1 Biological Nitrogen Fixation 184
11.3.2 Nitrification 185
11.3.3 Denitrification 186
11.4 Other Minerals 187
11.4.1 Phosphorus 187
11.4.2 Sulfur 188
11.5 Future Perspectives 189
Acknowledgments 190
References 190

11.1 INTRODUCTION

Microorganisms drive nutrient cycles in soil, and without this key activity, many essential elements would not be available to plants. Conversely, without the input of carbon and energy, primarily from photosynthesis by green plants, soil would consist mainly of mineral particles produced by the weathering of rocks. Residues from plant, animal, and microbial activity provide organic components, making soils fertile and binding together mineral particles into aggregates that, with the associated pore spaces, confer structure to the matrix that supports terrestrial life. In addition to providing a substrate for plants, soil also hosts a complex food web of microorganisms, microfauna, and mesofauna.

Soils host large and diverse microbial communities, which comprise an estimated 10^9 bacterial and archaeal cells per gram belonging to 10^4–10^6 operational taxonomic units (OTUs) per gram in temperate soils. Fungal biomass, as estimated from phospholipid fatty acid biomarkers, is often, but not always, <4% of the bacterial biomass. This is even lower when calculated on the basis of protein or RNA biomarkers, with correspondingly fewer OTUs detected (Dassen et al. 2017). As outlined in Chapter 1, the microscale physicochemical variability of soil creates multiple microenvironments that are at the basis of the multitude of niches required to host this high diversity. The soil microbiome is involved in many nutrient transformations, cycling essential elements between abiotic and biotic pools. The spatial separation of microsites enables parallel evolution of multiple lineages in any soil. In general, microbial transformations of nutrients can be divided into those that are undertaken by many diverse organisms (categorized as "broad" processes) and those that are more specific or "narrow," performed by defined groups of specialists (Schimel and Schaeffer 2012).

In common with all life on Earth, the microbial cells in soil are composed of >96% hydrogen, oxygen, carbon, and nitrogen. The microbial nutrient cycles that support this soil life include the fixation of C and N from the atmosphere and the degradation (or mineralization) of complex molecules containing C, N, and other elements into organic residues. Moreover, the solubilization of soil minerals to provide other essential elements, including the macronutrients P, K, S, Ca, and Mg, and a range of trace element micronutrients (B, Cl, Cu, Fe, Mn, Mo, and Zn), is important. Although the natural weathering of rocks and biological nitrogen fixation (BNF) have supported plant growth in natural ecosystems, agriculture that feeds the growing human population requires extra inputs of N and the other macro- and micronutrients, to compensate for the resources that are removed from the soil by harvesting of crops or grazing by livestock. However, oversupply of N, and also P, can result in environmental damage, including greenhouse gas (GHG) production and eutrophication of water bodies. A better understanding of the microbiology that underpins soil nutrient cycling will aid the management of nutrients and suggest mitigation strategies, leading to more sustainable agricultural practices, as discussed in Section 11.5.

11.2 THE SOIL CARBON CYCLE

The soil food web recycles nutrients such as carbon. Respiration of carbonaceous substances returns C to the atmosphere as CO_2, whereas other gases and solutes are also lost from the soil due to microbial activity. C, N, and the soil macro- and micronutrients listed previously are relocated by the large-scale removal of plant material (roaming herds of herbivores, agricultural practices). They are also released from plant residues in soil by aboveground grazing along with the activity of mesofauna and microorganisms in soil. The major sources, sinks, and outflows for C in soils are described in Figure 11.1.

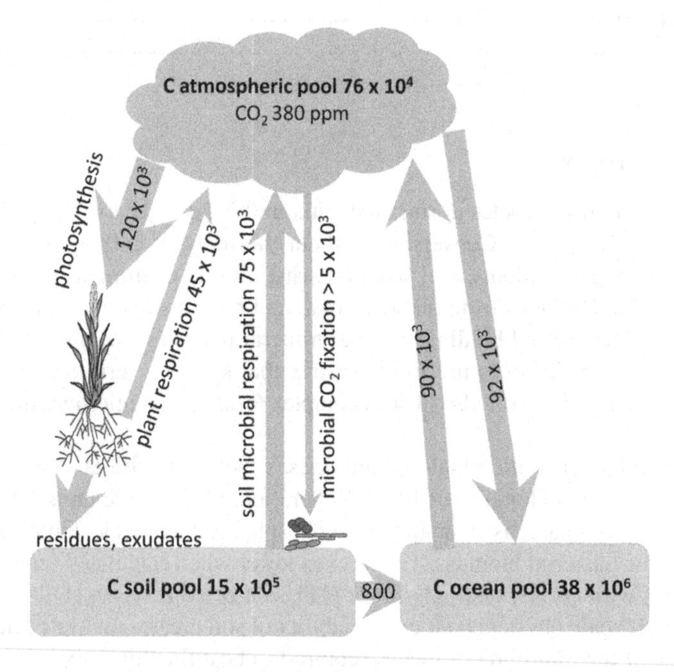

FIGURE 11.1 The global C cycle. Fluxes in Tg (teragram) per year; pools in Tg; cited sources updated from IPCC 2007; only the major C fluxes and pools are shown (Schlesinger and Andrews 2000). Most organic C in soil is derived from atmospheric CO_2, photosynthetically fixed by green plants with a smaller contribution from soil microorganisms (Miltner et al. 2005). Currently, CO_2 respired by soil organisms and plants is balanced by photosynthetic CO_2 fixation, although losses from soil may increase with global warming (Lal 2008). C is also lost by leaching, ultimately to the oceans.

11.2.1 Organic Carbon Cycling

Microorganisms play a major role in breaking down plant and animal residues and are also an important reservoir of organic C in soil. However, soil also holds a large reservoir of C-rich organic matter, which is to some extent degraded and may "condense" over time to form humus, a recalcitrant C reservoir, and other molecules that are physically protected from microbial degradation in the soil matrix and on mineral surfaces (Schimel and Schaeffer 2012). In agricultural systems, inversion tillage (ploughing) makes the organic C more available to soil microorganisms; in undisturbed soils (where anoxia can develop), organic C tends to accumulate.

Although soil organic C recycling is a ubiquitous function, not all members of the soil microbiome may be involved to a similar extent. Instead, particular groups of microorganisms are thought to play key roles, especially those producing extracellular enzymes. Soil fungi are known to have major roles in the decomposition of plant residues containing lignin, cellulose, and hemicellulose, and their hyphal growth enables an efficient colonization of plant litter and woody residues. The role of soil fungi is further discussed in Chapter 5. In addition, many groups of bacteria, including members of the Proteobacteria, Actinomycetes, Firmicutes, and Bacteroidetes, have also been shown to produce enzymes capable of degrading these plant polymers (Lopez-Mondejar et al. 2016, Nyyssonen et al. 2013). The increasing use of metagenomics to analyze soil microbiomes (see Chapters 13 and 14) is providing better insights into the range of taxa and catabolic potential in soil, including the identification, isolation, and cloning of functional genes (Jacquiod et al. 2014, Nyyssonen et al. 2013).

11.2.2 CO_2 Fixation

Microbial activity involving the decomposition of organic C results in the emission of CO_2 from soils. This soil respiration is commonly used as a proxy measure of microbial biomass, and it is estimated that soil microorganisms emit 75×10^3 Tg of CO_2 per year globally (Schlesinger and Andrews 2000). However, some soil bacteria and archaea can fix CO_2 autotrophically—rates up to 5% of respired CO_2 have been reported (Miltner et al. 2005). The Calvin–Benson–Bassham cycle, responsible for photosynthesis in green plants as well as microorganisms, is the predominant pathway for CO_2 incorporation into soil microorganisms. Ribulose-1,5-bisphosphate carboxylase/oxygenase (RubisCO) is the key enzyme responsible for the reduction of CO_2. Although the majority of soil carbon is derived from green plants, surface crusts of Cyanobacteria are important in arid regions and phototrophs belonging to other bacterial phyla, including the Proteobacteria, Chlorobi, Chloroflexi, Firmicutes, Acidobacteria, and Gemmatimonadetes, also contribute fixed C to the soil carbon reservoir. Other soil bacteria carry the RubisCO large-subunit gene *cbbL* and fix CO_2 during autotrophic growth, although C fixation mechanisms are also known to exist in heterotrophic bacteria and archaea (Berg 2011). Carbon monoxide (CO) is present in relatively low concentrations compared to CO_2 in soil. Up to 20% of this CO can be oxidized by Proteobacteria and Actinomycetes that carry the CO dehydrogenase gene *coxL* (Lynn et al. 2017). Quantitative polymerase chain reaction (PCR) on metagenomic soil DNA for *cbbL* and *coxL* indicated that both genes are relatively abundant, with abundances of 0.1–1.3% of the 16S rRNA gene copies present, depending on soil type (Lynn et al. 2017). This indicates that both CO_2 fixation and CO oxidation play important roles in the soil C cycle.

11.2.3 The Methane Cycle

Methane (CH_4) is a major GHG, more potent than CO_2. It is generated in wet, organic C-rich soils during anaerobic processes, including fermentation by bacteria and anaerobic respiration by a specific group of archaea, denoted as methanogenic archaea (Levine et al. 2011). The most common methanogenesis pathway uses CO_2 as a substrate but there are other mechanisms involving methylated compounds or small organic acids such as acetate (Liu and Whitman 2008). Wetlands, including rice paddies, generate around 160 Tg CH_4 per year, which is a major input of methane into the atmosphere.

This is more than the other major methane source, ruminant animals, which produce 100 Tg CH_4 per year. Conversely, methanotrophic Proteobacteria in soil are estimated to re-oxidize 30 Tg of CH_4 per year (Levine et al. 2011). These methanotrophs are specialist α- and γ-Proteobacteria that produce the enzyme methane monooxygenase. The enzyme is also found in some Verrucomicrobia, but these bacteria have not been shown to actively oxidize CH_4, in contrast to the Proteobacteria (Esson et al. 2016). Methanogenesis, on the one hand, and methane oxidation, on the other, are increasingly important, as CH_4 emission rates rise globally, as a consequence of global warming and thawing permafrost. Control measures such as draining rice paddies and managing soil organic matter incorporation into the soil are required but have not yet had great impact (Singh et al. 2010).

11.3 THE NITROGEN CYCLE

Nitrogen is an essential component of living organisms, comprising 1%–4% of all living cells. In the form of N_2, it makes up 78% of Earth's atmosphere. Bacteria and archaea that possess the nitrogenase enzyme can reduce the triple bond in atmospheric N_2 to form ammonia, which can subsequently be assimilated and enter the food web (Dixon and Kahn 2004). No other organisms can access the atmospheric N_2 directly, although it is an essential component of amino acids, nucleic acids, and other molecules such as chlorophyll. Prior to the introduction of man-made fertilizers, the only other natural route for atmospheric N_2 to reach life on the Earth was via lightning, which is estimated to generate 5 Tg per year of nitrous oxides. Soil microorganisms maintain supplies of bioavailable N for plants, whether inputs are from N fixation or from organic fertilizers, and they modulate the N availability from inorganic fertilizers. Excess N, whether from animal excreta or from fertilizer applications, can be lost from the soil by physicochemical routes (volatilization, leaching) or via microbially mediated nitrification and denitrification (Figure 11.2).

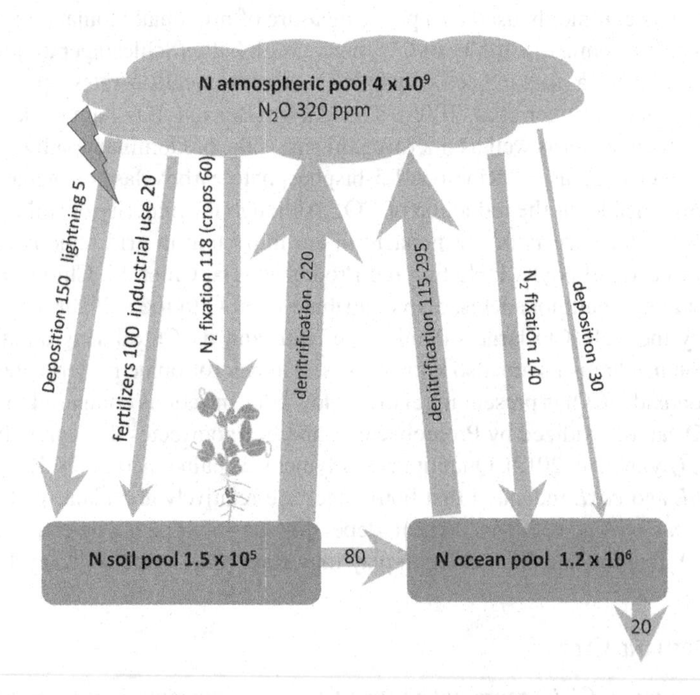

FIGURE 11.2 The global N cycle. Fluxes in Tg per year (Fowler et al. 2013); pools in Tg (Sorai et al. 2007). In addition to biological nitrogen fixation (BNF) in soil by bacteria and archaea, and atmospheric deposition, lightning generates 5 Tg per year nitrous oxides (Shepon et al. 2007). Manures recycle about 90 Tg annually, almost equaling the contribution of N fertilizers manufactured by the Haber–Bosch process (Bouwman et al. 2013). Soil N leaching to water bodies and the oceans causes problems of eutrophication; some of this is eventually buried in sediments.

Atmospheric N_2 enters the N cycle via nitrogen fixation. The microbial N cycle is usually considered to start and finish with N_2 (Figure 11.3). Its major steps, that is, N fixation, nitrification, and denitrification, are described as follows. Nitrite, an intermediate in several reactions in the N cycle, is both reactive and toxic, and microorganisms have a variety of methods for processing it by either oxidation or reduction. Some of the processes shown in Figure 11.3 play minor roles in soil. For instance, during nitrification, hydroxylamine can spontaneously decompose to form N_2O. In addition, dissimilatory nitrate reduction to ammonium (DNRA) by certain bacteria may have a substantial role in environments that are rich in organic carbon with relatively low nitrate, resulting in N conserved as ammonia (Sanford et al. 2012). Various fungi break down organic N to produce nitrate rather than ammonia, a process known as "heterotrophic nitrification." Some bacteria can nitrify ammonia to nitrate and subsequently reduce it to N_2, notably the α-Proteobacterium *Paracoccus denitrificans* (Crossman et al. 1997).

The mineralization of organic N by fungi and heterotrophic bacteria is a "broad" function, which is an important part of the food web, especially in natural ecosystems. As discussed in Chapter 5, the root-colonizing arbuscular mycorrhizal fungi (AMF) form an intimate mutualistic symbiotic association with many plants. They are best known as improving phosphate nutrition of their host (for further details, see Section 11.4.1), but they also transfer N from soil organic matter, nitrate, and ammonia to their host plants. AMF are not host specific and the hyphal network can transfer N between different plant species. The conversion of organic forms of N to ammonia by microbial ammonification, "mineralization," and urease activity are broad functions that are found in many microorganisms, whereas DNRA and the anaerobic oxidation of ammonia in a reaction with nitrite (ANAMMOX) are confined to specific bacteria. ANAMMOX occurs in anoxic conditions and is carried out by as yet uncultured Planctomycetes (Strous et al. 2006). In anaerobic sludges and

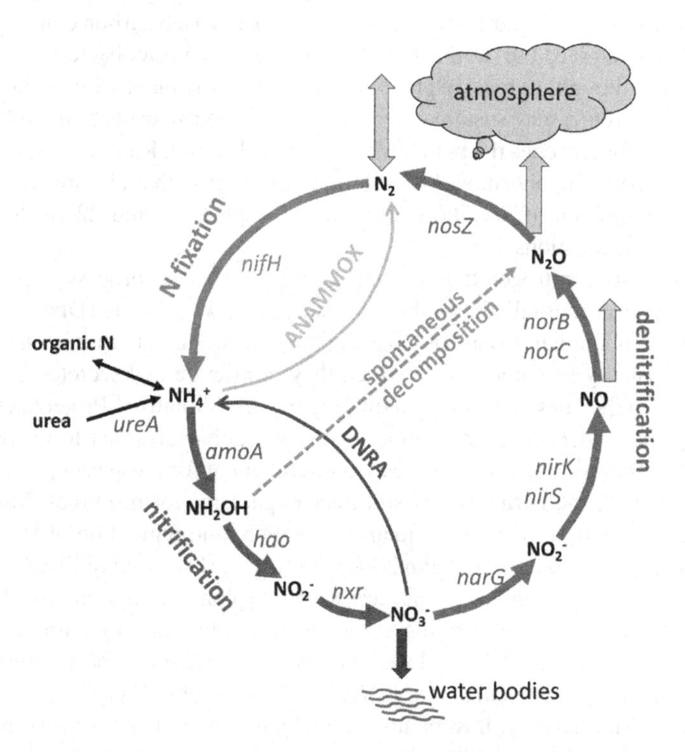

FIGURE 11.3 (See color insert) The biological N cycle. Broad gray arrows indicate gasses taken up or evolved; other arrows show the processes of N_2 fixation, nitrification (oxidation of ammonia—NH_3 to nitrite—NO_2^- via hydroxylamine—NH_2OH), and denitrification (reduction of nitrate—NO_3^- to nitrous oxide—N_2O and N_2 via nitrite and nitric oxide—NO). The majority of ammonia in soil is present as the ammonium ion—NH_4^+.

marine systems, this process plays an important role in returning N_2 to the atmosphere. It may cause significant N fertilizer losses in paddy fields; however, its relevance for aerobic soil and plant nutrition remains to be demonstrated.

Finally, the recently described comammox bacteria (Daims et al. 2015) perform all the steps in nitrification, in contrast to the two separate steps undertaken by either ammonia oxidizers or nitrite oxidizers, as classically recognized (further discussed in Section 11.3.2). The possibility of the existence of comammox bacteria had been predicted before the actual discovery of comammox *Nitrospira* sp. in 2015. However, before this finding, only bacteria and archaea that oxidize ammonia to nitrite, and different groups of bacteria that subsequently oxidize nitrite to nitrate, had been identified.

11.3.1 Biological Nitrogen Fixation

Plants are unable to fix atmospheric N_2. Only the nitrogenase enzyme that occurs in particular bacteria and archaea can reduce N_2 to NH_3 (Dixon and Kahn 2004). The gene for this key enzyme, *nifH*, is found across members of the dominant bacterial phyla in soil: the α-, β-, and/or γ-Proteobacteria (Gaby and Buckley 2011). However, although *nifH* and other *nif* genes have been found in members of many bacterial and archaeal phyla, functional N fixation (diazotrophy) has not been demonstrated in all of these. Currently, it is not known how many different microorganisms effectively fix nitrogen in our soils. The process is of key importance in soils (that are often nitrogen limited), as the end result of the energy-intensive fixation process, ammonia, is incorporated into the soil organic matter pool. This process, releasing nitrogenous products, is carried out by many organisms.

Photosynthetic cyanobacteria are the predominant diazotrophs in many aquatic systems (Gaby and Buckley 2011) and also play roles in agricultural soils (Zhalnina et al. 2013). They are important constituents of the photosynthetic crusts that form on bare soil. Other free-living diazotrophs especially in rhizosphere soil benefit indirectly (by receiving energy-rich carbon compounds) from plant photosynthesis. For example, the well-known *Azotobacter* (γ-Proteobacteria) and *Azospirillum* (α-Proteobacteria) are thought to receive photosynthates from the plant while being active as nitrogen fixers. These two groups may sustain other members of the rhizosphere microbiome. However, the amount of fixed N that reaches the plant directly may be limited. Rather, most host plant benefits are thought to arise from phytohormonal stimulation of root growth and improved stress tolerance conferred by the microbiome (Carvalho et al. 2014). Chapters 10 and 22 further examine such bacterium–host plant interactions.

As indicated previously, nitrogen fixation is an energy-dependent process. It is often performed by endophytic bacteria in mutualistic symbiotic associations with plants (Dixon and Kahn 2004). There are two major functional groups of diazotrophic endosymbionts that induce their host plant to form structures, usually root nodules, in which they proliferate and secrete significant amounts of N during plant growth. These are rhizobia (mainly α-, occasionally β-Proteobacteria) that nodulate leguminous plants and *Frankia*, a member of the Actinobacteria that forms nodules on some shrubs and trees. The nodules provide a protective microhabitat for these groups, where they do not need to compete with the plethora of other soil microorganisms for resources. Moreover, the system for host plant recognition, root entry, nodule formation, and regulation of N fixation is highly developed and specialized (Dixon and Kahn 2004). As further discussed in Chapter 22, *Rhizobium*, *Bradyrhizobium*, and *Mesorhizobium* species are widely exploited in agriculture. These organisms are applied to crops grown in soils lacking compatible rhizobia, and they are also important for many grassland forage species such as clover. Specific rhizobia increase in numbers in the soil following their application as inoculants (e.g., *Bradyrhizobium* for soybeans, typically applied to large tracts of land). Alternatively, host plants are nodulated by indigenous rhizobial strains, such as found in the cultivation of clovers. Together, the input of N to soils by free-living and endophytic diazotrophic bacteria equals the contribution of N fertilizers (Figure 11.2); around half is due to leguminous crops. The plants nodulated by *Frankia* (e.g., *Casuarina* sp.) often grow on marginal land and are important as pioneer species. Another association can be found in the Cyanobacteria, which are involved in several plant symbioses. For instance, the water fern *Azolla* lives in association with the cyanobacterium *Anabaena*, and the angiosperm *Gunnera* and various cycads with *Nostoc* and with fungi forming lichens.

N fixed by non-nodulating endophytic bacteria, such as *Gluconacetobacter* (α-Proteobacteria), *Herbaspirillum* (β-Proteobacteria), and *Enterobacter* (γ-Proteobacteria), which inhabit plant intercellular spaces and xylem, is not usually released until cell death. With the exception of sugarcane, there are still few reliable reports of plants receiving detectable levels of N from diazotrophic endophytes (Carvalho et al. 2014).

11.3.2 Nitrification

Ammonia (NH_3), the product of BNF, may leak from root nodules, and it is also released from organic matter, including plant residues and animal wastes, by many different groups of microorganisms. In addition, ammonium nitrate (NH_4NO_3) is a common fertilizer of agricultural land in Europe, ammonium (NH_4^+), being more surface bound and slower acting than nitrate (NO_3^-). The latter molecule, being less bound to soil particles, is more mobile than ammonia in soil. This increases the probability of contact with plant roots but also of losses by leaching. Excess NO_3^- is known to cause eutrophication of water bodies. In cases where urea (NH_2CONH_2) is the most used N fertilizer (in much of the world and increasingly in Europe), it is rapidly hydrolyzed to NH_4^+ by the many soil microorganisms that possess the enzyme urease. Oxidation of NH_3 to nitrite (NO_2^-) via hydroxylamine (NH_2OH) is performed by specific groups of autotrophic bacteria and archaea, namely, ammonia-oxidizing bacteria (AOB) and ammonia-oxidizing archaea (AOA) (Prosser and Nichol 2012). Further oxidation of nitrite to nitrate, as illustrated in Figure 11.4, is performed by a few groups of nitrite-oxidizing bacteria (NOB). These include *Nitrobacter* (α-Proteobacteria), *Nitrolancea hollandica* (Chloroflexi), and a deep-branching phylum, Nitrospirae. The recently discovered comammox Nitrospirae that perform complete ammonia oxidation from NH_3 to NO_3^- have not been shown to be active in soil to date.

The relative contribution of the different NOB groups to nitrite oxidation in soil is not known, but the process does not seem to be a rate-limiting step for nitrification. Evidence for the latter statement comes from the fact that nitrite is never observed to accumulate in ammonia-fertilized soils unless specific inhibitors are added. In contrast, the activity of ammonia oxidizers appears to regulate the N cycle in agricultural soils. Thus, nitrification inhibitors that reduce AOB activity are increasingly applied together with fertilizers and manures (Monteny et al. 2006) to limit N losses.

Although ammonia monooxygenase (AMO) catalyzes the oxidation of ammonia to hydroxylamine in both AOA and AOB, the enzymes and gene sequences are distinct and the corresponding *amo* genes can be readily differentiated by diagnostic PCR primers and in metagenomic DNA sequence analyses. Spurring a competing process, nitrite reductase genes are present in both AOA and AOB;

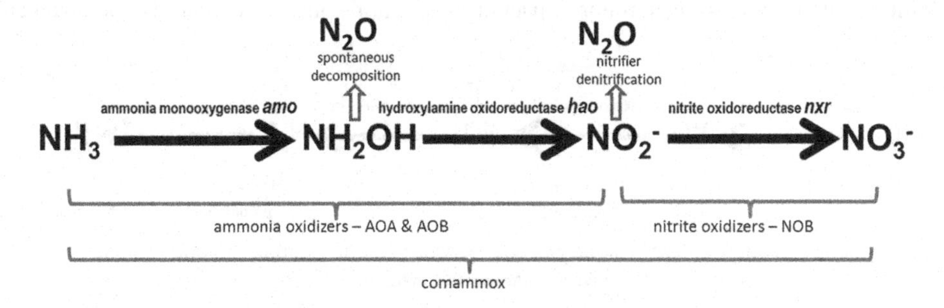

FIGURE 11.4 Nitrification. The solid arrows indicate the process carried out by the different nitrifier groups; the relevant gene names encoding the key enzymes for each step are shown in italics; hollow arrows indicate the points where the GHG nitrous oxide may be produced.

these have been shown to produce the GHG nitrous oxide by "nitrifier denitrification" in aerobic conditions. The second stage of ammonia oxidation, the oxidation of hydroxylamine to nitrite, is performed by hydroxylamine oxidoreductase (HAO) in AOB. No gene comparable to *hao* has been detected in the AOA, and hence, it is assumed that AOA use an alternative mechanism (Walker et al. 2010). In most soils, AOA are reported to be more abundant than AOB (Leininger et al. 2006), but there is conflicting evidence with respect to the relative contribution of each group to nitrification. Therefore, it is still unknown whether or not AOB and AOA are functionally interchangeable.

However, AOA are presumed to be more successful than AOB in low-input unfertilized and forest soils, despite having lower cellular rates of ammonium oxidation as they require less energy and NH_3 to survive (Martens-Habbena et al. 2009, Tourna et al. 2011). In addition, they are also more abundant in (often slightly acidic) agricultural soils (Zhalnina et al. 2013), which may indicate an effect of pH. The equilibrium between NH_3 and NH_4^+ shifts towards NH_4^+ at low pH, limiting substrate availability and growth of AOB, even when N inputs are relatively high. This may explain why AOA are reportedly more abundant in acidic soils (Prosser and Nichol 2012). The most abundant AOB in soil are the β-Proteobacteria *Nitrosospira* and *Nitrosomonas.* These are obligate chemoautotrophs that rely on CO_2 fixation for C and energy. They have been shown to increase in abundance with increasing N fertilizer inputs (Mendum and Hirsch 2002). The first AOA isolated from soil, *Nitrososphaera viennensis*, revealed mixotrophic growth, being able to use pyruvate as a carbon source (Tourna et al. 2011). This was taken to imply that some other abundant soil AOA will have similar properties.

The ability to use organic C and low levels of NH_3 may give AOA an advantage over AOB in soils where N inputs are sporadic. However, it does not predict that AOA would be more active in nitrification. Nitrification inhibitors that act on the AMO enzyme are known to inhibit the activity of AOB and to slow the production of NO_3^-, and there is little evidence that AOA are inhibited, unless much higher inhibitor concentrations are used (Shen et al. 2013).

11.3.3 Denitrification

Denitrification returns most of the N that is deposited on, applied to or fixed in, soil to the atmosphere. It is a major cause of fertilizer N losses from soil and of GHG emissions (Figures 11.2, 11.3, and 11.5). Denitrification can be considered as a "broad" function. At least 5% of all soil bacteria can denitrify (Philippot et al. 2007), using nitrite/nitrate and nitrous oxide as terminal electron acceptors in anaerobic respiration, ultimately producing N_2 to complete the N cycle (Figure 11.3). This process is thought to provide a competitive advantage for those soil bacteria that are facultative anaerobes, as they can switch to denitrification in waterlogged (anoxic) conditions when N and organic matter are abundant. Fungi that contain a bacterial-type nitrite reductase gene in their mitochondria can also denitrify. The less-studied process DNRA (shown in Figure 11.3) has nitrite as an intermediate and N_2O as a by-product. Although it is considered to be important in particular soil conditions, it may have only a minor impact in agricultural soils, as compared to denitrification.

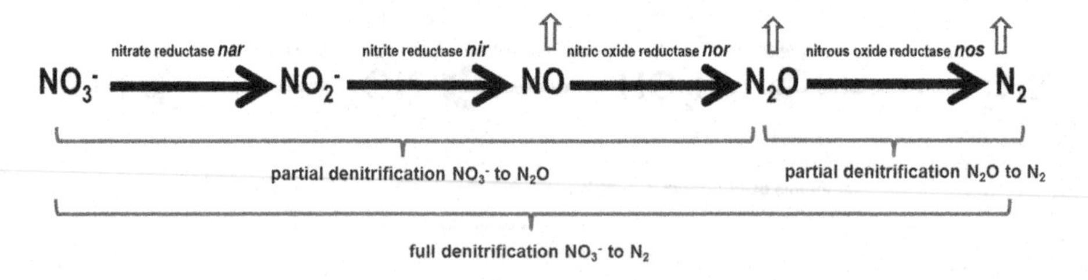

FIGURE 11.5 Denitrification. The solid arrows indicate the steps and the key enzymes involved; the gene names are shown in italics. Hollow arrows indicate gaseous products. The activities of different functional groups are described in the main text.

There are two alternate genes for nitrite reductase, *nirK* and *nirS* (Jones et al. 2008). These occur in closely related species but appear to be incompatible in the same cell, although they are functionally interchangeable (Glockner et al. 1993). The large number of yet-to-be-cultured soil bacteria makes it difficult to predict the actual relative abundance of the alternate *nir* genes, although a survey of environmental bacterial isolates reported that the cytochrome cd_1 variant of nitrite reductase (encoded by *nirS*) was more common than the Cu-dependent *nirK* (Coyne et al. 1989). The factors that influence denitrification, including anoxia and NO_3^- availability, are well known, and the tendency of manures and organic fertilizers to increase denitrification activity compared to mineral fertilizers is well documented. However, how this relates to denitrifier community structure in soil is unclear. Nevertheless, there is evidence that denitrifier distributions are influenced by soil and cultivation conditions and that denitrifier activity is related to *nir* gene abundance (Clark et al. 2012, Hallin et al. 2009).

The final step in the N cycle—returning N_2 to the atmosphere—is the reduction of nitrous oxide by nitrous oxide reductase, encoded by *nosZ* (Figure 11.5). Fewer than 70% of the sequenced denitrifier genomes have been shown to contain *nosZ* (Jones et al. 2008). Even if present, this gene is not always expressed, resulting in emissions of damaging N_2O rather than of innocuous N_2. Recently, it has been reported that some bacteria and archaea contain only the terminal part of the denitrification pathway, with a functional *nosZ* but no *nir* or *nor* genes (Jones et al. 2014, Sanford et al. 2012). This includes the soil bacterium *Anaeromyxobacter* (δ-Proteobacteria), which is responsible for DNRA. In addition to managing drainage, organic matter, and fertilizer inputs in soil to minimize denitrification, maintaining conditions that promote the presence and activity of microorganisms with a functional *nosZ* could be important for reducing GHG emissions.

11.4 OTHER MINERALS

While the atmosphere provides C and N to the soil community, and these, along with other essential elements, are cycled in the soil food web, other key nutrients may be depleted in soil. This refers in particular to P in agricultural systems, where it is removed from the system as crops are harvested (Bouwman et al. 2013). The extent of the availability or depletion of essential elements such as P depends on the underlying soil mineralogy, climatic factors, and previous plant cover and soil management. Fertilizers containing P (in addition to N, K, S, Ca, Mg, and the micronutrients listed in Section 11.1) are usually required to maintain crop yields. Soil microorganisms are involved in the "weathering" of rocks to release soluble elements from the constituent minerals and in solubilizing mineral fertilizers. For example, K is one of the most abundant elements in soil minerals, but it may not be available for plants without the solubilizing activity of soil microorganisms (Etesami et al. 2017).

11.4.1 PHOSPHORUS

Globally, there are limited supplies of mineral rock phosphate available as fertilizer for agricultural land, which may last only another 50 years. This is possibly a problem for intensive agriculture rather than for soil microbial communities, which require much lower concentrations and can access and recycle most of their needs within soil. Although phosphate is incorporated into plants and recycled in manures and residues (Figure 11.6), some is lost to water bodies or removed from the system in plant products (Bouwman et al. 2013). Plants cannot take up organic forms of P. Moreover, recycling organic residues is important in natural ecosystems as well as for sustainable agriculture. Phosphate released from organic residues by microbial activity can become firmly bound to soil minerals (Al and Fe oxides) in acidic soils and Ca-containing minerals in calcareous soils. It is then no longer available to plants and has limited availability to soil microorganisms. This sequestered P is only slowly released by microbial activity, sometimes in conjunction with chemical reactions in soil and acidification by root exudates (Richardson and Simpson 2011). Nevertheless, pollution from excess

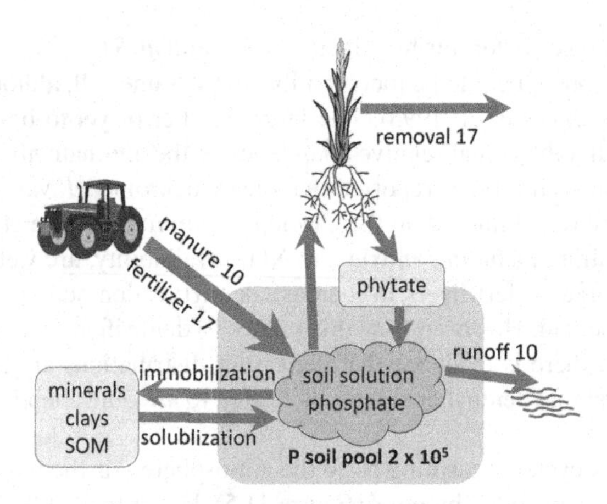

FIGURE 11.6 The P cycle in cropland. Major fluxes in Tg per year and pools in Tg (Yuan et al. 2018). Phosphate P is incorporated into plants and soil microorganisms, recycled in plant residues and manures (phytates and other phosphate polymers in plant material are broken down by microbial activity). P-containing plant residues are removed during harvest; phosphate in solution and particulate forms are lost to water bodies. Phosphate is immobilized on soil minerals and organic matter and released by microbial activity; some is irreversibly bound.

soluble phosphates in animal wastes and fertilizers, and soil particles carrying sequestered P, can cause severe eutrophication in receiving water bodies.

Both plants and microorganisms produce organic anions and phosphatases that solubilize mineral phosphates. Microorganisms can also secrete nucleases, releasing phosphate from nucleic acids, and phytases that degrade relatively stable plant-derived inositol phosphates (phytates). The latter is a major P storage compound of plants and a component of plant residues (Jones 1998). Inoculation of plants and of soil with phosphate-solubilizing bacteria and fungi has been proposed since the 1930s, but the extent to which their application has any significant effects on the P nutrition of plants in field conditions is controversial (Gyaneshwar et al. 2002). In natural ecosystems, there is a long-standing disagreement with respect to the contribution of microorganisms to plant P nutrition, with estimates in the literature ranging from 0% to 90%. There is, however, acceptance of the importance of mycorrhizal fungi to plant P nutrition and health. As discussed in Chapter 5, several groups of filamentous soil fungi form mutualistic symbiotic associations with plant roots in >80% of angiosperm genera. In natural ecosystems, the majority of plants are found to be mycorrhizal with notable exception of members of *Brassica* and related families. However, agricultural cultivation reduces infection levels, in part because tillage breaks up mycelial networks in soil. Also, plants replete in P and N may be less susceptible to colonization by AMF, although there is considerable variation between species.

11.4.2 Sulfur

Sulfur is an essential component of amino acids. It is also a component of many minerals and does not appear to limit microbial growth in soil although some crop plants may face deficiencies. Following the industrial revolution, until the end of the 20th century, considerable sulfur dioxide gas (SO_2) from burning of fossil fuels was deposited on land. When atmospheric SO_2 levels fell, the growth of some crops was found to be limited, and S is now part of many fertilizer formulations. A simplified global S cycle is shown in Figure 11.7.

Specialist anaerobic soil bacteria and archaea from several phyla (including δ-Proteobacteria, Firmicutes, and Nitrospirae) can reduce sulfate (SO_4^{2-}) to sulfide via sulfite (SO_3^{2-}) in a dissimilatory reaction, with simple organic compounds or H_2 acting as electron donors. Typically, it occurs

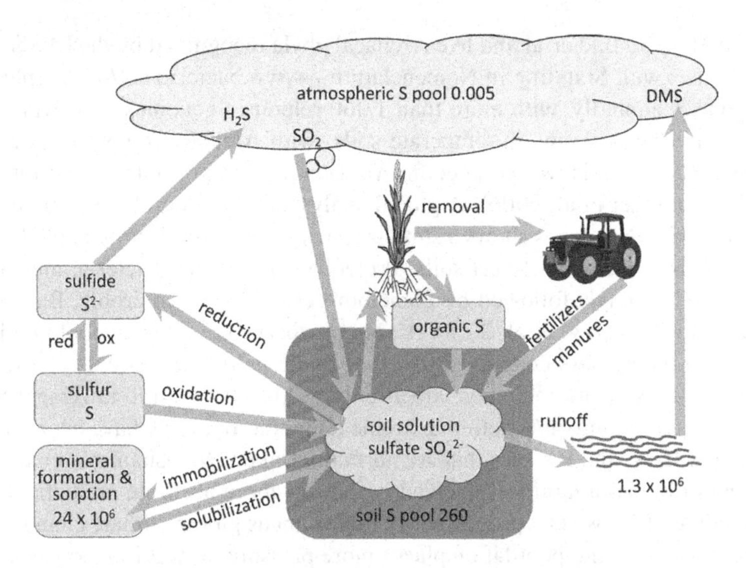

FIGURE 11.7 The global S cycle. Major pools in Tg (Plante 2007) The largest global S reserves are in the lithosphere, released to the soil pool through weathering; some S is returned to minerals through microbially mediated reactions with Fe, Mg, or Ca. Sulfate S is incorporated into plants and soil microorganisms, recycled in plant residues and manures. Organic S, particulates, and sulfates in water bodies and oceans result in a significant proportion of global S emissions due to the action of marine microorganisms generating dimethyl sulfoxide (DMS), estimated at around 25 Tg S per year. The major sources of atmospheric sulfur dioxide (SO_2) are volcanic activity and burning of fossil fuels; the latter was estimated to provide 70 Tg S per year during the peak emission years of the 1970s. During this time, acid rain was a significant problem; subsequently, technological advances have decreased SO_2 emissions significantly. Hydrogen sulfide (H_2S) is generated by microbial reduction of SO_4^{2-} and S, especially in anoxic environments: atmospheric DMS, H_2S, and SO_2 are oxidized to form SO_4^{2-}.

in waterlogged soils (including rice paddies) and sediments and results in the familiar "rotten egg" odor when such soils are disturbed. Sulfide is oxidized to elemental S in soil by anaerobic autotrophic bacteria, which have pigments for photosynthesis: the green sulfur bacteria belonging to the phylum Chlorobi and purple sulfur bacteria from the γ-Proteobacteria. Non-photosynthetic microaerophilic β- and γ-Proteobacteria, *Thiobacillus* and *Beggiatoa*, can oxidize sulfide using simple organic compounds.

Elemental S is oxidized, under aerobic conditions, to sulfate by a range of heterotrophic bacteria, archaea, and fungi in soils, in contrast to extreme systems such as hydrothermal vents, where specialist groups of chemolithotrophs are active (Plante 2007). However, it is likely that more microbial groups involved in S cycling in soil remain to be discovered, possibly related to the more exotic extremophiles that are thought to represent some of the oldest life forms on Earth.

11.5 FUTURE PERSPECTIVES

The modern approaches to studying the living soil now provide information on microbial communities in much greater depth than was available in the past. For example, soil metagenomics describes the genetic composition, transcriptomics reveals which genes are expressed, and proteomics identifies gene products in the soil microbiome. Key developments in soil metagenomics, metatranscriptomics, and metaproteomics are discussed in Chapters 14–16. These powerful techniques should make it easier to understand the synergistic action of organisms with both defined "narrow" functions and those with broad activities that—together—give the soil microbiome its resilience in the face of change.

By 2016, there were 30 Bacterial and five Archaeal phyla recognized by the LPSN website (List of Prokaryotic names with Standing in Nomenclature—www.bacterio.net/-classifphyla.html) and the number increases annually, with more than 1,000 reference genomes now available. Bacteria are the dominant microorganisms in temperate soils, with Archaea comprising <2% of the cell numbers (Bates et al. 2011). However, specific Archaea have important roles in nitrification and methanogenesis (Leininger et al. 2006). A global analysis of soil microbiomes based on the relative abundance of 16S rRNA genes shows a similar distribution of the dominant phylotypes in most sites, the exception being tropical forest soils. Proteobacteria, Actinobacteria, and Acidobacteria were the most abundant phyla followed by Planctomycetes, Verrucomicrobia, Bacteroidetes, and Firmicutes (Delgado-Baquerizo et al. 2018). A core phylotype community could be identified for most situations, and the authors concluded that they could predict the effects of various soil and climatic factors on the soil microbiomes, albeit only at high taxonomic rankings. Some factors could lead to the loss of organism/function: extreme temperature events have been shown to result in the decoupling of C, N, and P cycles that are normally linked (Mooshammer et al. 2017). A better understanding of the communities involved will allow more accurate predictions and suggest possible interventions. This will be progressively important as global climate change coupled with the needs of an ever-increasing population places more pressure on soil functions and agricultural sustainability.

The full range of bacteria, archaea, fungi, protists, microfauna, and mesofauna involved in nutrient cycling in all soil types and conditions may never be fully described. However, it is known that consortia of individual OTUs possessing a range of complementary functions are required and that alternative OTUs and functions may substitute existing ones if conditions change. An example—described in Section 11.3.2—is the dynamics between archaeal and bacteria ammonia oxidizers in soil. Another example, described in Section 11.3.3, is the existence of two alternate nitrate reductases in denitrifying bacteria. Thus, from the large diversity of microorganisms in soil, there is a capacity for the selection of appropriately adapted microbiomes as the environment changes.

Although many soil microorganisms have been proposed as soil inoculants to enhance plant nutrient acquisition, it is often difficult for inoculants to compete with the existing soil microbiome and the results of soil inoculations are often unreliable (Gyaneshwar et al. 2002). This issue is further explored in Chapters 4 and 22. Managing agricultural soils to minimize the adverse effects on the microbiome and to select for beneficial active groups, for example, by optimizing pH, organic matter, and moisture, may be a more efficient way of supporting sustainable agriculture.

ACKNOWLEDGMENTS

Rothamsted Research receives strategic funding from the Biology and Biological Sciences Research Council (BBSRC). The author acknowledges the funding from the BBSRC program BB/E/C/0005196; projects BB/L0258868/1, BB/NO13468/1, and NERC Project NE/MO16978/1 and thanks colleagues Ian Clark, Tim Mauchline and Andrew Neal for help in preparing the chapter.

REFERENCES

Bates, S.T., Berg-Lyons, D., Caporaso, J.G., Walters, W.A., Knight, R. and N. Fierer. 2011. Examining the global distribution of dominant archaeal populations in soil. *ISME J* 5, 908–917.

Berg, I.A. 2011. Ecological aspects of the distribution of different autotrophic CO_2 fixation pathways. *Appl Environ Microbiol* 77, 1925–1936.

Bouwman, L., Goldewijk, K.K., van der Hoek, K.W. et al. 2013. Exploring global changes in nitrogen and phosphorus cycles in agriculture induced by livestock production over the 1900–2050 period. *Proc Natl Acad Sci USA* 110, 20882–20887.

Carvalho, T.L., Balsemao-Pires, E., Saraiva, R.M., Ferreira, P.C. and A.S. Hemerly. 2014. Nitrogen signalling in plant interactions with associative and endophytic diazotrophic bacteria. *J Exp Bot* 65, 5631–5642.

Clark, I.M., Buchkina, N., Jhurreea, D., Goulding, K.W. and P.R. Hirsch. 2012. Impacts of nitrogen application rates on the activity and diversity of denitrifying bacteria in the Broadbalk wheat experiment. *Philos Trans R Soc Lond B Biol Sci* 367, 1235–1244.

Coyne, M.S., Arunakumari, A., Averill, B.A. and J.M. Tiedje. 1989. Immunological identification and distribution of dissimilatory heme cd1 and nonheme copper nitrite reductases in denitrifying bacteria. *Appl Environ Microbiol* 55, 2924–2931.

Crossman, L.C., Moir, J.W., Enticknap, J.J., Richardson, D.J. and S. Spiro. 1997. Heterologous expression of heterotrophic nitrification genes. *Microbiology* 143, 3775–3783.

Daims, H., Lebedeva, E.V., Pjevac, P. et al. 2015. Complete nitrification by Nitrospira bacteria. *Nature* 528, 504–509.

Dassen, S., Cortois, R., Martens, H. et al. 2017. Differential responses of soil bacteria, fungi, archaea and protists to plant species richness and plant functional group identity. *Mol Ecol* 26, 4085–4098.

Delgado-Baquerizo, M., Oliverio, A.M., Brewer, T.E. et al. 2018. A global atlas of the dominant bacteria found in soil. *Science* 359, 320–325.

Dixon, R. and Kahn, D. 2004. Genetic regulation of biological nitrogen fixation. *Nat Rev Microbiol* 2, 621–631.

Esson, K.C., Lin, X.J., Kumaresan, D., Chanton, J.P., Murrell, J.C. and J.E. Kostka. 2016. Alpha- and Gammaproteobacterial methanotrophs codominate the active methane-oxidizing communities in an acidic boreal peat bog. *Appl Environ Microbiol* 82, 2363–2371.

Etesami, H., Emami, S. and H.A. Alikhani. 2017. Potassium solubilizing bacteria (KSB): mechanisms, promotion of plant growth, and future prospects – a review. *J Soil Sci Plant Nut* 17, 897–911.

Fowler, D., Pyle, J.A., Raven, J.A. and M.A. Sutton. 2013. The global nitrogen cycle in the twenty-first century: introduction. *Philos Trans R Soc Lond B Biol Sci* 368, 20130165.

Gaby, J.C. and Buckley, D.H. 2011. A global census of nitrogenase diversity. *Environ Microbiol* 13, 1790–1799.

Glockner, A.B., Jungst, A. and W.G. Zumft. 1993. Copper-containing nitrite reductase from *Pseudomonas aureofaciens* is functional in a mutationally cytochrome Cd1-free background (*nirS*$^-$) of *Pseudomonas stutzeri*. *Arch Microbiol* 160, 18–26.

Gyaneshwar, P., Kumar, G.N., Parekh, L.J. and P.S. Poole. 2002. Role of soil microorganisms in improving P nutrition of plants. *Plant Soil* 245, 83–93.

Hallin, S., Jones, C.M., Schloter, M. and L. Philippot. 2009. Relationship between N-cycling communities and ecosystem functioning in a 50-year-old fertilization experiment. *ISME J* 3, 597–605.

Jacquiod, S., Demaneche, S., Franqueville, L. et al. 2014. Characterization of new bacterial catabolic genes and mobile genetic elements by high throughput genetic screening of a soil metagenomic library. *J Biotechnol* 190, 18–29.

Jones, C.M., Stres, B., Rosenquist, M. and S. Hallin. 2008. Phylogenetic analysis of nitrite, nitric oxide, and nitrous oxide respiratory enzymes reveal a complex evolutionary history for denitrification. *Mol Biol Evol* 25, 1955–1966.

Jones, C.M., Spor, A., Brennan, F.P. et al. 2014. Recently identified microbial guild mediates soil N_2O sink capacity. *Nat Clim Change* 4, 801–805.

Jones, D.L. 1998. Organic acids in the rhizosphere – a critical review. *Plant Soil* 205, 25–44.

Lal, R. 2008. Carbon sequestration. *Philos Trans R Soc Lond B Biol Sci* 363, 815–830.

Leininger, S., Urich, T., Schloter, M. et al. 2006. Archaea predominate among ammonia-oxidizing prokaryotes in soils. *Nature* 442, 806–809.

Levine, U.Y., Teal, T.K., Robertson, G.P. and T.M. Schmidt. 2011. Agriculture's impact on microbial diversity and associated fluxes of carbon dioxide and methane. *ISME J* 5, 1683–1691.

Liu, Y.C. and Whitman, W.B. 2008. Metabolic, phylogenetic, and ecological diversity of the methanogenic archaea. In: *Incredible Anaerobes: From Physiology to Genomics to Fuels*, (eds.) J. Wiegel, R.J. Maier, and M.W.W. Adams, Blackwell on behalf of the New York Academy of Sciences, Boston, MA, 1125, pp. 171–189.

Lopez-Mondejar, R., Zuhlke, D., Becher, D., Riedel, K. and P. Baldrian. 2016. Cellulose and hemicellulose decomposition by forest soil bacteria proceeds by the action of structurally variable enzymatic systems. *Sci Rep* 6, 25279.

Lynn, T.M., Ge, T.D., Yuan, H.Z. et al. 2017. Soil carbon-fixation rates and associated bacterial diversity and abundance in three natural ecosystems. *Microb Ecol* 73, 645–657.

Martens-Habbena, W., Berube, P.M., Urakawa, H., de la Torre, J.R. and D.A. Stahl. 2009. Ammonia oxidation kinetics determine niche separation of nitrifying Archaea and Bacteria. *Nature* 461, 976–979.

Mendum, T.A. and P.R. Hirsch. 2002. Changes in the population structure of beta-group autotrophic ammonia oxidising bacteria in arable soils in response to agricultural practice, *Soil Biol Biochem* 34, 1479–1485.

Miltner, A., Kopinke, F.D., Kindler, R., Selesi, D.E., Hartmann, A. and M. Kastner. 2005. Non-phototrophic CO_2 fixation by soil microorganisms. *Plant Soil* 269, 193–203.

Monteny, G.J., Bannink, A. and D. Chadwick. 2006. Greenhouse gas abatement strategies for animal husbandry. *Agric Ecosyst Environ* 112, 163–170.

Mooshammer, M., Hofhansl, F., Frank, A.H. et al. 2017. Decoupling of microbial carbon, nitrogen, and phosphorus cycling in response to extreme temperature events. *Sci Adv* 3(5), e160278.

Nyyssonen, M., Tran, H.M., Karaoz, U. et al. 2013. Coupled high-throughput functional screening and next generation sequencing for identification of plant polymer decomposing enzymes in metagenomic libraries, *Front Microbiol* 4, 282.

Philippot, L., Hallin, S. and M. Schloter. 2007. Ecology of denitrifying prokaryotes in agricultural soil. *Adv Agron* 96, 249–305.

Plante, A.F. 2007. Soil biogeochemical cycling of inorganic nutrients and metals. In: *Soil Microbiology, Ecology and Biochemistry*, (ed.) E.A. Paul, Academic Press, Oxford, UK, pp. 389–432.

Prosser, J.I. and G.W. Nicol. 2012. Archaeal and bacterial ammonia-oxidisers in soil: the quest for niche specialisation and differentiation. *Trends Microbiol* 20, 523–531.

Richardson, A.E. and R.J. Simpson. 2011. Soil microorganisms mediating phosphorus availability. *Plant Physiol* 156, 989–996.

Sanford, R.A., Wagner, D.D., Wu, Q. et al. 2012. Unexpected nondenitrifier nitrous oxide reductase gene diversity and abundance in soils. *Proc Natl Acad Sci USA* 109, 19709–19714.

Schimel, J.P. and S.M. Schaeffer. 2012. Microbial control over carbon cycling in soil. *Front Microbiol* 3, 348.

Schlesinger, W.H. and J.A. Andrews. 2000. Soil respiration and the global carbon cycle. *Biogeochemistry* 48, 7–20.

Shen, T., Stieglmeier, M., Dai, J., Urich, T. and C. Schleper. 2013. Responses of the terrestrial ammonia-oxidizing archaeon *Ca.* Nitrososphaera viennensis and the ammonia-oxidizing bacterium *Nitrosospira multiformis* to nitrification inhibitors. *FEMS Microbiol Lett* 344, 121–129.

Shepon, A., Gildor, H., Labrador, L.J. et al. 2007. Global reactive nitrogen deposition from lightning NO_x. *J Geophys Res Atmos* 112, D06304.

Singh, B.K., Bardgett, R.D., Smith, P. and D.S. Reay. 2010. Microorganisms and climate change: terrestrial feedbacks and mitigation options. *Nat Rev Microbiol* 8, 779–790.

Sorai, M., Yoshida, N. and M. Ishikawa. 2007. Biogeochemical simulation of nitrous oxide cycle based on the major nitrogen processes. *J Geophys Res Biogeosci* 112, G01006.

Strous, M., Pelletier, E., Mangenot, S. et al. 2006. Deciphering the evolution and metabolism of an anammox bacterium from a community genome. *Nature* 440, 790–794.

Tourna, M., Stieglmeier, M., Spang, A. et al. 2011. *Nitrososphaera viennensis*, an ammonia oxidizing archaeon from soil. *Proc Natl Acad Sci USA* 108, 8420–8425.

Walker, C.B., de la Torre, J.R., Klotz, M.G. et al. 2010. *Nitrosopumilus maritimus* genome reveals unique mechanisms for nitrification and autotrophy in globally distributed marine crenarchaea. *Proc Natl Acad Sci USA* 107, 8818–8823.

Yuan, Z., Jiang, S., Sheng, H. et al. 2018. Human perturbation of the global phosphorus cycle: changes and consequences. *Environ Sci Technol* 52, 2438–2450.

Zhalnina, K., de Quadros, P.D., Gano, K.A. et al. 2013. *Ca.* Nitrososphaera and *Bradyrhizobium* are inversely correlated and related to agricultural practices in long-term field experiments, *Front Microbiol* 4, 104.

Section II

Methods Chapters

12 Methods to Determine Bacterial Abundance, Localization, and General Metabolic Activity in Soil

Lise Bonnichsen, Nanna Bygvraa Svenningsen,
Mette Haubjerg Nicolaisen, and Ole Nybroe
University of Copenhagen

CONTENTS

12.1 Introduction 196
12.2 Detection Methods Based on Extraction of Nucleic Acids 197
12.2.1 Nucleic Acids as Targets for Detection Methods 197
12.2.1.1 Extraction of DNA from Soil 197
12.2.2 Hybridization-Based Detection of DNA Extracted from Soil 199
12.2.3 PCR-Based Detection of DNA Extracted from Soil 199
12.2.3.1 Quantitative PCR 200
12.2.3.2 Digital Droplet PCR 201
12.3 Direct Detection Methods Involving Microscale or Single-Cell Analysis 203
12.3.1 Methods for Extraction of Bacterial Cells from Soil 203
12.3.2 Instrumentation Required for Microscale or Single-Cell Analysis 204
12.3.2.1 Epifluorescence Microscopy (EFM) 204
12.3.2.2 Flow Cytometry 205
12.3.3 Methods for Direct Detection 206
12.3.3.1 Determination of Total Bacterial (Direct) Counts 206
12.3.3.2 Fluorescence *In Situ* Hybridization (FISH) 206
12.3.3.3 Marker Genes for Direct Detection of Specific Bacteria in Soil 207
12.3.3.4 ImmunoFluorescence (IF; Fluorescent Antibody) Methods 207
12.3.4 Determining Bacterial Viability and Activity 208
12.3.4.1 The Concepts of Bacterial Viability and Activity 208
12.3.4.2 Assays Determining Bacterial Viability 209
12.3.4.3 Assays Determining Bacterial Activity: Membrane Integrity and Potential 210
12.4 Detection and Enumeration of Soil Bacteria by Cultivation-Dependent Methods 210
12.5 Conclusions and Outlook 211
References 212

12.1 INTRODUCTION

In this chapter, we present the methods used to detect specific bacteria in the soil environment. This toolbox becomes relevant when

1. The abundance of functional groups of bacteria that contribute to global biogeochemical cycles is to be quantified (see Chapters 11 and 20).
2. The occurrence of bacteria that are threats to human, animal, or plant health is to be monitored (See Chapter 21 and 22).
3. Specific bacterial strains are to be detected in the soil environment.

The latter is relevant in environmental biotechnology where bacterial strains are introduced into the soil or rhizosphere to improve plant health (see Chapter 22) or to degrade pollutants in soil bioremediation (see Chapter 23).

Any detection method needs a target or marker molecule, or trait in or on the organism of interest. An intrinsic marker is a "non-introduced DNA sequence or natural phenotype that serves as a signature for a particular organism or group of organisms." Intrinsic markers could be a specific nucleic acid sequence for polymerase chain reaction (PCR) detection, or a unique cell surface structure that can be detected by immunochemical assays. Alternatively, cells can be tagged with a marker gene defined as "a DNA sequence, introduced into an organism, which confers a distinct genotype or phenotype to enable monitoring in a given environment." The most commonly used marker genes encode autofluorescent proteins or enzymes involved in the emission of bioluminescence.

While *in vitro* detection is standard, the properties of the soil environment represent a challenge to the application of detection methods. Soils are structurally complex, and humic acids as well as clay minerals inhibit enzymatic reactions that are part of molecular detection assays. Further, bacteria in soil typically adhere to the surfaces of minerals and organic matter. These particles can mask the bacterial cells, making direct microscopy observations difficult. As the bacteria are difficult to release from their association with soil particles, even the efficiency of extraction-based methods is crucial. Despite these challenges, continuous methodological developments have enabled an expansion of methods usable for investigating microbial life in the soil habitat.

Here, the initial focus will be on molecular detection methods targeting DNA extracted from soil samples (see Figure 12.1 for an outline). Subsequently, we turn to methods for the direct detection of bacteria at the single-cell or microscale level with or without previous extraction of the cells from the soil, and we address how bacterial cells can be extracted from soils. We emphasize whether a method can be conveniently coupled to assays demonstrating metabolic activity or viability of the target cells. Finally, we deal briefly with cultivation-dependent techniques that remain important for

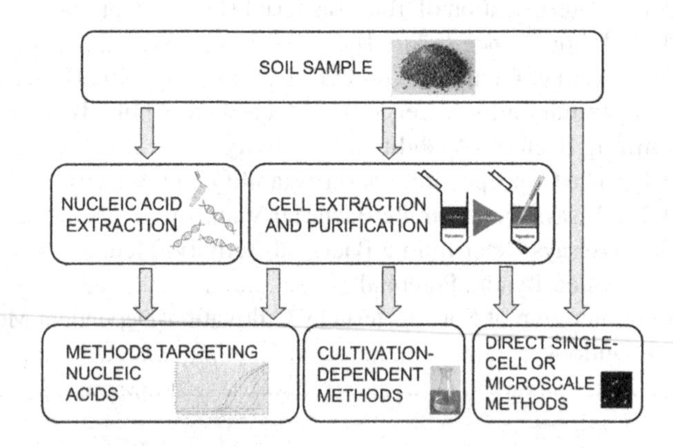

FIGURE 12.1 Conceptual overview of approaches available for the determination of abundance, localization, and general metabolic activity of bacteria in soil.

the detection of soil bacteria. Throughout the text, we briefly present the principles of the selected methods and include examples of their use in the bulk soil or in the rhizosphere/rhizoplane.

12.2 DETECTION METHODS BASED ON EXTRACTION OF NUCLEIC ACIDS

12.2.1 Nucleic Acids as Targets for Detection Methods

Molecular detection methods can target either DNA or RNA, depending on the research objective. DNA primarily provides information on the presence of bacteria harboring a specific target sequence in their genome, whereas the detection of RNA molecules provides a measure of expressed genes (mRNA) and cellular growth and activity (ribosomal RNA), respectively (see Chapter 15). DNA has been the choice for the majority of published nucleic acid-based detection methods applied to soil due to its high stability compared to RNA. Hence, DNA-targeted techniques will be the focus of the subsequent sections. Especially, the gene encoding the ribosomal small subunit RNA (16S rRNA) has been extensively used as a target, as this gene is currently used as the principal taxonomic marker for prokaryotes (see Chapter 3). Detection of functional genes that are exclusively found in specific functional groups of bacteria (e.g., the ammonia monooxygenase gene *amoA* from ammonia-oxidizing bacteria) (Sterngren et al. 2015) may constitute an attractive alternative to the 16S rRNA gene-targeted approach. This is particularly relevant when aiming at coupling DNA-based detection of bacteria to specific biogeochemical processes or other soil functions.

12.2.1.1 Extraction of DNA from Soil

There are two general approaches to consider for DNA extraction from soil: the direct extraction procedure where cells are lysed within the soil matrix and the indirect extraction procedure where cells are extracted from the soil matrix prior to lysis. Direct extraction procedures generally provide the highest DNA yields, and several easy-to-use commercial kits relying on this approach are now available. These provide both high quantity and quality of the extracted DNA for most soil types (e.g., extraction and purification kits from Qbiogene, Aurora, Ohio, and Mo Bio Laboratories, Solana Beach, California). Hence, the direct isolation procedure is currently the method of choice for studies on soil microbiomes (Sterngren et al. 2015). However, contamination, in particular with humic acids, is still a problem for soils with high clay or organic matter contents. The extraction of DNA from these soils may benefit from the indirect procedure. This approach is highly dependent on efficient, unbiased separation of microbial cells from soil aggregates (see Section 12.3.1 for details), but has the advantage of minimizing the co-extraction of enzyme inhibitors such as humic acids, and of providing high-molecular-weight DNA.

The direct and indirect extraction procedures require an efficient cell lysis to release the nucleic acids. Three main protocols for cell lysis are currently applied separately or in combination and depend on (1) enzymatic, (2) chemical, or (3) physical disruption. The most common enzymatic treatment involves digestion with lysozyme, followed by the chemical lysis of cells by the addition of surfactants such as sodium dodecyl sulfate, which dissolve the hydrophobic components of the cell wall. Both enzymatic and chemical cell lyses are often combined with the physical treatment of the sample to improve the lysis efficiency. Incorporation of physical treatments such as freezing–thawing and freezing–boiling cycles or mechanical disruption by bead beating in the cell lysis procedure leads to highly efficient lysis of the whole bacterial community. In particular, extended bead beating might enable the detection of DNA from difficult-to-lyse forms, such as the spores of Gram-positive bacteria (Sergeant et al. 2012). However, the application of harsh mechanical treatments for prolonged times can lead to extensive shearing of the released DNA and thereby lower the quality of the DNA extract for subsequent analyses.

After cell lysis, the crude DNA extracts are purified to remove cell debris and soil constituents including proteins and inhibitory humic substances—before the subsequent analyses involving enzymatic reactions (e.g., PCR) can be carried out. This is especially relevant for the direct lysis protocols, which provide extracts with high contents of contaminating substances after cell lysis. Most commercial kits use solid-phase purification on silica membranes in the presence of a high concentration of chaotropic salts. For noncommercial protocols, a range of different purification methods has been proposed. Phenol–chloroform extraction/purification has been extensively used. However, this procedure is currently uniquely used when co-extraction of RNA together with the DNA is desirable from soil samples (Paulin et al. 2011). It generally gives high yields and acceptable purity in soil systems, but the toxicity of the chemicals calls for alternatives. The addition of polyvinyl polypyrrolidone (PVPP) and cetyl trimethyl ammonium bromide to the soil before the extraction has been suggested as a means to limit the co-extraction of inhibitors, but the use of PVPP might lead to significant DNA losses. Other methods of purification involve ion-exchange chromatography and gel filtration, as well as CsCl density gradient ultracentrifugation. The latter method is efficient in separating proteins and other cell material from the DNA fraction, whereas humic acids will disperse throughout the gradient, and thus, a second purification step is often required. A different approach for obtaining pure extracts of target DNA is to use specific "capture probes" that can be linked to magnetic beads in a procedure referred to as magnetic capture/hybridization. The procedure relies on the specific binding of a probe to single-stranded target DNA and subsequent recovery of specifically bound DNA from the solution by magnetic capture.

Regardless of the extraction protocol used, specific soil parameters such as low pH and high clay content can cause problems for the efficient extraction of high-quality DNA. Several protocol improvements have been published to overcome these challenges, including the additions of sorption site competitors such as short fragments of nontarget DNA (Paulin et al. 2011) or skimmed milk (Takada-Hoshino and Matsumoto 2004). In addition to improving nucleic acid recovery from high-clay soils, adsorption site competitors also improve the recovery rates from low-biomass soils (Paulin et al. 2013; Figure 12.2).

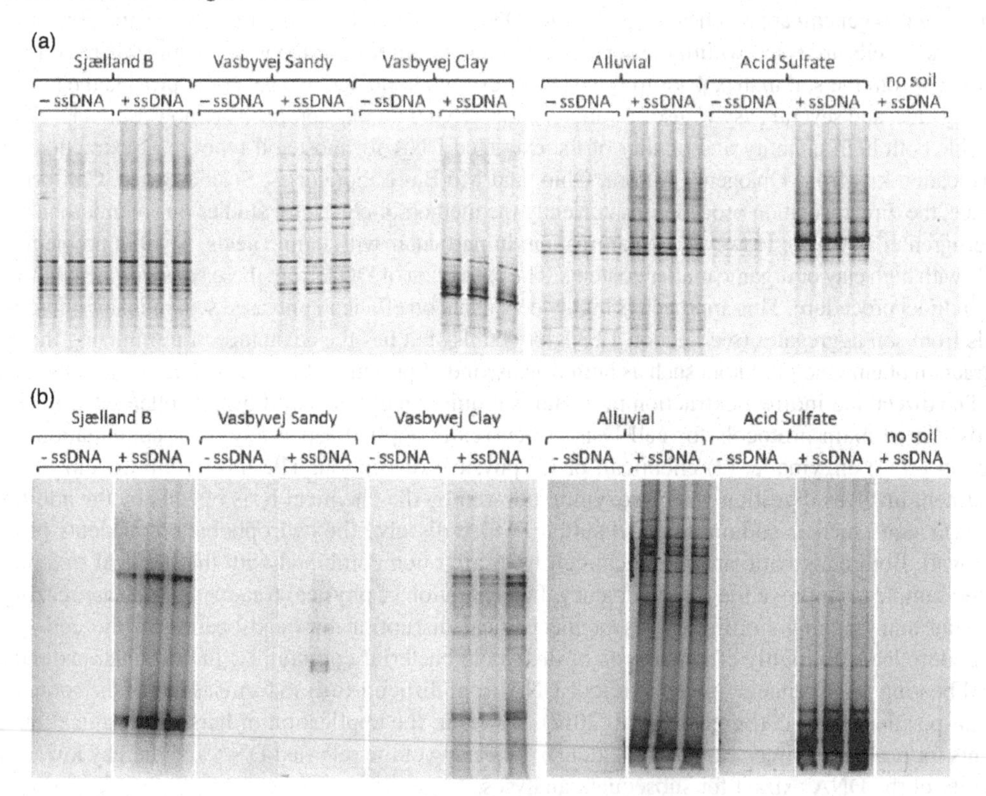

FIGURE 12.2 PCR-denaturing gradient gel electrophoresis (DGGE) profiles of (a) 16S rRNA genes and (b) 16S rRNA gene transcripts from low-biomass and high adsorption soils without (–) and with (+) the addition of 10 g kg^{-1} of salmon sperm DNA (ssDNA). Extracts of 10 g kg^{-1} ssDNA without soil were included as a control (Paulin et al. 2013).

Due to the high complexity of soils, the methods to extract soil DNA are continuously evaluated and compared. A standardized "ISO-11063 Soil quality method" has been published in 2010 (Philippot et al. 2010). However, only 2 years later, a modification of this method—enabling improved DNA extraction—was published (Plassart et al. 2012). This underlines the importance of the development of methods for optimized soil DNA extraction in our hunt for the best approaches for assessment of soil microbiome dynamics. Hence, studies relying on soil nucleic acid extraction should consider the prior knowledge on the soil sampled, as well as the requested concentration and quality of the recovered DNA, before the exact extraction procedure is chosen.

12.2.2 Hybridization-Based Detection of DNA Extracted from Soil

A first approach to the analysis of soil microbiome DNA is provided by hybridization to DNA microarrays. Such microarrays for microbial community analysis have been classified into three categories:

1. Community genome arrays
2. Oligonucleotide microarrays (rRNA based; phylochips, phylogenetic oligonucleotide arrays)
3. Functional gene arrays

Independent of the approach, microarrays are constructed by placing DNA probes, which can be either purified PCR products or synthesized oligonucleotides, on a solid surface (e.g., glass slide), at a density of hundreds to thousands of probes per cm^2. Subsequently, nucleic acids extracted from the soil sample are fluorescently labeled and hybridizations carried out. A hybridization signal is then acquired by laser scanning detection. Inhibitory substances in the nucleic acid extract can influence the hybridization signal. Moreover, even single-base-pair mismatches have significant effects on hybridization sensitivity and specificity. Despite these caveats, microarray technology has developed rapidly, holding great potential for the studies of microbial community structure and function in soil (see Asuming-Brempong, 2012, for a comprehensive review). Especially, the phylochip that targets 16S rRNA genes has gained momentum. A case in point is given by the study of Mendes et al. (2011), who identified the key microbial taxa that are involved in fungal pathogen suppression in disease-suppressive soils.

12.2.3 PCR-Based Detection of DNA Extracted from Soil

PCR is a sensitive method for the detection and quantification of specific genes, and the organisms harboring them, in any soil sample from which relatively pure DNA extracts can be obtained. During PCR, large quantities of a specific gene fragment are synthesized during repeated cycles of denaturation, primer annealing, and DNA polymerase-catalyzed elongation of the strands. The specificity of PCR-based detection primarily depends on the primer DNA sequences used and can be intentionally tailored to target DNA sequences representing various taxonomic levels (e.g., from strains to domains). Usually, phylogenetically based detection is based on the 16S rRNA gene (see Chapter 3). However, for the detection of specific species or strains, unique target DNA sequences may be identified by typing PCR analyses or by *in silico* analyses of genome sequences. Furthermore, particular functional genes are often also of interest to detect, for instance, organisms involved in the degradation of pollutants, biogeochemical cycling processes, or plant protection.

PCR methods may be categorized as either nonquantitative or quantitative, depending on whether they allow for the quantification of target DNA copy numbers in a sample or not. Quantitative PCR (qPCR) offers excellent sensitivity and is able to detect and quantify even rare populations of bacteria (e.g., specific pathogenic bacteria in environmental samples). However, as is true for all PCR-based assays, potential biases associated with nucleic acid extraction, primer specificity, and general PCR conditions may occur and should be carefully evaluated. Target gene numbers may not always be directly translated into cell numbers, as the cell-specific target gene copy number may vary between different bacteria. In addition, it is important to have in mind that PCR analysis does not distinguish between living or dead cells or DNA freely present in the soil (see Section 3.4 for the methods used to assess the viability or activity of bacterial cells).

In Sections 12.2.3.1 and 12.2.3.2, we present the methodology of qPCR, in its classical form (also referred to as real-time PCR) and in the form of digital droplet PCR (ddPCR), which includes a technological overlay to the classical qPCR. qPCR is widely used in soil microbiome assessments and can also be used for the detection of RNA after a reverse transcription step (see Chapter 15).

12.2.3.1 Quantitative PCR

Currently, qPCR constitutes the state-of-the-art method for the quantification of taxonomic and functional marker genes in soil. It detects the product formation in real time during the process of PCR amplification and is thus not based on endpoint detection. The formation of double-stranded (ds) DNA is proportional to a fluorescence signal that is detected directly after each amplification cycle. qPCR requires a specialized thermocycler equipped with a fluorescence detector (Smith and Osborn 2009). The most commonly used reporter system relies on the fluorescent dye SYBR Green that binds to the minor groove of the dsDNA, and increases fluorescence along with the amplicon number accumulation during PCR. Primer specificity is crucial when relying on SYBR Green-based detection systems, as the lack of specificity will lead to the formation of nontarget—next to target—products that are both detected by the general dye. To confirm that the fluorescent signal is generated only from the target of interest, a post-PCR dissociation curve analysis is often carried out. Initial screenings should, however, also include the verification of primer specificity via gel electrophoresis.

A more specific qPCR approach is enabled by the use of Taqman probes (Figure 12.3). Taqman probes are non-extendable dual-labeled fluorogenic hybridization probes, which will bind to target sequences in a similar manner to the primers. The probe is labeled with a reporter dye (e.g., 6-carboxyfluorescein) at one end and a quencher dye (e.g., 6-carboxy-tetramethyl-rhodamine) at the other. As long as the probe is intact, the fluorescence of the reporter dye is quenched by the quencher dye, and no fluorescence is detectable. Taqman probe quantification relies on the 5′ nuclease activity of the DNA polymerase (e.g., *Taq* polymerase), which causes the degradation of the probe as elongation proceeds. This process releases the reporter dye into solution where its fluorescence is no longer quenched, and therefore, fluorescence increases the proportional to the product formation.

The fluorescence signal (whether obtained by SYBR Green staining or degradation of Taqman probes) is plotted against the cycle number, and amplification plots are thus obtained (Figure 12.3). Based on the standard deviation determined from the data points collected during the first PCR cycles (the baseline), a threshold line (usually ten times the standard deviation of the baseline) is determined, and the cycle at which the amplification curve crosses the threshold line, defined as C_T, is used to estimate the number of targets in comparison with an external standard curve (e.g., a serial dilution of known concentrations) (Figure 12.3).

In studies on soil, quantification often relies on a standard curve that is made from the known copy numbers of the target gene in clean "soil-free" solutions. However, for application to the soil environment, it is preferable to use standard curves made by the addition of increasing numbers of cells containing the target gene to soil, prior to DNA extraction. This is because the PCR amplification efficiency may decrease due to co-extracted PCR inhibitors.

The major advantages of the qPCR procedure are the wide dynamic range (up to eight orders of magnitude) and the reliability of the assay (given the knowledge of the full amplification profile), which is of great value when working with nucleic acids extracted from complex soil samples.

qPCR is being increasingly used to study the abundance of bacteria in soil, at both taxonomic and functional levels. It has been of paramount importance for expanding the understanding of microbial processes in soil systems. In a study by Bælum et al. (2008), qPCR was used to monitor the population dynamics of indigenous phenoxyacetic acid degraders in soil by targeting the

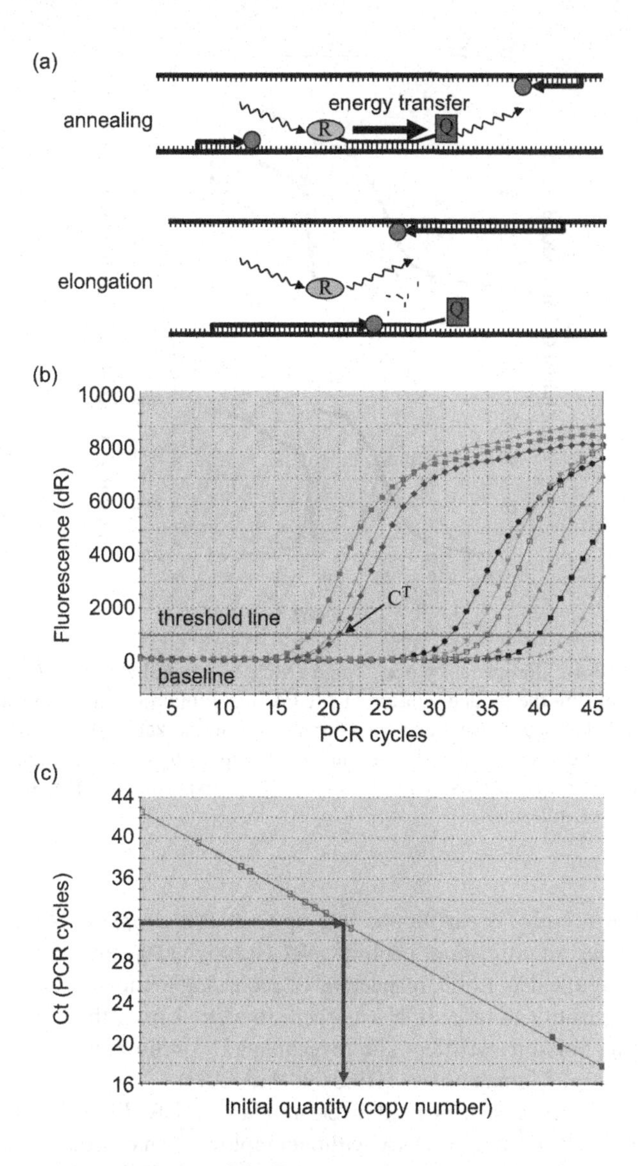

FIGURE 12.3 (a) Schematic presentation of the Taqman probe approach. R: reference dye; Q: quencher dye; circles: Taq polymerase; bended arrows: primer binding. (b) Amplification plot (see text for explanation). (c) Standard curve (see text for explanation).

catabolic gene *tfdA* that is involved in the degradation of phenoxyacetic acid herbicides. Soil amended with the phenoxyacetic acid compound 2-methyl-4-chlorophenoxy acetic acid (MCPA) was used as the experimental platform to study the biodegradation process. The population size of bacteria harboring the *tfdA* gene increased upon confrontation with, and degradation of, MCPA. At the same time, expression of the *tfdA* gene was detected by coupling the qPCR to a prior reverse transcription step (using the total extracted nucleic acids) (see Figure 12.4). Currently, qPCR is the method of choice for the quantification of presence and expression of specific genes in soil (see Chapter 15).

12.2.3.2 Digital Droplet PCR

ddPCR is the latest recently developed DNA quantification technology, which has great potential of use in soil (Hindson et al. 2011). The method is based on water–oil emulsion droplet technology,

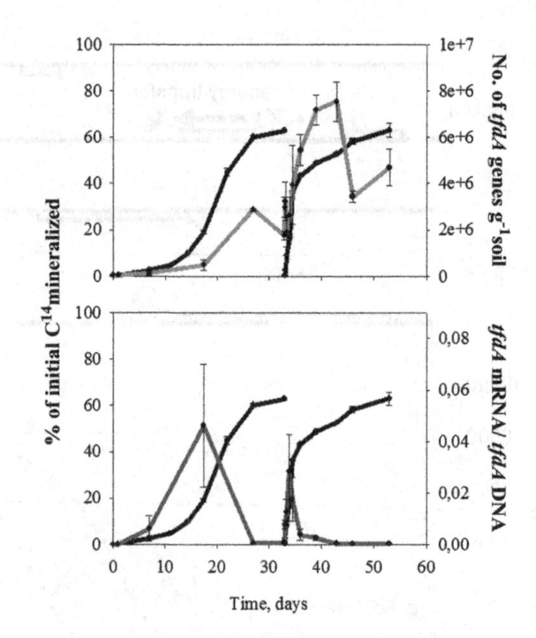

FIGURE 12.4 2-Methyl-4-chlorophenoxy acetic acid (MCPA) mineralization, *tfdA* gene expression, and *tfdA* gene enumeration during the biodegradation of MCPA in soil microcosms. The soil was enriched with MCPA on days 0 and 32. The figure illustrates the cumulative mineralization data of MCPA (left axis, black lines), *tfdA* mRNA quantities expressed as DNA equivalents (upper figure, right axis, gray lines), and *tfdA* DNA quantities (lower figure, right axis, gray lines) during the experiment. Error bars represent standard errors of the mean for soil triplicates.

in which a single PCR mixture is partitioned into up to 20,000 droplets. The PCR occurs in each individual droplet. After amplification, each droplet is assigned as either "template positive" or "template negative" (digitized as 1 or 0, respectively), according to the threshold fluorescence signal intensity. The initial quantity of target DNA is then calculated from the numbers of positive and negative droplets using Poisson statistics. The reagents and workflows used for ddPCR are akin to those used in conventional Taqman assays (Hindson et al. 2011).

There are several advantages of ddPCR over standard qPCR. First of all, ddPCR allows to quantify the absolute concentration of DNA without employing an external standard curve, which can be laborious and time consuming to prepare. Since the ddPCR provides an endpoint measurement and relies on binary detection (either positive or negative), it is less sensitive to PCR inhibitors that are commonly present in environmental samples. Quantification of DNA by ddPCR relies on thousands of data points rather than a single result, which provides great statistical power to the analyses. One of the major drawbacks of ddPCR compared to qPCR is a lower detection range, since the system can only enumerate up to about 2×10^5 copies using 20,000 droplets, which is much less than the capability of qPCR (usually up to 10^8 copies) (Hindson et al. 2011). As a consequence, a pre-run of samples has to be included in the ddPCR protocol in order to evaluate the optimal dilution factor for samples with diverse quantities of target genes. Moreover, ddPCR is relatively expensive, laborious, and time-consuming, since it encompasses several steps, such as droplet generation and droplet reading (6 versus 3 h operation time for ddPCR and qPCR, respectively) (Kim et al. 2014).

So far, ddPCR has been sparsely used for assessments in soil. However, in a study by Kim et al. (2014), ddPCR was compared to qPCR for its ability to quantify *Cupriavidus* strain MBT14 and *Sphingopyxis* strain MD2 in soil. Standard curve analyses on tenfold dilution series showed that both qPCR and ddPCR exhibited excellent linearity ($R^2=1.00$) and PCR efficiency (≥92%) across the detection ranges. However, ddPCR was more sensitive in detecting low quantities of DNA than qPCR (tenfold lower detection limit). Analysis of the temporal changes of the two bacterial strains in soil by the two methods showed an excellent quantitative agreement. Therefore, ddPCR is a promising tool for monitoring specific bacterial populations in soil.

12.3 DIRECT DETECTION METHODS INVOLVING MICROSCALE OR SINGLE-CELL ANALYSIS

In the heterogeneous soil environment, different bacterial cells may be metabolically active, dormant, forming spores, or dead, even if they are separated by only a few micrometers (see Chapters 1 and 9 for considerations that underpin this contention). Thus, phenotypic as well as genotypic variability among individual cells or groups of cells is expected. Soil microbiologists often have specific interests in investigating hot spot environments (e.g., worm casts or decaying organic material) or (hot spot) gradient environments (e.g., the rhizosphere, the mycosphere, or the spermosphere), where spatial heterogeneity may be large. Given this heterogeneity, microscale or single-cell analyses of the distribution and activity of specific organisms are often required. For example, investigations at this fine level of resolution are fundamental to the identification of the microhabitats supporting survival or activity of introduced cell populations, and for the studies of microbial interactions in the soil (see Chapter 9). Importantly, the PCR-based techniques dealt with in Section 12.2.3 cannot provide such detailed spatial analyses as they typically rely on the amplification of DNA extracted from soil samples that exceed the small microhabitat size (e.g., typically 0.5–1 g).

12.3.1 Methods for Extraction of Bacterial Cells from Soil

Microscale analyses of bacterial cells colonizing surfaces in the soil, for example, soil colloids, plant roots, or fungal hyphae, are typically carried out for samples that are as intact as possible, as subsequent microscopic analyses can then reveal their localization and any spatial relationships between different cells (see Section 12.3.2.1).

However, for many other approaches, in particular those relying on downstream single-cell analysis by flow cytometry (see Section 12.3.2.2), a first prerequisite is to obtain the intact cells detached from soil particles that otherwise might interfere with the detection. In soil, most bacterial cells are attached to the surfaces of organic and inorganic particles via hydrogen bonds, van der Waals forces, and electrostatic interactions. Additionally, extracellular polymeric substances (EPSs) excreted by the cells form strong frameworks that aid in cell adhesion and promote dense aggregation of soil particles. Hence, the initial step of cell extraction requires both efficient dispersion of soil aggregates and breaking of the bonds between the cells and soil particles. Dispersion of soil and detachment of bacterial cells can be conducted by physical methods, either alone or in combination with chemical ones. The most common physical dispersion techniques include mild sonication and homogenization of soil by vortexing, shaking, or blending with water or a buffered ion solution. The extraction efficiency increases with the intensity of the treatment, however, at the expense of survival rates of the more sensitive bacterial cells (Liu et al. 2010). Adding chelating agents (e.g., sodium pyrophosphate) in the physical treatment can increase the recovery of bacterial cells by weakening the bonds between cells and soil particles, whereas detergents (e.g., Tween/Triton) will dissolve the EPS layers. Chemical treatments are, in particular, effective for the recovery of cells that are strongly associated with soil surfaces, which is relevant for many indigenous microbes, especially slow-growing ones. They are also useful for extracting bacterial inoculants from soil. The optimal combination of cell extraction methods will, nevertheless, depend on the physical and chemical characteristics of the soil. If a specific inoculant strain is extracted from soil, it will further depend on the cell surface charge and hydrophobicity. Optimization of dispersion protocols considering these factors therefore needs to be carried out in order to obtain the highest possible cell recovery.

The second step in the procedure is purification and concentration of the detached cells from soil debris. For some microscale or single-cell analyses (such as microscopy), detached cells can be examined directly without further treatment. However, autofluorescence from soil organic matter may interfere with epifluorescence microscopic (EFM) detection. For high-throughput single-cell analysis by flow cytometry, clean samples devoid of particles that can block the instrument or interfere with data acquisition are required. Bressan et al. (2015) have recently succeeded in reliably quantifying bacteria in non-purified soil samples by flow cytometry, even in the presence of a high soil particle background. However, purification using density gradient centrifugation with a solution of a higher density than the bacterial cells (e.g., Nycodenz) is routinely used in most cell extraction protocols. After soil dispersion and a short low-speed centrifugation step that separates the cells that detached from soil particles on the basis of sedimentation rate, cell suspensions are placed upon a layer of Nycodenz or another high-density suspension. High-speed centrifugation is performed for 10–90 min, resulting in a layer of bacterial cells remaining above the high-density solution, whereas soil particles (containing nondetached bacteria) pass through the layer. The original protocol employed a Nycodenz solution with a density of 1.3 g mL^{-1}; however, more recent protocols recommend solutions with even higher densities to better capture cells with high densities.

Purification protocols have been optimized to increase the recovery of bacterial cells up to 70%–90%. However, recovery rates are highly dependent on specific soil properties and on how tightly bacterial cells are associated with soil surfaces. Thus, lower recoveries are often obtained (Liu et al. 2010). Recently, Eichorst et al. (2015) developed a protocol by which ~10^9 cells per gram of dry weight soil were recovered from two different soils. It was possible to obtain the intact cells in suspensions that were clean and sufficiently concentrated to enable nanoscale secondary ion mass spectrometry (NanoSIMS). This allowed the analysis of activity (nitrogen fixation and utilization of cellulose) of individual microbial cells.

Not all bacterial cells in soil are recovered with equal efficiencies using the Nycodenz technique. Using 16S rRNA gene amplicon sequencing, studies have shown that the composition of Nycodenz extracted bacterial communities differs from that of the native soil communities prior to extraction. The procedure may even increase the detectability of certain taxonomic groups (Holmsgaard et al. 2011, Eichorst et al. 2015). Hence, Nycodenz extraction has drawbacks for the studies of total microbial soil diversity. Yet, it is useful when focusing on population changes, or quantification of cells with a certain property by flow cytometry (see Section 12.3.2.2.), if the cells of interest belong to genera that are well recovered (Holmsgaard et al. 2011).

12.3.2 Instrumentation Required for Microscale or Single-Cell Analysis

12.3.2.1 Epifluorescence Microscopy (EFM)

EFM is a classical technique for the enumeration of total fluorescent bacterial populations in soil samples, in particular when physical interactions are of interest. Different light sources (Hg, Hg/Xe lamps) and filter combinations are used in the microscope, depending on the specific fluorescent dye (fluorochrome) or fluorescent marker gene used. Major limitations include the need for a high cell number, typically above 10^5 cells per gram of soil, and a tedious manual counting procedure. Analysis of rhizosphere samples may also be problematic due to the physical dimensions of the root specimen and the limited depth of field in conventional EFM. Finally, autofluorescence from soil organic material may cause problems for EFM.

Confocal laser scanning microscopy (CLSM) is a powerful technique for qualitative analyses of cell distributions in soil, for example, the rhizosphere or soil aggregates. CLSM is based on a laser-supported, computerized microscope that allows for three-dimensional digital imaging of the specimen. Recent studies have employed CLSM for the qualitative analysis of distribution of microbes in soil, especially in rhizosphere samples. Thus, colonization by different plant growth-promoting rhizobacteria of the potato rhizosphere was investigated using CLMS (Krzyzanowska et al. 2012). Potato seed tubers were inoculated with green fluorescent protein (GFP)-tagged strains of *Bacillus subtilis*, *Ochrobactrum* sp., and *Pseudomonas* sp. After 4 weeks of plant growth, the roots with adhering soil were subjected to enrichment in selective media before CLMS analysis with a broad-field-of-view device. This setup enabled the specific monitoring of the colonization patterns of the inoculant bacteria across the potato rhizosphere (Figure 12.5).

12.3.2.2 Flow Cytometry

Flow cytometry and accompanying cell sorting is a technique that provides information at the single-cell level in a fast high-throughput manner. In the flow cytometer, cells from a suspension pass through a laser in a hydrodynamically focused line. A detector records the scattering of light caused by the laser beam hitting the cells one by one, which gives information about the size and granularity of each single cell. In addition, emission from cell-associated fluorescent molecules is captured and recorded by a varying number of filters and detectors, thereby generating specific multiparameter data for each cell. The latter enables close monitoring of heterogeneity within a bacterial population in soil.

Staining a cell suspension with general fluorescent nucleic acid stains such as SYBR Green provides a fast quantification of the numbers of bacterial cells in a background of nonfluorescent particles (Bressan et al. 2015). The use of one fluorescent dye for simple detection can be combined with other stains (e.g., viability stains) (see Section 12.3.4), giving information about the physiological status of the cells. Thus, subpopulations of cells showing different responses to, for instance, stress conditions in the soil can be identified (Nielsen et al. 2009).

Survival of specific inoculant cell populations in soil can be assessed by flow cytometry after tagging the cells with a fluorescent marker such as the GFP (see Section 12.3.3.3). Additionally, fluorescently labeled bioreporter cells extracted from soil can be tracked, allowing to the presence or absence of potential heterogeneity in the bacterial population upon confrontation with certain stimuli or conditions in soil. For example, Burmølle et al. (2005) could demonstrate the production of quorum-sensing (QS) signals among indigenous soil bacteria by the flow cytometric detection of fluorescence from a QS-responsive bioreporter strain that had been introduced into the soil.

Fluorescence-activated cell sorting (FACS) is an additional capability of flow cytometry, where cells showing specific fluorescent properties are separated from the remaining cells after passing a laser beam. The sorted cells can subsequently be subjected to molecular analyses such as sequencing to reveal their identity (Shintani et al. 2014).

By FACS in combination with fluorescence *in situ* hybridization (FISH; see Section 12.3.3.2) for single-cell sorting, it is possible to detect and isolate rare uncultivated bacteria from soil with the potential for whole-genome amplification and sequencing. By using this methodology in combination with enrichment on a soil substrate membrane system, Ferrari et al. (2012) were able to

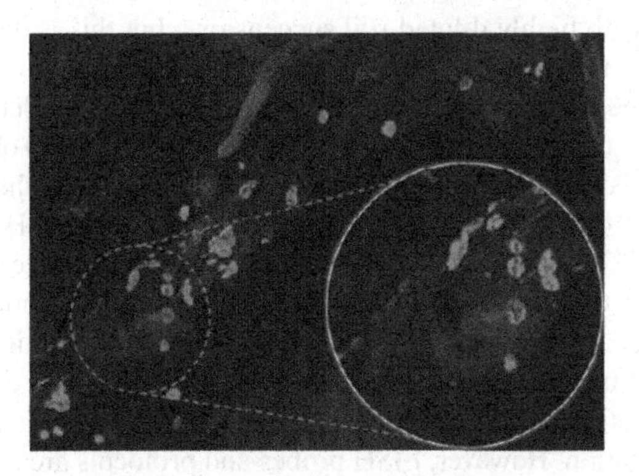

FIGURE 12.5 (See color insert) Colonization of the potato rhizosphere by GFP-tagged *Ochrobactrum* sp. A4. Images were obtained with a Leica TCS LSI Macro confocal laser microscope. Whitish (green in color insert) pseudocolor indicates the position of GFP-expressing minicolonies (Krzyzanowska et al. 2012).

sort bacteria from Candidate Division TM7 from a complex microbial soil community. The high speed and throughput of so-called FACS-based single-cell genomics compared to other methods for single-cell isolation without cultivation (e.g., optical tweezing and laser-capture microdissection) has made this technique an emergent area in soil microbiology (Rinke et al. 2014).

12.3.3 Methods for Direct Detection

12.3.3.1 Determination of Total Bacterial (Direct) Counts

The total number of bacterial cells in soil is typically determined by direct microscopic counting after staining with DNA-binding fluorochromes. The introduction of direct counting procedures in environmental microbiology revealed the phenomenon known as "the great plate count anomaly," that is, the observation that cell counts from environmental samples obtained by cultivation on agar media are orders of magnitude lower than those observed under the microscope. Bacterial total count procedures use DNA-specific dyes such as SYBR Green, acridine orange, or 4′,6-diamidino-2-phenylindole. Such direct total counts can show up to hundredfold differences between soil samples. Hence, not all soils harbor the common text book reference number of 10^9 bacterial cells per gram.

A new method for bacterial total cell quantification, based on SYBR Green staining and cell enumeration by flow cytometry, seems to offer a rapid alternative to the aforementioned methods (Bressan et al. 2015). Importantly, this method provides cell counts that correlate with bacterial rRNA gene numbers as assessed by qPCR. In addition, the method works for cells obtained by a simple extraction procedure, hence avoiding the bias that may be introduced by the application of Nycodenz density gradient protocols as mentioned in Section 12.3.1.

12.3.3.2 Fluorescence *In Situ* Hybridization (FISH)

FISH is based on the hybridization of ribosomal RNA with fluorescent oligonucleotide probes for the detection of single cells in complex samples. FISH has been used to detect and quantify bacteria from various aquatic environments based on probes targeting specific phylogenetic groups, ranging from the phylum to the species level. The use of FISH in soil systems is still limited, primarily due to technical challenges. However, good progress has been obtained in rhizosphere studies (see Chapter 10), where improved detection has been obtained by catalyzed reporter deposition-FISH, using an enzymatic cascade to amplify the signal (Eickhorst and Tippkötter 2008).

Some major challenges of FISH applied to soil are comparable to those mentioned for immunofluorescence (IF) staining (Section 12.3.3.4). For example, the masking of signal by particles and the high background fluorescence that may be encountered are key issues. One way to circumvent the problems is to work with highly diluted soil suspensions, but this is only relevant when targeting dominant members of the soil bacterial community. As reliable enumeration of cells requires hundreds of cells to be counted, diluting samples in order to avoid interference from soil particles might thus require tedious counting. In addition, Gough and Stahl (2003) observed that the number of bacterial cells counted per microscopic field did not adequately reflect the dilution level, probably due to masking of the cells. Hence, it may be necessary to perform FISH on cells extracted from soil particles (see Section 12.3.1) (Bertaux et al. 2007). Unfortunately, the cell extraction approaches impede the detection of a connection between microhabitat and local microbiome. The standard FISH approach also suffers from target cells not being detectable due to insufficient permeabilization (as entry of the used oligonucleotide probe may be blocked or due to low ribosome contents in the target cells). Suboptimal and unknown specificity of the used oligonucleotide probes also constitutes a serious limitation. However, FISH probes and protocols are continuously being modified as part of an iterative process to improve probe specificity and the overall performance of the method, even in difficult environments such as soil. The Probe Database (www.microbial-ecology.net/probebase/) constitutes a useful resource for conducting FISH-based studies.

12.3.3.3 Marker Genes for Direct Detection of Specific Bacteria in Soil

Insertion of marker genes into bacterial genomes represents a useful method for the detection and quantification of specific bacteria in soil. Marker genes are frequently delivered to recipient cells on plasmid vectors that are somewhat stably maintained in natural environments. However, they have also been inserted into bacterial chromosomes by site-specific transposon mutagenesis, which ensures a higher stability of the tag.

The most widely used fluorescent marker genes encode autofluorescent proteins that emit green (GFP) or red (dsRed or mCherry) fluorescence. These signals ensure the specific detection in soil and rhizosphere samples. The fluorescence is often sufficient for single-cell detection and does not depend on cellular energy or added substrates. Moreover, the autofluorescent proteins are stable in the tagged cells, which leads to an increased fluorescence and hence lower detection limit with time. Multiple autofluorescent proteins can be applied simultaneously for the visualization of different populations of bacteria in complex soil or soil–plant habitats. The combination of green and red proteins (e.g., GFP and mCherry) is suitable, since the excitation and emission spectra of these proteins are well separated. There are even more colors available as GFP derivatives that emit different colors such as blue, cyan, or yellow have been designed—though they frequently emit fluorescence with a lower intensity than that of GFP. Monitoring the fate and localization of bacteria tagged with fluorescent marker genes is nowadays routine in soil and rhizosphere studies (see Rochat et al. 2010). Fluorescent marker genes are often also exploited in bioreporter (biosensor) constructions, where they are fused to an environmentally regulated promoter of a gene of interest. Bioreporters (biosensors) are organisms that respond to specific signals in a system, which can subsequently be detected by detection devices. Thus, local conditions can be monitored, for instance, the level of particular nutrients or toxic compounds in a soil or rhizosphere setting. Expression of a bioreporter can be monitored simultaneously with the constitutive marker of a given strain. This was exploited in a study by Rochat et al. (2010) who assessed the colonization of roots of various cereal cultivars by the biocontrol strain *Pseudomonas fluorescens* CHA0 (with a constitutive mCherry marker) using EFM. At the same time, the expression of genes encoding the biosynthesis of antifungal metabolites was monitored by bioreporters fused to GFP. This setup allowed the authors to elegantly study the biocontrol pseudomonad–crop interaction at a single-cell level.

The bioluminescence markers also encompass the bacterial luciferase (*lux*) and the eukaryotic luciferase (*luc*) genes. When these genes are expressed, they catalyze bioluminescent reactions that depend on the metabolic energy reserves of the tagged bacteria. Hence, these markers can—with some right—be considered as "activity assays" due to the close coupling of bioluminescence and activity state. For *in situ* analyses, the specific and sensitive detection of bioluminescence at the microscale level is possible by charge-coupled device-enhanced microscopy. The resolution of the technique typically allows the detection of bacterial microcolonies.

The technology behind bioluminescence markers is older than that of fluorescent marker technology, but the former markers remain useful due to the possibility of determining the activity of tagged cells in a system. In the study by Maraha et al. (2004), a *Pseudomonas* strain was tagged with a marker gene cassette encoding GFP as well as bacterial luciferase. In a cell population introduced into soil, a decline in luciferase activity (bioluminescence) was found, which agreed well with an increase in cells with damaged membranes (see Section 12.3.4.3), indicating that a large fraction of the cell population became inactive or died.

12.3.3.4 ImmunoFluorescence (IF; Fluorescent Antibody) Methods

Antibodies enable the detection of specific bacteria by binding to antigenic determinants, which can be components of the flagellum or the cell wall of the target cells. For Gram-negative bacteria, the O-side chains of the outer membrane lipopolysaccharides are useful targets. Antibodies raised against O-side chains provide specificity at the subspecies level, but not necessarily at the strain level. However, when tracking a bacterial inoculum, it is straightforward to include the necessary controls to ascertain assay specificity in a given soil or rhizosphere sample. IF staining, also known as the fluorescent antibody (FA) technique, employs antibodies coupled to fluorescent compounds so that the tagged cells can be visualized by fluorescence microscopy or flow cytometry.

The method is useful for studies on bacterial inoculants under field conditions where the introduction of cells tagged with marker genes is not feasible due to regulatory constraints (Troxler et al. 2012). It also enables visualizing bacterial inoculants associated with the surfaces of plant roots or fungal hyphae as shown in Figure 12.6. When applying FA methods to soil and rhizosphere, one should be aware that the antibodies might bind nonspecifically to soil particles or root surfaces and that soil and plant materials might be autofluorescent. It is possible to optimize the stringency of the immunoassay, to rely on morphological criteria when identifying target cells, or to rely on CLSM to minimize these problems. Basically, the IF technique does not distinguish between active and inactive bacterial cells. However, it can be conveniently coupled to assays that monitor the viability or activity of individual cells as exemplified in Section 12.3.4.2.

12.3.4 Determining Bacterial Viability and Activity

12.3.4.1 The Concepts of Bacterial Viability and Activity

The methods described in Sections 12.3.3.1–12.3.3.4 are all useful for the detection of cells at the microscale and single-cell levels in soil. They detect all cells and do not distinguish between living (viable) and dead cells. Classical methods for determining viable bacteria are based on their ability to form colonies on an agar medium, and thereby satisfy the definition of bacterial viability, as cells able to perform cell divisions. However, the cultivation-based techniques severely underestimate the number of viable bacterial cells for several reasons. First, it is impossible to define growth media supporting cell divisions by all members of diverse bacterial communities in soil. Second, even bacteria that normally can be cultivated may enter a physiological state where they lose their ability to form colonies, yet maintain other functions indicative of an active metabolism.

A cell in the latter state is referred to as viable but nonculturable (VBNC), which is defined as "a cell which is metabolically active, while being incapable of undergoing the cellular divisions required for growth in or on a medium normally supporting growth of that cell." The significance of the VBNC state has been debated, but it may represent a survival strategy used by stressed bacterial cells, in which growth is arrested while tolerance to environmental insults is increased. Recently, the intriguing proposition has been made that there might be a connection between antibiotic-resistant persister bacterial cells and cells in the VBNC state; see recent review by Ayrapetyan et al. (2015). Importantly, VBNC cells are thought to be able to resuscitate when the environmental stressor inducing the condition is removed or reversed. This has consequences, for example, in relation to the

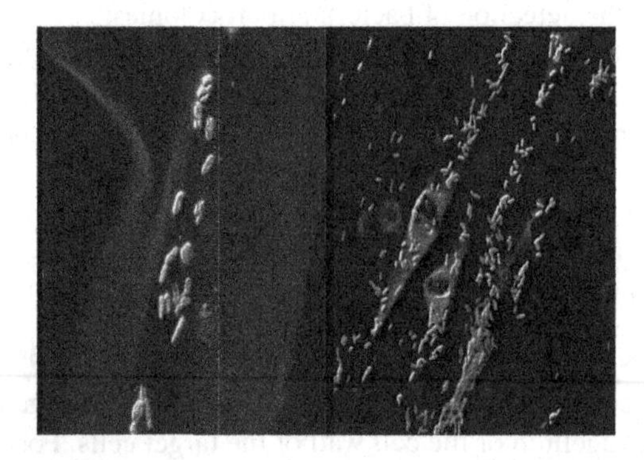

FIGURE 12.6 **(See color insert)** Fluorescence images obtained by confocal laser scanning microscopy (CLSM) of an IF-labeled *Pseudomonas fluorescens* strain (DF57) colonizing barley roots. (The figure was kindly provided by Dr. Michael Hansen, Department of Plant and Environmental Sciences, University of Copenhagen, Copenhagen, Denmark.)

monitoring of pathogenic bacteria, which could remain undetected in a soil by the classical cultivation methods, but maintain their capacity to cause infections when reentering a host. By analogy, even the population sizes of active bacteria performing beneficial tasks in the soil environment, such as bacterial inoculants for plant growth promotion, could be severely underestimated by cultivation methods.

Several methods are available for the cultivation-independent assessment of bacterial viability or metabolic activity at the single-cell level. A few of these methods actually demonstrate the viability (cell divisions) such as the microcolony assay and the direct viable count (DVC) methods described in Section 12.3.4.2. Others are based on fluorescent viability probes or fluorogenic enzyme substrates, determining different aspects of bacterial activity rather than viability (Hammes et al. 2011). A limited number of these methods are routinely used to study soil microbiomes, and these are presented in Sections 12.3.4.2 and 12.3.4.3.

Facing the diverse criteria for viability/activity, it becomes obvious that we merely obtain operational definitions of viable/active versus dead/inactive cells by the available assays. As underlined by Hammes et al. (2011), there is no single method that can unequivocally distinguish between viable/active and dead/inactive cells. However, it can be argued that an intact and polarized cytoplasmic membrane is essential, whereas other manifestations of activity such as different enzyme reactions are less essential. A few studies have applied a battery of viability/activity assays to bacterial cultures subjected to stress. A stepwise loss of activities was observed, in that the loss of membrane integrity occurs later than, for example, that of membrane potential and culturability. Hence, for the studies of bacterial activity in the complex soil environment, it is recommended to apply several complementary methods in a coordinated fashion. Importantly, it is advisable to test and optimize each method for the organism under study, as cell physiology and cell accessibility to the viability stains may differ hugely between bacteria (Hammes et al. 2011).

12.3.4.2 Assays Determining Bacterial Viability

As stated in Section 12.3.4.1, viable bacterial cells are traditionally defined as those cells able to perform cell divisions. Only two methods, the microcolony assay and the DVC procedure, distinguish between viable and nonviable cells according to this definition. These assays are conceptually related to the classical cultivation techniques that detect bacteria able to divide repeatedly and form visible colonies on growth media.

In the microcolony assay, viable cells are identified by their ability to perform one or a limited number of cell divisions when placed on a growth medium. As the initial cell divisions that lead to the formation of microcolonies can be inspected directly by microscopy, it is possible to determine whether an individual cell is nonviable (nondividing) at the time of sampling or viable, but stops dividing during the incubation period. The DVC method is a related method, in which viable cells respond to added nutrients, frequently yeast extract, by increasing their biomass under conditions where their cell divisions are inhibited by the addition of an antibiotic such as nalidixic acid or ciprofloxacin that blocks DNA replication. Active cells, which would have divided if antibiotics were not included in the growth medium, become elongated in response to the added nutrients. The larger size of the viable cells as compared to nonviable cells can be determined by fluorescence microscopy or by flow cytometry if the cells are appropriately tagged.

In a classical study, Winding et al. (1994) demonstrated that the number of viable indigenous soil bacteria, as determined by the microcolony assay (the ability to perform one or a few cell divisions), was up to hundredfold higher than the number of culturable soil bacteria as determined by conventional plating (the ability to perform continuous cell divisions). Interestingly, the difference narrowed down with increasing incubation time on the growth medium, underlining the significance of slow-growing bacteria in soil. Also, for investigating the fate of bacterial strains introduced into the soil environment, the microcolony and DVC assays can be useful. For example, Mascher et al. (2014) investigated the influence of soil pH on survival of *Pseudomonas protegens* CHA0 in soil. They showed that low soil pH reduced the culturability and increased the subpopulation of CHA0 cells in the VBNC state by using the DVC assay combined with cell detection by IF microscopy.

12.3.4.3 Assays Determining Bacterial Activity: Membrane Integrity and Potential

The assays described in this paragraph focus on the aspects of cellular activity that are thought to be indicative for living cells. Two properties of bacterial cell membranes are frequently probed by such assays: membrane integrity and membrane potential. In pure culture studies, stressed bacteria tend to lose membrane potential faster than membrane integrity. Membrane potential dyes (dyes that accumulate in the cytoplasm of cells with a transmembrane electrochemical potential) have, to the best of our knowledge, not yet been applied to soil microbiome studies, whereas the assays for membrane integrity are more widely used.

Cells with intact membranes can be distinguished from those with damaged membranes by staining with two DNA-binding stains, SYTO-9 and propidium iodide (PI). SYTO-9 is a green fluorescent stain that stains the DNA of bacterial cells irrespective of membrane integrity. By contrast, PI is a red fluorescent nucleic acid dye that is excluded from the active cells with intact membrane function. Because the fluorescence from PI suppresses that from SYTO-9, the cells with damaged membranes appear with red fluorescence, whereas the active ones stain green. However, it should be noted that the distinction between green and red fluorescence may be blurred. The commercially available Live/Dead BacLight kit uses SYTO-9/PI double staining, but alternatives may be used. For instance, SYBR Green may substitute SYTO-9 and SYTOX might be used instead of PI. The Live/Dead staining protocol has an unfortunate name, as indicated by Hammes et al. (2011), as this assay actually does NOT strictly distinguish between living and dead cells. However, it can be assumed that Live/Dead staining provides a rather conservative estimate of the number of active cells in a sample.

An example of the application of the Live/Dead BacLight staining protocol to soil examined ca. 2,400-year-old permafrost samples. The study showed that 26% of the total cells were active according to the membrane integrity criterion, whereas minor fractions of the bacterial cells formed microcolonies or could be cultivated under aerobic or anaerobic conditions (Hansen et al. 2007). In more conventional soils, the percentage of cells with intact membranes is often higher.

12.4 DETECTION AND ENUMERATION OF SOIL BACTERIA BY CULTIVATION-DEPENDENT METHODS

Despite the generally low culturability of soil microorganisms, a total count of colony-forming units can be preferable. The significant improvements that have been made to media that enable the growth of soil microorganisms (see Chapter 18) make this approach viable, even though the direct molecular detection of soil microbes has become the state of the art. When the cultivation methods are used to monitor specific bacterial groups or inoculant strains, it is necessary that they enable specific detection of the target bacterium in the presence of a large number of diverse nontarget organisms. Selective media can be of use for the enumeration of bacteria belonging to a given group. For example, pseudomonads can be recovered on Gould's S1 medium, or derived selective soil extract media (see Riber et al. 2014). However, frequently specific markers become important to improve the specificity at which the bacterial group is detected. For specific enumeration of individual bacterial strains, these markers represent an absolute requirement. Intrinsic markers can be useful as targets for immunochemical or molecular detection of colonies appearing on a growth medium. For example, both genus- and strain-specific antibodies have been used to detect pseudomonads in soil and rhizosphere using a colony blot assay. In this assay, colony material is transferred to a filter, which serves as the solid phase for a subsequent enzyme-linked immunoassay. An alternative immunochemical approach is given by IF colony staining, in which specific colonies developing in an agar growth medium are revealed by an FA.

Moreover, several different marker genes can even be introduced into bacteria, as mentioned in Section 12.3.3.3. For the cultivation-dependent methods, the most useful marker genes are those that provide the tagged cells with a selectable phenotype, such as resistance to antibiotics or heavy metals.

12.5 CONCLUSIONS AND OUTLOOK

In this chapter, we have addressed the toolbox that is available for monitoring the population dynamics and activity/viability of specific bacteria, ranging from individual strains to broader taxonomic or functional groups, in soil. Although this panel of methods has developed rapidly, there are still major challenges to consider. For example, the molecular methods applied to analyses of soil microbiomes rely on prior DNA extraction, and current procedures have different efficiencies for different bacterial groups. Moreover, even methods for extracting bacterial cells provide a biased representation of bacteria in soil. Undoubtedly, method development will continue in the years to come, but it is important to realize that benchmarks for this development are hard to define, for example, due to the immense variation in soil properties at a small scale (Nesme et al. 2016).

To address the future utility of the approaches outlined in the foregoing, we need to turn to the current developments in soil microbiome analyses. During the last decade, the area has been transformed by the massive characterization of soil community DNA by amplicon sequencing and metagenomics approaches. These approaches have provided novel insights into the community structures, phylogenetic diversities, and functional potential within these microbiomes (see Chapters 3 and 14). However, we are still far from an exhaustive representative depiction of who is who in the soil (Nesme et al. 2016). The sheer complexity and diversity of populations within the microbiomes that has been uncovered is profound, but it is important to realize that the methods that are currently used to decipher microbiomes integrate information from relatively large soil samples and hence do not provide microscale or single-cell information. The need to address the aspects of soil heterogeneity in studies on microbiome behavior and effects calls for single-cell resolution analysis of soil microbiomes. Improved protocols for flow cytometry and advanced bioimaging will be important in the years to come.

Furthermore, we still need to move on from the descriptive studies on microbiome composition, and make attempts to link specific microorganisms to specific functional roles in the soil systems studied. Most often, this aim requires to bring these organisms into culture, so that their physiology can be investigated. Interestingly, current research tends to indicate that a larger proportion of soil microbiomes can be brought into culture than had previously been assumed (see Chapter 18). Hence, the continued development of improved cultivation protocols will continue to be on the research agenda. For selected microorganisms, we can apply small-scale or single-cell detection systems to allow investigations of relevant soil microhabitats, and use functional assays to reveal their ecophysiology under conditions as close to the *in situ* ones as possible (van Elsas and Boersma 2011).

Methods that provide the researcher with additional information about the function of target cells will also have to be further developed and refined. This could, for instance, be achieved by complementing rRNA-targeted with mRNA-targeted FISH staining, so that the expression of key metabolic functions can be determined within narrower or broader taxonomic groups of bacteria. Parallel approaches could be the application of the microautoradiography-FISH approach to soil bacteria or to develop FISH-NanoSIMS methods for simultaneous analysis of cell distribution and function. Finally, reporter bacteria, which sense their immediate surroundings by a changed expression of specific, introduced reporter genes, also hold potential for a more thorough understanding of the functions expressed by individual bacteria in soil.

Most detection methods and activity assays will, if used alone, provide a biased estimate of the population sizes and activity levels of a target bacterial group or bacterial isolate. It is therefore noteworthy that researchers need to use a sufficient number of samples and experimental replicates to allow robust conclusions to be made. As stated above, we need an even tighter combination of cultivation-independent and cultivation-based approaches. Hence, several complementary methods should preferentially be used as part of a polyphasic approach. Only in this way may we ultimately obtain a comprehensive understanding of the microbes inhabiting a soil environment, which continues to constitute a great frontier for discovery and bioprospecting.

REFERENCES

Asuming-Brempong, S. 2012. Microarray technology and its applicability in soil science – A short review. *Open J Soil Sci* 2, 333–340.

Ayrapetyan, M., Williams, T.C. and J.D. Oliver. 2015. Bridging the gap between viable but non-culturable and antibiotic persistent bacteria. *Trends Microbiol* 23, 7–13.

Bertaux, J., Gloger, U., Schmid, M., Hartmann, A. and S. Scheu. 2007. Routine fluorescence in situ hybridization in soil. *J Microbiol Methods* 69, 451–460.

Bressan, M., Trinsoutrot Gattin, I., Desaire, S., Castel, L., Gangneux, C. and K. Laval. 2015. A rapid flow cytometry method to assess bacterial abundance in agricultural soil. *Appl Soil Ecol* 88, 60–68.

Burmølle, M., Hestbjerg Hansen, L. and S.J. Sørensen. 2005. Use of a whole-cell biosensor and flow cytometry to detect AHL production by an indigenous soil community during decomposition of litter. *Microb Ecol* 50, 221–229.

Bælum, J., Nicolaisen, M.H., Holben, W.E., Strobel, B.W., Sørensen, J. and C.S. Jacobsen. 2008. Direct analysis of *tfd*A gene expression by indigenous bacteria in phenoxy acid amended agricultural soil. *ISME J* 2, 677–687.

Eichorst, S.A., Strasser, F., Woyke, T., Schintlmeister, A., Wagner, M. and D. Woebken. 2015. Advancements in the application of nanoSIMS and Raman microspectroscopy to investigate the activity of microbial cells in soils. *FEMS Microbiol Ecol* 91, fiv106.

Eickhorst, T. and R. Tippkötter. 2008. Improved detection of soil microorganisms using fluorescence in situ hybridization (FISH) and catalyzed reporter deposition (CARD-FISH). *Soil Biol Biochem* 40, 1883–1891.

Ferrari, B.C., Winsley, T.J., Bergquist, P.L. and J. van Dorst. 2012. Flow cytometry in environmental microbiology: a rapid approach for the isolation of single cells for advanced molecular biology analysis. *Methods Mol Biol* 881, 3–26.

Gough, H.L. and D.A. Stahl. 2003. Optimization of direct cell counting in sediment. *J Microbiol Methods* 52, 39–46.

Hammes, F., Berney, M. and T. Egli. 2011. Cultivation-independent assessment of bacterial viability. *Adv Biochem Eng Biotechnol* 124, 123–150.

Hansen, A.A, Herbert, R.A., Mikkelsen, K. et al. 2007. Viability, diversity and composition of the bacterial community in a high Arctic permafrost soil from Spitsbergen, Northern Norway. *Environ Microbiol* 9, 2870–2884.

Hindson, B.J., Ness, K.D., Masquelier, D.A. et al. 2011. High-throughput droplet digital PCR system for absolute quantitation of DNA copy number. *Analyt Chem* 83, 8604–8610.

Holmsgaard, P.N., Norman, A., Hede, S.C. et al. 2011. Bias in bacterial diversity as a result of Nycodenz extraction from bulk soil. *Soil Biol Biochem* 43, 2152–2159.

Kim, T.G., Jeong, S.Y. and K.S. Cho. 2014. Comparison of droplet digital PCR and quantitative real-time PCR for examining population dynamics of bacteria in soil. *Appl Microbiol Biotechnol* 98, 6105–6113.

Krzyzanowska, D., Obuchowski, M., Bikowski, M., Rychlowski, M. and S. Jafra. 2012. Colonization of potato rhizosphere by GFP-tagged *Bacillus subtilis* MB73/2, *Pseudomonas* sp. P482 and *Ochrobactrum* sp. A44 shown on large sections of roots using enrichment sample preparation and Confocal Laser Scanning Microscopy. *Sensors* 12, 17608–17619.

Liu, J., Li, J., Feng, L., Cao, H. and Z. Cui. 2010. An improved method for extracting bacteria from soil for high molecular weight DNA recovery and BAC library construction. *J Microbiol* 48, 728–733.

Maraha, N., Backman, A. and J.K. Jansson. 2004. Monitoring physiological status of GFP-tagged *Pseudomonas fluorescens* SBW25 under different nutrient conditions and in soil by flow cytometry. *FEMS Microbiol Ecol* 51, 123–132.

Mascher, F., Hase, C., Bouffaud, M.-L., Défago, G. and Y. Moënne-Loccoz. 2014. Cell culturability of *Pseudomonas protegens* CHA0 depends on soil pH. *FEMS Microbiol Ecol* 87, 441–450.

Mendes, R., Kruijt, M., de Bruijn, I. et al. 2011. Deciphering the rhizosphere microbiome for disease-suppressive bacteria. *Science* 332, 1097–1100.

Nesme, J., Achouak, W., Agathos, S.N. et al. 2016. Back to the future of soil metagenomics. *Front Microbiol* 7, 1–5.

Nielsen, T., Harder, O., Sjøholm, R. and J. Sørensen. 2009. Multiple physiological states of a *Pseudomonas fluorescens* DR54 biocontrol inoculant monitored by a new flow cytometry protocol. *FEMS Microbiol Ecol* 67, 479–490.

Paulin, M.M., Nicolaisen, M.H., Jacobsen, C.S., Louise, A., Sørensen, J. and J. Bælum. 2013. Improving Griffith's protocol for co-extraction of microbial DNA and RNA in adsorptive soils. *Soil Biol Biochem* 63, 37–49.

Paulin, M.M., Nicolaisen, M.H. and J. Sørensen. 2011. (R,S)-dichlorprop herbicide in agricultural soil induces proliferation and expression of multiple dioxygenase-encoding genes in the indigenous microbial community. *Environ Microbiol* 13, 1513–1523.

Philippot, L., Abbate, C., Bispo, A. et al. 2010. Soil microbial diversity: an ISO standard for soil DNA extraction. *J Soils Sediments* 10, 1344–1345.

Plassart, P., Terrat, S., Thomson, B. et al. 2012. Evaluation of the ISO standard 11063 DNA extraction procedure for assessing soil microbial abundance and community structure. *PLoS One* 7, 1–8.

Riber, L., Poulsen, P.H.B., Al-Soud, W.A. et al. 2014. Exploring the immediate and long-term impact on bacterial communities in soil amended with animal and urban organic waste fertilizers using pyrosequencing and screening for horizontal transfer of antibiotic resistance. *FEMS Microbiol Ecol* 90, 206–224.

Rinke, C., Lee, J., Nath, N. et al. 2014. Obtaining genomes from uncultivated environmental microorganisms using FACS-based single-cell genomics. *Nat Protoc* 9, 1038–1048.

Rochat, L., Péchy-Tarr, M., Baehler, E., Maurhofer, M. and C. Keel. 2010. Combination of fluorescent reporters for simultaneous monitoring of root colonization and antifungal gene expression by a biocontrol pseudomonad on cereals with flow cytometry. *Mol Plant Microbe Interact* 23, 949–961.

Sergeant, M.J., Constantinidou, C., Cogan, T., Penn, C.W. and M.J. Pallen. 2012. High-throughput sequencing of 16S rRNA gene amplicons: effects of extraction procedure, primer length and annealing temperature. *PLoS One* 7, 1–10.

Shintani, M., Matsui, K., Inoue, J.I. et al. 2014. Single-cell analyses revealed transfer ranges of IncP-1, IncP-7, and IncP-9 plasmids in a soil bacterial community. *Appl Environ Microbiol* 80, 138–145.

Smith, C.J. and A.M. Osborn. 2009. Advantages and limitations of quantitative PCR (q-PCR)-based approaches in microbial ecology. *FEMS Microbiol Ecol* 67, 6–20.

Sterngren, A.E., Hallin, S. and P. Bengtson. 2015. Archaeal ammonia oxidizers dominate in numbers, but bacteria drive gross nitrification in N-amended grassland soil. *Front Microbiol* 6, 1–8.

Takada-Hoshino, Y. and N. Matsumoto. 2004. An improved DNA extraction method using skim milk from soils that strongly adsorb DNA. *Microbes Environ* 19, 13–19.

Troxler, J., Svercel, M., Natsch, A. et al. 2012. Persistence of a biocontrol *Pseudomonas* inoculant as high populations of culturable and non-culturable cells in 200-cm-deep soil profiles. *Soil Biol Biochem* 44, 122–129.

van Elsas, J.D. and F.G.H. Boersma. 2011. A review of molecular methods to study the microbiota of soil and the mycosphere. *Eur J Soil Biol* 47, 77–87.

Winding, A., Binnerup, S.J. and J. Sørensen. 1994. Viability of indigenous soil bacteria assayed by respiratory activity and growth. *Appl Environ Microbiol* 60, 2869–2275.

13 Soil Microbiome Data Analysis

Francisco Dini-Andreote
Netherlands Institute of Ecology (NIOO-KNAW)

CONTENTS

13.1 Introduction 215
13.2 Profiling the Diversity of Soil Microbiomes 216
13.2.1 Selection of Molecular Markers 216
13.2.1.1 The Case of Bacteria and Archaea 216
13.2.1.2 The Case of Fungi 216
13.3 Soil Microbiome Data Analysis 217
13.3.1 Initial Data Processing: From Raw Data to Operational Taxonomic Units 218
13.3.2 From OTU Table to Diversity Metrics 219
13.3.2.1 Data Normalization 219
13.3.2.2 Microbiome Diversity Analysis 220
13.3.3 Correlation Analysis 220
13.4 Concluding Remarks 222
References 225

13.1 INTRODUCTION

The study of soil microbiomes has experienced major breakthroughs in the past two decades, mostly as a result of the continuing advances in molecular sequencing technologies. These technological developments have enabled to reach an increased resolution in the assessment of the microbial constituents of soil, that is, the soil microbiome. Most importantly, the application of DNA sequencing technologies in soil microbiome studies circumvents the limitations of organismal culturability, caused by the so-called great plate count anomaly (GPCA). As further discussed in Chapter 18, the GPCA refers to the notion that only a small fraction of the total bacterial cells in a soil sample (about 1%–5%) can be retrieved by culturing under standardized laboratory conditions (Staley and Konopka 1985). The application of molecular methods starts with nucleic acid extraction from a given (or collection of) soil sample(s), followed by amplification and high-throughput sequencing of a specific gene or DNA region that is used as a molecular marker. For instance, the bacterial and archaeal 16S ribosomal RNA (rRNA) gene, the fungal 18S rRNA gene, or the internal transcribed spacer 1 and 2 (ITS1 and ITS2) regions. In this chapter, a focus is placed on key aspects of amplicon sequencing and data analysis, as it is currently applied to the profiling of soil microbiomes. As this is a rapidly progressing area of research, the analytical methods are addressed with emphasis on their fundaments rather than up-to-date discussions on the methodological specificities and developments. This chapter thus provides a basis for understanding the analytical metrics that are routinely applied in soil microbiome studies, serving as a reference for researchers and students seeking information about the fundamental aspects of bioinformatics and statistical methods in soil microbiome data analysis.

13.2 PROFILING THE DIVERSITY OF SOIL MICROBIOMES

13.2.1 Selection of Molecular Markers

Perhaps one of the most troubling themes in soil microbial ecology pertains to the precise description of the distribution of microbial species across samples, and their variations in both space and time. The complexity of the soil system, coupled to our inherent inability to access the large proportion of microbes by classical culture-dependent methods, makes such cataloging of species distributions a daunting task. However, the use of molecular markers that enable operational species discrimination provides a way to circumvent this limitation. By definition, a *molecular marker* is a DNA sequence that can be unambiguously assigned to a particular taxon for phylogenetic profiling, or to a function for trait-based profiling (see also Chapter 12 for an outline of different markers). Importantly, the molecular marker needs to fulfill a set of requirements, as follows: (1) it must be present in every member of a microbiome, (2) it should consistently differ only between individuals with distinct genomes, and (3) it should possess a proportional degree of distinctiveness that matches the (evolutionary) distance between the microbial types that are considered (i.e., on the basis of their genomes). In the current literature, several examples of molecular markers are provided, all with specific advantages and disadvantages (e.g., Vos et al. 2012). Here, a focus is placed on markers that are currently broadly used to profile the taxonomic diversity of members of soil microbiomes.

13.2.1.1 The Case of Bacteria and Archaea

Since the pioneering studies of Carl Woese and colleagues (Woese and Fox 1977, Balch et al. 1977, Fox et al. 1980), the 16S rRNA gene [also called small subunit rRNA (ssu-rRNA)] has been widely adopted as the main phylogenetic marker for studies of bacteria and archaea. Across the known bacteria and archaea, this gene sequence is about 1,550 bp long and is composed of both variable and conserved regions. The conserved regions are frequently used as sites that can be efficiently targeted by PCR primers, whereas the variable ones provide interspecific polymorphisms that allow distinguishing among distinct microbial taxa. The full-length 16S rRNA gene encompasses nine hypervariable regions (referred to as V1–V9). Given the constraints of the currently available high-throughput sequencing platforms with respect to read length (extending several hundreds of bases), it is not common practice in microbiome studies to sequence the full-length 16S rRNA genes of the microbes present in a given sample. Rather, partial sequences are used. As such, the V1–V3 (targeted by primer set 27F/338R) and V4 (primer set 515F/806R) regions are commonly used as molecular markers (Caporaso et al. 2012), with the V4 one being mostly preferred for the currently popular Illumina sequencing platforms. The argument in favor of the V4 region is built on the understanding that this region reflects the full 16S rRNA polymorphism quite well, whereas it distinguishes a broad range of taxa within the bacteria and archaea (Kuczynski et al. 2012, Walters et al. 2015). In contrast, the V1–V3 regions are known to be biased away from some of the archaea, having a rather low molecular resolution that limits the differentiation across a few distinct bacterial taxa.

Moreover, once a suite of sequences is produced from a soil sample, it is commonly compared to predefined sequences in reference databases. Hence, it is of critical importance that such databases are properly curated. With respect to bacterial and archaeal 16S rRNA gene sequences, the most used databases are GreenGenes (DeSantis et al. 2006), the Ribosomal Database Project (RDP; Cole et al. 2009), and Silva (Pruesse et al. 2007) (Table 13.1). These databases contain well-categorized information of partial and full-length 16S rRNA sequences of both cultured and uncultured bacteria and archaea. Moreover, some of these reference databases provide phylogenetic trees against which unknown sequences can be compared and phylogenetically allocated, with a particular level of accuracy.

13.2.1.2 The Case of Fungi

Profiling the taxonomic composition of fungal communities in soil is, in practical terms, similar to the methodology commonly used for bacteria. However, the difference lies in the lack of a standardized and universally accepted molecular marker for fungi. Although several molecular marker

TABLE 13.1
Software Packages and Database Resources Dedicated to Microbiome Data Analysis

Software Packages	Explanation	Website
QIIME	Bioinformatics package for marker gene data	http://qiime.org/
Mothur	Bioinformatics package for marker gene data	www.mothur.org/
Database Resources	**Explanation**	**Website**
SILVA	rRNA sequence data	www.arb-silva.de/
RDP	rRNA sequences and tools for sequencing analysis	http://rdp.cme.msu.edu/
GreenGenes	16S rRNA gene database and tools	http://greengenes.lbl.gov/
UNITE	Fungal ITS sequence database and analytical tools	https://unite.ut.ee/
MaarjAM	18S rRNA, 28S rRNA, and ITS databases for arbuscular mycorrhiza fungi	http://maarjam.botany.ut.ee/
Other Pipelines	**Explanation**	**Website**
VAMPS	Visualization and microbiome analysis tools	https://vamps.mbl.edu/
MetaAmp	Web interface pipeline for microbiome analysis	http://ebg.ucalgary.ca/metaamp/

options exist for fungi (e.g., the 18S ssu-rRNA and/or the 28S large subunit rRNA), ITS regions located between the 18S and 5.8S subunits (ITS1) and between the 5.8S and 28S ones (ITS2) are generally preferred for obtaining effective taxonomic resolution. The use of primer sets targeting either the ITS1 or ITS2 (or both) regions comes with differential amplification biases of specific fungal lineages (for a detailed discussion on the topic, see Lindahl et al. 2013). Moreover, the choice of a specific marker has to take into account the availability of reference databases, against which obtained sequences will be compared. With respect to this, ITS sequences have by far the most robust database for fungi (Begerow et al. 2010). It is worth mentioning that, although useful for taxon separation, both ITS regions are highly variable in terms of nucleotide composition and size, which hampers the proper phylogenetic assessment of organisms at higher taxonomic ranks, that is, at the levels of "family" and "order." As such, in studies where the estimation of phylogenetic distances across distantly related fungal groups is required, the use of 18S rRNA and/or 28S rRNA genes as molecular markers provides a viable option, being both genes more conserved than the fungal ITS regions. The use of either 18S rRNA or 28S rRNA genes allows us to align sequences across distantly related taxa, thus providing a tool for phylogenetic reconstruction and taxonomic assessment at high taxonomic ranks.

The database UNITE (Abarenkov et al. 2010) is currently broadly used as a reference for ITS sequence-based analysis of fungal sequences. The UNITE website (https://unite.ut.ee/index.php) contains a set of useful tools for ITS sequence annotation and a collection of pre-formatted datasets for command-based software packages. Moreover, the website provides a full list of available primer sets for ITS1 and ITS2 amplifications, with a detailed description of the target regions and fragment lengths, and remarks on primer advantages and limitations across distinct fungal lineages.

13.3 SOIL MICROBIOME DATA ANALYSIS

Given the fact that soil microbiomes are diverse and complex (see Chapter 3) datasets generated from soil samples that report on microbiome composition are usually large. This requires computer-assisted analysis methods. Here, the analytical aspects of (soil) microbiome data analysis are discussed with a focus on bacteria, as these are by far the best-classified group of microbes. Similar analytical methods may be applied to other microbial groups (such as archaea and fungi), with minor changes and specific adjustments. Microbiome data analysis commonly uses a combination of bioinformatics, statistical, and computational tools. The most used software packages for such a purpose are QIIME (Caporaso et al. 2010) and Mothur (Schloss et al. 2009) (see Table 13.1 for additional references). Both software packages are open source and have been consistently updated and supported by the developers. Here, we examine the workflow that is commonly used in the QIIME package (Figure 13.1). However, it is important to note that many of the aspects that will be presented are common to both platforms.

13.3.1 Initial Data Processing: From Raw Data to Operational Taxonomic Units

The analysis of amplicon sequencing data aims at processing the raw sequence data (the output of a sequencer platform) up to the point where information on microbial community composition, ecological indices, and comparative ecological metrics can be calculated and inferred across a set of samples. The raw sequence data are commonly provided as one batch, in which the reads constitute the results of sequencing of the various sample types, each one carrying a specific barcode. The initial processing of these raw amplicon data encompasses essential steps that improve the quality of the assessment. The first step pertains to so-called *quality filtering*, focusing on sequencing errors (which vary according to the sequencer platform), as well as the removal of primer sequences that are

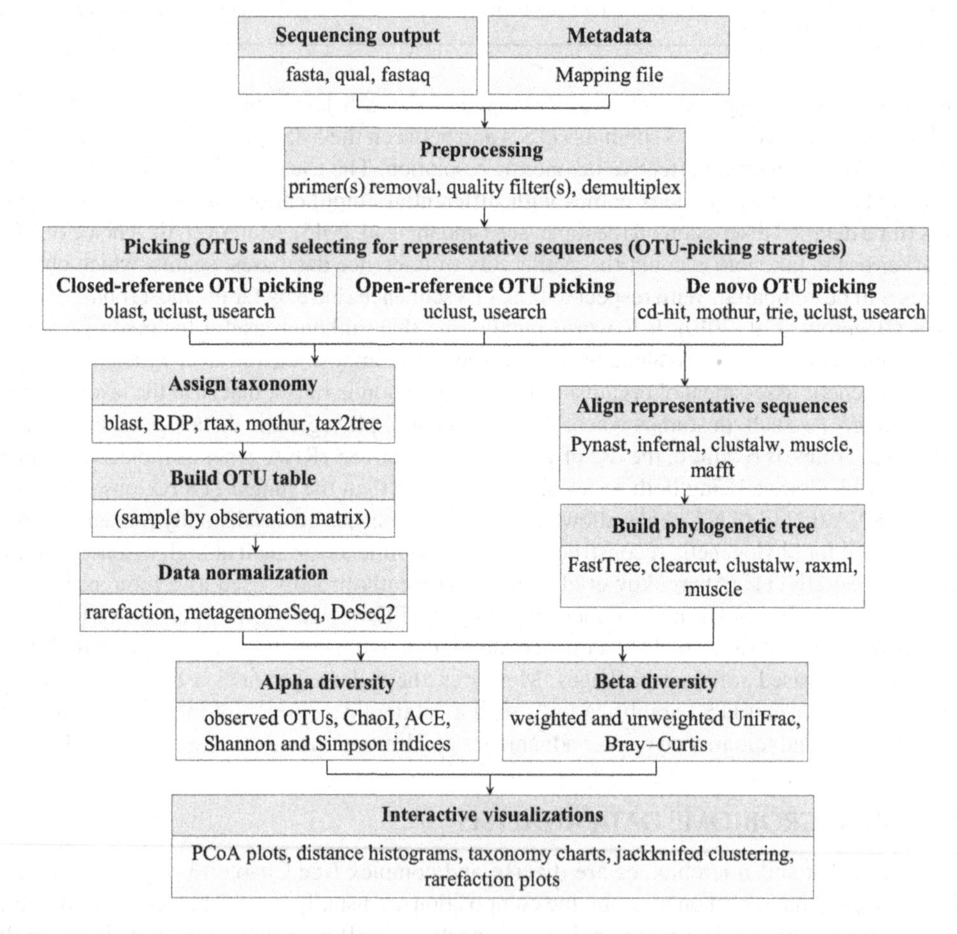

FIGURE 13.1 Schematic workflow of soil microbiome analysis based on amplicon sequencing data using the QIIME toolkit. The workflow is presented in the sequential order of analytical processing as discussed in the main text in detail. Note that not all algorithm variations are fully discussed; rather, the figure aims at providing an overview of the analytical procedure. (Figure modified from Navas-Molina et al. 2013)

located at the edge(s) of the reads. This is followed by *demultiplexing*, a step in which reads are allocated to the samples based on their respective barcodes. The resulting filtered reads are then *binned* based on their similarity to each other. This results in clusters of reads, which are referred to as operational taxonomic units (OTUs). This strategy is commonly used, as each OTU ideally represents an actual bacterial species. Here, the commonly accepted and broadly used threshold for binning of OTUs based on bacterial and archaeal 16S rRNA gene sequences is 97% of nucleotide identity.

A suite of different algorithms can be used to cluster reads into OTUs (e.g., cd-hit, usearch, uclust). Most importantly, there are three—fundamentally distinct—ways in which the clustering can be performed. These are described as being either *de novo*, closed-reference, or open-reference *OTU picking*. In generic terms, these approaches differ with respect to how centroids ("seed" sequences) are defined to initiate the binning process. That is, the centroids are either defined internally based solely on the sequences that are being clustered (*de novo*), predefined based on a known database of cluster centroids (closed-reference), or defined based on a combination of the former two (open-reference). It is important to note that each of these approaches has intrinsic benefits and drawbacks, and hence, a careful consideration of the method according to each study type is required.

De novo OTU picking: In this method, the trimmed reads are aligned against one another, and those that share greater than a predetermined threshold of similarity (e.g., 97%) are defined as belonging to the same OTU. There are several variations and parameters used in picking OTUs by *de novo* binning, for example, the proper determination of the number of alignments to be performed in order to ensure the assignment of a read to a specific OTU, or to define an OTU. The advantage of this approach is that a preexisting collection of reference sequences prior to clustering is not required, which is useful in the case of a new (or uncommon) marker gene.

Closed-reference OTU picking: In this method, the trimmed reads are aligned against a predefined cluster of centroids (a reference database). Sequences that match the reference centroids are kept and allocated in their respective OTUs, whereas those that do not have a representative centroid in the database are excluded. The main advantages of this method are the resulting high-quality taxonomic assignments of the OTUs (as it is based on the reference centroids) and the high-quality phylogenetic tree that is generated. However, the drawback is that this method excludes potentially "novel" sequences from the analysis. Thus, it restricts the data to the well-known categorized sequences in the databases. It is expected that this limitation will rapidly decrease over time, as the used reference databases increase in coverage of the extant soil microbial diversity.

Open-reference OTU picking: This method combines both previously mentioned OTU picking strategies. For that, sequences are initially clustered against a reference database in a closed-reference OTU picking process. In a second step, sequences that fail to match the reference database are clustered *de novo* in a serial process. Open-reference OTU picking offers advantages over both the *de novo* and closed-reference strategies. Since it includes *de novo* OTU picking of the sequences that fail to hit the reference database, all sequences are clustered, so analyses are not restricted to already-known OTUs. However, because the *de novo* clustering process is run serially, it can be computationally slow for large datasets or datasets with a substantial number of sequences that fail to find hits in the reference database.

13.3.2 From OTU Table to Diversity Metrics

13.3.2.1 Data Normalization

A common outcome of amplicon sequencing projects is that the number of sequences obtained is uneven across samples. That is, the number of sequences acquired in a sequencing run varies across samples due to technical rather than biological reasons. These differences in sequencing depth are known to affect diversity estimates. There are statistical methods that allow dealing with such variations. The most commonly applied method is rarefaction, in which an equal number of sequences is randomly selected from each sample. The number of sequences retained is usually the sequence count of the sample with the lowest (and acceptable) number of sequences. This sequence count is defined during the analytical procedure in a way that reflects a balance between retaining as many sequences as possible without excluding too many low-sequence-number samples. Despite being broadly used, rarefaction of amplicon data for normalization purposes has been recently criticized. Benchmark data analyses have shown that rarefaction can introduce errors in the analyses, resulting in inappropriate detection of differentially abundant OTUs across samples. It was even depicted as an inefficient method in the statistical sense (McMurdie and Holmes 2014). Alternative methods for data normalization include those based on variance stabilization. In brief, the implementation of these methods includes, for example, the "metagenomeSeq" and "DESeq2" software packages. However, proper validation of such methods across different datasets and case studies is still lacking. Importantly, to make proper use of these methods, it is necessary to address a larger number of replicates per sample (i.e., sample per category) in any given experimental design. Moreover, as these methods assume normal distributions of the data, and are therefore subjected to parametric tests (See Chapter 19), it is necessary to carefully evaluate the dataset prior to normalization procedures. Also, both methods are computationally demanding, in particular for large datasets containing hundreds of samples with thousands of sequences per sample.

13.3.2.2 Microbiome Diversity Analysis

Analysis of microbiome diversity is typically based on the calculation of within—(i.e., α) and between—(i.e., β) diversities (Table 13.2). α-Diversity metrics act as a statistical and ecological summary of the individual communities. The absolute value of community richness is defined as the number of "observed OTUs." To calculate α-diversity, several estimators of community richness are commonly used, including the ChaoI and abundance-based coverage estimator (ACE) measures. Conversely, several estimators of community diversity (which take into account both OTU richness and evenness) are used, as exemplified by the Shannon and Simpson indices. It is important to note that these metrics are estimators of the number and distribution of taxa in a community based on predetermined sequencing depth. As such, increasing the number of sequenced taxa allows to produce increasingly precise estimates of total community diversity. In soil microbiome studies, these values are often displayed in a comparative manner across samples (i.e., communities). Additionally, when comparing community diversity between samples (β-diversity), ecological matrices are used to calculate a similarity score between each pair of communities. The β-diversity metrics provide a measure of the degree to which samples differ from one another, in a more precise and robust way than by looking at the composition and α-diversity metrics of the individual samples. There are several ways by which β-diversity can be calculated. The different metrics are either quantitative (taking into account sequence abundance, e.g., Bray–Curtis and weighted UniFrac) or qualitative (considering only presence/absence of sequences, e.g., unweighted UniFrac). Moreover, these can be categorized as phylogeny-based metrics (e.g., weighted and unweighted UniFrac distances) and non-phylogenetic metrics (e.g., Bray–Curtis). The use of these metrics is often followed by sample clustering or dimensionality reduction (e.g., by principal components analysis). A set of different metrics exists to estimate and calculate α- and β-diversities, each one serving to reveal different aspects of the soil microbiome. The choice of the method should be based on the detailed examination of the question posed and the advantages and intrinsic limitations of each metric (or complementary use of many).

13.3.3 Correlation Analysis

An emerging approach in soil microbiome analysis is the examination of "correlation networks" to make inferences with respect to the potential biological or biochemical relationships between the data in a dataset. Essentially, correlation networks consist of *nodes* (defined OTUs or features, e.g., physicochemical data) and *edges* (positive or negative correlational connections between the nodes). Despite the promise of the approach, the proper measurement of these relationships in a dataset and the further visualization of the complexity of the potential interactions across features are still computationally challenging. First, soil microbiomes have immense complexity (up to more than

TABLE 13.2
List of α- and β-Diversity Metrics Commonly Used in Soil Microbiome Analysis

α-diversity		
Richness		
Observed OTUs		Total number of OTUs in a sample
Chaol	$S_1 = S_{obs} + \frac{F_1^2}{2F_2}$	S_{obs} is the number of OTUs in the sample F_1 is the number of singletons in the sample (OTUs with a single occurrence in the sample)
ACE	$S_{ace} = S_{abund} + \frac{S_{rare}}{C_{ace}} + \frac{F_1}{C_{ace}}\gamma_{ace}^2$	S_{abund} are the OTUs that occur more than ten times in the sample S_{rare} are the OTUs that occur ten times or less in the sample C_{ace} is the sample ACE: $C_{ace} = 1 - \frac{F_1}{N_{rare}}$, where $N_{rare} = \sum_{i=1}^{10} iF_i$ F_1 is the number of singletons (OTUs with a single occurrence in the sample) γ_{ace} is the estimated coefficient of variation for F_1: $\gamma_{ace}^2 = \max\left[\frac{S_{rare}\sum_{i=1}^{10} i(i-1)F_i}{C_{ace}(N_{rare})(N_{rare}-1)}\right]$
Diversity		
Shannon	$H' = -\sum_{i=1}^{S}(p_i \ln(p_i))$	S is the number of OTUs in the sample p_i is the proportion of the total sample represented by the ith OTU
Simpson	$D = \sum_{i=1}^{S} p_i^2$	S is the number of OTUs in the sample p_i is the proportion of the total sample represented by the ith OTU
β-Diversity		
Bray–Curtis	$BC_{ij} = \frac{S_i + S_j - 2C_{ij}}{S_i + S_j}$	S_i and S_j are the number of OTUs in the samples i and j, respectively C_{ij} is the total number of OTUs at the location with the fewest species
Weighted UniFrac	$u = \frac{\sum_{i=1}^{N} l_i \left\lvert \frac{A_i}{A_T} - \frac{B_i}{B_T} \right\rvert}{\sum_{j=1}^{S} L_j}$	N is the number of nodes in the tree S is the number of sequences represented by the tree l_i is the branch length between node i and its parent L_j is the total branch length from the root to the tip of the tree for sequence j A_i and B_i are the number of sequences from samples A and B that descend from the node A_T and B_T are the total number of sequences from samples A and B
Unweighted UniFrac	$W = \frac{\sum_{i=1}^{N} l_i \lvert A_i + B_i \rvert}{\sum_{i=1}^{N} l_i \max(A_i, B_i)}$	N is the number of nodes in the tree l_i is the branch length between node i and its parent A_i and B_i are the indicators equal to 0 or 1 as descendants of node i are absent or present in communities A and B, respectively

10,000 OTUs per unit soil), which implies a huge number of possible positive/negative two-feature interactions. This complexity scales up when taking into account that microorganisms in the soil often interact with two, three, or more features. Second, the common complexity of the dataset may harbor diverse relationship types (e.g., linear, exponential, or periodic), and this requires proper statistical tests to effectively detect them (Table 13.3). And, third, sequencing data provide information

TABLE 13.3
List of Available Coefficients for Correlation Analysis

Naive Correlation Coefficients	Correlation Type
Bray–Curtis	Abundance similarity
Pearson's correlation coefficient	Linear relationships
Spearman's correlation coefficient	Rank relationships

on taxon abundance based on a fixed set of sequences, which implies that relative, rather than absolute, values are taken into account for the analysis. This introduces the problem of using relative—or compositional—data in statistical procedures (Lovell et al. 2010).

Several software programs have been developed to calculate and visualize correlation networks, which are specifically optimized to correct for different aspects of correlation analysis in microbiome data. For example, SparCC (Sparse Correlations for Compositional data) (Friedman and Alm 2012) is designed to deal with compositional data, and it is essentially based on Aitchison's log-ratio analysis. In addition, CoNet (Correlation Networks) (Faust et al. 2012) uses an ensemble method with a ReBoot procedure for *P*-value computation to retrieve information from different correlational metrics. Local similarity analysis (Ruan et al. 2006) is designed to build correlation networks from time-series data, as it is optimized to detect nonlinear, time-sensitive relationships. Finally, maximal information coefficient analysis (Reshef et al. 2011) is a nonparametric method that captures a wide range of associations without limitations to specific metrics and relation types (linear or exponential), whereas molecular ecological network analysis (Deng et al. 2012) adapts random matrix theory from physics to microbiome data, thus being optimized to be robust to noise and arbitrary significance thresholds.

In generic terms, the level of performance and the limitations of these methods are still under debate in particular, with respect to their analytical sensitivity, specificity, precision, and noise. As such, most of these analytical procedures should be used with caution, with respect to the analytical capacity that allows providing interpretable results. However, it can be safely posited that the use of correlation networks in soil microbiome data analysis provides an interesting direction for prospective hypothesis generation and/or design of experiments.

13.4 CONCLUDING REMARKS

Microbiome data analysis is a rapidly evolving field. As DNA sequencing technologies advance, bioinformatics and analytical tools are constantly being changed and optimized. It is noticeable that, in the past few years, the analytical aspects of amplicon sequencing analysis have progressed towards great maturity. For example, the current use of established pipelines for data analysis allows for standardized experimental and bioinformatics procedures that validate the robustness of the methods applied to answer distinct biological questions of interest in soil microbiome studies. These include the use of standardized concepts (e.g., the 97% nucleotide identity cutoff used for OTU definition) and databases for taxon annotation and curation (e.g., the GreenGenes and Silva databases). However, it is important to realize that these are not yet fully standardized procedures. For example, recent advances have shown advantages of using refined algorithms that allow finer taxonomic resolution at the level of exact sequence variants (rather than OTUs) to categorize organismal taxa (e.g., UNOISE3, DADA2, and Deblur) (Callahan et al. 2017). As such, follow-up procedures are expected to emerge in the near future, with the focus on improving analytical precision, reproducibility, and comprehensiveness in soil microbiome data analysis.

Another important analytical step consists of integrating the aforementioned pipelines with the R environment for further statistical analyses. In this sense, the raw output data of the workflow for microbiome analysis presented in this chapter may serve as the basis for further analysis and data exploration. For example, this may include the use of multivariate data analysis, machine learning algorithms for data prediction, the integration of microbiome data with ecological models, and computational simulations. Several R packages that integrate and extend particular aspects of the analysis are available (e.g., Phyloseq, microbiome R package, ggplot2, and Vegan). These are just a few examples that illustrate how to strengthen the examination of data in soil microbiome analyses.

Of critical importance to soil microbiome analyses, and irrespective of the methodological idiosyncrasies in the data analysis, the following aspects are of utmost importance:

i. The proper delineation of the study (i.e., replicability, reducing confounding factors)
ii. Caution with sample type selection and collection

With respect to the latter, it is critical to pre-establish a protocol for the proper collection of soil samples. Current studies in the literature report variations in soil microbiomes at the continental (Fierer et al. 2006) as well as soil microaggregate scales (Vos et al. 2013). The definition of what is the "optimum" soil sample is still under debate and must be carefully considered with respect to each study design and aim. In addition, there is also an increasing awareness with respect to the use of standardized molecular and bioinformatics methods. Given the rapid advances in this field, some progress has been made towards standardizing the protocols recommended for soil microbiome studies (e.g., the Earth Microbiome Project, www.earthmicrobiome.org/). One key issue is metadata collection: it is critical to keep track of all metadata that can be measured or collected at the sampling time. It is of critical importance that the methods used in data analysis are both documented and shared with the scientific community (including versions of the software used and lists of commands). Finally, safe and accessible data deposits are crucial; optimally, data should be processed and deposited in public databases. For the sake of clarity, public databases should work towards standardized data and metadata formats. This last aspect must connect each individual researcher to the larger scientific community, thus prompting the next generation of analyses that can profit from the multitude of databases, allowing the development of further improved analytical procedures in soil microbiome data analysis. Box 13.1 provides an example of these molecular methods, applied to determine the diversity of bacteriomes across sites along a chronosequence of land formation.

BOX 13.1 CASE STUDY: SUCCESSION IN THE SOIL MICROBIOME ALONG A CHRONOSEQUENCE (DINI-ANDREOTE ET AL, 2014)

Study system: The island of Schiermonnikoog (N53°30′ E6°10′; The Netherlands) presents a well-documented soil successional gradient. This so-called soil chronosequence is formed by progressive and continuous increments in the base of elevation of the soil caused by the deposition of particles carried out by wind and frequent flooding events. From the west to the east, this ecosystem has developed along ~8 km encompassing over a century of land formation. The chronosequence has been properly validated with respect to the progressive land formation, and macroflora and fauna establishment, by a combination of data gathered from topographic maps, aerial photographs, and ecological monitoring of permanent plots of over 20 years.

Bacterial α-diversity along the soil chronosequence: In studying the composition of bacterial communities in the system, it was shown that the trend in bacterial α-diversity was opposite to that observed for plant communities. That is, while a progressively higher plant community diversity was recorded along the chronosequence, that of bacteria decreased. This counterintuitive result stands in sharp contrast with the *Whittaker conundrum* "diversity begets diversity" (meaning that distinct communities within an ecosystem tend to develop towards a gradual increase in species complexity).

Bacterial β-diversity along the soil chronosequence: Interestingly, the bacterial communities along the chronosequence also showed different community "turnover." That is, the highest β-diversity was found at the initial successional stages, with the rate of community turnover progressively decreasing as succession proceeded. Most of the variation in community turnover was explained by the temporal variability in soil chemistry nutrients as well as immigration, next to the higher amplitude of variation in environmental conditions (salinity, oxygen, and temperature) at the poorly vegetated sites. As succession proceeded, the buffering effects of soil, as well as plants, became more dominant, reducing the amplitudes of variation and community turnover.

Evidence for temporal niche partitioning: By reconstructing co-occurrence networks of bacterial communities at each successional stage, a progressive decrease in network complexity, as indicated by a reduction in network edges, towards the intermediate and later stages of succession was found. Thus, it was suggested that the daily fluctuations in oxygen, temperature, and water saturation/salinity caused by the tides led to temporally driven niche partitioning. This phenomenon, combined with high immigration as well as colonization rates, was suggested to elicit high phylotype coexistence at the early successional sites. On the other end, in the late successional stages, less dynamic communities were assembled, with low complexity in network co-occurrence and relatively lower diversity.

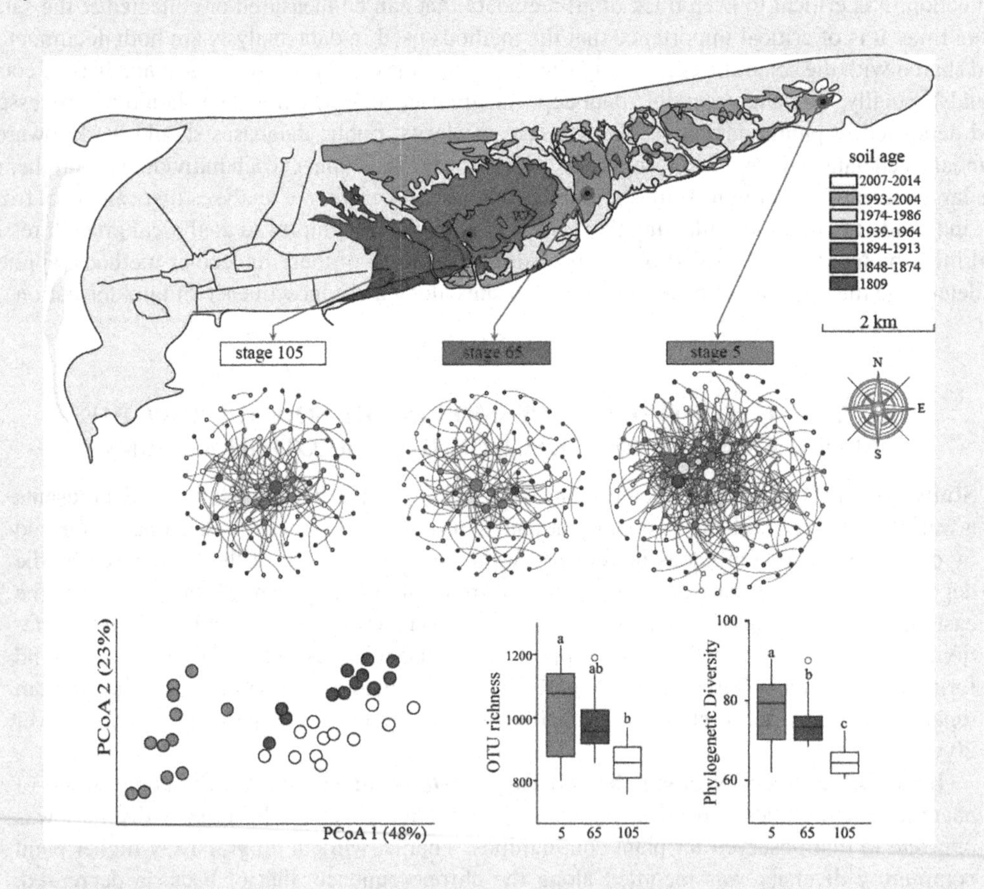

A general overview of the Schiermonnikoog soil chronosequence. Stages are shown in years of soil development (i.e., 5, 65, and 105 years). Plots display OTU co-occurrence networks per stage, α-diversity, and β-diversity of bacterial communities in the system.

REFERENCES

Abarenkov, K., Nilsson, R.H., Larsson, K.-H. et al. 2010. The UNITE database for molecular identification of fungi – recent updates and future perspectives. *New Phytol* 186, 281–285.

Balch, W.E., Magrum, L.J., Fox, G.E., Wolfe, R.S. and C.R. Woese. 1977. An ancient divergence among the bacteria. *J Mol Evol* 9, 305–311.

Begerow, D., Nilsson, H., Unterseher, M. and W. Maier. 2010. Current state and perspectives of fungal DNA barcoding and rapid identification procedures. *Appl Microbiol Biotechnol* 87, 99–108.

Callahan, B.J., McMurdie, P.J. and S.P. Holmes. 2017. Exact sequence variants should replace operational taxonomic units in marker-gene data analysis. *ISME J* 11, 2639–2643.

Caporaso, J.G., Kuczynski, J., Stombaugh, J. et al. 2010. QIIME allows analysis of high-throughput community sequencing data. *Nat Methods* 7, 335–336.

Caporaso, J.G., Lauber, C.L., Walters, W.A. et al. 2012. Ultra-high-throughput microbial community analysis on the Illumina HiSeq and MiSeq platforms. *ISME J* 6, 1621–1624.

Cole, J.R., Wang, Q., Cardenas, E. et al. 2009. The Ribosomal Database Project: improved alignments and new tools for rRNA analysis. *Nucl Acids Res* 37, D141–D145.

Deng, Y., Jiang, Y.H., Yang, Y. et al. 2012. Molecular ecological network analyses. *BMC Bioinf* 13, 113.

DeSantis, T.Z., Hugenholtz, P., Larsen, N. et al. 2006. Greengenes, a chimera-checked 16S rRNA gene database and workbench compatible with ARB. *Appl Environ Microbiol* 72, 5069–5072.

Dini-Andreote, F., Pereira e Silva, M.C., Triadó-Margarit, X. et al. 2014. Dynamics of bacterial community succession in a salt marsh chronosequence: evidences for temporal niche partitioning. *ISME J* 8, 1989–2001.

Faust, K., Sathirapongsasuti, J.F., Izard, J. et al. 2012. Microbial co-occurrence relationships in the human microbiome. *PLoS Comput Biol* 8, e1002606.

Fierer, N. and R.B. Jackson. 2006. The diversity and biogeography of soil bacterial communities. *Proc Natl Acad Sci USA* 103, 626–631.

Fox, G.E., Stackebrandt, E., Hespell, R.B. et al. 1980. The phylogeny of prokaryotes. *Science* 209, 457–463.

Friedman, J. and E.J. Alm. 2012. Inferring correlation networks from genomic survey data. *PLoS Comput Biol* 8, 1002687.

Kuczynski, J., Lauber, C., Walters, W.A. et al. 2012. Experimental and analytical tools for studying the human microbiome. *Nat Rev Genet* 13, 47–58.

Lindahl, B.D., Nilsson, R.H., Tedersoo, L. et al. 2013. Fungal community analysis by high-throughput sequencing of amplified markers – a user's guide. *New Phytol* 199, 288–299.

Lovell, D., Müller, W., Taylor, J., Zwart, A. and C. Helliwell. 2010. Caution! Compositions! Technical report and companion software (publication-technical). *Technical Report EP10994*, CSIRO.

McMurdie, P.J. and S. Holmes. 2014. Waste not, want not: why rarefying microbiome data is inadmissible. *PLoS Comput Biol* 10, e1003531.

Navas-Molina, J.A., Peralta-Sánchez, J.M., González, A. et al. 2013. Advancing our understanding of the human microbiome using QIIME. In: E.F. de Long (ed.), *Microbial Metagenomics, Metatranscriptomics, and Metaproteomics*, Elsevier, London, pp. 371–344.

Pruesse, E., Quast, C., Knittel, K. et al. 2007. SILVA: a comprehensive online resource for quality checked and aligned ribosomal RNA sequence data compatible with ARB. *Nucl Acids Res* 35, 7188–7196.

Reshef, D.N., Reshef, Y.A., Finucane, H.K. et al. 2011. Detecting novel associations in large data sets. *Science* 334, 1518–1524.

Ruan, Q., Dutta, D., Schwalbach, M.S. et al. 2006. Local similarity analysis reveals unique associations among marine bacterioplankton species and environmental factors. *Bioinformatics* 22, 2532–2538.

Schloss, P.D., Westcott, S.L., Ryabin, T. et al. 2009. Introducing Mothur: open-source, platform-independent, community-supported software for describing and comparing microbial communities. *Appl Environ Microbiol* 75, 7537–7541.

Staley, J.T. and A. Konopka. 1985. Measurement of *in situ* activities of non-photosynthetic microorganisms in aquatic and terrestrial habitats. *Ann Rev Microbiol* 39, 321–346.

Vos, M., Quince, C., Pijl, A.S., Hollander, M. and G.A. Kowalchuk. 2012. A comparison of *rpoB* and 16S rRNA as markers in pyrosequencing studies of bacterial diversity. *PLoS One* 7, e30600.

Vos, M., Wolf, A.B., Jennings, S.J. and G.A. Kowalchuk. 2013. Micro-scale determinants of bacterial diversity in soil. *FEMS Microbiol Rev* 37, 936–954.

Walters, W.A., Hyde, E.R., Berg-Lyons, D. et al. 2015. Improved bacterial 16S rRNA gene (V4 and V4–5) and fungal internal transcribed spacer marker gene primers for microbial community surveys. *mSystems* 1, e00009–e00015.

Woese, C.R. and G.E. Fox. 1977. Phylogenetic structure of the prokaryotic domain: the primary kingdoms. *Proc Natl Acad Sci USA* 74, 5088–5090.

14 Soil Metagenomics
Deciphering the Soil Microbial Gene Pool

Sara Sjöling
Södertörn University

Jan Dirk van Elsas
University of Groningen

Francisco Dini Andreote
Netherlands Institute of Ecology (NIOO-KNAW)

Jorge L. Mazza Rodrigues
University of California

CONTENTS

14.1 Introduction 227
14.2 Principles of Soil Metagenomics 228
14.3 Soil Microbiome DNA Isolation 229
14.4 Direct Sequencing Approaches 230
14.4.1 Direct Sequencing Techniques 230
14.4.2 Key Considerations for Successful Metagenomics-Based Analysis of Soil 231
14.4.3 Analysis of Metagenomics Data 232
14.4.4 Construction of Genomes from Metagenomes 232
14.4.5 Statistical Significance of Metagenomic Profiling 233
14.5 Targeted Gene Discovery and Bioprospecting 233
14.5.1 Metagenome Libraries, Vectors, and Hosts 235
14.5.2 Screening of Metagenome Libraries 236
14.5.2.1 Sequence-Based Screening 236
14.5.2.2 Functional Screening 236
14.6 Soil Metagenomics to Assess and Understand Microbial Ecology and Evolution 237
14.7 Examples of Successful Metagenomics-Based Studies of Soils 238
14.8 Commercialization of Activity-Based Metagenomic Products 239
14.9 Future Contribution of Soil Metagenomics to Different Scientific Aims 240
Acknowledgments 241
References 241

14.1 INTRODUCTION

In previous chapters, we learned that the traditional culture-dependent methods to study soil microorganisms poorly represents the extant microbial diversity in soil. Depending on the soil type, as little as about 1% of the microbial cells present is amenable to studies through cultivation, leaving up to 99% of the total community unexplored. Thus, advanced culture-independent methods to unravel microbial diversity, on the basis of molecular markers that are present in the microbiome DNA, have demonstrated the existence of an immense and unexplored diversity of microorganisms in soil. Many of these microorganisms were shown to have no known cultured representatives to date. The analysis of sequences of the gene encoding the RNA present in the small subunit of the ribosome [i.e., the 16S or 18S ribosomal RNA (rRNA) genes, used as a universal molecular marker for, respectively, bacteria and fungi; see Chapter 3] has suggested that there are commonly several thousands of different microbial species (expressed either as exact sequence variants or as operational taxonomic units per gram of soil: Thompson et al. 2017). However, these diverse organisms remain unexplored with respect to their genomic characteristics.

A remedy to the lack of information on the extent of microbial diversity present in soils consists in unlocking the collective genomes of the microbial community as a whole, that is, to access the so-called soil community *metagenome*. Such a soil metagenome contains much more genetic information than what can be found in the culturable subset of organisms. It provides an enormous, untapped and novel gene pool that enables studies on the ecological functioning and evolution of soil microorganisms. Moreover, it also serves as a basis for the identification of previously unknown soil microbial species, biochemical pathways, as well as genes and biomolecules for biotechnological exploration and utilization.

Soil metagenomics is defined as the analysis of—or the access to—the collective genomes that are present in soil microbiomes. It was for a better understanding of soil functioning, for example, with respect to how environmental change impacts the soil ecosystem, as well as a demand for compounds of agricultural, pharmaceutical, or other industrial interest.

Given the rapid development of this field, with the radical development of DNA sequencing and other metagenomics tools, there are various routes to the metagenomics-based analysis of soil samples. Generally, it can be achieved by two divergent approaches:

1. Direct shotgun sequencing of the soil microbiome DNA
2. Cloning of DNA fragments into bacterial or fungal vectors for the construction of metagenomic libraries, followed by activity- and/or sequence-based selection and analysis of gene products from library clones

The success of soil metagenomics—genomics on a grand scale—thus depends on the successful isolation of soil microbial DNA, cloning technologies, the recently developed mass sequencing technologies, as well as bioinformatics, statistical analyses, high-throughput screening, and biomolecule production platforms. The aim is to detect, isolate, explore, and/or exploit selected target genes or derived molecules (gene products) from the soil microbiome. In addition, metagenomics can be used to generate information on the functional capacity of individual microbial species as well as the whole community in the soil environment. This may serve to ascertain the broader role of microorganisms in the soil ecosystem and/or to establish microbial properties that determine soil quality (Vestergaard et al. 2017).

Here, we examine the state of the art of soil metagenomics approaches, addressing both the technical achievements and challenges, and the progress made with respect to our understanding of soil microbial diversity and function.

14.2 PRINCIPLES OF SOIL METAGENOMICS

Soil metagenomics provides a unique access route to the genes present in the genomes of uncultured soil organisms, offering a broad spectrum of parameters of soil microbial communities, ranging from single genes to metabolic pathways and/or to entire genome reconstructions. No prior knowledge of gene sequences is required, and multiple screening methodologies can be applied. However, soil metagenomics is also subjected to substantial technical constraints. First, biases in the metagenome data resulting from (1) DNA extraction inefficiencies, in particular due to the difficulties of obtaining high-quality DNA representative of the community/population sampled, and (2) cloning and/or sequencing inefficiencies (due to sensitivities of the respective cloning/sequencing tools to soil compounds such as humic substances) may pose significant challenges (for a careful review of the intricacies associated with working with soil DNA and RNA, see Chapters 12 and 15).

Furthermore, both the direct shotgun sequencing and the clone library construction strategies only allow examination of the "tip of the iceberg" of the soil microbiome, which may leave the "rare biosphere" inaccessible (Delmont et al. 2015). This also depends on predefined sequencing and cloning depths. In many cases, the sequencing of DNA obtained from high-microbial-diversity soils will be stochastic, meaning that whereas dominant genes will invariably show up, rare genes may be haphazardly present at low and erratic frequencies. This poses particular difficulties for evaluating functional gene detection and isolation (Gabor et al. 2003).

In the cloning approach, current host/vector systems, although useful, may not be compatible with particular genes, which precludes the detection of these genes by expression analysis. This problem may be overcome through a sequence-based approach by synthetically matching the sequence of the cloned gene with the bacterial host's codon usage, as recently shown by Maruthamuthu et al. (2017). It is also of critical importance that one takes into account whether bacterial or eukaryotic DNA is cloned and accessed, as this may guide vector choice and thus facilitate success in (heterologous) gene expression. Currently, a large suite of bacterial host/vector systems is available, enabling broader expression-based screening than before. The successful development of large-scale fungal expression platforms (for secondary metabolites produced from biosynthetic gene clusters) has significantly promoted the cloning of the relevant (highly demanded) clusters (Clevenger et al. 2017). Robotics-based heterologous expression screening using efficient scoring pipelines may now allow the identification of novel metabolites from full-length biosynthetic gene clusters present in diverse fungal species (Clevenger et al. 2017). The success of metagenomics analyses is, in certain cases, even independent of the size or representative nature of a library, as an inherently biased library may still yield the gene products of interest.

The different steps in the metagenomics-based analysis and exploration of soil microbiomes are shown in Figure 14.1 and will be detailed in the following sections. They include the direct isolation and purification of high-quality DNA from soil samples without prior enrichment of the organisms, followed by either direct sequencing or cloning of the DNA into a metagenomic library and subsequent analysis of the cloned fragments using sequence-based or activity-based screening approaches.

14.3 SOIL MICROBIOME DNA ISOLATION

Chapters 12 and 13 provide discussions on soil DNA isolation. As we have seen, the protocols for the isolation of DNA from soil microbiomes are typically divided into direct and indirect extraction methods, and the choice for an extraction method will depend on the project goals. When using direct methods, soil microorganisms are lysed directly in the soil matrix and total microbiome DNA is then obtained. Indirect methods aim to dislodge microbial cells from the soil particles and organic matter to which they are attached. A homogenizer is often used to increase extraction yields, which causes the genomic DNA to fragment. This is not a problem for short-read direct sequencing approaches. However, if the goal is to identify complete biosynthetic pathways, for example, in a polyketide synthase cluster, a method that yields larger genomic DNA fragments, for example, of >50 kb on average, should be considered. The challenge is the construction of a useful genomic library from high-molecular-weight soil DNA. An alternative is the physical separation of cells from soil using density gradient centrifugation or differential centrifugation, followed by immobilization of the soil-extracted cells in low-melting-point agarose plugs and *in situ* lysis with protease and lysozyme. The genomic DNA can then either be digested with rare-cutting restriction

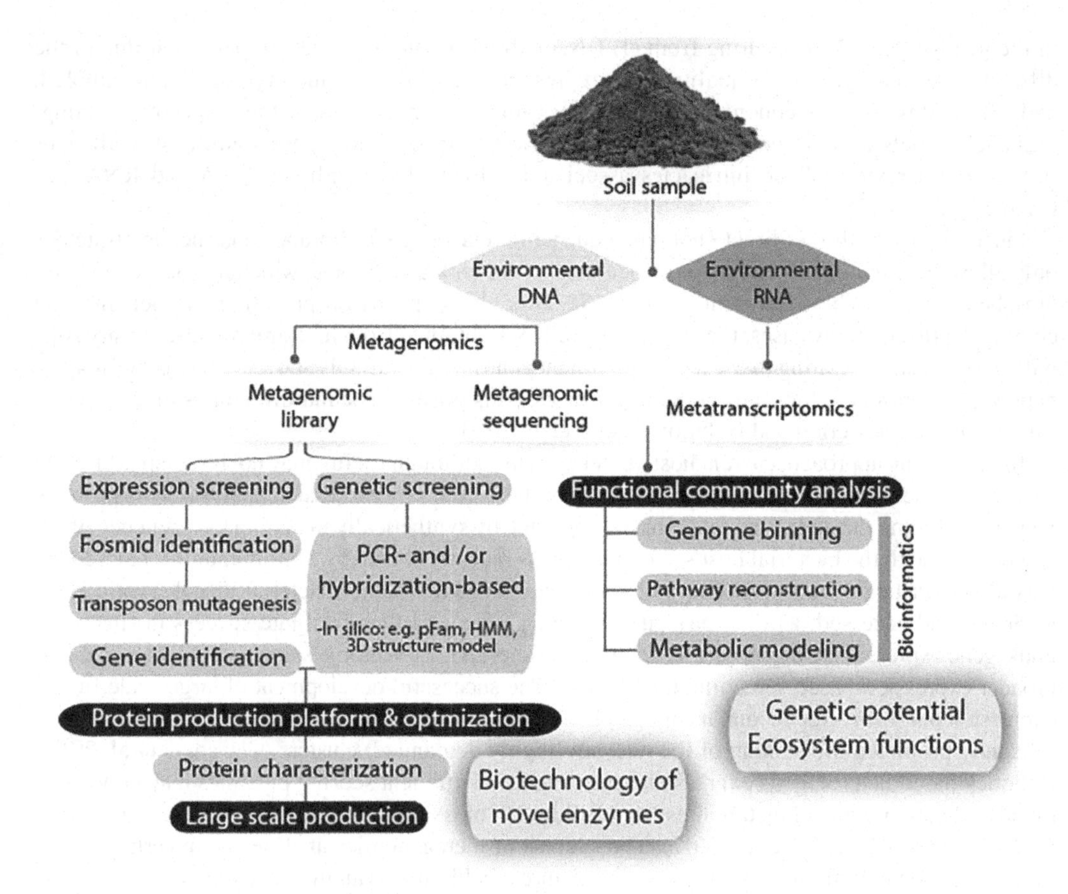

FIGURE 14.1 The principles of soil metagenomics. Explanation: HMM: sequence analysis using profile hidden Markov Models (see Section 14.4.3); pFam: database of protein families; 3D: threedimensional.

enzymes or end-repaired before separation by pulsed-field gel electrophoresis. The latter step is advantageous, as it enables the purification of cellular DNA from soil contaminants while minimizing DNA fragmentation. Soils frequently contain high concentrations of polyphenolic compounds such as humic acids, which tend to co-purify with the DNA, thus inhibiting restriction enzymes and cloning procedures.

14.4 DIRECT SEQUENCING APPROACHES

14.4.1 Direct Sequencing Techniques

Following the initial work (see Rondon et al. 2000, Tyson et al. 2004, Tringe et al. 2005), direct shotgun sequencing of soil-derived DNA has skyrocketed over the past 10–15 years. This is likely due to the development of efficient and fast soil DNA isolation kits, the rapid reduction in sequencing costs, and the increasing availability of user-friendly bioinformatics tools. To date, close to a thousand studies have been published on soil metagenomes and significantly more on soil microbiomes. Global soil microbiome surveys, for example, the Earth Microbiome Project (www.earthmicrobiome.com), TerraGenome (www.terragenome.org), the Brazilian Microbiome Project (www.brmicrobiome.org), the China Soil Microbiome Initiative (http://english.issas.cas.cn/), EcoFINDERS (http://ecofinders.dmu.dk/), and the MicroBlitz citizen science project (www.microblitz.com.au/), recently have explored and deciphered the taxonomic and functional diversity of microbiomes across disparate soil types. Nonetheless, the degree to which data from these endeavors reflect the original soil community compositions is still unknown, given the biases in soil DNA isolation methods and data analyses. Current sequencing techniques utilized in metagenomics include third-generation sequencing and may range from "Illumina HiSeq" (Illumina Inc., San Diego, California) to long-read single-molecule real-time chemistry (Pacific Biosciences of California Inc., Menlo Park, California). The latter procedure is less amenable to misassembly of sequence reads. Remarkably, compared with early shotgun sequencing tools, a portable field platform, that is, the MinION (Oxford Nanopore, Oxford, UK) that was recently developed, has shown promising results (Edwards et al. 2016).

14.4.2 Key Considerations for Successful Metagenomics-Based Analysis of Soil

Although the theoretical framework of a soil microbiome study may be well designed, there are practical considerations that need attention to achieve successful metagenomics-based analyses. In the light of the highly diverse microbial communities observed in soils, researchers have asked how much sequencing should be allotted to a single soil sample to yield a scientifically robust data set (Rodriguez and Konstantinidis 2014). This is not only a scientific question (Is the proposed sequencing depth enough for species coverage within my sample of interest?) but also a technical/pragmatic question (What are the costs of my metagenomics project?). These questions are not easily answered, but a few considerations prior to the start of any metagenomics project may help to guide priorities and achievable goals:

1. **Target group (or species) of interest**: Soil microorganisms have a range of genome sizes, varying from about 1.8 Mb in some archaea to 13 Mb in myxobacteria, with an estimated average genome size of 4.7 Mb (Raes et al. 2007). Prior knowledge of the genome size of a particular microbial group will help with estimates of the requirements of metagenomics approaches.
2. **Abundance of a microbial population**: Different microbial taxa have different relative abundances depending on soil conditions. While the contribution of Proteobacteria can reach up to 40% of the total bacteriome in a given soil, members of the subphylum δ-Proteobacteria may comprise only up to 2% of this total (Janssen et al. 2006). The rarest populations may even consist of less than a hundred individuals in a given sample. Profiling of the microbiome through amplicon sequencing on the basis of the 16S rRNA marker gene can greatly assist in this step (see also Chapters 3 and 13). One caveat associated with the use of amplicon sequencing to estimate the microbial diversity of soil is the potential bias of so-called universal primers. A novel approach, which combines the addition of a poly(A) tail to sequences followed by reverse transcription, was shown to double the number of sequences available in rRNA databases (Karst et al. 2018). The approach may be particularly suitable for rare soil biosphere microorganisms, such as soil archaea and other low-abundance taxa.
3. **Sequence coverage**: High-quality draft microbial genomes require about tenfold coverage per nucleotide. It is assumed that this target has yet to be obtained in soils, where an estimated 50 Tb of sequence would be required to describe all (dominant) genomes present in a single gram (Howe et al. 2014). As sequencing technologies almost continually increase output, this issue will become less important in the near future, in particular with the development of increasing overlaps in the same DNA region being sequenced.

Given the above challenges, the use of key sequencing simulator software, such as MetaSim (Richter et al. 2008) or CONCOCT (Alneberg et al. 2014) can be advantageous, as it allows to estimate the number of reads needed for a metagenome and the number of samples that are required to be sequenced to achieve sufficient coverage. Sample coverage has two aspects in the context of work on soil metagenomes: one is related to the mean number of times that a particular nucleotide is sequenced, as a confirmation that reads will overlap, facilitating assembly (e.g., threefold coverage means that a nucleotide was sequenced three times), and the other one is statistical coverage, a measure that can provide confidence intervals for the observed results. Both issues should be carefully considered during the design phase of any metagenomics project.

14.4.3 Analysis of Metagenomics Data

To gain access to the genetic and biological information within a metagenomics dataset, the millions or billions of sequence reads (which have been quality-checked, trimmed, filtered, and sorted) are often assembled into larger genome fragments, so-called contigs (i.e., contiguous sequences). Numerous routes, or pipelines, for sequence preparation and analysis, including assembly, have been recently developed (for a list of software packages and evaluation of their performance, please see Sczyrba et al. 2017). Approaches applied for metagenomic short-read (e.g., paired-end or mate-paired) assembly are usually *de novo* assembly strategies. Alternatively, comparative assembly procedures can be used, which take advantage of existing genome information. In the assembly approach, the contiguous sequences (*contigs*) provide access to coding regions that allow better prediction of functional capacities or pathways. As mentioned, population size, the size of the dataset or sequence depth and coverage, together with nucleotide composition and frequencies and the occurrence of conserved genomic regions, may collectively affect the quality of the assembly. There are contrasting estimates with respect to the depths needed for successful assembly/binning. Whereas Howe et al. (2014) posited that depths of up to 50 Tb are required to recover the genomes of the majority of species in soil, Hultman et al. (2015) reasoned that 100 Gb per sample would be necessary for successful binning of the reads, based on the assumption that samples were from average soils. Ultimately, the research aims will dictate the required number and coverage of the metagenomic reads. For instance, if the reconstruction of major metabolic pathways is the goal, lower than "complete" sequencing depth may suffice (Vestergaard et al. 2017; and references therein). As the delivered metagenomics-based datasets become larger, reaching terabase dimensions, the suitability of common genome assemblers becomes limited, raising the need for specific powerful metagenome assemblers.

For the analyses, different algorithms can be applied according to specific metagenomic datasets, for example:

1. Use of fixed-length *k-mers* for *de novo* assemblies—*k-mers* are DNA fragments of a defined size that are used to create so-called de Bruijn graphs that represent the overlaps between the sequences. This approach is particularly useful for high-coverage, short reads, which can be used for *de novo* assembly projects.
2. Gene-targeted assembly—The approach uses multiple sequence alignments from a gene family to build a hidden Markov model (HMM) that guides the assembly (Wang et al. 2015, Li et al. 2017). While the method was not designed for novel gene discoveries, it can facilitate the assembly of operons with well-known functions, for example, methanogenesis, steps in the nitrogen and sulfur cycles.

14.4.4 Construction of Genomes from Metagenomes

With the advent of high-throughput sequencing and improved assembly algorithms, we are now able to perform cultivation-independent recovery of complete genomes from metagenomes, also termed *genome-resolved metagenomics generating metagenome-assembled genomes* (MAGs) (Parks et al. 2017). Genome sequence data produced from metagenomics may also help to design selective isolation strategies to cultivate organisms based on the presence, absence, or distribution of particular metabolic capacities. Some of the early reconstructed near-complete genomes of a shotgun metagenome are reported in the Acid Mine Drainage study (Tyson et al. 2004). From the reconstruction of a few genomes from the low-diversity extreme subsurface habitat, recent results (by the same researcher) show the reconstruction of several thousands of near-complete genomes; these are now available in publicly accessible metagenome datasets (Parks et al. 2017). Such approaches hold the potential of large-scale recovery of complete or near-complete genomes, that are otherwise primarily obtained through often tedious single-cell genomics. These genomes, however, may consist of mergers of a suite of highly similar genomes that co-occur in the sample and emerged as a result of local divergence. They are obtained by "binning" the assembled contigs with similar sequence compositions, depth of coverage across one or more related samples, and taxonomic affiliations. Several different approaches can be used to filter, order, and sort sequence reads. Thus, several parameters are used to create the genomic "bins," for example, the tetranucleotide and *k-mer* frequencies and coverage of sequences. Numerous bioinformatics-based methods and algorithms are currently available that exploit sources of information to effectively retrieve genomes from metagenomics data. See Albertsen et al. (2013) for some proposed pipelines, with developed variants CONCOCT (Alneberg et al. 2014) and MetaBAT (Kang et al. 2015). Of critical importance is the ongoing efforts to evaluate the effectiveness of the different approaches. A detailed review on the performance of methods, tools, and benchmarking analytical aspects can be found in Sczyrba et al. (2017).

An excellent example of the power of soil metagenomics can be found with the reconstruction of 17 novel bacterial genomes from a 540 Mb metagenome constructed from soil (Delmont et al. 2015). Soil enrichment techniques followed by metagenomics were able to capture the genetic information of microorganisms that were thought to be rare (0.0001%), including new taxa. The genomes of these rare members would likely have remained unassembled without *in situ* enrichment even after extraordinary sequencing depth and bioinformatics efforts. Moreover, a new standard that specifies the minimum information needed about a MAG (Bowers et al. 2017) has been proposed, including information on assembly quality and estimates of genome completeness and the maximum accepted rate of genomic contamination.

14.4.5 Statistical Significance of Metagenomic Profiling

A common goal of projects that apply in-depth sequencing of metagenomes is to identify biologically relevant results through taxonomic and/or functional alterations between treatments or conditions. Although biologically important functions may not always be statistically significant, careful evaluation of sampling size and variation will help with data interpretation. A particularly attractive option for researchers starting their first metagenomics project is to take advantage of software with user-friendly graphical interfaces, such as Statistical Analyses of Metagenomic Profiles (STAMP) (Parks and Beiko 2010). Sequence reads that were annotated through MG-RAST (Meyer et al. 2008), MEGAN (Huson et al. 2009), or IMG/M (Markowitz et al. 2008) can be tabulated and used as inputs to calculate confidence intervals for variation observed in taxonomic and functional profiles.

An alternative method that allows to evaluate the statistical significance of metagenomics-based results is to normalize samples of different sequencing depths and calculate the variation in the proportion of annotated genes for specific biogeochemical functions. This alternative, however, requires the metagenomics data to be tested for normal distribution, as not all sequences of interest will be present in every sample. If necessary, the data can be transformed to establish normality prior to the statistical analyses that are performed. A discussion of normal versus non-normal data distributions and the methods used for testing and processing are given in Chapter 19.

14.5 TARGETED GENE DISCOVERY AND BIOPROSPECTING

There is a demand for novel and sustainably produced biomolecules for pharmaceutical, biotechnological, and other industrial applications, such as in medicine and agricultural biocontrol. Therefore, screening of soil microbiome genomic contents for targeted gene discovery is an important aspect of metagenomics. For example, a research team from the Rockefeller University in New York discovered a completely new class of antibiotics that are potentially powerful against multiresistant "superbugs" (Hover et al. 2018). This is described in more detail in Box 14.1. Hereunder, we examine the approaches, challenges, and pitfalls of such metagenomics-based gene discoveries.

BOX 14.1 A METAGENOMICS-BASED STRATEGY FOR THE DISCOVERY OF A NEW CLASS OF CALCIUM-DEPENDENT ANTIBIOTICS FROM THE GLOBAL SOIL MICROBIOME

Many potentially powerful microbial natural products are hidden in the global soil microbiome. Unlocking this resource of chemistries encoded by the uncultured majority is highly attractive for society but also challenging. Through an innovative approach and by combining targeted genetic screening with activity-based screening, a new class of antibiotics was discovered by metagenomic screening of over 2,000 soil samples from across the United States (Hover et al., 2018). A team led by S. Brady (The Rockefeller University, New York) developed the culture-independent natural product discovery platform that involves sequencing, bioinformatics analysis, and heterologous expression of biosynthetic gene clusters from metagenomes. The team specifically targeted calcium-dependent antibiotics because individual family members had been shown to have discrete modes of action, targeting either cell wall biosynthesis or cell membrane integrity. The team hypothesized that the conserved Asp-X-Asp-Gly calcium-binding motif might be indicative of a broader collection of uncharacterized antibiotics (a). One distinct eDNA-specific clade was found in 19% of the metagenomes (b). Through heterologous production of metabolites biosynthesized by non-ribosomal peptide synthetases (NRPSs) and metabolite profiling, they identified a new class of calcium-dependent antibiotics, named the malacidins (metagenomic acidic lipopeptide antibiotic-cidins) (c). Excitingly, the malacidins show activity against multidrug-resistant pathogens and sterilized methicillin-resistant *Staphylococcus aureus* skin infections in animal wound models.

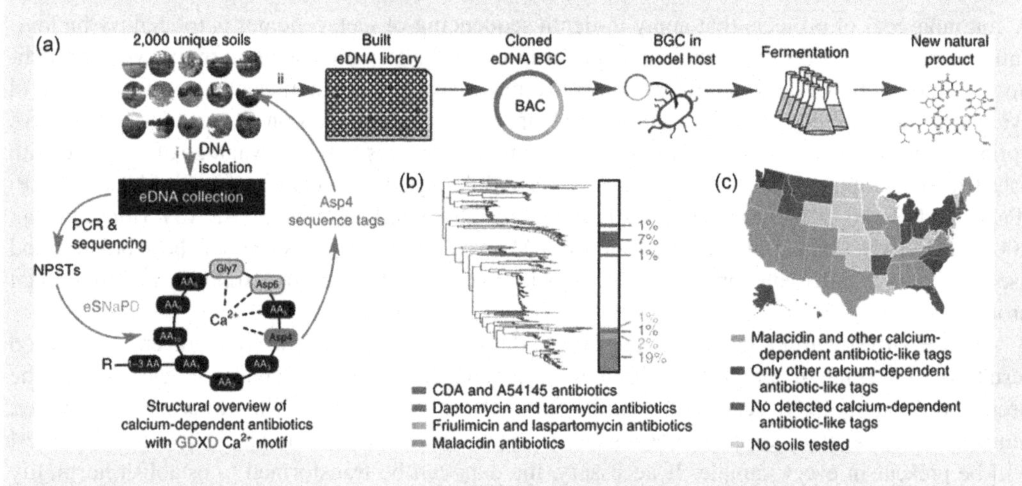

(See color insert)

This image and figure legend were reproduced from the article: Hover, B.M., Kim, S.H., Katz, M. et al. 2018. Culture-independent discovery of the malacidins as calcium-dependent antibiotics with activity against multidrug-resistant Gram-positive pathogens. *Nat Microbiol*, February 12, 2018, Springer Nature ISSN 2058-5276 (online) according to the Creative Commons license (http://creativecommons.org/licenses/by/4.0/) and with permission from Sean F. Brady and *Nature Microbiology*, Springer Nature Publishing.

14.5.1 Metagenome Libraries, Vectors, and Hosts

A sampled metagenome can be "stored" and "handled" using metagenome libraries. These are convenient for functional metagenomics using activity-based screening but are also amenable to sequence-based screening.

The selection of a suitable vector–host system to prepare a clone library from soil DNA has to take into account the targeted activity, the screening strategy, as well as the size of the DNA fragments to be cloned. Requirements for successful heterologous expression systems include the existence of compatible genetic elements, either encoded by the vector or the cloned DNA fragment, for successful transcription and translation of the targeted genes, or NRPS production of biomolecules. Correct transport, sorting, and folding by the host system are also necessary if the target is a protein, in activity-based screenings.

Vectors: For expression screening, typically libraries with small inserts (up to a few kilobases) will be constructed using plasmid-based vectors. Moreover, libraries with medium inserts (up to several tens of kilobases) will be based on cosmid, fosmid, or phage λ (lambda)-based vectors; and libraries with larger inserts (up to 100–200 kb) on bacterial artificial chromosome (BAC) or fungal artificial chromosome (FAC) vectors. In addition, the stability of the cloned DNA is best maintained in vectors of which the copy number can be regulated. *Escherichia coli*-based systems are commonly used since the promoter of an *E. coli* vector is often inducible to higher copy number and sufficient to drive the expression of small genes. A large suite of small or copy control medium-sized insert vectors has been used. Thus, screening for expression of soil-derived genes has allowed the identification of novel chitinases (Hjort et al. 2014, Cretoiu et al. 2015), amylases, agarase, dioxygenase, esterase/lipase, acetyl hexosaminidases, fucosidases, xylosidases, glucosidases, and endoglucanases (for a review, see Berini et al. 2017a).

Large-insert vectors, that is, BAC or FAC vectors, increase the probability of cloning whole genes of larger sizes, of operons encoding complete pathways, such as antibiotic biosynthesis clusters, and of indigenous promoter regions, improving the chances of gene expression (Wang et al. 2014). Currently, large DNA fragment cloning can also be outsourced to companies, for example, using FAC or BAC library preparation (Intact Genomics Inc., St. Louis, Michigan). Vector innovations that may be advantageous include the fusion of the gene target to a leader sequence for extracellular secretion and the use of promoters that drive low basal expression and tight regulation of expression levels (e.g., using a tetracycline resistance promoter) (T-REx™ System; Thermo Fisher Scientific, Waltham, MA).

Nowadays, the demand for improved metagenomic library systems and increased screening efficiency has led to the development of novel platforms with unprecedented capacities. Examples include the large-scale adapted system for sequence-based screening and barcoding in antibiotics and secondary metabolite analysis (anti-SMASH system; Blin et al. 2013). Other novelties are the SMART-BAC/-FAC vectors (Clevenger et al. 2017) and the ultra-high-throughput screening method using a pooling strategy (e.g., BAC Sudoku; Varigen Biosciences, Madison, Wisconsin) and microfluidic nanopore systems that allow the detection of biomolecules (Clevenger et al. 2017).

Shuttle vectors have been developed for heterologous DNA expression in more than one host (Lewin et al. 2017), since the use of alternative hosts with different genetic and metabolic systems may increase the chance of expression of heterologous gene products encoded by the metagenomic DNA. For example, the pPAC-S2 vector is an *E. coli–Streptomyces* artificial chromosome shuttle vector that has been used to clone and express soil DNA of up to 100 kb. The use of the SuperBAC1 vector may also increase hit rates, as it has been modified for a range of alternative hosts, including species of *Bacillus*, *Enterococcus*, *Listeria*, *Staphylococcus*, and *Streptococcus*.

Hosts: The selection of hosts is best guided by the efficiency of vector transformation as well as the stability of the vector. For expression screening, the host cell must also have the genetic expression machinery that is capable of expressing the encoded protein and exporting it out of the cell without background expression of the function to be screened. Several commercially available bacterial hosts can aid in the screens—including mutations for endogenous proteases and the co-expression of chaperones or disulfide isomerases to aid in heterologous protein folding. *Streptomyces* has been shown to express a significantly higher number of metagenomics-derived genes in comparison with *E. coli*, which contains only half of the RNA polymerase sigma factors contained in *Streptomyces* (Katz et al. 2016). Other hosts include *Agrobacterium*, *Rhodococcus*, *Pseudomonas putida*, *Bacillus subtilis*, *Burkholderia glumae*, *Rhodobacter capsulatus*, *Sinorhizobium meliloti*, *Thermus thermophilus*, and *Gluconobacter oxydans*. A few archaeal systems have also been used, for example, *Sulfolobus solfataricus* (Liebl et al. 2014). Heterologous expression in eukaryotic systems requires the construction of a cDNA library from environmental mRNA, as outlined in Marmeisse et al. (2017). Similarly to archaeal genes, eukaryotic genes cloned into *E. coli* are generally not efficiently expressed, requiring a eukaryotic host such as *Saccharomyces*, *Trichoderma*, *Pichia*, *Aspergillus*, or *Kluyveromyces*.

14.5.2 Screening of Metagenome Libraries

As discussed previously, either sequence-based and/or function- or activity-based screenings can be used for examining soil metagenome libraries. Sequence-based screening may include random, end, and transposon-facilitated sequencing polymerase chain reaction (PCR) amplification of target genes and hybridization with oligonucleotide probes. Functional (expression) screening includes complementation, immunodetection, and detection of enzyme activity or of other bioactive properties. In addition to allowing gene and bio-compound discovery and phylogenetic identification, metagenomic DNA libraries can also provide relevant information for particular metabolic pathway reconstructions. They can also serve as tools in functional genomics for the preparation of microarrays to monitor the distribution, gene expression, and environmental responses of uncultured microorganisms.

14.5.2.1 Sequence-Based Screening

The genetic screening approaches include direct sequencing of cloned fragments, so-called *in silico* mining, or gene-targeted analyses, via probes or degenerate primers of conserved sequence regions of interest. Conserved DNA sequences can be used to construct HMMs, which can be used in the design of PCR primers or hybridization probes in robotics-based high-throughput screenings for detection of the sequences of interest. Enrichment of specific environmental nucleotide sequences was first developed by a commercial company, as biopanning. Nowadays, large genomic regions harboring gene clusters of interest have been explored using target capture probes combined with sequencing of the enriched genomic fragments (Denonfoux et al. 2013). However, the gene-targeted approach imposes a bias towards the recovery of known sequences. Combined with activity-based screening, this route may, however, function as a pre-sieving step to resolve some of the screening bottlenecks in cases of low hit rates of clones detected by activity. Heterologous expression of genes enriched through such pre-sieving on the basis of homology has resulted in the discovery of novel enzymes as well as classes of putative antibiotics from soil (Cretoiu et al. 2015, Hover et al. 2018).

With the fast development of the gene-targeted approach through ultra-high-throughput screening and new bioinformatics tools for mining complete sequenced genomes for secondary metabolite biosynthetic clusters (e.g., anti-SMASH, Blin et al. 2013), the potential for the identification of novel natural products, such as antibiotics (Hover et al. 2018), is promising (Box 14.1).

14.5.2.2 Functional Screening

Functional or activity-based screening is achieved through selection for a phenotype, which can be an enzymatic activity or a metabolite produced by the library clones. Thus, genes that encode enzymes (or compounds) with low identities to known sequences can be identified, for example, colorimetrically, fluorimetrically, spectrophotometrically, antibacterially, or by complementation (see Chow and Streit 2016). Examples of heterologously expressed enzymes identified from soil metagenome libraries are numerous (for reviews of recently identified enzymes and compounds, see Berini et al. 2017a and Katz et al. 2016, respectively).

The success of activity-based screening depends on the efficiency of the heterologous expression. Hit rates are typically low since functional screening often results in low frequencies of recombinant clones that are able to express the selected characteristic in a functional form and at detectable levels (Sjöling et al. 2007). This is problematic, even with the application of robotics. The phenotypic selection has conventionally been performed on solid media (such as agar plates), on a membrane or in liquid medium on microtiter plates or by flow cytometry-assisted cell sorting, as mentioned. Today, it is understood that efficient ultra-high-throughput systems are essential as enough DNA "sequence space" is to be covered. For example, monodispersed cells using water-in-oil droplets, dispersed in microfluidic chips and exposed to ultra-high-throughput micro- or pico-fluidic cell sorting (Ufarté et al. 2016), were used to identify genes for novel hydrolases in pooled metagenome libraries (Colin et al. 2015).

As outlined in Sections 14.4 and 14.5, enrichment of the target sequence is sometimes used, as in the case of sequences of rare organisms (Ufarté et al. 2016). An extension of the enrichment approach is stable isotope probing (SIP) (e.g., using a ^{13}C-labeled substrate; see Chapter 17) or bromodeoxyuridine (BrdU)-labeling of DNA or RNA, respectively. SIP identifies the metabolically active microbial populations (Hernández et al. 2017). Product-induced gene expression uses a reporter assay with the green fluorescent protein (Uchiyama and Miyazaki 2010) and metabolite-regulated expression to identify the target gene.

The prospect of accessing novel biologically active compounds from metagenome libraries is bright, with the recent development of micro-fluidic high-throughput screening. Hundreds of metagenome libraries prepared from soil DNA have been successfully screened for bioactive compounds and/or enzyme activities. However, few of the identified compounds have as yet become commercially available (Berini et al. 2017a). One technical bottleneck in achieving commercially valuable metagenome-sourced enzymes is reaching sufficiently high yields in protein production. Modification and optimization of a protein production platform may require at least as much effort as the initial metagenomic screening, depending on the respective protein. A strategy to reduce problems with inclusion bodies and to obtain a suitable yield was recently demonstrated for a soil-derived chitinase with biocontrol activity of a crop pest (see Berini et al. 2017b).

14.6 SOIL METAGENOMICS TO ASSESS AND UNDERSTAND MICROBIAL ECOLOGY AND EVOLUTION

Understanding the interactions between different soil organisms and their surrounding environment, that is, linking microbial communities to their ecological roles and functions, is an important task in soil science (Allen and Banfield 2005). Metagenomics provides a robust platform to assess soil microbial ecology questions, which include both the phenomenological and mechanistic understanding of microbial population interactions, biogeochemical activities, and general soil functioning. Analysis of microbial genomes across environmental gradients can yield clues about differences in the fitness of the organisms harboring these genomes and resolve the genetic and metabolic capacities of communities and their functions in the soil (Allen and Banfield 2005). For example, Ji and coworkers propose that the main sources of energy and carbon that support Antarctic soil communities are (soil) atmospheric H_2, CO_2, and CO, suggesting that such atmospheric energy sources provide an alternative basis for ecosystem function to solar or geological energy sources (Ji et al. 2017). Stable nitrogen and carbon analysis or substrate labeling integrated with metagenomics will be helpful to enhance our understanding of the metabolic activity of such soil populations. If microbiome genomic data analysis is integrated with concurrent gene expression studies, the determination of how microbe-driven functions are distributed among different members of the communities through space and time becomes possible (Faust et al. 2015). With strategic sampling schemes (Kuzyakov and Blagodatskaya 2015), biologically and ecologically relevant information can thus be obtained from comparative metagenomics (Vestergaard et al. 2017). Metagenomics may also enable the assessment of evolutionary processes, including natural selection, on the basis of the analysis of nucleotide substitutions and horizontal gene transfer events (Allen and Banfield 2005; see also Chapter 7). Studies focusing on the origin and drivers of the richness of soil microbiomes are still scarce (Prosser 2015). Comparative metagenomics, that is, the comparison of sequence data within and among microbiomes of spatial heterogeneity and at different scales, may reveal drivers behind the assembly, dispersal, stability, and evolutionary patterns and processes that lead to genome diversification and speciation (Allen and Banfield 2005, Tringe et al. 2005). Thus, comparative metagenomics aids in our understanding of these processes, allowing to predict the features of the soil environments sampled, such as the available energy sources and/or the extent of pollution (Tringe et al. 2005). Through integrating comparative metagenomics with other "meta-omics" approaches (metaproteomics, metatranscriptomics, metabonomics, and metaphenomics), a better understanding of the links between soil microbial taxonomic and functional diversity, activity, regulation, and ecosystem functioning can be reached.

14.7 EXAMPLES OF SUCCESSFUL METAGENOMICS-BASED STUDIES OF SOILS

To date, numerous metagenomics studies have been performed in soil, with the majority being untargeted. Here, we provide details of three studies that may serve as benchmarks for future-targeted soil metagenomic studies:

1. **Genome reconstruction from rare microorganisms**: A combination of *in situ* enrichment techniques for soils of a well-studied experimental site, Rothamsted Research, Harpenden, and metagenomics provided enough sequencing information for the *de novo* assembly of 17 draft bacterial genomes, 11 of which being without any close relatives. Some of the assembled genomes had an estimated "natural" population abundance of <0.0001% of the original soil microbiome (Delmont et al. 2015). The genomes of these rare members would likely have remained unassembled without *in situ* enrichment, even after extraordinary sequencing depth and bioinformatics efforts.
2. **Recreating ecosystem services for preagricultural soils**: Highly productive biomes, such as grasslands, have a dynamic relationship with belowground biogeochemical processes carried out by soil microbiomes. Metagenomics-based analyses of the last pristine prairies in the United States have revealed that the biogeographical patterns were largely driven by the abundance of members of the phylum Verrucomicrobia (Fierer et al. 2013), being directly correlated with genes for carbohydrate metabolism. Land use changes caused by agriculture may have caused functional alterations in these microbiomes, with direct implications for ecosystem productivity. Thus, metagenomics may guide restoration efforts of degraded soil resources.
3. **Regulation of ecosystem-scale process in a changing world**: Thawing of permafrost can result in carbon losses through greenhouse gas emissions, with positive feedback to Earth's climate. Using a combination of isotopic signature mass discrimination for CH_4 production and consumption, and metagenomics, McCalley et al. (2014) observed the CH_4 emissions that were associated with changes from hydrogenotrophic to acetoclastic methanogenesis. Metagenomic sequencing of partially thawed permafrost samples allowed for the recovery

of a near-complete genome of an archaeal microorganism, that was named *Candidatus* Methanoflorens stordalenmirensis (Mondav et al. 2013). Population abundance of the methanogen was posited to be a key predictor of CH_4 isotopic shifts. Incorporation of microbial community abundance and function into Earth system models may help us to improve predictions of climate change in the future. For instance, the incorporation of CH_4 processes (methanotrophy versus methanogenesis) and functional group attributes (abundance, composition, and activity) have been recommended to remediate uncertainties of the Community Earth System Model.

14.8 COMMERCIALIZATION OF ACTIVITY-BASED METAGENOMIC PRODUCTS

The evolution of microorganisms over billions of years has resulted in immense genetic diversity, providing physiological adaptations to allow survival in virtually every environmental niche on the Earth. Soil biotopes include many of these environments, allowing the survival of *psychrophilic*, *thermophilic*, *acidophilic*, *alkaliphilic*, *halophilic*, and other *extremophilic* organisms.

Microbial genomes represent large gene pools and thus constitute rich resources for agricultural, pharmaceutical, and industrial biotechnology. A single microbial genome harbors several thousands of protein-encoding genes, of which many might have wide commercial potential. The world's commercial enzyme market (estimated at over US$1 billion per annum) and the (multibillion dollar) antibiotic market have both developed from the exploitation of single genomes, typically through the cultivation of soil microbial isolates. The soil metagenome most likely contains a wealth of exploitable genes that await discovery. An excerpt from this functional richness, produced by enrichment cultures followed by metagenomics, is presented in the study described in Box 14.2. As in numerous previous studies, the authors discovered three novel enzymes that, in conjunction, were able to improve the hydrolysis of wheat straw by a commercially available enzyme preparation. Numerous (start-up and other) companies are active within this market sector. The discovery of a novel enzyme or bioactive compound is only the first step in a long and expensive developmental path, during which a very large proportion of primary "hits" are discarded for reasons of efficacy, toxicity, allergenicity, yield, and/or production costs.

BOX 14.2 METAGENOMICS-AIDED DISCOVERY OF NOVEL GLYCOSYL HYDROLASES THAT PROMOTE WHEAT STRAW HYDROLYSIS

Using soil-based enrichment cultures grown on wheat straw, Maruthamuthu and van Elsas (2017) identified, purified, and characterized the genes for several novel thermo-alkaliphilic enzymes that were active in the hydrolysis of hemicellulose. Cloning of three of these genes, denoted as P1, P5, and P6, enabled the production of the respective enzymes in *E. coli*. The main catalytic activities of the three proteins were characterized as beta-galactosidase (P1) and alpha-glucosidase (P5 and P6). Following purification, the activities of the three enzymes in the hydrolysis of, in particular, the hemicellulose moiety of steam-exploded wheat straw were confirmed (Maruthamuthu 2017). In one revealing experiment, the use of the three enzymes, in varying combinations, in a commercial enzyme preparation to boost the hydrolysis of steam-exploded wheat straw resulted in strongly enhanced production of the reducing sugars arabinose, glucose, and xylose (see the following figure). Enrichment culture-based metagenomics approaches offer excellent possibilities for further exploration of the (selectable) enzymatic potential present in soil microbiomes.

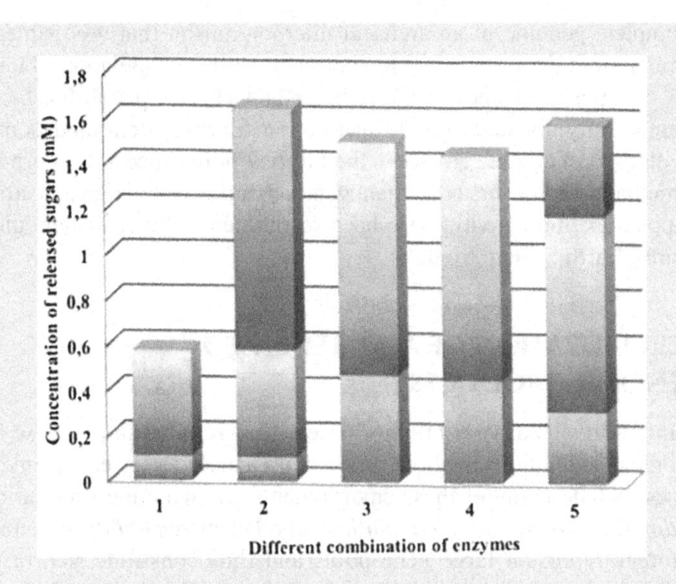

The figure depicts the amounts of reducing sugars released: from steam-exploded wheat straw by varying combinations of enzymes, including three novel ones (P1, P5 and P6) produced from metagenomic libraries. Explanation: (1) control (commercial enzyme mix without any extra added enzymes); (2) biodegradative enzymes P1, P5, P6 added; (3) P1, P5 added; (4) P1, P6 added; (5) P5, P6 added. Bars: lower gray: arabinose, shaded darker gray (second from bottom): glucose, whitish (third from bottom): xylose, dark gray (top parts, only in bars 2 and 5): unknown disaccharide.

14.9 FUTURE CONTRIBUTION OF SOIL METAGENOMICS TO DIFFERENT SCIENTIFIC AIMS

Although we are only starting to unlock the true diversity of soil microbiome systems, metagenomic approaches have already contributed with many novel scientific advances. What will be the future? First, we foresee that soil microbiome studies will be performed with multiple metagenomics samples. This requires new algorithms that aim to handle the assembly of progressively larger datasets and to model the relative contribution of soil factors to ecosystem outcomes (e.g., microbial responses to permafrost thaw, the stoichiometry of the soil C cycle; Mackelprang et al. 2011, Hartman et al. 2017). Computational methods developed in other disciplines will assist in this development. A case in point is offered by network theory, which is used in social and computer sciences to measure the degree of co-occurrence among microbial taxa and their genes (Faust and Raes 2012). Network studies on soil microbiomes have been limited to the taxon level, but there is no impediment to its application to functions and processes in soil.

At the technical level, the per-base sequencing price has dropped substantially whereas sequence length has increased. This will increase our capability of detecting changes in soil microbial communities at spatially and temporally explicit scales. A combination of sequencing platforms will also improve the assembly of metagenomics data. Currently, sequencing is already being distributed between Illumina paired-ended reads and long PacBio reads. The former will provide accuracy, whereas the latter facilitate mapping onto long scaffolds. Finally, soil metagenomics provides a unique opportunity to understand how information encoded in the genomes drives the aforementioned large-scale biogeochemical cycles. We have yet to identify and understand how the soil virome regulates microbial abundance in bulk soils and the rhizosphere (Pratama and van Elsas 2018; see Chapter 6). Moreover, there is a need to characterize the role of CRISPR-Cas systems as a driver of microbial community structures (Burstein et al. 2017) and to define the scales of functional biogeography for soil microbes (Violle et al. 2014). The latter topic provides a bridge to the Earth sciences.

ACKNOWLEDGMENTS

Research in the Soil Ecogenomics Laboratory at the University of California, Davis, Davis, California, is supported by the National Science Foundation—Dimensions of Biodiversity (DEB 1442214), NSF-FAPESP (2014-50320-5), by the Agriculture and Food Research Initiative Competitive Grant 2009-35319-05186 from the U.S. Department of Agriculture–National Institute of Food and Agriculture, and by the U.S. Department of Energy Joint Genome Institute, a DOE Office of Science User Facility, supported by the Office of Science of the U.S. Department of Energy under Contract No. DE-AC02-05CH11231.

The Baltic and Eastern Europe Foundation Project 3150-3.1.1-2017 (SS), the Office of Science Sabbatical Grant at Södertörn University (SS).

REFERENCES

Albertsen, M., Hugenholtz, P., Skarshewski, A., Nielsen, K.L., Tyson, G.W. and P.H. Nielsen. 2013. Genome sequences of rare, uncultured bacteria by differential coverage binning of multiple metagenomes. *Nat Biotechnol* 31, 533–538.

Allen, E.E. and J.F. Banfield. 2005. Community genomics in microbial ecology and evolution. *Nat Rev Microbiol* 3, 489–498.

Alneberg, J., Bjarnason, B.S., De Bruijn, I. et al. 2014. Binning metagenomic contigs by coverage and composition. *Nat Methods* 11, 1144–1146.

Berini, F., Casciello, C., Marcone, G.L. and F. Marinelli. 2017a. Metagenomics: novel enzymes from non-culturable microbes. *FEMS Microbiol Letts* 364, fnx211.

Berini, F., Presti, I., Beltrametti, F. et al. 2017b. Production and characterization of a novel antifungal chitinase identified by functional screening of a suppressive-soil metagenome. *Microb Cell Fact* 16, 16.

Blin, K., Medema, M.H., Kazempour, D. et al. 2013. AntiSMASH 2.0—a versatile platform for genome mining of secondary metabolite producers. *Nucl Acids Res* 41, W204–W212.

Bowers, R.M., Kyrpides, N.C., Stepanauskas, R. et al. 2017. Minimum information about a single amplified genome (MISAG) and a metagenome-assembled genome (MIMAG) of bacteria and archaea. *Nat Biotechnol* 35, 725–731.

Burstein, D., Harrington, L.B., Strutt, S.C. et al. 2017. New CRISPR-Cas systems from uncultivated microbes. *Nature* 542, 237–241.

Chow, J. and W. Streit. 2016. Screening of enzymes: novel screening technologies to exploit non cultivated microbes for biotechnolgy. In: Hilterhause L., Liese A., Kettling U., Antranikian G. (eds.), *Applied Biocatalysis: From Fundamental Science to Industrial Applications.* Wiley-VCH Verlag GmbH & Co Weinheim, Germany.

Clevenger, K.D., Bok, J.W., Ye, R. et al. 2017. A scalable platform to identify fungal secondary metabolites and their gene clusters. *Nat Chem Biol* 13, 895–901.

Colin, P.Y., Kintses, B., Gielen, F. et al. 2015. Ultrahigh-throughput discovery of promiscuous enzymes by picodroplet functional metagenomics. *Nat Commun* 6, 10008.

Cretoiu, M.S., Berini, F., Kielak, A.M., Marinelli, F. and J.D. van Elsas. 2015. A novel salt-tolerant chitobiosidase discovered by genetic screening of a metagenomic library derived from chitin-amended agricultural soil. *Appl Microbiol Biotechnol* 99, 8199–8215.

Delmont, T.O., Eren, A.M., Maccario, L. et al. 2015. Reconstructing rare soil microbial genomes using *in situ* enrichments and metagenomics. *Front Microbiol* 6, 358.

Denonfoux, J., Parisot, N., Dugat-Bony, E. et al. 2013. Gene capture coupled to high-throughput sequencing as a strategy for targeted metagenome exploration. *DNA Res* 20, 185.

Edwards, A., Debbonaire, A.R., Sattler, B., Mur, L.A.J. and A.J. Hodson. 2016. Extreme metagenomics using nanopore DNA sequencing: a field report from Svalbard, 78 N. *BioRxiv.* doi:10.1101/073965.

Faust, K., Lahti, L., Gonze, D., de Vos, W.M. and J. Raes. 2015. Metagenomics meets time series analysis: unraveling microbial community dynamics. *Curr Opin Microbiol* 25, 56–66.

Faust, K. and J. Raes. 2012. Microbial interactions: from networks to models. *Nat Rev Microbiol* 10, 538–550.

Fierer, N., Ladau, J., Clemente, J.C. et al. 2013. Reconstructing the microbial diversity and function of pre-agricultural tallgrass prairie soils in the United States. *Science* 342, 621–624.

Gabor, E.M., de Vries, E.J. and D.B. Janssen. 2003. Efficient recovery of environmental DNA for expression cloning by indirect extraction methods. *FEMS Microbiol Ecol* 44, 153–163.

Hartman, W.H., Ye, R., Horwath, W.R. and S.G. Tringe. 2017. A genomic perspective on stoichiometric regulation of soil carbon cycling. *ISME J* 11, 2652–2665.

Hernández, M., Neufeld, J.D. and M.G. Dumont. 2017. Enhancing functional metagenomics of complex microbial communities using stable isotopes. In: Charles T., Liles M., Sessitsch A. (eds.), *Functional Metagenomics: Tools and Applications*. Springer, Cham.

Hjort, K., Presti, I., Elväng, A., Marinelli, F. and S. Sjöling. 2014. Bacterial chitinase with phytopathogen control capacity from suppressive soil revealed by functional metagenomics. *Appl Microbiol Biotechnol* 98, 2819–2828.

Hover, B.M., Kim, S.H., Katz, M. et al. 2018. Culture-independent discovery of the malacidins as calcium-dependent antibiotics with activity against multidrug-resistant Gram-positive pathogens. *Nat Microbiol* 3, 415–422.

Howe, A.C., Jansson, J.K., Malfatti, S.A., Tringe, S.G., Tiedje, J.M. and C.T. Brown. 2014. Tackling soil diversity with the assembly of large complex metagenomes. *Proc Natl Acad Sci USA* 111, 4904–4909.

Hultman, J., Waldrop, M.P., Mackelprang, R. et al. 2015. Multi-omics of permafrost, active layer and thermokarst bog soil microbiomes. *Nature* 521, 208–212.

Huson, D., Richter, D., Mitra, S., Auch, A. and S. Schuster. 2009. Methods for comparative metagenomics. *BMC Bioinf* 10, S1–S12.

Janssen, P.H. 2006. Identifying the dominant soil bacterial taxa in libraries of 16S rRNA and 16S rRNA genes. *Appl Environ Microbiol* 72, 1719–1729.

Ji, M., Greening, C., Vanwonterghem, I. et al. 2017. Atmospheric trace gases support primary production in Antarctic desert surface soil. *Nature* 552, 400–403.

Kang, D.D., Froula, J., Egan, R. and Z. Wang. 2015. MetaBAT, an efficient tool for accurately reconstructing single genomes from complex microbial communities. *PeerJ* 3, e1165.

Karst, S.M., Dueholm, M.S., McIlroy, S.J., Kirkegaard, R.H., Nielsen, P.H. and M. Albertsen. 2018. Retrieval of a million high-quality, full-length microbial 16S and 18S rRNA gene sequences without primer bias. *Nat Biotechnol* 36, 190–195.

Katz, M., Hover, B.M. and S.F.J. Brady. 2016. Culture-independent discovery of natural products from soil metagenomes. *Ind Microbiol Biotechnol* 43, 129–141.

Kuzyakov, Y. and E. Blagodatskaya. 2015. Microbial hotspots and hot moments in soil: concept & review. *Soil Biol Biochem* 83, 184–199.

Lewin, A., Lale, R. and A. Wentzel. 2017. Expression platforms for functional metagenomics: emerging technology options beyond *Escherichia coli*. In: Charles T., Liles M., Sessitsch A. (eds.), *Functional Metagenomics: Tools and Applications*. Springer, Cham.

Li, D., Huang, Y., Leung, C.-M., Luo, R., Ting, H.-F. and T.-W. Lam. 2017. MegaGTA: a sensitive and accurate metagenomic gene-targeted assembler using iterative de Bruijn graphs. *BMC Bioinf* 18, 67–75.

Liebl, W., Angelov, A., Juergensen, J., Chow, J., Loeschke, A. and T. Drepper. 2014. Alternative hosts for functional (meta)genome analysis. *Appl Microbiol Biotechnol* 98, 8099–8109.

Mackelprang, R., Waldrop, M.P., and DeAngelis, K.M. 2011. Metagenomic analysis of a permafrost microbial community reveals a rapid response to thaw. *Nature* 480, 368–371.

Markowitz, V.M., Ivanova, N.N., Szeto, E. et al. 2008. IMG/M: a data management and analysis system for metagenomes. *Nucleic Acids Res* 36, D534–D538.

Marmeisse, R., Kellner, H., Fraissinet-Tachet, L. and P. Luis. 2017. Discovering protein-coding genes from the environment: time for the Eukaryotes? *Trends Biotechnol* 35, 824–835.

Maruthamuthu, M. 2017. Novel bacterial enzymes for plant biomass degradation discovered by meta-omics approaches. *PhD thesis*, University of Groningen, Groningen, Netherlands

Maruthamuthu, M. and J.D. van Elsas. 2017. Molecular cloning, expression, and characterization of four novel thermo-alkaliphilic enzymes retrieved from a metagenomics library. *Biotechnol Biofuels* 10, 142.

Maruthamuthu, M., Jiménez, D.J. and J.D. van Elsas. 2017. Characterization of a furan aldehyde-tolerant β-xylosidase/α-arabinosidase obtained through a synthetic metagenomics approach. *J Appl Microbiol* 123, 145–158.

McCalley, C.K., Woodcroft, B.J., Hodgkins, S.B. et al. 2014. Methane dynamics regulated by microbial community response to permafrost thaw. *Nature* 514, 478–481.

Meyer, F., Paarmann, D., D'Souza, M. et al. 2008. The metagenomics rast server—a public resource for the automatic phylogenetic and functional analysis of metagenomes. *BMC Bioinf* 9, 386.

Mondav, R., Woodcroft, B.J., Kin, E.-H. et al. 2013. Discovery of a novel methanogen prevalent in thawing permafrost. *Nat Commun* 5, 3212.

Parks, D.H. and R.G. Beiko. 2010. Identifying biologically relevant differences between metagenomic communities. *Bioinformatics* 26, 715–721.

Parks, D.H., Rinke, C., Chuvochina, M. et al. 2017. Recovery of nearly 8,000 metagenome-assembled genomes substantially expands the tree of life. *Nat Microbiol* 2, 1533–1542.

Pratama, A.A. and J.D. van Elsas. 2018. The 'neglected' soil virome – potential role and impact. *Trends Microbiol* 1533, 1–14.

Prosser, J. 2015. Dispersing misconceptions and identifying opportunities for the use of "omics' in soil microbiology. *Nat Rev Microbiol* 13, 439–446.

Raes, J., Korbel, J.O., Lercher, M.J., von Mering, C. and P. Bork. 2007. Prediction of effective genome size in metagenomic samples. *Genome Biol* 8, R10.

Richter, D.C., Ott, F., Auch, A.F., Schmid, R. and D.H. Huson. 2008. Metasim—a sequencing simulator for genomics and metagenomics. *PLoS One* 3, e3373.

Rodriguez, L.M. and K.T. Konstantinidis. 2014. Estimating coverage in metagenomics data sets and why it matters. *ISME J* 8, 2349–2351.

Rondon, M.R., August, P.R., Bettermann, A.D. et al. 2000. Cloning the soil metagenome: a strategy for accessing the genetic and functional diversity of uncultured microorganisms. *Appl Environ Microbiol* 66, 2541–2547.

Sczyrba, A., Hofmann, P., Belmann, P. et al. 2017. Critical assessment of metagenome interpretation—a benchmark of computational metagenomics software. *Nat Methods*, 14, 1063–1071.

Sjöling, S., Stafford, W. and D.A. Cowan. 2007. Soil metagenomics: exploring and exploiting the soil microbial gene pool. In: van Elsas J.D., Jansson J., Trevors J.T. (eds.), *Modern Soil Microbiology 2*. CRC Press, Boca Raton, FL, pp. 409–434.

Thompson, L.R., Sanders, J.G., McDonald, D. et al. 2017. A communal catalogue reveals Earth's multiscale microbial diversity. *Nature* 551, 457–463.

Tringe, S.G., von Mering, C., Kobayashi, A. et al. 2005. Comparative metagenomics of microbial communities. *Science* 308, 554–557.

Tyson, G.W., Chapman, J., Hugenholtz, P. et al. 2004. Community structure and metabolism through reconstruction of microbial genomes from the environment. *Nature* 428, 37–43.

Uchiyama, T. and K. Miyazaki. 2010. Product-induced gene expression, a product-responsive reporter assay used to screen metagenomic libraries for enzyme-encoding genes. *Appl Environ Microbiol* 76, 7029–7035.

Ufarté, L., Bozonnet, S., Laville, E. et al. 2016. Functional metagenomics: construction and high-throughput screening of fosmid libraries for discovery of novel carbohydrate-active enzymes. *Methods Mol Biol* 1399, 257–271.

Vestergaard, G., Schulz, S., Scholer, A. and M. Schloter. 2017. Making big data smart – how to use metagenomics to understand soil quality. *Biol Fertil Soils* 53, 479–484.

Violle, C., Reich, P.B., Pacala, S.W., Enquist, B.J. and J. Kattge. 2014. The emergence and promise of functional biogeography. *Proc Natl Acad Sci USA* 111, 13690–13696.

Wang, L., Nasrin, S., Liles, M.R. and Z. Yu. 2014. Use of bacterial artificial chromosomes in metagenomic studies, overview. In: Nelson K.E. (ed.), *Encyclopedia of Metagenomics*. Springer, Boston, MA, pp. 1–12.

Wang, Q., Fish, J.A., Gilman, M. et al. 2015. Xander: employing a novel method for efficient gene-targeted metagenomic assembly. *Microbiome* 3, 32.

15 Analysis of Transcriptomes to Assess Expression and Activity Patterns of the Soil Microbiome

Stefanie Schulz, Anne Schöler, and Michael Schloter
Helmholtz Zentrum München

Shilpi Sharma
Indian Institute of Technology

CONTENTS

15.1 Introduction 245
15.1.1 Microbial RNA—An Activity Parameter for Soil Microbial Communities? 245
15.1.2 Challenges in Transcript Analyses 246
15.2 mRNA from Soils—Extraction and Processing 247
15.2.1 Sampling and Storage of Samples 247
15.2.2 Extraction Protocols 248
15.2.3 Purification 249
15.2.4 Quantification of RNA 250
15.2.5 Reverse Transcription 250
15.3 mRNA from Soils—Downstream Analyses 251
15.3.1 PCR-Based Approaches 251
15.3.2 Microarrays 252
15.3.3 Metatranscriptomes 252
15.4 Conclusions and Outlook 253
References 255

15.1 INTRODUCTION

15.1.1 MICROBIAL RNA—AN ACTIVITY PARAMETER FOR SOIL MICROBIAL COMMUNITIES?

As described in previous chapters, the first molecular studies of soil microbiomes used gene probe technology (hybridization) to screen for the presence/absence of specific genes in DNA extracted from soil. The ability to detect specific DNA sequences directly from a soil community has since proven to be a useful tool to improve our understanding of particular soil processes. See Chapter 12 for a treatise of soil DNA extraction methods. Even though these studies provide information in respect of the genotype of members of the microbial community, they cannot be used to determine whether the specific genes detected are actually expressed within that community. The presence of specific genes in the soil microbiome provides information on the soil's functional potential, while actual soil functioning depends on the local expression of those genes. Expression of genes is regulated by many factors, with control of transcriptional initiation being the major point of regulation, at least for the prokaryotes. Hence, the mere presence of a particular gene in a soil microbiome does not necessarily imply its expression under the conditions studied.

One step forward towards understanding the microbial activity of soil is offered by the analysis of ribosomal RNA (rRNA), since the amount of ribosomes (and their rRNA) per cell was found to be about proportional to the growth of several bacteria when in pure culture (Ramos et al. 2000). However, this correlation between the activity and the rRNA content of a cell is only true for so-called *r*-strategists (see Chapter 4), and it is certainly not a general concept of bacteria. Furthermore, the correlation between microbial rRNA levels and activity is not always straightforward, since it is possible for ribosomes to be transcribed at different rates in cells in one and the same microbial population (Hopkins and McFarlane 2000). Moreover, information on the overall activity of a specific soil population should not be regarded as an indication of a specific function. An increased rRNA level may only indicate an overall higher growth rate but not provide any information about the expression of any particular gene. Often, a specific population has the potential to carry out different enzymatic processes—one genotype is thus linked to several phenotypes. A key approach in studying specific microbial activities of soil is thus to investigate the expression of genes by the analysis of messenger RNA (mRNA) transcripts. The usefulness of such transcription analysis is strongly affected by the short half-life of mRNA, which is sometimes on the order of minutes. Regulation at the transcriptional level almost immediately affects the rate of protein synthesis. Hence, the detection of mRNA provides a strong indication of specific gene expression at the time of sampling, which can be used to improve our mechanistic understanding on the interactions between microbes as well as the interplay of soil biota with their abiotic environment.

However, in the case of eukaryotes, the relationship between mRNA synthesis and expression of the gene is not straightforward. Whereas transcription and translation are coupled in prokaryotes, in eukaryotes these processes are spatially and temporally separated. Eukaryotic mRNA is at times stored as mRNA–protein complexes until the translational signal is received. The process of eukaryotic gene expression involves a diverse number of steps, including transcription, RNA processing, transport, translation, and posttranslational modifications. These processes are not covered extensively in this chapter.

15.1.2 Challenges in Transcript Analyses

Reliable analysis of mRNA from soil is still considered a challenge. The short half-life, also called half-time, of the mRNA provides a major challenge, as the temporal dynamics of mRNA in soils are far larger compared to DNA. Thus, the analysis of transcriptional pattern requires more sampling effort compared to DNA. Whereas for DNA-based analysis, often seasonal changes are followed, for mRNA analysis, changes occur even between day and night (Baraniyan et al. 2018). Also, one sampling time point might reflect a snapshot, which makes only sense if it is integrated into a dense network of other sampling time points. Furthermore, the half-time of transcripts differs not only between microorganisms but also between genes of the same microorganism. It is obvious that, for example, transcripts from genes coding for toxins are highly sensitively regulated compared to transcripts from housekeeping genes, which have longer half-times. On the other hand, much like for DNA, also for RNA, we are facing a huge heterogeneity in patterns over soil spaces, mainly when hot spots such as rhizosphere, mycosphere, detritusphere, or drilosphere are compared to bulk soil samples. New sampling schemes and concepts are required to cover the huge temporal dynamics and pronounced heterogeneity in space in terms of local gene expression patterns (Petersen and Esbensen 2005).

In addition, mRNA extraction from soil *per se* is challenging due to the matrix of surfaces that are capable of binding nucleic acids, mostly clays and organic materials, next to a range of compounds, mostly humic materials that inhibit enzymatic reactions such as reverse transcription (RT) or DNA extension like in the polymerase chain reaction (PCR). Finally, the ubiquitous presence of RNase in RNA preparations is a very challenging obstacle.

It should be noticed that if gene abundance pattern should be linked to transcript rates, there is a need for co-extraction of DNA and RNA as the different types of bias related to different nucleic acid extraction protocols are then avoided.

Despite the difficulties mentioned previously, the successful extraction of total RNA from soils and sediments has been reported more than two decades ago (Fleming et al. 1998). Most of the early work on soil transcriptomes was done following the introduction of a dense cell population into soil. Later on, reliable mRNA isolation and analysis from non-inoculated soils were successfully carried out; overall RNA yields of up to 56.1 μg g^{-1} soil have been reported (Hurt et al. 2001). A generalized scheme of soil sampling, and RNA extraction, processing, and analysis is shown in Figure 15.1. Whereas all of the steps in the procedure appear to be rather straightforward, technical as well as intrinsic challenges still hamper progress in this area. In the following section, general issues of mRNA extraction and analysis will be discussed, with an emphasis on the state of the art of the technology.

15.2 mRNA FROM SOILS—EXTRACTION AND PROCESSING

15.2.1 Sampling and Storage of Samples

As outlined in Chapter 1, soils are vertically and horizontally structured ecosystems, which are composed of a multitude of different microhabitats comprising diverse physical, chemical, and biological properties. The degree of heterogeneity strongly depends on (1) the sampled

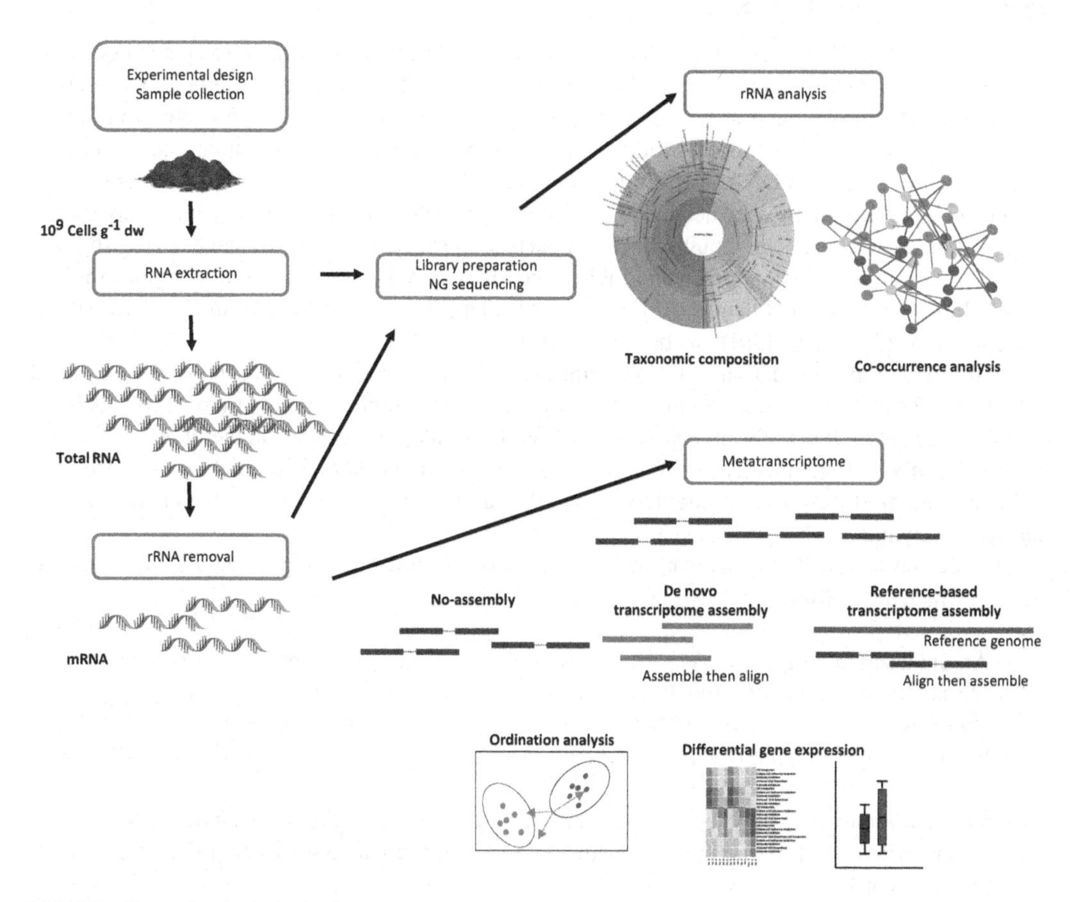

FIGURE 15.1 Analytical pipeline to analyze activity patterns of the soil microbiome. *de novo* assembly: assembly without the aid of a sequence matrix.

compartment, for example, the rhizosphere may appear as less heterogeneous compared to the bulk soil (Hinsinger et al. 2009); (2) the soil texture, which strongly influences aggregate formation and also nucleic acid extraction efficiency; (3) the aboveground diversity and plant coverage; (4) season; and (5) specific site characteristics, such as slope, degree of shading, groundwater levels, and other factors (Rhee et al. 2004). To analyze mRNA from soil, a sampling schema must be chosen that reflects these heterogeneities but also takes the strong dynamics of microbial transcriptomes across time into account, as discussed in the foregoing paragraph.

For instance, several authors immediately extract RNA following soil sampling. However, this is often not practical, especially when field samples or many replicates are to be investigated. Today, a standard approach is to store soil samples before proceeding with transport to the laboratory and extraction of the RNA. Here, it is essential to shock-freeze soil in liquid nitrogen or dry ice at the time of sampling, in order to minimize RNA degradation and to maintain mRNA integrity. Despite the fact that mRNA of specific biodegradation genes from soil has been successfully detected, even when samples were stored at –20°C (Urich et al. 2008) or lyophilized (Sharma et al. 2006), storage of samples at –80°C is widely used to minimize RNA degradation (Baraniyan et al. 2018). This approach has also recently been implemented into the ISO guidelines for soil analysis (ISO 18512). The described procedure (including the snap-freezing step) often results in higher RNA yields than the extraction procedure from fresh samples. This yield increase may be due to additional cell disruptions caused by the freezing and subsequent thawing steps.

15.2.2 Extraction Protocols

One of the most cited protocols for DNA/RNA co-extraction from soil, which was published about two decades ago (Griffiths et al. 2000), has served as a basis for a number of authors, who subsequently made modifications for their particular use (e.g., Töwe et al. 2011). Approaches to produce rRNA were even published earlier (Moran et al. 1993). As RNA (or DNA) isolation—at some stage of the protocol—included DNA or RNA digestion, respectively, by DNase or RNase, further purification and separation steps were needed in the protocol. The later steps, in consequence, may induce selective ranges of error, even though the initial parts of the protocol incite the same biases. In other efforts, a protocol for the co-extraction of RNA and DNA based on precipitation with polyethylene glycol (PEG) (Arbeli and Fuentes 2007) and subsequent RNA separation from DNA via silica-based columns (Tsai et al. 1991) has been developed.

To minimize the degradation of RNA during the extraction procedure, special precautions need to be taken. These include treating all solutions with RNase inhibitor, diethyl pyrocarbonate, heating all glassware at 180°C for at least 4 h, and keeping RNA preparations on ice unless otherwise indicated. Although one ideally uses one protocol for the extraction of RNA and DNA, biases can still occur due to differences in interactions of DNA and RNA molecules with the environmental matrices (Newman et al. 2016).

An ideal procedure for recovering mRNA from environmental samples should meet several criteria, which are outlined as:

i. An unbiased and high recovery efficiency so that the final transcriptome is representative of the total expression profile of the naturally occurring microbial community
ii. Minimum degradation of mRNA during the protocol to enable the synthesis of full-length complementary DNA (cDNA) for community gene library construction and gene cloning
iii. Efficient removal of inhibitors, such as humic acids, for subsequent molecular analysis
iv. Simple protocol of extraction and purification so that the whole recovery process is rapid and inexpensive
v. Be robust, reproducible and reliable

Although all procedures published to date use steps that release bacterial cells from the soil matrix and subsequently lyse them, followed by purification step(s), a number of different approaches have been reported, which are summarized in Table 15.1. Some protocols use acidic conditions in order to remove proteins, lipids, and DNA by phase separation using phenol and chloroform, leaving RNA in the liquid phase. Alternatively, the contaminating compounds can be removed by enzymatic digestion.

15.2.3 Purification

The extraction of mRNA (similar to DNA) from soil requires the removal of inhibitors of subsequent enzymatic reactions, such as RT and PCR. Hexadecyl methylammonium bromide (CTAB) and polyvinylpolypyrrolidone (PVPP) in extraction buffers have been used to complex and remove contaminants from soil RNA (Wang et al. 2012). Jacobsen and Holben (2007) described a protocol that allows to purify RNA with a hybridization probe conjugated to paramagnetic beads. In another protocol, a combined low-pH extraction (pH 5.0)/Q-Sepharose chromatography approach was used (Mettel el al. 2010). The removal efficiency of humic acids was 94%–98% for all soils tested.

Besides these protocols, a large number of kit-based solutions are available. Luis et al. (2005) used a commercially available kit (RNeasy Plant Mini kit; Qiagen, Hilden, Germany) to purify RNA extracted from soil before the RT step. They successfully showed that there was no inhibition of the cDNA synthesis step. Besides Qiagen, there is a number of other manufacturers who provide similar solutions (e.g., Omega-Bio-Tek, Norcross, GA; Nippon Gene, Toyama, Japan; MO BIO (Qiagen Life Sciences); Norgen, Thorold, ON, Canada). As the exact mode of action of most kits is

TABLE 15.1
Different Sample Storage, and RNA Extraction and Purification Protocols Used for mRNA Analysis

Storage	Lysis and Extraction	Precipitation and/or Purification	References
Fresh extraction	Freeze-drying in liquid N_2 followed by grinding; extraction with guanidine thiocyanate, phenol, chloroform	Sequential purification with Sephadex G-75, QIAeasy (Qiagen), and PVPP columns; ethanol precipitation	Kolb and Stacheter (2013)
Fresh extraction	Lysis with silica beads and extraction buffer [CTAB, 1,4-dithio-DL-threitol (DTT), sodium phosphate buffer, NaCl, EDTA]; phenol–chloroform extraction	Precipitation with polyethylene glycol (PEG)-NaCl	Bogan et al. (1996)
−20°C	Extraction with guanidine thiocyanate, sarcosyl, and mercaptoethanol with shaking	Phenol–chloroform extraction and ethanol precipitation	Rhee et al. (2004)
−40°C	Freeze-drying in liquid N_2 (with sterile sand) and grinding; extraction with CTAB, Tris-HCl, NaCl, EDTA, sodium dodecyl sulfate (SDS), and phenol–chloroform	Precipitation with isopropyl alcohol; purification with Sephadex G-75 resin slurry	Hesse et al. (2015)
−80°C	Lysis with glass beads and diatomaceous earth in extraction buffer (SDS, phenol)	Precipitation with ethanol–sodium acetate; purification using RNeasy Plant Mini Kit (Qiagen)	Hopkins and MacFarlane (2000)
−80°C	Incubation with extraction buffer (proteinase K and lysozyme, SDS, CTAB) followed by freeze–thaw cycles; phenol-chloroform extraction	Precipitation with isopropanol–sodium acetate	Mendum et al. (1998)

unknown and the composition changes on a regular basis, it is recommended to use "handmade" protocols, mainly if long time series are planned, to ensure comparability of the results.

After purification of the extracted nucleic acids, contaminating DNA needs to be removed from the solution using DNase. The degree of removal of the DNA needs to be checked, for example, by performing a PCR using universal bacterial primers, before RT is commenced.

15.2.4 Quantification of RNA

The traditional method for determining total RNA concentrations in cell suspensions is by UV spectroscopy ($A_{260/280}$). The aromatic ring structure of the purine and pyrimidine moieties that make up the nucleoside bases of DNA and RNA is responsible for the absorbance of UV light at 260 nm. Recently, the use of fluorescent dyes that bind to RNA has proven to be more accurate for determining RNA concentrations in single cells and mixed populations. The performance of this technique is dependent on the availability of fluorochromic dyes with high sensitivity and high resistance to photo-bleaching. Numerous fluorescent dyes, including magdala red, hypocrellin A, thiazole orange homodimer, oxazole yellow homodimer, and ethidium bromide, have been used for the quantification of RNA. RiboGreen (Molecular Probes, Eugene, Oregon) is the most sensitive fluorescent dye that is currently available for RNA quantification in solution, allowing concentrations as low as 1 ng mL^{-1} to be quantified with a spectrofluorometer.

15.2.5 Reverse Transcription

Different methods that have been used to quantify specific mRNA molecules extracted from environmental samples include ribonuclease (RNase) protection assays, Northern blotting, and RT of the RNA into cDNA, which is subsequently used as a template for PCR or direct sequencing. The RNase protection assay is an extremely sensitive technique for the quantification of specific RNAs in solution. The sensitivity of the assay is derived from the use of a complementary *in vitro* transcript probe which is radiolabeled to high specific activity. The probe and target RNA are hybridized in solution, after which the mixture is diluted and treated with RNase to degrade all remaining single-stranded RNA. The hybridized portion of the probe will be protected from digestion and can be visualized via electrophoresis of the mixture on a denaturing polyacrylamide gel followed by autoradiography. Northern blotting involves the separation of RNA by size via electrophoresis in an agarose gel under denaturing conditions. The RNA is then transferred to a membrane, cross-linked, and hybridized with a labeled probe. Northern blotting is the only method that will provide information regarding transcript size and integrity of RNA, whereas RNase protection assays are the easiest way to simultaneously examine multiple messengers.

During RT, RNA is converted into single-stranded cDNA, which is then used as a template for PCR or direct sequencing. RT-PCR is more rapid and sensitive, and can be more specific than Northern blot analysis, but quantification can be difficult because many sources of variation exist, including template concentration and amplification efficiency. In the following sections, RT and its derived techniques, and their applications to soil microbiology, will be described in further detail, as these techniques provide great potential to assess gene expression in soil. RT has revolutionized gene expression analysis mainly when combined with direct sequencing (metatranscriptomics). It is theoretically possible to detect transcripts from any gene regardless of the abundance of the specific mRNAs, unless there is no expression. However, for RT-PCR, the initial step in is the production of a single-stranded cDNA copy of the RNA using the retroviral reverse transcriptase enzyme followed by the amplification by PCR. The reaction can be performed in either one-step or two-step formats, with cDNA synthesis and PCR being separated in the latter. Two-step RT-PCR is popular and useful for detecting multiple messages from the same sample, whereas one-step RT-PCR is more advantageous when processing multiple samples, as carryover contamination is minimized. In bacterial systems, the RT reaction is usually primed with gene-specific primers.

For metatranscriptomics, random hexamers for cDNA synthesis from soil RNA were used (Sharma et al. 2004). The use of such random primers in cDNA synthesis produces a suite of randomly synthesized cDNA molecules, which are to be analyzed for the presence of multiple genes. The analysis of multiple sequences following random RT is also thought to minimize the biases of using gene-specific primers (due to high sequence heterogeneity, a common feature in environmental samples).

The most commonly used enzymes for cDNA synthesis from RNA extracted from soil are avian myeloblastosis virus and Moloney murine leukemia virus reverse transcriptases. Generally, the former enzyme is better suited for the synthesis of short cDNAs, whereas the latter will better generate longer molecules. Unfortunately, few data are available on the specific inhibition of these reverse transcriptases or of enzymes with both reverse transcriptase and DNA polymerase activity, such as r*Tth* DNA polymerase. Hence, for each study in soil, selection of the enzyme will still need to be done on a case-by-case basis.

15.3 mRNA FROM SOILS—DOWNSTREAM ANALYSES

Figure 15.1 shows a number of approaches that analyze the mRNA produced from soil microbiomes. These are divided into sequencing-based approaches (no assembly, *de novo* transcriptome assembly, and reference-based transcriptome assembly), ordination analysis, and gene expression measurements. For the latter, RT-PCR approaches have been developed that allow to quantify the number of copies of particular messengers, as outlined in the following.

15.3.1 PCR-Based Approaches

The coupling of RT-PCR with fluorescence techniques and advanced technology capable of automated detection and quantification of mRNAs has led to the development of new technologies that have dramatically changed gene expression studies. These include advanced quantitative versions of RT-PCR such as competitive RT-PCR (cRT-PCR) and quantitative real-time RT-PCR (qRT-PCR, also denoted as RT-qPCR). In cRT-PCR, known amounts of an internal standard are co-amplified in one reaction with a sequence of interest, allowing the expression levels of the gene under investigation to be determined. Theoretically, there is a quantitative relationship between the amount of starting target and the amount of PCR product at any given cycle number. qRT-PCR detects the accumulation of amplicons during the reaction. Many studies have been performed in the past to prove the activity of single genes using the extracted RNA and qRT-PCR, and in fact, any available primer system suitable for the quantification of a particular gene by qPCR can also be applied for measuring transcription rates of the same gene by qRT-PCR. Thus, a large number of assays has been described in the literature focusing on the transcription rates of genes catalyzing the turnover of carbonaceous compounds, for example, naphthalene dioxygenase (Wilson et al. 1999) or lignin peroxidase (Bogan et al. 1996) in soils or driving the transformation of nutrients such as N (Pierre et al. 2017), as well as genes involved in the degradation of pesticides (Pesaro et al. 2004). Also, genes involved in respiration and redox processes have been analyzed (Bürgmann et al. 2003, Sharma et al. 2006). Details on the use of qPCR-based approaches are given in Chapter 12.

In principle, many of the primer systems developed for qPCR can also be used for the assessment of the diversity of the active microflora with respect to a particular trait. This approach just requires an appropriate length of the fragment amplified by the used primers (>250 bases). Overall, the number of studies on the diversity of microbes involved in a specific transformation with marker genes that amplify the gene for the respective enzyme has been increasing in the last years. Thus, steps in major nutrient cycles, with a focus on N transformations (Zani et al. 2000, Nogales et al. 2002, Sharma et al. 2005), have been addressed. This approach requires a stable pipeline for bioinformatic analyses of the sequenced amplicon libraries. Major issues related to amplicon sequencing are described in Chapter 13.

The expression of particular genes can be estimated by quantifying the level of their transcripts. This approach can work well if the regulation of the gene studied is understood, and mRNA levels can be correlated specifically with the measurements of activity. The combination of "baiting" (labeling) key functional groups with isotopically labeled substrates followed by the recovery of labeled DNA, RNA, or lipid molecules has led to significant breakthroughs in the quest to link microbial function to microbial identity (Manefield et al. 2002, Wellington et al. 2003). This approach involves the incorporation of stable-isotope-labeled substrates into cellular biomarkers that can be used to identify the organisms that assimilated the substrates [stable-isotope probing (SIP); see for details Chapter 17]. rRNA has been suggested to constitute a more responsive biomarker for SIP compared to DNA because of the higher rates of rRNA synthesis than those of DNA.

15.3.2 Microarrays

Some years ago, microarray hybridization analysis has been considered to represent a major advance in the semiquantitative analysis of a large number of genes in parallel, which could be also useful for RNA-based studies to assess activity pattern. Chapter 12 discusses microarray experiments in some more detail. In brief, mRNA from a given tissue or cell line is used to generate a labeled sample, sometimes referred to as a target, which is hybridized—in parallel—to a large number of DNA sequences that are immobilized onto a solid surface in an ordered array. Recently, a 50-mer-based oligonucleotide microarray was assayed as a tool to effectively monitor biodegrading populations (Rhee et al. 2004). This array has been further developed, and the recent version is based on more than 20,000 known genes and pathways involved in nutrient cycling, biodegradation, and metal resistance. When determining its sensitivity, it was found that the detection limit of mRNA was the equivalent messenger load of ~1.3×10^7–5.0×10^7 cells, depending on the genes and the experimental conditions. Roughly 40- to 100-fold enhanced gene expression was observed in naphthalene-enriched cells grown on pyruvate.

15.3.3 Metatranscriptomes

Figure 15.1 gives a generic impression of how to produce and analyze a metatranscriptome from a soil microbiome. Metatranscriptomics allows the nontargeted analysis of transcripts produced in a complex microbial community at any point in time and space. It follows—in principle—the same procedures described for metagenomes (Vestergaard et al. 2017). See Chapter 14 for details of the procedures used in metagenomics of soil. In soil metatranscriptomics, following separation of the RNA from other compounds including DNA, elimination of the rRNA from the total RNA is needed. In most cases, this rRNA accounts for >90% of the total RNA extracted from soil. Furthermore, in studies of plant-associated microbiomes, depletion of the often dominating host plant RNA is needed, which can be achieved by making use of the poly-A tails of plant-derived mRNA. Reverse transcription of thus purified mRNA will yield cDNA ready for sequencing. However, compared to amplicon-based sequencing, direct mRNA-based sequencing requires high amounts of high-quality cDNA, with 500 pg to 50 ng often being necessary; this critically depends on the procedure or kit used for library preparation. Unfortunately, the amount of mRNA obtained from bulk soil is often limited. Hence, so far mostly hot spots for microbial activity, such as the rhizosphere or drilosphere, have been studied with respect to metatranscriptomics (Urich et al. 2008, Garoutte et al. 2016). A number of studies has focused on the response patterns of the root microbiome to pesticide application or to biotic stressors (Cabral et al. 2016, Yergeau et al. 2018, Gonzalez et al. 2018). Also, shifts in activity patterns of fungal and bacterial communities towards N deposition in two maple forests were investigated (Hesse et al. 2015). All of these studies indicated that a large number of expressed genes in these systems is involved in host–microbe interactions (Cao et al. 2015, Zhang et al. 2017). Furthermore, a study in forest soils during litter degradation revealed a stratified grouping of active microorganisms related to soil depth and compartment (Baldrian et al. 2012). This key study identified some important ecological parameters, as further illustrated in Box 15.1. The study points to a

BOX 15.1 ACTIVE MICROBIAL COMMUNITIES IN FOREST LITTER IDENTIFIED

In a hallmark study, Baldrian et al. (2012) showed that the active microbial communities in a forest soil were different from the overall picture of diversity. That is, only specific parts of the sampled microbiome communities turned out to be active in the system. On the basis of data obtained with advanced molecular tools, the researchers showed that, whereas the microbiome diversities based on DNA versus RNA assessments were quite comparable between the two approaches, the microbiome structures differed strongly. Several highly active taxa, especially fungal ones, showed low abundances or even complete absence in the DNA-based analyses, whereas these were major contributors to the communities detected by the RNA-based analyses. The results thus indicated that (fungal) species that are apparently present in low abundance in the (forest) soil microbiome can make important contributions to the local decomposition processes.

Interestingly, the fungal activities were more 'clustered' than the bacterial ones. The data indicated that the diversities and absolute numbers of expressed fungal gene sequences were higher in the litter horizon that in the organic soil layers. This was mainly true for key genes involved in the breakdown of litter material and included a number of important *exocellulases*. The authors also found differences in spatial heterogeneity patterns between the bacterial and fungal functions.

way forward in soil metatranscriptome studies, as it highlights the "physiological stratification" in the microbiomes that are active in a forest bed. However, most studies did not account for the spatiotemporal scale at which microorganisms act (short to intermediate distances, short to large time spans). Moreover, to achieve proper data interpretation, a link to data from assembled metagenomes from the same location is often needed (White et al. 2016).

15.4 CONCLUSIONS AND OUTLOOK

The analysis of gene expression in soil microbiomes via a suite of methods—as described in this chapter—is still subjected to an array of considerable limitations. These limitations are related to both technical constraints and constraints imposed by the nature of soil microbiome gene expression, which is highly context and scale dependent.

On the positive side, the positive detection of mRNA from soil microbiomes is a clear indicator of the occurrence of (overall) gene expression. However, the levels of mRNA that are found may not always equate to the actual activity of the targeted genes, because of the possibility of posttranscriptional modification of the RNA and/or posttranslational modifications of the proteins that are synthesized. The levels of gene expression within microorganisms in soil are controlled at multiple levels, including the following:

1. Promoter strength
2. mRNA half-life
3. Ribosome binding site

Bacterial mRNAs have very short half-lives, ranging from <1 to 10–20 min. Several endo- and exonucleases carry out different reactions during the maturation phase of pre-mRNAs, and these reactions are strongly related to cellular physiological status. Moreover, secreted enzymes (e.g., cellulases) may bind to clay particles or to the soil matrix and retain activity, while the cells producing them may no longer express the mRNA. In such cases, the failure to detect specific gene transcripts may not necessarily translate into the absence of (current) enzymatic activity. However, despite this inherent lack of synchronicity, several studies in soil (Bogan et al. 1996, Baldrian et al. 2012) have successfully demonstrated correlations between the levels of specific transcripts and other parameters of the specific activity, such as transformation rates and/or protein levels.

Overall, our current ability to analyze the transcriptomes of whole microbiomes represents an important tool in studies of their general metabolic status. As summarized in Table 15.2, any one of the molecular detection methods that has been developed to achieve this aim has inherent advantages and disadvantages. Analysis of the collective mRNA is theoretically possible for a suite of naturally occurring soil microbiomes, including bulk soil. However, for reasons of methodological efficiency, it has been most successfully applied to microbiomes from soil activity hot spots, such as rhizospheres and forest litter.

On another notice, to understand the functioning of soil microbiomes at a fine level, gene expression studies will need to account for the microorganisms themselves, both in space (e.g., the micrometer to millimeter level) and time (e.g., the minute to hour/day level), as it is at these spatio-temporal scales that most processes in soil are being regulated. Such fine-scale-level studies should

TABLE 15.2
Comparison of Various Techniques for Soil mRNA Analysis

Technique	Advantages	Disadvantages
RT-PCR	Rapid, sensitive, easy	Semiquantitative because of the kinetics of PCR product accumulation, dynamic range only thousandfold
cRT-PCR	Rapid, sensitive, easy, quantitative	Post-PCR processing of PCR products needed
Quantitative (real–time) RT-PCR	Wide dynamic range (up to 10^7 fold), high sensitivity (~5 copies), high precision, no post-PCR steps, high-throughput, multiplexing possible, no radioactivity	PCR product increases exponentially, variation increases with cycle number, overlap of emission spectra, expensive instrument and reagents
TaqMan RT-PCR	Extremely sensitive, specific, multiplex detection, quantitative, probes can be applied in real-time PCR	Additional expensive fluorophore probe needed, PCR product increases exponentially, variation increases with cycle number, expensive instrument and reagents
SYBR® Green RT-PCR	No requirement of design and synthesis of fluorescently labeled target-specific probes, inexpensive	No distinction of non-sequence specific products (e.g., primer dimers) so optimization of primer design and PCR assay conditions needed
cDNA array	Rapid, multiple gene analysis, high throughput, sensitive, quantitative, no amplification biases	Expensive equipment, diverse target sequences in environmental samples, humic acids inhibit hybridization
FISH-MAR	Rapid, high efficiency of hybridization and detection, insight into spatial niches	Prior information needed for design of special fluorescent probes, radioactivity involved, dependent on the permeability of cells and accessibility of the target
RNA-SIP	High sensitivity, enables coupling of structure and function	Discrimination by microbes against heavy isotope
Metatranscriptomics	Complete picture of transcribed genes at a given time point	Well-developed bioinformatics needed; reference metagenomes often useful, depletion of rRNA and eukaryotic RNA

RT-PCR, reverse transcription-PCR; cRT-PCR, competitive DNA RT-PCR; Real-time RT-PCR, real-time monitored RT-PCR; SYBR® Green, a specific dye detecting PCR amplicons; cDNA array, hybridization-based detection of target RNA molecules using a cDNA microarray; FISH-MAR, fluorescence *in situ* hybridization—molecular amplification; RNA-SIP, RNA-based stable-isotope probing (Chapter 17).

accompany broader-scale studies that are most commonly being performed, as these provide overall assessment of the processes being studied.

Finally, mRNA-based analyses have the potential to elucidate the resilience of soil microbiomes to perturbations, such as pollutant input and climate change. For instance, the influence of soil storage-related stress on the resilience (and resistance) of different microbiological soil characteristics (including 16S rRNA gene diversity) was successfully studied by assessing the impact of a single severe drying–rewetting cycle and by monitoring the recovery from this event for 34 days (Pesaro et al. 2004). Such an approach can be further extended to the mRNA level by analyzing the expression of genes that are responsive to the particular perturbation.

REFERENCES

Arbeli, Z. and C.L. Fuentes. 2007. Improved purification and PCR amplification of DNA from environmental samples. *FEMS Microbiol Lett* 272, 269–275.

Baldrian, P., Kolařík, M., Stursová, M. et al. 2012. Active and total microbial communities in forest soil are largely different and highly stratified during decomposition. *ISME J* 6, 248–258.

Baraniyan, D., Nannipieri, P., Kublik, S., Vestergaard, G., Schloter, M. and A. Schöler. 2018. The impact of the diurnal cycle on the microbial transcriptome in the rhizosphere of barley. *Microb Ecol*. doi:10.1007/s00248-017-1101-0.

Bogan, B.W., Schoenike, B., Lamar, R.T. and D. Cullen. 1996. Manganese peroxidase mRNA and enzyme activity levels during bioremediation of polycyclic aromatic hydrocarbon-contaminated soil with *Phanerochaete chrysosporium*. *Appl Environ Microbiol* 62, 2381–2386.

Bürgmann, H., Widmer, F., Sigler, W.V. and J. Zeyer. 2003. mRNA extraction and reverse transcription-PCR protocol for detection of *nifH* gene expression by *Azotobacter vinelandii* in soil. *Appl Environ Microbiol* 6, 1928–1935.

Cabral, L., Júnior, G.V., Pereira de Sousa, S.T. et al. 2016. Anthropogenic impact on mangrove sediments triggers differential responses in the heavy metals and antibiotic resistomes of microbial communities. *Environ Poll* 216, 460–469.

Cao, H.X., Schmutzer, T., Scholz, U., Pecinka, A., Schubert, İ. and G.T. Vu. 2015. Metatranscriptome analysis reveals host-microbiome interactions in traps of carnivorous Genlisea species. *Front Microbiol* 14, 526.

Fleming, J.T., Yao, W.H. and G. Sayler. 1998. Optimization of differential display of prokaryotic mRNA: Application to pure culture and soil microcosms. *Appl Environ Microbiol* 64, 3698–3706.

Garoutte, A., Cardenas, E., Tiedje, J. and A. Howe. 2016. Methodologies for probing the metatranscriptome of grassland soil. *J Microbiol Methods* 131, 122–129.

Gonzalez, E., Pitre, F.E., Pagé, A.P. et al. 2018. Trees, fungi and bacteria: Tripartite metatranscriptomics of a root microbiome responding to soil contamination. *Microbiome* 6, 53.

Griffiths, R.I., Whiteley, A.S., O'Donnell, A.G. and M.J. Bailey. 2000. Rapid method for coextraction of DNA and RNA from natural environments for analysis of ribosomal DNA- and rRNA-based microbial community composition. *Appl Environ Microbiol* 66, 5488–5491.

Hesse, C.N., Mueller, R.C., Vuyisich, M. et al. 2015. Forest floor community metatranscriptomes identify fungal and bacterial responses to N deposition in two maple forests. *Front Microbiol* 23, 337.

Hinsinger, P., Bengough, A.G., Vetterlein, D. and I.M. Young. 2009. Rhizosphere: Biophysics, biogeochemistry and ecological relevance. *Plant Soil* 321, 117–152.

Hopkins, M.J. and G.T. MacFarlane. 2000. Evaluation of 16S rRNA and cellular fatty acid profiles as markers of human intestinal bacterial growth using the chemostat. *J Appl Microbiol* 89, 668–677.

Hurt, R.A., Qiu, X.Y., Wu, L.Y. et al. 2001. Simultaneous recovery of RNA and DNA from soils and sediments. *Appl Environ Microbiol* 67, 4495–4503.

Jacobsen, C.S. and W.E. Holben. 2007. Quantification of mRNA in *Salmonella* sp. seeded soil and chicken manure using magnetic capture hybridization RT-PCR. *J Microbiol Methods* 69, 315–321.

Kolb, S. and A. Stacheter. 2013. Prerequisites for amplicon pyrosequencing of microbial methanol utilizers in the environment. *Front Microbiol* 5, 268.

Luis, P., Kellner, H., Martin, F. and F. Buscot. 2005. A molecular method to evaluate basidiomycete laccase gene expression in forest soils. *Geoderma* 128, 18–27.

Manefield, M., Whiteley, A.S., Griffiths, R.I. and M.J. Bailey. 2002. RNA stable isotope probing, a novel means of linking microbial community function to phylogeny. *Appl Environ Microbiol* 68, 5367–5373.

Mendum, T.A., Sockett, R.E. and P.R. Hirsch. 1998. The detection of Gram-negative bacterial mRNA from soil by RT-PCR. *FEMS Microbiol Lett* 164, 369–373.

Mettel, C., Kim, Y, Shrestha, P.M. and W. Liesack. 2010. Extraction of mRNA from soil. *Appl Environ Microbiol* 76, 5995–6000.

Moran, M.A., Torsvik, V.L., Torsvik, T. and R.E. Hodson. 1993. Direct extraction and purification of ribosomal RNA for ecological studies. *Appl Environ Microbiol* 59, 915–918.

Newman, M.M., Lorenz, N., Hoilett, N. et al. 2016. Changes in rhizosphere bacterial gene expression following glyphosate treatment. *Sci Total Environ* 15, 32–41.

Nogales, B., Timmis, K.N., Nedwell, D.B. and A.M. Osborn. 2002. Detection and diversity of expressed denitrification genes in estuarine sediments after reverse transcription-PCR amplification from mRNA. *Appl Environ Microbiol* 68, 5017–5025.

Pesaro, M., Nicollier, G., Zeyer, J. and F. Widmer. 2004. Impact of soil drying-rewetting stress on microbial communities and activities and on degradation of two crop protection products. *Appl Environ Microbiol* 70, 2577–2587.

Petersen L. and K.H. Esbensen. 2005. Representative process sampling for reliable data analysis—a tutorial. *J Chemom* 19, 625–647.

Pierre, S., Hewson, I., Sparkks, J. et al. 2017. Ammonia oxidizer populations vary with nitrogen cycling across a tropical montane men annual temperature gradient. *Ecology* 98, 1996–1977.

Ramos, C., Mølbak, L. and S. Molin. 2000. Bacterial activity in the rhizosphere analyzed at the single-cell level by monitoring ribosome contents and synthesis rates. *Appl Environ Microbiol* 66, 801–809.

Rhee, S.K., Liu, X., Wu, L., Chong, S.C., Wan, X. and J. Zhou. 2004. Detection of genes involved in biodegradation and biotransformation in microbial communities by using 50-Mer oligonucleotide microarrays. *Appl Environ Microbiol* 70, 4303–4317.

Sharma, S., Aneja, M., Mayer, J., Munch, J.C. and M. Schloter. 2005. Diversity of transcripts of nitrite reductase genes (*nirK* and *nirS*) in rhizospheres of grain legumes *Appl Environ Microbiol* 71, 2001–2007.

Sharma, S., Aneja, M., Mayer, J., Schloter, M. and J.C. Munch. 2004. RNA fingerprinting of microbial community in the rhizosphere soil of grain legumes. *FEMS Microbiol Lett* 240, 181–186.

Sharma, S., Szele, Z., Schilling, R., Munch, J.C. and M. Schloter. 2006. Influence of freeze-thaw on the structure and function of microbial communities in soil. *Appl Environ Microbiol* 72, 2148–2154.

Töwe, S., Wallisch, S., Bannert, A. et al. 2011. Improved protocol for the simultaneous extraction and column-based separation of DNA and RNA from different soils. *J Microbiol Methods* 84, 406–412.

Tsai, Y.L., Park, M.J. and B.H. Olson. 1991. Rapid method for direct extraction of mRNA from seeded soils. *Appl Environ Microbiol* 57, 765–768.

Urich, T., Lanze, A., Qi, J., Huson, D.H. and C. Schleper. 2008. Simultaneous assessment of soil microbial community structure and function through analysis of the meta-transcriptome. *PLoS One* 3, e2527.

Vestergaard, G., Schulz, S., Schöler, A. and M. Schloter. 2017. Making big data smart—how to use metagenomics to understand soil quality. *Biol Fertil Soils* 53, 479–484.

Wang, Y., Hayatsu, M. and T. Fujh. 2012. Extraction of bacterial RNA from soil: Challenges and solutions. *Microbes Environ* 27, 111–121.

Wellington, E.M.H., Berry, A. and M. Krsek. 2003. Resolving functional diversity in relation to microbial community structure in soil: Exploiting genomics and stable isotope probing. *Curr Opin Microbiol* 6, 295–301.

White, R.A., Bottos, E.M., Roy Chowdhury, T. et al. 2016. Long-read sequencing facilitates assembly and genomic binning from complex soil metagenomes. *mSystems* 1(3).

Wilson, M., Bakermans, C. and E. Madsen. 1999. In situ, real-time catabolic gene expression. Extraction and characterization of naphthalene dioxygenase mRNA transcripts from groundwater. *Appl Env Microbiol* 65, 80–87.

Yergeau, E., Tremblay, J., Joly, S. et al. 2018. Soil contamination alters the willow root and rhizosphere metatranscriptome and the root-rhizosphere interactome. *ISME J* 12, 869–884.

Zani, S., Mellon, M.T., Collier, J.L. and J.P. Zehr. 2000. Expression of *nifH* genes in natural microbial assemblages in lake George, New York, detected by reverse transcriptase PCR. *Appl Environ Microbiol* 66, 3119–3124.

Zhang, Y., Xu, J., Riera, N., Jin, T., Li, J. and N. Wang. 2017. Huanglongbing impairs the rhizosphere-to-rhizoplane enrichment process of the citrus root-associated microbiome. *Microbiome* 10, 97.

16 Metaproteomics of Soil Microbial Communities

Paolo Nannipieri, L. Giagnoni, and G. Renella
University of Firenze

CONTENTS

16.1 Introduction 257
16.2 Environmental Proteomics: Definitions, Technology, Challenges, and Perspectives 258
16.2.1 Definitions and Developments 258
16.2.2 Approaches in Environmental Proteomics 259
16.3 Soil Metaproteomics 260
16.3.1 Proteomics Studies in Model Soil and Laboratory Systems 260
16.3.2 Proteomics Studies in Soil: The State of the Art and the Technical Challenges 261
16.3.3 Conceptual Challenges: Protein Distribution in the Soil Matrix, Soil Proteomics, and Soil Metaproteomics 264
16.4 Conclusions and Prospects for Future Research 265
References 267

16.1 INTRODUCTION

Soil functionality depends mainly on the activity of microbial communities. Microbial properties are more sensitive to changes in management practices or environmental conditions than chemical and physical properties of soil (Nannipieri et al. 2017). Both general (e.g., soil respiration) and specific (e.g., enzyme activities) methods can be used to determine soil microbial activities. In addition, transformations of key nutrients can be quantified in soil by using isotope methods (Nannipieri et al. 2017). Given the importance, molecular studies of microbial diversity are popular (see Chapters 3 and 13–15). However, the soil metabolic activity can only be ascertained by detecting the synthesized proteins (Nesatyy and Suter 2007). Rocca et al. (2014) conducted a meta-analysis of the relationships between the abundances of genes, transcripts, and the corresponding process rates, on the basis of 59 in 415 studies where this was possible. They concluded that gene abundance cannot *a priori* be taken as a proxy of the corresponding process rate, based on the fact that only a few studies showed a correlation between the two parameters. Proteins are better markers of biological function than nucleic acids because they perform metabolic reactions (e.g., enzymes) and are involved in regulatory cascades. Therefore, they provide more direct information on microbial activity than genes, at the level of abundance (DNA) or expression (mRNA). However, the translation of mRNA to proteins is not the final step in microbial functionality, since proteins can be regulated or modified following biosynthesis and before expressing their activity (Nesatyy and Suter 2007). Thus, proteomics, which involves the detection of the proteins expressed by the genome of an organism, can yield insights in the activity of the target organism (Greaves and Haystead 2002). In addition, proteomics studies are not based on prior assumptions as to which genes are important, and are therefore less restrictive than metagenomics and metatranscriptomics approaches. The term "metaproteomics" was coined to describe the determination of the most abundant proteins expressed in environmental samples (Nesatyy and Suter 2007). Generally, a complete metaproteomics analysis involves the following main steps: sampling, handling and storing the collected samples, protein extraction, protein purification and separation, proteolytic digestion, mass spectrometric analysis, database searches, and finally, data interpretation (Siggins et al. 2012). As shown below, proteolytic digestion can follow protein purification and precede separation of peptides produced by the digestion.

Soil metaproteomics can improve our knowledge on the response of soil microorganisms to environmental changes and on the interactions among microorganisms, between microorganisms and micro-, meso- and macrofauna, and between microorganisms and plants. Other expected outcomes from this research encompass the identification of functional indicators for evaluation of soil quality (Ogunseitan 2006, Renella et al. 2014, 2015). Bastida and Jehmlich (2016) examined the application of soil metaproteomics for studying the impact of deforestation, agricultural practices, restoration of semiarid soils, biodegradation of pollutants, and microbial dynamics in permafrost, providing information about functional–phylogenetic relationships, also in the context of resilience and resistance in soil.

The aim of this chapter is to discuss the state of the art of soil proteomics by underpinning the present methodological and conceptual challenges and the potentialities for improving the determination of microbial activities as main drivers of soil functionality. First, the development of analytical tools in proteomics studies will be examined, in order to foster improved applications in soil. Then, proteomics-based studies of soil microbiomes in microcosm and mesocosm systems will be addressed. Moreover, we will discuss proteomics of soil samples collected under extreme environmental conditions which are hypothesized to have a more limited microbial functionality than other soils, also in the context of a historical perspective. Finally, the different locations of proteins in the soil matrix will be addressed. This will allow to distinguish different soil proteomics approaches, focusing either on proteins reflecting the actual response of microorganisms to stresses, changes in environmental conditions, and soil management or on stabilized extracellular proteins that may be indicative of past microbial events. In this context, the distinction between soil proteomics and soil metaproteomics is important, since the former term concerns the protein expression profile under a specific set of biotic and abiotic conditions at the sampling time, whereas the latter term also concerns extraction of all proteins, including those not expressed by microbial activity at the sampling time.

16.2 ENVIRONMENTAL PROTEOMICS: DEFINITIONS, TECHNOLOGY, CHALLENGES, AND PERSPECTIVES

16.2.1 Definitions and Developments

In 1993, Ogunseitan initiated environmental proteomics by studying the proteins extracted from a suite of environmental samples, including soil. Then—in 1995—Witzmann et al. applied a novel technique, that is, two-dimensional (2D) electrophoresis mapping, to find protein biomarkers of human and rodent hepatic stresses induced by toxic chemicals. In 1999, the same authors pioneered the application of mass spectrometry to identify proteins in environmental proteome samples. After these seminal publications, the field of environmental proteomics gradually matured, with contributions studying selected proteins and additional studies aiming to improve methodologies (Renella et al. 2014). In 2002, Craig and Collins applied such techniques in a search for protein markers in human settlements in archeological sites. After these early studies, most soil proteomics studies aimed to increase protein extraction yields from soil samples for inventory purposes rather than to address the expression of proteins of key processes (Renella et al. 2015). New definitions were introduced according to the type of soil proteomics study. Thus *exoproteomics* concerns the study of extracellular proteins released (actively or by leakage) by microbial cells for nutrient uptake, motility, cell attachment, defense, communication, and antagonism. Extracellular proteins can be obtained by removing cells from the culture medium before protein extraction (Armengaud et al. 2012, Christie-Oleza et al. 2015). Analysis of peptides/

proteins with molecular masses lower than 15 kDa but higher than 0.5 kDa has been denominated *peptidomics* (Nesatyy and Suter 2007).

16.2.2 Approaches in Environmental Proteomics

Both bottom-up and top-down approaches can be used in environmental (meta)proteomics. In the former approach, purified proteins or complex mixtures of proteins are trypsin-digested into peptides, which are separated [usually by liquid chromatography (LC)] before the determination of proteins by a mass spectrometric (MS) or by tandem MS (MS/MS) analysis. In top-down proteomics, intact proteins are loaded into an MS, and protein ions are generated by electrospray mass spectrometry (ESI) and subjected to gas-phase fragmentation (Nesatyy and Suter 2007). The main limitation of bottom-up proteomics is that generally low coverage of the protein sequences is obtained, as only a small and variable fraction of the total peptide population of a protein can be recovered. This limitation is important considering that each open reading frame can give rise to several protein isoforms. The top-down approach is often better than the bottom-up approach for two reasons:

i. MS analysis of intact proteins that have not been cleaved allows determining structural protein characteristics or posttranslational modifications that are mostly destroyed in the bottom-up approach.
ii. Elimination of protein digestion results in significant time savings (Nesatyy and Suter 2007).

Despite these advantages, top-down proteomics is less used than bottom-up proteomics due to serious technical problems. First, proteins are generally more complicated to handle than small peptides (generated in the bottom-up approach), which is mainly due to their poor solubility. For example, some high-molecular-weight (HMW) proteins require a detergent such as sodium dodecyl sulfate (SDS) for their solubilization, and this detergent is not compatible with the use of ESI. Second, the sensitivity of MS is more suited for detection of peptides than that of proteins. Finally, detection of proteins with molecular weights higher than 50 kDa is not possible due to problems in destroying their tertiary structure. For these reasons, top-down proteomics is only used for studying single proteins or low-to-moderate complexity protein mixtures (Nesatyy and Suter 2007).

Two-dimensional electrophoresis separates proteins according to their isoelectric point and molecular weight. Proteins are stained for visualization, after which protein spots can be excised and proteins solubilized, digested by trypsin, and the resulting peptides determined by MS or tandem MS. LC, either one- or two-dimensional, can be used to separate peptides resulting from the trypsin digestion (Nesatyy and Suter 2007). Usually, low-abundance proteins and proteins with extreme isoelectric points (acid or basic) or extreme molecular weights (very low or very high) are not separated by this method. Also, hydrophobic proteins are excluded in the separation (Keller and Hettich 2009). One example of 2D LC is the running of acidified peptides first by a multiple-step-gradient separation through strong cation-exchange chromatography followed by a high-performance reversed chromatography of the eluates. Three-dimensional LC-MS/MS technology (involving three types of LC in sequence) can give a high separation power of complex protein mixtures (Wei et al. 2005).

High mass-accuracy tandem spectrometers (MS/MS) are increasingly used for protein analyses (Keller and Hettich 2009). The first MS determines the relative mass of each peptide, and then, each isolated peptide is dissociated by collision with a neutral gas before determination of its mass-to-charge ratio (Keller and Hettich 2009). Advancements in proteomics have paralleled the development of high-efficiency peptide ionization methods in MS. Nowadays, different MS techniques, such as matrix-assisted laser desorption–ionization (MALDI) and MALDI combined with analysis of time of flight (MALDI-TOF), are used. Fragmentation of peptides can be obtained with electrospray source (ESI-MS/MS) (Keller and Hettich 2009). The use of the ultra-high-field Orbitrap analyzer has improved the speed, resolution, and sensitivity of detection of peptides (Scheltema et al. 2014). Techniques such as "selected multiple reaction monitoring" or "multiple reaction monitoring" can detect up to 100 proteins simultaneously (Picotti et al. 2010).

Moreover, stable-isotope probing (SIP; see also Chapter 17) used in metaproteomics can assess the proteins of metabolically active microorganisms (Jehmlich et al. 2016). In SIP metaproteomics, proteins synthesized by active microbial cells are labeled by using substrates carrying a stable isotope (e.g., ^{13}C-glucose, ^{15}N-NH_4^+, labeled amino acids). MS applied to the resulting microbiome samples can determine the mass shift due to the presence of the stable isotope, which yields an indirect evaluation of the metabolic activity. The number of studies using ^{13}C is much higher than that of those using ^{15}N, as C is present in proteins in higher amounts than N, and, thus, its use will give a higher mass shift than the use of ^{15}N (Jehmlich et al. 2016). The sensitivity of SIP-based detection of ^{13}C-labeled proteins is comparable to that of similarly labeled phospholipid fatty acids. It is much higher than that of nucleic acids-based SIP, which requires cell replication for incorporation and thus incubation times longer than those required for labeling proteins. This incurs the risk of cross-feeding. It has been estimated that detection of labeled proteins is already possible at 1% incorporation of the stable isotope, whereas the threshold for nucleic acids-based SIP is at least 30% (Jehmlich et al. 2016).

The final step in soil proteomics studies is the identification of proteins from MS readings using a proteome database. Unfortunately, the current gap between annotated protein information and genomic information limits proteomics studies as compared to genomics ones. This problem is particularly critical in environmental proteomics due to the insufficient annotation of the proteomes of environmental microorganisms (Renella et al. 2015). Proteins predicted from fully sequenced genomes are identified by computational scoring, which involves matching peptides from proteins expressed by genes with those measured. In the case of unsequenced species, proteins can be identified from peptides measured by MS in two ways: (1) matching these peptides with those in databases and (2) subjecting them to a computational method that predicts protein structure (Keller and Hettich 2009). A further challenge in protein identification by MS is caused by posttranslational modifications. The present databases should be improved including such modifications (e.g., by phosphorylation, glycosylation, or truncation). These modifications often change protein function and, if detected, provide further functional information on microbial metabolic responses in the environment.

16.3 SOIL METAPROTEOMICS

16.3.1 Proteomics Studies in Model Soil and Laboratory Systems

Given the complexity of soil and its immense microbial diversity and biomass (Nannipieri et al. 2017), it is problematic to study the full plethora of soil microbial functions by expressed proteins. Thus, the initial proteomic studies on microbiome processes or interactions have been carried out using simplified systems, such as mixtures of surface-reactive soil particles inoculated with single soil bacteria or simple consortia. Next to information on expressed proteins, pertaining to interactions of target microorganism with surface-reactive soil particles or with other microbes or plant roots, information on the efficiency of the analytical methods can be produced. Here, we discuss such experiments on the interaction of soil microorganisms with surface-reactive particles. However, caution is required to extrapolate the results obtained with these model soil systems to "real-world" soil.

Giagnoni et al. (2011) showed that surface-reactive particles affect the (extracted) proteome in experiments performed with the soil bacterium *Cupriavidus metallidurans* CH34 in model soils (mixtures of sand, goethite, humic acids, kaolinite, or montmorillonite). Reduced numbers of generally low-quality proteins were found—when using phosphate buffer (also containing SDS and a protease inhibitor cocktail) extraction—upon increases in the montmorillonite content of the soils. The highest yields were found in the artificial soils with kaolinite. Two-dimensional electrophoresis of the model soil extracts showed that some proteins present in the control (cell extract) could not be found. Further MS-based analyses showed that the proteomes of the culture resembled those extracted from the sand and kaolinite model soils, but not those from the montmorillonite one (Giagnoni et al. 2012). In addition, protein recovery from all microcosms decreased upon prolonged incubation. Thus, the presence of surface-reactive particles affects, qualitatively and quantitatively, the efficiency of extraction of proteins from soil. Interactions with soil organic matter (SOM) also affect protein identifications from soil. Arenella et al. (2014) showed that interactions of myoglobin, α-glucosidase, and β-glucosidase with soil humic acids changed the conformation of these proteins (particularly that of myoglobin), thus affecting successive trypsin digestions and the production of tryptic peptides to be determined by MS.

Several Gram-negative bacteria adopt the viable but non-culturable (VBNC) status as a survival strategy when subjected to stresses (see Chapter 4; Giagnoni et al. 2018). Recent studies on the proteome of *C. metallidurans* CH34 in the aforementioned model soil showed that proteins related to cell shape and protein synthesis were downregulated during the transition from growth to the VBNC state in response to water and nutrient limitations (Giagnoni et al. 2018). Reversion from the VBNC to the culturable state—by adding gluconate as a nutrient source—showed the restoration of proteins involved in biosynthesis and energy generation pathways. In contrast, water addition caused the upregulation of only six proteins, which were involved in transport, translation, and protein folding. These results agree with those by Bastida et al. (2016), who observed that proteins involved in replication and nucleic acid integrity and metabolism, especially those of Proteobacteria, are more abundant in soil with higher dissolved organic carbon (DOC) than in that with lower DOC values.

To address the potential microbially driven mineralization of organic P by extracellular phosphatases (found in rhizosphere soil in response to low P concentrations), the extracellular proteins released from three plant-associated *Pseudomonas* strains (*P. putida* BIRD-1, *P. fluorescens* SBW25, and *P. stutzeri* DSM4166) were studied *in vitro*. These showed different patterns, revealing, in addition to alkaline phosphomonoesterase (of which the synthesis occurs in response to P limitation), phosphotriesterases, putative phosphonate transporters, and other outer membrane proteins (Lidbury et al. 2016).

In other experiments, the *in vitro* proteomics of a Gram-positive bacterium (*Bacillus atrophaeus*) and a Gram-negative bacterium (*P. putida*), both active in bioremediation and inhabitants of rhizosphere soil, was assessed. Proteomes were affected when antagonistic compounds were added to culture media or when these bacterial species interacted between them (Chignell et al. 2018). These results confirm that the proteome of individual bacterial cells in soil is affected not only by the environmental conditions but also by the presence of other species.

16.3.2 Proteomics Studies in Soil: The State of the Art and the Technical Challenges

Early soil proteomics studies addressed either specifically expressed proteins or subsets of proteins, for example, those expressed in response to pollutants in soil (Ogunseitan et al. 2006, Nannipieri et al. 2006). In an early study, proteins extracted from an Entisol with phosphate buffer set at pH 6.0 and separated by sodium dodecyl sulfate–polyacrylamide gel electrophoresis (SDS–PAGE) gave several bands that ranged in size from 35 to 68 kDa. One of these bands was identified as a homologue of a thermostable cellulase produced by the thermophilic fungi *Humicola insolens* and *H. grisea* (Muranase et al. 2003). In another study, it was found that Cd pollution in soil decreased the amount of extracted proteins but increased that of small proteins, which were, unfortunately, not identified (Singleton et al. 2003). These early studies give a snapshot of what might be possible in soil proteomics but also hinted at extreme challenges posed to researchers, as discussed next.

Proteins can be extracted from soil by both direct and indirect (Table 16.1) extractions. In the former method, cells are lysed in the soil, and proteins are extracted, whereas in the latter method, cells are first separated from soil particles and then lysed (Maron et al. 2007, Nannipieri 2006,

TABLE 16.1
Soil (Meta)Proteomics Studies Based on the Extraction of Proteins via Direct or Indirect Methods

Sequential Extraction	Results	Soil Type	References
	Direct Methods		
Citrate buffer + SDS, phenol followed by precipitation with methanol	About 630 identified proteins	Rhizosphere soils from rice, sugarcane, and tobacco fields	Wang et al. (2010)
Citrate buffer, SDS, phenol	Only one- and two-dimensional profiles	Arable soil	Chen et al. (2009)
Citrate and Tris buffer + SDS at pH 8.5	38 proteins expressed differently by comparing rhizosphere and bulk soil	Sugarcane rhizosphere soils	Lin et al. (2013)
Citrate and Tris buffer + SDS at 8.5	Bacterial and fungal proteins involved in the tomato resistance to *Fusarium*	*Fusarium*-infected rhizosphere soil sampled under tomato	Manikandan et al. (2017)
NaOH and phenol	12 identified proteins including 1,2-dioxygenase and chaperone and membrane proteins	Compost-treated soil	Benndorf et al. (2007)
Four different protein extraction methods	The type and number of the identified proteins depended on the extraction method	Forest and potting soils	Keiblinger et al. (2012)
	Indirect Methods		
Separation of microbiomes from soil particles by density gradient centrifugation. Separation of proteins from pellets by Tris-HCl buffer (pH 8.5) + sucrose + EDTA + KCl + 2-mercaptoethanol; final protein extraction with phenol	Protein patterns of toluene-exposed soils differed from those of glucose-exposed soils. Only small subsets of expressed proteins were identified	• Native prairie soil covered by mixed C_3 and C_4 plants • Soil with herbaceous crops	Taylor and Williams (2010), Williams et al. (2010)
	Multi-omics Approach		
Metagenomic, metatranscriptomic, and proteomic (SDS lysis-based) method with trichloroacetic acid precipitation	Many detected proteins (most involved in housekeeping functions) and cold-shock proteins detected in the permafrost	Permafrost and thermokarst bog soils	Huitman et al. (2015)

Ogunseitan 2006, Renella et al. 2014). Generally, the direct methods give the highest extraction yields, but are less specific for microbial cells than the indirect ones (Renella et al. 2014). Moreover, the extracts produced by the direct methods often contain substances (e.g., humic substances) that interfere with subsequent analytical steps, and this makes purification complex. Chourey et al. (2010) compared direct protein extraction (involving the use of SDS Tris-HCl buffer at pH 8.5 with additives promoting protein desorption from soil) with indirect extraction, using soil inoculated with *P. putida* or *Arthrobacter chlorophenolicus*. Each method allowed the identification of more than 500 unique proteins, with the direct method providing deeper data than the indirect one. Cytoplasmic membrane, outer membrane, and periplasmic, intra-, and extracellular (next to unknown) proteins, mostly trackable to the introduced bacteria, were observed. In contrast, Taylor and Williams (2010) found that direct extraction applied to soil and successive separation of proteins by SDS–PAGE gave unresolved bands, whereas this problem did not occur with the indirect extraction method.

Unfortunately, several authors of recent soil proteomics studies have ignored the extensive bibliography on enzyme extraction from soil. The reasons may be twofold: (1) a too superficial knowledge of the complexity of soil and (2) the fact that most "old" bibliography on enzyme extraction from soil is not available to libraries. In early work, it was shown that alkaline solutions (NaOH, $Na_2P_2O_4$) for extracting organic matter will give high enzyme/protein extraction yields from soil (Nannipieri et al. 1996). Increasing the pH of the extraction solution increases the amount of proteins extracted from soil, but protein separation by SDS–PAGE patterns was found to give fuzzy bands and tailing (Muranase et al. 2003). The extracts obtained with these solutions, even with $Na_2P_2O_4$ near neutrality, are dark due to high concentrations of humic molecules. As mentioned above, several purification steps are needed to obtain pure proteins (Nannipieri et al. 1996). Schulze et al. (2005) separated humic acids and proteins in forest leachate by gel filtration, and separated proteins by SDS–PAGE. Using the National Center for Biotechnology Information (NCBI) protein and taxonomy database, they found bacterial proteins accounted for 50% of the total detected proteins. The forest leachate contained fungal, bacterial, and animal proteins in amounts greater than those found in a (control) leachate of a peat bog. In the latter, 78% of total proteins were from bacteria (involved in transferring one C residue, acyl groups, N-containing groups or P-containing groups). In addition, peroxidases were only present in the forest leachate, whereas transferases were present in the peat bog leachate. These differences may reflect the different microbial activities in the two sampled habitats, probably due to the presence of different microbial communities.

An NaOH and phenol solution was used by Benndorf at al. (2007) to extract proteins from an organic rich soil packed in a column. Only 12 proteins were identified, among which a 1,2-dioxygenase and some chaperone and membrane proteins. Considering that NaOH is one of the most effective SOM extracting solutions, the low number of identified proteins underlines that effective purification of the extracted humic molecules from soil can markedly decrease the number of identified proteins. Therefore, an effective method in soil proteomics should also consider the loss of proteins after the purification step following soil extraction.

The protein extraction step in soil metaproteomics is very critical. Ideally, the quality and quantity of the extracted proteins would be representative of those present in the sample (Nannipieri 2006, Ogunseitan 2006, Renella et al. 2014, Taylor and Williams 2010). In a hallmark study by Keiblinger et al. (2012), different proteins were extracted from a forest and a potting soil (50% white peat, 25% clay, and 25% pumice) using four different methods, with a method involving SDS-phenol extraction giving the highest number of extracted proteins. However, the protein recovery rate was only around 50% as determined by spiking the two sterilized soils with proteins from *Pectobacterium carotovorum* and *Aspergillus nidulans.* Differences in functional proteins were more marked among the four extraction methods with the potting than with the forest soil. Most of the identified proteins were presumably involved in translational processes, followed by post-translational modification, protein turnover and chaperones, and energy production and conversion.

Completeness of soil proteomes and technical solutions—Despite the technical challenges, it is important to underline that soil proteomics can detect specific proteins nowadays, as discussed above. Soil protein N constitutes one of the N pools in holistic studies aiming to quantify the N dynamics in the soil–plant system. By detecting specific proteins, soil proteomics allows to open this "box" by detecting proteins with specific functions; this is needed to foster our understanding of soil functionality. For example, Huitman et al. (2015) detected cold-shock proteins in permafrost soil, probably reflecting the physiological response of soil microorganisms to freezing conditions, as discussed below. However, a key finding of all studies has been that the resulting proteomes were generally incomplete, revealing the presence only of abundant proteins that "survived" the extraction protocols as detectable entities. The technical challenges are related to the high numbers of proteins that can be expressed from active microorganisms in soil and their persistence, which also depends on the type of soil. Obviously, the choice of the extraction (direct or indirect) method is also related to the soil type. With an organic matter rich soil, the indirect method may be preferred over the direct one, so as to overcome the problems of humic material extraction. This is because several purification steps are needed to obtain pure proteins in cases of strong humic matter contamination (Nannipieri et al. 1996). Such strong purification steps markedly reduce the number of proteins that can be identified. In addition, as mentioned above, protein molecules show a wide range of chemical properties, being either hydrophilic or hydrophobic, having either low or high molecular weight, next to high or low isoelectric points. Consequently, it is difficult to extract all proteins once they are released into soil with the direct extraction method. Given the binding of many proteins to protein adsorption sites in soil (as discussed in the foregoing), one may include blockers of protein adsorption sites in the direct extraction protocols. For example, the extraction of hydrophobic proteins from soil was found to be increased by blocking such adsorbing sites with bovine serum albumin and Triton-X prior to microbial lysis (Fornasier and Margon 2007).

Combined omics—Combining metaproteomics with the other omics approaches can give a more complete picture of activity and species diversity than the use of a single omics approach. Particularly, *proteogenomics* (the link between genomics and proteomics) with cross-checks of the identified proteins with their respective genes will discover new genes, validate predicted genes, correct translational starts, identify posttranslational modifications and give functional information on proteins. These outcomes give insights in environmental and physiological conditions affecting protein synthesis (Bland and Armengaud 2014). Amplicon sequencing combined with metaproteomics using ^{15}N-labeled tobacco plants showed that bacteria (but not fungi) were involved in the short-term assimilation of plant N (Starke et al. 2016). As far as we know, this is the first report showing the combined use of metagenomics and metaproteomics based on protein-SIP to track the fate of added organic N. A multi-omics approach combined with measurements of iron reduction, sulfate and nitrate utilization, denitrification, methane oxidation, and trace gas (CO_2, N_2O, CH_4) production was conducted on permafrost soil cores and a young thermokarst bog (Huitman et al. 2015). Metagenomics- and metatranscriptomics-based analyses identified members of the Actinobacteria, Acidobacteria, and Proteobacteria as active organisms. Most of the identified proteins matched the ones predicted from the metagenomics data; however, many were unknown. In addition to many proteins involved in housekeeping functions, cold-shock proteins were abundantly found, which was consistent with the need of the organisms to adapt to frozen conditions.

Understanding the proteomic responses of known microorganisms is useful as it gives information on the relationship between microbial presence, local conditions, and soil functions (Giagnoni et al. 2018).

16.3.3 Conceptual Challenges: Protein Distribution in the Soil Matrix, Soil Proteomics, and Soil Metaproteomics

The complexity of the soil system is, generally, not considered in soil proteome studies. In particular, the distribution of proteins in soil is neglected. Most of the total N in soils is present as organic N (usually 95%–99% of the total), with, on average, 4% of organic N being present as microbial biomass N (Nannipieri et al. 2006). Since not all microbial N is present as proteins and/or peptides, the intracellular protein N is less than 4% of the total organic N (Figure 16.1). It is well established that most protein N and peptide N is present extracellularly in soil, accounting for 30%–50% of organic N. The extracellular proteins and peptides can be present as free, loosely adsorbed or loosely bound, tightly bound, or even entrapped molecules (Nannipieri and Paul 2009). Microorganisms may actively release extracellular proteins, for example enzymes to degrade HMW substrates, or proteins and peptides can be leaked after cell lysis. The protein molecules can be used as a nutrient source by active microbial cells. If bound to soil surfaces, the binding can be weak (when molecules can easily be solubilized by water), strong (due to hydrogen bonds or electrostatic interactions), or strong due to the formation of covalent bonds (Nannipieri et al. 1996). Both the extracellular and intracellular proteins are key target in soil (meta)proteomics studies (Figure 16.1).

In terms of what proteomes tell us about soil function, extracted proteins may not always reflect the microbial activity at the time of sampling, as adsorption/desorption processes differ. The finding of stabilized extracellular proteins in soil may give insights into their properties (for instance,

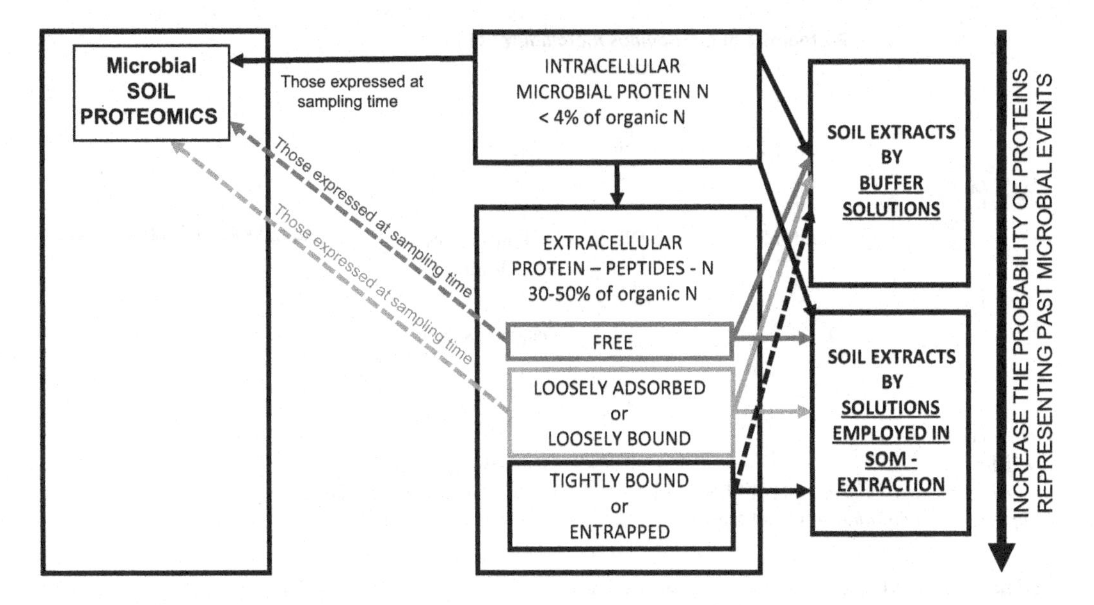

FIGURE 16.1 Protein distribution in the soil matrix as related to soil proteomics and soil metaproteomics.

hydrophobic versus hydrophilic proteins, and glycoproteins) and thus on the mechanisms (hydrophobic or electrostatic interactions, and hydrogen bonds) responsible for their stabilization. Such stabilized extracellular proteins may give insights in past microbial events.

The number of proteins that is required for detecting and evaluating microbial activities in soil is another challenge. Considering that several thousands of microbial species may inhabit 1 g of soil and that each microbial species can synthesize a few thousand proteins, the diversity of proteins can be very high. The number depends on the active microbial species, which in hot spots, such as rhizosphere soil, can reach a hundred thousand and even millions of proteins. Such numbers are much higher than those of the proteins extracted from the soil so far. Therefore, by considering the high microbial diversity of soil, current techniques are not yet adequate for resolving the soil metaproteome.

Moreover, indirect protein extraction methods applied to soil may be biased, as proteomes can change during the separation of microbial cells from soil particles. The microbial proteome is sensitive to changes as shown for *C. metallidurans* cells 1 day after introduction into a model soil (Giagnoni et al. 2012; see also Figure 16.2) Sequential extraction from soil may be a possible strategy to separate intracellular, free extracellular, and loosely adsorbed proteins from the extracellular tightly bound/entrapped ones (Figure 16.1). Extraction of proteins from soil should always include a control based on inoculation of the sterilized soil with a microbial species whose proteome is known, as done by Keiblinger et al. (2012).

16.4 CONCLUSIONS AND PROSPECTS FOR FUTURE RESEARCH

Despite extensive past research efforts and technological improvements, soil proteomics still shows challenges and pitfalls. The number of proteins can be theoretically very high in soil. Even if we would be able to extract all of these proteins, their determination by current techniques is not possible. Moreover, the chemical and physical properties of proteins are diverse, and for this reason, a plethora of different extraction protocols has been proposed. The genomes and (predicted) proteomes of an as-yet-limited number of soil organisms have been covered, and hence the databases of environmental nucleic acid and protein sequences are incomplete. This picture is made even more complex by the dependency of gene expression in soil on environmental factors and by the high genetic variability even across members of the same taxon (Nesatyy and Suter 2007). There is also a number of technological problems that prevents us from being all-encompassing in our drive to understand the full soil metaproteome.

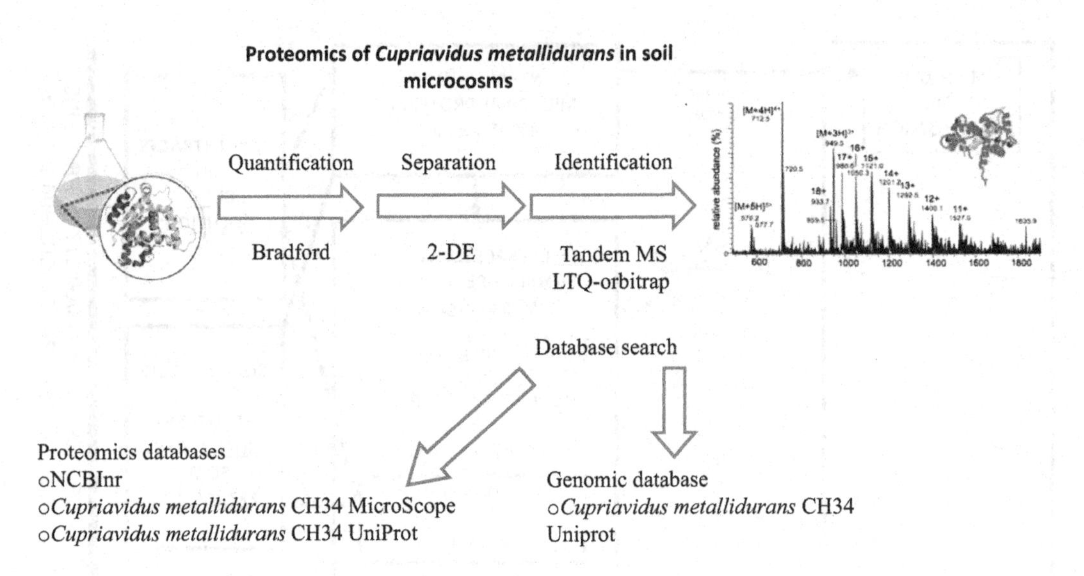

FIGURE 16.2 Metaproteomics of *Cupriavidus metallidurans* in soil-mimicking microcosms. Cells of *C. metallidurans* harvested during the exponential growth phase were introduced into microcosms consisting of quartz sand, kaolinite, and montmorillonite alone, or mixed together with goethite and humic acids (Giagnoni et al. 2012). Proteins were extracted from cells or from microcosms by phosphate buffer, purified to remove humic acids, and separated by two-dimensional gel electrophoresis. After staining the gels, spots were manually excised, then destained, and dehydrated, and proteins were subjected to trypsin digestion. The digested proteins were analyzed by MALDI-TOF MS and mass fingerprinting searches done in the NCBInr and Swiss-PROT databases (Giagnoni et al. 2012).

Accurate MS is not universally available and the application of this and other advanced technique(s) is still limited to specific laboratories. Current MS-based techniques are not even able to characterize the overall proteome of a single organism, let alone that of a soil microbiome. Hence, incomplete proteome coverage ensues, which is often accompanied by problems of irreproducible results. Finally, proteomics still gives a static snapshot of the local microbiome constitution in the soil, and temporally-explicit proteomic characterizations are highly needed to allow assessments of developments. Only the assessment of temporal trends will allow a better understanding of the target biological process (Nesatyy and Suter 2007). Therefore, future research should use time courses in combination with optimized proteomics protocols, in order to foster our understanding of gene expression in the soil environment. Soil proteomics should consider the intracellular and free extracellular proteins, reflecting the microbial activities at the moment of soil sampling. On the other hand, soil metaproteomics will deal with all soil proteins, including those extracellularly present in soil. This latter study may give insights on past microbial events and on the mechanisms responsible for the stabilization of proteins in soil. All soil metaproteomics studies performed to date have indicated that the qualitative and quantitative composition of proteomes will depend on the extracting method used as well as the soil type. Proteins synthesized by organisms in response to pollutants or other stresses can be used as biomarkers, but caution is needed since the response of an organism under laboratory conditions can be different from that of the same organism in the soil (Barea and Gomez-Ariza 2006). The soil composition, in terms of SOM and clay contents, should be considered, as humic substances and clays can interfere with soil proteomic analysis, hampering protein recovery or identification by MS (Arenella et al. 2014, Giagnoni et al. 2011, 2012). Studies in soil proteomics have almost exclusively been conducted with gram-scale samples of soil, thus not considering the importance of spatial scales in soil (discussed in Chapters 1 and 9). Another expected outcome from soil proteomics research concerns the identification of functional indicators to be used to assess soil quality. Recently, Schloter et al. (2018) discussed the need to include molecular indicators in a good framework for soil quality assessment.

Finally, it is important to underline that model studies are important to resolve some of the present limitations in soil proteomics. In addition, coupling soil metagenomics, transcriptomics, and proteomics studies can give insights in the regulation responses of specific proteins and also validate the detection of proteins identified with low confidence parameters.

REFERENCES

Arenella, M., Giagnoni, L., Masciandaro, G., Ceccanti, B., Nannipieri, P. and G. Renella. 2014. Interactions between proteins and humic substances affect protein identification by mass spectrometry. *Biol Fertil Soils* 50, 447–445.

Armengaud, J., Christie-Oleza, J.A., Clair, G., Malard, V. and C. Duport. 2012. Exoproteomics: Exploring the world around biological systems. *Expert Rev Proteomics* 9, 561–575.

Barea, J.L. and J.L. Gomez-Ariza. 2006. Environmental proteomics and metallomics. *Proteomics* 6, S51–S62.

Bastida, F. and N. Jehmlich. 2016. It's all about functionality: How can metatranscriptomics helps us to discuss the attributes of ecological relevance in soil? *J Proteomics* 144, 159–161.

Bastida, F., Torres, I., Moreno, J.L. et al. 2016. The active microbial diversity drives ecosystem multifunctionality and is physiologically related to carbon availability in Mediterranean semi-arid soils. *Mol Ecol* 25, 4660–4673.

Benndorf, D., Balcke, G.U., Harms, H. and M. von Bergen. 2007. Functional metaproteome analysis of protein extracts from contaminated soil and ground water. *ISME J* 1, 224–234.

Bland, C. and J. Armengaud. 2014. Proteogenomics: A new integrative approach for a better description of protein diversity found in soil microflora. In *Omics in Soil Science*, P. Nannipieri, G. Pietramellara, and G. Renella (eds.), Custer Academic Press: Norfolk, UK, 139–162.

Chen, S., Rillig, M. and W. Wang. 2009. Improving soil protein extraction for metaproteome analysis and glomalin-related soil protein detection. *Proteomics* 9, 4970–4973.

Chignell, J.F., Park, S., Lacerda, C.M.R., De Long, S.K. and K.F. Reardon. 2018. Label-free proteomics of a defined, binary co-culture reveals diversity of competitive responses between members of a model soil microbial system. *Microb Ecol* 75, 701–719.

Chourey, K., Jansson, J.K., Verberkmoes, N. et al. 2010. Direct lysis/protein extraction for soil metaproteomics. *J Proteome Res* 9, 6615–6622.

Christie-Oleza, J.A., Armengaud, J., Guerin, P. and D.J. Scanlan. 2015. Functional distinctness in the exoproteomes of marine *Synechococcus*. *Environ Microbiol* 17, 3781–3794.

Craig, O.E. and M.J. Collins. 2002. The removal of protein from mineral surface: Implications for residue analysis of archeological materials. *J Archaeolog Sci* 29, 1077–1082.

Fornasier, F. and J. Margon. 2007. Bovine serum albumin and Triton X-100 greatly increase phosphomonoesterases and arylsulphatase extraction yield from soil. *Soil Biol Biochem* 39, 2682–2684.

Giagnoni, L., Arenella, M., Galardi, E., Nannipieri, P. and G. Renella. 2018. The bacterial culturability and viable but not culturable (VBNC) state studied by proteomic approach using an artificial soil. *Soil Biol Biochem* 118, 51–58.

Giagnoni, L., Magherini, F., Landi, L. et al. 2011. Extraction of microbial proteome from soil: Potential and limitations assessed through a model study. *Eur J Soil Sci* 62, 74–81.

Giagnoni, L., Magherini, F., Landi, L. et al. 2012. Solid phases effects on the proteomic analysis of Cupriavidus *metallidurans* CH34. *Biol Fertil Soils* 48, 425–433.

Greaves, P.R. and T.A.J. Haystead. 2002. Molecular biologist's guide to proteomics. *Microbiol Mol Biol Rev* 66, 39–63.

Huitman, J., Waldrop, M.P., Mackelprang, R. et al. 2015. Multi-omics of permafrost, active layer and thermokarst bog soil microbiomes. *Nature* 521, 208–212.

Jehmlich, N., Vogt, C., Lunsmann, V., Richnow, H.H. and M. von Bergen. 2016. Protein-SIP in environmental studies. *Curr Opin Biotechnol* 41, 26–33.

Keiblinger, K.M., Wilharititz, I.C., Schneider, T. et al. 2012. Soil metaproteomics-Comparative evaluation of protein extraction protocols. *Soil Biol Biochem* 54, 14–24.

Keller, M. and R. Hettich. 2009. Environmental proteomics: A paradigm shift in characterizing microbial activities at the molecular level. *Microbiol Mol Biol Rev* 73, 62–70.

Lidbury, I.D.E.A., Murphy, A.R.J., Scanlan, D.J. et al. 2016. Comparative genomic, proteomic and exoproteomic analyses of three *Pseudomonas* strains reveals novel insights into phosphorus scavenging capabilities of soil bacteria. *Environ Microbiol* 18, 3535–3549.

Lin, W., Wu, L., Lin, S. et al. 2013. Metaproteomic analysis of ratoon sugarcane rhizospheric soil. *BMC Microbiol* 13, 135–148.

Manikandan, R., Karthikenyan, G. and T. Raguchander. 2017. Soil proteomics for exploitation of microbial diversity in *Fusarium* wilt infected and healthy rhizosphere soils of tomato. *Physiol Mol Plant Pathol* 100, 185–193.

Maron, P-A., Ranjard, L., Mougel, C. and P. Lemanceau. 2007. Metaproteomics: A new approach for studying functional microbial ecology. *Microb Ecol* 53, 486–493.

Muranase, A., Yoneda, M., Ueno, R. and K. Yonebayashi. 2003. Isolation of extracellular protein from greenhouse soil. *Soil Biol Biochem* 35, 733–736.

Nannipieri, P. 2006. Role of stabilized enzymes in microbial ecology and enzyme extraction from soil with potential applications in soil proteomics. In *Nucleic Acids and Proteins in Soil*, P. Nannipieri, and K. Smalla (eds.), Springer: Berlin, Germany, 75–94.

Nannipieri, P., Ascher-Jenull, J., Ceccherini, M.T., Giagnoni, L., Pietramellara, G. and G. Renella. 2017. Reflections on microbial diversity and soil functions. *Eur J Soil Sci* 68, 1–26.

Nannipieri, P. and E.A. Paul. 2009. The chemical and functional characterization of soil N and its biotic components. *Soil Biol Biochem* 41, 2357–2369.

Nannipieri, P., Sequi, P. and P. Fusi. 1996. Humus and enzyme activity. In *Humic Substances in Terrestrial Ecosystem*, A. Piccolo (ed.), Elsevier: Amsterdam, The Netherlands, 293–328.

Nesatyy, V.J. and M.J.-F. Suter. 2007. Proteomics for the analysis of environmental stress response organisms. *Environ Sci Technol* 41, 6891–6900.

Ogunseitan, O.A. 1993. Direct extraction of proteins from environmental samples. *J Microbiol Methods* 17, 273–281.

Ogunseitan, O.A. 2006. Soil proteomics: Extraction and analysis of proteins from soils. In *Nucleic Acids and Proteins in Soil*, P. Nannipieri, and K. Smalla (eds.), Springer: Berlin, Germany, 95–115.

Picotti, R., Rinner, O., Stallmach, R. et al. 2010. High-throughput generation of selected reaction-monitoring assays for proteins and proteomes. *Nat Methods* 7, 43–46.

Renella, G., Giagnoni, L., Arenella, M. and P. Nannipieri. 2014. Soil proteomics. In *Omics in Soil Science*, P. Nannipieri, G. Pietramellara, and G. Renella (eds.), Custer Academic Press: Norfolk, UK, 95–125.

Renella, G., Ogunseitan, O., Giagnoni, L. and M. Arenella. 2015. Environmental proteomics: A long march in the pedosphere. *Soil Biol Biochem* 69, 34–37.

Rocca, J.D., Hall, E.K., Lennon, J.T. et al. 2014. Relationship between protein-encoding gene abundance and corresponding process are commonly assumed yet rarely observed. *ISME J* 9, 1693–1699.

Scheltema, R.A., Hauschild, J.P., Lange, O. et al. 2014. The Q Exactive HF, a benchtop mass spectrometer with a prefilter high-performance quadrupole and an ultra-high-field Orbitrap analyser. *Mol Cell Proteomics* 13, 3698–3708.

Schloter, M., Nannipieri, P., Sorensen, S.J. and J.D. van Elsas. 2018. Microbial indicators for soil quality-A perspective. *Biol Fertil Soils* 54, 1–10.

Schulze, W.X., Gleixner, G., Kaiser, K., Guggenberger, G., Mann, M. and E.-D. Schulze. 2005. A proteomic fingerprint of dissolved organic carbon and soil particles. *Oecologia* 142, 335–343.

Siggins, A., Gunnigle, E. and F. Abram. 2012. Exploring mixed microbial community functioning: Recent advances in metaproteomics. *FEMS Microbiol Ecol* 80, 265–280.

Singleton, I., Merrington, G., Colvan, S. and J.S. Delahunty. 2003. The potential soil protein-based methods to indicate metal contamination. *Appl Soil Ecol* 23, 23–52.

Starke, R., Kermer, R., Ullmann-Zeunert, L. et al. 2016. Bacteria dominate the short-term assimilation of plant-derived N in soil. *Soil Biol Biochem* 96, 30–38.

Taylor, E.-B. and M.A. Williams. 2010. Microbial protein in soil: Influence of extraction method and C amendment on extraction and recovery. *Microb Ecol* 59, 390–399.

Wang, H.-B., Zhang, Z.-X., Li, H. et al. 2010. Characterization of metaproteomics in crop rhizospheric soil. *J Proteome Res* 10, 932–940.

Wei, J., Sun, J., Yu, W. et al. 2005. Global proteome discovery using on line three-dimensional LC-MS/MS. *J Proteome Res* 4, 801–808.

Williams, M.A, Williams, M.A. and H.P. Mula. 2010. Metaproteomic characterization of a soil microbial community following carbon amendment. *Soil Biol Biochem* 42, 1148–1156.

Witzmann, F.A., Fultz, C.D., Grant, R.A., Wright, L.S., Kornguth, S.E. and F.L. Siegel. 1999. Regional protein alteration in rat kidneys induced by lead exposure. *Electrophoresis* 20, 943–951.

Witzmann, F.A., Fultz, C.D. and J. Lipscomb. 1995. Comparative 2-D electrophoretic mapping of human and rodent hepatic stress proteins as potential biomarkers. *Appl Theor Electrophor* 5, 113–117.

17 Stable Isotope Probing—Detection of Active Microbes in Soil

Marie E. Kroeger and Klaus Nüsslein
University of Massachusetts Amherst

CONTENTS

17.1 Introduction 269
17.1.1 What Is Stable Isotope Probing (SIP)? 269
17.1.2 How Is SIP Used? 270
17.1.3 What Are the Typical Applications of SIP in Soil? 270
17.2 Types of SIP 272
17.2.1 Experimental Design 272
17.2.1.1 Choice of Target Molecule 272
17.2.1.2 Choice of the Stable Isotope 273
17.2.1.3 Choice of Incubation Technique 273
17.2.2 Carbon Cycling Studies 275
17.2.3 Nitrogen Cycling Studies 276
17.2.4 RNA-SIP in Soil Research 287
17.2.5 PLFA-SIP in Soil Research 287
17.2.6 Protein-SIP in Soil Research 288
17.2.7 Raman-SIP in Soil Research 288
17.3 Analyzing SIP High-Throughput Sequencing Data 288
17.4 Standards for Reporting SIP Studies 289
17.5 Conclusion and Outlook 289
References 290

17.1 INTRODUCTION

17.1.1 What Is Stable Isotope Probing (SIP)?

Soil is the most diverse habitat on Earth, and the active microorganisms in soil drive essential global biogeochemical cycles. Understanding the diversity of active populations in combination with their function in soil remains a substantial challenge to understand and utilize microbial processes *in situ*. One method to overcome this challenge is Stable Isotope Probing (SIP, Dumont and Murrell 2005, Neufeld et al. 2007, Richnow and Lueders 2016).

SIP is a molecular technique that supplies substrates labeled with the heavier and stable (non-radioactive) isotopes of common elements to microorganisms. When they consume such labeled substrates, actively growing cells can assimilate the heavy isotopes and incorporate these into the building blocks of new cell components. A specific group of active microorganisms can be targeted by labeling any of its molecular building blocks such as DNA, RNA, proteins, or lipids of entire cells. These biological markers that are now labeled with heavier stable isotopes can be selectively partitioned by density gradient centrifugation. Then, further processing will allow the identification of a microbe with active metabolism from which this biomarker originated. The power of SIP lies in the ability to link the biochemical pathways and metabolic activity of a microbial community with its phylogenetic composition.

This chapter details the pros and cons of all current techniques of SIP in soil systems and assists the reader in making a focused decision as to which application to employ. A brief discussion to exemplify the need for such an overview follows. Most commonly, DNA is used as a biomarker with ^{13}C as the isotopic label (although ^{2}H, ^{18}O, or ^{15}N could also be used). RNA, however, is a more responsive biomarker for SIP experiments, because the rates of RNA synthesis are much higher than those of DNA synthesis. Moreover, they are a function of copy number which represents cellular activity rather than replication like DNA synthesis (Manefield et al. 2002). SIP of mRNA is more sensitive than that of DNA, since the label is rapidly incorporated into mRNA and is independent of cellular replication. This potentially allows to include the labeling of microorganisms that are not multiplying (Pratscher et al. 2011). The benefit of using mRNA-SIP compared to DNA-SIP or rRNA-SIP is that it can identify the organisms involved in assimilating a substrate and determine the genes being expressed during the process of assimilation. In addition to the selection of the SIP method and the choice of the target molecule, other decisions have to be made in the strategic design of a SIP experiment. The type(s) of stable isotope to be used within a substrate molecule has to be chosen, the type and duration of incubation with the labeled substrate has to be decided on, and internal controls for label incorporation efficiency or sensitivity of target molecule separation and detection have to be determined. Considering all the decisions that have to be made, the goal of this chapter is to provide an overview and guide the reader to the most appropriate type of SIP to identify active microbes in a particular soil sample. Figure 17.1 depicts the general workflow of common SIP techniques.

17.1.2 How Is SIP Used?

Chemical elements occur in the form of isotopes. An isotope is a variation of the same chemical element in which the number of neutrons in the nucleus is different. For example, the chemical element carbon exists as three different isotopes: ^{12}C, ^{13}C, and ^{14}C. The first two are stable isotopes, whereas ^{14}C is radioactive. Their chemical properties are the same. Most carbon in the biosphere is ^{12}C, because biological systems prefer the lighter isotope. Similarly, nitrogen exists as two stable isotopes, mostly as the lighter isotope ^{14}N, which is again preferred by biological systems, and the less common ^{15}N. The preference for the lighter isotope by biological systems is true also for other chemical elements.

For the study of biological systems, stable isotopes of carbon, hydrogen, nitrogen, oxygen, and sulfur have been used in the forms of ^{13}C, ^{2}H, ^{15}N, ^{18}O, and $^{34/36}S$. Isotopically labeled substrates applied to study metabolic processing are assimilated by microbial cells, and the isotopic label is incorporated into a variety of biomarkers such as carbohydrates, proteins, lipids, nucleic acids, and metabolites.

17.1.3 What Are the Typical Applications of SIP in Soil?

With the extensive diversity of microorganisms in most soil habitats, it is important to distinguish metabolically active or growing populations from those that are dormant or dead. Soil microbial ecologists aim to capture abundance, diversity, and reaction plasticity across time and across the heterogeneity of overlapping physical–chemical gradients in soil. SIP is a powerful way to link metabolic function to identity, and it can be employed with the goal to evaluate the microbial impact on a given soil system. Different applications for the SIP technique enable a variety of answers. While DNA-SIP directly addresses actively growing populations, RNA-SIP focuses on the physiological activity of target populations. In many studies, both techniques are complemented with

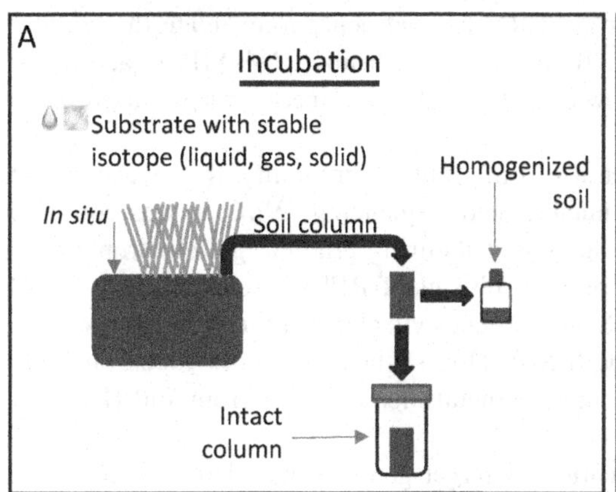

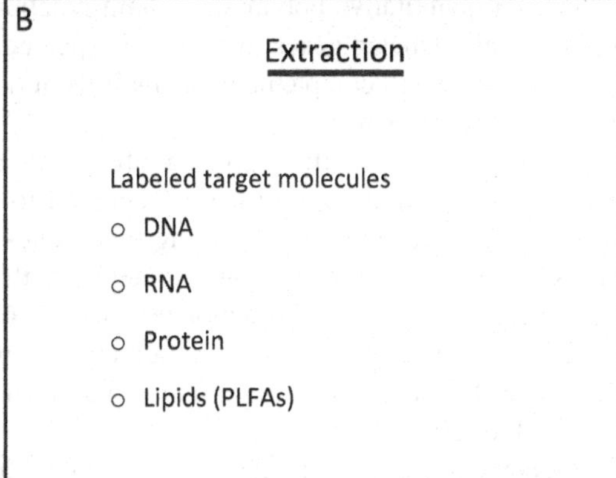

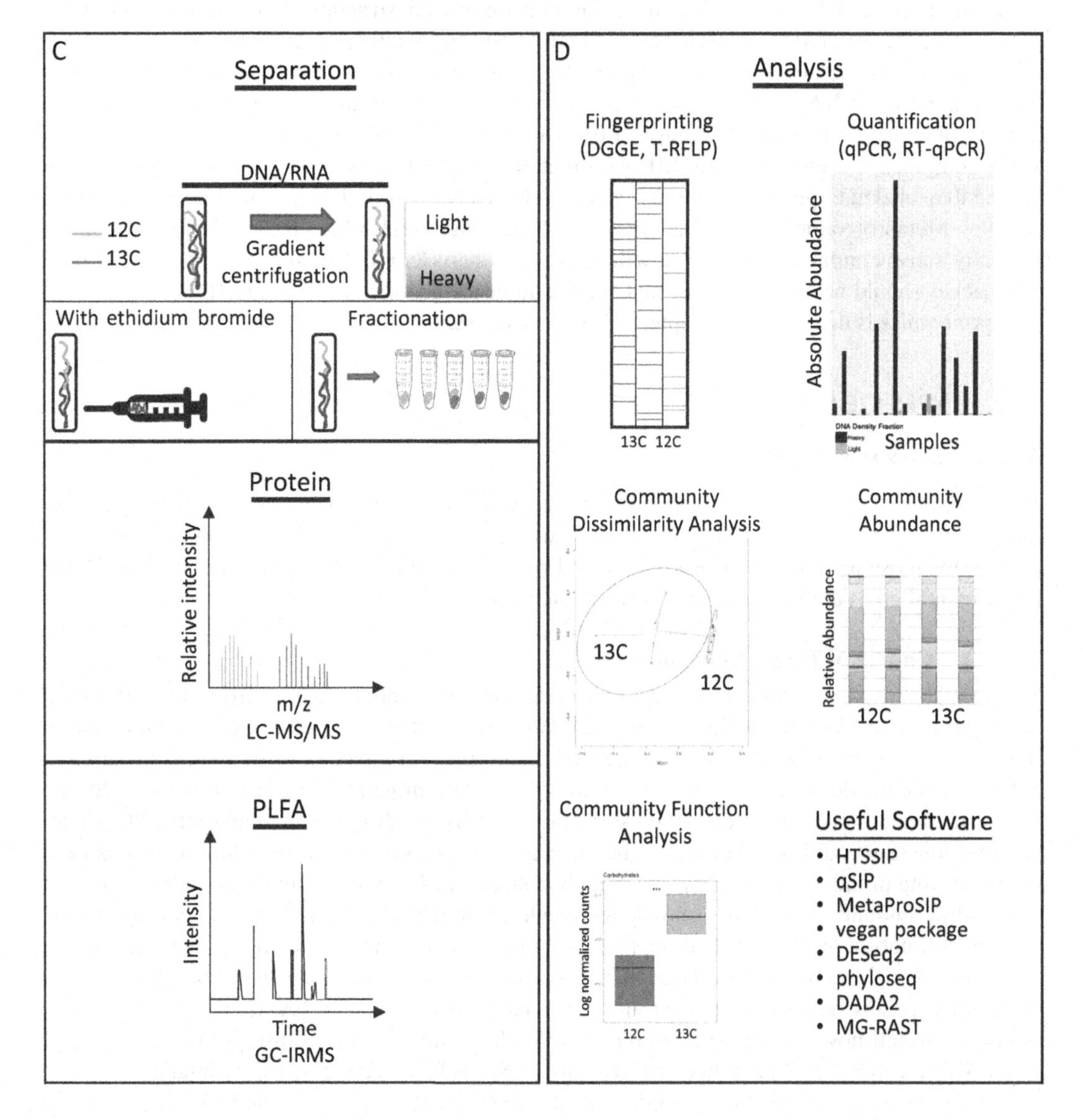

FIGURE 17.1 General workflow of SIP processing steps.

additional quantitative polymerase chain reaction (qPCR) of target genes to evaluate the number of potentially labeled genes that can be expected. To evaluate specifically RNA-SIP, an additional RT-qPCR step will complement the analysis of how efficiently label was incorporated into the total number of target genes.

For the analysis of the active members within a soil microbial community, the measurement of *in situ* microbial activity can be coupled to nucleic acid sequencing. While it was common to first use molecular fingerprinting techniques such as denaturing gradient gel electrophoresis (DGGE) or terminal restriction fragment length polymorphisms (T-RFLP) (Schwartz 2007), and then direct sequencing of functional genes or 16S rRNA genes via clone libraries, it is now common to combine high-throughput sequencing with SIP. This sequencing could focus on large amplicon databases, or even on entire metagenomes or metatranscriptomes from soil (Dumont et al. 2006, 2013).

Metagenomes, in particular, allow a quantification of target genes or populations, given that sufficient DNA can be isolated for metagenomic library construction. Furthermore, individual genomes can be inferred from metagenomes by first binning and then assembling genes that belong to one population genome (see also Chapter 14). Metatranscriptomes based on the study of stable-isotope-labeled mRNA molecules provide a powerful snapshot in time of the potential metabolic activity of associated enzymes. A case in point is given in a study by Dumont et al. (2013). This study combined, for the first time, SIP with metatranscriptomics in a lake sediment and demonstrated that labeled target transcriptomes can be selectively recovered from complex environmental samples. Metaproteomics in combination with SIP labeling, which allows the identification of metabolically active community members, will be the next step (Jehmlich et al. 2016), and multi-omics approaches should be expected after that. The state of the art of soil metatranscriptomics and soil metaproteomics is discussed in Chapters 15 and 16, respectively.

17.2 TYPES OF SIP

17.2.1 Experimental Design

When designing an experiment to investigate the active microorganisms in soil, there are many factors to consider. What molecule should be extracted from soil after labeling, what isotope will provide the most quantitative incorporation, and which incubation technique is most suitable for the experimental goals are some of the dominant questions.

17.2.1.1 Choice of Target Molecule

The first choice to be made is what type of molecule one wants to extract from the soil: DNA, RNA, protein, and/or phospholipid fatty acids (PLFAs). With each of these extracted molecules, there are certain benefits and limitations. PLFAs were the first molecules that were considered in SIP experiments. For example, Mohanty et al. (2006) determined the effect of nitrogen addition on the methane-consuming activity of soil microbes. By labeling the methane with ^{13}C, diversity and function could be assessed by the analyses of PLFAs extracted from soil microcosms at different time points during incubation. The benefits of PLFA-SIP are that it is highly sensitive, detects low amounts of label, does not depend on high-throughput sequencing, and is quantitative. A major limitation of PLFA-SIP, though, is that one cannot get the same phylogenetic resolution as achievable in other methods. PLFA-SIP provides a coarse analysis of how lipids from different taxonomic groups respond to a substrate, but—due to overlaps in PLFA across species—there is no way to tell how specific species respond. With the advent of high-throughput sequencing, DNA-SIP and RNA-SIP have become feasible. DNA-SIP is dependent on cellular replication, which means longer incubation periods are necessary for the stable isotope to be incorporated into the microorganisms' DNA. On the other hand, RNA-SIP is dependent on transcription,

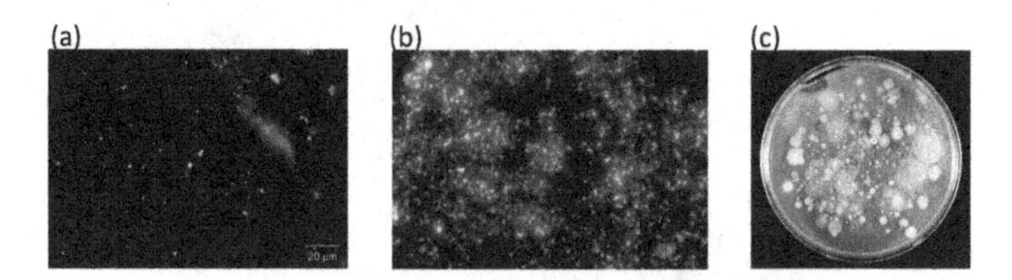

FIGURE 4.1 Bacterial cells in soil. (a) Microscopic impression of bacterial cells in soil. Soil extract was directly observed using FISH. Note the scattered occurrence of cells (yellow-green). (b) Cells extracted from soil stained with 4′-6-diamidino-2-phenylindole (DAPI) and examined under a fluorescence microscope. DAPI stains cells by interacting with their DNA. Note the great abundance and diversity of appearances of the cells. (c) Bacterial colonies from soil growing on a general growth agar. Note the diversity of the morphologies of the emerged colonies. (Courtesy of P. Hirsch, Rothamsted, Harpenden, UK.)

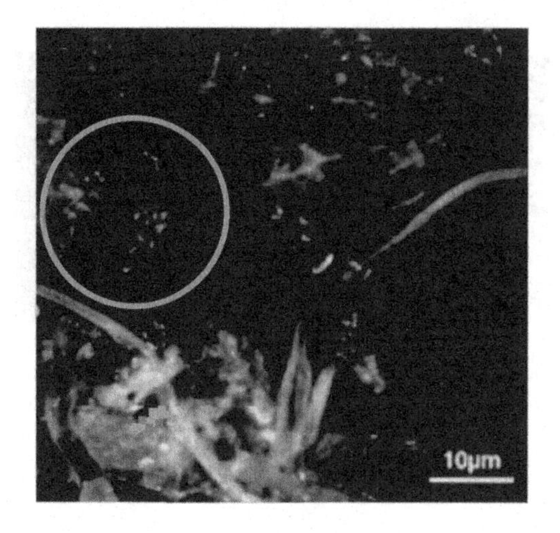

FIGURE 4.2 Detection of uncultured *Holophaga* sp. by fluorescent in situ hybridization (FISH) of soil. Blue (here greyish): *Holophaga* sp. cells stained using a specific ("IROG1") probe.

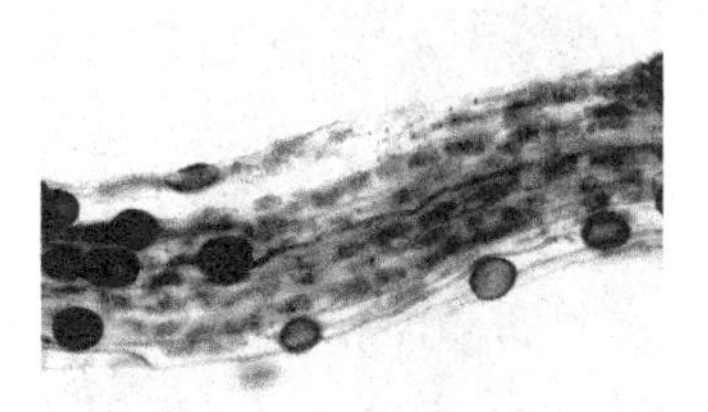

FIGURE 5.2 *Trifolium repens* root colonized by the AM fungus *Claroideoglomus* (formerly *Glomus*) *claroideum*—note vesicles and arbuscules.

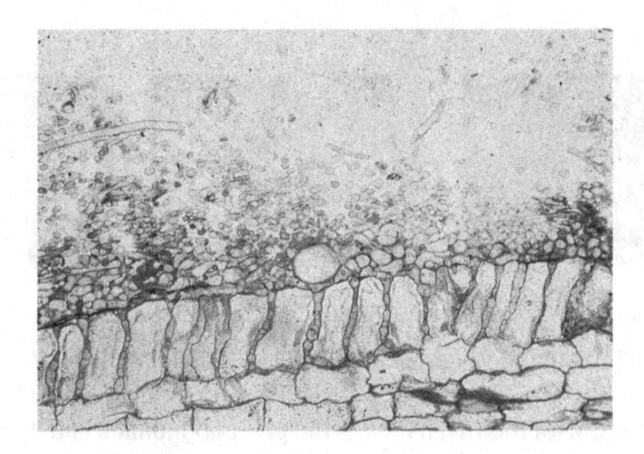

FIGURE 5.3 Longitudinal section of *Betula pendula* root showing colonization by the ectomycorrhizal fungus *Paxillus involutus* including the mantle and Hartig net.

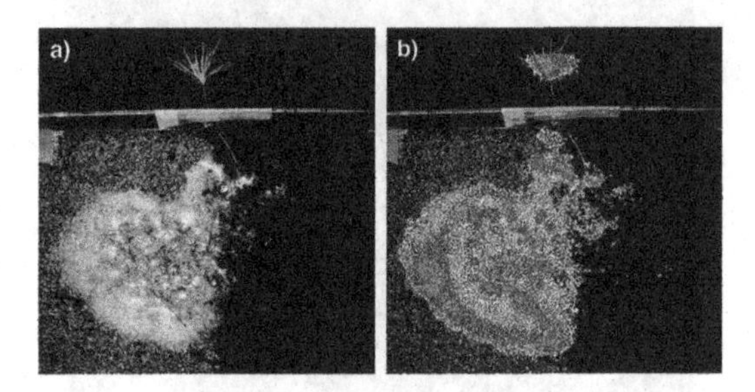

FIGURE 5.4 (a and b) Colonization of different substrates by extraradical mycelium of ectomycorrhizal fungus *Hebeloma crustuliniforme* showing active carbon allocation to E-horizon material after 2 months of hyphal growth.

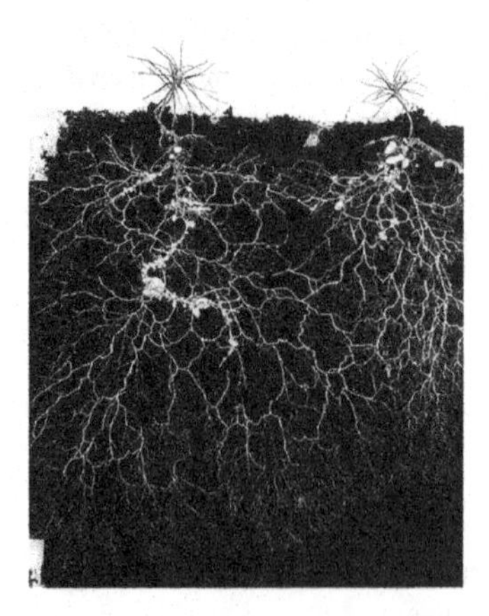

FIGURE 5.5 Extraradical mycelium of the ectomycorrhizal fungus *Suillus bovinus* colonizing *Pinus sylvestris* seedlings.

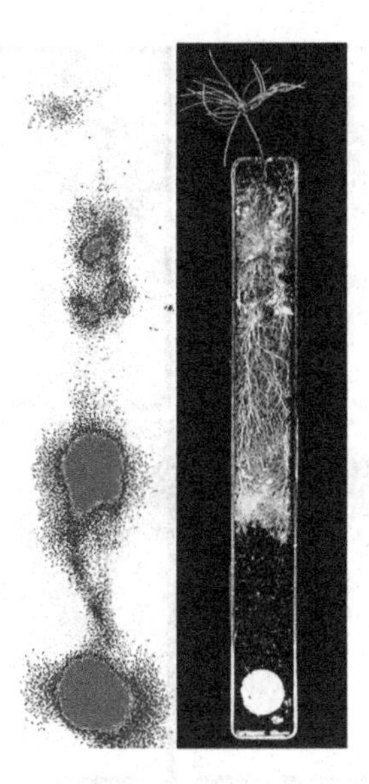

FIGURE 5.8 Interactions between ectomycorrhizal fungus *S. bovinus* and the saprotrophic fungus *H. fasciculare*, showing the capture of ^{32}P from the saprotroph by the mycorrhizal fungus.

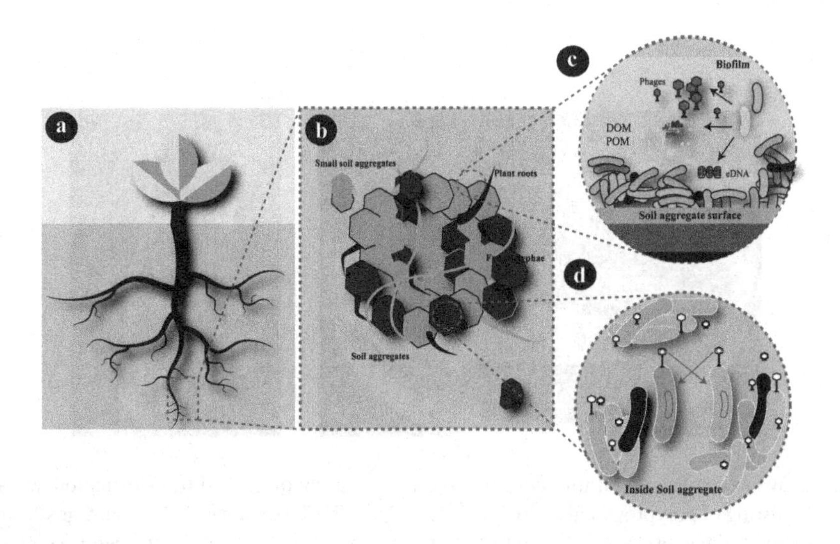

FIGURE 6.1 Role of viruses in soil microbiomes. (a) The heterogeneous nature of soil incites the presence of a great number of diverse organisms, as multiple microhabitats, including soil hot spots, are created. (b) Soil aggregates consist of microaggregates, soil organic matter, biotic binding agents (plant roots, fungal hyphae), water, and microbes. Thus, niches exist for bacteria, archaea, protozoa, nematodes, earthworms, and arthropods. (c) On surfaces, biofilms can emerge; in these, local lysis caused by phage activity introduces channels, in which low-molecular-weight compounds and extracellular DNA (eDNA) emerge. Both contribute to the dynamic development of the biofilm. (d) Soil aggregates in hot spots promote spatial isolation for bacteria, allowing separate evolution. Within a soil aggregate, pressure due to nutrient depletion may incite the excision of phages from host genomes. As a consequence, soil bacterial evolution is spurred. (Modified from Pratama and van Elsas (2017a,b)).

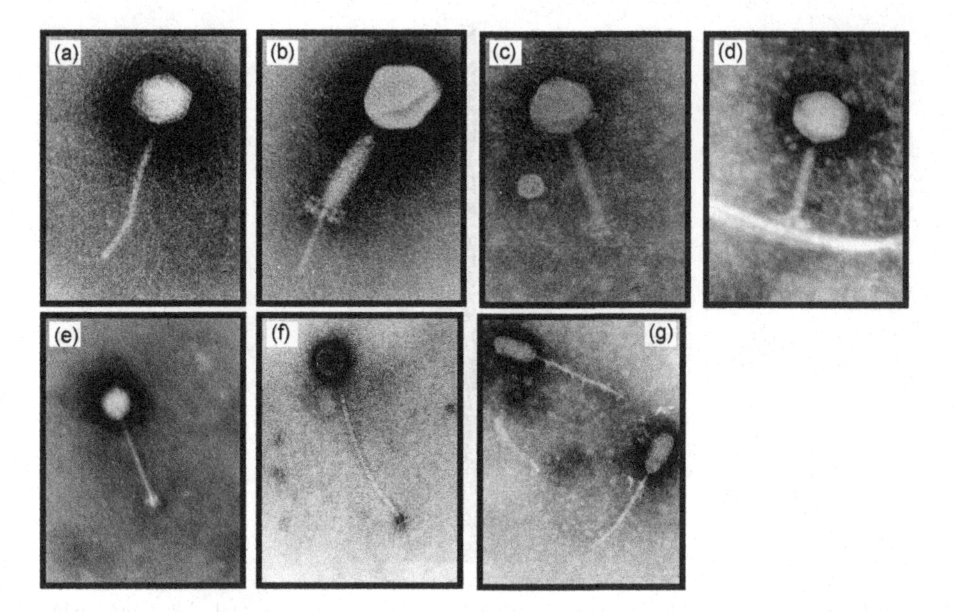

FIGURE 6.2 Diversity of soil phages associated with *Bacillus subtilis* and *Bacillus megaterium* (observed by TEM—staining with 2% uranyl oxalate). The phages were isolated from Brazilian soils using *B. subtilis* and *B. megaterium* isolated from the same soils as the hosts. (a) *B. subtilis* F6 phage FS2, magnification 258,000×; (b) *B. subtilis* F6 phage FS7, magnification 226,000×; (c) *B. subtilis* F6 phage IS1, magnification 170,000×; (d) *B. subtilis* F6 phage GS1, magnification 179,000×; (e) *B. megaterium* F4 phage MJ3, magnification 240,000×; (f) *B. megaterium* F4 phage MJ1, magnification 194,000×; (g) *B. megaterium* F4 phage MJ4, magnification 149,000×.

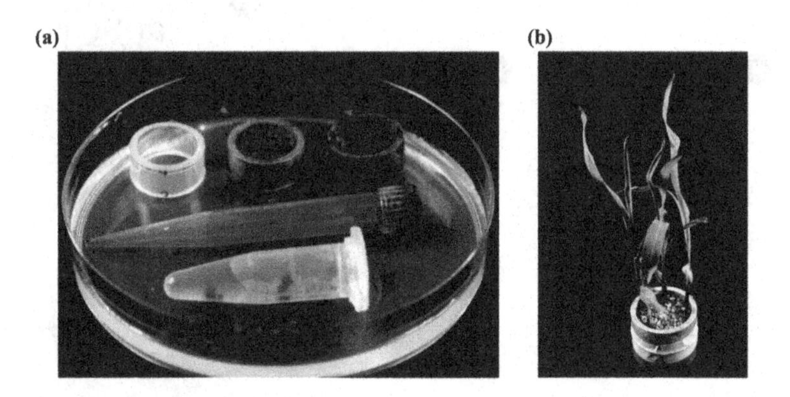

FIGURE 7.2 Two examples of soil microcosms used for the study of natural transformation. (a) Small soil microcosms consisting of polypropylene cylinders filled with soil. The number of bacterial transformants arising after the addition of bacteria and DNA to the soil is determined after sampling the central core of the soil using the wide end of a pipette tip and transferring the soil portion to an Eppendorf tube for suspension, dilution, and plating on transformant-selective agar plates. (b) Soil microcosm consisting of two plastic cylinders divided with a fine nylon mesh to allow bacterial activity and HGT to be investigated at the root–soil interface (rhizosphere). Various plant species can be germinated and grown in the upper cylinder, and the resulting soil–root interface on the lower cylinder was sampled and transferred to a dilution tube for suspension, dilution, and plating on transformant-selective agar plates.

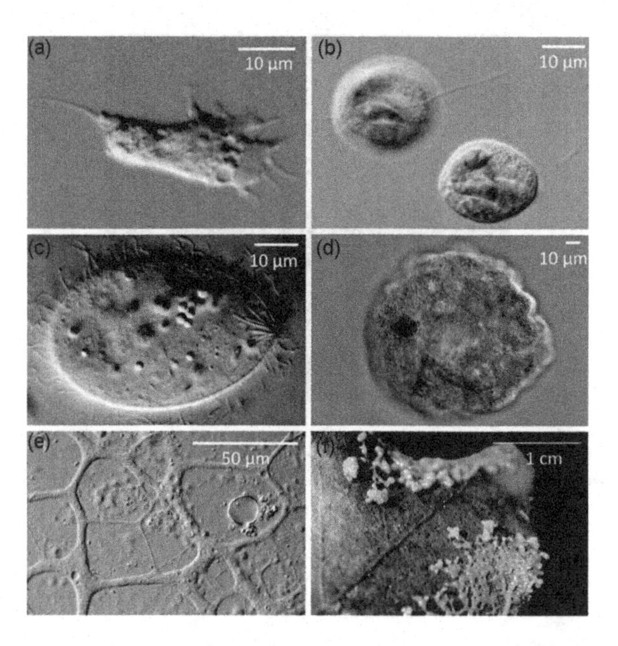

FIGURE 8.2 Examples of protistan morphological diversity. (a) *Acanthamoeba*, a very common soil amoeba with "spiky" pseudopods. (b) *Cercomonas*, a common amoeboflagellate, with ingested green algae (*Characium* sp.) cells in food vacuoles. (c) *Parenchelys terricola*, a predatory ciliate with extrusomes at the apical end. (d) *Amphizonella* (Amoebozoa), a shell-bearing amoeba with purple pigments and various ingested eukaryotes. (e) A multinucleate variosean amoeba (Amoebozoa) forming a network. (f) A multinucleate giant amoeba of Myxomycetes of the order Physarales (Amoebozoa) on a dead leaf, starting to form fruiting bodies. (Photographs by the authors.)

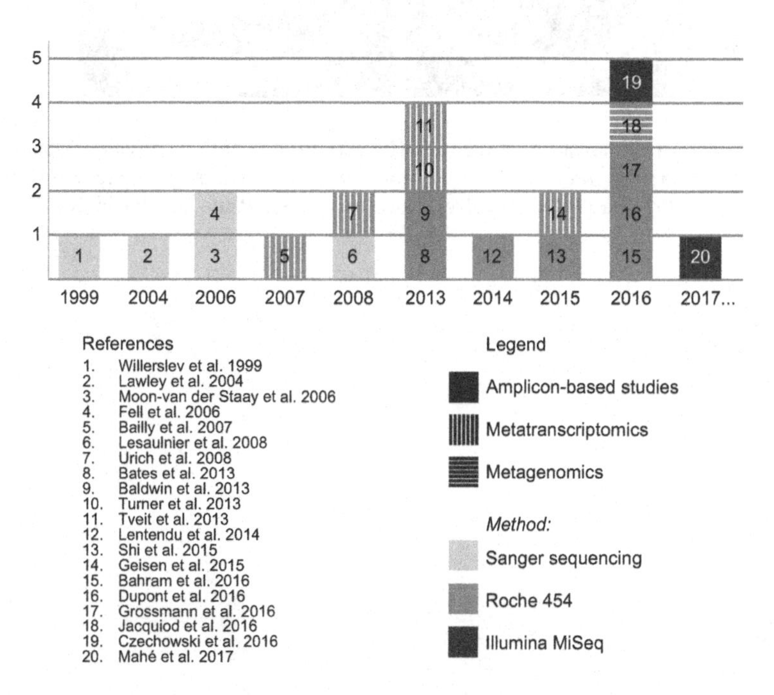

FIGURE 8.3 Summary of the molecular environmental studies targeting protists, per year, and indicating the method.

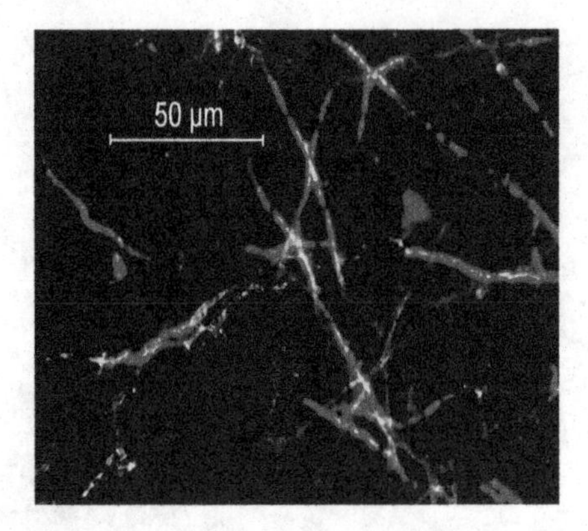

FIGURE 9.3 Biofilm formation by the soil bacterium *Paraburkholderia terrae* on the soil fungus *Trichoderma asperellum*. Yellow-green (here whitish): fungal tissue. Red (here grayish): bacteria adhering to fungal tissue. (Courtesy of Dr. D. Tazetdinova, University of Kazan, Russia.)

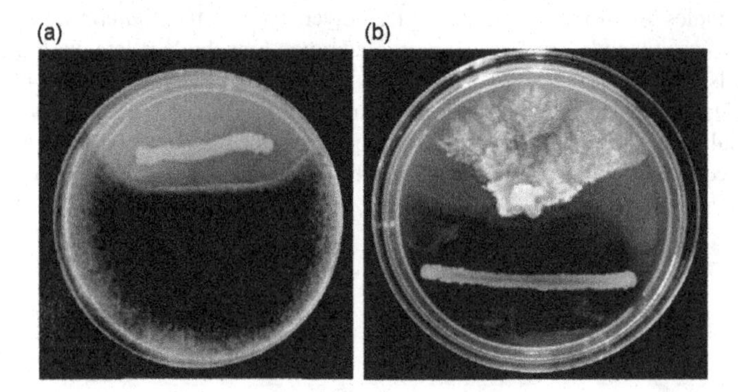

FIGURE 9.4 Antagonistic interaction between a soil bacterium and soil fungi revealed on dual-culture plates. Inhibition of (a) *Ceratocystis paradoxa* and (b) *Phytophthora parasitica* by the bacterium *Paraburkholderia seminalis*. (Courtesy of Priscila J.R.O. Gonçalves, University of São Paulo, São Paulo, Brazil.)

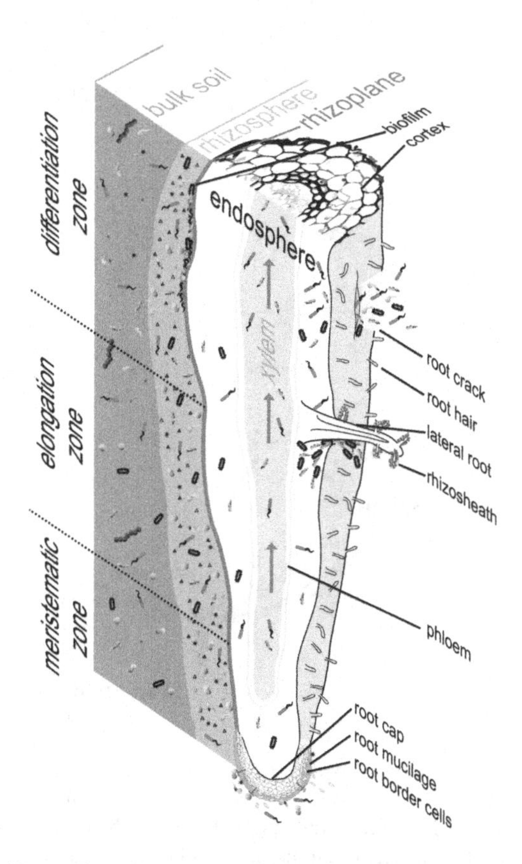

FIGURE 10.2 Schematic illustration of bacterial colonization and distribution in different root microhabitats. Gradual shifts in microbiota composition occur in root-associated habitats from the bulk soil to the root endosphere. Root cracks and sites of lateral roots are among the hot spots of bacterial colonization. Only microorganisms with special traits (e.g., cell wall-degrading enzymes) can enter root tissues. Arrows indicate the translocation of microorganisms, here bacteria, inside the vascular tissue. Root border cells originating from the root cap may be involved in defense against pathogens. The mucilage (shown in light green) secreted from the root and microorganisms leads to the formation of the rhizosheath, which enhances nutrient uptake and water-holding capacity of the root. The bacteria with the ability to form a biofilm and attach to the surface are good colonizers of the rhizoplane.

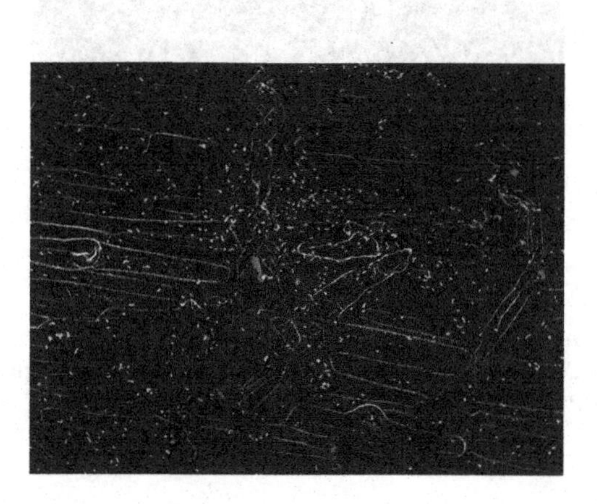

FIGURE 10.3 Confocal microscopy image of a maize root surface gray (maize root) and whitish (beneficial bacterium). (Photograph credit: Stéphane Compant.)

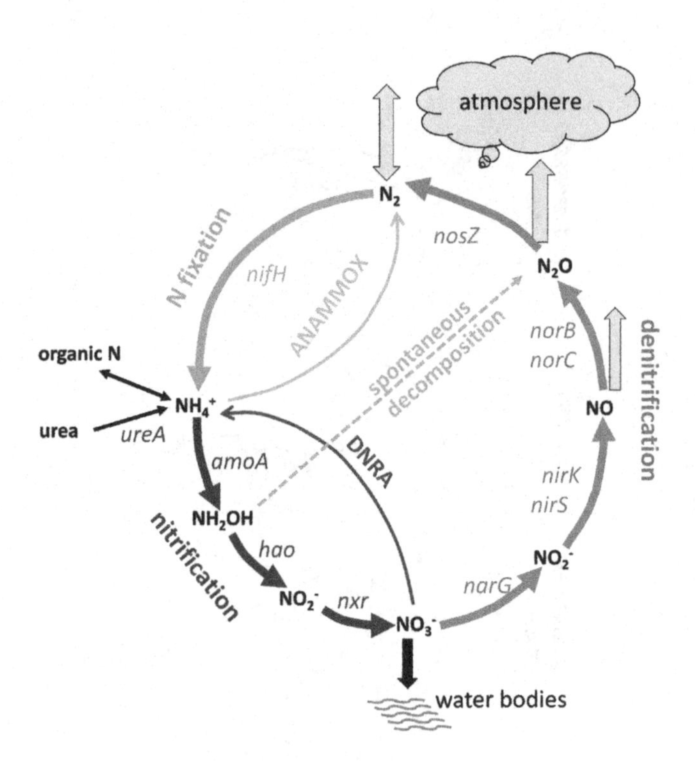

FIGURE 11.3 The biological N cycle. Broad gray arrows indicate gasses taken up or evolved; other arrows show the processes of N_2 fixation, nitrification (oxidation of ammonia—NH_3 to nitrite—NO_2^- via hydroxylamine—NH_2OH), and denitrification (reduction of nitrate—NO_3^- to nitrous oxide—N_2O and N_2 via nitrite and nitric oxide—NO). The majority of ammonia in soil is present as the ammonium ion—NH_4^+.

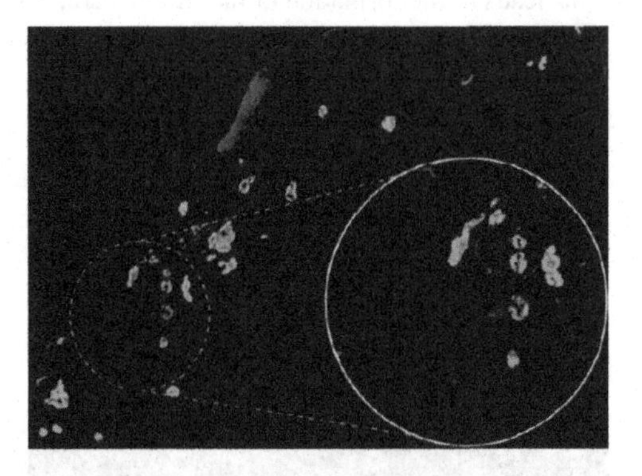

FIGURE 12.5 Colonization of the potato rhizosphere by GFP-tagged *Ochrobactrum* sp. A4. Images were obtained with a Leica TCS LSI Macro confocal laser microscope. Whitish (green in color insert) pseudocolor indicates the position of GFP-expressing minicolonies (Krzyzanowska et al. 2012).

FIGURE 12.6 Fluorescence images obtained by confocal laser scanning microscopy (CLSM) of an IF-labeled *Pseudomonas fluorescens* strain (DF57) colonizing barley roots. (The figure was kindly provided by Dr. Michael Hansen, Department of Plant and Environmental Sciences, University of Copenhagen, Copenhagen, Denmark.)

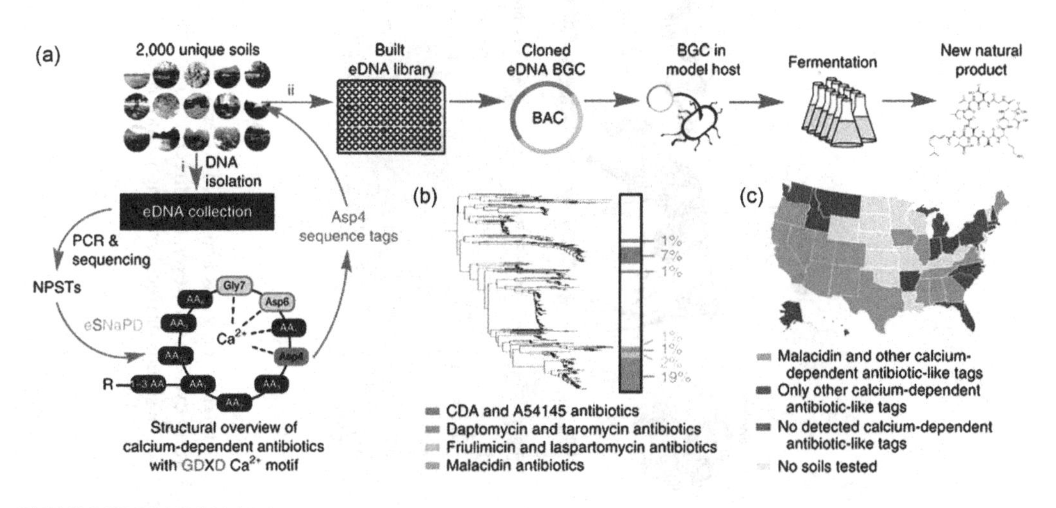

FIGURE FROM BOX 14.1.

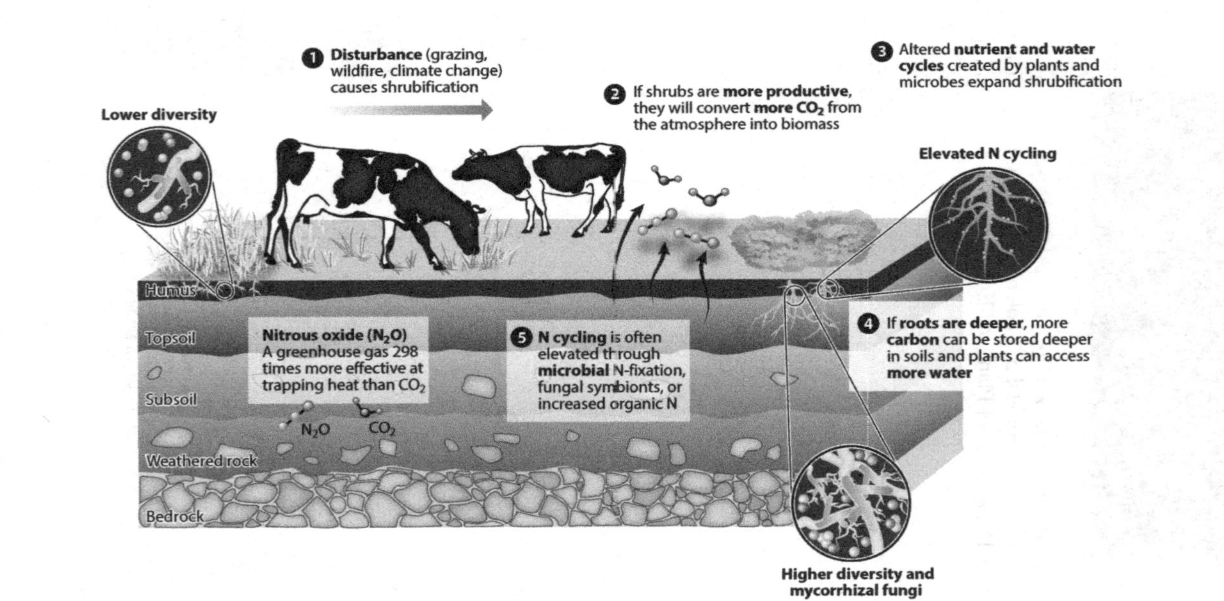

FIGURE 20.1 Impact of press disturbance on grass and shrub cover. Increases in shrubs or trees relative to grasses typically induce changes in microbial communities and in carbon, nitrogen, and water cycling. However, the directionality of these changes varies based upon the ecosystem. As shown for drier climates, carbon and nutrients are typically higher below the shrubs or trees, creating islands of fertility relative to the surrounding areas.

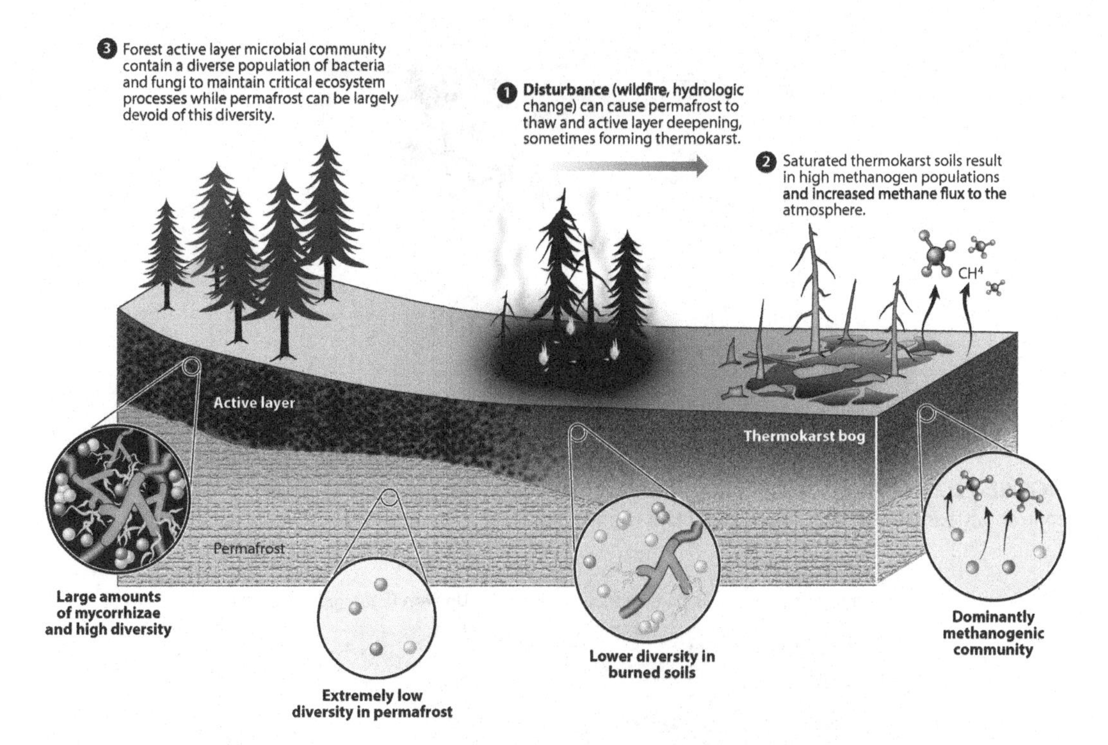
3 Forest active layer microbial community contain a diverse population of bacteria and fungi to maintain critical ecosystem processes while permafrost can be largely devoid of this diversity.
1 Disturbance (wildfire, hydrologic change) can cause permafrost to thaw and active layer deepening, sometimes forming thermokarst.
2 Saturated thermokarst soils result in high methanogen populations and increased methane flux to the atmosphere.
CH4
Active layer
Thermokarst bog
Permafrost
Large amounts of mycorrhizae and high diversity
Extremely low diversity in permafrost
Lower diversity in burned soils
Dominantly methanogenic community

FIGURE 20.2 Characterization of the microbial response to permafrost thawing. The diversity of soil microbial communities is affected by wildfire but is even more dramatically impacted by permafrost thawing and the formation of thermokarst bogs. Under these circumstances, the microbial community rapidly shifts to a methanogen dominated microbial community. Permafrost soils generally have lower biomass and diversity relative to active layer soils, and it is unclear how much impact permafrost microorganisms have on biogeochemical processes once thawed.

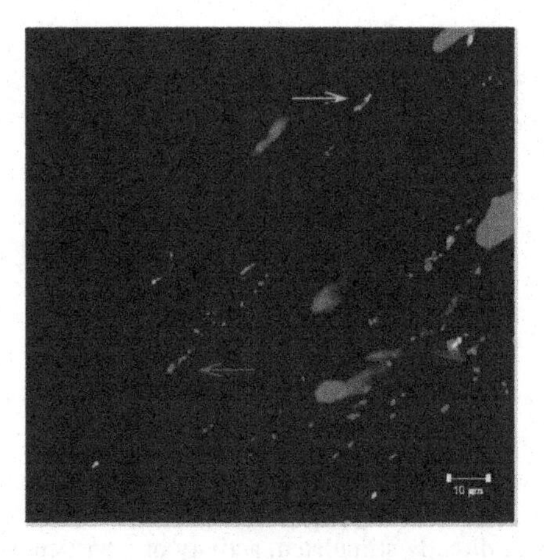
10 µm

FIGURE 22.1 Confocal laser scanning microscope image of *Pseudomonas fluorescens* 92rkG5 cells expressing green fluorescent protein and stained with propidium iodide. Green (here whitish) cells (green (here white) arrow) are alive, whereas the red (here grayish) cells (red (here gray) arrow) are dead.

FIGURE 22.2 Effects of the bacterial endophyte *Pseudomonas migulae* 8R6 on periwinkle: uninoculated control (left) and bacterially inoculated plant (right).

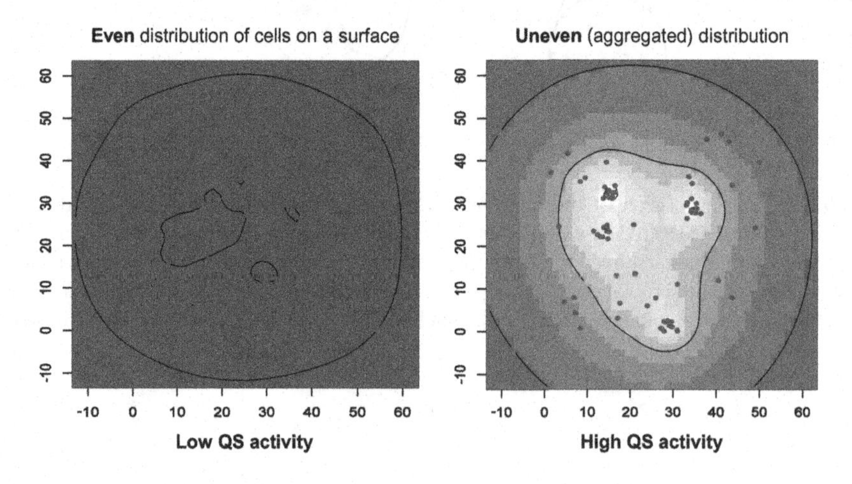

FIGURE FROM BOX 22.1.

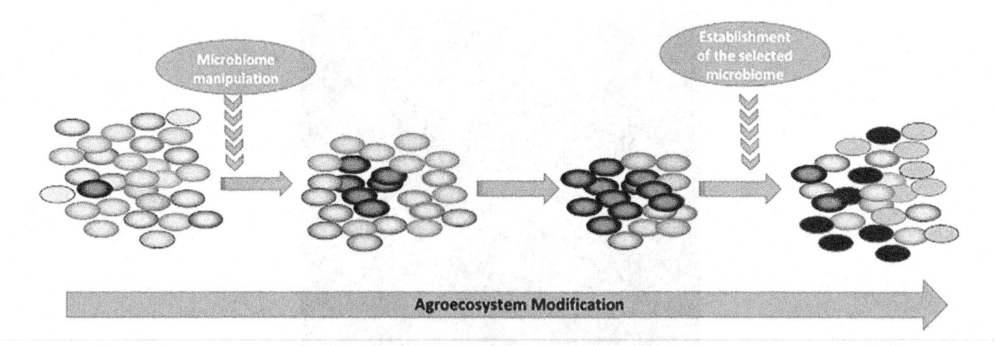

FIGURE 25.3 Representation of modifications in the soil microbiome after soil modulation events. The soil management strategy used is an important selection factor for target species and functions, or for the "core microbiome." In addition to being directly stimulated, activity of a particular community member may also generate adequate substrate for growth of other populations. The strategy applied should consider the specific soil microbiome changes, next to soil physicochemical data.

which allows for shorter incubation times, but the instability of the RNA makes this method more challenging. Protein-SIP has become more popular in recent years due to the increased efficiency of protein extraction from soil. The main benefit of protein-SIP is that it is directly related to changes in the activity of microorganisms instead of using replication or transcription as a proxy for activity. There are many challenges with protein-SIP though. Although protein extraction from soil has improved, it is still a bottleneck and, similar to RNA, there is the problem of uncontrolled protein degradation in soil (Keiblinger et al. 2012). Therefore, RNA- and protein-SIP provide snapshots of organismal activity in soil but may not provide the full picture. Furthermore, for RNA- and protein-SIP, a seeding experiment of the targeted soil sample with a previously labeled type strain is an excellent way to evaluate the detection limit of the experimental setup. Rangel-Castro et al. (2005) studied the effect of liming of soil on the structure of the rhizosphere microbial community metabolizing root exudates using $^{13}CO_2$ pulse labeling field grasses, followed by RNA-SIP analysis of the soil. To determine the lower detection limits, they seeded the experimental system with ^{13}C-labeled *Escherichia coli* and *Pseudomonas fluorescens* cells, and found a detection limit between 10^5 and 10^6 cells per gram of soil. In all SIP experiments, an added challenge is to optimize incubation time and substrate concentration. The goal is always to balance the minimization of dilution of the added isotope through the turnover of biomass within the incubated sample, while at the same time confirming the levels of enriched target molecules within the range of detection.

All types of SIP experiments come with pros and cons, which makes it essential to think about the specific research question and soil type to determine what method is optimal to answer the question. Some researchers combine these methods to get a better understanding of their system, but this comes with a serious cost. Table 17.1 shows an overview of some examples of SIP studies in soil.

17.2.1.2 Choice of the Stable Isotope

A choice that coincides with deciding on the method of SIP is what stable isotope to use, hence, a choice out of ^2H (deuterium), ^{13}C, ^{18}O, or ^{15}N. These isotopes are incorporated in varying amounts into nucleic acids and/or other cellular macromolecules, have different weights, and are found in different molecules. For example, if the research question is "how does the 'active' methane-consuming community differ between forest and an adjacent pasture," then the best stable isotope to use is ^{13}carbon (in methane). The logic behind this is that methanotrophs oxidize methane (CH_4), so the only choices are carbon and hydrogen for stable isotopes. Since carbon is much heavier than hydrogen and more likely to be assimilated, it makes separation of "light" and "heavy" molecules easier. Another example would be to differentiate between heterotrophic and autotrophic metabolisms. Here, a double-SIP approach with both inorganic ^{13}carbon and ^{2}hydrogen (as deuterated water) would be appropriate (Kellermann et al. 2012).

17.2.1.3 Choice of Incubation Technique

After deciding on the molecule to be extracted and the stable isotope substrate to be used, the next important question to address is what type of incubation is best suited for a SIP experiment. There are four currently used options: homogenized soil, soil slurry, intact soil cores, and in-field incubations. Depending on the research question, the process being studied, and which labeled substrate is used, the best incubation type can vary and should be adapted to the goals of the study (Figure 17.2). This includes also the cost of SIP methods, which are affected by the high cost of isotopically labeled substrates or the need to custom-synthesize them, laborious experimental procedures, or lengthy downstream sequence analyses.

There are detailed published protocols for all types of SIP, e.g. Dunford et al. (2010) for DNA-SIP, Whiteley et al. (2007) for RNA-SIP, Jehmlich et al. (2010) for protein-SIP, and Quideau et al. (2016) for PLFA-SIP.

TABLE 17.1
Overview of Different SIP Techniques and Their Applications in Soil

SIP Technique	Experimental Detail	Pros	Cons	Outcome	References
DNA-SIP	*Goal:* Phylogeny *Target:* Amplicons (16S rRNA, functional genes, fingerprinting) *Goal:* Total community genome *Target:* Metagenomics	• High taxonomic resolution • Inferred metabolic potential • Correlate metabolism with identity • Genomic DNA available for further analyses (e.g., targeted genome assembly)	• Potential cross-feeding • Time consuming • Low throughput • Issues separating due to GC content • Limited commercial availability of labeled substrates	Investigated the active methylotrophic community in a forest soil. Found that the predominant active methylotrophs in this acidic forest soil were most closely related to α-Proteobacteria and Acidobacteria.	Radajewski et al. (2000)
RNA-SIP	*Goal:* Phylogeny of community expression profile *Target:* rRNA *Goal:* Active metabolism of community *Target:* Metatranscriptome	• Higher sensitivity than DNA-SIP • Short incubation time • Detect changes quickly • Correlate metabolism with taxonomic identity • Labeling independent of cell replication	• RNA degrades quickly • Difficult to extract • Potential cross-feeding • Limited availability of labeled substrates • For mRNA: low amounts of target	^{13}C-labeled methanol was added to oxic soil. After 6 days, labeled Methylobacteriaceae could be detected. After longer periods, the label could also be detected in soil fungi and protozoa, revealing carbon-processing trophic networks.	Lueders et al. (2004)
PLFA	*Goal:* Specific microbial activity, metabolic function, lipid biosynthesis, and carbon flow *Target:* Membrane lipids (hopanoids for bacteria, isoprenoid-based molecules for archaea, and steroids for eukaryotes)	• Sensitive • Low cost • Highly quantitative	• Low taxonomic resolution • Requires isotope ratio mass spectrometry	Soil microbial populations responsible for toluene degradation were identified by ^{13}C incorporation into selected PLFAs. After 5 days of incubation with ^{13}C-toluene, 96% of the ^{13}C was detected in only 16 of the total 59 PLFAs. 85% of the labeled PLFAs were identical to those in a toluene-metabolizing bacterium isolated from the same soil.	Hanson et al. (1999)

(*Continued*)

TABLE 17.1 (*Continued*)
Overview of Different SIP Techniques and Their Applications in Soil

SIP Technique	Experimental Detail	Pros	Cons	Outcome	References
Protein	*Goal:* Currently active metabolic pathways *Target:* Protein extract, metaproteomics	• Functional information • Phylogeny directly linked to activity • Quantitative • Short incubation time (minimum incorporation required) • Highly sensitive • Labeling independent of cell replication	• Low throughput and high cost • Difficult extraction from soil • Limited databases • Degradation in soil • Little taxonomic information • Special instrumentation required	16S and 18S rDNA gene profiling and functional metaproteomics were employed to characterize the composition of a soil microbial community assimilating ^{15}N-labeled plant-derived organic matter. Tracking the flow of N within the soil microbial community showed that mostly bacteria (not fungi) were involved in the short-term assimilation of plant-derived N.	Starke et al. (2016)
Raman	*Goal:* Functions of single microorganisms in their natural habitat *Target:* Single-cell probing with stable isotopes	• Single-cell analysis	• Cell extraction from soils prior to analysis • Special instrumentation required	N_2-fixing bacteria in both artificial and complex soil communities were discerned and imaged at the single-cell level. In addition, the extent of N_2 fixation of diverse soil bacteria in the soil microbiome could be quantified.	Cui et al. (2018)

17.2.2 Carbon Cycling Studies

Almost all DNA-SIP studies in soil have used ^{13}C, with a focus on either methane oxidation or the metabolism of plant-derived carbon. The first study to demonstrate the potential of DNA-SIP as a tool in soil investigated the active methylotrophic community in a forest soil (Radajewski et al. 2000). This study found that the predominant active methylotrophs in this acidic forest soil were most closely related to α-Proteobacteria and Acidobacteria. After the technique had been established, many research groups began utilizing it to understand the active cycling of carbon in soil (Table 17.2). Due to methane and methanol having only one carbon atom, these substrates are ideal for carbon-SIP. Many groups have investigated active methanotrophy in soil, since it is essential to reducing emissions of methane, a potent greenhouse gas (Cébron et al. 2007a, 2007b, Dumont et al. 2006, Gupta et al. 2012, Leng et al. 2015). These studies used homogenized soil, which alters the gas diffusivity, and therefore oxygen and methane availability. One reason why many researchers use homogenized soil is to decrease the incubation time. However, this type of incubation is far from environmental conditions, and therefore, activity may not be comparable to *in situ* soil conditions. Recent research into the active carbon cycling in soil has focused on the metabolism of complex carbon sources such as cellulose and polycyclic aromatic hydrocarbons (PAHs). Pepe-Ranney et al. (2016) used ^{13}C-cellulose and ^{13}C-xylose to investigate what microorganisms are actively metabolizing plant-derived carbon in agricultural soil. Overall, this study found that active cellulose

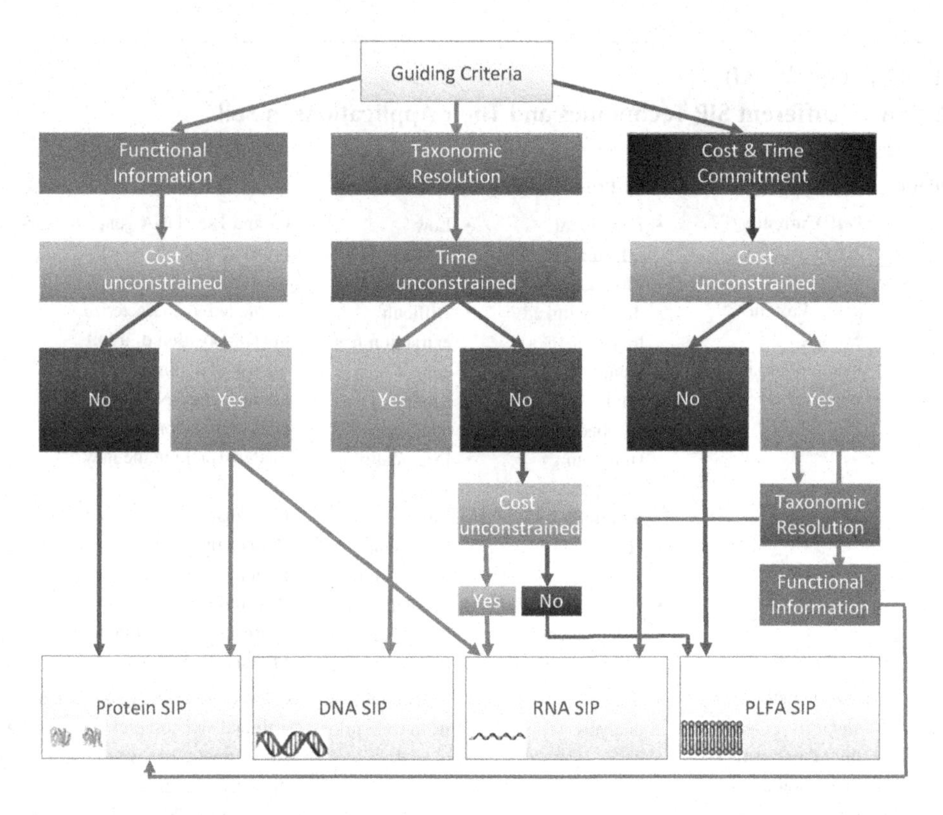

FIGURE 17.2 Decision tree for SIP analyses in soil.

metabolism is carried out by members of the Verrucomicrobia, Chloroflexi, Bacteroidetes, and Planctomycetes, with many of these active cellulolytic microorganisms being uncultured. Another investigation into the metabolism of plant-derived carbon in soil found that Clostridia were the main active taxon responsible for consuming rice callus carbon (Lee et al. 2017). The metabolism of PAHs is of interest because these are toxic, being found in many soils and sediments. In 2015, Rodgers-Vieira and colleagues investigated the active metabolism of an oxygenated PAH, anthraquinone, next to anthracene, in soil. This study discovered that PAHs and oxygenated PAHs are not metabolized by the same microbes. Microorganisms most closely related to the genera *Phenylobacterium* and *Sphingomonas* degraded anthraquinone, whereas uncultivated microorganisms not related to these groups consumed anthracene. A study by Song et al. (2016) focused on identifying PAH degraders from a natural soil environment unlike the previous study, which used PAH-contaminated soil. Rather, Song et al. used phenanthrene, anthracene, and fluoranthene labeled with ^{13}C to identify the microbes capable of consuming these compounds. Each PAH had a different group responsible for its metabolism, with phenanthrene, anthracene, and fluoranthene being consumed by Sphingomonads, *Rhodanobacter*, and Acidobacteria, respectively. These studies are beginning to unravel the complex carbon cycling in soil that is essential to mitigating climate change, bioremediation, and understanding carbon storage (see also Chapters 20 and 23).

17.2.3 Nitrogen Cycling Studies

The first investigation in soil using a ^{15}N substrate was conducted in 2005 by Cadisch and colleagues. This study provided the foundation for future work in soil by demonstrating the level of incorporation necessary to separate heavy and light DNA, which resulted in ideally >50% for clear separation. Even though the separation of nitrogen-labeled DNA is challenging, many groups

TABLE 17.2
Overview of SIP Applications in Soil Microbiome Studies

SIP Type	Goal	Substrate	Heavy Isotope	Sample	Incubation Conditions	Paper
DNA	Identify methylotrophs in soil	Methanol	Carbon	Oak forest soil	Sieved, air-dried, 10 g in serum vial, 44 days	Radajewski et al. (2000)
	Identify methanotrophs	Methane	Carbon	Peat soil	10 g of fresh soil in serum vial, 40 days	Morris et al. (2002)
	Identify methanotrophs and how nutrient amendment impacts activity	Methane	Carbon	Wytham soil	5 g of dried, sieved soil in serum vial either in NMS*, DI water, or adjusted to natural moisture content	Cébron et al. (2007a)
	Identify methanotrophs in landfill cover soil	Methane	Carbon	Landfill cover soil	5 g of dried, sieved soil in serum vial either in NMS or in DI water	Cébron et al. (2007b)
	Determine the microorganisms involved in herbicide degradation	2,4-Dichloro-phenoxyacetic acid	Carbon	Agricultural field	Sieved, 10 days	Cupples and Sims (2007)
	Investigate phenol-degrading microorganisms	Phenol	Carbon	Collamer silt loam soil	Field based, 0 or 11 daily doses	DeRito et al. (2005), DeRito and Madsen (2009)
	Determine the active methanotrophs in forest soil and their diversity	Methane	Carbon	Gisburn forest soil	Soil slurry in ANMS* medium in serum vial	Dumont et al. (2006)
	Determine the microorganisms associated with plants and plant exudates	Carbon dioxide	Carbon	Wheat, maize, rape, and barrel clover grown in eutric cambisol soil	Harvested soil after 36 days	Haichar el Zahar et al. (2008)
	Identify the active ammonia-oxidizing microorganisms in agricultural soil	Carbon dioxide	Carbon	Maize agricultural soil previously fertilized	Sieved, 60% WHC*, serum bottles, fertilized every week with $(NH_4)_2SO_4$	Jia and Conrad (2009)
	Identify the nitrogen-incorporating, hydrocarbon-degrading microorganisms	Monoammonium phosphate	Nitrogen	JP-8 fuel-contaminated soil	Soil slurry in ANMS medium in serum vial	Bell et al. (2011)

(*Continued*)

TABLE 17.2 (*Continued*)
Overview of SIP Applications in Soil Microbiome Studies

SIP Type	Goal	Substrate	Heavy Isotope	Sample	Incubation Conditions	Paper
	Identify the active methanotrophs from nutrient-poor and nutrient-rich peatland ecosystems	Methane	Carbon	Peatland soil	Composited, homogenized, 5 g incubated in serum vial for 120 days	Gupta et al. (2012)
	Identify the microorganisms metabolizing rice callus as a proxy for rice straw	Rice callus	Carbon	Paddy field	Sieved, 55% WHC, 10 g of soil in test tubes for 3–56 days	Lee et al. (2011)
	Identify the microorganisms degrading rice callus as a proxy for root cap cells	Rice callus	Carbon	Paddy field	Sieved, 55% WHC, 10 g of soil in test tubes for 3–56 days	Li et al. (2011)
	Identify the microorganisms involved in nitrification	Carbon dioxide	Carbon	Agricultural soil	60% WHC, 6 g in serum bottles for 3, 7, or 28 days	Xia et al. (2011)
	Identify the ammonia-oxidizing microorganisms in agricultural soil	Carbon dioxide	Carbon	Agricultural soil	10 g of sieved soil in serum bottles for 0, 14, or 28 days	Zhang et al. (2010)
	Determine the contributions of AOA and AOB to autotrophic ammonia oxidation in acidic agricultural soil	Carbon dioxide	Carbon	Agricultural soil	10 g of sieved soil in serum bottles with and without DCD* for 0, 15, and 30 days	Zhang et al. (2012)
	Identify the active nitrogen-fixing microorganisms	Dinitrogen	Nitrogen	Caldwell field soil (Williamson fine sandy loams)	10 g of sieved soil was incubated for 28 days	Buckley et al. (2007)
	Identify the dimethyl sulfide (DMS)-degrading microorganisms in soil and lake sediments	DMS	Carbon	Brassica crop field soil and Tocil Lake sediment	1 g of wet soil or lake sediment incubated over a time course of approx. 20 days	Eyice et al. (2015)

(*Continued*)

TABLE 17.2 (*Continued*)
Overview of SIP Applications in Soil Microbiome Studies

SIP Type	Goal	Substrate	Heavy Isotope	Sample	Incubation Conditions	Paper
	Identify the microbes responsible for plant-derived carbon metabolism, specifically xylose and cellulose in agricultural soil	Cellulose and xylose	Carbon	Agricultural field soil	Sieved, homogenized, 10 g incubated for 2 weeks without substrate then with substrate for 1, 3, 7, 14, and 30 days	Pepe-Ranney et al. (2016)
	Identify the microorganisms responsible for plant-derived carbon metabolism in paddy field soil	Rice callus cells	Carbon	Paddy field soil	1 kg of sieved soil was flooded with DI water for 14 days and preincubated under different conditions. Then, 10 g of soil was taken and incubated with substrate for 8 weeks	Lee et al. (2017)
	Identify the microorganisms capable of metabolizing PAHs in forest soil	PAHS: anthracene, phenanthrene, and fluoranthene	Carbon	Forest soil	3 g of homogenized and sieved soil was incubated in 10 mL of phosphate-buffered mineral medium with labeled substrate for 3, 9, and 14 days	Song et al. (2016)
	Identify the microorganisms metabolizing anthracene and anthraquinone from a contaminated soil	Anthraquinone and anthracene	Carbon	PAH-contaminated soil	1 g of air-dried and sieved soil was preincubated, while shaking for 2 days in reactor buffer, and then incubated with substrate for 20 days	Rodgers-Vieira et al. (2015)

(*Continued*)

TABLE 17.2 (*Continued*)
Overview of SIP Applications in Soil Microbiome Studies

SIP Type	Goal	Substrate	Heavy Isotope	Sample	Incubation Conditions	Paper
	Understand the role of archaea in N_2O reduction as a function of carbon and nitrogen availability	Ammonium sulfate	Nitrogen	Paddy field soil	Soil was sieved and homogenized, incubated under submerged conditions for 14 days without substrate, and then incubated for 1, 5, 10, 20, 30, and 60 days	Cucu et al. (2017)
	Determine the active ammonia-oxidizing archaea in acidic soil	Carbon dioxide and urea	Carbon	Acidic agricultural soil	10 g of sieved soil at 60% WHC was incubated for 0, 28, and 56 days	Wang et al. (2014)
	Determine the impact of different organic matter on anammox bacteria in paddy soil	Carbon dioxide	Carbon	Paddy field soil	25 g of soil	Zhang et al. (2018)
	Identify the active microorganisms degrading pentachlorophenol (PCP) in paddy soil	PCP	Carbon	Paddy field soil	5 g of soil in a medium was incubated for 10, 20, 30, 40, 50, and 60 days	Tong et al. (2015)
RNA	Identify propionate-degrading microorganisms under methanogenic conditions	Propionate	Carbon	Rice field soil	20 g of air-dried, sieved soil in 20 mL sterile, anoxic water in serum vial for 2 days, 3 weeks, or 7 weeks	Lueders et al. (2004)
	Identify the methanogens in rice rhizosphere	Carbon dioxide	Carbon	Rice field soil	Rice plants in microcosms, pulse labeled, 7 days	Lu and Conrad (2005)
	Determine which is capable of degrading PCP in the bacterial community	PCP	Carbon	Grassland soil	10 g of sieved soil in 28 mL glass bottle for 0, 4, 7, 14, 21, 35, 49, or 63 days	Mahmood et al. (2005)

(*Continued*)

TABLE 17.2 (*Continued*)
Overview of SIP Applications in Soil Microbiome Studies

SIP Type	Goal	Substrate	Heavy Isotope	Sample	Incubation Conditions	Paper
	Assess the response of soil bacteria to carbon enrichment	Glucose or acetate	Carbon	Agricultural soil	15×10cm soil core stored for 26days with and without earthworms, then 5g of bulk and cast samples was taken and incubated with glubose or acetate for 5–24h	Monard et al. (2008)
	Investigate the microbial food web related to methane	Methane	Carbon	Wetland rice soil	Air-dried, sieved, ground to <2mm, 14g of soil on 10μm-thick PTFE* membrane	Murase and Frenzel (2007)
	Determine how methanotrophs respond to fertilization	Methane	Carbon	Paddy field soil	Air-dried, ground to <2mm, soil slurry made with 1g of dry soil and 5g of sterile distilled (DI) water in 28mL pressure tubes. Samples were shaken and incubated for 10days	Noll et al. (2008)
	Determine how liming soil impacts carbon turnover by the active microbial community	Carbon dioxide	Carbon	Upland grassland soil	Pulse-labeled soil and vegetation in the field, then extracted 5×15cm soil cores after 3, 6, 12, 24h and 5days	Rangel-Castro et al. (2005)
	Identify the active acetate-assimilating, iron-reducing microorganisms from a long-term fertilized soil	Acetate	Carbon	Paddy field soil	Air-dried, sieved, soil slurry made with 1:1 ratio of soil to sterile DI water, aliquots were placed in serum vials and preincubated for 21days, then incubated with heavy isotope for 0, 0.5, 1, 1.5, 2, 3, and 4days	Ding et al. (2015)

(*Continued*)

TABLE 17.2 (*Continued*)
Overview of SIP Applications in Soil Microbiome Studies

SIP Type	Goal	Substrate	Heavy Isotope	Sample	Incubation Conditions	Paper
	Identify the microorganisms assimilating plant-based carbon in the root and rhizosphere soil	Carbon dioxide	Carbon	Oilseed rape soil	Following plant labeling with CO_2, plants were harvested at 24 and 72 h, and 7 and 14 days	Gkarmiri et al. (2017)
	Identify the active acetate-assimilating microorganisms that reduce ferric oxides	Acetate	Carbon	Paddy field soil	Air-dried, sieved, soil slurry made with 1:1 ratio of soil to sterile DI water, aliquots were placed in serum vials and preincubated for 21 days and then incubated with heavy isotope for 0, 8, 16, 24, 48, and 72 h	Hori et al. (2010)
	Identify the active acetate-assimilating microorganisms under methanogenic conditions	Acetate	Carbon	Paddy field soil	Air-dried, sieved, soil slurry made with 1:1 ratio of soil to sterile DI water, aliquots were placed in serum vials and preincubated for 30 days, then incubated with heavy isotope for 0, 3, 6, and 9 days	Hori et al. (2007)
	Understand how biomass carbon is mineralized and assimilated in soil	*E. coli* biomass labeled with glucose	Carbon	Agricultural soil	2 kg of soil in 5.7 L bioreactor, incubated for 1 h and 1, 2, 4, and 8 weeks with 1.3×10^8 cells per gram dry weight	Lueders et al. (2006)
	Understand how the frequency of burning alters the functional diversity of cellulolytic soil fungi	Cellulose	Carbon	Wet sclerophyll forest	Sieved, adjusted to 55% WHC, 30 g of soil in 1 L jars with sterile moist paper towel incubated for 35 days	Bastias et al. (2009)

(*Continued*)

TABLE 17.2 (*Continued*)
Overview of SIP Applications in Soil Microbiome Studies

SIP Type	Goal	Substrate	Heavy Isotope	Sample	Incubation Conditions	Paper
DNA/RNA	Identify the active acetate metabolizing microbes under methanogenic thermal conditions	Acetate	Carbon	Rice field soil	Air-dried, sieved, soil slurries were made using 700 g of dry soil, 0.35 g of dry ground rice straw, and 700 mL of anoxic sterile water. The slurries were incubated for 1, 2, 4, 7, 9, 11, 15, 17, 21, 25, 28, 32, and 67 days	Liu and Conrad (2010)
DNA/RNA	Identify the microorganisms actively assimilating plant-derived carbon	Wheat biomass labeled with carbon dioxide	Carbon	Calcareous silty-clay soil	10 g of sieved soil in 150 mL flasks, adjusted to 80% WHC and incubated for 0, 3, 7, 14, and 28 days	Bernard et al. (2007)
DNA/RNA	Understand the interaction between earthworms and soil microbial community, especially as it relates to methane oxidation	Methane	Carbon	Landfill cover soil	Air-dried, sieved, post-earthworm incubation 5 g of soil was put in vials and incubated for 7 days with substrate	Héry et al. (2008)
DNA/RNA	Identify the microorganisms capable of reducing arsenate in sediments amended with acetate	Sodium acetate	Carbon	Fine-grained sediment	20 g of sediment was added to 40 mL of artificial groundwater, incubated for 0, 4, 8, 12, 16, 20, and 30 days	Lear et al. (2007)
DNA/RNA	Understand the carbon dioxide assimilation as it related to nitrification, specifically ammonia oxidation	Carbon dioxide	Carbon	Agricultural soil	10 g of air-dried, sieved, homogenized soil was incubated at 60% WHC, with different amounts of ammonium for 0–12 weeks	Pratscher et al. (2011)

(*Continued*)

TABLE 17.2 (*Continued*)
Overview of SIP Applications in Soil Microbiome Studies

SIP Type	Goal	Substrate	Heavy Isotope	Sample	Incubation Conditions	Paper
DNA/RNA	Characterize the bacterial diversity of active soil organisms	Water	Oxygen	Brandt silty clay loam soil	10 g of dry homogenized soil in 50 mL conical tube with 1 mL of heavy water and 2 mL of sterile d H_2O incubated for 38 days	Rettedal and Brözel (2015)
DNA/RNA	Identify the active diazotrophs	Dinitrogen	Nitrogen	Dystric cambisol soil	3 g of sieved and homogenized soil was incubated in artificial root exudate mixture for 3, 7, and 21 days	Angel et al. (2018)
DNA/RNA	Identify methanotrophs from a rice field that respond to acetate amendments	Methane	Carbon	Rice field soil	10 g of sieved, air-dried soil was preincubated for 6 days and then incubated with substrate for 0, 21, and 39 days	Leng et al. (2015)
PLFA/mRNA	Active methanotroph community in acidic peatlands	Methane	Carbon	Acidic, oligotrophic, ombrogenous blanket peat	30 cm deep cores were sliced into 5 cm tall discs, and 5 g of soil from each disc was homogenized and incubated for 65 days	Chen et al. (2008)
PLFA	Role of microbial community in the stability of cellulose- and lignin-derived carbon in semiarid soils	Lignin, cellulose	Carbon	Semiarid soil after intensive agricultural use, Calcaric Regosol and Haplic Calcisol	100 g of dry homogenized soil from top 15 cm for long-term incubations (1, 4, 20, 60, and 120 days), and 1 g of soil for substrate mineralization experiments (1, 4, 10, 20, and 30 days)	Torres et al. (2014)

(*Continued*)

TABLE 17.2 (*Continued*)
Overview of SIP Applications in Soil Microbiome Studies

SIP Type	Goal	Substrate	Heavy Isotope	Sample	Incubation Conditions	Paper
PLFA/DNA	Determine the active microorganisms in mofette soil (i.e., high CO_2 environment and highly acidic)	Carbon dioxide	Carbon	Wetland with mofettes	8 g mofette soil was incubated for 0, 5, 14, and 28 days	Beulig et al. (2015)
PLFA	Understand the methane-derived carbon transport and sequestration in the soil food web	Methane	Carbon	Landfill cover soil	Sieved, 20 g incubated in petri dish in a flow through incubation chamber for 0, 3, 6, 9, 12, 18, 21, 27, 38, 50, 53, 56, 65, 85, 117, 154 days	Maxfield et al. (2006)
PLFA	Position-specific ^{13}C-labeling allows insight into carbon utilization by various microbial groups	Amino acids	Carbon	Agriculturally used loamy Luvisol	*In situ* application of substrates and soil sampling after 3 and 10 days. Extraction of 15 g subsets each	Apostel et al. (2013)
Protein	Tracking the assimilation of plant-derived N by genetic profiling and functional metaproteomics showed that bacterial populations assimilate plant-derived N faster than fungi	Leaf litter (tobacco was grown for 10 days with ^{15}N-NO^{3-})	Nitrogen	Typic Ariudoll	50 g of homogenized soil was mixed with 0.5 g of shredded plant material, and mesocosms were sampled during a 14-day incubation (after 0.1, 0.8, 1.9, 2.9, 5.0, 7.9, and 13.9 days)	Starke et al. (2016)
	Follow the fate of N in soil by measuring assimilation of compound-specific N by the soil microbial biomass	Ammonium, nitrate, glutamate	Nitrogen	Stagni-Vertic Cambisol	10 g of homogenized soil was preincubated at 50% water holding capacity for 4 days, and then incubated for 1.5, 3, 6, and 12 h and 1, 2, 4, 8, 16, and 32 days	Charteris et al. (2016)

(*Continued*)

TABLE 17.2 (*Continued*)
Overview of SIP Applications in Soil Microbiome Studies

SIP Type	Goal	Substrate	Heavy Isotope	Sample	Incubation Conditions	Paper
RAMAN	^{15}N-labeled substrate by the soil microbial biomass	$^{15}N_2$ gas	Nitrogen	Grassland soil	2 g of homogenized soil was amended with 500 μL of 0.5 M glucose solution and incubated at room temperature (~25°C) under low-light conditions for 12 days with an 80% $^{15}N_2$ atmosphere	Cui et al. (2018)
DNA/ functional metagenomics	Active-yet-uncultivated soil microorganisms and the diversity and activity of their glycoside hydrolases	Glucose, arabinose, cellobiose, xylose, cellulose	Carbon	Soils from arctic tundra, temperate rainforest, and agricultural soil	Soil samples (top 10 cm) were preincubated for 2 weeks to minimize readily available carbon. 10 g of homogenized soil was then incubated for up to 6 weeks	Verastegui et al. (2014)

* NMS, nitrate mineral salts; ANMS, ammonium nitrate mineral salts; WHC, water holding capacity; DCD, Dicyandiamide; PTFE, polytetrafluoroethylene.

have used DNA-SIP in soil to address questions related to nitrification. Understanding nitrification in agricultural soil is important, since these soils are generally nitrogen-limited unless supplemented with fertilizer. Thus, Xia et al. (2011) investigated the active ammonia-oxidizing microbial community in such a soil. In this agricultural soil from China, ammonia-oxidizing bacteria were found to dominate the active community, specifically *Nitrosospira* and *Nitrosomonas* spp. Recently, Cucu et al. (2017) investigated how archaea are involved in nitrous oxide reduction in paddy field soil. Archaea were found to play an important role in the last step of paddy soil denitrification. However, when rice straw (i.e., carbon) was added, the archaeal *nosZ* gene decreased compared to the treatment without rice straw. The authors attributed these results to archaea assimilating NH_4^+ when rice straw was not present. Another study investigated nitrification, in particular archaeal ammonia oxidation, in acidic soil (Wang et al. 2014). Uncultured microorganisms from within the *Nitrososphaera* cluster were found to constitute the active ammonia-oxidizing archaea under acidic conditions in soil. These had previously only been found in neutral environments, leading to an expansion of this group's habitats. Another important part of nitrogen cycling in soil is nitrogen fixation. Buckley et al. (2007) investigated this process by N-based SIP and found that three taxonomic groups were labeled: *Rhodoplanes*, unclassified β-proteobacteria, and unclassified Actinobacteria. Angel et al. (2018) also sought to identify the active diazotrophs, in this case in a beech forest soil instead of a field soil, and found that the active diazotrophs were members of the classes Clostridia and Bacilli. With continued technical improvements, there will likely be more researchers using ^{15}N in SIP studies to improve our understanding of nitrogen cycling in soil (Cui et al. 2018).

17.2.4 RNA-SIP in Soil Research

RNA-SIP has been mainly utilized in soil microbiology in order to understand carbon cycling. Lueders et al. (2004) first described a fractionation method instead of the use of ethidium bromide for gradient analysis in their RNA-SIP study in rice field soil. The aim was to determine what microorganisms were responsible for propionate degradation under methanogenic conditions in this soil (Table 17.2). Instead of collecting large amounts of DNA that had been stained with ethidium bromide in cesium chloride gradients, Lueders et al. gained enhanced sensitivity by first fractionating the gradients into 14 equal fractions and then quantifying the distribution of total nucleic acids and rRNA throughout all gradient fractions by fluorometry and real-time PCR. Most RNA-SIP studies, due to their independence from replication and quick label incorporation rates, have been conducted on a short timescale (i.e., less than 2 weeks), while many DNA-SIP incubations lasted a minimum of 1 month. The latter approach allowed the assessment of how fertilization impacts methanotrophs (Noll et al. 2008). Nitrogen availability impacts the two known types of methanotrophs (type I and type II) differently (Hanson and Hanson 1996). Noll et al. extended this concept and found that two type I methanotrophs, *Methylocaldum* and *Methylomicrobium*, assimilated methane under fertilization conditions, with the overall diversity of methanotrophs being low under high-nitrogen conditions. RNA-SIP is not dependent on DNA replication, and hence, it is possible to get expression data from studies in mesocosms or in the field. Gkarmiri et al. (2017) recently did a pulse-labeling experiment on oilseed rape (*Brassica napus*) to identify the microorganisms associated with the root and rhizosphere involved in plant-derived carbon metabolism. *Streptomycetes*, *Rhizobium*, *Clonostachys*, and *Fusarium* were found to assimilate plant-based carbon quickly. Multiple studies investigated the active acetate-assimilating microorganisms involved in iron reduction in paddy field soil (Hori et al. 2007, Hori et al. 2010, Ding et al. 2015). The studies by Hori and colleagues found that *Geobacter* sp., known iron-reducers, and *Anaeromyxobacter* sp. actively assimilated acetate and reduced iron oxides under methanogenic or anoxic conditions. Ding et al. (2015) recently found that long-term nitrogen fertilization promotes iron reduction and that *Geobacter* spp., along with microbes related to Proteobacteria and Firmicutes, are responsible for acetate assimilation coupled to iron reduction. It is becoming more common to use a combination of RNA- and DNA-SIP to get a more complete understanding of the active carbon and nitrogen cycling in soils under varying conditions. Since the availability of oxygen is essential to methane oxidation, Héry et al. (2008) investigated how earthworms impact this process in landfill cover soil. Earthworms did not shift the active communities with both type-I and type-II methanotrophs present, but did stimulate either growth or activity. The combination of DNA- and RNA-SIP has also been used to characterize the diversity of active soil microorganisms using $^{18}O\text{-}H_2O$. The authors found that a majority of the microbial community had metabolically active representatives (Rettedal and Brözel 2015). As costs decrease, using such combinations of SIP approaches will likely become more common.

17.2.5 PLFA-SIP in Soil Research

Microbial membrane lipids confer another biomarker that can be used to identify active microbial community members by incorporation of isotopic labels from assimilated labeled substrates. Common SIP targets are the PLFAs (Maxfield et al. 2006). However, lipid-SIP has also been applied to a suite of other microbial lipid types such as hopanoids, steroids, and isoprenoid-based molecules (Wegener et al. 2016). Because the phylogenetic resolution of microbial lipids is low, lipid-targeting SIP methods should always be paired with nucleic acid-based analyses such as 16S rRNA phylogenies. The PLFA-SIP approach can be refined to determine the growth kinetics *in vitro* and to identify the size of a targeted population (Maxfield et al. 2006). Recently, a combined approach using both inorganic carbon and deuterated water (D_2O) [called dual-SIP] was introduced, allowing the simultaneous analysis of heterotrophic and autotrophic carbon fixation without altering the substrate availability of the microbial community present (Kellermann et al. 2012, Wegener et al. 2016).

17.2.6 Protein-SIP in Soil Research

In protein-SIP, SIP techniques are matched with mass spectrometry. The incorporation of the isotopic label into proteins can be used to interpret how much of the labeled substrate was originally assimilated. The addition of mass spectrometry provides phylogenetic information from the peptide sequence analyses. This enables metaproteomic approaches with the advantage that the incorporation of stable isotopes can be determined with high sensitivity (Jehmlich et al. 2010). In addition to identifying active community members, elemental fluxes can also be followed quantitatively (Lünsmann et al. 2016). A drawback of the application of this method in soil is that proteins have to be isolated efficiently prior to analyses, and quantitative isolation of proteins from soil is not trivial (Keiblinger et al. 2012).

As in other SIP methods, the soil microbial community is exposed to a labeled substrate, and following cultivation, proteins are extracted, fractionated, and separated. The proteins are now proteolytically digested and identified by mass spectrometry combined with liquid chromatography (together, denoted as LC-MS). The identification of labeled proteins is challenged by both the limited database of proteins and the mass shift resulting from partial labeling by incorporation of varying amounts of heavier stable isotopes under natural conditions. However, the development of multiple software approaches allows for a more streamlined analysis (e.g., SIPPER, Slysz et al. 2014), and metaproteomics approaches can now be analyzed using MetaProSIP (Sachsenberg et al. 2015). The identification of proteins is greatly supported by the availability of a parallel metagenome analysis.

17.2.7 Raman-SIP in Soil Research

Raman spectroscopy allows the nondestructive identification of individual microbial cells by creating a wave spectrum for each cell type. If combined with SIP, Raman spectroscopy enables the direct observation of functional properties of individual microbial community members (Wang et al. 2016), which has been successfully applied for soil (Eichorst et al. 2015). However, for this single-cell analysis, soil bacterial cells have to be extracted from the soil environment prior to analysis. An important advantage of Raman-SIP is that cells can be cultivated after analysis or used in downstream molecular analyses. Recently, Cui et al. (2018) reported the use of single-cell Raman to study N_2-fixing bacteria in complex ecosystems.

17.3 ANALYZING SIP HIGH-THROUGHPUT SEQUENCING DATA

When DNA-SIP and RNA-SIP were initially developed, labeled populations were identified by targeting their genes using fingerprinting techniques such as DGGE and T-RFLP, or by directly sequencing marker genes such as 16S rRNA or functional genes. Today, fingerprinting techniques are used only to confirm sufficient isotope labeling (Gutierrez et al. 2013), while nucleic acids are analyzed by modern high-throughput sequencing methods. This enables the more comprehensive analysis of large amplicon datasets, or metagenomes and metatranscriptomes.

There are many methods for the complex analysis of datasets to analyze DNA- and RNA-SIP data to determine what taxa or genes are enriched in the "heavy" compared to "light" DNA/RNA fractions. Until relatively recently, the majority of SIP studies were qualitative, using fingerprinting techniques such as DGGE, automated ribosomal intergenic spacer analysis, T-RFLP, or gene cloning to identify the taxa found in the "heavy" compared to "light" DNA/RNA. Occasionally, a study would use qPCR or quantitative reverse transcription-PCR (qRT-PCR) to quantify the taxa. Now, DNA-SIP and RNA-SIP can be combined with high-throughput sequencing, which allows for the generation of massive amounts of sequencing data. Note that if one wants a quantitative analysis of these sequencing data, then be sure to sequence all collected fractions with DNA, as described by Hungate et al. (2015).

Like most sequencing data, the data are unlikely to be normally distributed. Therefore, it is important to select the proper statistical methods. First, one needs to annotate the sequences with a program such as DADA2 for 16S rRNA or MG-RAST for metagenomes/metatranscriptomes. Second, the resulting table of operational taxonomic units can be offered to "phyloseq" or "count data" in the DESeq2 program, respectively. Following the phyloseq analysis, the publicly available software package HTSSIP can be used. This recently developed software combines several SIP data analysis methods in order to accurately map *in situ* metabolic processes from soil with the microbial taxa (Youngblut et al. 2018). In addition to visualization methods such as nonmetric multidimensional scaling plots, HTSSIP also allows to identify the taxa that significantly incorporated the stable isotope and possibly quantify the incorporation depending on the setup. If the diversity and richness in the samples are to be determined, an "R" package like vegan can be used (Oksanen et al. 2017). For protein-SIP data analysis, there is a program called MetaProSIP (Sachsenberg et al. 2015).

A novel way to analyze nucleic acids labeled with ^{13}C stable isotopes was recently reported by Wilhelm et al. (2014). The use of ultra-high-performance LC-MS enabled a highly sensitive method that quantifies label incorporation into DNA or RNA. Having this method and the necessary instrumentation available will also minimize the expenses for isotopically labeled substrates.

Due to the expense of SIP methods (e.g., the common combination of DNA-SIP with high-throughput sequencing), the accuracy of DNA-SIP methods is difficult to assess. Recently, the toolset SIPSim was developed, which simulates complex DNA-SIP datasets and evaluates their analytical accuracy prior to incurring expenses (Youngblut and Buckley 2017).

17.4 STANDARDS FOR REPORTING SIP STUDIES

As the number of SIP studies in soils increases, it is crucial to follow a standard for the reporting of minimally required information with respect to the experimental data and the appropriate metadata. We suggest that the following minimum information is included in every report on SIP applied to soil:

- Experimental goal
- Soil metadata
- Soil sampling protocol and storage
- Type of SIP
- Choice of stable isotope label and substrate
- Target molecule and extraction protocol
- Type of sample incubation
- Incubation time
- Refeeding schedule with labeled substrate
- Ultracentrifugation rotor and full protocol
- Fractionation technique
- Quality control of target molecule extraction and purification
- Molecular analysis of target molecule
- Software packages employed to model SIP
- Software package employed to analyze post-SIP data

17.5 CONCLUSION AND OUTLOOK

The use of SIP approaches in tracer experiments with labeled substrates has significantly improved the understanding of microbial processes and interactions in complex soil microbial systems.

Further advances in target molecule separation combined with integrated methods (e.g., double-SIP labeling in combination with metagenomics and next-generation nucleic acid analyses) will further expand our understanding of selected microbiome-driven processes and functions in soils. Tracing active microbial populations that feed on specific carbon- or nitrogen-containing substrates will yield deeper insights into the combined flow of carbon and nitrogen in the soil network. The approach might also enable a better understanding of the plant–microbe interplay in the rhizosphere, either as a beneficial coexistence or as a competitive interaction for nutrients. In the near future, the metagenome-assisted assembly of microbial genomes from SIP-targeted nucleic acids will permit the identification of multiple members of the soil microbiome that are involved in a targeted function, even if these are members of the rare biosphere. Further technical improvements will continue to make SIP techniques into essential and even more powerful tools for soil microbiome investigations.

REFERENCES

Angel, R., Panhölzl, C., Gabriel, R., et al. 2018. Application of stable-isotope labelling techniques for the detection of active diazotrophs. *Environ Microbiol* 20, 44–61.

Apostel, C., Dippold, M., Glaser, B. and Y. Kuzyakov. 2013. Biochemical pathways of amino acids in soil: Assessment by position-specific labeling and ^{13}C-PLFA analysis. *Soil Biol Biochem* 67, 31–40.

Bastias, B.A., Anderson, I.C., Rangel-Castro, J.I., Parkin, P.I., Prosser, J.I. and J.W. Cairney. 2009. Influence of repeated prescribed burning on incorporation of 13C from cellulose by forest soil fungi as determined by RNA stable isotope probing. *Soil Biol Biochem* 41, 467–472.

Bell, T.H., Yergeau, E., Martineau, C., Juck, D., Whyte, L.G. and C.W. Greer. 2011. Identification of nitrogen-incorporating bacteria in petroleum-contaminated arctic soils by using ^{15}N-DNA-based stable isotope probing and pyrosequencing. *Appl Environ Microbiol* 77, 4163–4171.

Bernard, L., Mougel, C., Maron, P.A., et al. 2007. Dynamics and identification of soil microbial populations actively assimilating carbon from ^{13}C-labelled wheat residue as estimated by DNA-and RNA-SIP techniques. *Environ Microbiol* 9, 752–764.

Beulig, F., Heuer, V.B., Akob, D.M., et al. 2015. Carbon flow from volcanic CO_2 into soil microbial communities of a wetland mofette. *ISME J* 9, 746–752.

Buckley, D.H., Huangyutitham, V., Hsu, S.F. and T.A. Nelson. 2007. Stable isotope probing with $^{15}N_2$ reveals novel noncultivated diazotrophs in soil. *Appl Environ Microbiol* 73, 3196–3204.

Cébron, A., Bodrossy, L., Stralis-Pavese, N., et al. 2007a. Nutrient amendments in soil DNA stable isotope probing experiments reduce the observed methanotroph diversity. *Appl Environ Microbiol* 73, 798–807.

Cébron, A., Bodrossy, L., Chen, Y., et al. 2007b. Identity of active methanotrophs in landfill cover soil as revealed by DNA-stable isotope probing. *FEMS Microbiol Ecol* 62, 12–23.

Charteris, A.F., Knowles, T.D.J., Michaelides, K. and R.P. Evershed. 2016. Compound-specific amino acid ^{15}N stable isotope probing of nitrogen assimilation by the soil microbial biomass using gas chromatography/combustion/isotope ratio mass spectrometry. *Rapid Commun Mass Spectrom* 30, 1846–1856.

Chen, Y., Dumont, M.G., McNamara, N.P., et al. 2008. Diversity of the active methanotrophic community in acidic peatlands as assessed by mRNA and SIP-PLFA analyses. *Environ Microbiol* 10, 446–459.

Cucu, M.A., Marhan, S., Said-Pullicino, D., Celi, L., Kandeler, E. and F. Rasche. 2017. Resource driven community dynamics of NH_4^+ assimilating and N_2O reducing archaea in a temperate paddy soil. *Pedobiologia* 62, 16–27.

Cui, L., Yang, K., Li, H.Z., et al. 2018. Functional single-cell approach to probing nitrogen-fixing bacteria in soil communities by resonance Raman spectroscopy with $^{15}N_2$ labeling. *Anal Chem* 90, 5082–5089.

Cupples, A.M. and G.K. Sims. 2007. Identification of in situ 2, 4-dichlorophenoxyacetic acid-degrading soil microorganisms using DNA-stable isotope probing. *Soil Biol Biochem* 39, 232–238.

DeRito, C.M. and E.L. Madsen. 2009. Stable isotope probing reveals *Trichosporon* yeast to be active in situ in soil phenol metabolism. *ISME J* 3, 477–482.

DeRito, C.M., Pumphrey, G.M. and E.L. Madsen. 2005. Use of field-based stable isotope probing to identify adapted populations and track carbon flow through a phenol-degrading soil microbial community. *Appl Environ Microbiol* 71, 7858–7865.

Ding, L.J., Su, J.Q., Xu, H.J., Jia, Z.J. and Y.G. Zhu. 2015. Long-term nitrogen fertilization of paddy soil shifts iron-reducing microbial community revealed by RNA-^{13}C-acetate probing coupled with pyrosequencing. *ISME J* 9, 721–729.

Dumont, M.G. and J.C. Murrell. 2005. Stable isotope probing—linking microbial identity to function. *Nat Rev Microbiol* 3, 499–504.

Dunford, E.A. and J.D. Neufeld. 2010. DNA stable-isotope probing (DNA-SIP). *J Visual Exp*, 42, 1–6.

Dumont, M.G., Pommerenke, B. and P. Casper. 2013. Using stable isotope probing to obtain a targeted metatranscriptome of aerobic methanotrophs in lake sediment. *Environ Microbiol Rep* 5, 757–764.

Dumont, M.G., Radajewski, S.M., Miguez, C.B., McDonald, I.R. and J.C. Murrell. 2006. Identification of a complete methane monooxygenase operon from soil by combining stable isotope probing and metagenomic analysis. *Environ Microbiol* 8, 1240–1250.

Eichorst, S.A., Strasser, F., Woyke, T., Schintlmeister, A., Wagner, M. and D. Woebken. 2015. Advancements in the application of NanoSIMS and Raman microspectroscopy to investigate the activity of microbial cells in soils. *FEMS Microbiol Ecol* 91, fiv106.

Eyice, Ö., Namura, M., Chen, Y., Mead, A., Samavedam, S. and H. Schäfer. 2015. SIP metagenomics identifies uncultivated Methylophilaceae as dimethylsulphide degrading bacteria in soil and lake sediment. *ISME J* 9, 2336.

Gkarmiri, K., Mahmood, S., Ekblad, A., Alström, S., Högberg, N. and R. Finlay. 2017. Identifying the active microbiome associated with roots and rhizosphere soil of oilseed rape. *Appl Environ Microbiol* 83, e01938-17.

Gupta, V., Smemo, K.A., Yavitt, J.B. and N. Basiliko. 2012. Active methanotrophs in two contrasting North American peatland ecosystems revealed using DNA-SIP. *Microb Ecol* 63, 438–445.

Gutierrez, T., Singleton, D.R., Berry, D., Yang, T., Aitken, M.D. and A. Teske. 2013. Hydrocarbon-degrading bacteria enriched by the Deepwater Horizon oil spill identified by cultivation and DNA-SIP. *ISME J* 7, 2091–2104.

Haichar el Zahar, F., Marol, C., Berge, O., et al. 2008. Plant host habitat and root exudates shape soil bacterial community structure. *ISME J* 2, 1221–1230.

Hanson, R.S. and T.E. Hanson. 1996. Methanotrophic bacteria. *Microbiol Rev* 60, 439–471.

Hanson, J.R., Macalady, J.L., Harris, D. and K.W. Scow. 1999. Linking toluene degradation with specific microbial populations in soil. *Appl Environ Microbiol* 65, 5403–5408.

Héry, M., Singer, A.C., Kumaresan, D., et al. 2008. Effect of earthworms on the community structure of active methanotrophic bacteria in a landfill cover soil. *ISME J* 2, 92–104.

Hori, T., Noll, M., Igarashi, Y., Friedrich, M.W. and R. Conrad. 2007. Identification of acetate-assimilating microorganisms under methanogenic conditions in anoxic rice field soil by comparative stable isotope probing of RNA. *Appl Environ Microbiol* 73, 101–109.

Hori, T., Müller, A., Igarashi, Y., Conrad, R. and M.W. Friedrich. 2010. Identification of iron-reducing microorganisms in anoxic rice paddy soil by ^{13}C-acetate probing. *ISME J* 42, 267–277.

Hungate, B.A., Mau, R.L., Schwartz, E., et al. 2015. Quantitative microbial ecology through stable isotope probing. *Appl Environ Microbiol* 81, 7570–7581.

Jehmlich, N., Schmidt, F., Taubert, M., et al. 2010. Protein-based stable isotope probing. *Nat Protoc* 5, 1957–1966.

Jehmlich, N., Vogt, C., Lünsmann, V., Richnow, H.H. and M. von Bergen. 2016. Protein-SIP in environmental studies. *Curr Opin Biotechnol* 41, 26–33.

Jia, Z. and R. Conrad. 2009. Bacteria rather than Archaea dominate microbial ammonia oxidation in an agricultural soil. *Environ Microbiol* 11, 1658–1671.

Keiblinger, K.M., Wilhartitz, I.C., Schneider, T., et al. 2012. Soil metaproteomics—comparative evaluation of protein extraction protocols. *Soil Biol Biochem* 54, 14–24.

Kellermann, M.Y., Wegener, G., Elvert, M., et al. 2012. Autotrophy as a predominant mode of carbon fixation in anaerobic methane-oxidizing microbial communities. *Proc Nat Acad Sci USA* 109, 19321–19326.

Lear, G., Song, B., Gault, A.G., Polya, D.A. and J.R. Lloyd. 2007. Molecular analysis of arsenate-reducing bacteria within Cambodian sediments following amendment with acetate. *Appl Environ Microbiol* 73, 1041–1048.

Lee, C.G., Watanabe, T. and S. Asakawa. 2017. Bacterial community incorporating carbon derived from plant residue in an anoxic non-rhizosphere soil estimated by DNA-SIP analysis. *J Soils Sediments* 17, 1084–1091.

Lee, C.G., Watanabe, T., Sato, Y., Murase, J., Asakawa, S. and M. Kimura. 2011. Bacterial populations assimilating carbon from ^{13}C-labeled plant residue in soil: Analysis by a DNA-SIP approach. *Soil Biol Biochem* 43, 814–822.

Leng, L., Chang, J., Geng, K. and K. Ma. 2015. Uncultivated *Methylocystis* species in paddy soil include facultative methanotrophs that utilize acetate. *Microb Ecol* 70, 88–96.

Li, Y., Lee, C.G., Watanabe, T., Murase, J., Asakawa, S. and M. Kimura. 2011. Identification of microbial communities that assimilate substrate from root cap cells in an aerobic soil using a DNA-SIP approach. *Soil Biol Biochem* 43, 1928–1935.

Liu, F. and R. Conrad. 2010. Thermoanaerobacteriaceae oxidize acetate in methanogenic rice field soil at 50°C. *Environ Microbiol* 12, 2341–2354.

Lu, Y. and R. Conrad. 2005. In situ stable isotope probing of methanogenic archaea in the rice rhizosphere. *Science* 309, 1088–1090.

Lueders, T., Kindler, R., Miltner, A., Friedrich, M.W. and M. Kaestner. 2006. Identification of bacterial micropredators distinctively active in a soil microbial food web. *Appl Environ Microbiol* 72, 5342–5348.

Lueders, T., Wagner, B., Claus, P. and M.W. Friedrich. 2004. Stable isotope probing of rRNA and DNA reveals a dynamic methylotroph community and trophic interactions with fungi and protozoa in oxic rice field soil. *Environ Microbiol* 6, 60–72.

Lünsmann, V., Kappelmeyer, U., Benndorf, R., et al. 2016. In situ protein-SIP highlights Burkholderiaceae as key players degrading toluene by para ring hydroxylation in a constructed wetland model. *Environ Microbiol* 18, 1176–1184.

Mahmood, S., Paton, G.I. and J.I. Prosser. 2005. Cultivation-independent in situ molecular analysis of bacteria involved in degradation of pentachlorophenol in soil. *Environ Microbiol* 7, 1349–1360.

Manefield, M., Whiteley, A.S., Griffiths, R.I. and M.R. Bailey. 2002. RNA stable isotope probing, a novel means of linking microbial community function to phylogeny. *Appl Environ Microbiol* 68, 5367–5373.

Maxfield, P.J., Hornibrook, E.R.C. and R.P. Evershed. 2006. Estimating high-affinity methanotrophic bacterial biomass, growth and turnover in soil by phospholipid fatty acid 13C labeling. *Appl Environ Microbiol* 72, 3901–3907.

Mohanty, S.R., Bodelier, P.L., Floris, V. and R. Conrad. 2006. Differential effects of nitrogenous fertilizers on methane-consuming microbes in rice field and forest soils. *Appl Environ Microbiol* 72, 1346–1354.

Monard, C., Binet, F. and P. Vandenkoornhuyse. 2008. Short-term response of soil bacteria to carbon enrichment in different soil microsites. *Appl Environ Microbiol* 74, 5589–5592.

Morris, S.A., Radajewski, S., Willison, T.W. and J.C. Murrell. 2002. Identification of the functionally active methanotroph population in a peat soil microcosm by stable-isotope probing. *Appl Environ Microbiol* 68, 1446–1453.

Murase, J. and P. Frenzel. 2007. A methane-driven microbial food web in a wetland rice soil. *Environ Microbiol* 9, 3025–3034.

Neufeld, J.D., Vohra, J., Dumont, M.G., Lueders, T., Manefield, M., Friedrich, M.W. and J.C. Murrell. 2007. DNA stable-isotope probing. *Nat Protoc* 2, 860–866.

Noll, M., Frenzel, P. and R. Conrad. 2008. Selective stimulation of type I methanotrophs in a rice paddy soil by urea fertilization revealed by RNA-based stable isotope probing. *FEMS Microbiol Ecol* 65, 125–132.

Oksanen, J.F., Blanchet, G., Friendly, M., et al. 2017. Vegan: Community Ecology Package. R package version 2.4-5. https://CRAN.R-project.org/package=vegan.

Pepe-Ranney, C., Koechli, C., Potrafka, R., et al. 2016. Non-cyanobacterial diazotrophs mediate dinitrogen fixation in biological soil crusts during early crust formation. *ISME J* 10, 287–298.

Pratscher, J., Dumont, M.G. and R. Conrad. 2011. Ammonia oxidation coupled to CO_2 fixation by archaea and bacteria in an agricultural soil. *Proc Nat Acad Sci USA* 108, 4170–4175.

Quideau, S.A., McIntosh, A.C.S., Norris, C.E., Lloret, E., Swallow, M.J.B. and K. Hannam. 2016. Extraction and analysis of microbial phospholipid fatty acids in soils. *J Visual Exp* 114, e54360.

Radajewski, S., Ineson, P., Parekh, N.R. and J.C. Murrell. 2000. Stable-isotope probing as a tool in microbial ecology. *Nature* 403, 646–649.

Rangel-Castro, J.I., Killham, K., Ostle, N., et al. 2005. Stable isotope probing analysis of the influence of liming on root exudate utilization by soil microorganisms. *Environ Microbiol* 7, 828–838.

Rettedal, E.A. and V.S. Brözel. 2015. Characterizing the diversity of active bacteria in soil by comprehensive stable isotope probing of DNA and RNA with $H_2{}^{18}O$. *Microbiol Open* 4, 208–219.

Richnow, H.H. and T. Lueders. 2016. Editorial overview: Probing environmental processes and microbiome functions using stable isotopes as smart tracers in analytical biotechnology. *Curr Opin Biotechnol* 41, iv–vii.

Rodgers-Vieira, E.A., Zhang, Z., Adrion, A.C., Gold, A. and M.D. Aitken. 2015. Identification of anthraquinone-degrading bacteria in soil contaminated with polycyclic aromatic hydrocarbons. *Appl Environ Microbiol* 81, 3775–3781.

Sachsenberg, T., Herbst, F.A., Taubert, M., et al. 2015. MetaProSIP: Automated inference of stable isotope incorporation rates in proteins for functional metaproteomics. *J Proteome Res* 14, 619–627.

Schwartz, E. 2007. Characterization of growing microorganisms in soil by stable isotope probing with $H_2{}^{18}O$. *Appl Environ Microbiol* 73, 2541–2546.

Slysz, G.W., Steinke, L., Ward, D.M., et al. 2014. Automated data extraction from in situ protein-stable isotope probing studies. *J Proteome Res* 13, 1200–1210.

Song, M., Jiang, L., Zhang, D., et al. 2016. Bacteria capable of degrading anthracene, phenanthrene and fluoranthene as revealed by DNA based stable-isotope probing in a forest soil. *J Hazard Mater* 308, 50–57.

Starke, R., Kermer, R., Ullmann-Zeunert, L., et al. 2016. Bacteria dominate the short-term assimilation of plant-derived N in soil. *Soil Biol Biochem* 96, 30–38.

Tong, H., Liu, C., Li, F., Luo, C., Chen, M. and M. Hu. 2015. The key microorganisms for anaerobic degradation of pentachlorophenol in paddy soil as revealed by stable isotope probing. *J Hazard Mater* 298, 252–260.

Torres, I.F., Bastida, F., Hernández, T., Bombach, P., Richnow, H.H. and C. García. 2014. The role of lignin and cellulose in the carbon-cycling of degraded soils under semiarid climate and their relation to microbial biomass. *Soil Biol Biochem* 75, 152–160.

Verastegui, Y., Cheng, J., Engel, K., et al. 2014. Multisubstrate isotope labeling and metagenomic analysis of active soil bacterial communities. *MBio* 5, e01157-14.

Wang, Y., Huang, W.E., Cui, L., M. Wagner. 2016. Single cell stable isotope probing in microbiology using Raman microspectroscopy. *Curr Opin Biotechnol* 41, 34–42.

Wang, B., Zheng, Y., Huang, R., et al. 2014. Active ammonia oxidizers in an acidic soil are phylogenetically closely related to neutrophilic archaeon. *Appl Environ Microbiol* 80, 1684–1691.

Wegener, G., Kellermann, M.Y. and M. Elvert. 2016. Tracking activity and function of microorganisms by stable isotope probing of membrane lipids. *Curr Opin Biotechnol* 41, 43–52.

Whiteley, A.S., Thomson, B., Lueders, T. and M. Manefield. 2007. RNA stable-isotope probing. *Nat Protoc* 2, 838–844.

Wilhelm, R., Szeitz, A., Klassen, T.L. and W.W. Mohn. 2014. Sensitive, efficient quantitation of ^{13}C-enriched Nucleic Acids via Ultrahigh-Performance Liquid Chromatography–Tandem Mass Spectrometry for Applications in Stable Isotope Probing. *Appl Environ Microbiol* 80, 7206–7211.

Xia, W., Zhang, C., Zeng, X., et al. 2011. Autotrophic growth of nitrifying community in an agricultural soil. *ISME J* 5, 1226.

Youngblut, N.D., Barnett, S.E. and D.H. Buckley. 2018. HTSSIP: An R package for analysis of high throughput sequencing data from nucleic acid stable isotope probing (SIP) experiments. *PLoS One* 13, e0189616.

Youngblut, N.D. and D.H. Buckley. 2017. Evaluating the accuracy of DNA stable isotope probing. *bioRxiv*, 138719. doi:10.1101/138719.

Zhang, Q., Gu, C., Zhou, H., Liang, Y., Zhao, Y. and H. Di. 2018. Alterations in anaerobic ammonium oxidation of paddy soil following organic carbon treatment estimated using ^{13}C-DNA stable isotope probing. *Appl Microbiol Biotechnol* 102, 1407–1416.

Zhang, L.M., Hu, H.W., Shen, J.P. and J.Z. He. 2012. Ammonia-oxidizing archaea have more important role than ammonia-oxidizing bacteria in ammonia oxidation of strongly acidic soils. *ISME J* 6, 1032–1039.

Zhang, L.M., Offre, P.R., He, J.Z., Verhamme, D.T., Nicol, G.W. and J.I. Prosser. 2010. Autotrophic ammonia oxidation by soil thaumarchaea. *Proc Nat Acad Sci USA* 107, 17240–17245.

18 Isolation of Uncultured Bacteria

Ulisses Nunes da Rocha
Helmholtz Centre for Environmental Research—UFZ

CONTENTS

18.1 Introduction—The Great Plate Count Anomaly 295
18.2 Uncultured Bacteria—The Unseen Majority 296
18.3 Growth Media 297
18.3.1 Strategies in the Development of Growth Media 297
18.3.2 The Use of Different Media and Growth Conditions 300
18.3.2.1 Dilution of Traditional Media 301
18.3.2.2 Use of Oxidation-Protective Agents 301
18.3.2.3 Addition of Micronutrients and Use of Different Atmospheric Compositions 301
18.4 How Far Have We Come with the Novel Isolation Approaches? 302
18.5 Outlook and Prospects 303
References 304

18.1 INTRODUCTION—THE GREAT PLATE COUNT ANOMALY

A large discrepancy between the bacterial numbers obtained from direct microscopic cell counts and those calculated from viable plate counts, being the latter up to three orders of magnitude lower, was observed and described almost 90 years ago (Razumov 1932) in studies on aquatic habitats. A more recent illustration of this anomaly was provided ca. 30 years ago in studies in both terrestrial and aquatic habitats (Staley and Konopka 1985). The phenomenon has been denominated the *great plate count anomaly*. The question arises: why does this discrepancy happen? As an easy explanation, it might be cogitated that most bacterial cells from the natural systems might not be able to grow under laboratory conditions and are thus "non-culturable." Nevertheless, bacterial cells could be alive ("viable") under field conditions; that is, they could be perfectly adapted to their environmental niche (Staley and Konopka 1985). If the explanation is correct, then it is important to understand under what conditions such bacteria might be active and increase their population sizes, on the basis of knowledge gained from the natural systems. One might actually surmise that most intact bacterial cells in natural systems are actually alive and in principle capable of activity, and that their non-culturability is due to a lack of precision in laboratory cultivation procedures. This leads us to the question "Which factors are crucial for environmental (soil) bacteria in order to grow under laboratory conditions?"

As a result of our limited knowledge of the conditions that prevail in soil microhabitats, the traditional cultivation techniques by definition cannot mimic these. The growth conditions in such sites are characterized by (adequate amounts of) growth substrates as well as micronutrients, next to adequate levels of pH, temperature, and moisture. In addition, spatiotemporal gradient-like conditions may be required, with in particular temporal gradients being difficult to mimic. Other factors relate to the gaseous conditions of the atmosphere the organism is exposed to, such as specific CO_2:O_2 ratio's (Stevenson et al. 2004). In addition, some bacteria may need particular signal molecules to initiate cell division (Nichols et al. 2008). Alternatively, bacteria may grow only when they are in mixtures or consortia, together with other microorganisms (Ettwig et al. 2008), as explored in the work of Chen et al (2014). Finally, there may be conditions applied in cultivation efforts that even inhibit growth, such as excess substrate levels. And, it cannot be excluded that bacteria residing in a soil system are injured (so, intrinsically non-culturable and probably nonviable); this may result in a permanent loss of culturability (Carini et al. 2017).

Pioneering studies performed about 15 years ago showed that enhancement of the culturability of environmental bacteria was possible when conditions that are better-tuned to the habitat were applied, for soil (Janssen et al. 2002) as well as freshwater habitats (Bruns et al. 2003). In these studies, the key factors found to be relevant for the improved cultivation were (1) reduction of nutrient availability, (2) prolonged incubation times, and (3) reduction of oxidative stress by the addition of protective agents. Big steps in the recovery of cultured organisms from soils may thus be taken by fairly simple modifications of existing culturing protocols. However, we do not know to what extent these improvements will work for a major part of the extant bacterial diversity, and so a large part of the soil bacterial diversity remains cryptic to date (Hug et al. 2016).

Simple modifications of existing protocols have already resulted in the ready isolation (in axenic culture) of rare bacterial groups. For example, avoiding autoclaving of phosphate together with agar was shown to enhance the colony formation by bacteria on agar plates (Kamagata and Tamaki 2005). However, in spite of advances made in recent years in isolation research, a major part of the extant bacterial diversity in soil still remains cryptic. This chapter examines the state of the art in our understanding of the triggers that lead to successful cultivation of bacteria—and archaea—from soil.

18.2 UNCULTURED BACTERIA—THE UNSEEN MAJORITY

Prior to the introduction of modern "omics" techniques in analyses of soil bacterial communities, investigations of bacterial diversity were hampered by the need to dispose of pure cultures of the relevant extant bacteria. Use of amplicon sequencing of soil samples has resulted in a large fraction of 16S rRNA gene sequence data in public databases (such as the Ribosomal Database Project; Cole et al. 2014) consisting of sequences derived from direct molecular (cultivation-independent) data. Cultivation efforts have not kept pace with the ever-expanding sequence databases.

In this respect, it is difficult to estimate how far we are from bridging the gap between the soil DNA-based picture of the extant bacterial diversity and the cultured diversity. According to a recent assessment, representatives of each of 54 of the 96 bacterial phyla known are yet to be recovered in pure culture (Hug et al. 2016). For instance, only recently representatives of species from well-characterized taxonomic groups, such as slow-growing *Bradyrhizobium* spp., have been recovered in pure culture (Zhalnina et al. 2018).

The Ribosomal Database Project (Cole et al. 2014) (release 11, update 5, September 30, 2016; http://rdp.cme.msu.edu/index.jsp) is used here to pinpoint examples of bacterial groups that dominate sequence databases yet are sparsely present in isolate collections. The following dominant soil bacterial groups had more than 95% of their 16S rRNA gene sequences identified from cultivation-independent studies: Armatimonadetes (previously *Candidatus* phylum OP10; 1.3% of sequences associated with cultured strains), *Candidatus* phylum Saccharibacteria (previously TM7; 0.9%), Gemmatimonadetes (1.2%), Acidobacteria (1.4%), Caldiserica (previously OP5; 1.7%), Chloroflexi (0.9%), Verrucomicrobia (2.3%), and Planctomycetes (3.0%) (Figure 18.1). This analysis may be slightly biased due to the different methods that were used to produce the sequences, but it gives a robust impression of the state of the art at this point in time. Overall, a sizeable fraction of the bacterial phyla still either remains completely uncultured or encompasses just a few cultured representatives (Solden et al. 2016).

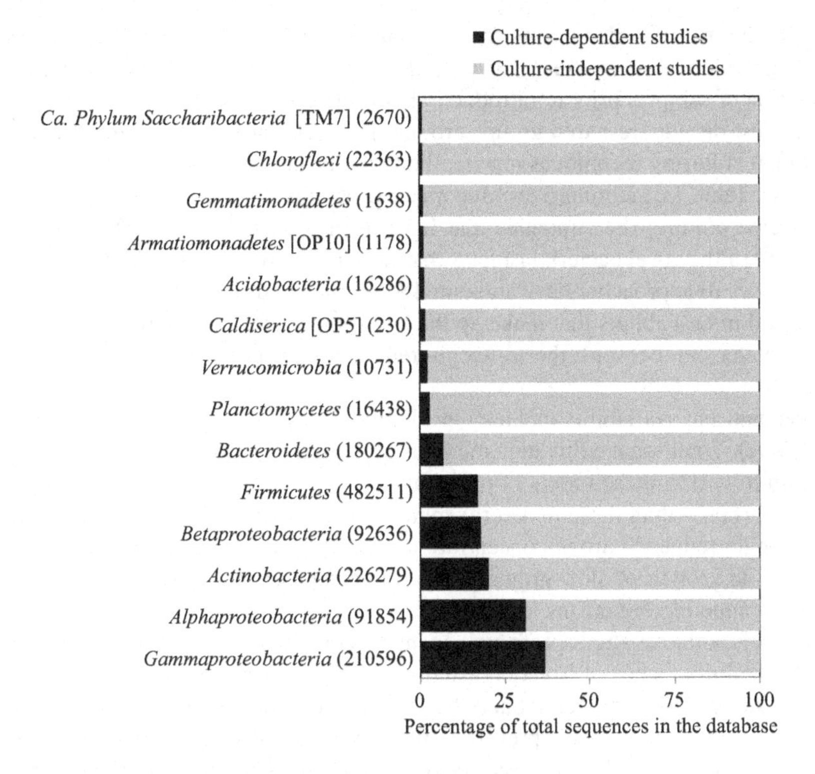

FIGURE 18.1 Percentage of 16S rRNA gene sequences from bacterial groups common to the rhizosphere and obtained via culture-dependent and culture-independent approaches. All sequences (at least 1,200 bp in length), derived from different ecosystems, were obtained from the Ribosomal Database Project (Cole et al. 2014), release 11 update 5 (last accession, 15 July 2018). Numbers between parentheses represent the total number of sequences per bacterial group found in the Ribosomal Database Project. Codes between brackets represent taxa before culturable isolates were recovered.

18.3 GROWTH MEDIA

18.3.1 Strategies in the Development of Growth Media

In order to remain alive and grow, organisms in soil need to acquire energy and nutrients from their surrounding environment. Traditional microbiology teaches us that each type/species of organism has its own nutritional requirements and that the diversity of metabolic and ecological types in nature is truly daunting. To cultivate these diverse organisms in the laboratory, scientists have formulated a variety of different growth media that supply the nutrient and energy needs of specific types of organisms. For example, the use of specific liquid media for isolation often results in enrichment cultures. Although not yet growing in axenic culture, the target organism may tend to occur in enhanced abundance. Sequential use of the same selective liquid medium may result in highly enriched target microbes that can subsequently be placed on solidified medium for purification (Kessel et al. 2015). Growth media are often liquid, as they serve the purpose of enrichment, yet can be made "solid" by the addition of solidifying agents, such as agar or gellan gum. The advantage of using solid media is that the stress imposed by inter-organism competition is reduced, and so single populations may better grow out in colonies for further purification and isolation. The approach has been traditionally useful and is still being used for the recovery of novel isolates in pure culture (Pascual et al. 2016).

Traditionally, the first steps of bacterial isolation (both in liquid and solid media) consist of the following: (1) extraction of cells from their native environment; (2) disassembly of cellular aggregates (by the use of a blender, mechanical shaking or mortar and pestle); (3) if present, the removal of large abiotic particles (e.g., for soil, filtration or dispersal by using pyrophosphate); and (4) serial dilution of samples prior to introduction into the growth media.

During the last decade, research groups around the world have studied how small adjustments of the traditional culturing techniques may facilitate the isolation of hitherto "rare" bacteria in culture collections. Table 18.1 summarizes nine major problems involved in the recovery of bacterial isolates in axenic culture, the strategies that have been used to circumvent these, and the taxa (at phylum level) of the novel bacterial isolates that were recovered using these different approaches. In a generic sense, all approaches have attempted to mimic conditions that were presumed to occur *in situ* in the soil microhabitats that make up the environment for local soil bacteria. The rationale behind five strategies to recover hitherto-uncultured bacteria is given as follows:

- Reduced nutrient availability and long incubation times, as originally proposed by Janssen et al. (2002). Traditional cultivation methods are based on media that contain much higher (often up to two to three orders of magnitude) nutrient concentrations—next to different types of carbon—than those in natural environments. It has been hypothesized that media with strongly reduced carbon concentrations, together with long incubation periods, may facilitate the growth of slow-growing (e.g., oligotrophic) rare organisms, as that of fast-growing competing organisms is restricted.
- Addition of oxidative stress-protective agents, as originally proposed by Stevenson et al. (2004). The initial steps of isolation usually encompass separation of bacterial cells from their environmental microhabitat and their introduction in a completely new environment (the cultivation medium). This procedure may damage cell surfaces or yield conditions of oxidative stress that inhibit their subsequent growth in laboratory conditions. The use of oxidation-protective agents may protect cells from damage caused by free radicals.
- Use of alternative solidifying agents, as originally proposed by Kamagata and Tamaki (2005). The most used solidifying agent for bacterial growth media is agar-agar. This solidifying agent is extracted from a group of red-to-purple marine algae, and it has been shown to be toxic for a large number of bacterial species. The use of alternative solidifying agents, such as gellan gum (produced by particular bacterial species), has allowed the isolation of several hitherto-uncultured bacterial groups. Thus, in recent years, members of the Acidobacteria (de Castro et al. 2013, Campanharo et al. 2016), Verrucomicrobia (Rygaard et al. 2017, Zhalnina et al. 2018), and Bacteroidetes (Nunes da Rocha et al. 2015) have been obtained in axenic culture.
- Use of increased CO_2 tension during incubation (Nunes da Rocha et al. 2015). Traditional cultivation techniques use incubations under atmospheric conditions that are optimal for aerobic organisms (e.g., about 18% O_2, 80% N_2, and the atmospheric level of CO_2, i.e., about 0.3%). Microbial habitats in soil often exhibit different atmospheric conditions. For example, CO_2 tensions as high as 5% may occur, in particular microhabitats in soil, whereas in others, O_2 may be (slightly to severely) reduced from 18% to below 1%.
- Use of educated-guess-based differences in media compositions, as exemplified by Rocha et al. (2010) and Zhalnina et al. (2018). Small changes in culture conditions were shown to allow the selection of different subparts of the microbial diversity that is present in soil environments. In detail, multiple media were used with different solidifying agents, carbon sources (levels and types), and incubation times, in the context of high and low CO_2 tensions, as described earlier.

Given these issues, it is recommended that (1) researchers use educated guesses with respect to the physiologies of the organismal types they aim to isolate, and (2) in cases of random (untargeted) isolations, they use several media and incubation conditions with a given sample, in order to cultivate the widest possible diversity from it. Particularly when one is interested in rare microorganisms, one

TABLE 18.1
Nine Factors Hampering Bacterial Growth upon Isolation from Natural Environments and Proposed Approaches to Improve Culturability

Factors that Limit Growth of the Hitherto-Uncultured Bacteria	Proposed Approaches to Improve Culturability	Phylogeny of Novel Isolates That Were Obtained	Environments from Where Novel Isolates Were Obtained	References
1. Inability to grow at high nutrient concentrations and overgrowth by faster growing species	• Reduce nutrient availability in growth medium • Apply a longer incubation time	Actinobacteria, α-proteobacteria, Acidobacteria, Verrucomicrobia, Armatimonadetes, Planctomycetes	• Antarctic soil • Grassland rhizosphere • Aquatic rhizoplane • Freshwater • Water treatment recycling tank	Tamaki et al. (2011), Lage and Bondoso (2012), Pulschen et al. (2017), Zhalnina et al. (2018)
2. Media selectiveness for particular groups of organisms	• Development of media that select for a different or a broader spectrum of microorganism • Cyclic cultivation	β-proteobacteria, Bacteroidetes, Acidobacteria, *Actinobacteria*, *Firmicutes*	• Wastewater • Bulk soil	Dorofeev et al. (2014), Campanharo et al. (2016), Lycus et al. (2017)
3. Compounds inhibiting bacterial growth	• Diffusion or dilution of growth-inhibiting compounds, for example, by making use of diffusion chambers • Application of alternative solidifying agents, for example, gellan gum instead of agar	Acidobacteria, Verrucomicrobia, Bacteroidetes, *Nitrospira*, unclassified Bacteria, *Armatimonadetes*, Gemmatimonadetes	• Bulk soil • Intertidal samples • Wastewater treatment plants • Activated sludge • Nickel-contaminated soil	Aoi et al. (2009), Tamaki et al. (2011), de Castro et al. (2013), Remenár et al. (2015), Pascual et al. (2016)
4. Syntrophic growth requirements for growth factors produced by other microorganisms	• Incubation of environmental samples in diffusion chamber • Addition of signaling molecules • Addition of syntrophic bacteria in enrichment cultures • Adding host extracts • *In situ* colonization	Bacteroidetes, α-proteobacteria, γ-proteobacteria, Verrucomicrobia, Gemmatimonadetes, Chloroflexi, Planctomycetes	• Marine sponge • Seawater • Lichen • Bulk soil • Submarine hot spring • Biofilm community of macroalgae	Lage and Bondoso (2012), Nunoura et al. (2013), Steinert et al. (2014), Biosca et al. (2016), Pascual et al. (2016), Rygaard et al. (2017)
5. Gas tension during incubation	• Increase of CO_2 during incubation • Aeration of headspace • *In situ* colonization	Bacteroidetes, *Caldiserica*, Chloroflexi	• Bulk soil • Hot spring • Submarine hot spring	Mori et al. (2009), Nunoura et al. (2013), Takahashi and Aoyagi (2018)
6. Low abundance in environmental samples	• Single-cell detection combined with microbial cultivation • High-throughput dilution to extinction cultivation • Micro-petri dish	Cyanobacteria, Firmicutes, Actinobacteria, α-proteobacteria	• Bioreactor • Polycyclic aromatic contaminated soil	Ingham et al. (2007), Abalde-Cela et al. (2015), Jiang et al. (2016)

(*Continued*)

TABLE 18.1 (*Continued*)
Nine Factors Hampering Bacterial Growth upon Isolation from Natural Environments and Proposed Approaches to Improve Culturability

Factors that Limit Growth of the Hitherto-Uncultured Bacteria	Proposed Approaches to Improve Culturability	Phylogeny of Novel Isolates That Were Obtained	Environments from Where Novel Isolates Were Obtained	References
7. Formation of colonies undetectable by the unarmed eye	• Colony identification by fluorescent *in situ* hybridization (FISH) in microtiter plates • Colony identification by hybridization on nylon membranes • Microcolony cultivation	Bacteroidetes, Ca. Phylum Saccharibacteria	• Bulk soil • Seawater	Ferrari et al. (2005), Tandogan et al. (2014), Berdy et al. (2017)
8. Restriction of bacterial growth upon the presence of reactive oxygen species	• Reduction of oxidative stress by addition of oxidation-protective agents • Growth under anaerobic conditions	Bacteroidetes, Firmicutes, Chloroflexi	• Biological soil crust • Submarine hot spring	Nunoura et al. (2013), Nunes da Rocha et al. (2015)
9. Lack of ability to grow under specific circumstances, that is, growth on solid or in a liquid substrate	• Screening for isolates in both liquid and in solid media • Cell size contractions	Bacteroidetes	• Seawater	Tandogan et al. (2014)

Source: Adapted and updated from Nunes da Rocha et al. (2009).

should randomly pick and purify a broad range of colonies from the plates. As colony morphologies may be little discriminative, grouping isolates by colony morphology may be difficult, and random colony picking and analysis may be required.

18.3.2 The Use of Different Media and Growth Conditions

Here, we give examples of media and growth conditions that have been used to recover novel bacteria in axenic culture (the novel bacterial taxa recovered by the different strategies presented in Table 18.1 are discussed in Section 18.4). R2A (Greenberg et al. 1985) is a generic isolation medium that has been often used to isolate bacteria from environmental samples. This medium is composed of complex carbon sources and nitrogen sources (proteose peptone, casamino acids, yeast extract and soluble starch, dipotassium phosphate, magnesium sulfate, and sodium pyruvate). In most cases, when preparing this medium, researchers use tap or distilled water, in which cases the medium may still contain minor compounds coming from the water. This phenomenon increases in relevance as more dilute forms of R2A (in the same water) are used. An immense diversity of bacterial species can grow on R2A or tenfold diluted R2A (0.1 × R2A), and novel and rare species have been recovered using the latter medium. To determine if a novel medium or incubation condition is responsible for the recovery of novel species, it is suggested to compare its efficacy of isolation to that of the standard medium.

18.3.2.1 Dilution of Traditional Media

The use of differentially diluted traditional agar media, like R2A, may recover different pools of organisms. For example, 1:20 and 1:100 dilutions of R2A medium were found to incite restriction of the growth rates of renowned fast-growing organisms and concomitant reduction of competition among the emerging colonies (Zhalnina et al. 2018). In two studies, The colony-forming unit counts from soil environments plated on 0.1 × R2A were statistically similar to those observed in 0.05 and 0.01 × R2A (Nunes da Rocha et al. 2015, Zhalnina et al. 2018). Nevertheless, the organisms recovered on 0.1 × R2A were mostly "fast" growers that were affiliated with Actinobacteria, Firmicutes, and (*alpha*, *beta*, and *gamma*) Proteobacteria. Differently, those recovered from 0.05 and 0.01 × R2A many times were affiliated with "difficult-to-recover" bacterial species, such as members of the Acidobacteria, Verrucomicrobia, and Bacteroidetes, next to several slow-growing Proteobacteria (Nunes da Rocha et al. 2015, Zhalnina et al. 2018).

18.3.2.2 Use of Oxidation-Protective Agents

It is hypothesized that any of the sample processing steps—as outlined in Section 18.3.1—may damage cell surfaces physically or chemically, which makes the cells susceptible to the action of free radicals. To circumvent this phenomenon, diverse oxidative-protective agents have been added to culture media. As a case in point, catalase was added to a nutrient-limited medium from which several representatives of Subdivision 1 of the Verrucomicrobia were recovered (Rocha et al. 2010). This bacterial phylum is highly abundant in several different ecosystems (e.g., soil, water, and the human intestinal microbiome). However, only 230 isolates affiliated with Verrucomicrobia have so far been recovered in axenic culture (Figure 18.1), which classifies this bacterial phylum as "generally recalcitrant to cultivation."

18.3.2.3 Addition of Micronutrients and Use of Different Atmospheric Compositions

As different bacteria have different nutritional requirements in order to grow, variation of nutrient level and range in the growth media may be successful to broaden the scope of isolation. A medium coined "VXG" (vitamins-xylan-gellan gum), derived from VL55 (Janssen et al. 2002), was based on low concentrations of xylan (0.5 g L^{-1}), used gellan gum instead of agar, and supplied a broad spectrum of trace elements and vitamins. The use of the trace elements and vitamins was thought to satisfy the specific nutritional requirements of particular members of the Acidobacteria (de Castro et al. 2013) and Verrucomicrobia (Zhalnina et al. 2018). The use of VXG coupled to incubation of the medium in high CO_2 tensions has allowed the recovery of several representatives of the Bacteroidetes (Nunes da Rocha et al. 2015). In a generic sense, several bacterial groups recalcitrant to culturing have been recovered in media supplemented with micronutrients and/or incubated under different atmospheric conditions. Until now, no study attempted to understand which micronutrients (or thereof combination) and how specific atmospheric compositions affected the recovery of microorganisms. A solution might lie in the gas phase to which bacteria are exposed. In a seminal study by Greening et al. (2015), representatives of the genus *Pyrinomonas* (Acidobacteria Group 4) were found to survive and grow under low carbon availability by scavenging picomolar concentrations of hydrogen gas from the atmosphere (Greening et al. 2015). Molecular hydrogen is hypothesized to be—for certain soil bacterial types—a key energy source, if present at adequate concentrations. Thus, this key finding may suggest that a strong emphasis should be placed on specific atmosphere (hydrogen)-related conditions. In this way, potentially other representatives of Acidobacteria or of other cultivation-recalcitrant bacterial phyla may become culturable.

18.4 HOW FAR HAVE WE COME WITH THE NOVEL ISOLATION APPROACHES?

The large-scale use of advanced molecular ecology techniques coupled to bioinformatics analyses has increased the soil bacterial diversity that is now known to man (Hug et al. 2016). As described in the foregoing, in 2016 representatives of over 42 bacterial phyla were yet to be recovered in pure culture (Hug et al. 2016). There is an increased perception of the need to obtain cultures of key environmental bacteria, as only with cultures, the functioning of simple to complex systems, including the interactions therein, can be fully understood. Hence, the isolation of defined or pure cultures is of utmost importance, and it may be guided by the following recommendations based on "genome readings" (Gutleben et al. 2018):

- Test multi-omics-based ecological hypotheses by deciphering microbial physiology.
- Improve and curate database annotations.
- Develop novel applications.

In a broad sense, however, isolation strategies can be divided into untargeted and targeted approaches. Untargeted approaches may be defined as those approaches in which researchers attempt to unveil the largest possible diversity from a given sample, whereas targeted approaches focus on the isolation of a specific group of organisms or a specific organism from that sample. There is a fundamental issue here, namely that the physiological needs of unknown (rare) microbes are per definition mostly unknown. Thus, untargeted approaches have been responsible for the recovery of most novel species of Acidobacteria (Tanaka et al. 2017), Verrucomicrobia (Zhalnina et al. 2018), Chloroflexi (Nunoura et al. 2013), Gemmatimonadetes (Pascual et al. 2016), *Caldiserica* (Mori et al. 2009), Planctomycetes (Lage and Bondoso 2012), and Bacteroidetes (Steinert et al. 2014, Remenár et al. 2015). In contrast, targeted isolations, in cases no genomic information is available (see Section 18.5), are limited to prior knowledge on the physiologies of the targeted bacteria or media that by chance are selective for particular groups. This strategy has been used to recover novel isolates from the Acidobacteria (de Castro et al. 2013, Campanharo et al. 2016) and Planctomycetes (Lage and Bondoso 2012). Furthermore, a targeted isolation strategy was successfully used to recover two representatives of the Nitrospirae from a sand bed that were capable of complete ammonia oxidation (Kessel et al. 2015). These organisms were recovered in enrichment cultures and characterized using a combination of culture-dependent and culture-independent approaches. Similarly, isolation strategies based on enrichment cultures have recovered several Nitrosomonadaceae species (affiliated with the β-proteobacteria) that generally perform the first steps of nitrification, that is, oxidizing ammonia via hydroxylamine to nitrite, controlling the subsequent oxidation (by bacteria) of nitrite to nitrate (Prosser et al. 2014).

Basic physiological traits (such as growth rate, biomass composition, and secondary metabolism) are difficult—or impossible—to define from culture-independent studies (Gutleben et al. 2018). Here, we will examine the Acidobacteria as an example of how axenic cultures are important to understand the ecology and physiology of soil bacteria. This bacterial phylum is among of the most widespread and abundant ones on the planet (Kielak et al. 2016). A first isolate was obtained in 2002 (Janssen et al. 2002), and to date, 230 16S rRNA gene sequences of bacterial isolates are present in the Ribosomal Database Project (release 11, update 5, last visited on July 15, 2018). Cultivation of axenic and mixed cultures, followed by the characterization of genomes, together with analyses of multi-omics data from environmental samples, has shown that the Acidobacteria are as metabolically diverse as Proteobacteria (Kielak et al. 2016). This daunting diversity leaves us with questions with respect to the ultimate aim of culturing all extant soil bacterial diversity. What are the main issues that need to be further addressed to diminish the soil culturability gap related to the *great plate count anomaly*? And, which strategies should be further explored or extended?

18.5 OUTLOOK AND PROSPECTS

It has been estimated that about 10^{12} bacterial species inhabit our planet (Locey and Lennon 2016). We are by far incomplete in our approaches to cover all extant diversity in soil, and it appears to be beyond our current capacities to culture all. A fundamental aspect emerges here and that is the bacterial species concept. Remaining controversial, a debate on the concepts and definitions of bacterial species has been carried out by generations of scientists. In the light of potentially frequent—or evolutionarily relevant—horizontal gene transfer events, bacterial species are thought to exist as clusters of genetic and ecological types (Shapiro et al. 2016). This may mean that with a single population in soil, many divergent "types" exist that may be on their way to further divergence and, ultimately, speciation. How can we deal with this phenomenon? Single-cell analyses offer a step forward, but at the same time pose obvious limits. In any case, it is necessary to increase the throughput (and biomass production for downstream analysis) of our current isolation strategies. However, as we increase our throughput and thus generate diverse bacterial growth, we will increasingly require some "clear-cutting sward" that enables us to draw lines as to what to sample and analyze and what not to do.

Another major issue is the hypothesis that many hitherto-uncultured bacteria resist cultivation because of their dependency on one another or on other (micro)organisms (Rocha et al. 2010). Such interdependency poses problems to cultivation, as it is virtually impossible to mimic all subtle changes that may occur in microbial agglomerates in biofilms and other environments that set the stage for bacterial growth. Novel cultivation approaches (see the following) will have to be developed to take the next steps in this issue. Several strategies have been proposed to deal with the interdependency and other issues (see also Table 18.1). Attempts to achieve high-throughput isolation techniques have developed micro-Petri dishes (Ingham et al. 2007), microdroplet platforms (Abalde-Cela et al. 2015), and isolation of microorganisms in submicrometer constrictions (Tandogan et al. 2014). The necessity of (as-yet-unknown nutrients and growth factors) produced by other organism has been studied by the use of diffusion chambers (Steinert et al. 2014, Remenár et al. 2015, Pascual et al. 2016, Berdy et al. 2017). Isolation approaches based on diffusion chambers have been responsible for the recovery of several α-proteobacteria (Steinert et al. 2014), β-proteobacteria (Remenár et al. 2015), γ-proteobacteria (Steinert et al. 2014, Remenár et al. 2015), Bacteroidetes (Steinert et al. 2014), and extremely rare Gemmatimonadetes (Pascual et al. 2016). Nevertheless, three main drawbacks have been highlighted to this approach (Berdy et al. 2017). First, some bacterial species require not only growth factors but also require close proximity to their synergistic partners. This is the case for, for example, hydrogen-producing bacteria and methanogenic archaea (Berdy et al. 2017). The thermodynamics of their synergy would be compromised if they were placed at different sides of a membrane. Second, diffusion chamber membranes are often susceptible to drying out, requiring high moisture contents in the environment for the success of the isolation. Third, the high moisture contents required for the procedure may cause anoxic conditions below the diffusion chambers, creating unnatural growth conditions in these. As the limitations of the techniques have been well studied, diffusion chambers offer a large number of opportunities to recover novel hitherto-uncultured bacteria.

Moreover, attempts to mimic the constantly changing conditions that spur microbial growth have used two strategies. The first strategy has been named "cyclic culture," where cultivation under cyclically varying conditions comparable to the duration of the cell cycle is used (Dorofeev et al. 2014). The second strategy uses programmable and automated micro-chemostats to vary the growth of single cells to small aggregates (Kim et al. 2017). More studies that merge culture-dependent with advanced culture-independent approaches should be developed. The combination of these approaches will shed light as to which ecological niches, putative roles, functions, and modes of interactions are present in the yet-to-be cultured bacteria.

REFERENCES

Abalde-Cela, S., Gould, A., Liu, X., Kazamia, E., Smith, A.G. and C. Abell. 2015. High-throughput detection of ethanol-producing cyanobacteria in a microdroplet platform. *J R Soc Interface* 12, 20150216.

Aoi, Y., Kinoshita, T., Hata, T., Ohta, H., Obokata, H. and S. Tsuneda. 2009. Hollow-fiber membrane chamber as a device for in situ environmental cultivation. *Appl Environ Microbiol* 75, 3826–3833.

Berdy, B., Spoering, A.L., Ling, L.L. and S.S. Epstein. 2017. *In situ* cultivation of previously uncultivable microorganisms using the Ichip. *Nat Protoc* 12, 2232–2242.

Biosca, E.G., Flores, R., Santander, R.D., Díez-Gil, J.L. and E. Barreno. 2016. Innovative approaches using lichen enriched media to improve isolation and culturability of lichen associated bacteria. *PLoS One* 11, e0160328.

Bruns, A., Nübel, U., Cypionka, H. and J. Overmann. 2003. Effect of signal compounds and incubation conditions on the culturability of freshwater bacterioplankton. *Appl Environ Microbiol* 69, 1980–1989.

Campanharo, J.C., Kielak, A.M., Castellane, T.C.L., Kuramae, E.E. and E.G.M. Lemos. 2016. Optimized medium culture for Acidobacteria Subdivision 1 strains. *FEMS Microbiol Lett* 363.

Carini, P., Marsden, P.J., Leff, J.W., Morgan, E.E., Strickland, M.S. and N. Fierer. 2017. Relic DNA is abundant in soil and obscures estimates of soil microbial diversity. *Nat Microbiol* 2, 16242.

de Castro, V.H.L., Schroeder, L.F., Quirino, B.F., Kruger, R.H. and C.C. Barreto. 2013. Acidobacteria from oligotrophic soil from the Cerrado can grow in a wide range of carbon source concentrations. *Can J Microbiol* 59, 746–753.

Chen, Y.-C., Cheng, Y.H., Kim, H.S., Ingram, P.N., Nor, J.E. and E. Yoon. 2014. Paired single cell co-culture microenvironment isolated by two-phase flow with continuous nutrient renewal. *Lab Chip* 14, 2941–2947.

Cole, J.R., Wang, Q., Fish, J.A. et al. 2014. Ribosomal database project: Data and tools for high throughput RRNA analysis. *Nucleic Acids Res* 42 (Database issue), D633–D642.

Dorofeev, A.G., Grigor'eva, N.V., Kozlov, M.N., Kevbrina, M.V., Aseeva, V.G. and Y.A. Nikolaev. 2014. Approaches to cultivation of 'nonculturable' bacteria: Cyclic cultures. *Microbiology* 83, 450–461.

Ettwig, K.F., Seigo, S., Katinka, T. et al. 2008. Denitrifying bacteria anaerobically oxidize methane in the absence of archaea. *Environ Microbiol* 10, 3164–3173.

Ferrari, B.C., Binnerup, S.J. and M. Gillings. 2005. Microcolony cultivation on a soil substrate membrane system selects for previously uncultured soil bacteria. *Appl Environ Microbiol* 71, 8714–8720.

Greenberg, A.E., Clesceri, L.S. and A.D. Eaton. 1985. *Standard Methods for the Examination of Water and Wastewater*, 16th ed., American Public Health Association: Washington, DC.

Greening, C., Carere, C.R., Rushton-Green, R. et al. 2015. Persistance of the dominant soil phylum Acidobacteria by trace gas scavenging. *Proc Natl Acad Sci USA* 112, 10497–10502.

Gutleben, J., De Mares, M.C., van Elsas, J.F., Smidt, H., Overmann, J. and D. Sipkema. 2018. The multi-omics promise in context: From sequence to microbial isolate. *Crit Rev Microbiol* 44, 212–229.

Hug, L.A., Baker, B.J., Anantharaman, K. et al. 2016. A new view of the tree of life. *Nat Microbiol* 1 (5), 16048.

Ingham, C.J., Sprenkels, A., Bomer, J. et al. 2007. The micro-petri dish, a million-well growth chip for the culture and high-throughput screening of microorganisms. *Proc Nat Acad Sci USA* 104, 18217–18222.

Janssen, P.H., Yates, P.S., Grinton, B.E., Taylor, P.M. and M. Sait. 2002. Improved culturability of soil bacteria and isolation in pure culture of novel members of the divisions acidobacteria, actinobacteria, proteobacteria, and verrucomicrobia. *Appl Environ Microbiol* 68, 2391–2396.

Jiang, C.-Y., Dong, L., Zhao, J.-K. et al. 2016. High-throughput single-cell cultivation on microfluidic streak plates. *Appl Environ Microbiol* 82, 2210–2218.

Kamagata, Y. and H. Tamaki. 2005. Cultivation of uncultured fastidious microbes. *Microbes Environ* 20, 85–91.

Kessel, M.A.H.J., Speth, D.R., Albertsen, M. et al. 2015. Complete nitrification by a single microorganism. *Nature* 528, 555–559.

Kielak, A.M., Barreto, C.C., Kowalchuk, G.A., Van Veen, J.A., and E.E. Kuramae. 2016. The ecology of *Acidobacteria*: Moving beyond genes and genomes. *Front Microbiol* 7. doi:10.3389/fmicb.2016.00744.

Kim, M., Bae, J. and T. Kim. 2017. Long-term and programmable bacterial subculture in completely automated microchemostats. *Anal Chem* 89, 9676–9684.

Lage, O.M. and J. Bondoso. 2012. Bringing *Planctomycetes* into pure culture. *Front Microbiol* 3. doi:10.3389/fmicb.2012.00405.

Locey, K.J. and J.T. Lennon. 2016. Scaling laws predict global microbial diversity. *Proc Nat Acad Sci USA*, 113, 5970–5975.

Lycus, P., Bøthun, K.L., Bergaust, L. et al. 2017. Phenotypic and genotypic richness of denitrifiers revealed by a novel isolation strategy. *ISME J* 11, 2219–2232.

Mori, K., Yamaguchi, K., Sakiyama, Y. et al. 2009. Caldisericum Exile Gen. Nov., Sp. Nov., an anaerobic, thermophilic, filamentous bacterium of a novel bacterial phylum, Caldiserica Phyl. Nov., Originally Called the Candidate Phylum OP5, and Description of Caldisericaceae Fam. Nov., Caldisericales Ord. Nov. and Caldisericia Classis Nov. *Int J Syst Evol Microbiol* 59, 2894–2898.

Nichols, D., Lewis, K., Orjala, J. et al. 2008. Short peptide induces an 'uncultivable' microorganism to grow in vitro. *Appl Environ Microbiol* 74, 4889–4897.

Nunes da Rocha, U., Cadillo-Quiroz, H., Karaoz, U. et al. 2015. Isolation of a significant fraction of non-phototroph diversity from a desert biological soil crust. *Front Microbiol* 6, 277.

Nunes da Rocha, U., Van Overbeek, L., and J.D. van Elsas. 2009. Exploration of hitherto-uncultured bacteria from the rhizosphere. *FEMS Microbiol Ecol* 69, 313–328.

Nunoura, T., Hirai, M., Miyazaki, M. et al. 2013. Isolation and characterization of a thermophilic, obligately anaerobic and heterotrophic marine *Chloroflexi* bacterium from a *Chloroflexi*-dominated microbial community associated with a Japanese shallow hydrothermal system, and proposal for *Thermomarinilinea Lacunofontalis* Gen. Nov., Sp. Nov. *Microbes Environ* 28, 228–235.

Pascual, J., García-López, M., Bills, G.F. and O. Genilloud. 2016. *Longimicrobium terrae* Gen. Nov., Sp. Nov., an oligotrophic bacterium of the under-represented Phylum *Gemmatimonadetes* isolated through a system of miniaturized diffusion chambers. *Int J Syst Evol Microbiol* 66, 1976–1985.

Prosser, J.I., Head, I.M., and L.Y. Stein. 2014. The family Nitrosomonadaceae. In *The Prokaryotes: Alphaproteobacteria and Betaproteobacteria*, 901–918.

Pulschen, A.A., Bendia, A.G., Fricker, A.D., Pellizari, V.H., Galante, D. and F. Rodrigues. 2017. Isolation of uncultured bacteria from Antarctica using long incubation periods and low nutritional media. *Front Microbiol* 8, 1346.

Razumov, A.S. 1932. Direct count of bacteria in water. Its comparison with the Koch method. *Microbiology* 1, 131–146.

Remenár, M., Karelová, E., Harichová, J., Zámocký, M., Kamlárová, A. and P. Ferianc. 2015. Isolation of previously uncultivable bacteria from a nickel contaminated soil using a diffusion-chamber-based approach. *Appl Soil Ecol* 95, 115–127.

Rocha, U.N., Andreote, F.D., Azevedo, J.L., van Elsas, J.D. and L.S. van Overbeek. 2010. Cultivation of hitherto-uncultured bacteria belonging to *Verrucomicrobia* Subdivision 1 from the potato (*Solanum Tuberosum* L.) rhizosphere. *J Soils Sediments* 10, 326–339.

Rygaard, A.M., Thøgersen, M.S., Nielsen, K.F., Gram, L. and M. Bentzon-Tilia. 2017. Effects of gelling agent and extracellular signaling molecules on the culturability of marine bacteria. *Appl Environ Microbiol* 83, e00243-17.

Shapiro, B.J., Leducq, J.B. and J. Mallet. 2016. What Is speciation? *PLoS Genet* 12, e1005860.

Solden, L., Lloyd, K. and K. Wrighton. 2016. The bright side of microbial dark matter: Lessons learned from the uncultivated majority. *Curr Opin Microbiol* 31, 217–226.

Staley, J.T. and A. Konopka. 1985. Measurement of in situ activities of nonphotosynthetic microorganisms in aquatic and terrestrial habitats. *Ann Rev Microbiol* 39, 321–346.

Steinert, G., Whitfield, S., Taylor, M.W., Thoms, C. and P.J. Schupp. 2014. Application of diffusion growth chambers for the cultivtion of marine sponge-associated bacteria. *Mar Biotechnol* 16, 594–603.

Stevenson, B.S., Eichorst, S.A., Wertz, J.T., Schmidt, T.M. and J.A. Breznak. 2004. New strategies for cultivation and detection of previously uncultured microbes. *Appl Environ Microbiol* 70, 4748–4755.

Takahashi, M. and H. Aoyagi. 2018. Effect of intermittent opening of breathable culture plugs and aeration of headspace on the structure of microbial communities in shake-flask culture. *J Biosci Bioeng* 126, 96–101.

Tamaki, H., Tanaka, Y., Matsuzawa, H. et al. 2011. *Armatimonas rosea* Gen. Nov., Sp. Nov., of a novel bacterial phylum, *Armatimonadetes* Phyl. Nov., formally called the Candidate phylum OP10. *Int J Syst Evol Microbiol* 61, 1442–1447.

Tanaka, Y., Matsuzawa, H., Tamaki, H. et al. 2017. Isolation of novel bacteria including rarely cultivated phyla, *Acidobacteria* and *Verrucomicrobia*, from the roots of emergent plants by simple culturing method. *Microbes Environ* 32, 288–292.

Tandogan, N., Abadian, P.N., Epstein, S., Aoi, Y. and E.D. Goluch. 2014. Isolation of microorganisms using sub-micrometer constrictions. *PLoS One* 9, e101429.

Zhalnina, K., Louie, K.B., Hao, Z. et al. 2018. Dynamic root exudate chemistry and microbial substrate preferences drive patterns in rhizosphere microbial community assembly. *Nat Microbiol* 3, 470–480.

19 Statistical Analyses of Microbiological and Environmental Data

*Alexander V. Semenov**
Incotec Group BV

CONTENTS

19.1 Introduction 307
19.2 Quantifying Natural Phenomena 308
  19.2.1 Variables and Measurement Scales 308
  19.2.2 Descriptive Statistics 310
  19.2.3 Graphical Descriptions 313
19.3 Hypothesis Testing 314
  19.3.1 Distributions of Sample Data and Populations 317
  19.3.2 Null and Alternative Hypotheses 317
  19.3.3 Parametric and Nonparametric Analyses 318
19.4 Multivariate Statistics, Classification, and Hypothesis Generation 320
  19.4.1 Ordination 321
  19.4.2 Cluster Analysis 322
  19.4.3 Phylogenetics and Parsimony Analyses 323
  19.4.4 Substantiating a Classification 324
19.5 Modern Machine Learning Approaches 325
19.6 Conclusions 326
References 327

19.1 INTRODUCTION

The complexity/randomness of soil systems poses a formidable challenge to the design, analysis, and interpretation of microbiological experiments. Natural variation in the abundances and activities of microorganisms, and the metabolites and enzymes they produce, can be tremendous even under the most controlled conditions. The temporal and spatial scales over which control can be exerted are often much larger than those within which microbes exist. For example, the abundance of specific bacterial populations may vary by over several orders of magnitude depending on the availability of the resources that are heterogeneously distributed in the soil. Such a difference in scale requires scientists to use assays of ever-increasing sensitivity and throughput in order to capture the patterns of natural microbial phenomena and their influences on "macrobiological" systems. One hallmark of modern soil microbiological methods is the tremendous amount of data these generate. Whether it is DNA or RNA sequence data or the amount of carbon present in a soil sample, high-throughput analytical methods are expanding

* Modified from: McSpadden-Gardener, Modern Soil Microbiology II, 2007.

our database on soil microbes and their surroundings. To make sense of this ever-increasing amount of information, numerous statistical tools have been developed and applied. In this chapter, we review the classical quantitative methods and their application in soil science as well as advanced highly promising approaches such as machine learning and variations of artificial intelligence (e.g., deep learning). More thorough descriptions of the mathematics underlying the various statistical procedures described here are provided elsewhere, several of which are referenced. Here, key concepts and definitions related to the quantitative analyses of data will be presented. Additionally, specific examples will be used to illustrate the type of statistics that can be used to characterize microbes, their activities, and their impacts on the environment.

19.2 QUANTIFYING NATURAL PHENOMENA

The scientific method formalizes and structures experiential learning. In soil microbiology, observations of the natural environment lead to questions that focus on the elements of, and interactions among, microbes and their environment (Yates et al. 2016). Such questions require that measurements are taken on one or more components of the experimental system. Recent texts describe a range of experimental methods (as discussed in Chapters 12–18), which are useful to the soil microbiologist, all of which yield numerical data on soil systems that must be analyzed. The types of data obtained by an investigator will determine the types of analyses that can be properly performed. Initially, graphs and descriptive statistics can be used to summarize large amounts of data. Examinations of these summaries can lead to refinements of experimental design that may improve the likelihood of making meaningful descriptions and interpretations from the data obtained.

A major goal of any scientific investigation is to produce an accurate description and understanding of natural phenomena. To develop that description, we begin by gathering data. The data are obtained by categorizing and/or measuring one or more sets of subjects that are thought to represent those subjects in nature. Because it is often difficult to observe patterns in raw data sets, two ways of summarizing the information are used. First, one can make numerical summaries of the data, especially of the central tendency, degree of variation, and relationships between data sets. These *descriptive statistics* can then be used to make inferences about the populations from which the samples were taken. Second, statistical methods, including machine learning algorithms, can be used to find natural patterns in data that generate insight and help to better understand the behavior of complex systems such as soil microbiomes.

19.2.1 Variables and Measurement Scales

A variable is any feature of an experimental subject that can be measured. For example, it may be the number of cells recovered from a soil core, the amount of metabolite produced in a test tube, or the rate of metabolic activity observed under some set of natural conditions. In any experimental system, there are many variables that can be measured, but only a few are typically examined in detail. Which variables are to be measured depends on the question being asked and the resources available to make measurements.

Measurements are almost always made on samples. A sample is a representative portion of the whole population or environment about which we would like to make an inference. Chapter 3 further discusses the intricacies of sampling of soil. A sample needs to be selected or prepared at random from the larger population which it is supposed to represent. While this is rarely possible in an absolute sense, efforts should be made to draw one's experimental sample at random with regard to the phenomenon being studied. Under rare circumstances, one may study a whole population, but this is only the case when all members can be isolated and measured. More commonly, questions in soil microbiology focus on making inferences about populations or environments that are much larger than can be practically measured. For example, scientists will define the types and numbers of soil cores, cell cultures, or nutrients under the environmental conditions to be studied. The methods they use to measure a characteristic of their sample will constrain the scale on which that variable is measured.

Measurement scales may be *cardinal*, *ordinal*, or *nominal*. These may also be referred to as interval/ratio, rank, and categorical, respectively. Cardinal scales have evenly defined intervals. For example, the temperature scale is evenly defined in intervals called degrees. In contrast, ordinal scales are numerically ranked but may differ in amount between different levels. Such scales extend across a predetermined number of levels (N) from least (e.g., 0 or 1) to greatest (N). Nominal scales are simply qualitative descriptions of sample subsets, such as the names of microbial species or the identity of DNA sequences. Binary data are a type of nominal data for which there are only two categories, such as the presence or absence of a specific trait. Most biologists are comfortable with cardinal and nominal scales, because—most commonly—experimental variables related to identity, mass, energy, and motion are measured on these scales. However, it is not always possible to obtain cardinal measurements.

Nonlinear responses in the isolation and/or detection of experimental subjects prevent cardinal scales from being fully applied. These types of responses occur at the limits of detection of measuring devices and at those points when new properties emerge in an experimental system. A classic example of this is the measurement of bacterial cell density in a spectrophotometer. At intermediate cell densities (e.g., log 5 to log 7 cells per mL), light absorbance is proportional to cell concentration. However, at low cell densities, most spectrophotometers are not sensitive enough to detect changes in light absorbance. Conversely, at high cell densities, linear increases in light absorbance are not proportional to cell density because time- and density-dependent changes in bacterial physiology result in altered chemistry of the bacterial capsule and culture medium. Such nonlinear responses are actually quite common in the realm of modern soil microbiology and associated disciplines. They are widely known to occur in the recovery and amplification efficiency of DNA from different soils. Real-time polymerase chain reaction (PCR) techniques often involve calibrations with known standards, but such standards are not considered to be absolute if reaction conditions vary with the experimental conditions used. And, while not yet systematically investigated, it is reasonable to assume that the hybridization efficiency of fluorescent probes will not always follow simple kinetics, particularly *in situ*. In related disciplines, ordinal scales are already commonly used. For example, in plant pathology, disease ratings are typically defined with the lowest level (e.g., 0) indicating no signs of disease, the highest level (e.g., 5) indicating plant death after a specified amount of time, and the intervening levels defined by the amount of damage observed. Since disease progress is not linear in time and the relationships between visual measures of disease severity and plant health are complex, an ordinal scale is most appropriate. That is to say, the intervals of disease severity between levels 1 and 2 are not necessarily the same as the intervals between 3 and 4, and 4 and 5. Thus, in analyzing data from measurements involving complex reactions or interactions, it will most often be appropriate to use ordinal scales and their associated nonparametric statistics (see Section 19.2.2).

Variables can be measured at multiple levels. The choice of scale and level of measurement one undertakes depends on the scientific question, the methods used to acquire the data, and the amount of resources available to pursue the research. Generally, cardinal data are more informative than ordinal ones, which are more informative than nominal ones. And, coarse scales (i.e., those with few measured intervals) lump subjects together, and fine scales (i.e., those with many possible levels) differentiate subjects. Measurement scales may also be described as *continuous* (e.g., temperature, pH) or *discrete*, that is, with intervals being limited to a fixed, definable unit (e.g., cell numbers). The scales at which variables are measured determine which descriptive statistics and which inferential statistical procedures can be appropriately applied. Numerical and graphical approaches can be used to describe a sample data set. The numerical approach involves calculating descriptive statistics, and the graphical approach involves plotting data or summaries thereof. Details of these two approaches are given in Sections 19.2.2 and 19.2.3.

19.2.2 Descriptive Statistics

Tables 19.1–19.3 and Figure 19.1 are relevant to the topic of descriptive statistics. In common parlance, the terms statistic and statistics refer interchangeably to numerical descriptions (described here) and their analyses (described in Sections 19.2 and 19.3). However, in order to understand the natural world, it is important to separate the two concepts. Mathematically, a statistic is simply a number. That number may be a direct measurement or a numerical summary of multiple measurements. Because all variables worth measuring fluctuate in time and/or space, modern soil microbiologists must make multiple measurements to describe and interpret natural phenomena. Descriptive statistics is used to summarize the numerical information obtained from multiple measurements made on a sample or set of samples. These summaries may describe different numerical aspects of a data set, including *central tendency*, *dispersion*, *association*, and *diversity*.

The most commonly reported descriptive statistics approach focuses on summarizing the distribution of sample values (Table 19.1). Common measures of central tendency are the sample *mean* (i.e., arithmetic average) and *median* (the middlemost value), though in large data sets, it is sometimes useful to describe the *modes* (most common values). Dispersion, also called data scatter,

TABLE 19.1
Commonly Used Sample Statistics Used as Parameter Estimates

Characteristic	Measure	Abbreviation[a]	References
Central tendency	Mean	$\bar{x}$	
	Median	med	Fry et al. (1993)
	Mode	—	Fry et al. (1993)
Dispersion	Range	—	Fry et al. (1993)
	Interquartile range	Q1–Q3	Fry et al. (1993)
	Sample variance	s^2	Fry et al. (1993)
	Standard deviation	s	Fry et al. (1993)
	Standard error of the mean	SEM	Fry et al. (1993)

[a] Note that abbreviation conventions can vary by text.

TABLE 19.2
Commonly Used Descriptive Statistics of Bivariate and Multivariate Associations

Type	Descriptor	Abbreviation[a]	Reference
Similarity	Pearson's normal correlation	r	Sheskin (2011)
	Kendall's rank correlation	τ	Sheskin (2011)
	Jaccard	S_J	Fry et al. (1993)
	Simple matching	S_{SM}	Fry et al. (1993)
	Sorenson's	S_S	Fry et al. (1993)
	Gower's	S_G	Fry et al. (1993)
Dissimilarity	Euclidean distance	d_{AB}	Fry et al. (1993)
	Manhattan distance	D_{AB}	Fry et al. (1993)
	Mahalanobis distance	D^2	Ellison (2001)
	Percentage dissimilarity	PD_{AB}	Ellison (2001)
Regression	Linear slope	a	Sheskin (2011)
	Intercept	b	Sheskin (2011)

[a] Note that abbreviation conventions can vary by text.

TABLE 19.3
Several Useful Diversity Statistics

Type	Descriptor	Abbreviation[a]	Reference
Richness	Number of defined groups	N_0, g	McGarigal et al. (2000)
	MRR-like (median rank regression) estimators	S_{Chao1}, S_{ACE}	Grunwald et al. (2003)
Evenness	Relative distribution of all groups	E_1, E_5	McGarigal et al. (2000)
Diversity	Shannon's index	H'	McGarigal et al. 2000)
	Hill's index	N_1	McGarigal et al. (2000)
	Stoddart and Taylor's index	G, N_2	McGarigal et al. (2000)
	Total genetic variation	θ	Martin et al. (2002)

[a] Note that abbreviation conventions can vary by text.

can be described in many different ways. Students are often introduced to the *standard deviation*, *variance*, and sums of squares in introductory courses on statistics. All of these statistics are mathematical measures related to the difference of sample values from the sample mean. They assume a normal distribution of possible sample values in a population. However, other measures of dispersion are more generally useful in describing data scatter. The *range* describes the difference between the minimum and maximum values in a data set. *Quantiles*, such as percentiles, deciles, and quartiles, are the values represented by a fixed ranking among the data. These extend from the 1st percentile, through the 50th percentile (which is equal to the median), and upwards to the 100th percentile (which is the maximum value). Analogous to the full range, interquartile ranges define the difference between the 75th percentile and 25th percentile values.

When two or more variables are measured, it is possible to describe the degree of resemblance that they have (Table 19.2). The most commonly used resemblance measures are correlation statistics. Their calculated values range from −1 to +1, with the magnitude indicating the relative strength of the association between or among variables. The *bivariate correlation coefficient*, *r*, first defined by Pearson, describes the degree to which the sample data conform to a simple linear relationship between two or more variables measured on a continuous cardinal scale. However, if measurements are made on ordinal or nominal scales, other descriptive statistics are more appropriate (Sheskin 2011).

It is not appropriate to use the abovementioned correlation statistics when the relationship between variables does not appear to follow a simple relationship, that is, a straight line for cardinal data sets, or a continual rise or fall for ordinal data sets. In such cases, *linear regression* is more appropriate. Different linear regression models can be used to more precisely define the relationships between two or more variables. These are all derived from the general model:

$$y = b + a_1x_1 + a_2x_2 + a_3x_3 + \cdots + a_nx_n,$$

where y is the value of the response variable, x is the value of the n different predictor variables, a is the coefficient for each of the n predictors, and b is the intercept.

There are no limits to the degree of mathematical sophistication that can be applied to fitting a data set. For example, the coefficients may each be equations in and of themselves, the predictor variables may be elevated to different powers, or there may be multiple response variables. However, the simple goal of defining such models is to mathematically predict the value of one or more response variables from one or more predictor variables. The values of the coefficients and the order of the predictor variables derived by fitting the data to different models are, like the bivariate measures noted previously, simplified numerical summaries of the data and should not be confused with the subsequent evaluations relative to a statistical hypothesis (see Section 19.3).

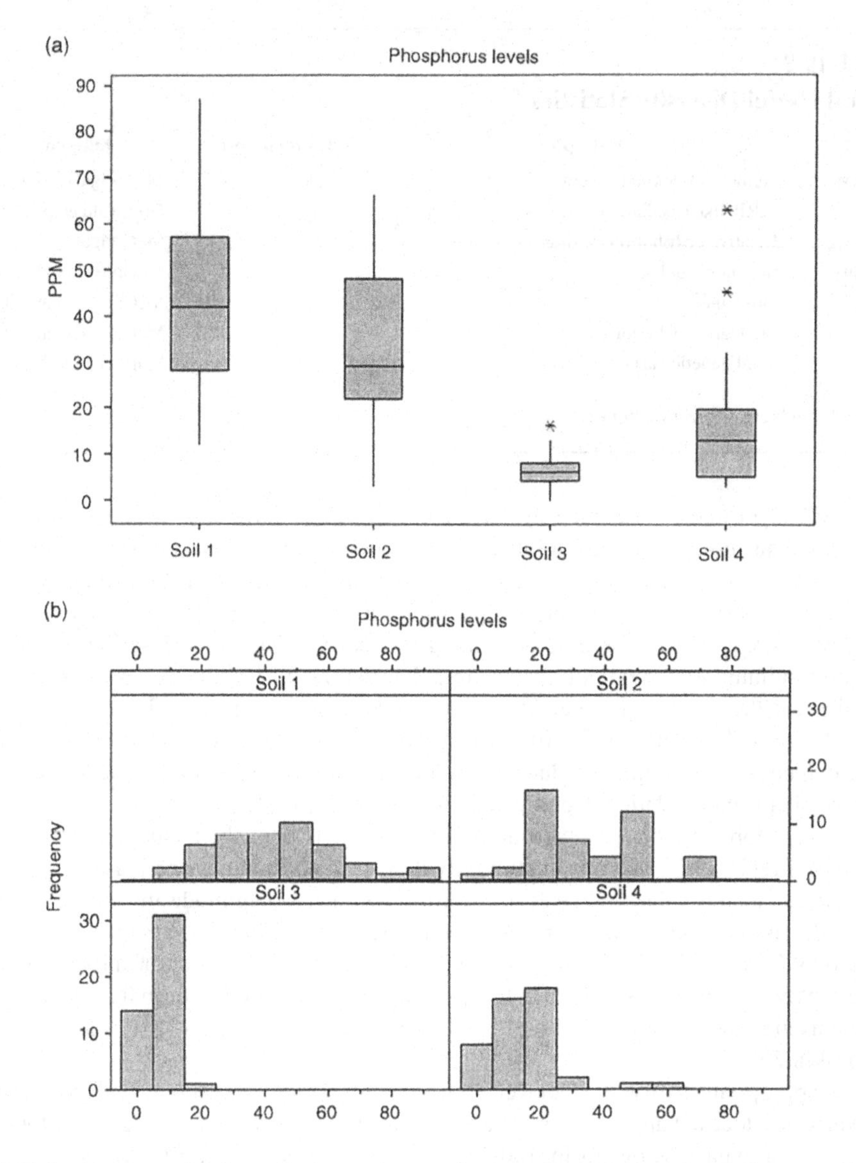

FIGURE 19.1 Graphs useful for preliminary characterization of sample data. In this instance, the amount of phosphorus present in multiple soil cores taken from four different soils is represented. (a) Box plots provide information on the median (center line), interquartile range (circumscribed by the upper and lower box boundaries), and outlying data points (indicated by lines and/or asterisks beyond the box boundaries). (b) Histograms provide information on the distribution of sample values, with sample values plotted on the *x*-axis, and the frequency of observations of each sample value is plotted on the *y*-axis. Given sufficient sample size, histograms indicate the distribution of values in the populations from which the samples were drawn. Here, the sample values obtained from Soils 3 and 4 appear to have skewed, non-normal distributions. Data were plotted using Minitab release 14.0.

When large amounts of data are available, machine learning approaches typically perform better in relating inputs to outputs of interest. This is also the case for soil microbiology, where the underlying processes are often too complex to model via classical statistical models. One of the major advantages of using machine learning approaches for biologists is the opportunity to discover patterns while simultaneously looking at a combination of factors and their interactions, instead of analyzing the influence of each variable individually. This has been a major limitation in the past because the high dimensionality of the data sets (especially, in the era of next-generation sequencing) makes these difficult to analyze through standard statistical models.

Another way to deal with sample data is to mathematically describe the diversity of subjects obtained in a study. Again, there are multiple measures that can be used to summarize the abundance and relative distribution of those subjects (Table 19.3). The first of these involves counting the number of taxonomically distinct species (*richness*). Second, the relative abundance of each species within a sample can be estimated using different measures of *evenness*. Mathematical combinations of these two measures, that is, total number of species and relative abundance, are captured under the denominator diversity. To describe it, Shannon's index has been commonly used, particularly in the older literature on soil microbes. However, analyses have indicated the limitations of such older indices and pointed towards more useful estimators of microbial diversity (Grunwald et al. 2003, Hughes et al. 2001, Martin 2002). Some of these are particularly useful in community analyses, because samples of microbial populations are typically quite small relative to the total diversity present. One measure of genetic diversity, θ, is also used for calculating an index of population structure across different spatial scales called F_{ST}. The interpretation of these descriptive statistics is complicated by the fact that both the term "species" and the context in which they are being observed must be precisely defined. Most often, the subjects are specific DNA sequences, and inferences about the organisms are made from those sequences. Differences in nucleic acid recovery rates from different microbes (e.g., bacteria and fungi) can influence such diversity estimates. Given these challenges, diversity measures describing soil microbial populations and their metabolites should be considered as relative (not absolute) measures and are, perhaps, best compared across studies on ordinal scales.

19.2.3 Graphical Descriptions

Graphs are used to visually summarize complex data sets using shapes and colors to represent quantities and lines and lettering to emphasize patterns (Unwin 2015) (see Figure 19.1). The types of graphs that are commonly used have evolved with the development of new techniques and software. However, in microbiological studies, several graphical techniques have been used for over a century and retain their usefulness even in an era of rapidly evolving technologies. *Scatter plots* show data points or measures of central tendency with associated measures of data scatter, for example, the standard deviation. *Bar graphs* use elevation to describe a central measure and may also be accompanied by error bars or lettering to indicate degrees of variation. In some instances, bar graphs are used to indicate the presence or absence of particular molecules, as in high-performance liquid chromatography and mass spectroscopy data. In *line graphs*, points are connected to indicate trends in *y*-axis values. Data from gels, such as those containing multiple bands of DNA, are now commonly converted into line graphs in order to readily visualize the relative abundance of different DNA products in different samples. *Box plots* display the median surrounded by a rectangle generally encompassing the interquartile range. Additional data may be indicated as "whiskers" to an extended range or distinct outlying data points. *Histograms* are special bar graphs indicating the distribution of sampled values for a given variable. Box plots and histograms are particularly useful for preliminary assessments of sample data sets, especially when the number of measurements for a single variable is large (i.e., >30) (Figure 19.1). This is because they allow for a rapid evaluation of the distribution of the data and convenient comparisons across samples. Such exploratory examinations can be particularly useful for evaluating which statistical tests might be appropriate and predicting the likelihood that observed differences will be both statistically and phenomenologically significant.

More recently, complex graphical representations have emerged as a response to the tremendous amount of data generated by the new microbiological methods. Often, these large data sets include the measurements of multiple variables across multiple cases. To summarize such complex data sets, a variety of shapes and forms may be used, but sequence alignments, ordination plots, dendrograms, and concentric circle plots are most frequently encountered in the microbiology literature. Sequence alignments share characteristics with graphs, especially when symbols are used to

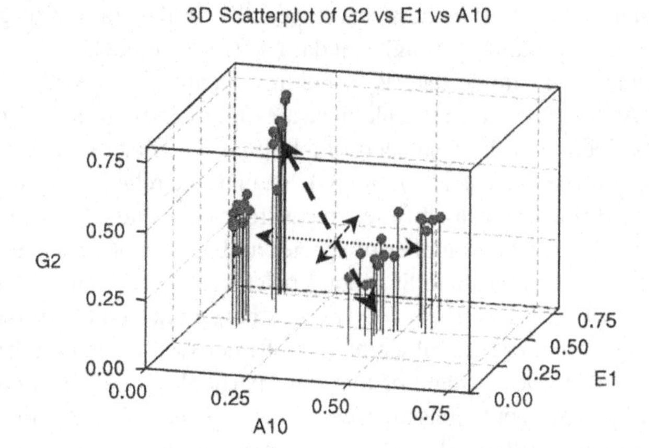

FIGURE 19.2 Mathematical construction of multivariate space and the basis for ordination analyses. Each axis in multivariate space represents a scale upon which a single variable is measured (solid lines). In this instance, the three axes represent the bacterial growth, measured as light absorbance units, on three different carbon sources in wells A10, G2, and El of a BIOLOG assay plate. Ordination axes (dashed lines) are drawn across data points in the directions of greatest data scatter. Mathematical descriptions of multivariate data sets with *N* variables, and described by an equal number of principal components, are mathematical extrapolations of these diagrams in *N*-dimensional space. Data were plotted using Minitab release 14.0.

indicate sequence identity or similarity. Most striking is the use of dots to indicate identity so that sequence differences are made more visible. Ordination plots are simply scatter plots of multivariate data (e.g., Figure 19.2) on axes defined by various complex algorithms. The scales of the different axes are all relative to thc data set and the algorithms used for plotting, and they are rarely defined in detail. Dendrograms (e.g., Figure 19.3) are line plots that connect subjects, such as sequences or organisms, in such a way as to indicate various types of relationships. In general, the shorter the lines between subjects, the more closely related they are. However, as with the ordination plots, the relative distances are reflections of complex algorithms, and they are only meaningfully interpreted within the context of a specific data set (see Section 19.4 for more information). Concentric circle graphs are now widely used to describe genome contents and compare the genomes of different organisms. Different concentric circles indicate unique or homologous sequences (both predicted and defined), as well as their positions and orientations relative to a defined standard. Furthermore, microarray data have blurred the line between primary data presentation and graphical representation by virtue of the use of color enhancements.

The purpose of graphs and tables is to succinctly communicate the salient features of a data set. Common errors are to oversimplify (e.g., by not including a measure of data scatter), overcomplicate (e.g., inclusion of both raw data and summary points or the inclusion of too many categories), or minimize the contrast between data sets that are predicted to differ. Such errors can obscure the conclusions drawn by the investigator and should be avoided. It is also important to remember that, regardless of whether they are presented numerically or graphically, descriptive statistics refer only to the measurements made on the sample, and only by inference do they apply to populations from which the samples were taken. Simply put, sample statistics can be calculated, but population parameters can only be estimated. Thus, if a sample is not representative of the whole population, inferences based on sample statistics will not be sound.

19.3 HYPOTHESIS TESTING

Descriptions of data lead to questions concerning comparisons between and among sample groups. Simple comparisons of descriptive measures are not sufficient to establish experimental observations as "facts." Because all experiments are performed on samples of the larger environment, inferences about the experimental systems are limited by the methods used to obtain data from those samples and the degree of replication performed. Investigators need to establish a level of confidence in the

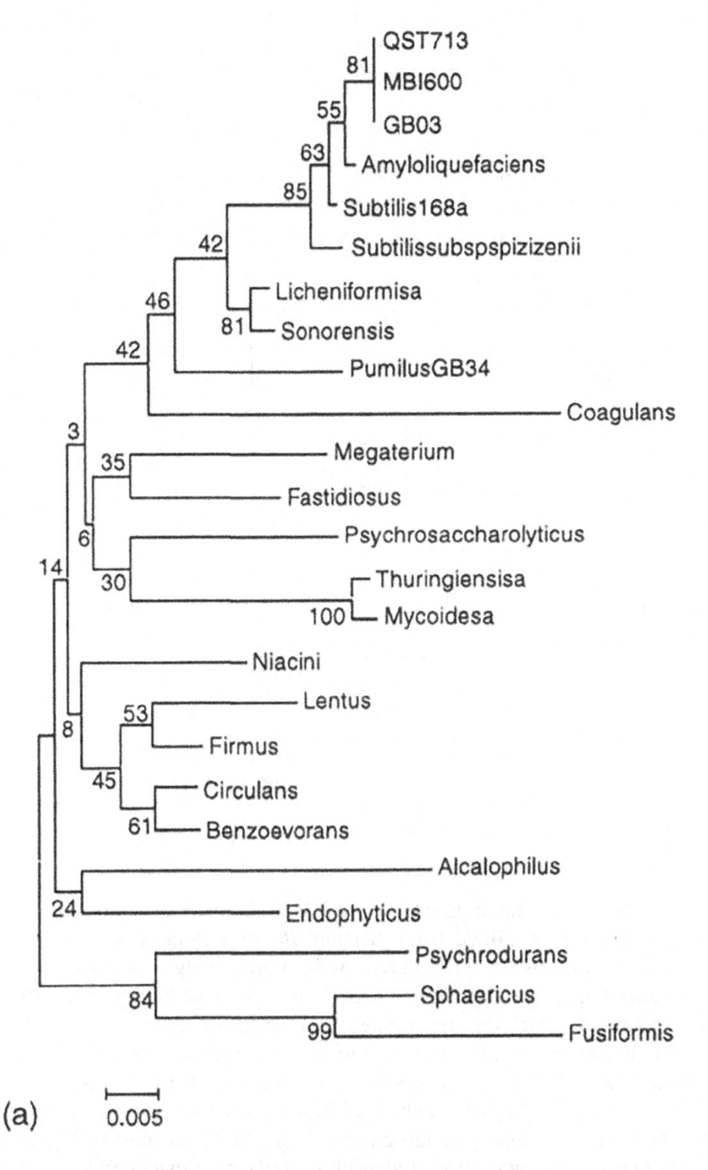

FIGURE 19.3 Dendrograms displaying evolutionarily significant clustering. An 850-bp portion of 16S gene sequences amplified from different *Bacillus* isolates was analyzed. Sequences were aligned with CLUSTAL X, and phylogenetic analyses were subsequently performed using MEGA release 3.0. Dendrograms were generated using either the neighbor-joining (a) or the maximum parsimony (b) algorithms. The scale in the first case indicates the magnitude of sequence differences measured on the dendrogram in a horizontal direction. No such scale is provided in the second case, where the focus is on modeling the patterns of descent and the actual variation currently present is not of interest. Differences in tree topology indicate that the use of two different, but widely used, approaches can result in different models of association. Additionally, the bootstrap support (ranging from 1% to 100%) for most branch events differed in the dendrograms generated using the two different algorithms, indicating that confidence in different models of association would vary depending on the approach used.

(Continued)

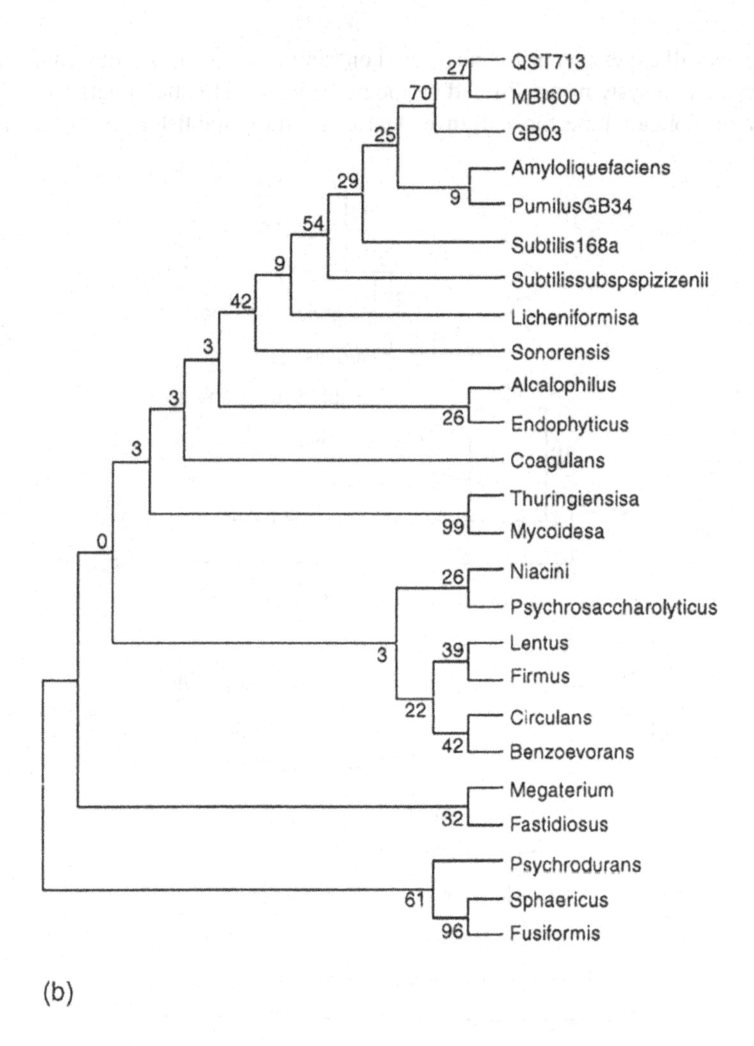

FIGURE 19.3 (CONTINUED) Dendrograms displaying evolutionarily significant clustering. An 850-bp portion of 16S gene sequences amplified from different *Bacillus* isolates was analyzed. Sequences were aligned with CLUSTAL X, and phylogenetic analyses were subsequently performed using MEGA release 3.0. Dendrograms were generated using either the neighbor-joining (a) or the maximum parsimony (b) algorithms. The scale in the first case indicates the magnitude of sequence differences measured on the dendrogram in a horizontal direction. No such scale is provided in the second case, where the focus is on modeling the patterns of descent and the actual variation currently present is not of interest. Differences in tree topology indicate that the use of two different, but widely used, approaches can result in different models of association. Additionally, the bootstrap support (ranging from 1% to 100%) for most branch events differed in the dendrograms generated using the two different algorithms, indicating that confidence in different models of association would vary depending on the approach used.

measurements that they obtain. To do so, they restate the experimental hypothesis in mathematical terms such that the values of certain descriptive statistics (e.g., the mean and variance) are compared to some mathematical model. Descriptions of, or assumptions about, the distribution of sample data determine the appropriate choice of statistical test. Because the exact population distributions of measured variables are rarely known, one can only make inferences about populations from sampled subjects. Statistical tests allow calculations of the degree of certainty with which one can assert that the patterns observed in sample data represent patterns observable in the larger populations. Here, we review the conceptual basis underlying statistical hypothesis testing and how to determine which statistical procedures are appropriate for the analysis of a specific data set.

19.3.1 Distributions of Sample Data and Populations

Samples are typically assumed to be representative of the population from which they are derived. For that to be true, they must be taken at random, that is, without an obvious and confounding bias with respect to the phenomenon being studied. The scope of sampling is decided by the context of the question or research hypothesis tested. For example, if one hypothesizes that nitrogen fixation occurs to a greater extent in non-tilled soils than that in conventionally tilled soils, then one would try to find soil sites that were roughly identical in all characteristics except tillage. If multiple independent samples are taken, the distribution of the sample values, for example, nitrogen fixation (or, for instance, denitrification) rate per soil core, can be assessed and inferences about the larger population, for example, all soil in the field plot, can be made. Because each sample leads to one or more quantifiable observations, the greater the sample size, n, the more confidence one can have in the nature of the observations being made. If the sample size is relatively large (e.g., $n > 100$), then fairly strong inferences about the population distribution can be made and tested. In contrast, if sample sizes are small (i.e., $n < 30$), few robust assessments can be made and several assumptions about the nature of the sample and the population from which it was derived must be made. Histograms and box plots can provide a description of sample distributions from which inferences can be made, but tests of specific statistical hypotheses must be conducted in order to evaluate those inferences.

19.3.2 Null and Alternative Hypotheses

When one conducts empirical research, one generates a hypothesis that can be supported or refuted by the data gathered. To assess the degree of certainty that the research hypotheses are true or false, equivalent mathematical statements need to be constructed. Such statistical hypotheses restate a research hypothesis into two (or more) contrasting models. Specifically, a *null hypothesis* and *alternative hypothesis* are formed using the quantifiable elements of the research hypothesis. The null hypothesis is a statement reflecting a negative result in an experiment, for example, that no differences were observed following an experimental treatment. In contrast, the alternative hypothesis reflects a positive result, for example, where the predicted response to an experimental treatment is observed. The null hypothesis is generally written as H_0 and the alternative as H_1. The alternative hypothesis may be considered directional, where *a* value > or < that of a fixed quantity is specified, or nondirectional, where *a* value that is simply not equal to that fixed quantity is specified.

To evaluate the alternative hypothesis, mathematical equations are used to calculate the value of a test statistic and compare that value to a probability distribution defined for that test statistic by the experimental design. A test statistic is a value that is calculated from sample data using specific formulas. It is used to determine the probability that the data collected in an experiment support the specified null hypothesis, given that the assumptions underlying the specific formula are met. For interpretation, the associated probabilities, or *P*-values, of the test statistic are most informative. With the ready availability of desktop computers and statistical software, calculation of test statistics and their associated probabilities has become automatic, so further discussion on the mathematics of specific test statistics will not be included here. Those interested in learning the mathematical underpinnings of the most commonly used test statistics can find them in several references (Sheskin 2011, Whitlock and Schluter 2014). The *P*-value calculated for any given test statistic provides the basis for rejecting the null hypothesis. Simply put, the lower the *P*-value, the more likely it is that the null hypothesis is false and should be rejected in favor of the alternative hypothesis. However, the converse is not strictly true. High *P*-values should not be taken as evidence in support of the null hypothesis, but simply as the absence of evidence for the alternative hypothesis.

What constitutes a statistically significant *P*-value is arbitrarily decided upon by the investigator. The rigor of that decision and its phenomenological significance are evaluated by colleagues who review the results. Most students are taught to reject the null hypothesis when $P < 0.05$, but this "rule of thumb" is not necessarily useful in all cases. In some instances, natural variation may be large and sampling intensity insufficient to achieve a particular level of statistical significance. This is often the case in agricultural studies, where practical constraints typically limit the number of field replicates to four per treatment and *P*-values of 0.10 are considered to be adequate for reporting yield responses. Mathematically, the sample size and amount of natural variation in a measured variable determine the level of significance that can be achieved in a given study. Table 19.4 shows this type of relationship for a simple comparison of two samples of equal size. In studies in which the observed differences in sample means are a fraction of the standard deviation, *s*, sample sizes must be quite large in order to achieve a high degree of confidence in the reproducibility of a result. An awareness of this relationship can help an investigator use the results of small, preliminary studies to more effectively design and execute larger experiments. As a practical matter, it is most important to report the experimental design, sample size, choice of statistical test, and the *P*-value obtained, so that the appropriateness of the analysis and the degree of confidence in the result can be fairly evaluated (Semenov et al. 2013).

19.3.3 Parametric and Nonparametric Analyses

Broadly speaking, there are two types of statistical tests, *parametric* and *nonparametric*, which differ in the assumptions they make about the distribution of the measured variable in the underlying population (Sheskin 2011). Parametric tests encompass linear regression and the associated models that analyze the variance within and between subjects. The different analyses of variance (ANOVA) of single-variable experiments described are special cases of approaches known as *generalized linear models* and *linear mixed models* analysis (McCulloch and Searle 2001). These approaches are computationally intensive, and the use of computers has thus facilitated their widespread application. In contrast, most nonparametric tests are computationally relatively simple and are less dependent on mathematical assumptions (Shah and Madden 2004).

All tests assume that samples are randomly and independently selected from larger populations. With the exception of a few nonparametric tests, most also assume that the underlying population distributions are homogeneous across sample sets that are to be compared. However, only parametric tests assume that the measured variable is normally distributed in the population from which the sample was drawn. These assumptions can be tested for any given sample set given a sufficient sample size (Table 19.5). If a low *P*-value is generated using a statistical test for normality (e.g., the Shapiro–Wilk test), parametric tests on the raw data should not be performed. In such a case, either the data need to be transformed mathematically to a scale for which no significant divergence from normality is detected, or the data should be analyzed using a nonparametric test. From a practical

TABLE 19.4
Minimum Sample Sizes Required for Detecting Statistically Significant Differences with Means Estimated from Equally Sized Samples

One-Tailed Test	$P = 0.05$	$P = 0.025$	$P = 0.01$
Two-Tailed Test	$P = 0.10$	$P = 0.05$	$P = 0.02$
	Difference in Means		
5s	2	2	3
2s	3	4	5
s	7	9	13
0.5s	23	32	46
0.2s	~135	~195	~270

TABLE 19.5
Tests of the Common Assumptions of Univariate Statistical Tests

Assumption	Test	References
Randomness	Wald–Wolfowitz	Bromham (2016)
Homogeneity of variance	Levene	McCulloch and Searle (2001)
	Bartlett	Fry et al. (1993)
	Cochran	Fry et al. (1993)
	Hartley	Fry et al. (1993)
Normality	Kolmogorov–Smirnov	Bromham (2016)
	Lilliefors	Bromham (2016)
	Shapiro–Wilk	Bromham (2016)

Where more than one test is listed, selection will be dependent on the type and amount of sample data being analyzed.

standpoint, randomness and normality cannot be robustly tested if the number of data points is small, $n<30$. Thus, in experiments with small sample sizes, the investigator must assume a random sample coming from a normally distributed population or simply a random sample. While many investigators choose to make both assumptions, it is more conservative to make only the latter assumption and rely more heavily on nonparametric statistics.

Because most variables of interest to soil microbiological questions are often not normally distributed, investigators must either transform their data by numerical conversion prior to running parametric statistical tests, apply machine learning approaches in which normality of the data distribution is not a requirement, or rely directly on the corresponding nonparametric statistical tests. While there is nothing inherently wrong with data value transformation from a mathematical point of view, it does present a problem for interpretation. For example, because cultured bacterial cell counts are obtained from 10-fold serial dilutions, they are often logarithmically transformed to $\log(x + 1)$ cell counts and reported as log colony-forming units (CFUs) per unit soil mass. While it is well understood that a CFU may be one or more bacterial cell giving rise to a colony, it is not at all clear what the basic unit of measurement, that is, a logarithm of a CFU, is in a physical sense. This would be equally true for other mathematical transformations such as the square root and arcsine.

In order to perform a statistical test, a specific value called a test statistic must be calculated with the appropriate formula. The most common parametric and nonparametric test statistics are listed in Table 19.6. Larger values of the test statistics indicate a higher probability that the null hypothesis is false. The precise relationship between the value of a test statistic and its associated P-value varies with the experimental design, so emphasis should be placed only on the more interpretable P-values. These are inferential tests that allow one to determine the probability that the data obtained support the null hypothesis. If the assumptions of these tests are not met, then the calculated P-values will be inaccurate. The degree of that inaccuracy will be dependent on the nature of the violation and is not easily predicted. Some tests allow for the comparison of multiple samples taken from different treatments, while others are designed to determine the associates between and among the measured variables. The choice of which test to use depends on the experimental design, the assumptions being made about the samples, and the nature of the variable being measured.

To illustrate a typical statistical analysis, let us consider an example where one is interested in comparing the relative abundance of six different species of bacteria inhabiting soil following an oil spill (i.e., contaminated vs uncontaminated soil). Five samples can be taken from each soil, and the abundance of each cell type measured by real-time PCR with an appropriate set of controls. A data matrix is thus generated with five values for each of six different species occurring in the two

TABLE 19.6
Some Equivalent Parametric and Nonparametric Test Statistics

Type of Test	Parametric	Nonparametric
	Comparisons	
One and two samples	Student's t	Wilcoxon T
		Mann–Whitney U
Multiple samples	Fisher's F	Kruskal–Wallis H
		Friedman's S
Multiple range tests	Tukey's honestly significant difference (HSD)	Dunn's multiple comparison test
	Fisher's least significant difference (LSD)	
	Dunnett's test	
	Associations	
Correlation	Pearson's r	Kendall's τ
Regression	Linear regression	Monotonic regression
		Ordinal logistic regression

different soils. The homogeneity of variance among our two samples (i.e., six tests total, with the sample variance s^2 calculated from two sets each of five measurements) is then tested. If the P-value of these tests is uniformly high, for example, $P > 0.10$, then there is no strong evidence for the violation of this assumption. However, with six tests, there is a chance that the variance of one of the six species sampled varies significantly. What to do? Here is where subjective judgment comes in. That one set should be analyzed and discussed separately or tossed out of the analysis altogether. The quality of the data is not in question, merely their fitness for statistical analyses. As long as any violation is reported along with the results of the subsequent comparisons, then others will recognize that the calculated P-values may not be precise. Now, in order to further justify the use of a parametric test, we must either assume normality and randomness or accept convention and assume normality of the log-transformed values. Alternatively, we could choose to use nonparametric tests. Here, one could use a general linear model approach, where this design could also be called a one-way ANOVA (or its nonparametric analog, the Kruskal–Wallis test). If a significant F (or H) test statistic is generated, then a multiple range test should be performed to determine how the abundance of each of the six species compares to one another. Alternatively, six different t tests (or Mann–Whitney tests) could be performed.

The choice between the two approaches described previously depends on one's definition of an experiment. Is this one test examining the effects of the experimental treatment on six different species or six independent tests of the influence of the experimental treatment? While such decisions should generally be made *a priori*, they are often decided *a posteriori*, after the preliminary analyses are made. Again, the calculations of these test statistics and their associated P-values are nowadays trivial with the advent of widely available statistical software (Sheskin 2011). For most parametric tests, similar nonparametric tests can be performed; however, the obtained P-values will rarely be exactly the same.

19.4 MULTIVARIATE STATISTICS, CLASSIFICATION, AND HYPOTHESIS GENERATION

In many studies, univariate statistical tests will be sufficient for investigation, but in some instances, multivariate analyses are preferable. Some *multivariate techniques* (i.e., multi-way ANOVA, multivariate regression, discriminant analyses, and canonical correlation analyses), including machine learning approaches (support vector machines, nearest neighbor, decision trees, and neural networks), test hypotheses analogous to their univariate counterparts. However, such approaches make mathematical assumptions about the independence, normality and homogeneity of variance of the measured variables that are rarely substantiated in complex, interconnected systems such as the soil. While such techniques can be useful, careful attention to the underlying assumptions and the methods used to validate them in a multivariate data set are required. A more thorough discussion of the usefulness and limitations of such techniques can be found elsewhere (McGarigal et al. 2000).

The multivariate analyses most commonly conducted in microbiology are *classification* methods, useful for hypothesis generation, not for hypothesis testing in the statistical sense (i.e., they do not typically yield interpretable *P*-values). There are two distinct approaches to classification. The first process delineates groups based solely on present-day attributes of the measured system. Because this taxonomic approach was developed prior to the molecular biology "revolution" and based on phenotypic data, it was first called *phenetics*. The approach has also been referred to as *numerical taxonomy* or *taxometrics*, and it has been applied to classifying abiotic variables in the natural environment as well as living organisms and their constituents. Specific mathematical relationships present in a multivariate data set are first defined using resemblance statistics. Then, various algorithms are used to summarize those relationships, and the outputs are tabulated and projected graphically. These methods are most useful for associating groups of subjects or variables with nonobvious relationships and for identifying the variables that, in combination, lead to such associations. The second approach to classification focuses on describing evolutionary pathways. This approach includes various algorithms described under the branches of science called *biosystematics*, *phylogenetics*, *phyletics*, and *cladistics*. In this approach, the different algorithms used imply various models of evolutionary change, and those that result in the most concise descriptions are generally favored as the most likely explanations. This approach to multivariate classification is useful for making inferences about the order, timing, and number of changes leading to the set of character states observed in the present day. While some outputs generated from these two approaches to classification may look similar (i.e., dendrograms), they have different meanings and interpretations.

19.4.1 Ordination

Ordination techniques, for example, *principal components analyses*, *correspondence analysis*, and *factor analysis*, use a procedure called Eigen analysis to reduce the complexity of a data matrix. While there are numerous approaches to ordination, distinguished by the use of different correlation coefficients and algorithms to identify the principal components, they all share common elements of data processing. Following the generation of a data matrix, a secondary matrix of correlation, covariance, or dissimilarity values is calculated for either variables or subjects. From this matrix, specific values are calculated that allow a simplified description of the relationships among variables (or subjects) to be presented graphically and numerically. Simply put, ordination analyses make the associations among subjects or variables visible along newly defined axes, called principal components, which represent the paths through multivariate space with the greatest variance.

To better understand ordination, consider a plot of data along three variable axes. In Figure 19.2, each axis represents a single variable, that is, bacterial growth rate on L-arabinose (*x*-axis), hydroxy-L-proline (*y*-axis), and *p*-hydroxyphenylacetic acid (*z*-axis). Each test was performed on 30 different subjects, that is, bacterial isolates, and the results were plotted for each in the three-dimensional space shown. From this graph, it is clear that the isolates show a range of responses on all three substrates, and the data reveal an elongated "cloud" in a direction that is skewed relative to all three axes. The first principal component, drawn as a dashed line, describes the path of the greatest variance across the data set and represents a composite of the three measured variables (growth rates on the three substrates). The second principal component is defined perpendicularly to the first one and represents another composite variable, with a different spectrum of affiliation for each of the three components defined in this study. The last principal component is defined perpendicularly to the first two in three-dimensional space and will represent the path of minimal variance. There are always as many ordination axes as variables, but the contribution of each to the classification will decrease unevenly. Graphical projections of the data relative to the ordination axes are then used to identify distinct subsets of subjects. Points that are plotted close to one another represent subjects that are more related to one another than those plotted further away, within the context of the data set being analyzed. In this case, the use of ordination and simple plots both indicate that there are two or three subsets of bacterial strains among the test group. Though impossible to fully visualize in a single graph, it is easy to mathematically expand this approach to studies with more than three variables and this is where ordination is of the greatest analytical utility. Because they encompass the greatest amount of variation, plots of the first few components reveal the strongest mathematical relationships among subjects. This process is particularly useful when the number of variables is large, because graphical projections in more than three dimensions are difficult to interpret. Furthermore, the same process can be applied to classifying variables, as long as the sample of subjects is a representative in some defensible way.

The basic interpretation of ordination outputs is composed of two steps. Most statistical software programs provide key information useful for interpreting ordination outputs. First, the amount of variation in the data set encompassed by each principal component is used to evaluate the relative strength of the associations identified. If a relatively high percentage of variance is encompassed by the first two or three components, then graphical interpretations are straightforward. That is to say, groups of data points that are separated from other groups of data points along these axes probably represent distinct subsets of the data. If multiple data points exist for cases (or variables) known or thought to be identical, then confidence intervals for apparent groupings can be generated. And, while there are no hard and fast rules for determining how many components to consider, one should probably be cautious of drawing conclusions from ordination plots when the amount of variance encompassed by the projected components is <50%. Second, the associations of the measured variables (or cases) and the principal components are used to determine which are interrelated and are most discriminatory. These associations may be given as factor loadings and/or correlation coefficients, depending on the software used. Those variables (or subjects) with loadings of large magnitudes, positive or negative, contribute most to the definition of the principal components and to the structuring of the data. The importance of the identified variables (or subjects, if one is classifying variables) can be subsequently investigated by generating specific, testable hypotheses about their contributions to structuring the system under investigation. Effective tests of those hypotheses, however, will require the acquisition of additional data, unless the separation of data subsets is large across the first two component axes.

19.4.2 Cluster Analysis

Cluster analyses approach classification in a completely different way to ordination. The primary approach to clustering used in biology is called *hierarchical* or *agglomerative clustering*. As with ordination analyses, multiple synonyms exist in the literature, so one will see such techniques also referred to as *sequential agglomerative hierarchical and nonoverlapping clustering* and *polythetic agglomerative hierarchical clustering* (McGarigal et al. 2000, Whitlock and Schluter 2014). Following the generation of a resemblance matrix, an algorithm is used to sequentially fuse units, of either subjects or variables, together, until all have been linked together as part of the classification. Many fusion algorithms exist, and the most commonly reported ones include single, complete, median centroid, and "unweighted pair group method with arithmetic mean" linkage algorithms. The results of these linkages are often tabulated as agglomeration schedules, but they are most indicated by dendrograms, which are graphical displays reflecting the ordered associations of subjects or variables, whichever is being classified. Most statistical software programs that perform hierarchical cluster analyses will provide scaled trees based on the values present in the resemblance matrix. This provides an indication of the relative closeness of the associations within and between different clusters depicted in a dendrogram. The lengths of the scaled lines on dendrograms are indicative of the degree of similarity among joined units of subjects or variables given the associations calculated across variables or subjects, respectively, in the resemblance matrix. However, it is important to recognize that the scale used is only relevant in an absolute sense to the data set being analyzed, and it should not be used to quantitatively describe the relationships among the entities being classified. Like the choice of resemblance measure, the choice of linkage algorithm can substantially influence the outcome of a classification. Because no linkage algorithm can be phenomenologically linked to a biological process, there is no *a priori* "best" linkage algorithm to use. As with ordination analyses, it is often worthwhile to reanalyze the data set with different resemblance measures and linkage algorithms in order to better substantiate the relationships presented in a dendrogram.

19.4.3 Phylogenetics and Parsimony Analyses

Classifications with evolutionary significance are analyzed using algorithms that are distinct from clustering and ordination analyses (Bromham 2016). While phenetic clustering relies on the application of an algorithm to generate an ordered representation of increasing resemblance in the data, phylogenetic clustering uses an algorithm *and* an optimality criterion to decide on the "most likely" stepwise path of change (unrooted), sometimes from a hypothesized "original" state (rooted). Phylogenetic models all assume that character state differences among subjects arose in a stepwise fashion with some defined constraints. Such models have historically focused on phenotypic data, with the number of character states limited by the breadth of the study. However, with the general availability of DNA sequencing, evolutionary studies of genes and genomes are increasingly more common. The general assumptions of phylogenetic analyses are vertical heritability, independent state changes, and lack of homoplasy. Vertical heritability refers to traits being passed from one generation to the next, for example, mother to daughter cells. For state changes to be independent, they must not result in a significant interaction with other traits being analyzed. And homoplasy refers to the presence of similar traits or sequences evolving from disparate ones through processes such as convergent and parallel evolution and unselected reversions. While these assumptions are almost always true in multicellular organisms, the occurrence of lateral DNA transfer, as well as the existence of mutator genes and the rapid and often uneven generation times in most bacteria, can greatly complicate the analyses. In Chapters 3 and 7, we examined these evolutionary aspects of bacteria.

The general approach to evolutionary analysis involves several steps. First, *meaningful* characters and character states need to be chosen for measurement. With most studies, this includes DNA sequence data, as the fundamental structures and activities of cells are encoded in the genome. Then, matrices of binary and/or multistate (e.g., DNA, RNA, or protein sequence) data are generated for analysis. One or more algorithms are then used to organize and subsequently define the possible paths of progression to current character states of all cases, generally from a single undefined progenitor. Because multiple paths of progression can be readily defined, additional criteria are used to identify the optimal path(s) for the development of current character states.

In general, there are two mathematically distinct approaches to defining these optimal paths. First, the *maximum parsimony* approach minimizes the number of changes required to explain existing data. Outputs represent the shortest hypothetical pathway of changes and imply a "most likely" evolutionary path. This is mathematically expedient as well as a reasonable expression of the philosophical assertion known as Occam's razor (i.e., the simplest explanations are most often the most correct ones). Different parsimony programs make different assumptions about the types of state changes that are possible. For example, Fitch and Wagner parsimony assumes that all state changes are freely reversible and equally probable. In contrast, Dollo parsimony assumes that state changes are reversible, but loss of a character is more likely than its addition. And, the Camin–Sokal parsimony algorithm assumes that state changes are not reversible, so character states once appearing are fixed in all lines of descent. Second, the *maximum-likelihood* approach estimates a parameter according to a predetermined model of processes. Outputs of such analyses represent the pathway of changes that is "most consistent" with the predetermined model. Like Fitch and Wagner parsimony, such programs generally assume that state changes are freely reversible and not dependent on previous character states. However, different models can be defined based on the types of substitutions allowed (e.g., specific transitions and transversions) and the relative frequency of different substitutions (e.g., transitions more frequent than transversions).

In most cases, multiple models with a minimal number of state changes (i.e., most parsimonious trees) will be identified. To determine which of these contrasting models is "best," the trees are compared using different approaches. Exhaustive algorithms and the computationally least intensive branch and bound algorithms can be used to identify the most parsimonious trees. Multiple options are available in different software packages, such as MEGA (Kumar et al. 2004). These algorithms allow for the construction of an evolutionary model that can be graphically represented. This may consist of a simple *cladogram* describing the order in which evolutionary changes are modeled. Or, more frequently, the results of phylogenetic analyses are presented as scaled dendrograms, where the distance between endpoints is a measure of evolutionary distance (Figure 19.3). As with all multivariate analyses, the use of different algorithms may affect the topology of the dendrograms depending on the composition of the data set. For example, in a phylogenetic analysis of different *Bacillus* strains, the branch paths between *B. subtilis* and *B. mycoides*, and the statistical support for such events, vary depending on the algorithm used (Figure 19.3).

19.4.4 Substantiating a Classification

Classification methods are used to generate testable hypotheses. While well-characterized subjects may be inserted into data sets for analysis, groups containing these "controls" should not be considered to represent quantified proofs of identification in a multivariate output, such as a dendrogram or ordination plot. Rather, the associations revealed can be validated *only* within the context of the supplied data. There are several ways in which this can be achieved.

Bootstrapping is a common technique to determine the confidence intervals for any statistic. In phylogenetic analyses, it is commonly used to determine the confidence (likelihood) that specific branching events occurred. In ordination analyses, one can calculate and average the correlations among pairs of easily circumscribed subjects based on the variables or the principal components. With some programs, *co-phenetic correlation* values can be calculated to compare the original resemblance matrices to those calculated from dendrograms generated by different hierarchical clustering algorithms. The dendrograms generating the highest co-phenetic correlations can be considered better models than those with much lower values. Because the methods described earlier are all exploratory, comparisons among classifications generated with different resemblance measures, algorithms, and complementary methods (e.g., clustering and ordination) can yield more reliable indications of the most consistent patterns of association among subsets of the data. Additionally, replication of measurements within a data set and the use of independent samplings can be used to better define the boundaries of distinct groups. For example, the use of independent measurements of the same subjects can be used to define the level of significant clustering in a dendrogram. In a cluster analysis of strains for carbon source utilization, different lines were drawn in the dendrograms at the minimum similarity level achieved by all or the large majority of the replicated samples. Such lines lead to robust and statistically supported descriptions of group identity. This approach can be used for any taxometric analysis, and it was first used to successfully classify a large collection of *Pseudomonas fluorescens* strains (see McSpadden-Gardener et al. 2000). The advantage of this method is that it unambiguously indicates the degree to which a classification will be reproducible no matter how much empirical noise exists in a data set. This is useful for ensuring that the classification of cases can be reproduced in other laboratories regardless of the technical difficulty of a given analytical procedure.

19.5 MODERN MACHINE LEARNING APPROACHES

When classical statistical models cannot handle highly dimensional and complex data or the accuracy of the conclusions (e.g., prediction of changes of soil microbial community structures over time) is too low, modern machine learning approaches offer good alternatives. Often spectacular results can be obtained. Such approaches can generalize and learn trends and/or patterns from an available data set, and on the basis of this, they can perform automatic decision-making without programming explicit rules (Figure 19.4).

Following several ground-breaking successes in solving problems which were previously assumed to be impossible to model—ranging from computer vision (Greenspan et al. 2016) and predictive analytics to natural language processing (Hinton et al. 2012), machine learning approaches have been introduced into bioinformatics (genome annotation; Min et al. 2017), cell biology, and other biological scientific fields, including soil microbiology (Chang et al. 2017, Sunil and Zulfiqar 2018).

Machine learning can be divided into two types of techniques:

1. Supervised learning (trains a model on known input and output data so it has predicting capacities)
2. Unsupervised learning (finds hidden patterns or structures in large data sets).

Supervised machine learning approaches build models that make predictions based on evidence in the presence of uncertainty. They take a known set of input and output data and train a model to generate predictions for the response to new data. Supervised learning can be divided into classification and regression approaches. Classification approaches predict discrete responses and classify input data into categories, for example, whether a certain species is present in soil. On the other hand, regression approaches predict continuous responses, for example, fluctuations in the abundance of members of soil microbiomes.

Unsupervised learning approaches are built to find hidden patterns or structures in data sets. They are used to draw inferences from data sets consisting only of input data without output data. Clustering is one of the most common unsupervised learning approaches. Clustering is used to explore the data to find hidden patterns or groupings in data. Applications for clustering include gene sequence analysis (Libbrecht and Noble 2015).

Choosing the right algorithm for both types of machine learning approaches can be very challenging—there are many supervised and unsupervised machine learning algorithms (Figure 19.4), and each of them takes a different approach to train. Currently, there is no best method to find the right algorithm, and hence, this is often a trial-and-error approach. Even an experienced computational biologist will not be able to tell on first hand whether an algorithm is suitable for each particular case. However, selection of machine learning might depend on the size and type of the data, the expected results from the data, and how such results will be used. Experience in applying various machine learning approaches will help to identify highly flexible models which tend to overfit data by modeling minor variations, while simple models might have too many prior assumptions. There are thus always trade-offs between the complexity of the model and its speed and accuracy.

Current progress in machine learning methodology has resulted in robust software tools (packages and toolboxes available in R, MATLAB®, and Python) that make the application of machine learning approaches more accessible for researchers that do not already have years of experience in the area.

Finally, it is important to mention the state of the art of the deep learning approach, which currently demonstrates ground-breaking performance in various fields, including bioinformatics, and is

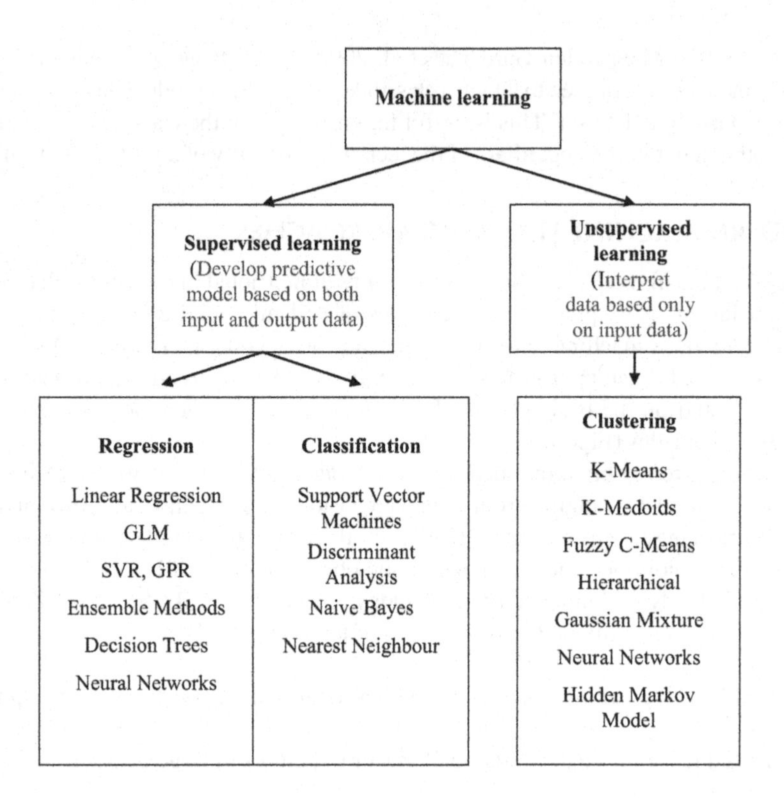

FIGURE 19.4 Selection of a machine learning algorithm based on research question and initial data available.

considered as an example of artificial intelligence (LeCun et al. 2015, Spencer et al. 2015). Inspired by the work flow of human brains, deep learning is built on an artificial neural network which consists of multiple nonlinear layers; thus, it can outperform even the most advanced machine learning approaches. The fundamental difference is that conventional machine learning approaches rely heavily on data representations (features) which are defined by experienced researches. Identification of such features for the given research is a difficult task that is often biased. Deep learning, a branch of machine learning, has overcome these limitations, by extracting simple features from data and combining them into complex features. Deep learning is a rapidly developing research area. Although there are several attempts to apply deep learning to ecological studies and bioinformatics (Min et al. 2017), its application in soil microbiome studies is still modest due to its complexity.

19.6 CONCLUSIONS

Modern soil microbiology encompasses descriptive, experimental, molecular and biotechnological studies that generate many different types of data. Analyses of both univariate and multivariate data will be used to successfully advance the field, but only if the appropriate tests and interpretations are applied. Exploratory data analysis involving the use of summary statistics, box plots, and histograms can be useful for identifying the likely patterns of variation. Statistical tests are used to evaluate the probability that the observed variation is significant given certain mathematical assumptions about the nature of the data analyzed. Moreover, machine learning approaches can reveal hidden patterns and predict the behavior of complex structures such as the parameters of soil microbiomes. Knowledge of these assumptions is a prerequisite for proper statistical analyses. In the absence of sufficient information on the distribution of a measured variable, it is recommended that nonparametric statistics be used. In studies involving the analysis of multivariate data sets, exceptional restraint should be applied to interpretations because, in most cases, the mathematical assumptions underlying the analyses are not met or arbitrarily specified by the user. In general, multivariate ordination and cluster analyses should only be used to classify subjects and variables. Such classifications can lead to the generation of new phenomenological models that can only be explicitly tested on the basis of new data. When properly used, the statistical methods described in this chapter can facilitate the description, testing, and development of new knowledge about soil microbiological systems.

REFERENCES

Bromham, L. 2016. *An Introduction to Molecular Evolution and Phylogenetics.* Oxford University Press: Oxford, UK.

Chang, H., Haudenshield, J.S., Bowen, C.R. and G.L. Hartman. 2017. Metagenome-wide association study and machine learning prediction of bulk soil microbiome and crop productivity. *Front Microbiol* 8, 519.

Ellison, A.M. 2001. Exploratory data analysis and graphical display. In: S.M. Scheiner and J. Gurevitch (eds.), *Design and Analysis of Ecological Experiments*, Chapman & Hall: London, UK, 37–62.

Fry, J.C. (ed.). 1993. *Biological Data Analysis: A Practical Approach.* IRL Press: Oxford, UK.

Greenspan, H., van Ginneken, B. and R.M. Summers. 2016. Deep learning in medical imaging: Overview and future promise of an exciting new technique. *IEEE Trans Med Imaging* 35, 1153–1159.

Grunwald, N.J., Goodwin, S.B., Milgroom, M.G. and W.E. Fry. 2003. Analysis of genotypic diversity data for populations of microorganisms. *Phytopathology* 93, 738–746.

Hinton, G., Deng, L. and D. Yu. 2012. Deep neural networks for acoustic modeling in speech recognition: The shared views of four research groups. *IEEE Signal Process Mag* 29, 82–97.

Hughes, J.B., Hellmann, J.J., Ricketts, T.H. and B.J.M. Bohannan. 2001. Counting the uncountable: Statistical approaches to estimating microbial diversity. *Appl Environ Microbiol* 67, 4399–4406.

Kumar, S., Tamura, K. and M. Nei. 2004. MEGA 3: Integrated software for molecular evolutionary genetics analysis and sequence alignment. *Briefings Bioinf* 5, 150–163.

LeCun, Y., Bengio, Y. and G. Hinton. 2015. Deep learning. *Nature* 521, 436–444.

Libbrecht, M.W. and W.S. Noble. 2015. Machine learning applications in genetics and genomics. *Nat Rev Genet* 16 (6), 321–332.

Martin, A.P. 2002. Phylogenetic approaches for describing and comparing diversity in microbial communities. *Appl Environ Microbiol* 68, 3673–3682.

McCulloch, C.E. and S.R. Searle. 2001. *Generalized, Linear, and Mixed Models.* John Wiley and Sons: New York.

McGarigal, K., Cushman, S. and S. Stafford. 2000. *Multivariate Statistics for Wildlife and Ecology Research.* Springer: New York.

McSpadden-Gardener, B.B. 2007. Statistical analyses of microbiological and environmental data. In: J.D. van Elsas, J.K. Jansson and J.T and Trevors (eds.), *Modern Soil Microbiology II*, CRC Press: New York, 553–584.

McSpadden-Gardener, B.B., Schroeder, K., Kalloger, S., Raaijmakers, J., Thomashow, L.S. and D.M. Weller. 2000. Genotypic and phenotypic diversity of *phlD*-containing *Pseudomonas* spp. isolated from the rhizosphere of wheat. *Appl Environ Microbiol* 66, 1939–1946.

Min, S., Lee, B. and S. Yoon. 2017. Deep learning in bioinformatics. *Briefings Bioinf* 18, 851–869.

Semenov, A.V., Van Elsas, J.D., Glandorf, D., Schilthuizen, M. and W.F. Boer. 2013. The use of statistical tools in field testing of putative effects of genetically modified plants on nontarget organisms. *Ecol Evol* 3, 2739–2750.

Shah, D.A. and L.V. Madden. 2004. Nonparametric analysis of ordinal data in designed factorial experiments. *Phytopathology* 94, 33–43.

Sheskin, D.J. 2011. *Handbook of Parametric and Nonparametric Statistical Procedures*, 5th ed., Chapman and Hall/CRC Press: Boca Raton, FL.

Spencer, M., Eickholt, J. and J. Cheng. 2015. A deep learning network approach to ab initio protein secondary structure prediction. *IEEE/ACM Trans Comput Biol Bioinf* 12, 103–112.

Sunil, K.J. and A. Zulfiqar. 2018. Soil microbial dynamics prediction using machine learning regression methods. *Comput Electron Agric* 147, 158–165.

Unwin, A. 2015. *Graphical Data Analysis with R (Chapman & Hall/CRC The R Series).* Chapman and Hall/CRC Press, Boca Raton, FL.

Whitlock, M.C. and D. Schluter. 2014. *The Analysis of Biological Data*, 2nd ed., W. H. Freeman and Company, New York, NY.

Yates, M.V., Nakatsu, C.H., Miller, R.V. and S.D. Pillai. 2016. *Manual of Environmental Microbiology*, 4th ed., ASM Press: Washington, DC.

REFERENCES

Section III

Applied Chapters

20 Soil Microbial Communities and Global Change

Mark P. Waldrop and Courtney Creamer
U.S. Geological Survey

CONTENTS

20.1 Introduction 331
20.2 Global Change: More Than Temperature 332
20.2.1 Vegetation Change and Microbial Communities 332
20.2.2 Wildfire Effects on Microbial Communities 334
20.2.3 Microbial Response to Permafrost Thaw 335
20.3 Methodological Revolutions and New Frontiers 337
20.4 Conclusions and Future Challenges 338
References 338

20.1 INTRODUCTION

Increasingly rapid changes are occurring in Earth's systems: from the atmosphere, to soil and water ecosystems and their biota. Atmospheric CO_2 is increasing to unprecedented levels, and more industrially fixed N is added to the environment than that from all natural processes combined (Vitousek 1994, 1997). Wildfires and droughts are increasing in frequency and magnitude, altering terrestrial ecosystems (Westerling et al. 2006, Kasischke et al. 2010, Dai 2012). As a result of these changes to many global biogeochemical cycles, the cycling of carbon and nutrients that sustain ecosystems is now considered to be human-dominated (Vitousek 1994). The changes are so dramatic and global in scope that the current epoch can be labelled the "Anthropocene" (Lewis and Maslin, 2016).

Soils and soil microbial communities mediate the biogeochemical processes that underly ecosystem-level changes. Understanding how microbial communities are responding to global changes will help us to understand how ecosystems will respond to, and potentially mediate, responses in a rapidly changing world, and improve our ability to model these processes and make predictions. Soil microbial communities are critical to several fundamental global change processes including N_2 fixation from the atmosphere; decomposition of plant tissues entering soils; mineralization of important elements including N, P, and S; formation of soil organic matter (SOM); greenhouse gas fluxes including CO_2, CH_4, and N_2O; mycorrhizal associations with plant communities that aid in water and nutrient uptake; and the cycling of elements such as Fe, As, and Hg. Soil microorganisms are the biological agents that decompose plant carbon (roots, root exudates, and plant litter) and produce stabilized SOM (Kleber et al. 2007). As discussed in previous chapters, the community of soil organisms contains an immense number of species of bacteria, archaea, fungi, nematodes, arthropods, and earthworms, although bacterial and fungal communities typically dominate biogeochemical activity. During the process of decomposition, microorganisms respire CO_2 back to the atmosphere and some of the carbon that flows through the microbial biomass becomes stabilized into SOM, the critical long-term carbon sink (Clemmensen et al. 2013, Lehmann and Kleber 2015). Microorganisms also produce and consume methane (CH_4), another important greenhouse gas with a greenhouse warming potential >25 times greater than CO_2. See also Chapter 11. Methane is produced by methanogens in water-saturated soils under anaerobic conditions; it can be oxidized both aerobically and anaerobically. Aerobic oxidation is more rapid than anaerobic oxidation, occurring near the surface of saturated soils where molecular oxygen is present as the terminal electron acceptor. When oxygen is not present, other molecules such as sulfate, ferric iron, nitrate, and nitrite are common electron acceptors. These are important for reducing methane emissions from a variety of habitats (Blazewicz et al. 2012). Besides being an important direct influence on the cycling of soil C, microorganisms also indirectly affect the carbon cycle by affecting the productivity of ecosystems, through mechanisms such as symbiotic relationships between mycorrhizae or N-fixing organisms and plants (Hoeksema et al. 2010, Hewitt et al. 2016), through nutrient mineralization, making inorganic N and P available for plant uptake, or through soil aggregation processes that increase soil stability (Jastrow et al. 1998, Ferrenberg et al. 2015).

In this chapter, we examine why soils and soil microbial communities are important for understanding impacts and feedbacks to global change. Furthermore, we will discuss the technological approaches and challenges that are at the frontiers of this research area. In so doing, it is first important to define the timing and magnitude of global change impacts on soil microbial communities because this affects how microbial communities will respond to, and mediate, these disturbances. Global change impacts on microbial communities can be categorized as *press* or *pulse disturbances* (Shade et al. 2012). Press disturbances are chronic and long term (although not necessarily of equal magnitude over time), whereas pulse disturbances are relatively short-term discrete events. Press disturbances can take a long time for microbial responses and feedbacks to occur. For example, press disturbances include higher temperatures and longer growing seasons, changes in precipitation patterns, and changes in atmospheric chemistry that, in turn, affect plant community productivity and composition. Pulse disturbances, on the other hand, may be accompanied by rapid changes in community composition towards organisms that can take most advantage of the new resource environment. Examples of pulse disturbances include wildfire, deforestation, flooding, hurricane damage, and draining of peatlands. Following pulse disturbances, the microbial community may return to its pre-disturbance level if the community and ecosystem are resilient (see also Chapter 3). Alternatively, if it does not return to the pre-disturbance level, it may enter an altered stable state, continuing to alter biogeochemical processes that provide feedback to plant communities for long periods of time (Shade et al. 2012). Climate change influences both the press disturbances of plant community change and the pulse disturbances of wildfire and permafrost thaw. We will focus on these examples from Western North America, where issues such as land cover change, wildfire, and permafrost thaw are some of the most observable global change impacts.

20.2 GLOBAL CHANGE: MORE THAN TEMPERATURE

Global increases in atmospheric temperature are among the most profound and concerning long-term changes affecting human society and will likely be so for centuries. Warming is occurring most rapidly at high latitudes (Hinzman et al. 2005, Soja et al. 2007, Schaefer et al. 2011), while decreases in precipitation, and resulting drought, are projected to be strongest in Western dryland ecosystems (Seager et al. 2007, Dai 2012). This will lead many of these areas to reach new ecosystem states over the next century (Westerling et al. 2006). In the Arctic, warmer and drier soils bring increased incidence of severe wildfire and increased permafrost thaw, due in part to the loss of the insulating forest floor (Turetsky et al. 2010). Changes in the temperature and precipitation regime can lead to a cascade of other ecosystem disturbances that are discussed in more detail below.

20.2.1 Vegetation Change and Microbial Communities

One of the most dramatic human-driven global changes is plant cover change (Vitousek 1994, Sala et al. 2000, Haberl et al. 2007). Plant cover has changed dramatically through land conversions, such as clearing of land for agriculture. For example, about 38% of the global ice-free land surface is under agriculture, including livestock grazing (Barger et al. 2011, Foley et al. 2012). This is projected to increase to 46% by 2050 (Tilman et al. 2011). Invasions by exotic plant species are affecting vast areas of the Western United States (Eviner et al. 2010). Plant communities will continue to change in response to other press disturbances such as elevated CO_2 levels, climate warming, and/or N deposition (Westerling et al. 2006, Barger et al. 2011). These press disturbances are expanding the global extent of shrublands (Stevens et al. 2016) at a rate of 0.5%–2% per year (Barger et al. 2011; Figure 20.1). Similarly, increased aridity, that is predicted for the Southwestern United States (Seager et al. 2007, Dai 2012), will likely decrease plant productivity, biomass and plant species richness (Brookshire and Weaver 2015), and microbial diversity and abundance (Maestre et al. 2015). Taken together, the distribution of vegetation in mid- and northern-latitude ecosystems will be substantially altered by the end of this century (Scheffer et al. 2001, Hinzman et al. 2005).

In the Arctic, increasing temperatures, longer growing seasons, and changing fire regimes are altering plant cover. This is most evident by the increasing abundance of shrubs in the Arctic tundra (Sturm et al. 2001, Hinzman et al. 2005) and the northward movement of the Arctic tree line (Soja et al. 2007). Wildfire and permafrost thaw will also rapidly change plant and microbial communities (Schuur et al. 2007, Johnstone et al. 2010, Mackelprang et al. 2011). Globally, plants are migrating northward at about 17 km per decade, and to higher elevations at about11–30 m per decade, to remain in their temperature optima (Lenoir et al. 2008, Chen et al. 2011). As a result, a pan-Arctic model scenario projects that at least half of the Arctic vegetation, across an area of about 37,000 ha, will shift to a different vegetation class by 2050 due to shrubification and northward migration of plants (Pearson 2013).

Microorganisms play an important role in mediating the changing composition of plant communities, due to their influence on plant fitness. It is increasingly recognized that plants and soil microbial communities are intimately linked to each other (see also Chapter 10). In some cases, soil microbes can influence plant function and establishment (Reinhart and Callaway 2004, Van der Heijden et al. 2008, Bennett et al. 2017). In general, soil microorganisms seem to be

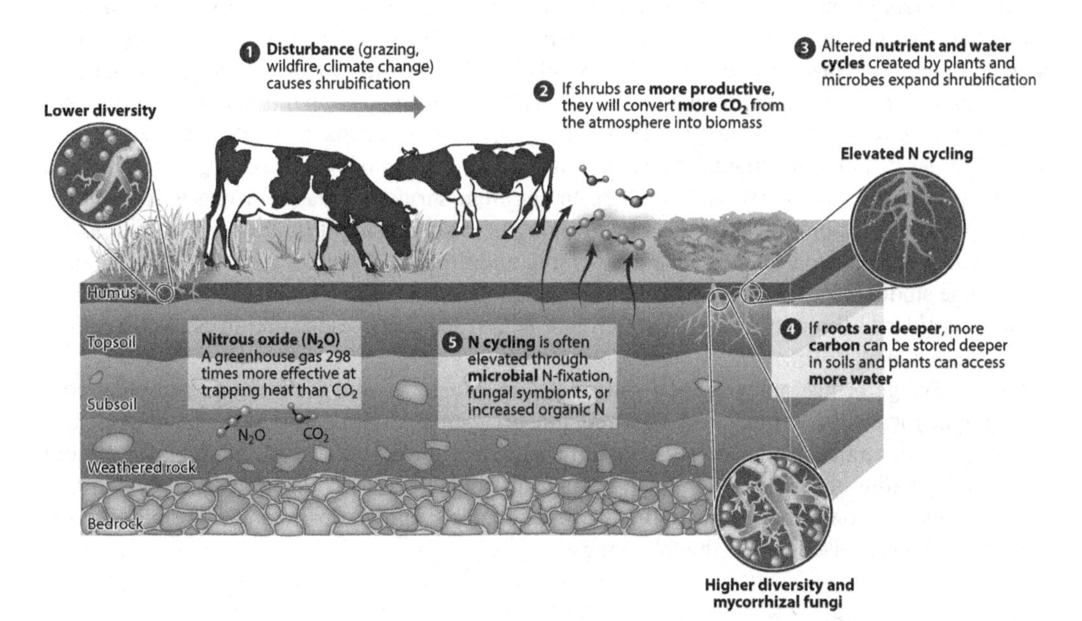

FIGURE 20.1 **(See color insert)** Impact of press disturbance on grass and shrub cover. Increases in shrubs or trees relative to grasses typically induce changes in microbial communities and in carbon, nitrogen, and water cycling. However, the directionality of these changes varies based upon the ecosystem. As shown for drier climates, carbon and nutrients are typically higher below the shrubs or trees, creating islands of fertility relative to the surrounding areas.

adapted to their particular climate regime (Waldrop and Firestone 2006), helping to maximize plant fitness when a plant is in the same climate regime. For example, a drought-adapted microbial community can improve the fitness of plants under drought conditions and a non-drought-adapted microbial community can improve plant fitness and productivity under high water conditions (Lau and Lennon 2012, Timmusk et al. 2014). The mechanism underlying the improvement of plant fitness is not well known, but it may be linked to changes in the microbial community that affect N availability, or to a changing abundance of mutualists or pathogens. Similarly, microbial communities can facilitate the establishment and/or the invasion of exotic plants (Reinhart and Callaway 2004, Inderjit and van der Putten 2010). This may occur through the presence or absence of pathogenic organisms or mutualistic fungi (Wolfe and Klironomos 2005, Hewitt et al. 2016), or of symbiotic microorganisms that improve nutrient acquisition (e.g., arbuscular mycoorhizal fungi [AMF], N-fixing *Rhizobia*). Leguminous shrubs, with N-fixing capability from symbiotic bacteria, have a competitive advantage over grasslands in dryland regions, thereby promoting invasive shrub expansion (Barger et al. 2011). Invasive species—such as *Bromus*—have reduced diversity and abundance of microbial populations in soils, potentially impacting the ability of native communities to resist invasion (Germino et al. 2016). Although several examples exist where microbes facilitate plant establishment, the *in situ* manipulation of microbial communities in order to alter plant–microbial feedbacks is difficult. The reason is that microbial dynamics in soil is complex (see also Chapter 3), success is seldom seen, and scalability is cost prohibitive. Only few studies have shown how manipulation of the microbial community in soil can be used to reduce the effects of plant invasion, although techniques such as the introduction of appropriate root symbionts or altering the timing of irrigation have been suggested (Sorensen et al. 2013, Germino et al. 2016) (for a discussion of soil/plant inoculants; see Chapter 22).

In Arctic systems, the interactions between plant and microbial biota will similarly drive the future responses of soil C cycling to climate change. Global warming facilitates the poleward movement of the forest–tundra boundary, furthering the establishment of shrubs in the Arctic, and changing the dominant forest species present (Sturm et al. 2001, Soja et al. 2007). Microbial communities may help to mediate these ecosystem transformations and biogeochemical processes resulting from plant community change. For example, the expansion of shrubs increases the phenolic concentrations of plant litter and soil solution, reducing the rates of decomposition through inhibitory effects on microbial enzymes (Bragazza 2012, Wang et al. 2015b). This results in higher dissolved nitrogen concentrations in soil, requiring fungal activity for the acquisition of N and creating a positive feedback to shrub productivity (Bragazza 2012). In addition, shrub-associated increases in phenolics and suppression of microbial activities can have the surprising result of limiting soil carbon losses, which would be expected to occur due to warming and drought (Wang et al. 2015b). On the other hand, some studies suggest that, by increasing the amount of C belowground, shrubs may increase the loss of older C through increasing microbial activity and decomposition of previously stable SOM (Walker et al. 2016).

Global change brings many press disturbances, in which decadal-scale changes in plant community dynamics will continue to be observed, that are mediated—in part—by the plant-associated microbial communities. However, pulse disturbances can quickly impact soil microbial communities, and depending upon the magnitude, timing, or frequency of the disturbance, they can have long-term impacts on system functioning. Two examples of pulse disturbances are wildfire and permafrost thaw as discussed in the following.

20.2.2 Wildfire Effects on Microbial Communities

Wildfire is a natural phenomenon that lies at the basis of the process of plant succession. Recovery and regrowth of vegetation after wildfire regenerates carbon and nutrient pools, such that long-term impacts on the ecosystems may be small (Hart et al. 2005). In the past, low-intensity fires acted by simply consuming a portion of the forest floor to create open parkland-type forests and/

or promote fire-adapted forest communities (Hart et al. 2005). Fire suppression activities over the last century have allowed forests to accumulate high levels of dead plant material, which has contributed to the more intense wildfires seen today (Hart et al. 2005). However, there is also compelling evidence that we are seeing more frequent, more intense wildfires and longer wildfire seasons due to warming, in both Alaska and Western temperate forests (Westerling et al. 2006, Kasischke et al. 2010). High-intensity wildfires with high plant mortality and oxidation of the surface organic horizon and/or high erosion rates can slow down postfire recovery and alter the magnitude and direction of forest succession and associated biogeochemical processes (Kashian et al. 2006, Johnstone et al. 2010).

The increasing intensities and frequencies of wildfires have the potential to alter plant and microbial community compositions to such an extent that the future trajectory of forests and their biogeochemical processes diverges from their historical norms (Johnstone et al. 2010, Kasischke et al. 2010, Westerling et al. 2006). Microbial communities are impacted by fire due to combustion of the surface organic horizons. However, more importantly, as fire frequency and intensity increase, communities may be impacted to such an extent that critical microbial processes such as decomposition and mycorrhizal relationships are reduced over decades (Holden et al. 2016). Intense wildfire reduces microbial biomass and respiration, with larger effects on Basidiomycetes and ectomycorrhizal fungi than on other (fungal) groups (Dooley and Treseder 2011, Holden et al. 2016). Increased fire severity also reduces fungal diversity in soils (Hewitt et al. 2013) and mycorrhizal interactions and seedling performance (Hewitt et al. 2016). This could impact soil C cycling in several ways. If decomposer abundance is reduced, this could result in a short-term negative feedback to climate change, with reduced rates of CO_2 fluxes to the atmosphere. If increasingly intense wildfire has a long-term deleterious effect on the populations of mycorrhizal fungi available to make association with plants, this may alter the composition and productivity of forests over the course of succession (Holden et al. 2016).

Carbon and nitrogen losses from soil resulting from wildfires can be substantial. Temperate and boreal forests can lose about one-quarter to one half of their soil C and N stocks after wildfire, and it takes one to two centuries for these soil C and N stocks to recover (Turetsky et al. 2010, Nave et al. 2011). In the boreal forest of Alaska, projections indicate that increasing intensity of fires, sometimes concomitant with loss of permafrost, has the potential to change large parts of the boreal forest to aspen-type parkland, with profound effects on wildlife and native communities which depend on those resources for their livelihood (Jolly et al. 2015). Fire impacts are expected to at least double this century (Flannigan et al. 2013), producing even greater pulse disturbances and alterations to biogeochemical processes and ecosystem structure and function.

20.2.3 Microbial Response to Permafrost Thaw

Permafrost covers about one-quarter of Earth's ice-free land area (Schaefer et al. 2011). Permafrost is perennially frozen soil, which can be several thousands to several hundreds of thousands years old. Widespread permafrost thaw is expected to occur over the coming century as a result of climate warming, particularly in central and southern Alaska, Canada, and Siberia (Schaefer et al. 2011). Permafrost thaw can be initiated by disturbance effects such as wildfire, which oxidizes the surface organic layer and reduces the insulating capacity of the organic forest floor, thus leading to warmer soils and deeper active layers (Jorgenson and Osterkamp 2005; see Figure 20.2). Diverted flows of water, as either groundwater or surface water, or simply the thermal power of water in lakes and wetlands, can also lead to new areas of thawing (Jorgenson and Osterkamp 2005). Permafrost thaw is particularly important because these soils contain more carbon than the atmosphere and terrestrial vegetation combined (Schuur et al. 2013). The potential release of permafrost C to the atmosphere could be as much as 100 pg of C over the next century (Schuur et al. 2013). Permafrost thaw may result in rapid losses of soil carbon and increased flux rates of CO_2 and CH_4 from microbial activity (Koven et al. 2011, O'Donnell et al. 2011). Although modeling studies have shown that, with

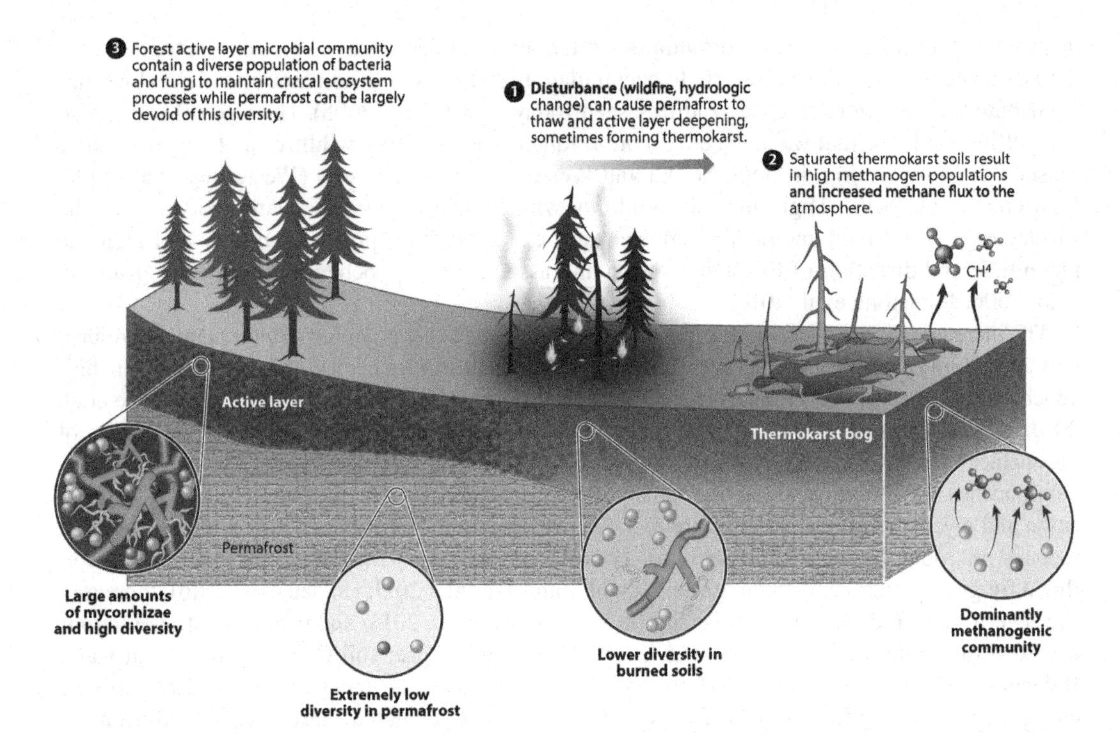
3 Forest active layer microbial community contain a diverse population of bacteria and fungi to maintain critical ecosystem processes while permafrost can be largely devoid of this diversity.
1 Disturbance (wildfire, hydrologic change) can cause permafrost to thaw and active layer deepening, sometimes forming thermokarst.
2 Saturated thermokarst soils result in high methanogen populations and increased methane flux to the atmosphere.
CH4
Active layer
Thermokarst bog
Permafrost
Large amounts of mycorrhizae and high diversity
Extremely low diversity in permafrost
Lower diversity in burned soils
Dominantly methanogenic community

FIGURE 20.2 (See color insert) Characterization of the microbial response to permafrost thawing. The diversity of soil microbial communities is affected by wildfire but is even more dramatically impacted by permafrost thawing and the formation of thermokarst bogs. Under these circumstances, the microbial community rapidly shifts to a methanogen dominated microbial community. Permafrost soils generally have lower biomass and diversity relative to active layer soils, and it is unclear how much impact permafrost microorganisms have on biogeochemical processes once thawed.

increasing temperatures, increases in plant productivity may compensate for this loss of C from soils, this compensation is expected to decline as permafrost thaw intensifies (Abbott et al. 2016).

Microbial communities in permafrost soils are particularly unique, because they are often reduced in diversity and abundance compared to unfrozen "active layer" soils that overlie them (Waldrop et al. 2010, Hultman et al. 2015). However, microbial communities have the ability to rapidly change their functional capacities after permafrost thaw (Mackelprang et al. 2011). If permafrost thaw results in more saturated soils, the ecology of methanogenic Archaea versus methane oxidizers becomes increasingly critical to understand the net flux of CH_4 to the atmosphere, and this balance can be strongly influenced by the type of vegetation that is present within the wetland (Turetsky et al. 2014). Another important aspect of Arctic systems is the unfrozen soil below the seasonal ice (called *taliks*) that maintains microbial activity over winter, which can be 6–9 months in many northern ecosystems. Microbial activities in taliks likely continue the process of decomposition and methanogenesis throughout the winter, contributing to the observed spring pulse of greenhouse gases observed in the Arctic (Raz-Yaseef et al. 2017).

The age of the permafrost layer may also have an important impact on microbial communities and rates of carbon cycling. It has been shown that Pleistocene-aged permafrost (often called Yedoma) contains high concentrations of organic carbon (Drake et al. 2015) yet tends to have a low abundance of methanogens (Rivkina et al. 2016). As a result, although high quantities of CO_2 can be rapidly released from Pleistocene soils (Drake et al. 2015) upon soil thawing, there can be limited production of CH_4 (Roy Chowdhury et al. 2014). Over the longer term, such CO_2 and CH_4 fluxes will likely depend strongly on other hydrological and ecological factors, such as soils becoming wetter or dryer, the future composition of the aboveground plant community, and its disturbance regime.

20.3 METHODOLOGICAL REVOLUTIONS AND NEW FRONTIERS

Rapid changes are occurring not only in the environment but also in scientific disciplines, particularly microbial ecology. We are currently experiencing a revolution in the field of microbial ecology that is far outpacing computation changes. The so-called Moore's law dictates that computational power doubles every 2 years. However, DNA sequencing has dramatically outpaced even improvements in computation in terms of increases in speed and decreases in cost (Hayden 2014). In the 1990s, DNA was difficult to extract from soil, and sequencing was expensive and throughput low. Today, we can sequence the genomes of entirely novel organisms from soil quickly (within weeks to months) and relatively cheaply (Mackelprang et al. 2011). See also Chapters 12–14 for details on these methods.

On the basis of the advanced molecular tools, we now have the ability to investigate how microorganisms respond to, and also influence, ecosystem processes in astonishing detail, down to the level of functional genes and transcripts (see Chapter 15). Such data will improve our ability to track, measure, and model the effects of global change, and therefore improve our ability to respond to, and mediate, environmental changes as they are occurring. However, as outlined in Chapter 15, the vast amount of information that is generated by the novel techniques can be overwhelming, and much of it may not pertain to processes at the ecosystem level. Methodological and experimental difficulties may thus preclude the finding of strong relationships between (meta)genomics and soil processes (Blazewicz et al. 2013). However, focused studies incorporating genomics, modeling, and field observations show ways in which this complexity can be distilled and used to improve biogeochemical models (Bouskill et al. 2012). In the following, we focus specifically on the development of a new era of carbon cycle models that incorporate some facets of the advanced information.

Historically, the chemical composition of C inputs into soil was thought to control carbon turnover, so that chemically complex (or so-called recalcitrant) biopolymers such as lignin would have slow decomposition rates and be selectively preserved in soils (Lehmann and Kleber 2015). This paradigm was supported by observations of changes in SOM chemistry during early stages of decomposition, particularly for litter or dissolved C components (Kalbitz et al. 2003, Manzoni and Porporato 2009). As a result, despite the fact that a number of soil C cycling models were developed (RothC and CENTURY are the primary examples), these all have a similar underlying structure: C inputs are divided into conceptual pools (typically three) based upon C chemical composition (e.g., lignin:N ratios). Each pool has a discrete and unchanging decomposition rate, where a proportion of C moves from the faster to the slower cycling pools as it decomposes (e.g., Parton et al. 1987, Manzoni and Porporato 2009). These models also incorporate the concept of physical protection of C that is adsorbed on soil minerals (Torn et al. 1997), typically by including soil texture. However, by including only metrics of chemical composition (and soil texture), microorganisms are not directly included in these models. Instead, their activity is represented implicitly through the first-order decay rates that control the decomposition of the conceptual soil C pools (Todd-Brown et al. 2011, Stevens et al. 2016). As a result, these models are controlled by carbon inputs, and the size of the microbial community has no impact on the size of the soil carbon pool. Despite this major limitation, these models have been able to predict large-scale distributions of soil C stocks relatively well, particularly in grassland and agricultural soils and under assumed steady-state conditions (Barger et al. 2011).

As our understanding has shifted to emphasize the importance of microorganisms on soil C turnover and stabilization, the ability of the traditional models to project soil C stocks under global change scenarios and in novel environments has come under increasing scrutiny. For example, many global Earth System Models (ESMs) do not agree with each other or with observational data on the distribution of soil C stocks, particularly in high-latitude soils (Koven et al. 2011, Todd-Brown et al. 2013). Moreover, these models do not produce similar projections of ecosystem or soil C to climate change suggesting either poor parameterization or incorrect model structure for modeling soil C decomposition (Manzoni and Porporato 2009, Todd-Brown et al. 2013). As these models are not reflecting our understanding of the importance of microbial activity for soil C cycling, the inclusion of properties of soil microbiomes that drive soil C cycling should improve model predictions. Here, the much larger question is, what properties of soil microbiomes (e.g., physiology, diversity, or functional metrics) are required to accurately model soil C cycling?

Some of the biggest improvements made to soil C models include microbial enzyme kinetics and the partitioning of carbon between microbial biomass and respiration. Microorganisms decompose soil C through the activity of extracellular enzymes, which are dependent on substrate concentrations (i.e., Michaelis–Menten kinetics) and temperature (i.e., Arrhenius kinetics), and both will be altered by climate change (Davidson and Janssens 2006, Soja et al. 2007). Microbial carbon use efficiency (CUE) or microbial growth efficiency, defined as the ratio of substrate C converted to microbial biomass or CO_2, is related to the functional roles of microbes as r- or K-strategists (Wieder et al. 2014; see Chapter 4) and is similarly sensitive to temperature (Frey et al. 2013). Both metrics have now been included, in various degrees of complexity, in microbially explicit soil C models. As predicted, these microbially explicit models, when coupled to ESMs, outperform traditional soil C models and improve the predictions of current soil C stocks (Wieder et al. 2013, Tang and Riley 2015). More importantly, the models project vastly divergent responses of soil C stocks to climate change (Wieder et al. 2013, Sulman et al. 2014). Interestingly, microbial CUE drives these differences in the response of soil C stocks to climate change, with either losses or gains of soil C projected with climate change depending on the response of CUE to climate change. Other microbial C cycle models that include soil enzymes (Wieder et al. 2014) microbial dormancy (Wang et al. 2015a), life history strategies (*r*- vs K-strategists; Wieder et al. 2014), density-dependent microbial turnover (Georgiou et al. 2017), and microbial social dynamics (Kaiser et al. 2015) have been created, and additional models, representing the further complexity of soil microbiome dynamics, are likely to develop. As soils constitute such large potential sinks (or sources) of atmospheric CO_2, without future constraining of this microbial response, we will be limited in our ability to predict whether soils will exert positive or negative feedbacks on climate change.

20.4 CONCLUSIONS AND FUTURE CHALLENGES

Microbial communities are critical for the Earth's biogeochemical cycles. As global change continues to affect soil microbial communities through press and pulse disturbances, understanding microbial responses will be critical to understanding how ecosystems respond. Microbial communities will likely be increasingly impacted by more frequent and intense wildfires and ecosystem changes resulting from permafrost thaw and plant invasions. But the ecosystem-level consequences of these shifts in microbial community composition and function are far from understood. The frontiers of microbial ecology lie not only in improved molecular technologies that allow us to examine microbial diversity in all of its wonder but also in asking the right questions and testing the correct hypotheses about the functioning of the system. This will enable to pinpoint and understand the intimate relationships between members of the microbial communities, as well as between these and the plant in its rooting zone. Moreover, we need to determine the emergent properties of microbial communities that allow us to understand what makes a microbial community resistant or resilient to disturbances such as posed by drought and wildfires. Finally, we need to question what level of functional detail is important to fine-tune the next generation of biogeochemical and ecological models that support predictions of the effects of global change.

REFERENCES

Abbott, B. W., J. B. Jones, E. A. G. Schuur, et al. 2016. Biomass offsets little or none of permafrost carbon release from soils, streams, and wildfire: An expert assessment. *Environ Res Lett* 11, 034014.

Barger, N. N., S. R. Archer, J. L. Campbell, C.-Y. Huang, J. A. Morton, and A. K. Knapp. 2011. Woody plant proliferation in North American drylands: A synthesis of impacts on ecosystem carbon balance. *J Geophys Res* 116, G00K07.

Bennett, J. A., H. Maherali, K. O. Reinhart, Y. Lekberg, M. M. Hart, and J. Klironomos. 2017. Plant-soil feedbacks and mycorrhizal type influence temperate forest population dynamics. *Science* 355, 181–184.

Blazewicz, S. J., R. L. Barnard, R. A. Daly, and M. K. Firestone. 2013. Evaluating rRNA as an indicator of microbial activity in environmental communities: Limitations and uses. *ISME J* 7, 2061–2068.

Blazewicz, S. J., D. G. Petersen, M. P. Waldrop, and M. K. Firestone. 2012. Anaerobic oxidation of methane in tropical and boreal soils: Ecological significance in terrestrial methane cycling. *J Geophys Res* 117, G02033.

Bouskill, N. J., J. Tang, W. J. Riley, and E. L. Brodie. 2012. Trait-based representation of biological nitrification: Model development, testing, and predicted community composition. *Front Microbiol* 3, 364.

Bragazza, L. 2012. Biogeochemical plant–soil microbe feedback in response to climate warming in peatlands. *Nat Clim Change* 3, 273–277.

Brookshire, E. N. J., and T. Weaver. 2015. Long-term decline in grassland productivity driven by increasing dryness. *Nat Commun* 6, 7148–7148.

Chen, I.-C., J. K. Hill, R. Ohlemüller, D. B. Roy, and C. D. Thomas. 2011. Rapid range shifts of species associated with high levels of climate warming. *Science* 333, 1024–1026.

Clemmensen, K. E., A. Bahr, O. Ovaskainen, et al. 2013. Roots and associated fungi drive long-term carbon sequestration in boreal forest. *Science* 339, 1615–1618.

Dai, A. 2012. Increasing drought under global warming in observations and models. *Nat Clim Change* 3, 52–58.

Davidson, E. A., and I. A. Janssens. 2006. Temperature sensitivity of soil carbon decomposition and feedbacks to climate change. *Nature* 440, 165–173.

Dooley, S. R., and K. K. Treseder. 2011. The effect of fire on microbial biomass: A meta-analysis of field studies. *Biogeochemistry* 109, 49–61.

Drake, T. W., K. P. Wickland, R. G. M. Spencer, D. M. McKnight, and R. G. Striegl. 2015. Ancient low–molecular-weight organic acids in permafrost fuel rapid carbon dioxide production upon thaw. *Proc Nat Acad Sci USA* 112, 13946–13951.

Eviner, V. T., S. A. Hoskinson, and C. V. Hawkes. 2010. Ecosystem impacts of exotic plants can feed back to increase invasion in western US rangelands. *Rangelands* 32, 21–31.

Ferrenberg, S., S. C. Reed, and J. Belnap. 2015. Climate change and physical disturbance cause similar community shifts in biological soil crusts. *Proc Nat Acad Sci USA* 112, 12116–12121.

Flannigan, M., A. S. Cantin, W. J. de Groot, M. Wotton, A. Newbery, and L. M. Gowman. 2013. Global wildland fire season severity in the 21st century. *For Ecol Manage* 294, 54–61.

Foley, J. A., N. Ramankutty, K. A. Brauman, et al. 2012. Solutions for a cultivated planet. *Nature* 478, 337–342.

Frey, S. D., J. Lee, J. M. Melillo, and J. Six. 2013. The temperature response of soil microbial efficiency and its feedback to climate. *Nat Clim Change* 3, 395–398.

Georgiou, K., R. Z. Abramoff, J. Harte, W. J. Riley, and M. S. Torn. 2017. Microbial community-level regulation explains soil carbon responses to long-term litter manipulations. *Nat Commun* 8, 1223.

Germino, M. J., J. Belnap, J. M. Stark, E. B. Allen, and B. M. Rau. 2016. Ecosystem impacts of exotic annual invaders in the genus bromus. In: *Exotic Brome-Grasses in Arid and Semiarid Ecosystems of the Western US*. Springer International Publishing: Cham, Switzerland, 61–95.

Haberl, H., K. H. Erb, F. Krausmann, et al. 2007. Quantifying and mapping the human appropriation of net primary production in earth's terrestrial ecosystems. *Proc Nat Acad Sci USA* 104, 12942–12947.

Hart, S. C., T. H. DeLuca, G. S. Newman, M. D. MacKenzie, and S. I. Boyle. 2005. Post-fire vegetative dynamics as drivers of microbial community structure and function in forest soils. *For Ecol Manage* 220, 166–184.

Hayden, E. C. 2014. Technology: The $1,000 genome. *Nature* 507, 294–295.

Hewitt, R. E., E. Bent, T. N. Hollingsworth, F. S. III. Chapin, and D. L. Taylor. 2013. Resilience of Arctic mycorrhizal fungal communities after wildfire facilitated by resprouting shrubs. *Ecoscience* 20, 296–310.

Hewitt, R. E., T. N. Hollingsworth, F. S. III. Chapin, and D. L. Taylor. 2016. Fire-severity effects on plant–fungal interactions after a novel tundra wildfire disturbance: Implications for Arctic shrub and tree migration. *BMC Ecol* 16(1), 1–11.

Hinzman, L. D., N. D. Bettez, W. R. Bolton, et al. 2005. Evidence and implications of recent climate change in northern Alaska and other Arctic regions. *Clim Change* 72, 251–298.

Hoeksema, J. D., V. B. Chaudhary, C. A. Gehring, et al. 2010. A meta-analysis of context-dependency in plant response to inoculation with mycorrhizal fungi. *Ecol Lett* 13, 394–407.

Holden, S. R., B. M. Rogers, K. K. Treseder, and J. T. Randerson. 2016. Fire severity influences the response of soil microbes to a boreal forest fire. *Environ Res Lett* 11, 1–10.

Hultman, J., M. P. Waldrop, R. Mackelprang, et al. 2015. Multi-omics of permafrost, active layer and thermokarst bog soil microbiomes. *Nature* 521, 208–212.

Inderjit and W. H. van der Putten. 2010. Impacts of soil microbial communities on exotic plant invasions. *Trends Ecol Evol* 25, 512–519.

Jastrow, J. D., R. M. Miller, and J. Lussenhop. 1998. Contributions of interacting biological mechanisms to soil aggregate stabilization in restored prairie. *Soil Biol Biochem* 30, 905–916.

Johnstone, J. F., F. S. Chapin, T. N. Hollingsworth, M. C. Mack, V. Romanovsky, and M. Turetsky. 2010. Fire, climate change, and forest resilience in interior Alaska. *Can J For Res* 40, 1302–1312.

Jolly, W. M., M. A. Cochrane, P. H. Freeborn, et al. 2015. Climate-induced variations in global wildfire danger from 1979 to 2013. *Nat Commun* 6, 7537–7537.

Jorgenson, M. T., and T. E. Osterkamp. 2005. Response of boreal ecosystems to varying modes of permafrost degradation. *Can J For Res* 35, 2100–2111.

Kaiser, C., O. Franklin, A. Richter, and U. Dieckmann. 2015. Social dynamics within decomposer communities lead to nitrogen retention and organic matter build-up in soils. *Nat Commun* 6, 8960–8960.

Kalbitz, K., D. Schwesig, J. Schmerwitz, et al. 2003. Changes in properties of soil-derived dissolved organic matter induced by biodegradation. *Soil Biol Biochem* 35, 1129–1142.

Kashian, D. M., W. H. Romme, D. B. Tinker, M. G. Turner, and M. G. Ryan. 2006. Carbon storage on landscapes with stand-replacing fires. *BioScience* 56, 598–606.

Kasischke, E. S., D. L. Verbyla, T. S. Rupp, et al. 2010. Alaska's changing fire regime—implications for the vulnerability of its boreal forests. *Can J For Res* 40, 1313–1324.

Kleber, M., P. Sollins, and R. Sutton. 2007. A conceptual model of organo-mineral interactions in soils: Self-assembly of organic molecular fragments into zonal structures on mineral surfaces. *Biogeochem* 85, 9–24.

Koven, C. D., B. Ringeval, P. Friedlingsteinet, et al. 2011. Permafrost carbon-climate feedbacks accelerate global warming. *Proc Nat Acad Sci USA* 108, 14769–14774.

Lau, J. A., and J. T. Lennon. 2012. Rapid responses of soil microorganisms improve plant fitness in novel environments. *Proc Nat Acad Sci USA* 109, 14058–14062.

Lehmann, J., and M. Kleber. 2015. The contentious nature of soil organic matter. *Nature* 528, 60–68.

Lenoir, J., J. C. Gégout, P. A. Marquet, P. de Ruffray, and H. Brisse. 2008. A significant upward shift in plant species optimum elevation during the 20th century. *Science* 320, 1768–1771.

Lewis, S. L., and M. A. Maslin. 2016. Defining the Anthropocene. *Nature* 519, 171–180.

Mackelprang, R., M. P. Waldrop, K. M. DeAngelis, et al. 2011. Metagenomic analysis of a permafrost microbial community reveals a rapid response to thaw. *Nature* 480, 368–371.

Maestre, F. T., M. Delgado-Baquerizo, T. C. Jeffries, et al. 2015. Increasing aridity reduces soil microbial diversity and abundance in global drylands. *Proc Nat Acad Sci USA* 112, 15684–15689.

Manzoni, S., and A. Porporato. 2009. Soil carbon and nitrogen mineralization: Theory and models across scales. *Soil Biol Biochem* 41, 1355–1379.

Nave, L. E., E. D. Vance, C. W. Swanston, and P. S. Curtis. 2011. Fire effects on temperate forest soil C and N storage. *Ecol Appl* 21, 1189–1201.

O'Donnell, J. A., M. T. Jorgenson, J. W. Harden, A. D. McGuire, M. Z. Kanevskiy, and K. P. Wickland. 2011. The effects of permafrost thaw on soil hydrologic, thermal, and carbon dynamics in an Alaskan peatland. *Ecosystems* 15, 213–229.

Parton, W. J., D. S. Schimel, C. V. Cole, and D. S. Ojima. 1987. Analysis of factors controlling soil organic matter levels in Great plains grasslands. *Soil Sci Soc Am J* 51, 1173–1179.

Pearson, R. G. 2013. Shifts in Arctic vegetation and associated feedbacks under climate change. *Nat Clim Change* 3, 673–677.

Raz-Yaseef, N., M. S. Torn, Y. Wu, et al. 2017. Large CO_2 and CH_4 emissions from polygonal tundra during spring thaw in northern Alaska. *Geophys Res Lett* 44, 504–513.

Reinhart, K. O., and R. M. Callaway. 2004. Soil biota facilitate exotic Acer invasions in Europe and North America. *Ecol Appl* 14, 1737–1745.

Rivkina, E., L. Petrovskaya, T. Vishnivetskaya, et al. 2016. Metagenomic analyses of the late Pleistocene permafrost – additional tools for reconstruction of environmental conditions. *Biogeosciences* 13, 2207–2219.

Roy Chowdhury, T., E. M. Herndon, T. J. Phelps, et al. 2014. Stoichiometry and temperature sensitivity of methanogenesis and CO_2 production from saturated polygonal tundra in Barrow, Alaska. *Global Change Biol* 21, 722–737.

Sala, O. E., F. S. Chapin, J. J. Armesto, et al. 2000. Global biodiversity scenarios for the year 2100. *Science* 287, 1770–1774.

Scheffer, M., S. Carpenter, J. A. Foley, C. Folke, and B. Walker. 2001. Catastrophic shifts in ecosystems. *Nature* 413, 591–596.

Schaefer, K., T. Zhang, L. Bruhwiler, and A. P. Barrett. 2011. Amount and timing of permafrost carbon release in response to climate warming. *Tellus B Chem Phys Meteorol* 63, 165–180.

Schuur, E. A. G., B. W. Abbott, W. B. Bowden, et al. 2013. Expert assessment of vulnerability of permafrost carbon to climate change. *Clim Change* 119, 359–374.

Schuur, E. A. G., K. G. Crummer, J. G. Vogel, and M. C. Mack. 2007. Plant species composition and productivity following permafrost thaw and thermokarst in Alaskan Tundra. *Ecosystems* 10, 280–292.

Seager, R., M. Ting, I. Held, et al. 2007. Model projections of an imminent transition to a more arid climate in Southwestern North America. *Science* 316, 1181–1184.

Shade, A., H. Peter, S. D. Allison, et al. 2012. Fundamentals of microbial community resistance and resilience. *Front Microbiol* 3, 417.

Soja, A. J., N. M. Tchebakova, N. H. F. French, et al. 2007. Climate-induced boreal forest change: Predictions versus current observations. *Global Planet Change* 56, 274–296.

Sorensen, P. O., M. J. Germino, and K. P. Feris. 2013. Microbial community responses to 17 years of altered precipitation are seasonally dependent and coupled to co-varying effects of water content on vegetation and soil C. *Soil Biol Biochem* 64, 155–163.

Stevens, N., C. E. R. Lehmann, B. P. Murphy, and G. Durigan. 2016. Savanna woody encroachment is widespread across three continents. *Global Change Biol* 23, 235–244.

Sulman, B. N., R. P. Phillips, A. C. Oishi, E. Shevliakova, and S. W. Pacala. 2014. Microbe-driven turnover offsets mineral-mediated storage of soil carbon under elevated CO_2. *Nat Clim Change* 4, 1099–1102.

Sturm, M., C. Racine, and K. Tape. 2001. Climate change. Increasing shrub abundance in the Arctic. *Nature* 411, 546–547.

Tang, J., and W. J. Riley. 2015. Weaker soil carbon-climate feedbacks resulting from microbial and abiotic interactions. *Nat Clim Change* 5, 56–60.

Tilman, D., C. Balzer, J. Hill, and B. L. Befort. 2011. Global food demand and the sustainable intensification of agriculture. *Proc Nat Acad Sci USA* 108, 20260–20264.

Timmusk, S., I. A. Abd El-Daim, L. Copolovici, et al. 2014. Drought-tolerance of wheat improved by rhizosphere bacteria from harsh environments: Enhanced biomass production and reduced emissions of stress volatiles. *PLoS One* 9, e96086.

Todd-Brown, K. E. O., F. M. Hopkins, S. N. Kivlin, J. M. Talbot, and S. D. Allison. 2011. A framework for representing microbial decomposition in coupled climate models. *Biogeochemistry* 109, 19–33.

Todd-Brown, K. E. O., J. T. Randerson, W. M. Post, et al. 2013. Causes of variation in soil carbon simulations from CMIP5 Earth system models and comparison with observations. *Biogeosciences* 10, 1717–1736.

Torn, M. S., S. E. Trumbore, O. A. Chadwick, and P. M. Vitousek. 1997. Mineral control of soil organic carbon storage and turnover. *Nature* 389, 170.

Turetsky, M. R., E. S. Kane, J. W. Harden, et al. 2010. Recent acceleration of biomass burning and carbon losses in Alaskan forests and peatlands. *Nat Geosci* 4, 27–31.

Turetsky, M. R., A. Kotowska, J. Bubier, et al. 2014. A synthesis of methane emissions from 71 northern, temperate, and subtropical wetlands. *Global Change Biol* 20, 2183–2197.

Van der Heijden, M. G. A., R. D. Bardgett, and N. M. van Straalen. 2008. The unseen majority: Soil microbes as drivers of plant diversity and productivity in terrestrial ecosystems. *Ecol Lett* 11, 296–310.

Vitousek, P. M. 1994. Beyond global warming: Ecology and global change. *Ecology* 75, 1861–1876.

Vitousek, P. M. 1997. Human domination of Earth's ecosystems. *Science* 277, 494–499.

Waldrop, M. P., and M. K. Firestone. 2006. Response of microbial community composition and function to soil climate change. *Microb Ecol* 52, 716–724.

Waldrop, M. P., K. P. Wickland, R. III. White, A. A. Berhe, J. W. Harden, and V. E. Romanovsky. 2010. Molecular investigations into a globally important carbon pool: Permafrost-protected carbon in Alaskan soils. *Global Change Biol* 16, 2543–2554.

Walker, T. N., M. H. Garnett, S. E. Ward, S. Oakley, R. D. Bardgett, and N. J. Ostle. 2016. Vascular plants promote ancient peatland carbon loss with climate warming. *Global Change Biol* 22, 1880–1889.

Wang, G., S. Jagadamma, M. A. Mayes, et al. 2015a. Microbial dormancy improves development and experimental validation of ecosystem model. *ISME J* 9, 226–237.

Wang, H., C. J. Richardson, and M. Ho. 2015b. Dual controls on carbon loss during drought in peatlands. *Nat Clim Change* 5, 584–587.

Westerling, A. L., H. G. Hidalgo, D. R. Cayan, and T. W. Swetnam. 2006. Warming and earlier spring increase western U.S. forest wildfire activity. *Science* 313, 940–943.

Wieder, W. R., G. B. Bonan, and S. D. Allison. 2013. Global soil carbon projections are improved by modelling microbial processes. *Nat Clim Change* 3, 909–912.

Wieder, W. R., A. S. Grandy, C. M. Kallenbach, and G. B. Bonan. 2014. Integrating microbial physiology and physio-chemical principles in soils with the MIcrobial-MIneral Carbon Stabilization (MIMICS) model. *Biogeosciences* 11, 3899–3917.

Wolfe, B. E., and J. N. Klironomos. 2005. Breaking new ground: Soil communities and exotic plant invasion. *BioScience* 55, 477–487.

21 Soil Suppressiveness to Plant Diseases

Christian Steinberg, Véronique Edel-Hermann, Claude Alabouvette, and Philippe Lemanceau
University of Bourgogne Franche-Comté

CONTENTS

21.1 Introduction 343
21.2 Assessment of Soil Infectious Potential and Soil Receptivity to Soilborne Diseases 344
21.3 Different Types of Disease Suppressiveness 346
21.4 Importance of Microbial Interactions in the Natural Suppression of Plant Diseases 347
21.4.1 Role of Biotic versus Abiotic Factors in Disease Suppression 347
21.4.2 General versus Specific Suppressiveness 348
21.4.3 Identification of Organisms Involved in Disease Suppression 348
21.4.4 Mechanisms Involved in Disease Suppression 349
21.5 Identification of Microorganisms, Genes, and Metabolites Involved in Disease Suppression 350
21.5.1 Early Studies and Importance of Diversity 350
21.5.2 High-Throughput Meta-Omics Techniques in the Context of Disease Suppression 351
21.5.3 Metaproteomics and Metabonomics in the Context of Disease Suppression 354
21.6 Soil Suppressiveness Revisited—Possible Spin-Offs 354
21.6.1 Isolation and Selection of Biocontrol Agents 354
21.6.2 Development of Methods to Characterize Phytosanitary Soil Quality 355
21.6.3 Agricultural Practices Modulate Disease Suppressiveness 355
References 357

21.1 INTRODUCTION

A plant disease results from the intimate interaction between a plant and a pathogen. Today, a great deal of research is devoted to the study of plant–pathogen interactions at the cellular and molecular level. These direct interactions are important but should not outshine the key roles of environmental factors, which influence these interactions and thereby disease incidence or severity. These indirect interactions are particularly important in the case of plant diseases caused by soilborne pathogens. The existence of soils that naturally suppress plant diseases (so-called *disease-suppressive soils*) provides an example of soil (biotic and/or abiotic) factors affecting the pathogen, the plant, or the interaction between the plant and the pathogen. In suppressive soils, disease incidence or severity commonly remains low in spite of the presence of the pathogen, a susceptible host plant and climatic conditions that would allow disease development (Cook and Baker 1983).

Studies of suppressive soils have led to the basic concept of "soil receptivity" to plant diseases, which is defined as the ability of soils to allow the expression of plant disease. Soil is not a neutral environment, in which pathogenic microorganisms interact freely with the roots of the host plant. On the contrary, soil can interfere in several ways with the relationships between and among microorganisms, pathogens, and plants. It can even modify the interactions among microorganisms themselves. Every soil has some potential for disease suppression, and this has to be considered as

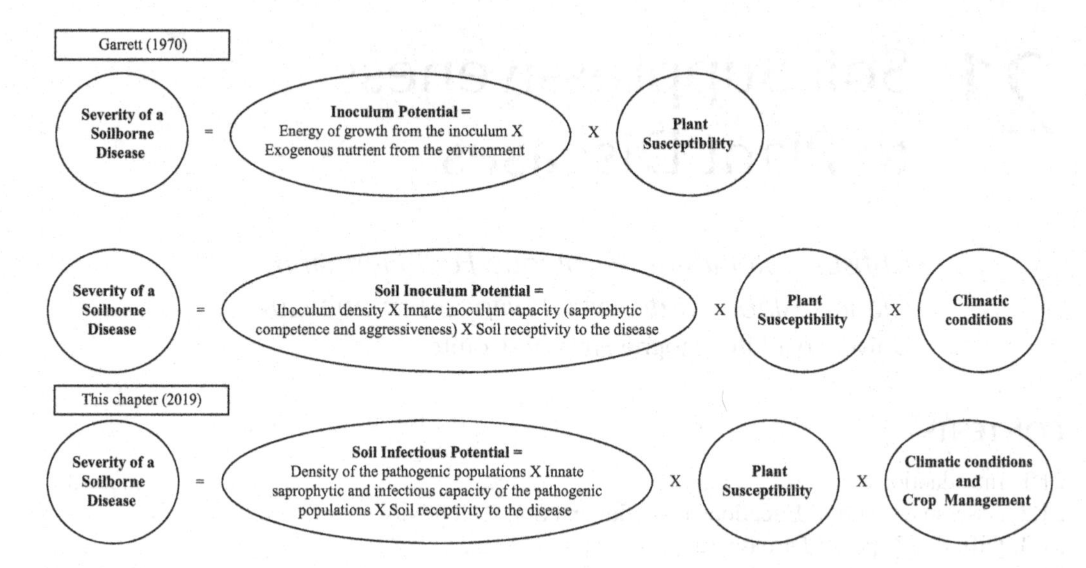

FIGURE 21.1 Parameters affecting the disease severity of soilborne diseases as proposed by Garrett (Garrett 1970) in the present chapter.

a continuum, from strongly suppressive to highly conducive states; a conducive soil—in contrast to a suppressive soil—thus is a soil that allows disease expression at a high level.

The idea of soil receptivity to diseases was already included in the "inoculum potential" concept defined by Garrett (1970) as "the energy of growth of a parasite available for infection of a host at the surface of the host organ to be infected." The concept states that the (pathogen) inoculum alone, although necessary, is not sufficient to explain the disease. Among the factors that affect the "energy of growth from the inoculum," Garrett pointed to "the collective effect of environmental conditions" and noted that "the endogenous nutrients of the inoculum might be augmented by exogenous nutrients from the environment." Three main factors may determine the soil inoculum potential:

1. The inoculum density, expressed as the number of pathogenic propagules per unit of soil (volume)
2. The inoculum capacity, which encompasses both the genetic and physiological capacities of the inoculum to infect the plant
3. The effect of the soil environment, which affects both the inoculum density and the inoculum capacity

The latter term is equivalent to what was described earlier as soil receptivity to diseases (see also Figure 21.1). Many authors consider the soil inoculum potential to be the key factor determining disease incidence or severity. Taking into account the diversity of pathogenic populations as described over the last years, we now would rather replace the term "inoculum"—which suggests that a specific pathogenic strain has been introduced—by "pathogenic populations." Similarly, the term "inoculum potential" in the definition of Garrett (1970) should be replaced by "infectious potential."

21.2 ASSESSMENT OF SOIL INFECTIOUS POTENTIAL AND SOIL RECEPTIVITY TO SOILBORNE DISEASES

In Garrett's definition of inoculum potential, the most important word is "energy." However, nobody knows to date how to measure the so-called pathogenic energy of the inoculum in the soil. A pragmatic solution to resolving this riddle is to test the infested soil with a highly susceptible plant cultivar under conditions favorable for disease expression. Quantification of the soil infectious potential can be achieved by "diluting" (mixing) naturally infested soil into a disinfested soil, growing the highly susceptible plant, and monitoring symptom appearance with time. After appropriate transformation, the correlation between the volume of naturally infested soil in the mixture and disease incidence can be calculated (using, e.g., regression analysis). As an example, this method applies to soil infested by *Pythium* spp. responsible for damping-off disease of cucumber. It allows the calculation of the infectious potential unit (IPU), which corresponds to the volume of soil necessary to induce the death of 50% of the plants (Figure 21.2). Based on the IPU value, the infectious potential of different soils can be compared and the soil infectious potential

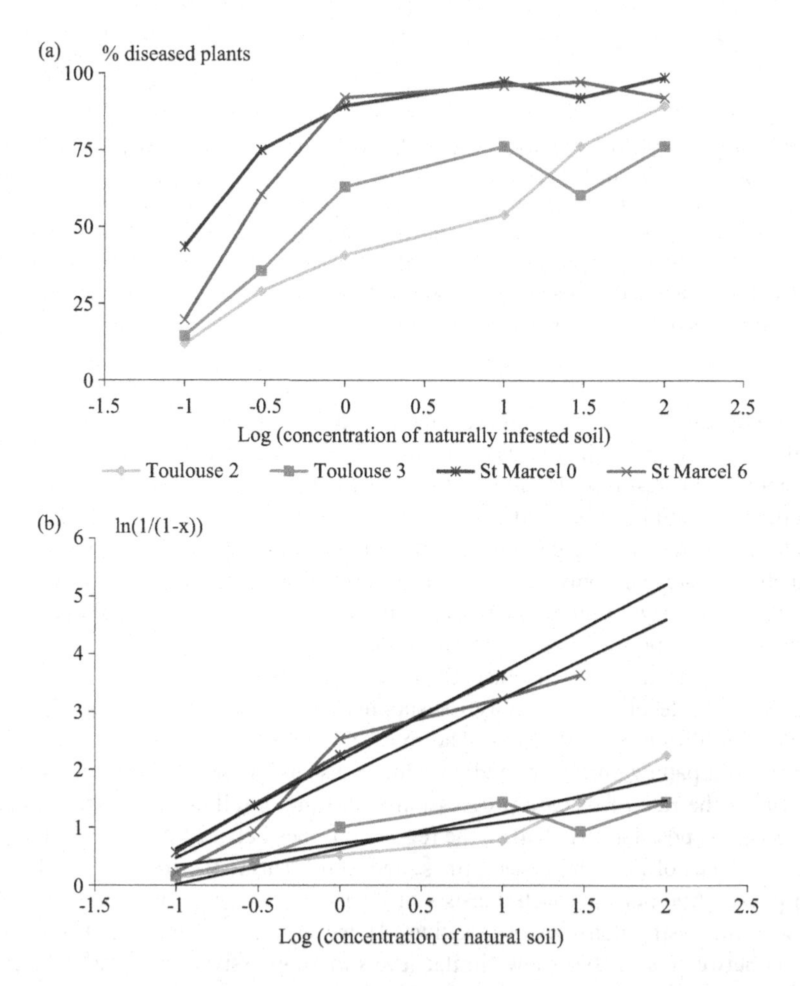

FIGURE 21.2 Methods of analysis of soil infectious potential to damping-off due to *Pythium* spp. (a) Measurement of damping-off severity in four soil samples naturally infested with *Pythium* spp. responsible for damping-off of cucumber (*Cucumis sativus*). The soil samples were diluted into steam-disinfected soil (ratios varying from 0.1% to 100%, v/v) and deposited at the crown of 6-day-old seedlings of cucumber previously grown in steamed potting mixture (three replicates/soil, five plants/replicate). The number of damped off plantlets was recorded 6 days later. (b) Data analysis. Linear regressions of transformed data allowed to calculate the number of IPU_{50} contained in 1 g of each of the natural soil samples. The samples from St. Marcel exhibited very high IPU_{50} (4.5 IPU_{50}/g soil) revealing a high soil infectious potential, while the soil samples from Toulouse had more than 100 times less IPU_{51}/g of soil (0.02 IPU_{51}/g soil) indicating a low soil infectious potential. To summarize, 0.2 and 500 g of soils of St Marcel and of Toulouse, respectively, are potentially able to induce mortality in 50% of the cucumber.

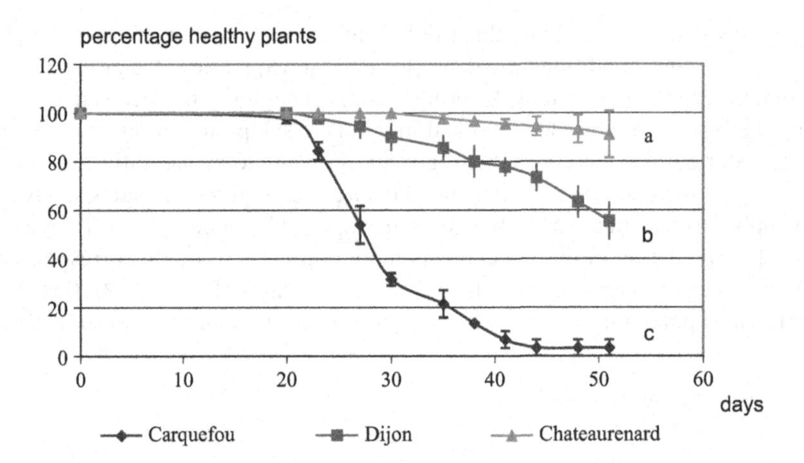

FIGURE 21.3 Suppressiveness to fusarium wilt of flax as assessed in three different soils. The soils were infested with 2,500 propagules of *F. oxysporum* f.sp. *lini*/g soil. Flax seeds (*Linum usitatissimum*) were cultivated in the infested soils for 53 days (three replicates/soil, 30 plants/replicate). Numbers of healthy and wilted plants were recorded twice a week. Areas under the disease progress curves (AUDPC) were measured and compared. The soil of Carquefou appeared as the soil that was most conducive to fusarium wilt, while the soil of Châteaurenard was qualified as a suppressive soil. The soil of Dijon, being in an intermediate situation, revealed that there is a continuum between suppressiveness and conduciveness to a given disease.

correlated with the disease severity observed in the field. The measurement of the infectious potential of a soil is valid only for the disease for which the bioassay was conducted and concerns the pathogenic populations responsible only for this disease, not for all soilborne diseases. Estimation of soil infectious potential can be used as a diagnostic tool to advise farmers whether or not a plant susceptible to the corresponding pathogenic populations can be safely cultivated in a given field. However, such estimates can only be made for diseases such as damping-off or root rot of which the symptoms are directly correlated with the activity of the pathogenic populations. No bioassay measures the infectious potential of soil for diseases in which the symptoms may appear long after the attack by pathogens, as is, for instance, the case for tracheomycoses.

Assessment of the level of receptivity (or suppressiveness) to disease of a given soil can be achieved following different approaches. The most common one consists of infesting soils with a given density of a pathogenic strain and growing a susceptible host plant under standard conditions, after which the rate of disease progression with time is followed. Using multiple infection transformations, it is possible to calculate the regression lines between disease incidence and time, to compare the slopes of these regression lines, and to determine the time required to reach 50% of diseased plants. Another approach consists of measuring the infectious potential of soil after infestation with increasing densities of inoculum. However, none of these methods allows for the differentiation between soils that show similar levels of suppressiveness. Thus, the best approach is to take into account the effects of both increasing densities of the pathogenic inoculum and time in relation to disease incidence. Different statistical procedures will allow the comparison of the receptivity level between soils. Thus, it is possible to calculate either the mean survival time of the plants or the area under the disease progress curve (AUDPC) to compare the levels of receptivity of several soils after variance analysis (Figure 21.3).

21.3 DIFFERENT TYPES OF DISEASE SUPPRESSIVENESS

Soils suppressive to diseases caused by the most important soilborne plant pathogens have been described all over the world. The latter include bacterial and fungal pathogens but also nematodes.

The large diversity of pathogens controlled by suppressive soils shows that soil suppressiveness is not a rare phenomenon.

Usually, disease suppressiveness of soil, as soil infectious potential, is specific to a given class of disease. This was illustrated for soils suppressive to fusarium wilts that were not suppressive to any other soilborne disease, not even to diseases caused by *Fusarium solani*. There is not a single mechanism of soil suppressiveness to diseases. Based on the concepts presented earlier, the absence of disease in a soil might be due to the following:

1. Absence of the pathogen
2. Presence of an avirulent pathogen
3. Effects of the environment that restrict either pathogen density or pathogenicity

When effects of the environment affect pathogen density, it would be preferable to qualify the soil as *pathogen suppressive* rather than *disease suppressive*. Baker and Cook (1974) described three types of suppressive soils: (1) the pathogen does not establish; (2) the pathogen establishes but fails to produce disease; and (3) the pathogen establishes and causes disease at first, but then, disease severity diminishes with continued cultivation of the same crop. The latter type of suppressiveness is illustrated by the well-known example of take-all decline (TAD) soils. TAD is defined as the spontaneous decrease of the incidence and severity of take-all disease, caused by *Gaeumannomyces graminis* in various crop plants. TAD occurs in wheat or other susceptible crops after one or more severe outbreaks of the disease (Cook 2003). The second type of suppressive soil is illustrated by fusarium wilt-suppressive soils. It has been well demonstrated that the pathogen is still present and virulent in these soils but that environmental conditions prevent disease expression (Alabouvette 1986). These two types of suppressive soils are among the most well-studied ones. For instance, TAD is acquired over time upon the monoculturing of susceptible host plants in the presence of the pathogen. On the other hand, suppressiveness to fusarium wilts does not seem to be related to any specific crop management regime, in spite of one report that describes the acquisition of suppressiveness to fusarium wilt after several monocultures of watermelon. Besides the fact that TAD is acquired, whereas suppressiveness to fusarium wilts is constitutive, suppressiveness to fusarium wilt is very stable over time, whereas TAD can be reduced or eliminated by breaking the monoculture with a nonhost crop. The stability of the suppressiveness to fusarium wilts points to a role of preexisting properties of the soil and associated microbiota, which explains why early studies focused on the relationship between disease incidence and soil type (Alabouvette et al. 1996).

The three types of disease suppressiveness described earlier underlay diverse mechanisms of disease suppression. Consequently, many studies have addressed the diverse biotic and abiotic interactions involved in disease suppressiveness.

21.4 IMPORTANCE OF MICROBIAL INTERACTIONS IN THE NATURAL SUPPRESSION OF PLANT DISEASES

21.4.1 Role of Biotic versus Abiotic Factors in Disease Suppression

From a theoretical point of view, either abiotic or biotic soil factors, or both, may be involved in the mechanisms of disease suppression in a given soil. The easiest way to differentiate biotic and abiotic factors is the application of soil disinfection, by which (active or living) soil microbiota is eliminated, and then to compare the level of disease suppressiveness of the disinfested to that of an untreated soil. The role of the soil microflora in the suppression of fusarium wilts was established by showing that the suppressive effect disappears upon soil disinfection by steam, methyl bromide, or γ-irradiation and that it is restored by mixing a small quantity of suppressive soil into heat-treated soil (Alabouvette 1986). The same strategy was used to demonstrate the microbial nature of the suppressiveness of soils to take-all and black root rot of tobacco (Stutz et al. 1986, Weller et al. 2002).

Even if these observations allow the conclusion that soil suppressiveness is often of microbial nature, there are examples where abiotic factors play a major role. This is the case with diseases caused by *Aphanomyces euteiches* and *Rhizoctonia solani*. Abiotic factors will always control, to some extent, the diversity and activity of soil microbiomes, and thus play a role in disease suppressiveness. For example, most soils suppressive to fusarium wilts are rich in smectite-type clays and have a pH higher than 7.0 (Alabouvette et al. 1996). Similarly, a clear relationship between soil type and suppressiveness to black root rot of tobacco has been established.

21.4.2 General versus Specific Suppressiveness

Natural disease suppressiveness is, thus, a complex phenomenon, resulting from the activity of various populations of microorganisms. Cook and Baker (1983) distinguished two types of mechanisms:

1. "General" suppressiveness
2. "Specific" suppressiveness

These two types of disease suppressiveness are thought to involve (1) the global microbiome acting as a type of nutrient sink (thereby antagonizing pathogens) and (2) specific populations of antagonists. Most fungal pathogens can survive in soil as resting structures (chlamydospores, microsclerotia, or oospores), which, in order to become active and cause a problem, have to germinate and grow saprophytically in soil towards the plant. During this saprophytic growth phase, these pathogens compete for nutrients, and nutrient competition is hence thought to be mainly responsible for the *fungistasis* that is sometimes observed, that is, the inhibition of spore germination in soil. The addition of a carbon source—such as glucose—was found to induce an increased germination of chlamydospores of *Fusarium oxysporum* in suppressive as well as in conducive soil (Alabouvette 1986). However, a higher concentration of added carbon was needed in the suppressive than in the conducive soil to induce the same level of chlamydospore germination and saprophytic growth of *Fusarium*. This suggests that fungistasis is expressed with greater intensity in suppressive than in conducive soils. Similarly, iron limitation can lead to decreased saprophytic growth of *F. oxysporum* (Lemanceau and Alabouvette 1993). Increased saprophytic growth of pathogenic *Fusarium*, resulting from the addition of carbon and iron, will thus lead to enhanced disease severity. This "facilitation" by adding growth-limiting compounds can make a suppressive soil conducive, and a conducive one even more conducive. In contrast, reduction of iron availability results in reduced saprophytic growth of pathogenic *F. oxysporum* and hence in increased disease suppressiveness in soils previously characterized as either conducive or suppressive. Interaction between carbon and iron competition makes the level of suppressiveness even more intense (Lemanceau and Alabouvette 1993). Thus, the intensity of competition for carbon appears to be directly related to the carbon sink associated with the microbial biomass and activity, thereby contributing to the general suppression. On top of that, the intensity of the competition for iron is determined by the physicochemical properties of the soils that are suppressive to fusarium wilts (high pH and $CaCO_3$ contents). Despite the importance of general disease suppressiveness, in some suppressive soils—including those to fusarium wilts—suppressiveness is disease specific. As general suppressiveness is theoretically based on aspecific mechanisms and is expected to act against any pathogen, this indicates that this form of suppressiveness cannot account for all disease suppression observed.

21.4.3 Identification of Organisms Involved in Disease Suppression

It is important to identify the microbial populations that are involved in specific disease suppression (Weller et al. 2002). Early strategies to identify microbial groups involved in suppressiveness consisted of subjecting suppressive soils to progressively increasing temperatures, for example, 50°C, 55°C, 60°C, 65°C, and 70°C. For soils suppressive both to fusarium wilts and to take-all, the suppressiveness was destroyed at 55°C–65°C. The next steps were as follows:

1. Identifying the microbial groups that were most affected by the temperature(s)
2. Isolating strains belonging to those groups from the suppressive soil prior to heat treatment
3. Reintroducing these strains into the heat-treated soil to restore disease suppressiveness

This strategy follows the classical *postulates of Koch*. Using this strategy, it turned out that fluorescent pseudomonads were involved in the natural suppressiveness of soils to black root disease of tobacco, fusarium wilts, and take-all disease (Stutz et al. 1986, Weller et al. 2002). Interestingly and in contrast to this, nonpathogenic *F. oxysporum* was also identified as playing a major role in the natural suppression of fusarium wilts (Alabouvette 1986). This observation stresses the limits of this strategy, which relates to the small number of microbial species that can be tested, to the lack of consideration of the diversity within these species and to the consideration of only culturable isolates. The newly developed molecular approaches, as discussed in Section 21.5 (see also Chapters 13 and 14), now identify taxa that are associated with disease suppression or even unravel the mechanisms involved. Examples are studies that revealed the production of particular metabolites by some of these taxa, and linking these to the disease suppression (Mendes et al. 2011, Sanguin et al. 2006, Siegel-Hertz et al. 2018). In a generic sense, they confirm that disease suppressiveness of soil often cannot be ascribed to a single bacterial or fungal taxon but is likely governed by microbial consortia.

21.4.4 Mechanisms Involved in Disease Suppression

With respect to "antagonism against disease," very diverse modes of action exist. These can be divided into two broad categories:

1. Microbial antagonism
2. Stimulation of the defense reaction of the host plant

Microbial antagonism has been discussed in detail in Chapter 9. Antagonism against a target pathogen will result in a reduction of the saprophytic growth of the latter and, consequently, in reduced infection of the plant and consequent disease severity. The antagonism can be based on parasitism, competition for nutrients, and/or antibiosis. Parasitism of plant–pathogenic fungi by viruses or virus-like particles can induce so-called *hypovirulence* (lowered virulence; see Chapter 6). On another notice, soil amoebae may perforate and destroy the spores of pathogenic fungi, and so reduce their populations. See Chapter 8 for a treatise of soil protists.

Competition for nutrients is a generalized phenomenon in soil, since, as we have seen, soil is grossly poor in readily available nutrients. Some microbial populations are particularly competitive for a given nutrient, and the use of this competitive saprophytic ability makes part of their antagonistic mode of action. For example, nonpathogenic *F. oxysporum*, which shows trophic characteristics similar to those of pathogenic *Fusarium*, efficiently competes with the pathogen for carbon in soil (Alabouvette et al. 1996). Also, fluorescent pseudomonads possess a very efficient iron uptake strategy based on the synthesis of siderophores that have an affinity for iron that is higher than that of the siderophores produced by various pathogenic fungi including *F. oxysporum* (Mazurier et al. 2009). In this way, these bacterial siderophores reduce the availability of iron to the pathogens. And, under these conditions, the pathogenic *F. oxysporum* even becomes more susceptible to carbon competition by a nonpathogenic *F. oxysporum* (Lemanceau and Alabouvette 1993).

Antibiosis exerted by particular bacteria and fungi has been discussed in Chapter 9. In the context of plant disease suppression, fluorescent pseudomonads are important, as they can produce a large diversity of antifungal secondary metabolites which are responsible for their biocontrol capacity (Gross and Loper 2009). These compounds include antibiotics, surfactants, and chitinolytic enzymes. Fungi such as *Trichoderma* and *Gliocladium* spp. also produce secondary metabolites that are toxic towards other fungi as well as enzymes responsible for cell wall degradation.

Most biological control agents, including fluorescent pseudomonads (Lemanceau and Alabouvette 1993) and nonpathogenic *F. oxysporum* (Aimé et al. 2013), can trigger plant defense reactions and induced systemic resistance. The relative importance of this mechanism in natural disease suppression is difficult to estimate but should not be neglected. Very recent studies point to a role of the endophytic microflora in the protection of plants and the stimulation of their immune defenses (Bacon and White 2016).

Antagonistic modes of action have been studied at both the cellular and molecular levels. They are now a part of ecological studies to demonstrate their involvement in natural soil suppressiveness. A case in point is given by the demonstration of antibiosis as the mechanism that causes TAD. The most effective *Pseudomonas* biocontrol strains isolated from TAD soils were shown to produce either 2,4-diacetylphloroglucinol (2,4-DAPG) or phenazine-1-carboxylic acid (PCA). Synthesis of these antibiotics in the rhizosphere and their involvement in the biocontrol activity of the bacterial strains were demonstrated (Mazurier et al. 2009, Weller et al. 2002). Using a specific polymerase chain reaction (PCR), the densities of the bacterial populations capable of producing either PCA or 2,4-DAPG were compared between TAD and conducive soils. Whereas PCA producers were not detected on the roots of wheat in either conducive or suppressive soils, 2,4-DAPG producers were present on wheat roots cultivated in suppressive soils at densities above the threshold required to control take-all but were below the threshold on roots in conducive soils (Raaijmakers and Weller 2001). Further evidence, including increased densities of 2,4-DAPG producers with wheat monoculture and a concomitant increased suppression of take-all, decrease of 2,4-DAPG producer densities with disruption of the wheat monoculture and concomitant elimination of the natural suppressiveness allowed to pinpoint 2,4-DAPG producers as major contributors to the TAD (Raaijmakers and Weller 2001). Such a conclusion could not be reached for PCA producers, despite the demonstration of the role of this metabolite in the control achieved by PCA produced in fluorescent pseudomonad strains, including *P. fluorescens* 2–79 (Weller et al. 2002).

21.5 IDENTIFICATION OF MICROORGANISMS, GENES, AND METABOLITES INVOLVED IN DISEASE SUPPRESSION

21.5.1 Early Studies and Importance of Diversity

Soils are major reservoirs of biodiversity, as discussed in Chapters 3 and 14. Soil microbiome biodiversity is currently increasingly accessible on the basis of the novel advanced omics methods (see Chapter 3). Thus, untargeted omics approaches that characterize the respective microbiomes offer great possibilities, also in the context of disease suppressiveness.

The general strategy to identify microbial populations involved in disease suppressiveness consists of comparing the structures and diversity of microbiomes across soils showing different levels of suppressiveness in order to identify those microbial groups and populations that are preferentially associated with suppressive soils. Thus, bioindicators of phytosanitary soil quality may be defined. Next to bacteria and fungi, oomycetes and, to a lesser extent, protozoa and nematodes are to be taken into consideration (Edel-Hermann et al. 2004, Waite et al. 2003). Another approach consists of subjecting a soil to treatments that affect the level of soil suppressiveness, either positively or negatively, and to search for shifts in the structures of the microbiomes as a result of the treatments.

Molecular fingerprintings and/or amplicon sequencing have been used as first steps in analyzing the structure and diversity of microbiomes in the context of disease suppression. In an early study, the structures of bacterial and fungal communities were compared by molecular fingerprinting, i.e. terminal restriction fragment length polymorphism (T-RFLP), of soils with suppressiveness levels modified by cultural practices. Organic amendments both enhanced the suppressiveness of the investigated soil to *Rhizoctonia solani* and induced major shifts in the bacterial and fungal community structures (Pérez-Piqueres et al. 2006). Another early study (Yin et al. 2003a,b) modified the level of suppressiveness of soil to *Heterodera schachtii* by introducing increasing amounts of a suppressive soil as well as by a biocidal treatment, namely, soil fumigation. An analysis of the rRNA genes associated with nematode cysts in the different treatments revealed a higher occurrence of defined bacteria (i.e. *Rhizobium* spp. and uncultured α-Proteobacteria) and fungi (i.e. *Dactylella oviparasitica*) in the highly suppressive soil. Quantitative PCR showed the consistent association of these groups with suppressiveness. Studying root rot of potato caused by *R. solani*, van Elsas et al. (2002) compared soils under different agricultural regimes showing different levels of suppression by using PCR-denaturing gradient gel electrophoresis analysis of the microbiomes. Both total and group-specific bacterial patterns showed higher diversities in the soils that exhibited higher pathogen suppression. In this case, there was a correlation between the level of suppressiveness and that of the *prnD* gene, as evidenced by quantitative PCR. The encoded protein, PrnD, is involved in the production of the antibiotic pyrrolnitrin, which is known to be able to suppress the growth of *R. solani*. In a more recent study, Cretoiu et al. (2013) revealed the selection of particular oxalobacteraceae and actinobacteria in conjunction with the enhancement of suppression of the pathogen *Verticillium dahliae* in a sandy field soil treated with chitin.

21.5.2 High-Throughput Meta-Omics Techniques in the Context of Disease Suppression

High-throughput sequencing can provide much more detail with respect to the taxa that are associated with suppressive soils (Gomez Exposito et al. 2017, Mendes et al. 2011, Rosenzweig et al. 2012, Siegel-Hertz et al. 2018). First, a comparative analysis of the fungomes and bacteriomes of fusarium wilt-suppressive and non-suppressive soils from the Châteaurenard region (France) was conducted (Siegel-Hertz et al. 2018). This analysis revealed the high complexity of disease suppression, as 10 fungal and 11 bacterial taxa were associated with the suppressive soil. Some, but not all, of these taxa harbor microorganisms known to produce metabolites that likely affect the development of *F. oxysporum*. None of them, however, is highly specific or completely biocidal and/or inhibitory to *F. oxysporum* infectious activity, or permanently active. On the other hand, the combination of these multiple mechanisms may exert a permanent "pressure" that only allows the pathogen to survive but limits its pathogenic behavior. Due to the multifactorial nature of this pressure linked to a diversity of mechanisms affecting its metabolism, *F. oxysporum* may not have developed means to circumvent the pressure, despite the remarkable adaptability that it has revealed in various abiotic conditions (Steinberg et al. 2016). Similarly, an analysis of the rhizosphere bacterial and archaeal communities associated with suppressiveness to *R. solani* in the Netherlands highlighted the predominance of a multitude of Proteobacteria, Firmicutes, and Actinobacteria, including more than 30,000 species (Mendes et al. 2011). Deeper investigation of the (γ-) Proteobacteria revealed that some of them were closely related to Pseudomonadaceae and harbored peptide synthetase genes similar to those encoding the toxic peptides syringomycin/syringopeptin in *Pseudomonas syringae* pv. *syringae*. Actually, these genes encoded a nine-amino-acid-chlorinated lipopeptide. The findings by these authors were trendsetting and are further described in Box 21.1. The mechanism they found, however, was probably complementary to the antagonistic activities that were exhibited by members of the Burkholderiaceae, Xanthomonadales, and Actinobacteria that were present in the same microbiome.

BOX 21.1 CONTRIBUTION OF META-OMICS TO THE UNDERSTANDING OF SOIL SUPPRESSIVENESS MECHANISMS

Although still enigmatic, the disease suppressiveness mechanisms of soil are beginning to be decrypted. Field observations and bioassays were the first systemic approaches highlighting the role of the soil microbiome in disease suppression. Microorganisms were isolated, and some (bacteria and/or fungi) had traits explaining their potential role in the process. However, the role of other organisms was often neglected. With the advent of the meta-omics tools, it is now possible to assess all biotic components of soil, addressing the system as one unit, in order to understand the mechanisms underlying disease suppression. Initial approaches based on molecular fingerprintings of diverse biomes have unfortunately remained descriptive. To highlight the mechanism involved in pathogen control, a focus must be placed on the dynamics of a given microbial group, *a priori* assuming its role in suppression. Recent systematic meta-omics approaches have provided explanatory notes to the mechanisms at play in disease suppression.

A hallmark study in this field was conducted by Mendes et al. (2011) on a soil that had become suppressive to *R. solani* disease in sugar beet. Using a 16S rRNA gene-based microarray, coupled to cloning, sequencing, and quantitative PCR, the authors first described the taxonomic diversity of bacterial and archaeal communities associated with the suppressive status, highlighting the prevalence of members of the Firmicutes (20%) and Proteobacteria (29%), the latter including γ-Proteobacteria, in particular Pseudomonadaceae. Special focus was given to this group, and *Pseudomonas* strains were isolated from the suppressive rhizosphere soil. The 16S rRNA genes of these isolates were similar to those of *Pseudomonas* prevalent in the suppressive soil. Random transposon mutagenesis of the *Pseudomonas* isolates active against *R. solani* AG4 generated mutants, one of which did not protect the sugar beet against *R. solani*, but in other aspects behaved similarly to the parent strain in the rhizosphere. The interrupted locus turned out to produce a lipopeptide coined thanamycin (encoded by a nonribosomal peptide synthetase gene cluster). The thanamycin, a monochlorinated lipopeptide, belongs to the syringomycin family of antifungal agents.

Although this study highlighted the role of one specific group of organisms, it is likely that other mechanisms corresponding to the metabolic activities of other microbial groups also participate in disease suppression. Thus, a novel study applied comparative metatranscriptomics-based analysis of the rhizosphere microbiome of wheat grown in soils that were either suppressive or conducive to *R. solani* AG8 (Hayden et al. 2018). The intricacies of soil metatranscriptomics are discussed in Chapter 15 of this book. In the Hayden et al. study, genes of 65 identified bacterial species were found to be significantly higher expressed in one or the other soil. The genes that were most expressed in the suppressive soil corresponded to a polyketide cyclase, a terpenoid biosynthesis backbone gene (*dxs*) and many cold shock proteins (*csp*). Surprisingly, the conducive soil was characterized by a greater expression of antibiotic production genes, for instance, a nonheme chloroperoxidase, which is involved in pyrrolnitrin synthesis, and phenazine biosynthesis family protein F and its transcriptional activator protein. The resultant metabolites are frequently associated with the control of pathogens in suppressive soils. Remarkably, none of the fungal species detected in the libraries showed differential expression between suppressive and conducive soils. This study identified antibiotic-type metabolites that may directly inhibit the development of the pathogen. The main step forward is that the physiology and metabolic functioning of the microbiome as a whole is highlighted, beyond the production of a simple antibiotic. Thus, the mechanisms behind general suppressiveness are addressed, including those behind fungistasis and/or competition, and others that may include signaling and molecular exchanges between microorganisms. Knowledge of these would enable to understand activities against more than just a single pathogen, and hence, this type of work opens perspectives that until now seemed inaccessible.

Metatranscriptomics approaches, based on the collective amount of messenger RNA produced by a microbiome, may identify the functional groups within soil microbiomes that contribute to disease suppression, as in Hayden et al. (2018). The study by Hayden et al. constitutes another hallmark study, in which advanced molecular methods fostered our understanding of the mechanisms of general disease suppressiveness (see Box 21.1). Thus, the metatranscriptomic analyses of wheat rhizosphere soil sampled from both a non-suppressive and an *R. solani* AG8 suppressive soil revealed taxa associated with either the suppressive or the conducive soil. Surprisingly, these taxa were different from those highlighted in the previous Dutch study. *Pseudomonas* and *Arthrobacter* spp. were dominant in the conducive soil, while *Stenotrophomonas* and *Buttiauxella* spp. dominated in the suppressive one. It is likely that one needs to assess the underlying organisms in more detail than just at the level of taxa, in order to understand the nature of the disease suppression as well as potential bioindicators of phytosanitary soil quality. Interestingly, besides a large set of genes encoding antibiotics such as pyrrolnitrin and phenazine, genes involved in detoxification of reactive oxygen species (ROS) were found in the conducive soil. Conversely, genes encoding polyketide cyclases, terpenoids, and cold shock proteins were found primarily in the suppressive soil. The abundance of the ROS detoxification genes in the conducive soil might be a response of the wheat roots to *R. solani* infection. The way in which the aforementioned metabolites interfere with *R. solani* pathogenicity has not been established. However, as in the case of the Dutch soil (Mendes et al. 2011), their direct impact on pathogen development should be tested under controlled conditions. These advanced analyses have shown that greater parts of the microbiome can be involved in disease suppression, either directly or indirectly. Analysis of co-occurrence networks (i.e., microorganisms that are not *a priori* involved but whose presence, even at low density, may be required) may highlight potential key organisms, yet causal relationships still need to be affirmed (Faust et al. 2017, Williams et al. 2014).

Next to metatranscriptomics, soil RNA extracts can also be investigated with microarrays consisting of either taxon-specific or functional oligonucleotide probes (Sanguin et al. 2006).

Another advanced method, stable-isotope probing (SIP; see Chapter 17)—using ^{13}C—allows to discern active from total microbial communities by assessing the incorporation of label into the cellular macromolecules DNA, RNA, or protein. SIP may allow another angle at the comparison of soils with different levels of suppressiveness, as key active groups are highlighted at the level of DNA or RNA.

Culture-independent methods can thus facilitate finding a link between soil suppressiveness and microbial diversity. However, these advances do not preclude the need for culturing of the organisms involved. Whether the identified microorganisms are merely passively "associated" with soil suppressiveness or are truly involved in the mechanisms of disease suppression is not *a priori* known from the direct omics approaches. Hence, the confirmation of the putative roles of specific microorganisms in suppressiveness will require their isolation from soil. The molecular and cultivation-based methods often detect different parts of the microbiomes. Therefore, the different methods should be considered as being complementary to each other, especially when the soil process investigated is as complex as soil suppressiveness (Garbeva et al. 2004).

21.5.3 Metaproteomics and Metabonomics in the Context of Disease Suppression

Current methodological progress in the characterization of the so-called soil *metaproteome* and *metabonome* (respectively, the collective proteins and metabolites present in the microbial community) opens up relevant functional approaches (Manikandan et al. 2017). Chapter 16 reveals the promises as well as the pitfalls of soil metaproteomics. Optimizations of extraction protocols of proteins as well as metabolites are available or underway (Goodacre et al. 2004), but soil metaproteomics is by no means a simple task. Next to the technical constraints (see Chapter 16), the information gained will not necessarily be correlated with the DNA (gene) or RNA (transcript) levels due to posttranscriptional regulation and possible proteins rearrangement, as well as differences in the stabilities of the respective molecules. However, it will be informative, since proteins preferentially associated with suppressive soils could be used as *bioindicators* of phytosanitary soil quality by immunological procedures, possibly allowing the targeting of novel functional genes by reverse genetics. In addition, comparison of the metabonomes in soils showing different levels of disease suppressiveness will identify the metabolites that are preferentially associated with suppressiveness. This requires the identification of the compound and of the producing microorganisms (Goodacre et al. 2004).

A polyphasic approach that is based on a combination of methods, including characterization of microbiomes by genotyping and cultivation, DNA arrays, gene expression, metagenomics, metaproteomics, and metabonomics assessments, should allow continued progress in the emerging understanding of the microorganisms, genes, and metabolites that are involved in natural suppressiveness of soil to soilborne diseases.

21.6 SOIL SUPPRESSIVENESS REVISITED—POSSIBLE SPIN-OFFS

Despite our still incomplete understanding of the microorganisms, genes and metabolites that are involved in, or associated with, disease-suppressive soils, studies on these soils have already provided data that are key to plant health. These are related to the following:

1. The isolation and selection of biological control agents (microbial inoculation)
2. The development of diagnostic methods for phytosanitary soil quality
3. The potential strategies to manage the microbial communities of soil in order to preserve and improve the level of disease suppressiveness (sustainable agriculture)

21.6.1 Isolation and Selection of Biocontrol Agents

The production and use of biocontrol agents is discussed in Chapter 22. We here address how disease-suppressive soils can be exploited as reservoirs of these agents. Rather than selecting antagonists at random from soils, selecting them from groups of microorganisms isolated from suppressive soils—more precisely among the groups involved in soil suppressiveness—might enhance the likelihood of success. This strategy was followed to isolate fluorescent pseudomonads from the aforementioned TAD soils (*P. fluorescens* 2–79, Q8r1–96), from soils suppressive to root rot of tobacco (*P. fluorescens* CHA0) or to fusarium wilts (*P. fluorescens* C7) (Alabouvette et al. 1993, Raaijmakers and Weller 2001, Stutz et al. 1986). Characterization of the metabolites and genes involved in the suppression of disease by the fluorescent pseudomonads from the suppressive soils also allowed the selection of strains based on the presence of these genes and their ability to synthesize the relevant metabolites. Other key organisms isolated from the fusarium wilt-suppressive soils were the aforementioned nonpathogenic *F. oxysporum* strains (Alabouvette 1986). One of these strains, denoted Fo47, when tested on large scale under glasshouse protected various crops against fusarium wilts (Alabouvette et al. 1993). A major limitation of the application of these biocontrol agents in soils is related to the difficulty to get them established in soil following introduction (see also Chapter 22), which possibly relates to a lack of "competence" (the ability to establish and persist in the system) of the inoculant.

21.6.2 Development of Methods to Characterize Phytosanitary Soil Quality

To date, there are no proven and reliable indicators of the phytosanitary quality of soils (Janvier et al. 2007). Robust omics data supported by statistics yield presumptive evidence for mechanisms that require support from bioassays to evaluate the soil infection potential. Thus, the populations that are preferentially associated with suppressive soil may eventually serve as the bioindicators. In most cases, however, suppression results from complex interactions between different microbial populations, and therefore, the approach may be inherently limited. However, natural suppressiveness to take-all disease was shown to be associated with the presence of 2,4-DAPG-producing fluorescent *Pseudomonas* spp. in various soils (Raaijmakers and Weller 2001). Such organisms might constitute excellent bioindicators of suppressiveness of soils to take-all; however, threshold density levels need to be carefully determined. The concepts and methods of phytosanitary soil quality determination (soil infectivity potential and soil receptivity), based on dedicated bioassays, are also useful, as they allow the integration of the complexity of the biotic and abiotic interactions.

The soil infectivity potential represents the risk of occurrence of a given soilborne disease in the test soil when cultivated with a susceptible cultivar. The methodology consists of the calculation of the IPU value (see Section 21.2), per gram of soil. The potential risk of occurrence of root rot of pea due to *A. euteiches* has thus been characterized (Moussart et al. 2009). In addition, sowing dates that would allow minimizing disease severity have been assessed. The IPU values analyzed across soils can be compared with those available in databases and data can be provided with respect to the risk of disease occurrence (high, medium, or nil). Because of the heterogeneous spatial distribution of microorganisms including pathogens, the number of samples collected within the field must be high enough to get a reliable evaluation of variability. Despite its relevance, this tool only provides an estimate of direct risk. On the other hand, a low infectivity potential may be ascribed to the complete absence or low density of the pathogen even when soil receptivity to the disease is high (Figure 21.1). In that specific case, soil contamination with aggressive pathogens, for instance by the introduction of infested plants, may cause a high disease severity because of the high soil receptivity (despite an initially low infectivity potential). Measurement of the infectious potential must thus be completed by measuring the receptivity of soil to the soilborne disease, which gives information on the potential risk of disease occurrence due to the accidental introduction of aggressive pathogenic populations.

21.6.3 Agricultural Practices Modulate Disease Suppressiveness

The effects of plant type, soil type, and soil management for soil microbiomes and suppressiveness to soilborne diseases have been stressed in many reviews (Garbeva et al. 2004, Larkin 2015). Further, the importance of the management of soil microbial community structure for disease suppression has been discussed (Garbeva et al. 2004, Latz et al. 2012, Mazzola 2007). Here, effects of agricultural practices, with special focus on crop rotation, intercropping, and organic amendments, on soil receptivity to soilborne diseases are discussed. In suppressive soils, the microbiomes, which can provide resilience (the ability to get back to an initial state) upon contamination with pathogenic populations, are often the sources for the buffering process. Low-input agricultural practices are hypothesized to preserve or favor microbiomes that are favorable to phytosanitary quality. An improved identification of microbial taxa but more surely of functional groups (gene clusters) that contribute to resilience or disease suppression will help growers to make relevant decisions with respect to their agricultural regimes (Janvier et al. 2007, Shtienberg 2013). Disease-suppressive soils can develop through crop rotation, intercropping, residue destruction, organic amendments, and tillage management practices, as well as through combinations of these regimes (Kinkel et al. 2011, Mazzola 2007).

The microflora associated with plant roots often differs in accordance with plant species (Garbeva et al. 2004, Latz et al. 2012, Lemanceau et al. 1995) and even cultivar. Thus, management of root-associated microbiomes through plant cultivation has been proposed, like adventitiously found in the wheat TAD soils. In that case, the ability of the plant to build up populations of 2,4-DAPG-producing fluorescent pseudomonads in the rhizosphere was wheat cultivar dependent. Thus, using a cultivar highly efficient in this selection process will allow a quick establishment of TAD (Mazzola 2007). However, the dominant 2,4-DAPG-producing pseudomonads in the wheat rhizosphere differed across cultivars, and consequently, the efficacy of take-all suppression varied (Weller et al. 2002). In a long-term grassland experiment, plant diversity was found to impact 2,4-DAPG and pyrrolnitrin producer communities and consequently soil suppressiveness to *R. solani* disease. The abundance of these organisms increased with plant diversity and that of the pyrrolnitrin producers also increased in the presence of grasses. In contrast, the abundances of both groups decreased under legume species. In turn, soil suppressiveness was maximal when the abundances of both groups were high (Latz et al. 2012). Similarly, long-term grassland versus a rotational cropping system revealed enhanced suppressiveness of the pathogen *R. solani* AG3 concomitantly with raised microbial diversity (Garbeva et al. 2004, van Elsas et al. 2002).

Mazzola (2007) reported the occurrence of suppressiveness to apple replant disease (caused by *Rhizoctonia*) in soil cropped to wheat monoculture. Thus, wheat cover cropping was recommended as a means to establish a disease-suppressive microbiome in apple orchards. Again, the efficacy of this measure differed in accordance with the wheat cultivar used and its ability to select specific fluorescent pseudomonads.

Depending on the persistence of the effect, microbial populations selected by a crop plant may improve the health of a forthcoming crop or be unable to do so. In the case of take-all suppression, unfortunately, the effect showed poor persistence, as the interruption of wheat monoculture reduced or abolished the capacity of pathogen suppression (Cook 2003).

Decomposition of organic matter resulting from plant residues will increase microbial biomass and microbial activities in soil, and therefore often enhance general disease suppression. Organic amendments have been used with success to increase the suppressiveness of soil to several diseases, including those caused by nematodes as well as those in farm truck and horticultural crops (Pérez-Piqueres et al. 2006, Postma and Schilder 2015). Control of soilborne plant pathogens was also improved using incorporation into soil of fresh organic matter in combination with a plastic cover and solar heating. Although modifications of microbiomes following such organic amendments have been reported, the stimulation of specific populations that actually exert antagonistic activity towards the pathogens has been difficult to prove (Edel-Hermann et al. 2004, Pérez-Piqueres et al. 2006). Besides the effect of introduction of plant residues on beneficial microflora through general and/or specific suppressiveness, it may also directly affect the pathogen. This is the case for Brassicaceae that produce glucosinolates and control various soilborne plant pathogens (Ren et al. 2018). Based on this property, Brassicaceae (cabbage, mustards, horseradish) may be cultivated as intermediate crops or enter a crop rotation, after which they are buried into the soil as so-called *green manures*. They thus act as either selective fungicides or fungistatics, limiting the development and activity of fungal populations, including pathogens of the forthcoming crop. Another example of organic amendments directly affecting pathogens is the control of *Allium* white rot due to *Sclerotium cepivorum* by incorporation of composted onion waste into soil. Roots of *Allium* species release alk(en)yl cysteine sulfoxides, which favor the germination of pathogen spores. When properly composted, onion waste including these compounds will trigger the germination of dormant sclerotia which, in the absence of roots of the host plant, will result in a decreased density of primary inoculum (Coventry et al. 2002). This strategy is not foolproof, since in some cases, the plant debris can preserve the pathogens. For instance, the incorporation of ground mustard actually reduced the inoculum density of *R. solani* AG2.2 and the disease severity on beet, but surviving sclerotia germinated after dissipation of glucosinolates using mustard fragments as a trophic source for their development (Friberg et al. 2009). Therefore, attention should be paid to the management of residues by burial through tillage or the promotion of rapid decomposition. In addition, modifying the structure of soil microbiomes through agricultural practices may make soils only temporarily suppressive to one or another disease (Raaijmakers and Mazzola 2016). Persistent suppressiveness, similar to the fusarium wilt suppressiveness of Châteaurenard soils, seems difficult to acquire. The balance reached by the soil microbiomes of the latter soil may result from a long natural evolutionary process; the suppressiveness of this soil is called "native" or stabilized. In the cases of "acquired" soil suppressiveness, the agricultural practices used to establish it may be still too recent to fix the microbiome structures in a sustainable way.

Finally, there is a major risk in soil disinfestation procedures, which can lead not only to partial destruction of the pathogen but also to that of the soil microflora involved in suppressiveness. Soil functionality depends on (structure of) the microbiome, and activity of microorganisms directly involved in a target function may only occur in the presence of the different components of this community, even without these components being directly involved in the function (Faust et al. 2017, Williams et al. 2014). Therefore, members of the microbial consortium responsible for soil suppressiveness might be identified by eroding diversity by serially diluting microbiomes of suppressive soils until these lose suppressiveness. The dilution level preceding the loss of suppressiveness would contain the minimum consortium necessary for controlling the pathogen. Although not yet realistic at the field scale, a kind of tailor-made core microbiome transfer therapy based on transferring the "hard core" of suppressive to conducive soil was also suggested, with reference to what has been achieved with gut microbiomes in medicine for human beings (Gopal et al. 2013).

Moreover, agricultural practices may yield variable results depending on the soil (and its history) in which tests are performed. The available literature shows that it is difficult to generalize and, so, to propose a standard protocol to improve suppressiveness, even for a given soilborne disease. In spite of great progress in the soil omics methods, causal relationships of microbiome changes to disease suppression have rarely been demonstrated. One problem is that taxa are unevenly present across soils, so it is difficult to promote those that are not present through agricultural practices. On the other hand, genes encoding functions that control the activities of plant pathogens can be present across many taxa (functional redundancy) across many soils. In these cases, the functional traits involved in the disease suppression mechanisms should thus be targeted. The relevance of agricultural practice could then be evaluated and quantified based on function (gene expression) rather than taxonomy within the microbiomes. Consequently, such trait-based assessments may yield risk indicators as well as decision support aids for agricultural practice for use in control of soilborne pathogens.

REFERENCES

Aimé, S., Alabouvette, C., Steinberg, C. and C. Olivain. 2013. The endophytic strain *Fusarium oxysporum* Fo47: A good candidate for priming the defense responses in tomato roots. *Mol Plant Microbe Interact* 26, 918–926.

Alabouvette, C. 1986. *Fusarium*-wilt suppressive soils from the Châteaurenard region: Review of a 10-year study. *Agronomie* 6, 273–284.

Alabouvette, C., Hoeper, H., Lemanceau, P. and C. Steinberg. 1996. Soil suppressiveness to diseases induced by soilborne plant pathogens. In: *Soil Biochemistry*, Vol. 9, G. Stotzky and J.M. Bollag (eds.), Marcel Dekker Inc.: New York, 371–413.

Alabouvette, C., Lemanceau, P. and C. Steinberg. 1993. Recent advances in the biological control of fusarium wilts. *Pestic Sci* 37, 365–373.

Bacon, C.W. and J.F. White. 2016. Functions, mechanisms and regulation of endophytic and epiphytic microbial communities of plants. *Symbiosis* 68, 87–98.

Baker, K.F. and R.J. Cook. 1974. *Biological Control of Plant Pathogens*, Freeman, W.H. (eds.), The American Phytopathological Society: San Francisco, CA.

Cook, R.J. 2003. Take-all of wheat. *Physiol Mol Plant Pathol* 62, 73–86.

Cook, R.J. and K.F. Baker. 1983. *The Nature and Practice of Biological Control of Plant Pathogens*, The American Phytopathological Society: St Paul, MN, 539.

Coventry, E., Noble, R., Mead, A. and J.M. Whipps. 2002. Control of allium white rot (*Sclerotium cepivorum*) with composted onion waste. *Soil Biol Biochem* 34, 1037–1045.

Cretoiu, M.S., Korthals, G.W., Visser, J.H.M. and J.D. van Elsas. 2013. Chitin amendment increases soil suppressiveness toward plant pathogens and modulates the actinobacterial and oxalobacteraceal communities in an experimental agricultural field. *Appl Environ Microbiol* 79, 1–11.

Edel-Hermann, V., Dreumont, C., Pérez-Piqueres, A. and C. Steinberg. 2004. Terminal restriction fragment length polymorphism analysis of ribosomal RNA genes to assess changes in fungal community structure in soils. *FEMS Microbiol Ecol* 47, 397–404.

Faust, K., Lima-Mendez, G., Lerat, J.S., et al. 2017. Cross-biome comparison of microbial association networks. *Front Microbiol* 6, 13.

Friberg, H., Edel-Hermann, V., Faivre, C., et al. 2009. Cause and duration of mustard incorporation effects on soil-borne plant pathogenic fungi. *Soil Biol Biochem* 41, 2075–2084.

Garbeva, P., van Veen, J.A. and J.D. van Elsas. 2004. Microbial diversity in soil: selection of microbial populations by plant and soil type and implications in soil suppressiveness. *Ann Rev Phytopathol* 42, 243–270.

Garrett, S.D. 1970. *Pathogenic Root-Infecting Fungi*, Cambridge University Press: London, 294.

Gomez Exposito, R., de Bruijn, I., Postma, J. and J.M. Raaijmakers. 2017. Current insights into the role of rhizosphere bacteria in disease suppressive soils. *Front Microbiol* 8, 252.

Goodacre, R., Vaidyanathan, S., Dunn, W.B., Harrigan, G.G. and D.B. Kell. 2004. Metabolomics by numbers: Acquiring and understanding global metabolite data. *Trends Biotechnol* 22, 245–252.

Gopal, M., Gupta, A. and G. Thomas. 2013. Bespoke microbiome therapy to manage plant diseases. *Front Microbiol* 4, 355.

Gross H. and J.E. Loper. 2009. Genomics of secondary metabolite production by pseudomonas spp. *Nat Prod Rep* 26, 1408–1446.

Hayden, H.L., Savin, K.W., Wadeson, J., Gupta, V. and P.M. Mele. 2018. Comparative metatranscriptomics of wheat rhizosphere microbiomes in disease suppressive and non-suppressive soils for *Rhizoctonia solani* AG8. *Front Microbiol* 9, 19.

Janvier, C., Villeneuve, F., Alabouvette, C., Edel-Hermann, V., Mateille, T. and C. Steinberg. 2007. Soil health through soil disease suppression: Which strategy from descriptors to indicators? *Soil Biol Biochem* 39, 1–23.

Kinkel, L.L., Bakker, M.G. and D.C. Schlatter. 2011. A coevolutionary framework for managing disease-suppressive soils. *Annu Rev Phytopathol* 49, 47–67.

Larkin, R.P. 2015. Soil health paradigms and implications for disease management. *Annu Rev Phytopathol* 53, 199–221.

Latz, E., Eisenhauer, N., Rall, B.C., et al. 2012. Plant diversity improves protection against soil-borne pathogens by fostering antagonistic bacterial communities. *J Ecol* 100, 597–604.

Lemanceau, P. and C. Alabouvette. 1993 Suppression of *fusarium*-wilts by fluorescent pseudomonads: Mechanisms and applications. *Bio Sci Technol* 3, 219–234.

Lemanceau, P., Corberand, T., Gardan, L., et al. 1995. Effect of two plant species, flax (*Linum usitatissimum* L.) and tomato (*Lycopersicon esculentum* Mill.), on the diversity of soilborne populations of fluorescent pseudomonads. *Appl Environ Microbiol* 61, 1004–1012.

Manikandan, R., Karthikeyan, G. and T. Raguchander. 2017. Soil proteomics for exploitation of microbial diversity in *fusarium* wilt infected and healthy rhizosphere soils of tomato. *Physiol Mol Plant Pathol* 100, 185–193.

Mazurier, S., Corberand, T., Lemanceau P. and J.M. Raaijmakers. 2009. Phenazine antibiotics produced by fluorescent pseudomonads contribute to natural soil suppressiveness to *fusarium* wilt. *ISME J* 3, 977–991.

Mazzola, M. 2007. Manipulation of rhizosphere bacterial communities to induce suppressive soils. *J Nematol* 39, 213–220.

Mendes, R., Kruijt, M., de Bruijn, I., et al. 2011. Deciphering the rhizosphere microbiome for disease-suppressive bacteria. *Science* 332, 1097–1100.

Moussart, A., Wicker, E., Le Delliou, B., et al. 2009. Spatial distribution of *Aphanomyces euteiches* inoculum in a naturally infested pea field. *Eur J Plant Pathol* 123, 153–158.

Pérez-Piqueres, A., Edel-Hermann, V., Alabouvette, C. and C. Steinberg. 2006. Response of soil microbial communities to compost amendments. *Soil Biol Biochem* 38, 460–470.

Postma, J. and M.T. Schilder. 2015. Enhancement of soil suppressiveness against *Rhizoctonia solani* in sugar beet by organic amendments. *Appl Soil Ecol* 94, 72–79.

Raaijmakers, J.M. and M. Mazzola. 2016. Soil immune responses. *Science* 352, 1392–1393.
Raaijmakers, J.M. and D.M. Weller. 2001. Exploiting genotypic diversity of 2,4-diacetylphloroglucinol-producing *Pseudomonas* spp.: Characterization of superior root-colonizing *P. fluorescens* strain Q8rl-96. *Appl Environ Microbiol* 67, 2545–2554.
Ren, G., Ma, Y., Guo, D., et al. 2018. Soil bacterial community was changed after Brassicaceous seed meal application for suppression of *fusarium* wilt on pepper. *Front Microbiol* 9, 185.
Rosenzweig, N., Tiedje, J.M., Quensen, J.F., Meng, Q.X. and J.J.J. Hao. 2012. Microbial communities associated with potato common scab-suppressive soil determined by pyrosequencing analyses. *Plant Dis* 96, 718–725.
Sanguin, H., Remenant, B., Dechesne, A., et al. 2006. Potential of a 16S rRNA-based taxonomic microarray for analyzing the rhizosphere effects of maize on *Agrobacterium* spp. and bacterial communities. *Appl Environ Microbiol* 72, 4302–4312.
Shtienberg, D. 2013. Will decision-support systems be widely used for the management of plant diseases? *Ann Rev Phytopathol* 51, 1–16.
Siegel-Hertz, K., Edel-Hermann, V., Chapelle, E., Terrat, S., Raaijmakers, J.M. and C. Steinberg. 2018. Comparative microbiome analysis of a *fusarium* wilt suppressive soil and a *fusarium* wilt conducive soil from the Châteaurenard Region. *Front Microbiol* 9, 16.
Steinberg, C., Lecomte, C., Alabouvette, C. and V. Edel-Hermann. 2016. Root interactions with non-pathogenic *Fusarium*. Hey *Fusarium oxysporum*, what do you do in life when you do not infect a plant? In: *Belowground Defence Strategies in Plants*, C.M. Vos and K. Kazan (eds.), Springer International Publishing: Switzerland, 281–299.
Stutz, E.W., Défago, G. and H. Kern. 1986. Naturally occurring fluorescent pseudomonads involved in the suppression of black root rot of tobacco. *Phytopathology* 76, 181–185.
van Elsas, J.D., Garbeva, P. and J.F. Salles. 2002. Effects of agronomical measures on the microbial diversity of soils as related to the suppression of soilborne plant pathogens. *Biodegradation* 13, 29–40.
Waite, I.S., O'Donnell, A.G., Harrison, A., et al. 2003. Design and evaluation of nematode 18S rDNA primers for PCR and denaturing gradient gel electrophoresis (DGGE) of soil community DNA. *Soil Biol Biochem* 35, 1165–1173.
Weller, D.M., Raaijmakers, J.M., Gardener, B.B.M. and L.S. Thomashow. 2002. Microbial populations responsible for specific soil suppressiveness to plant pathogens. *Ann Rev Phytopathol* 40, 309–348.
Williams, R.J., Howe, A. and K.S. Hofmockel. 2014. Demonstrating microbial co-occurrence pattern analyses within and between ecosystems. *Front Microbiol* 5, 358.
Yin B, Valinsky L, Gao XB, Becker JO, Borneman J, 2003a. Bacterial rRNA genes associated with soil suppressiveness against the plant-parasitic nematode *Heterodera schachtii. Appl. Environ Microbiol* 69, 1573–1580.
Yin, B., Valinsky, L., Gao, X.B., Becker, J.O. and J. Borneman. 2003b. Identification of fungal rDNA associated with soil suppressiveness against *Heterodera schachtii* using oligonucleotide fingerprinting. *Phytopathology* 93, 1006–1013.

22 Plant Growth-Promoting Bacteria in Agricultural and Stressed Soils

Elisa Gamalero
Università del Piemonte Orientale

Bernard R. Glick
University of Waterloo

CONTENTS

22.1 Introduction 361
22.1.1 Definitions 361
22.1.1.1 Rhizospheric versus Endophytic and Symbiotic Bacteria 361
22.1.2 Direct versus Indirect Mechanisms 363
22.1.3 Mechanisms Used by PGPB 363
22.1.3.1 Nitrogen Fixation 363
22.1.3.2 Phosphate Solubilization 365
22.1.3.3 Iron Sequestration 365
22.1.3.4 Synthesis of Phytohormones and/or Phytostimulators 367
22.1.3.5 Increasing Plant Tolerance through ACC Deaminase Synthesis 369
22.1.3.6 Stimulation of Plant Defense Pathways 370
22.1.3.7 Antibiotics and Hydrogen Cyanide 370
22.1.3.8 Cell Wall-Degrading Enzymes 371
22.1.3.9 Volatile Compounds 372
22.1.3.10 Bacteriophages 372
22.1.3.11 Quorum sensing and Quenching 373
22.2 Applications of Plant Growth-Promoting Bacteria 375
22.2.1 Uses to Improve Sustainability of Production 375
22.2.2 Critical Factors Determining the Outcome of PGPB Applications 375
22.2.3 Applications and Market Considerations 376
22.3 Conclusions and Perspectives 376
References 377

22.1 INTRODUCTION

22.1.1 Definitions

22.1.1.1 Rhizospheric versus Endophytic and Symbiotic Bacteria

In previous chapters, we learned that a "typical" soil may contain from several millions to several hundreds of millions microbial cells per gram of dry weight soil. In addition, the extent of soil microbial diversity can be extraordinary (see Chapter 3). However, this immense number of microorganisms is distributed in a heterogeneous manner in soils, with cells being localized mainly in nutrient-rich microhabitats representing favorable environments for growth and survival.

In 1904, Hiltner defined the rhizosphere as the thin (1–2 mm-thick) zone of soil surrounding plant roots that is influenced by the root system. Chapter 10 further provides a detailed description of the rhizosphere and its zones. Due to the release of a variety of organic compounds by the roots, the microbial densities and activities in the rhizosphere are often enhanced, so that it is considered a "hot spot" for microbial colonization and activities. Overall, members of this microbiome can be classified, according to their effects on the host plant, as beneficial, deleterious, or neutral. It has been estimated that about 2%–5% of the bacteria living in the rhizosphere possess physiological traits that can be involved in plant growth promotion and/or plant health improvement (Antoun and Prevost 2005). However, as a prerequisite to providing positive effects on the plant, plant growth-promoting bacteria (PGPB) need to be rhizosphere- and/or rhizoplane-competent, so the ability to colonize plants and survive in the rhizosphere and/or the rhizoplane is a first requirement for a good PGPB strain (Figure 22.1). Several factors, including chemotaxis, growth rate, quorum sensing (QS), amino acid and vitamin B1 syntheses, the *O*-antigen site of lipopolysaccharides, flagella, fimbriae, type IV pili, and siderophore production, are recognized as being involved in rhizosphere and rhizoplane competence (Compant et al. 2010).

Different PGPB interact with plants in different ways. Consequently, PGPB living in the rhizosphere or in spaces between root surface cells are considered extracellular or rhizospheric, whereas those living inside root tissues are intracellular, being called endophytes. Typically, endophytes may colonize different plant organs, with no external signs of infection or negative effects appearing on the host plant (Figure 22.2). They can occur in roots, stems, leaves, flowers, seeds, and fruits (Gamalero and Glick 2015, Glassner et al. 2015, Truyens et al. 2015).

Rhizobium spp. (and a limited number of other bacteria such as *Frankia* spp.) represent a key group of endophytic microorganisms that improve plant mineral nutrition mainly by nitrogen fixation. Following root colonization, they establish a mutualistic symbiosis with legume plants that leads to the formation of specific structures, that is, root nodules, where nitrogen fixation occurs via the enzyme nitrogenase. However, by moving from the root towards the shoot, such rhizobia can, to a limited extent, colonize internal root tissues of cereal crop plants. They thus can stimulate plant growth and grain yield independently of root nodule formation and nitrogen fixation.

PGPB act in varying ways. They can stimulate plant growth and yield, suppress disease development from soilborne pathogens, and increase plant tolerance to biotic and abiotic stresses.

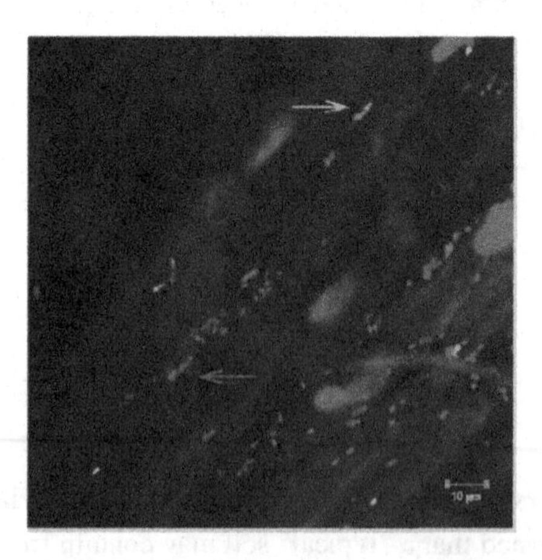

FIGURE 22.1 **(See color insert)** Confocal laser scanning microscope image of *Pseudomonas fluorescens* 92rkG5 cells expressing green fluorescent protein and stained with propidium iodide. Green (here whitish) cells (green (here white) arrow) are alive, whereas the red (here grayish) cells (red (here gray) arrow) are dead.

FIGURE 22.2 **(See color insert)** Effects of the bacterial endophyte *Pseudomonas migulae* 8R6 on periwinkle: uninoculated control (left) and bacterially inoculated plant (right).

Moreover, PGPB can improve the quality and nutritional value of seeds and fruits (Bona et al. 2015), thus contributing to an increase in food and feed yield and quality.

In a generic sense, PGPB can be exploited in agriculture either as biofertilizers or as pesticides, in environmental biotechnology for phytoremediation applications, and in food production as flavor and nutraceutical value enhancers. The mechanisms and the physiological traits used by PGPB are addressed in the following sections.

22.1.2 Direct versus Indirect Mechanisms

The positive effects caused by PGPB on plant growth may result from direct and/or indirect mechanisms (Figure 22.3). Direct mechanisms of plant growth promotion include growth stimulation by phytohormones and improvement of nutrient acquisition through processes such as atmospheric nitrogen fixation, iron sequestration (by siderophore synthesis), and phosphate solubilization. In addition, PGPB expressing 1-aminocyclopropane-1-carboxylate (ACC) deaminase, reducing ethylene synthesis in the plants, are able to enhance plant tolerance to a variety of environmental stresses.

Key indirect mechanisms of plant growth promotion are based on the suppression of soilborne diseases. This is achieved through a variety of mechanisms, for example, the production and release of antibiotics, cyanide, and extracellular lytic enzymes like chitinases (and cellulases) that hydrolyze fungal cell walls and volatile compounds. Moreover, parasitism by bacteriophages and induction of systemic resistance are major mechanisms. Many of these mechanisms are regulated in a density-dependent manner by cell–cell communication via signal molecule release, a phenomenon well known as QS. In this way, the degradation of the signal molecules synthesized by phytopathogenic organisms [quorum quenching (QQ)], represents another tool that can be exploited by PGPB involved in biocontrol.

22.1.3 Mechanisms Used by PGPB

22.1.3.1 Nitrogen Fixation

About 80% of the Earth's atmosphere consists of dinitrogen gas (N_2) that is unavailable to plants. To provide a sufficient amount of nitrogen to plants, large amounts of industrially produced N fertilizers are used in agriculture. However, from the perspective of sustainable agriculture, nitrogen fixation by microorganisms represents a key biological alternative. During biological nitrogen fixation, atmospheric N is converted into ammonia, through a cascade of reactions in which a two-component enzyme, known as nitrogenase, is important. The ammonia that is produced by the nitrogenase is

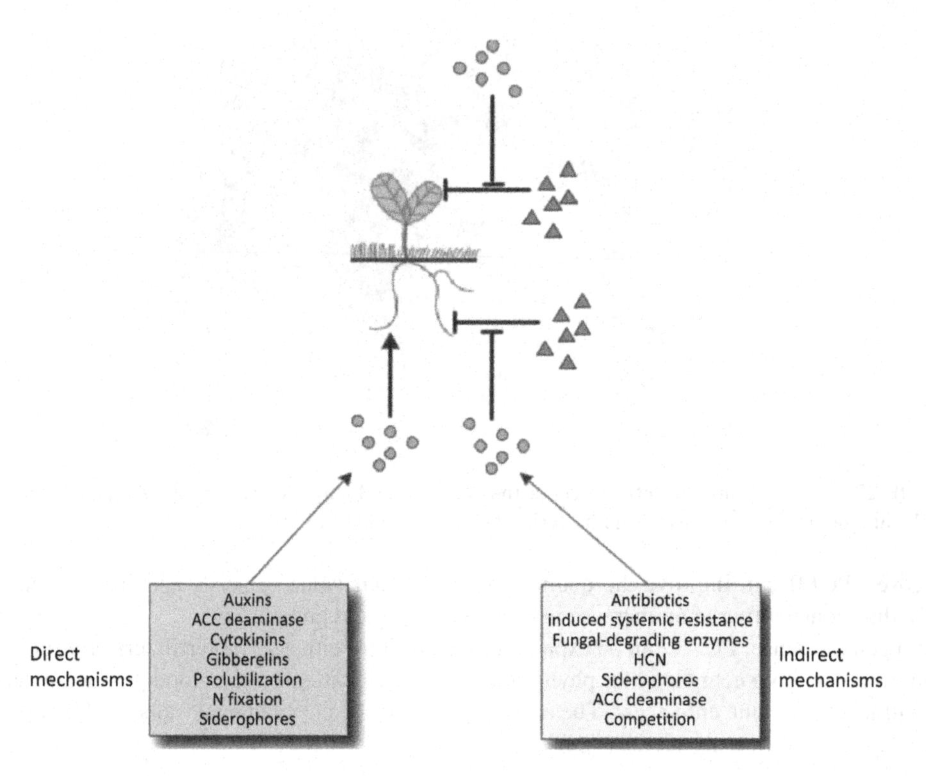

FIGURE 22.3 A schematic representation of the major mechanisms used by plant growth-promoting bacteria to stimulate the growth of plants. Arrows below indicate the effects on the plant system. Arrow above: positive effects. Inhibition arcs: negative effects (both belowground and aboveground).

assimilated by the organism and may end up in plants in symbioses, becoming an integral component of proteins, nucleic acids, and other biomolecules (for a detailed description of the root colonization process and nodule formation by rhizobia, see Chapter 11). Typically, each rhizobial species can interact with only a limited number of plant types and does not show interactions with plants other than its natural hosts. In addition to rhizobia, free-living nitrogen-fixing bacteria such as *Azospirillum*, *Azotobacter*, *Azoarcus*, and Cyanobacteria can provide ammonia to crops, although their contribution (10–160 kg N ha^{-1}) is typically lower than that provided by rhizobia to their host plants (13–360 kg N ha^{-1}).

A significant portion of the biologically fixed nitrogen is stored in plant tissues; however, the soil and the succeeding crops generally benefit from fixed N in the form of root and shoot residues (Bhattacharyya and Jha 2012). The symbiotic N fixation efficiency is dependent on the legume genotype, the rhizobial colonization efficiency, the chemical and physical parameters of the soil, and climatic factors. Besides the fixation of nitrogen in legumes, rhizobia are also able to synthesize other molecules involved in plant growth promotion, such as auxins, cytokinins, abscisic acid (ABA), siderophores, ACC deaminase, and vitamins (Gopala Krishnan et al. 2015). Moreover, they can suppress soilborne diseases via competition for nutrients, synthesis of antibiotics and hydrogen cyanide (HCN), release of lytic enzymes, and the production of siderophores (Gopala Krishnan et al. 2015, Bhattacharyya and Jha 2012). These physiological traits are strain specific. Co-inoculation of plants with rhizobia together with other plant-beneficial microorganisms such as mycorrhizal fungi (Ossler et al. 2015) or biocontrol agents/biofertilizers (Ahmad et al. 2013) can result in improved plant performance.

Rhizobia, given their good survival capabilities and huge PGP and protection potential, are excellent candidates for use as part of sustainable agriculture practices. In practice, formulations of rhizobial inoculants are used that lead to positive results. Unfortunately, such results are not always realized in open-field conditions.

22.1.3.2 Phosphate Solubilization

While many soils can contain large amounts of P, that is, 400–1,200 mg kg^{-1} of soil, the bioavailability of P is often limited, with the concentration of soluble P in soil being about ~1 mg kg^{-1} or less (Bhattacharyya and Jha 2012). The inorganic P in soil is often insoluble since it is typically bound to Fe, Al, and/or Ca, whereas organic phosphate is mainly present as phytic acid/phytate (Khan et al. 2007). Phytate is not bioavailable to plants, as plant roots generally produce only low levels of phytases, the enzymes that break down phytate. The low bioavailability of P in most soils is considered to be a key factor that limits plant growth. Therefore, P is usually added to soils as part of fertilizers that are produced through energy-intensive processes. However, the efficiency of utilization of the applied P rarely exceeds 30% as a result of its fixation and precipitation in soil. Moreover, in the long term, chemical P fertilizers may negatively affect beneficial microbial activity in the soil.

In light of the aforementioned considerations, bacteria able to solubilize and mineralize P, that is, phosphate-solubilizing bacteria (PSB), offer great promises for sustainable agriculture. The mechanism used by PSB for the solubilization of inorganic P consists in the synthesis of low-molecular-weight organic acids that lead to soil acidification and increased P solubilization (Khan et al. 2007). Moreover, some exopolysaccharides synthesized by PSB may have a role in the solubilization of tricalcium phosphates by binding free P and modulating the homeostasis of P solubilization. The mineralization of organic P occurs through the synthesis of phosphatases (phosphomonoesterase, phosphodiesterase, and phosphotriesterase) that cleave phosphoric esters. In addition, a number of different soil bacteria, that is, *Pseudomonas* sp., *Bacillus* sp., *Raoultella* sp., *Escherichia coli*, *Citrobacter braakii*, and *Enterobacter* sp., readily degrade phytate. Moreover, strains belonging to the genera *Azospirillum*, *Azotobacter*, *Pseudomonas*, *Bacillus*, *Rhizobium*, *Burkholderia*, *Enterobacter*, and *Streptomyces* have been recognized as relevant PSB, and both P solubilization and mineralization can occur in the same bacterial strain.

Frequently, there is no correlation between the density of PSB in soil and the amount of soluble P present in soil. Moreover, the role of phosphate solubilization in plant growth promotion is frequently overshadowed by other plant-beneficial traits expressed by PSB, such as auxin or siderophore synthesis. But, reassuringly, numerous reports have demonstrated the existence of a direct connection between phosphate solubilization activity and increased P levels in tissues of plants inoculated with PSB.

Moreover, it has been shown that inoculation of rice with PSB such as *Gluconacetobacter* sp. and *Burkholderia* sp. resulted in increased phosphorus content and uptake by plants. The highest nutrient uptake and plant yield were observed in rice grown in rock phosphate and inoculated with both microorganisms that revealed a positive synergistic interaction. These bacterial strains could be further developed into biofertilizers, after assessing their performance in open-field condition, either alone or as components of integrated nutrient management systems (Stephen et al. 2015).

Microorganisms able to solubilize phosphate have assisted in the phytoremediation of heavy metal-impacted soil (Ahemad 2015). When plants are grown in heavy metal-contaminated soils, their growth and physiological activities are impaired, and they become more susceptible to soilborne diseases. In this condition, their capacity to take up and tolerate heavy metals is depressed. Exploitation of PSB with multiple plant-beneficial activities and concurrent metal-detoxifying potentials may indirectly enhance phytoremediation efficiency by stimulating plant growth and improving their health even in such a stressful condition (Oteino et al. 2015).

22.1.3.3 Iron Sequestration

Fe deficiency is recognized as one of the major causes of yield limitation in agriculture. In aerobic soils, iron is mostly present as hydroxides, oxyhydroxides, and oxides, so that the amount of iron available for assimilation by living organisms is low, ranging from about 10^{-7} to 10^{-23} M at pH 3.5 and 8.5, respectively. Both microbes and plants have a high iron requirement (i.e., about 10^{-5}–10^{-7} and 10^{-4}–10^{-9} M, respectively). This need is accentuated in the rhizosphere, where strong competition for Fe occurs among plant roots, bacteria, and fungi (Mimmo et al. 2014).

Plant Fe deficiency impairs chloroplast development, affects the functioning of the photosynthetic process, and results in leaf chlorosis and in a general reduction in plant growth. Plants thus activate Fe-uptake mechanisms that use either of two strategies. Strategy I is found in dicotyledonous and non-gramineous monocotyledonous plants. It is based on pH reduction in the rhizosphere by the release of inorganic (H^+) and organic (e.g., organic acids, phenolics) substances. As a consequence of this rhizosphere acidification, ferric iron (Fe^{3+}) is reduced to its ferrous form (Fe^{2+}) that is transported inside root cells. Strategy II is typical of grasses and graminaceous plants. It is based on the synthesis and release of Fe^{3+} chelators (phytosiderophores) and on the uptake of the Fe–phytosiderophore complex inside the root cells.

The iron dynamics in the rhizosphere is regulated not only by the soil properties and the plant, but is also under the control of microbially produced compounds (Lemanceau et al. 2009). Iron starvation occurs when the concentration reaches about 10^{-6} M Fe. In this condition, bacterial cells are impaired in growth and show altered morphology, and the rates of RNA and DNA synthesis are lowered (Braun and Hantke 2011). At the same time, the synthesis of iron chelators (called siderophores) is promoted. These low-molecular-weight (<1 kDa) molecules have a high affinity for Fe^{3+} (K_a ranging from 10^{23} to 10^{52}). Once synthesized, siderophores are released from cells through an efflux pump. They then chelate Fe^{3+}, and the resulting ferri–siderophore complex is taken up via specific outer membrane transporters. Inside the bacterial cells, reductases convert Fe^{3+} to Fe^{2+} that has a lower affinity for the chelator. The complex is then destabilized, and iron is provided to the bacterial proteins. According to their chemical structure, siderophores are classified into catecholates, hydroxamates, and carboxylates. Siderophores containing at least two different iron-binding moieties are known as mixed ligands.

Bacterial siderophores are involved in both direct (stimulation of plant growth by improving iron nutrition) and indirect effects (stimulation of plant growth by phytopathogen inhibition) on plant growth. Plants may or may not be affected by the iron depletion induced by siderophores released by bacteria. Some plants can even take up and utilize bacterial ferri–siderophores. Positive effects on plant growth of siderophores synthesized by bacteria have been demonstrated by supplying radiolabeled ferri–siderophores to plants as a sole source of iron. Thus, the Fe–pyoverdine complex of *Pseudomonas fluorescens* C7 was efficiently taken up by *Arabidopsis thaliana* leading to an increase in iron content inside plant tissues and to improved plant growth. Moreover, the ferri–siderophore complexes of *P. fluorescens* as well as *Trichoderma asperellum* can behave as Fe donors for plants, overcoming Fe deficiency in hydroponic cultures (Nagata et al. 2013).

When a plant grows in a heavy metal-polluted soil, iron nutrition improvement by soil bacteria becomes even more important. Some, but not all, siderophores can bind magnesium, manganese, chromium (III), gallium (III), cadmium, copper, nickel, arsenic, lead, and zinc, next to radionuclides such as plutonium (IV). However, the affinity of these siderophores for metals other than iron is generally lower than the affinity for iron.

Liu et al. (2015) demonstrated that the purified siderophore produced by *Pseudomonas* PG12 boosted the As accumulation in the fronds of *Pteris vittata*, an arsenic-hyperaccumulating fern, cultivated in the presence of $FeAsO_4$. As a consequence of the enhanced arsenic accumulation and of the As-induced P uptake, the biomass of *P. vittata* increased significantly in the PG12 siderophore treatment. Thus, the *Pseudomonas* PG12 siderophore was effective in solubilizing As from $FeAsO_4$ and also assisted As translocation from the roots to the fronds.

Siderophores produced by PGPB limit iron nutrition for phytopathogenic fungi that (1) are unable to synthesize iron chelators, (2) release low amount of siderophores, or (3) produce siderophores with low iron affinity. Plant pathogen suppression by bacterial siderophores has recently been reported in the biocontrol of *Macrophomina phaseolina, Rhizoctonia solani, Pythium* spp., and *Fusarium* spp. (Ali and Vidhale 2012, Sulochana et al. 2014). Additionally, bacterial siderophores may promote induced systemic resistance (ISR) in plants, thus enhancing phytopathogen resistance (Aznar and Dellagi 2015). Induction of systemic resistance against *M. oryzae* infection was found in rice treated with a pseudobactin synthesized by *Pseudomonas fluorescens* WCS374r. Plant inoculation with a mutant of this bacterial strain that was unable to produce pseudobactin led to disease development. Moreover, the treatment of plant roots in hydroponics with pseudobactin restored disease suppression. Pseudobactin was hypothesized to trigger plant immunity by modulating the expression of the hormones jasmonic acid and ethylene.

Notwithstanding this clear outcome, the role of bacterial metabolites in plant ISR stimulation can be complex. Djavaheri et al. (2012) examined the role of pseudobactin374, salicylic acid, and pseudomonine produced by the PGPB *P. fluorescens* WCS374r in its colonization of *A. thaliana* roots and in ISR against *P. syringae* pv. *Tomato*. Inoculation of plants with the bacterial wild type and mutants disrupted in the production of these three metabolites demonstrated that their production by *P. fluorescens* WCS374r was not required for eliciting ISR in *Arabidopsis*.

22.1.3.4 Synthesis of Phytohormones and/or Phytostimulators

Auxins, cytokinins, gibberellins, ethylene, and ABA are phytohormones that have regulatory roles in plant growth and development. When their concentrations inside the plants change by environmental stresses, plant development is altered. As a consequence, phytohormones or hormone-like substances able to stimulate seed and tuber germination, root formation, or fruit ripening are frequently added to commercial biofertilizers.

Many PGPB are able to synthesize and modulate phytohormones under *in vitro* conditions, and these physiological traits are thought to be essential assets for new plant-beneficial strains. Production of the auxin indole-3-acetic acid (IAA) has been described in a wide range of soil bacteria. Auxins regulate many plant processes. These are cell extension and differentiation, seed and tuber germination, xylem and root development, lateral and adventitious root initiation, responses to light and gravity, florescence, and fructification. They also modulate photosynthesis, pigment formation, biosynthesis of various metabolites, and resistance to stressful conditions of plants.

Although several naturally occurring auxins have been described, IAA is the most studied auxin, and frequently auxin and IAA are considered as interchangeable terms. Different amounts of auxin inside plant tissues lead to diverse plant responses that are a function of the type of plant, the tissue involved, and the developmental stage of the plant. However, the plant endogenous auxin pool is affected by soil microorganisms able to synthesize this phytohormone. Thus, the concentration of endogenous auxin is the determinant of the (positive or negative) effect on plant growth of bacterial IAA. High levels of auxin boost ethylene synthesis in plants, leading to leaf and fruit abscission, inhibition of stem growth, and promotion of femaleness in dioecious flowers.

The frequency of occurrence of soil bacteria able to synthesize auxin is high, being estimated at levels as high as 80% of all the isolates. IAA synthesis by bacteria relies on six metabolic pathways, five of them based on tryptophan as the main IAA precursor. These pathways may be constitutively expressed or inducible and encoded by genomic or plasmid DNA. Nevertheless, they are classified—according to their intermediate compounds—as indole-3-acetamide, indole-3-pyruvate, tryptamine, tryptophan side-chain oxidase, indole-3-acetonitrile, and tryptophan independent (Duca et al. 2014).

It has been demonstrated that maize plants, when grown in the presence of L-tryptophan or the PGPB *Paraburkholderia* (formerly *Burkholderia*) *phytofirmans* PsJN, exhibited enhanced growth compared to untreated controls. In these experiments, the use of PsJN with plants treated with L-tryptophan (10^{-5} M) gave the best results in terms of plant height, photosynthesis, chlorophyll content, and root and shoot biomass. Thus, auxin synthesis induced by tryptophan in *P. phytofirmans* PsJN enhances maize plant growth and improves its physiology in terms of survival, root/shoot biomass, and nutrient content (Naveed et al. 2015).

Bacteria belonging to the genus *Azospirillum* have been reported to significantly promote the plant growth, presumably through a combination of nitrogen fixation and auxin synthesis. Commercial formulations of *Azospirillum* may increase plant yield by about 30% (Hungria et al. 2010), probably by modification of root architecture by increasing the number of lateral roots and root hairs. In recent work, Spaepen et al. (2014) indeed observed that inoculation of *Arabidopsis thaliana* with the IAA producer *Azospirillum brasilense* reduced the primary root length and increased the root branching degree and the number of root hairs. Mutants of *A. brasilense* impaired in auxin synthesis did not show such modifications of root architecture. This demonstrates the involvement of bacterial IAA in the modification of *Arabidopsis* root architecture. Plant transcriptomics analyses then revealed that genes involved in cell wall modification and root hair differentiation were overexpressed in the roots of plants inoculated with the wild-type (IAA-producing) *A. brasilense*, while the expression of these genes was not induced in plants inoculated with the mutant. Moreover, genes involved in DNA, RNA, and protein synthesis, as well as cell division and cell movement processes, were downregulated in plants inoculated with the mutant.

Cytokinins are N^6-substituted aminopurines that are widely distributed in higher plants, algae, and bacteria. Cytokinins play a key role in a wide range of physiological processes including stimulation of cell division and enlargement, tissue expansion, regulation of interruption of the quiescence of dormant buds, activation of seed germination, promotion of branching, accumulation of chlorophyll, leaf expansion, and delay of senescence. Moreover, these molecules regulate the expression of the gene that encodes expansin, which induces weakening of plant cell walls and facilitates cell expansion through turgor modulation. This, in turn, leads to modifications of plant cell size and shape.

The gene encoding the enzyme responsible for the synthesis of cytokinins was initially found in the phytopathogen *Agrobacterium tumefaciens*, and later in methylotrophic and methanotrophic bacteria, next to other phytopathogens such as *Pseudomonas savastanoi*, as well as PGPB including *Azotobacter*, *Azospirillum*, *Rhizobium*, *Bacillus*, and *Pseudomonas* isolated from different plant species.

Seed inoculation with bacteria able to synthesize cytokinins leads to a higher cytokinin content in plant tissues, with a concomitant influence on plant growth and development. Various environmental stresses—such as drought—can also modulate cytokinin levels. For example, when soil is drying, the cytokinin levels in plants decrease, leading to stomatal closure limiting foliar water loss. Liu et al. (2013) demonstrated that the leaves of *Platycladus orientalis*, cultivated under drought stress and inoculated with a strain of *Bacillus subtilis* was able to produce cytokinin, had higher relative water content and leaf water potential compared to those of uninoculated ones. Moreover, the leakage of electrolyte was reduced by the presence of *B. subtilis*. Thus, cytokinin-synthesizing bacteria may be beneficial for plants subjected to drought stress. The inoculation of plants exposed to drought stress with a bacterial strain able to produce cytokinin may support seedling growth by maintaining their water content.

Moreover, it has been recently reported that cytokinins synthesized by bacteria living on plant roots may manipulate the amino acid release by the latter, which attracts microbes and stimulates their growth. Wheat inoculation with *B. subtilis* IB-22 (that is able to produce zeatin-type cytokinins) leads to a 30% increase in the total amino acid concentration in the soil compared to untreated plants. In contrast, plants inoculated with the nonproducer *B. subtilis* IB-21 did not affect the amount of amino acids present in the soil (Kudoyarova et al. 2014). In this context, since *N*-deficient plants exude lower quantities of amino acids, synthesis of cytokinins by bacteria stimulates the release of amino acids by the plant, resulting in greater fitness for soil microbes especially in nutrient-poor soils, thereby facilitating an increase in both plant and microbial productivity.

Gibberellins are diterpenoid acids consisting of isoprene residues (generally with four rings). They were first identified as phytohormones involved in rice seedling disease caused by infection by the fungus *Fusarium moniliforme* (previously named *Gibberella fujikuroi*). Infection by the fungus reduces grain yield and induces abnormal development with increased stem elongation, inhibition of root growth, and leaf yellowing. Gibberellins are synthesized by plants, fungi, and bacteria. They modulate cell division and elongation and are involved in seed germination, stem elongation, determination of root hair abundance, flowering, fruit setting, and delay of senescence in a wide range of plant species. Early flowering, increased crop yield, and bigger fruit sizes are also expected in plants treated with this hormone.

Among bacteria, the ability to synthesize gibberellins-like substances was first described in *A. brasilense* and *Rhizobium* spp. Since the early discovery, gibberellin synthesis has been found in a range of bacterial genera, including *Azotobacter, Arthrobacter, Azospirillum, Promicromonospora, Pseudomonas, Bacillus, Acinetobacter, Flavobacterium, Micrococcus, Agrobacterium, Clostridium, Rhizobium, Burkholderia*, and *Xanthomonas* (Tudzynski et al. 2016). Plant growth promotion by gibberellin-producing PGPB has been reported by several researchers, and the positive effect observed on plant biomass is frequently correlated with an increased content of gibberellins in plant tissues.

Genome analysis of the biocontrol bacteria *Photorhabdus*, well known for its ability to invade and kill insects, revealed the presence of many genes responsible for the production of antimicrobial and insecticidal metabolites (Seo et al. 2012), siderophores, as well as the phytohormone IAA (Ullah et al. 2013). Interestingly, *Photorhabdus temperata* strain M1021 released several gibberellins (denoted GA_1, GA_3, GA_4, and GA_7). Once inoculated on rice plants, these molecules incited increased plant growth, as measured by plant length, chlorophyll content, and fresh and dry biomass (Ullah et al. 2014).Similarly, cucumber inoculation with *Acinetobacter calcoaceticus* SE370 increased the amount of endogenous GA1 and GA4 and their immediate precursors, with positive consequences for the growth of plants in both the absence and presence of salt stress, especially in terms of shoot biomass and chlorophyll content.

22.1.3.5 Increasing Plant Tolerance through ACC Deaminase Synthesis

ACC deaminase is an enzyme that cleaves ACC, the immediate precursor of ethylene in plants, releasing ammonia and α-ketobutyrate. As a result, the amount of ethylene that the plant can synthesize is lowered (Glick 2014). ACC deaminase is mainly produced by bacteria.

As we have seen in the previous sections, ethylene modulates key plant developmental stages, that is, seed germination, tissue differentiation, formation of root and shoot primordia, root branching and elongation, lateral bud development, flowering, flower senescence, fruit ripening and abscission, anthocyanin production, synthesis of volatile organic compounds (VOCs) (responsible for aroma formation in fruits), storage product hydrolysis, leaf senescence, and abscission (Glick 2014). During the establishment of symbioses, for instance with mycorrhiza and/or rhizobia, local increases of ethylene occur. In this context, bacteria able to synthesize ACC deaminase may facilitate symbiosis development (Gamalero et al. 2008, Glick 2014). Moreover, ethylene regulates plant responses to stress. A two-phase model was proposed to explain the observation that ethylene may alleviate and exacerbate the effects caused by pathogen infection. According to this model, a short time after the occurrence of stress, a small peak of ethylene is produced inside the plant. This first ethylene flush is generated from the existing ACC pool within the plant and may activate defensive plant genes (Stearns et al. 2012). In a second step, following the production of additional ACC within the plant, a much larger ethylene peak is observed. This peak acts as a signal to initiate processes such as senescence, chlorosis, and abscission, all of which are inhibitory to plant growth and survival. Thus, symptoms in plants faced with various environmental stresses are often more of a consequence of the second (large) peak of ethylene than from the direct effects of the stress itself. Based on this model, bacteria that synthesize ACC deaminase may decrease the stress-related damage to plants.

While ACC deaminase synthesis was first reported in *Pseudomonas* sp. ACP and in the yeast *Cyberlindnera saturnus*, it has now been found in numerous Gram-positive and Gram-negative bacteria (Glick 2014, Nascimento et al. 2014). Producer bacteria are common in most soils. Genes encoding ACC deaminase (the structural gene *acdS* and the regulatory gene *acdR*) have been found in many different soil bacteria belonging to the genera *Azospirillum, Rhizobium, Agrobacterium, Achromobacter, Burkholderia, Ralstonia, Paraburkholderia, Pseudomonas*, and *Enterobacter* (Nascimento et al. 2014).

In higher plants, ethylene is derived from *S*-adenosyl-methionine through the action of the enzyme ACC synthase. Especially when plants face environmental stresses, ethylene may reach concentrations that can be inhibitory to plant growth. Bacterial ACC deaminase that effectively modulates ethylene levels inside plant tissue represents one of the key bacterial traits supporting plant growth under stressed conditions (Glick 2014). In addition, ACC may also act as a signaling molecule in several plant processes, including root-to-shoot communication. Therefore, PGPB that synthesize ACC deaminase may decrease the extent of ACC signaling of specific plant functions such as the regulation of cell wall function. Thus, bacteria that have ACC deaminase facilitate plant growth under a variety of (ethylene-producing) environmental stresses, including flooding; pollution by organic toxicants and by heavy metals including nickel, lead, zinc, copper, cadmium, cobalt; salinity; drought; and bacterial and fungal phytopathogen infection and nematodes (Gamalero and Glick 2015).

22.1.3.6 Stimulation of Plant Defense Pathways

PGPB can promote plant health through stimulation of the plant's immune system. The plant-mediated immune response is called induced systemic resistance (ISR). It is expressed not only in the zone of induction, but also in distant plant organs. Since the early experiments, a body of literature has provided considerable insight into the mechanisms used by PGPB. In particular, organisms of the genera *Pseudomonas*, *Serratia*, and *Bacillus* have been found to boost plant defenses by ISR. Moreover, ISR may also be induced by plant-beneficial fungi, including arbuscular mycorrhizae.

ISR is driven by a complex network of interconnected signaling pathways in which plant hormones, mainly jasmonic acid/jasmonate and ethylene, play major regulatory roles (Pieterse et al. 2014). Besides ethylene and jasmonate produced by the plant, several bacterial molecules and components, such as the *O*-antigenic side chain of the bacterial outer membrane lipopolysaccharide, flagellar fractions, pyoverdine, 2,4-diacetyl phloroglucinol (2,4-DAPG), cyclic lipopeptide surfactants, and, in some instances, salicylic acid, have been implicated as signals that can induce systemic resistance (Pieterse et al. 2014).

All of these signaling molecules activate the transcription of defense-related plant genes that—in turn—lead to the construction of structural defense barriers. For example, Kumar et al. (2012) demonstrated that one of the mechanisms involved in protecting *A. thaliana* leaves from phytopathogen stress is the fast closure of the stomata induced by a *Bacillus subtilis* strain, FB17. This occurred only when the pathogen was present.

Plant signaling pathways were triggered by *B. amyloliquefaciens* strain FZB42 (via a unique cyclic lipopeptide) in lettuce infected with the phytopathogenic fungus *R. solani*. The expression of the gene for this lipopeptide was characterized through reverse transcription (RT)-PCR (Chowdhury et al. 2015). In healthy plants, inoculation with strain FZB42 led to increased expression of a (PR-1) pathogenesis protein, whereas in plants infected by *R. solani* a defensin was found. Overall, these results demonstrate that *B. amyloliquefaciens* FZB42 can boost plant defense responses in healthy plants via salicylic acid- and ethylene-triggered pathways, while in pathogen-infected plants a synergistic activation of the jasmonic acid and ethylene pathways, along with suppression of the salicylic acid induced pathway, occurs. Currently, the full complexity of the regulation of ISR has not been elaborated. However, a better understanding of this process is beginning to emerge from the experiments using plant proteomics and transcriptomics. Further exploration will enable us to understand how various PGPB and phytopathogens affect plant gene expression.

22.1.3.7 Antibiotics and Hydrogen Cyanide

The synthesis of antibiotics is most commonly associated with the ability of PGPB to suppress plant pathogen development (biocontrol). In particular, antibiotics produced by fluorescent pseudomonads have been widely studied. Antibiotics involved in soilborne disease suppression include phenazines, phloroglucinols, pyoluteorin, pyrrolnitrin, cyclic lipopeptides, and hydrogen cyanide. The main targets of these antibiotics are the electron transport chain (phenazines, pyrrolnitrin), metalloenzymes such as copper-containing cytochrome C oxidases (hydrogen cyanide), or cell membrane and zoospores (2,4-DAPG, biosurfactants). However, their precise modes of action are largely unknown.

Pseudomonas chlororaphis strain PA23 produces a number of metabolites, many of which may contribute to antagonism against fungal pathogens together with hydrogen cyanide, protease, lipase, and chitinase (Selin et al. 2010). Strain PA23 can protect canola against *Sclerotinia sclerotiorum*, wherein the synthesis of pyrrolnitrin is essential (Selin et al. 2010). Recently, the role of the pyrrolnitrin synthesized by *P. chlororaphis* PA23 in the suppression of the nematode *Caenorhabditis elegans* has been investigated (Munmun et al. 2015; Nandi et al. 2015). In this case, pyrrolnitrin reduced nematode egg hatching. In addition, pyrrolnitrin and HCN are highly toxic towards *C. elegans*. When strain PA23 and *C. elegans* were grown together, PA23 genes associated with biocontrol (i.e., *phzA*, *hcnA*, *phzR*, and *phzI* encoding phenazine and HCN production) were overexpressed. These results suggest that *P. chlororaphis* PA23 is able to sense the presence of *C. elegans* and can both repel and kill the nematodes.

2,4-DAPG produced by pseudomonads is considered to be an important factor involved in the inhibitory activities against many phytopathogens. Currently, a number of *P. fluorescens* strains synthesizing 2,4-DAPG have been reported to protect crops from a variety of soilborne fungal or bacterial pathogens. Zhou et al. (2014) constructed a mutant of *P. fluorescens* strain J2 that overexpressed 2,4-DAPG. It colonized the root system of tomato more efficiently and showed a higher level of suppression of *Ralstonia solanacearum*, than the wild-type strain. This suggests that increasing the amount of 2,4-DAPG (or other antibiotics) produced by biocontrol fluorescent pseudomonad strains represents a potential strategy for improving the biocontrol activity of those strains.

Phenazine, which is produced by pseudomonads as well as other soil bacteria, can inhibit the proliferation of phytopathogens. Thus, the *in vitro* growth as well as symptom production of the potato pathogen *Streptomyces scabies* by the PGPB *Pseudomonas* sp. LBUM223 can be reduced with basis in the synthesis of phenazine-1-carboxylic acid (PCA). Concomitantly, expression of the *S. scabies* gene *txtA* (encoding thaxtomin, a molecule that inhibits cellulose synthesis and then weakens cell walls) was repressed. Thus, under soil conditions, PCA released by *Pseudomonas* sp. LBUM223 reduces the virulence of *S. scabies* by repressing the synthesis of thaxtomin (Arsenault et al. 2013).

In addition to antibiotics, bacteria such as *Pseudomonas, Chromobacterium*, and *Rhizobium* can produce HCN (Munmun et al. 2015). This secondary metabolite inhibits electron transport and disrupts energy supply to the cells. Several workers have investigated its role in suppression of fungal pathogens, and in fact, the ability to release cyanide is one of the physiological traits that are now investigated in screening protocols aimed at finding new biocontrol strains. Such cyanogenic bacteria may show different behavior on different plant species. As an example, *P. fluorescens* WSM3455 inhibits the growth of wild radish weed root (*Raphanus raphanistrum*) and stimulates the development of subterranean clover root (*Trifolium subterraneum*) (Zdor 2015). This dual behavior, underlining the differences in plant sensitivity to HCN, has generated interest in the possible use of cyanogenic rhizobacteria in weed management schemes (Zdor 2015).

Typically, most PGPB strains that synthesize HCN produce only a low level of this compound. However, it is thought that HCN may act synergistically with bacterially encoded antibiotics with the HCN, augmenting the effectiveness of the antibiotic. This synergism might help to ensure that pathogens do not develop resistance to antibiotics.

22.1.3.8 Cell Wall-Degrading Enzymes

The synthesis of lytic enzymes is one of the many biocontrol mechanisms used by PGPB. Chitinases, cellulases, β-1,3glucanases, proteases, and lipases can lyse fungal cell walls and thereby limit phytopathogen populations. In particular, the degradation of the chitin (contained in many fungal cell walls) is crucial for biocontrol of both phytopathogenic fungi and insects (Nagpure et al. 2014), as no adverse side effects on humans and animals are expected. Chitinases are produced by a range of Gram-positive and Gram-negative bacteria. Among the Gram-positives, *Streptomyces* spp. and bacilli are key producers. As examples, chitinases produced by *Streptomyces* spp. and bacilli are commonly found to be active against pathogenic fungi such as *Candida* sp. and *R. solani*, reducing the incidence of stem canker and black scurf by 22.3% and 30%, respectively (Saber et al. 2015). Finally, the chitinase producer *Lysobacter capsici* YS1215, applied as a biocontrol agent of root-knot nematode *Meloidogyne incognita* in tomato, increased shoot biomass by strongly decreasing the disease severity.

22.1.3.9 Volatile Compounds

Bacteria produce a wide range of low-molecular-weight (<300 Da) VOCs that include alcohols, terpenes, esters, ketones, hydrocarbons, nitrogen-containing heterocycles, sulfur derivatives, and carboxylic acids (Audrain et al. 2015). Moreover, a single bacterial strain can synthesize many different compounds. The composition of these compounds can vary according to the available nutrients and oxygen, and the physiological state of the cells (Schmidt et al. 2015). After release, VOCs can spread over long distances and thus affect a large area. The emission of VOCs is one of the strategies by which bacteria interact with host plants and other microbial species via an intra- and interspecies dialogue (Kanchiswamy et al. 2015, Schmidt et al. 2015). Bacterial VOCs can inhibit the growth and differentiation of a suite of phytopathogenic fungi, that is, *Sclerotinia sclerotium, Phytophthora infestans, Fusarium oxysporum*, and *R. solani* (Elkahoui et al. 2015, Giorgio et al. 2015, Hunziker et al. 2015, Tenorio-Salgado et al. 2013, Wu et al. 2015). The different fungi have different sensitivities to the VOCs, and therefore, the efficacy of fungal inhibition depends on the bacterial–fungus combination, as well as local conditions.

Besides their antifungal activity, bacterial VOCs can also show nematicidal activity (NA). A suite of strains belonging to the genera *Alcaligenes, Pseudomonas, Proteus, Providencia*, and *Staphylococcus* synthesize VOCs that are active against *C. elegans* as well as *Meloidogyne incognita*. In a key study, seven bacterially produced VOCs responsible for NA were identified as acetophenone, S-methyl thiobutyrate, dimethyl disulfide, ethyl 3,3-dimethylacrylate, nonan-2-one, 1-methoxy-4-methylbenzene, and butyl isovalerate. Among them, *S*-methyl thiobutyrate showed a stronger NA than the chemical insecticide dimethyl disulfide (Xu et al. 2015).

The emission of VOCs by some bacteria can result in direct growth promotion in some plants, including *A. thaliana* and *Medicago truncatula* (Orozco-Mosqueda et al. 2013, Velazquez-Becerra et al. 2011). On the other hand, VOCs such as ammonia, dimethyl disulfide, HCN (Blom et al. 2011), and 3-phenylpropionic acid (produced by strains belonging to *Burkholderia, Chromobacterium, Pseudomonas, Serratia*, and *Stenotrophomonas*) can result in phytotoxicity and plant growth inhibition (Bailly and Weisskopf 2012). Taken together, these results highlight the involvement of the so-called bacterial volatilome in bacteria–bacteria, bacteria–fungi, and bacteria–plant interactions.

22.1.3.10 Bacteriophages

Chapter 6 discusses the role of viruses, in particular bacteriophages, in soil. Here, we examine the use of (lytic) bacteriophages in their role as controllers of population densities of phytopathogenic bacteria. Interest in bacteriophages as biocontrol agents is mainly related to human pathogens in the environment or foods. As a case in point, a preparation containing a bacteriophage lytic on (multiresistant) *P. aeruginosa*, denoted "Biophage-PA," is effective as a surface treatment for otitis. Moreover, it is well established that cell numbers of detrimental bacteria on surfaces can been reduced by the use of phages. This is relevant for treatment of pathogenic *Listeria, Salmonella*, and *E. coli* on food and hard surfaces. Thus, several phage-based systems have been approved for use on food products.

With respect to plant disease control, bacteriophages have been applied with success to control the plant pathogens *Agrobacterium tumefaciens, Burkholderia* sp., *Pectobacterium carotovorum, Erwinia amylovora, Pseudomonas syringae, Ralstonia solanacearum, S. scabies*, and *Xanthomonas* spp. (Frampton et al. 2012).

A key advantage of the use of bacteriophages as biocontrol agents is their specificity for target bacteria. Moreover, bacteriophages are self-limiting obligate parasites since they require specific hosts to replicate and do not persist for a long time in the absence of the host. However, a serious concern about phage therapy *in vivo* is the fact that phages are not lytic under all physiological conditions. Moreover, some phages may impart toxic properties to the bacterial host, thus enhancing their virulence, and hence, a careful scrutiny of this phenomenon is needed.

Presently, several bacteriophage-based biocontrol agents have received a license for use. For example, a range of products exists for the control of *Xanthomonas campestris* pv. *Vesicatoria,* which causes bacterial spot of tomatoes and peppers, and of *P. syringae* pv. *tomato*, which is the causative agent of bacterial speck on tomatoes. However, in the field, many phage preparations will work suboptimally as a consequence of their ultraviolet (UV) sensitivity. Hence, their application is typically done at dusk, when UV intensity is low. Another intrinsic problem is the persistence of a small number of bacteriophage-resistant pathogen cells. If these cells proliferate, they will dominate the system, ultimately nullifying the biocontrol effectiveness. To mitigate this problem, bacteriophages are generally delivered to plants as cocktails of different bacteriophages. This makes it less likely that bacteriophage-resistant mutants will develop, since each different bacteriophage binds to a unique site on the bacterial surface.

22.1.3.11 Quorum sensing and Quenching

As discussed in Chapter 9, QS is a form of intercellular communication, involving signaling molecules, used by some different bacteria in the environment. Here, we discuss it in the context of plant-associated bacteria and their mechanisms of action. Once bacterial cell densities attain a certain critical level, QS enables the cells to switch on particular gene sets that enable groups of bacteria to act in a coordinated manner. A suite of functions, mainly related to the expression of virulence factors in plant pathogens, are regulated by QS. Thus, antibiotic and siderophore synthesis, biofilm formation, motility, swarming, production of lytic enzymes, and Ti plasmid transfer (in *Agrobacterium tumefaciens*) are regulated by QS. In all systems, cell density and signal molecule levels increase proportionally. Once the two parameters reach a threshold level, cells start to behave differently, coordinating the expression of particular (e.g., virulence) traits.

In Gram-negative bacteria, QS communication generally occurs through *N*-acyl-homoserine lactones (AHLs) that possess acyl chains of different lengths. QS tools used by phytopathogens to express virulence are targets for the development of biocontrol strategies (Fetzner 2015). The consequent attenuation of a pathogen's virulence has been coined QQ. It may occur through inhibition of signal compound synthesis, degradation of the signal molecule, or alterations to the sensing of the latter. QQ may be widespread, as a large number of AHL-degrading bacteria, belonging to the Proteobacteria (*Agrobacterium, Bosea, Comamonas, Delftia, Ochrobactrum, Pseudomonas, Ralstonia, Sphingopyxis,* and *Variovorax*), *Actinobacteria,* and *Firmicutes* (*Arthrobacter, Bacillus, Rhodococcus,* and *Streptomyces*) have been isolated and characterized (Fetzner 2015). Interestingly, bacterial strains belonging to the same genus can be either AHL producers or degraders; in *Agrobacterium* and *Pseudomonas*, the same strain can sometimes both synthesize and cleave AHLs.

Although QQ offers fascinating perspectives, biocontrol by it does not remove a pathogen. Rather, the QQ action will reduce pathogen virulence. Consequently, the plant will be asymptomatic but contaminated with a pathogen, and so a reservoir for potential pathogen spread is present. Moreover, silencing of cell-to-cell communication is nonselective, so that nontarget functions (such as plant-beneficial traits [i.e., antifungal enzyme synthesis]) could also be impaired.

Remarkably, the tissue of various plants reveals responses to the QS signals that are secreted by plant-associated bacteria, as shown in a series of seminal papers. Box 22.1 gives an account of such responses. Plants possess mechanisms to "listen in" to bacterial cross talk in the phytosphere, and respond to it.

BOX 22.1 PLANT RESPONSES TO BACTERIAL QS COMPOUNDS

QS regulates the transcription of genes in a coordinated collective manner that is related to the density of the population and responds to habitat conditions. QS results in major phenotypic changes, optimizing the ecophysiological response. Therefore, QS has also been called "efficiency sensing." It is especially active in plant surface colonization processes, when bacteria switch to a plant-associated lifestyle. The activation of QS (especially the production of acyl homoserine lactones [AHLs] of Gram-negative bacteria) has been demonstrated *in situ* on root surfaces using AHL biosensors, identification of the produced AHLs (Rothballer et al. 2018), and modeling approaches (See Figure). During colonization, a "landscape" of AHL concentrations at the root surface is produced (lower mM range). QS is particularly involved in the activation of virulence genes of plant-pathogenic bacteria. With respect to symbiotic and beneficial endophytic bacteria, the functions related to successful interaction with the host plant are under QS control.

Lipophilic (hydrocarbon side chain with 12 and/or more C) and slightly hydrophilic (with side chain of 4–10 C) AHLs behave differently at the plant. Hydrophilic AHLs are taken up by roots and spread systemically (except for plants with high lactonase activity) (Sieper et al. 2014). Changes in enzymatic activities in roots and shoots were recorded in the presence of AHLs, leading to increased environmental fitness (Götz-Rösch et al. 2015). Lipophilic AHLs may interact with an uncharacterized membrane-located receptor, initiating a signaling cascade that primes an array of plant genes, optimizing the interaction with microbes. In the presence of pathogen-associated molecular patterns (PAMPs), increased expression of pathogen resistance genes, resulting in resistance, was reported (Schikora et al. 2011). In the absence of PAMPS, bacterial AHLs prime expression of developmental plant genes, leading to plant growth promotion and finally increased defense awareness (Schikora et al. 2016).

A QS response is effected through the autoinduction of AHLs in root-colonizing bacteria. The images show the result of mathematical simulation of AHL production using QS parameters of *P. aeruginosa* (courtesy of A. Hartmann, University of Munich, Germany). White to yellow color indicates highly induced QS response in microcolonies.

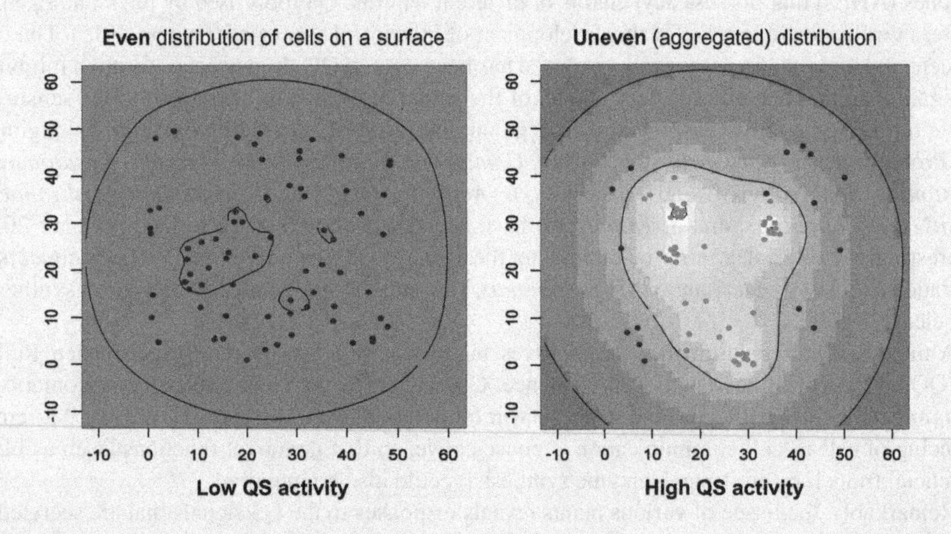

(See color insert)

22.2 APPLICATIONS OF PLANT GROWTH-PROMOTING BACTERIA

22.2.1 Uses to Improve Sustainability of Production

In the past few decades, consumers have increasingly demanded healthier and safer foods that are produced with low impacts on the environment. However, conventional crop cultivation is still based on external fertilizer and pesticide inputs that potentially impose adverse effects on human and environmental health. Thus, key research has focused on ecologically friendly biological alternatives to fertilizers and on soil management practices that reduce the use of chemicals in agriculture. A suite of approaches and organisms have been identified that may meet the requirements of sustainable agriculture, and this includes a range of beneficial bacteria next to fungi and other organisms. The consequent screening and testing of new PGPB has resulted in the production of numerous strains belonging to *Agrobacterium*, *Azospirillum*, *Azotobacter*, *Bacillus*, *Burkholderia*, *Delftia*, *Paenibacillus macerans*, *Pantoea agglomerans*, *Pseudomonas*, *Rhizobium*, and *Serratia* (Reed and Glick 2013). The diversity of mechanisms discussed in Section 22.1 applies to all of these strains, and each one of them has been formulated and marketed. However, the outcomes of the applications have been diverse, to date. For the most part, the detailed knowledge of the mechanisms used by each PGPB, as determined from laboratory and greenhouse studies, has served as a guide to predict the behavior and functioning of these strains at the plants in the field. However, in particular cases in the field, the outcomes have not met with the expectations, the reason being the unpredictability of field conditions that were potentially hostile to the inoculants. Hence, a key challenge has been to deliver a sufficient amount of active cells at the plant site where these were needed.

22.2.2 Critical Factors Determining the Outcome of PGPB Applications

As stated in the foregoing, the performance of the PGPB strains under field conditions has in particular cases been inconsistent, since these bacteria are highly dependent on conditions that reign in the soil/rhizosphere habitat to establish, survive, and function. Such conditions are largely dependent on the soil conditions, the soil management practice, and the host plant species and its physiology. Despite the fact that dozens of PGPB strains have already been successfully commercialized, a number of components in the soil/plant/inoculum system still remains to be optimized. It may appear that each inoculant strain has to encounter a suite of specific conditions in the recipient soil to be able to perform optimally. Thus, the main parameters to be considered when commercializing a PGPB strain (Reed and Glick 2013) are as follows:

i. Characterization of the strain's survival mechanisms in the respective soil/plant system
ii. Development of proper formulations in order to foster establishment and survival of strains
iii. Analysis of the inoculant's physiological traits that can impact plant growth promotion, biocontrol, or soil bioremediation
iv. Consistency among regulatory agencies in different countries regarding both rules for the deliberate release of PGPB in the environment and a wide range of perceived biological safety issues
v. Development of a knowledge base regarding the advantages and disadvantages of using PGPB rather than various chemicals
vi. Selection of strains that perform well under specific environmental conditions
vii. Development of new and more effective way of inoculating plants with selected bacterial strains under controlled or natural conditions (Bashan et al. 2014)
viii. The development of a better and more complete understanding of the interactions among various bacterial strains, the host plant, and other organisms living around or inside plant tissues

Overall, high rhizosphere competence, the ability to increase plant biomass, long-term survival on the plant, a wide range of plant-beneficial physiological traits, a lack of risk factors for human and environmental health, and high tolerance to environmental stresses encountered in soil/plant systems are essential characteristics that help to define new PGPB.

22.2.3 Applications and Market Considerations

While rhizobia constituted the first PGPB to be commercialized, in 1895 in both the United States and the United Kingdom, at the present time the market has (in addition to rhizobia) mainly turned to strains belonging to the genera *Azospirillum* (for their nitrogen-fixing capability), *Bacillus* (for their capability to solubilize phosphate or to behave as biocontrol agents), *Streptomyces* (for their ability to synthesize antibiotics and/or enzymes that are active in pathogen biocontrol), and *Pseudomonas* (for their potential use as both a biofertilizer and a biocontrol agent).

The economic impact of the commercialization of PGPB as biofertilizers or biocontrol agents is relevant. In previous assessments, it has been estimated that in developing countries such as Vietnam, the cost of N fertilization could be reduced up to 30-fold by replacing chemical fertilizers with various PGPB. Currently, the extent of biofertilizer use is limited across many countries, which is due—in part—to inappropriate formulations leading to low quality and shelf life of the inoculant. In this regard, as stated by Bashan et al. (2014), the parameters that need to be considered when formulating a commercial product are (in line with the parameters as stated earlier) the quality and the efficacy of the carriers, the mixture of the ingredients of the formulated products (quantities, conditions, ratio of mixtures), the improvements of the process involved in their production, the survival of the microorganism in the inoculant, the shelf life of the formulated product, the quality and the amount of additive to be introduced, and the development of commercial formulations based on mixtures of microorganisms. The latter requires an evaluation of the compatibility of the strains and a careful evaluation of their growth when they are cocultivated in the same bioreactor. Finally, the cost is a clear limiting parameter.

Unfortunately, several biofertilizers/biopesticides that are currently available on the market are ineffective and/or unreliable under field conditions. Product quality control is a major issue here. An evaluation of the purity of 65 commercial biofertilizers showed that 63% of these were contaminated by other bacterial strains (Herrmann and Lesueur 2013), while 40% did not contain the claimed strain, but only contaminants. Frequently, information provided on the label about the microorganisms and the carrier as well as the shelf life and the conservation of the product is limited or unclear. This highlights the necessity to develop and apply both quality control ("the process of measuring defined quality parameters of the inoculant") and quality assurance ("the overall evaluation that quality control procedures and techniques are achieving what they intend to achieve") measures during the formulation of biofertilizers and biopesticides. If this need is not carefully fulfilled, users such as farmers can lose their confidence in the use of biological products for agriculture.

22.3 CONCLUSIONS AND PERSPECTIVES

The world has seen a long history of efficient use of rhizobia in soils, which have been added in particular to foster soil nitrogen accumulation in nitrogen-poor soils. Moreover, in the past 30 years, researchers worldwide have gleaned an enormous amount of information on the mechanisms that are utilized by PGPB in their interaction with plants. However, there is still additional information that needs to be learned in terms of (i) understanding the detailed mechanisms that PGPB use to foster plant growth and protect against phytopathogens and (ii) elaborating how these organisms can be employed most effectively. Yet, on the basis of the existing knowledge, PGPB technology is ready to be developed on a larger scale than is currently applied. The large-scale use of PGPB in agriculture, horticulture, silviculture, and environmental cleanup is still in its infancy. Given the fact that microorganisms, in particular bacteria, have successfully interacted, often in beneficial ways, with plants for tens of millions of years, a reservoir of already-adapted organisms is available across soils. This can and will provide humanity with safe and efficacious alternatives to the continued widespread use of potentially problematic and dangerous agricultural chemicals. However, the frequent failure of organisms with proven biocontrol or biofertilizer capacity to act adequately in open-field situations is still alarming. The reasons for such failures often lie in the fact that an insufficient number of active cells are present at the plant site where the stimulus is required. As stated in the foregoing, such problems can be remediated by developing a proper and deeper understanding of the local limiting factors. The key critical success factors are, therefore, the selection of proper strains, the careful study of the local conditions that allow inoculant strain survival and activity, and the eventual development of strongly improved composite strain mixes. In this, the proper assessment of the survival and colonization capabilities of the developed PGPB is key. The possibilities of application of PGPB are vast, yet each application will probably require careful optimization with respect to success.

REFERENCES

Ahemad, M. 2015. Phosphate solubilising bacteria assisted phytoremediation of metalliferous soil: A review. *3 Biotech* 5, 111–121.

Ali, S.S. and N.N. Vidhale. 2012. Characterization and antifungal activity of siderophore produced by rhizospheric *Pseudomonas fluorescens* against fungal pathogen of soybean and groundnut. *Journal of Pure and Applied Microbiology* 6, 439–443.

Antoun, H. and D. Prévost. 2005. Ecology of plant growth promoting rhizobacteria. In: Siddiqui, Z.A. (ed.), *PGPR: Biocontrol and Biofertilization*. Springer, Dordrecht, pp. 1–38.

Arsenault, T., Goyer, C. and M. Filion. 2013. Phenazine production by *Pseudomonas* sp LBUM223 contributes to the biological control of potato common scab. *Phytopathology* 103, 995–1000.

Audrain, B., Farag, M.A., Ryu, C.-M. and J.M. Ghigo. 2015. Role of bacterial volatile compounds in bacterial biology. *FEMS Microbiology Review* 39, 222–233.

Aznar, A. and A. Dellagi. 2015. New insights into the role of siderophores as triggers of plant immunity: What can we learn from animals? *Journal of Experimental Botany* 21, 3001–3010.

Bailly, A. and L. Weisskopf. 2012. The modulating effect of bacterial volatiles on plant growth: Current knowledge and future challenges. *Plant Signalling Behaviour* 7, 79–85.

Bashan, Y., de-Bashan, L.E., Prabhu, S.R. and J.P. Hernandez. 2014. Advances in plant growth-promoting bacterial inoculant technology: Formulations and practical perspectives (1998–2013). *Plant and Soil* 378, 1–33.

Bhattacharyya, P.N. and D.K. Jha. 2012. Plant growth-promoting rhizobacteria (PGPR): Emergence in agriculture. *World Journal of Microbiology and Biotechnology* 28, 1327–1350.

Blom, D., Fabbri, C., Eberl, E. and L. Weisskopf. 2011. Volatile-mediated killing of *Arabidopsis thaliana* by bacterial volatiles is mainly due to hydrogen cyanide. *Applied and Environmental Microbiology* 77, 1000–1008.

Bona, E., Lingua, G., Manassero, P., Cantamessa, S., Marsano, F., Todeschini, V., Copetta, A., D'agostino, G., Massa, N., Avidano, L., Gamalero, E. and G. Berta. 2015. AM fungi and PGP pseudomonads increase flowering, fruit production, and vitamin content in strawberry grown at low nitrogen and phosphorus levels. *Mycorrhiza* 25, 181–193.

Braun, V. and K. Hantke. 2011. Recent insights into iron import by bacteria. *Current Opinion in Chemical Biology* 15, 328–334.

Chowdhury, S.P., Uhl, J., Grosch, R., Alquéres, S., Pittroff, S., Dietel, K., Schmitt-Kopplin, P., Borriss, R. and A. Hartmann. 2015. Cyclic lipopeptides of *Bacillus amyloliquefaciens* FZB42 subsp. *plantarum* colonizing the lettuce rhizosphere enhance plant defence responses towards the bottom rot pathogen *Rhizoctonia solani*. *Molecular Plant Microbe Interactions* 28, 984–995.

Compant, S., Clément, C. and A. Sessitsch. 2010. Plant growth-promoting bacteria in the rhizo- and endosphere of plants: Their role, colonization, mechanisms involved and prospects for utilization. *Soil Biology and Biochemistry* 42, 669–678.

Djavaheri, M., Mercado-Blanco, J., Versluis, C., Meyer, J.M., Van Loon, L.C. and P.A.H.M. Bakker. 2012. Iron-regulated metabolites produced by *Pseudomonas fluorescens* WCS374r are not required for eliciting induced systemic resistance against *Pseudomonas syringae* pv. *tomato* in *Arabidopsis. Microbiology Open* 1, 311–325.

Duca, D., Lorv, J., Patten, C.L., Rose, D. and B.R. Glick. 2014. Microbial indole-3-acetic acid and plant growth. *Antonie Van Leeuwenhoek* 106, 85–125.

Elkahoui, S., Djebali, N., Yaich, N., Azaiez, S., Hammami, M., Essid, R. and F. Limam. 2015. Antifungal activity of volatile compounds-producing *Pseudomonas* P2 strain against *Rhizoctonia solani. World Journal of Microbiology and Biotechnology* 31, 175–185.

Fetzner, S. 2015. Quorum quenching enzymes. *Journal of Biotechnology* 201, 2–14.

Frampton, R.A., Pitman, A.R. and P.C. Fineran. 2012. Advances in bacteriophage-mediated control of plant pathogens. *International Journal of Microbiology* 2012, 11, Article ID 326452. doi:10.1155/2012/326452

Gamalero, E. and B.R. Glick. 2015. Bacterial modulation of plant ethylene levels. *Plant Physiology* 169, 13–22.

Gamalero, E., Berta, G., Massa, N., Glick, B.R. and G. Lingua. 2008. Synergistic interactions between the ACC deaminase-producing bacterium *Pseudomonas putida* UW4 and the AM fungus *Gigaspora rosea* positively affect cucumber plant growth. *FEMS Microbiology Ecology* 64, 459–467.

Giorgio, A., De Stradis, A., Lo Cantore, P. and N.S. Iacobellis. 2015. Biocide effects of volatile organic compounds produced by potential biocontrol rhizobacteria on *Sclerotinia sclerotiorum. Frontiers in Microbiology* 6, Article ID 1056.

Glassner, H., Zchori-Fein, E., Compant, S., Sessitsch, A., Katzir, N., Portnoy, V. and S. Yaron. 2015. Characterization of endophytic bacteria from cucurbit fruits with potential benefits to agriculture in melons (*Cucumis melo* L.). *FEMS Microbiology Ecology* 91. doi:10.1093/femsec/fiv074

Glick, B.R. 2014. Bacteria with ACC deaminase can promote plant growth and help to feed the world. *Microbiological Research* 169, 30–39.

Gopala Krishnan, S., Sathya, R., Vijayabharathi, R., Varshney, R.K., Gowda, C.L.L. and L. Krishnamurthy. 2015. Plant growth promoting rhizobia: Challenges and opportunities. *Biotechnol* 5, 355–377.

Götz-Rösch, C., Riedel, T., Schmitt-Kopplin, P., Hartmann, A. and P. Schröder. 2015. Influence of bacterial *N*-acyl-homoserine lactones on growth parameters, pigments, antioxidative capacities and the xenobiotic phase II detoxification enzymes in barley and yam bean. *Frontiers in Plant Science* 6, 205.

Herrmann, L. and D. Lesueur. 2013. Challenges of formulation and quality of biofertilizers for successful inoculation. *Applied Microbiology and Biotechnology* 97, 8859–8873.

Hungria, M., Campo, R.J., Souza, E.M. and F.O. Pedrosa. 2010. Inoculation with selected strains of *Azospirillum brasilense* and *A. lipoferum* improves yields of maize and wheat in Brazil. *Plant and Soil* 3, 413–425.

Hunziker, L., Bonisch, D., Groenhagen, U., Bailly, A., Schulz, S. and L. Weisskopf. 2015. *Pseudomonas* strains naturally associated with potato plants produce volatiles with high potential for inhibition of *Phytophthora infestans. Applied and Environmental Microbiology* 81, 821–830.

Kanchiswamy, C.N., Malnoy, M. and M.E. Maffei. 2015. Chemical diversity of microbial volatiles and their potential for plant growth and productivity. *Frontiers in Plant Science* 6, 151.

Khan, M.S., Zaidi, A. and P.A. Wani. 2007. Role of phosphate-solubilizing microorganisms in sustainable agriculture – A review. *Agronomy and Sustainable Development* 27, 29–43.

Kudoyarova, G.R., Melentiev, A.I., Matìrtynenko, E.V., Timergalina, L.N., Arkhipova, T.N., Shendel, G.V., Kuzmina, L.Y., Dodd, I.C. and S.Y. Veselov. 2014. Cytokinin producing bacteria stimulate amino acid deposition by wheat roots. *Plant Physiology and Biochemistry* 83, 285–291.

Kumar, A.S., Lakshmanan, V., Caplan, J.L., Powell, D., Czymmek, K.J., Levia, D.F. and H.P. Bais. 2012. *Bacillus subtilis* restricts foliar pathogen entry through stomata. *Plant Journal* 72, 694–706.

Lemanceau, P., Bauer, P., Kraemer, S. and J.F. Briat. 2009. Iron dynamics in the rhizosphere as a case study for analyzing interactions between soils, plants and microbes. *Plant and Soil* 321, 513–535.

Liu, F., Xing, S., Ma, H., Du, Z. and B. Ma. 2013. Cytokinin-producing, plant growth-promoting rhizobacteria that confer resistance to drought stress in *Platycladus orientalis* container seedlings. *Applied Microbiology and Biotechnology* 97, 9155–9154.

Liu, X., Yang, G.M., Guan, D.X., Ghosh, P. and L.Q. Ma. 2015. Catecholate-siderophore produced by As-resistant bacterium effectively dissolved FeAsO4 and promoted *Pteris vittata* growth. *Environmental Pollution* 206, 376–381.

Mimmo, T., Del buono, D., Terzano, R., Tomasi, N., Vigani, G., Crecchio, C., Pinton, R., Zocchi, G. and S. Cesco. 2014. Rhizospheric organic compounds in the soil–microorganism–plant system: Their role in iron availability. *European Journal of Soil Science* 65, 629–642.

Munmun, N., Selin, C., Brassinga, A.K.C., Belmonte, M.F., Fernando, W.G.D., Loewen, P.C. and T.R. de Kievit. 2015. Pyrrolnitrin and hydrogen cyanide production by *Pseudomonas chlororaphis* strain PA23 exhibits nematicidal and repellent activity against *Caenorhabditis elegans*. *PloS One*. doi:10.1371/journal.pone.0123184

Nagata, T., Oobo, T. and O. Aozasa. 2013. Efficacy of a bacterial siderophore, pyoverdine, to supply iron to *Solanum lycopersicum* plants. *Journal of Bioscience and Bioengineering* 115, 686–690.

Nagpure, A., Choudhary, B. and R.K. Gupta. 2014. Chitinases: In agriculture and human healthcare. *Critical Reviews in Biotechnology* 34, 215–232.

Nandi, M., Selin, C., Brassinga, A.K.C., Belmonte, M.F., Fernando, W.G.D., Loewen, P.C. and T.R. De Kievit. 2015. Pyrrolnitrin and hydrogen cyanide production by *Pseudomonas chlororaphis* strain PA23 exhibits nematicidal and repellent activity against *Caenorhabditis elegans*. *PLoS One* 10, e0123184.

Nascimento, F.X., McConkey, B.J. and B.R. Glick. 2014. New insights into ACC deaminase phylogeny, evolution and evolutionary significance. *PLoS One* 9 (6), e99168.

Naveed, M., Qureshi, M.A., Zahir, Z.A., Hussain, M.B., Sessitsch, A. and B. Mitter. 2015. L-Tryptophan-dependent biosynthesis of indole-3-acetic acid (IAA) improves plant growth and colonization of maize by *Burkholderia phytofirmans* PsJN. *Annals of Microbiology* 65, 1381–1389.

Orozco-Mosqueda, M.D.C., Velázquez-Becerra, C., Macías-Rodríguez, L.I., Santoyo, G., Flores-Cortez, I., Alfaro-Cuevas, R. and E. Valencia-Cantero. 2013. *Arthrobacter agilis* UMCV2 induces iron acquisition in *Medicago truncatula* (strategy I plant) *in vitro* via dimethylhexadecylamine emission. *Plant and Soil* 362, 51–66.

Ossler, J., Zielinski, C.A. and K.D. Heath. 2015. Tripartite mutualism: Facilitation or trade-offs between rhizobial and mycorrhizal symbionts of legume hosts. *American Journal of Botany* 102, 1332–1341.

Oteino, N., Lally, R.D., Kiwanuka, S., Lloyd, A., Ryan, D., Germaine, K.J. and D.N. Dowling. 2015. Plant growth promotion induced by phosphate solubilizing endophytic *Pseudomonas* isolates. *Frontiers in Microbiology* 6, 745.

Pieterse, C.M., Zamioudis, C., Berendsen, R.L., Weller, D.M., Van Wees, S.C. and P.A. Bakker. 2014. Induced systemic resistance by beneficial microbes. *Annual Review of Phytopathology* 52, 347–375.

Reed, M.L.E. and B.R. Glick. 2013. Applications of plant growth-promoting bacteria for plant and soil systems. In: Gupta, V.K., Schmoll, M., Maki, M., Tuohy, M. and Mazutti, M.A. (eds.), *Applications of Microbial Engineering*, Taylor and Francis: Enfield, CT, pp. 181–229.

Rothballer, M., Uhl, J., Kunze, J., Schmitt-Kopplin, P. and A. Hartmann. 2018. Detection of the bacterial quorum sensing signaling molecules *N*-acyl-homoserine lactones and *N*-acyl-homoserine with an enzyme-linked immunosorbent assay (ELISA) and via ultrahigh performance liquid chromatography coupled to mass spectrometry (UPLC-MS). In: Leoni, L. and Rampioni, G. (eds.), *Quorum Sensing: Methods and Protocols*. Springer Series "Methods in Molecular Biology", Springer Nature: Switzerland, ISBN: 978-1-4939-7309-5.

Saber, W.I.A., Ghoneem, K.M., Al-Askar, A.A., Rashad, Y.M., Ali, A.A. and E.M. Rashad. 2015. Chitinase production by *Bacillus subtilis* ATCC 11774 and its effects on biocontrol of *Rhizoctonia* diseases of potato. *Acta Biologica Hungarica* 66, 436–448.

Schikora, A., Schenk, S.T. and A. Hartmann. 2016. Beneficial effects of bacteria-plant communication based on quorum sensing molecules. *Plant Molecular Biology* 90, 605–612.

Schikora, A., Schenk, S.T., Stein, E., Molitor, A., Zuccaro, A. and K.-H. Kogel. 2011. *N*-acyl-homoserine lactones confers resistance towards biotrophic and hemibiotrophic pathogens via altered activation of AtMPK6. *Plant Physiology* 157, 1407–1418.

Schmidt, R., Cordovez, V., De Boer, W., Raaijmakers, J. and P. Garbeva. 2015. Volatile affairs in microbial interactions. *ISME Journal* 9, 2329–2335.

Selin, C., Habibian, R., Poritsanos, N., Athukorala, S.N., Fernando, D. and T.R. de Kievit. 2010. Phenazines are not essential for *Pseudomonas chlororaphis* PA23 biocontrol of *Sclerotinia sclerotiorum*, but do play a role in biofilm formation. *FEMS Microbiology Ecology* 7, 73–83.

Seo, S., Lee, S., Hong, Y. and Y. Kim. 2012. Phospholipase A2 inhibitors synthesized by two entomopathogenic bacteria, *Xenorhabdus nematophila* and *Photorhabdus temperate* subsp. *temperata*. *Applied and Environmental Microbiology* 78, 3816–3823.

Sieper, T., Forczek, S., Matucha, M., Krämer, P., Hartmann, A. and P. Schröder. 2014. N-acyl-homoserine lactone uptake and systemic transport in barley rest upon active parts of the plant. *New Phytologist* 201, 545–555.

Spaepen, S., Bossuyt, S., Engelen, K., Marchal, K. and J. Vanderleyden. 2014. Phenotypical and molecular responses of *Arabidopsis thaliana* roots as a result of inoculation with the auxin-producing bacterium *Azospirillum brasilense*. *New Phytologist* 201, 850–861.

Stearns, J.C., Woody, O.Z., McConkey, B.J. and B.R. Glick. 2012. Effects of bacterial ACC deaminase on *Brassica napus* gene expression measured with an *Arabidopsis thaliana* microarray. *Molecular Plant-Microbes Interactions* 25, 668–676.

Stephen, J., Shabanamol, S., Rishad, K.S. and M.S. Jisha. 2015. Growth enhancement of rice (*Oryza sativa*) by phosphate solubilizing *Gluconacetobacter* sp. (MTCC 8368) and *Burkholderia* sp. (MTCC 8369) under greenhouse conditions. *3 Biotech* 5, 831–837.

Sulochana, M.B., Jayachandra, S.Y., Kumar, S.A., Parameshwar, A.B., Reddy, K.M. and A. Dayanand. 2014. Siderophore as a potential plant growth-promoting agent produced by *Pseudomonas aeruginosa* JAS-25. *Applied Biochemistry and Biotechnology* 174, 297–308.

Tenorio-Salgado, S., Tinoco, R., Vazquez-Duhalt, R., Caballero-Mellado, J. and E. Perez-Rueda. (2013). Identification of volatile compounds produced by the bacterium *Burkholderia tropica* that inhibit the growth of fungal pathogens. *Bioengineered* 4, 236–243.

Truyens, S., Weyens, N., Cuypers, A. and J. Vangronsveld. 2015. Bacterial seed endophytes: Genera, vertical transmission and interaction with plants. *Environmental Microbiology Reports* 7, 40–50.

Tudzynski, B., Studt, L. and Rojas, M.C. 2016. Gibberellins in fungi, bacteria and lower plants: Biosynthesis, function and evolution. *Annual Plant Reviews* 49. doi:10.1002/9781119210436.ch5

Ullah, I., Khan, A.R., Jung, B.K., Khan, A.L., Lee, I.J. and J.H. Shin. 2014. Gibberellins synthesized by the entomopathogenic bacterium, *Photorhabdus temperata* M1021 as one of the factors of rice plant growth promotion. *Journal of Plant Interactions* 9, 775–782.

Ullah, I., Khan, A., Park, G.-S., Lim, J.-H., Waqas, M., Lee, I.-J. and J.-H. Shin. 2013. Analysis of phytohormones and phosphate solubilization in *Photorhabdus* spp. *Food Science and Biotechnology* 22, 25–31.

Velazquez-Becerra, C., Macias-Rodriguez, L.I., Lopez-Bucio, J., Altamirano-Hernandez, J., Flores-Cortez, I. and E. Valencia-Cantero. 2011. A volatile organic compound analysis from *Arthrobacter* agilis identifies dimethylhexadecylamine, an amino-containing lipid modulating bacterial growth and *Medicago sativa* morphogenesis *in vitro*. *Plant and Soil* 339, 329–340.

Wu, Y.C., Yuan, J., Yaoyao, E., Raza, W., Shen, Q.R. and Q.W. Huang. 2015. Effects of volatile organic compounds from *Streptomyces albulus* NJZJSA2 on growth of two fungal pathogens. *Journal of Basic Microbiology* 55, 1104–1117.

Xu, Y.Y., Lu, H., Wang, X., Zhang, K.Q. and G.H. Li. 2015. Effect of volatile organic compounds from bacteria on nematodes. *Chemistry and Biodiversity* 12, 1415–1421.

Zdor, R.E. 2015. Bacterial cyanogenesis: Impact on biotic interactions. *Journal of Applied Microbiology* 118, 267–274.

Zhou, T.T., Li, C.Y., Chen, D., Wu, K., Shen, Q.R. and B. Shen. 2014. phlF(-) mutant of *Pseudomonas fluorescens* J2 improved 2,4-DAPG biosynthesis and biocontrol efficacy against tomato bacterial wilt. *Biological Control* 78, 1–8.

23 Biodegradation and Bioremediation of Organic Pollutants in Soil

Kam Tin Leung, Zi-Hua Jiang, and Nouf Almzene
Lakehead University (Thunder Bay Campus)

Kanavillil Nandakumar and Kurissery Sreekumari
Lakehead University (Orillia Campus)

Jack T. Trevors
University of Guelph

CONTENTS

23.1 Introduction 381
23.2 Biodegradation of Aliphatic and Aromatic Hydrocarbon Pollutants 383
23.2.1 Hydrocarbon Biodegradation 383
23.2.1.1 Aliphatic and Alicyclic Compounds 383
23.2.1.2 Aromatic Compounds 384
23.2.1.3 Hydrocarbon Biodegradation under Anaerobic Conditions 384
23.3 Biodegradation of Chlorinated Hydrocarbons 385
23.3.1 Biodegradation of Chlorinated Aliphatic Compounds 385
23.3.2 Biodegradation of Chlorinated Aromatic Compounds 386
23.3.2.1 Chlorinated Benzenes and Phenols 387
23.4 Biodegradation of Nitroaromatic Compounds 388
23.5 Biodegradation of Synthetic Organic Pollutants 389
23.5.1 Biodegradation of Pesticides 389
23.5.2 Biodegradation of Pharmaceuticals 391
23.5.3 Biodegradation of Plastics 393
23.6 Degradative Genes and Their Expression in Soil Microbial Communities 394
23.6.1 Activity-Based and Sequence-Based Metagenomics 394
23.6.2 Stable-Isotope Probing–Metagenomics 395
23.6.3 Expression of Degradative Genes in Soil 395
23.7 Bioremediation 396
23.7.1 Environmental Factors Affecting Bioremediation 399
23.8 Concluding Remarks 400
Acknowledgments 401
References 401

23.1 INTRODUCTION

Soil is one of the most important resources on the Earth, and the United Nations declared 2015 the International Year of Soils. Natural soil microbial communities possess immense species and metabolic diversity. The metabolic potential is essential for global nutrient cycling, facilitating the breakdown of diverse organic compounds in the environment. However, concerns with anthropogenic inputs into soil due to exploration and use of fossil fuels and the production of synthetic organic compounds, such as pesticides and pharmaceuticals, may exceed the biodegrading capability of microbial communities in soil and adversely affect soil health. Such compounds are released into the soil environment (accidentally or purposely) from either point or nonpoint source(s). In particular, many synthetic compounds have novel chemical structures that are recalcitrant to microbial degradation because microorganisms have not evolved to contain the enzymes to transform and degrade these compounds. Often these compounds are (1) halogenated, sulfonated, or substituted with nitro groups; (2) too toxic or present at inhibitory concentrations; and/or (3) unavailable to degradative microorganisms (due to low solubility in water or to being tightly sorbed to the soil matrix). Because of the persistent nature of some of these pollutants, they can accumulate in organisms at lower trophic levels and "biomagnify" in the higher trophic levels of the food chain. This problem was highlighted in the book *Silent Spring* by Rachel Carson (1964) for 1,1,1-trichloro-2,2-bis(*p*-chlorophenyl)ethane (DDT), a pesticide that accumulates in large predatory animals and birds to levels about four to six orders of magnitude higher than those in the environment.

Fortunately, many synthetic compounds are sufficiently similar to naturally occurring compounds. Given the right conditions and adequate time, they can be degraded (i.e., metabolized or co-metabolized) by microorganisms in the environment. The term "biodegradation" is used to describe a biological process that converts an organic compound to CO_2, H_2O, and/or inorganic salt(s) or, sometimes, to an intermediate or a dead-end product(s) that could be higher or lower in toxicity than the parent compound. To minimize ambiguity in terminology, the term "mineralization" describes the complete breakdown of organic compounds to CO_2, H_2O, and/or inorganic salt(s) that are present in the original compounds. *Bioremediation* refers to the application of the metabolic capacity of organisms to mineralize the pollutants or degrade them into less toxic or nontoxic substances in the environment.

Biodegradation of organic pollutants is a mechanism in which microorganisms use pollutants as sources of energy, carbon and/or other elements for growth, except for the special case of co-metabolism discussed in the following. The amount of carbon (C) assimilated into cells can be estimated by the difference between the initial amount of C in the substrate and the amount of C mineralized to CO_2. Depending on whether the process is aerobic or anaerobic, the concentration and reductive potential of the substrate, and the metabolic state of the degradative microorganisms, the percentage of assimilation can vary from $<10\%$ to $>90\%$ (Alexander 1999).

The term "co-metabolism" (sometimes referred to as fortuitous metabolism) describes the transformation of an organic compound by a microorganism that inadvertently carries out the reaction but is not capable of using the substrate for energy generation or growth. Co-metabolism can be beneficial or detrimental to the environment, depending on the chemical nature of intermediates and whether other species in the microbial community can mineralize the product(s) of the co-metabolic reaction. For example, parathion can be co-metabolized by *Pseudomonas stutzeri* to *p*-nitrophenol, which can be mineralized by *Pseudomonas aeruginosa* (Daughton and Hsieh 1977). However, the product(s) of a cometabolic reaction can be sometimes more toxic than the co-metabolic substrate (or secondary substrate). For instance, trichloroethene (TCE) can be co-metabolized by methanogens anaerobically to vinyl chloride, which is more carcinogenic than TCE (Freedman and Gossett 1989). Although not absolutely required, co-metabolism of the secondary substrate usually is dependent on the presence of a primary substrate that can support the growth of the microorganisms carrying out the reactions.

The success of bioremediation of organic pollutants at a contaminated site depends on several major factors. Besides the structural complexity of the pollutants and the nutritional and environmental (or physicochemical) conditions that a soil microbial community is exposed to, it also depends on the presence of microorganisms that possess the appropriate biodegradative enzymes. Because many bacteria in the environment are non-culturable, it is often not feasible to attempt to isolate these from a contaminated site and determine their biodegradative capability. However, a cultivation-independent metagenomics approach can provide opportunities to determine the diversity and biodegradative capability of a soil microbial community. This will improve the understanding of these and increase our success of bioremediation.

This chapter reviews some of the major microbial degradation pathways of organic pollutants based on the chemical structures. In addition, the biodegradation of synthetic compounds with specific functions such as pharmaceuticals, pesticides, and plastics is discussed. Finally, we focus on the principles of metagenomics, various approaches to bioremediation, and the major environmental factors that affect the success of bioremediation.

23.2 BIODEGRADATION OF ALIPHATIC AND AROMATIC HYDROCARBON POLLUTANTS

Petroleum hydrocarbons are among the most common sources of organic pollutions in the soil environment. They are mixtures of many compounds, such as alkanes, aromatics, asphaltenes, and resins. Among these, alkanes are the most readily biodegradable, followed by the low-molecular-weight (LMW) polyaromatic hydrocarbons (PAHs). Asphaltenes and resins are the most persistent to biodegradation. Many PAHs pose serious risks, as they have cytotoxic, mutagenic, and in some cases carcinogenic effects on animals and humans. In addition, components of gasoline such as benzene, toluene, ethylbenzene, and xylene (BTEX) are volatile mono-aromatic hydrocarbons that are major contributors to environmental pollution. For example, Jenkins et al. (2014) estimated that about 225,000 petroleum brownfield sites in the United States remained to be remediated.

23.2.1 Hydrocarbon Biodegradation

23.2.1.1 Aliphatic and Alicyclic Compounds

Complete degradation of aliphatic hydrocarbons results in the formation of carbon dioxide and water. Branched aliphatic hydrocarbons are relatively persistent to biodegradation, whereas medium-sized straight-chain compounds, such as C_{10}–C_{18} *n*-alkanes, are more readily degraded. The rate of degradation in soil also depends on the solubility of the compounds in water, and solubility decreases with an increase in the number of carbon atoms in the hydrocarbon chain. Significant enhancement of bioavailability of longer-chain organic compounds can be achieved by using surfactants or cosolvents (Becker and Seagren 2010).

Alkanes are saturated compounds, with a general formula C_nH_{2n+2}. They can be degraded via two biodegradation pathways. The most common pathway depends on monooxygenase enzymes specific for *n*-alkane compounds. This results in addition of an oxygen atom to the terminal methyl group, thereby producing an alcohol which is converted to an aldehyde and then to a fatty acid (Pathway 1, Figure 23.1). In the second pathway (Pathway 2, Figure 23.1), a dioxygenase enzyme acts on the terminal methyl group of an *n*-alkane compound, resulting in the addition of two oxygen atoms. This results in the formation of a peroxide compound that is converted to a fatty acid. The fatty acids resulting from either pathway are metabolized via a β-oxidation pathway, to form acetyl-CoA or propionyl-CoA. These are then metabolized via the tricarboxylic acid (TCA) cycle to CO_2 and H_2O.

Alkenes are degraded by many bacterial species. The biodegradation pathway of alkenes is similar to alkanes in that the degradation process starts with an oxidation of the compounds by a monooxygenase enzyme. Depending on the location of action, the end products could be either alcohol, when acted on terminal or subterminal methyl groups, or epoxides, when acted on the double bond. These compounds are further metabolized via the β-oxidation pathway, as described in the foregoing.

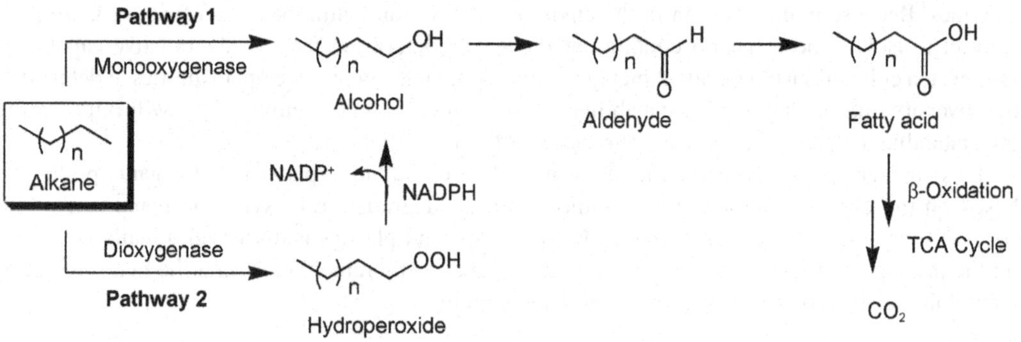

FIGURE 23.1 Biodegradation of alkanes.

Alicyclic hydrocarbons, such as cyclohexane, cyclopentane, and various cycloparaffins, comprise about 12% of the organic components of petroleum products. Among the different hydrocarbons, alicyclic compounds are the most resistant to biodegradation. Cyclohexane mineralization in bacteria is initiated by its oxidation to cyclohexanol by P450-dependent oxygenase systems. The cyclohexanol will then be oxidized to ε-caprolactone via cyclohexanone formation by a Baeyer–Villiger monooxygenation reaction. This lactone is split by an esterase (caprolactone hydrolase) to yield 6-hydroxyhexanoic acid, which is further oxidized to adipic acid and then to succinate. Succinate enters the TCA pathway to complete the mineralization process (Whitby 2010).

23.2.1.2 Aromatic Compounds

Aromatic hydrocarbons constitute major components of petroleum hydrocarbons. While some simple aromatic hydrocarbons (e.g., BTEX) are present in petroleum, naphthalene, with two aromatic rings, represents the simplest PAH. Many microorganisms, including representatives of bacteria, fungi, and algae, can degrade PAHs. Among the PAHs, the LMW ones (<3 rings) are more susceptible to microbial degradation than the high-molecular-weight (HMW) ones (≥3 rings). To date, no single microorganism has been reported that can mineralize HMW hydrocarbons, such as benzo(*a*)pyrene, as the sole source of energy, although transformations through co-metabolic activities have been reported. Low solubility is one reason for the low rate of biodegradation of these compounds in soil systems, in addition to problems due to production of toxic dead-end metabolites and/or a lack of co-metabolic substrates.

Aromatic hydrocarbons can be used as sole sources of carbon and energy by some bacteria. A key step required for their degradation is cleavage of the aromatic ring. This step is carried out by dioxygenase enzymes with molecular oxygen as the reactant. For example, in benzene catabolism, a dioxygenase facilitates the oxidation of benzene to benzene dihydrodiol, which is subsequently converted to catechol (Figure 23.2). Similarly in substituted aromatic compounds, such as toluene, ethylbenzene, or xylene isomers, the aromatic rings act as the initial sites of action of oxygenases, resulting in the formation of substituted catechols. Catechol (or substituted catechol intermediates from BTEX) will subsequently be catabolized by one of the following two pathways: (1) the ortho-cleavage (β-ketoadipate) pathway, in which the ring is cleaved between the two carbon atoms with hydroxyl groups, or (2) the meta-cleavage pathway, in which the ring splits between adjacent carbon atoms with and without a hydroxyl group. Breakdown metabolites such as acetate, succinate, pyruvate, or acetaldehyde subsequently enter the TCA cycle and are thus completely metabolized (Figure 23.2).

23.2.1.3 Hydrocarbon Biodegradation under Anaerobic Conditions

Hydrocarbons can also be degraded in the absence of oxygen by some microorganisms. Substrates such as Fe^{3+}, NO_3^-, Mn^{4+}, and SO_4^{2-} can act as the terminal electron acceptors (instead of oxygen)

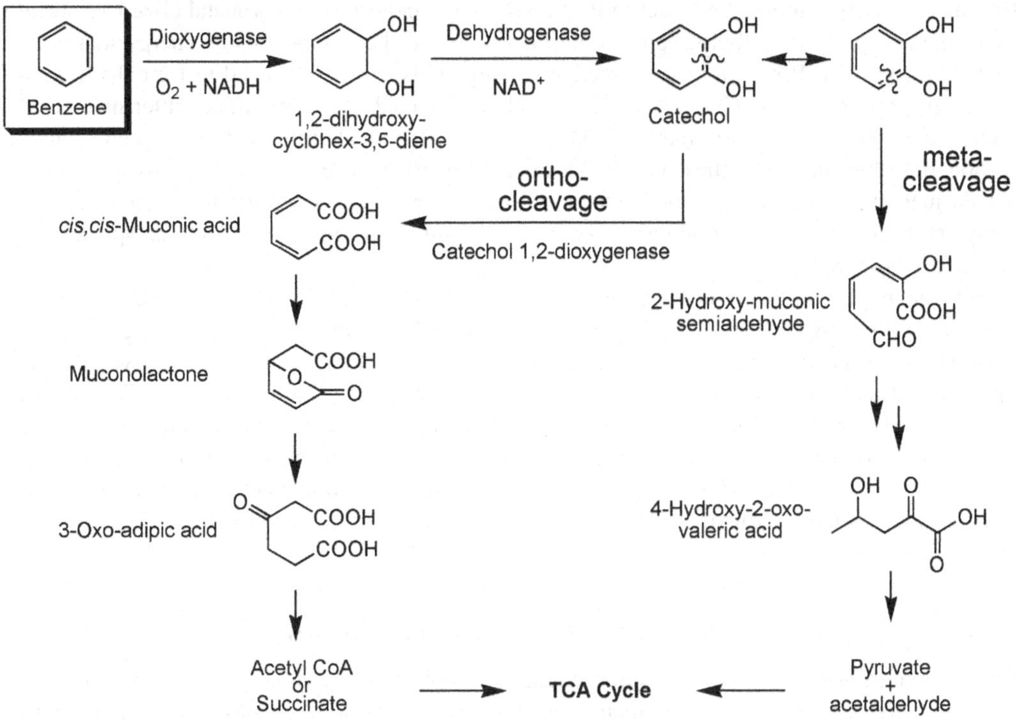

FIGURE 23.2 Biodegradation of benzene.

by undergoing reduction reactions or by methanogenesis (carbonates) (Becker and Seagren 2010). Sometimes, the pollutants themselves can be used as electron acceptors in the process. For example, BTEX molecules (benzene, toluene, ethylbenzene, xylene) can be metabolized anaerobically through a common intermediate, benzoyl-CoA. This benzoyl-CoA production is considered to be analogous to the production of catecholic ring compounds in the aerobic BTEX biodegradation (Gibson and Harwood 2002). Since the alternative electron acceptors are less efficient (i.e., have lower reduction potential) than oxygen to oxidize the compounds, growth under anaerobic conditions is not as efficient as that under aerobic conditions. While saturated aliphatic compounds are degraded slowly, those containing oxygen or unsaturated aliphatics are more readily degraded. The pathway for the degradation of unsaturated hydrocarbons involves the hydration of double bonds to alcohols which subsequently are converted to ketones or aldehydes and then to fatty acids that enter the common cellular catabolic routes for complete mineralization.

23.3 BIODEGRADATION OF CHLORINATED HYDROCARBONS

23.3.1 Biodegradation of Chlorinated Aliphatic Compounds

Mono- and dichloromethane, chloroform, tetrachloromethane, 1,2-dichloroethane, dichloroethylene, trichloroethylene (TCE), and tetrachloroethylene (or perchloroethene, PCE) are some of the major chlorinated aliphatic hydrocarbon (CAH) pollutants in many terrestrial environments. These compounds are toxic to humans and other animals. They are commonly used as solvents in some industries and contaminate some terrestrial and groundwater systems through spills and improper disposal.

Dechlorination is usually the first step of the microbial degradation processes. This can be achieved either by oxidation (or hydroxylation) or by reduction. CAHs with low numbers of chlorine substituents, such as mono- and di-chlorinated C1 compounds, are usually metabolized aerobically by methanotrophic or methylotrophic bacteria, which use the compounds as energy and carbon sources for growth. Both chlorinated methane compounds can be oxidized to formaldehyde and eventually broken down to CO_2 and water through formic acid. However, higher chlorinated CAHs such as chloroform, tetrachloromethane, TCE, and PCE are recalcitrant to aerobic dechlorination. Although dechlorination of these CAHs does occur under anaerobic conditions through reductive dechlorination (or dehalorespiration) where the CAHs are used as a terminal electron acceptor, complete dechlorination and mineralization of these compounds by a single bacterial species is not commonly observed.

For the highly chlorinated aliphatic organic solvents such as TCE and PCE, complete mineralization is usually achieved by co-metabolism under aerobic and/or anaerobic conditions. For instance, *Pseudomonas putida* F1 can dechlorinate and break down TCE to glyoxylate and formate fortuitously by toluene dioxygenase under aerobic conditions. Although *P. putida* F1 cannot metabolize the glyoxylate and formate, other aerobic soil bacteria will mineralize these compounds for energy and growth (Becker and Seagren 2010). Methane-utilizing microbial consortia can also degrade TCE aerobically by co-metabolism to products that can be used by other microorganisms in the consortia. Methane monooxygenase is involved in this TCE degradative process.

23.3.2 Biodegradation of Chlorinated Aromatic Compounds

Chlorinated aromatics, including chlorobenzenes, chlorobenzoates, chlorophenols, and polychlorinated biphenyls (PCBs), are major groups of halogenated aromatic pollutants in soil environments. Degradation of chloroaromatics can be either aerobic or anaerobic. The rate of aerobic decomposition usually decreases as the number of chlorine substituents increases, whereas the opposite is observed for the anaerobic decomposition process. One key reaction that leads to the complete breakdown of chloroaromatics is removal of some or all of the chlorine substituents, converting the molecules to chlorocatechol or catechol, respectively. Both intermediates can then be completely metabolized (or mineralized) through either the ortho- or meta-ring cleavage mechanism and the TCA cycle (Figure 23.3).

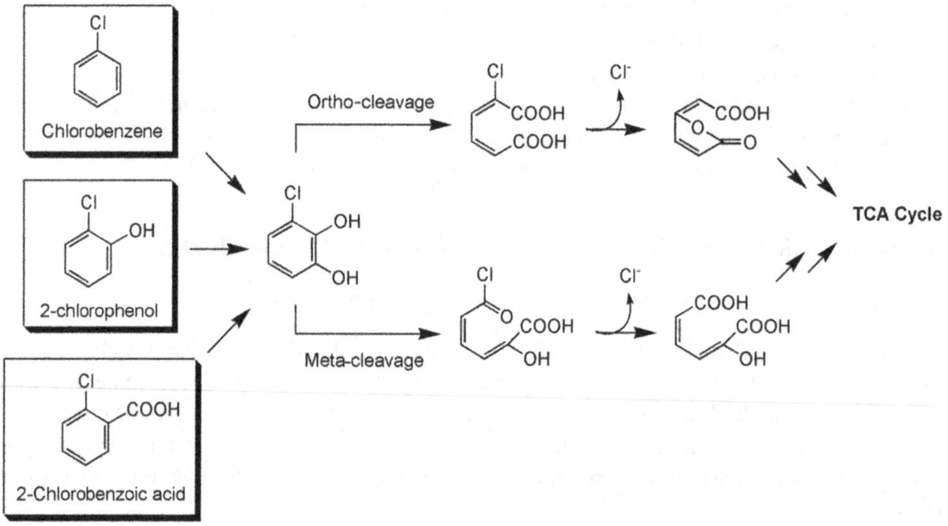

FIGURE 23.3 Biodegradation of chlorinated aromatic hydrocarbons.

23.3.2.1 Chlorinated Benzenes and Phenols

23.3.2.1.1 Polychlorinated Biphenyls

PCBs have been used in the past as hydraulic and dielectric fluids. Because of the lipophilic nature of these compounds, they concentrate in lipids of animals and can lead to detrimental bioaccumulation in the food chain. Although PCBs were phased out in most developed countries in the 1980s, they are still present in some terrestrial environments because of their recalcitrant nature.

PCBs consist of a biphenyl skeleton that contains one to ten chlorine substituents. There are 209 congeners of PCBs, and the number and position of the chlorine substituents on the compounds dictate the biodegradability of the congeners. Generally, PCBs are degraded by mixed microbial consortia. The highly chlorinated PCB congeners are preferentially dechlorinated under anaerobic conditions, converting them to lower chlorinated congeners (usually ≤3 chlorine substituents). The lower-chlorinated PCB congeners can be mineralized aerobically by particular microorganisms. *Dehalococcoides* species and some Proteobacteria (such as *Burkholderia*, *Pseudomonas*, and *Sphingomonas* spp.) are examples of anaerobic and aerobic degraders of PCBs, respectively.

For the highly chlorinated PCB congeners, biodegradation starts under anaerobic conditions. Reductive dehalogenases from the anaerobic PCB degraders convert the PCB congeners to lower chlorinated forms via dechlororespiration. However, the resulting molecules can only be degraded efficiently by aerobic PCB degraders. Under aerobic conditions, dioxygenases of the latter will usually hydroxylate the 2- and 3-positions of the lighter chlorinated ring. The hydroxylated ring will then be broken down by the meta-ring cleavage pathway and thus a PCB molecule is transformed to chlorobenzoate that will be completely degraded by aerobic PCB degraders and other chlorobenzene degraders in soil (Figure 23.4).

23.3.2.1.2 Chlorinated Phenols

Under aerobic conditions, mono- and dichlorophenols are most often reported to be degraded by an ortho-ring cleavage pathway (through a chlorocatechol intermediate), although alternative routes, such as via hydroxyquinol, have been reported. However, the degradation of tri-, tetra-, and pentachlorophenols (PCP) normally involves the formation of chloro-*p*-benzoquinones.

PCP is one of the most common chlorophenols in soils as it has been used worldwide as a biocide. Because of its toxicity to humans and animals, the use of PCP has been restricted in most developed countries. The elucidation of the aerobic degradation of PCP (Cai and Xun 2002) was mostly based on the study of *Sphingobium chlorophenolicum* (Figure 23.5). Anaerobic degradation of PCP

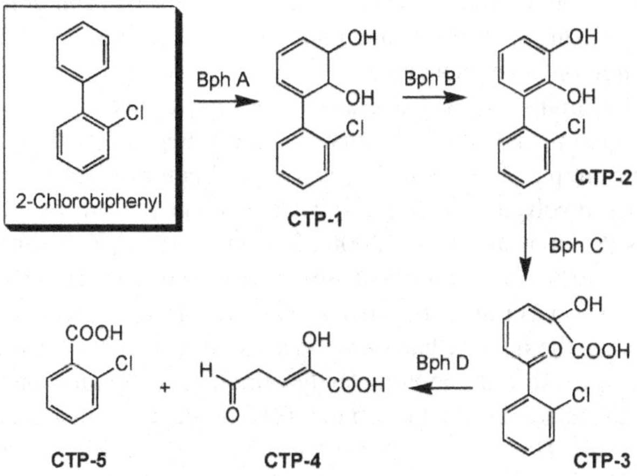

FIGURE 23.4 Biodegradation of 2-chlorobiphenyl.

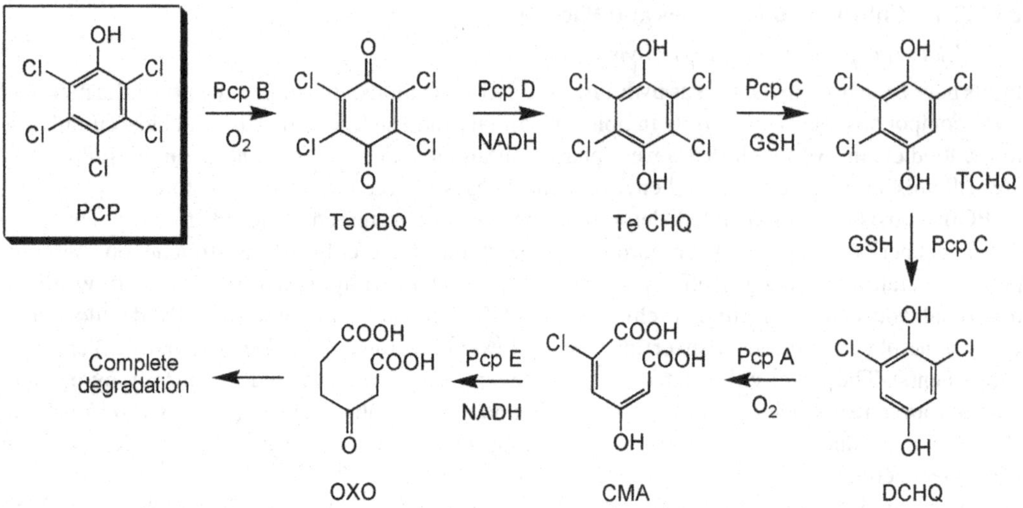

FIGURE 23.5 Aerobic degradation of pentachlorophenol. PCP, pentachlorophenol; TeCBQ, tetrachlorobenzoquinone; TeCHQ, tetrachloro-*p*-hydroquinone; TCHQ, 2,3,5-trichlorohydroquinone; DCHQ, 2,6-dichlorohydroquinone; CMA, 2-chloromaleylacetate; OXO, 3-oxoadipate; GSH, reduced glutathione; PcpB, PCP-4-monooxygenase; PcpD, TeCBQ reductase; PcpC, TeCHQ reductive dehydrogenase; PcpA, DCHQ 1,2-dioxygenase; PcpE, maleylacetate reductase.

or other chlorophenols, which can be found in methanogenic or sulfate-reducing environments, usually requires a consortium of microorganisms. Reductive dechlorination of PCP is a stepwise replacement of chlorine substituents with hydrogen. Although complete mineralization of PCP has been documented, incomplete degradation can result in the accumulation of less-chlorinated phenols (Lopez-Echartea et al. 2016).

23.4 BIODEGRADATION OF NITROAROMATIC COMPOUNDS

Nitroaromatic compounds are sometimes present in soil as intermediates of pesticides (e.g., parathion), explosives [e.g., trinitrotoluene (TNT) and picric acid], and azo dyes. Some of these compounds are the products of the incomplete combustion of gasoline and diesel fuel. Because the nitro group(s) of these compounds can be converted to toxic nitroso and/or hydroxylamino groups by intestinal microorganisms as well as some mammalian cells, nitroaromatic compounds can be highly toxic and/or carcinogenic to humans.

Nitroaromatic compounds can be degraded both anaerobically and aerobically in soil environments. Some mono- and di-nitroaromatic compounds are used as growth substrates by aerobic bacteria. For example, *Burkholderia cepacia* R34 can mineralize 2,4-dinitrotoluene, and genes for the enzymes involved (which evolved from mono- and di-oxygenases) exist in other degradative pathways (Symons and Bruce 2006). The nitro group(s) of nitroaromatic compounds can be removed, (1) as nitrite, by monooxygenases or dioxygenases; (2) as ammonia, via a hydroxylamine intermediate; or (3) as nitrite, by partial reduction of the aromatic ring. After elimination of the nitro substituent(s), the intermediates can be mineralized. However, higher nitro-substituted compounds such as TNT can only be completely mineralized by anaerobic bacterial consortia. Both the aerobic and the anaerobic degradation of TNT produce highly reactive intermediates that can covalently bind to soil. This may reduce the spread and toxicity of the contaminants. However, detailed information about the biochemical mechanisms of TNT degradation is limited (Symons and Bruce 2006).

23.5 BIODEGRADATION OF SYNTHETIC ORGANIC POLLUTANTS

23.5.1 Biodegradation of Pesticides

The worldwide use of pesticides is about two million tons per year, of which 45% is used in Europe, 25% in the United States, and 30% in the rest of the world (De et al. 2014). Due to their worldwide use, pesticides are key sources of organic pollutants in soils. In general, they can be degraded through chemical and/or biological processes. In soils, biodegradation is considered to constitute the primary mechanism. The mechanisms include oxidative, reductive, and hydrolytic processes. Oxidative processes involve hydroxylation, dealkylation, epoxidation, and sulfoxidation. These reactions are catalyzed by monooxygenase, dioxygenase, laccase, and peroxidase enzymes. Some pesticides are recalcitrant in soil, in particular those containing multiple chlorine substituents that are resistant to enzymatic breakdown. For instance, the half-life of DDT in terrestrial settings is about 15.6 years. Table 23.1 presents the current understanding of the biodegradation of some common pesticides (Box 23.1).

TABLE 23.1
The Biodegradation of Some Common Pesticides

Pesticide		Example of Degradative Microbe(s)	Products	References
Class	Example			
Organochloride	DDT (insecticide) 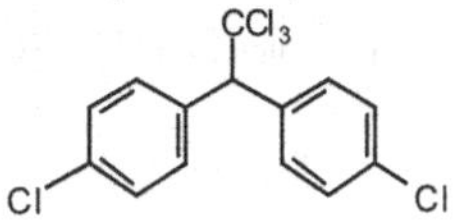 CCl_3 Cl Cl	Microbial consortium	[a]Completely metabolized	[b]EAWAG-BBD
	Lindane (insecticide) 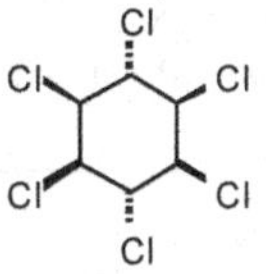 Cl Cl Cl Cl Cl Cl	*Sphingomonas paucimobilis* UT26	Completely metabolized	EAWAG-BBD
	Pentachlorophenol (fungicide) 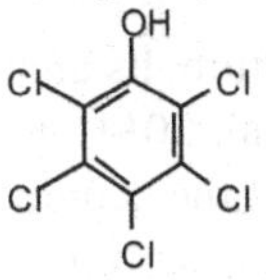 OH Cl Cl Cl Cl Cl	*Sphingobium chlorophenolicum* ATCC 39723	Completely metabolized	EAWAG-BBD
Organophosphate	Parathion (insecticide) 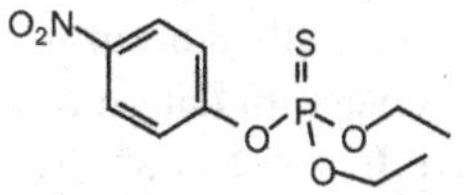 O_2N S P O O O	Microbial consortium or *Flavobacterium* sp. ATCC 27551	Completely metabolized	EAWAG-BBD
	Chlorpyrifos (insecticide) 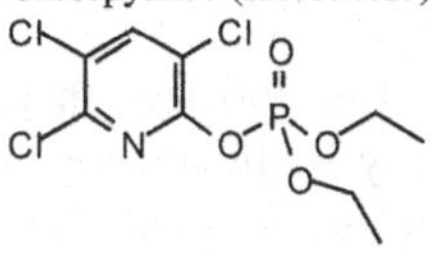 Cl Cl O Cl N O P O O	*Enterobacter* strain B-14	Completely metabolized	Singh et al. (2004)
	Glyphosate (herbicide) 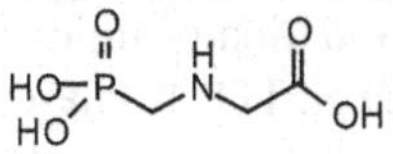 O H O HO P N OH HO	*Geobacillus caldoxylosilyticus* T20 and *Flavobacterium* sp.	Completely metabolized	EAWAG-BBD pathway prediction system

(Continued)

TABLE 23.1 (*Continued*)
The Biodegradation of Some Common Pesticides

Pesticide Class	Pesticide Example	Example of Degradative Microbe(s)	Products	References
Phenoxy alkanoic acid	2,4-D (insecticide)	*Alcaligenes eutrophus*, *B. cepacia*, and *Phanerochaete chrysosporium*	Completely metabolized	EAWAG-BBD
Carbamate	Carbofuran (insecticide)	Microbial consortium	3-(2-Hydroxy-2-methylpropyl) benzene-1,2-diol	EAWAG-BBD
Triazine	Atrazine (herbicide)	Microbial consortium	Cyanuric acid (it can be completely metabolized by other soil microorganisms)	EAWAG-BBD
Phenylurea	Isoproturon (herbicide)	*Sphingomonas* sp. Strain SRS2	Completely metabolized	Nielsen et al. (2015)

[a] Completely metabolized—the pesticide is completely broken down to CO_2, H_2O, and salt(s).
[b] EAWAG-BBD—the EAWAG Biocatalysis/Biodegradation Database (http://eawag-bbd.ethz.ch/index.html) (Gao et al. 2010).

BOX 23.1 CASE STUDY: IMIDACLOPRID BIODEGRADATION

Imidacloprid is one of the most widely used insecticides in agriculture. Its annual production was estimated to be about 20,000 tons in 2010 (Simon-Delso et al. 2014), and currently, it shares about 20% of the insecticide market. Imidacloprid belongs to the neonicotinoid family, in which there are currently seven major commercial products: acetamiprid, clothianidin, imidacloprid, nitenpyram, nithiazine, thiacloprid, and thiamethoxam. Neonicotinoid pesticides are a class of neuroactive substances that are structurally similar to nicotine. They act as agonists of insect nicotinic acetylcholine receptors. Overstimulation of these receptors causes paralysis and death to insects. Recent studies have hypothesized that neonicotinoid pesticides are toxic to bees and may be linked to the global decline of bee populations (Whitehorn et al. 2012).

The half-life of imidacloprid in soils ranges from 130 to 160 days in the absence of sunlight. Because imidacloprid can be degraded abiotically by hydrolysis and photolysis, the half-life can be shortened to about 100 days in humid subtropical climate. However, the half-life can be as long as 1,230 days under suboptimal conditions (Wagner 2016). A number of studies has been carried out to detail the biodegradation of imidacloprid in soil, and several transformation products have been identified (e.g., ITP-1–ITP-6, Figure 23.6) (Simon-Delso et al. 2014). Imidacloprid is readily biodegraded by metabolic attack at its

N-heterocyclylmethyl moiety, heterocyclic spacer, and *N*-nitroimine tip. Phase I metabolism is largely dependent on microsomal CYP450 isozymes and a nitroreductase. Although it is generally believed that neonicotinoids exhibit low toxicity to vertebrates, there have been growing concerns about the impact of imidacloprid and its biotransformation products on the environment and finally human health. More studies are warranted with respect to the environmental fate of imidacloprid and other neonicotinoids and their impact on the ecosystem and human health.

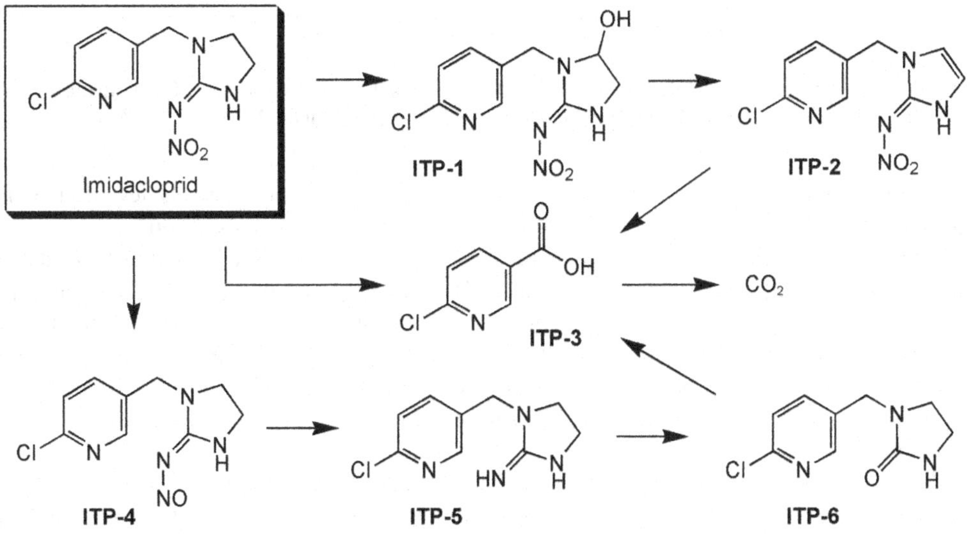

FIGURE 23.6 Biodegradation of imidacloprid in soil.

23.5.2 Biodegradation of Pharmaceuticals

Worldwide, the use of pharmaceuticals has increased in the past few decades, and it will continue to rise in the future with the development of new medicines for both human and veterinary purposes. The discharge of pharmaceuticals to the environment has become an issue of increasing concern due to the potential risks posed by the presence of these and derived substances in the environment. A few thousand pharmaceuticals, used as human and veterinary drugs, can reach the environment via various routes (Figure 23.7) (Mompelat et al. 2009). The use of veterinary pharmaceuticals in confined animal feeding operations alone has resulted in the annual discharge of some 3,000–27,000 tons of these via livestock manure into the environment (Song and Guo 2014).

Pharmaceuticals can be classified into 24 therapeutic classes, among which the following four classes are of major environmental concern: nonsteroid anti-inflammatory drugs, anticonvulsants, antibiotics, and lipid regulators (Mompelat et al. 2009). Pharmaceuticals, once in human/animal bodies, can undergo a set of biochemical reactions, such as oxidation and conjugation with polar groups to transform the parent compound into more polar molecules, facilitating their excretion from the body via urine or feces. They enter the environment either in their unchanged forms or as metabolites. Most pharmaceuticals are readily degraded in soil, as they are subjected to both biotic (biodegradation by microorganism) and abiotic (photolysis, hydrolysis, oxidation, reduction, complexation) transformations. During these transformation steps, breakdown (transformation) products may be produced that can be more toxic or more persistent in soil. Despite the widespread

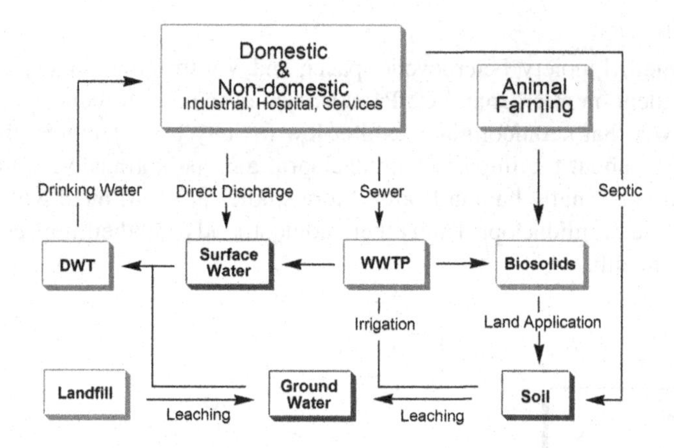

FIGURE 23.7 Origin of pharmaceuticals and routes of entry to the environment. DWT: drinking water treatment; WWTP: wastewater treatment plant. (Adapted from Mompelat et al. 2009.)

detection of different classes of pharmaceuticals in the environment, relatively few studies have detailed the biodegradation and transformation of specific pharmaceuticals in soil.

The antiepileptic drug carbamazepine is one of the most frequently detected human pharmaceuticals in wastewater effluents and biosolids. Li et al. (2013) reported that incubation of carbamazepine for 120 days under aerobic conditions in different soils resulted in less than 2% mineralization. A range of transformation products was detected, including acridine ('CTP-5', Figure 23.8). Since acridines are known to inhibit DNA repair and cell growth, and carbamazepine-10, 11-epoxide (CTP-1, Figure 23.8) is potentially toxic, it is important to further monitor the environmental fate of carbamazepine.

Sulfadiazine (4-amino-*N*-(2-pyrimidinel)benzene-sulfonamide) (Figure 23.9) is one of the sulfonamide antibiotics which are widely used in animal husbandry as well as in human medicine. Sulfadiazine is reported to be not readily biodegradable. Earlier studies showed that transformation products are generated primarily due to hydroxylation (STP-1), acetylation (STP-2), and formylation

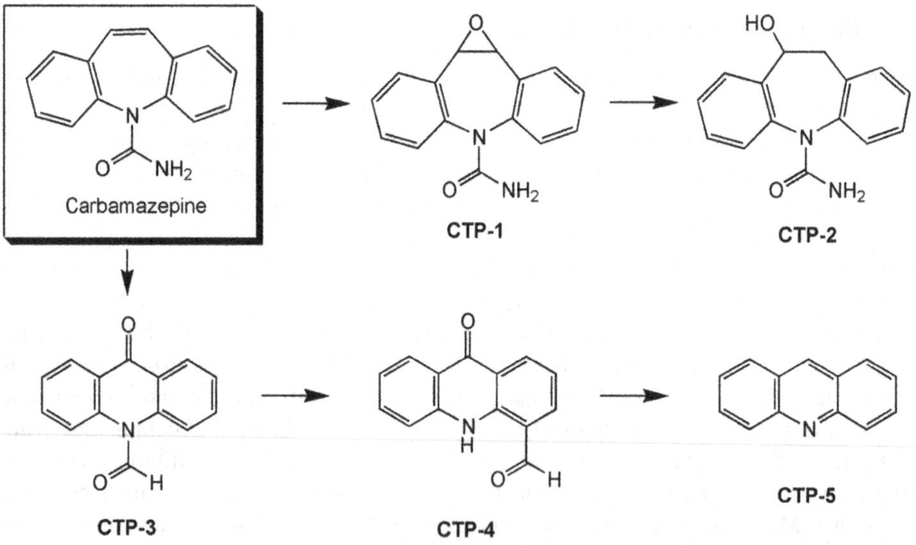

FIGURE 23.8 Proposed biodegradation pathways of carbamazepine in agricultural soils. (Adapted from Li et al. 2013.)

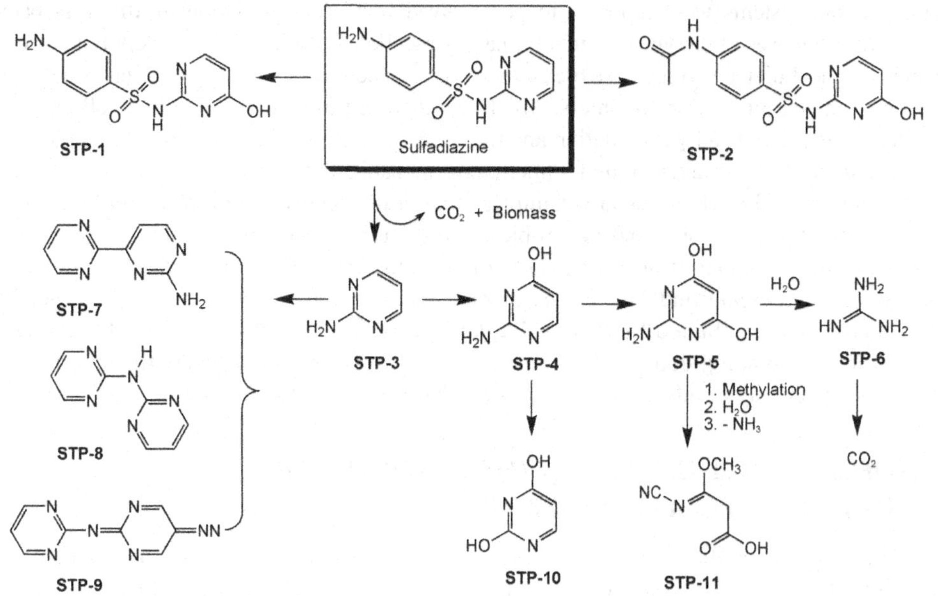

FIGURE 23.9 Biotransformation of sulfadiazine. (Adapted from Sittig et al. 2014 and Tappe et al. 2013.)

at either the aniline or the pyrimidine ring of sulfadiazine (Figure 23.9) (Sittig et al. 2014). Tappe et al. (2013) reported that *Microbacterium lacus* strain SDZm4 extensively mineralizes sulfadiazine while forming equimolar amounts of 2-aminopyrimidine (STP-3), which is considered to be a recalcitrant agent. Their studies resulted in the isolation of a *Terrabacter*-like bacterium, strain 2APm3, that was capable of degrading 2-aminopyrimidine (STP-3). Several transformation products (STP-4 to STP-11, Figure 23.9) were identified. It is hypothesized that a mixed culture of both species should be able to completely degrade sulfadiazine.

23.5.3 Biodegradation of Plastics

Plastic is usually a synthetic polymer, and its raw materials are generally extracted from crude oil, coal, and natural gas. In the recent past, the production and use of plastics have dramatically increased, which has elevated the issues of plastic waste disposal and pollution. According to some recent estimates, the global use of plastics is increasing annually by 12%, and the accumulation rate of plastic waste in the environment is about 25 million tons per year (Kaseem et al. 2012). Microorganisms such as bacteria, fungi, and algae are known to produce enzymes capable of degrading both natural and synthetic plastics. In the environment, plastic debris is grouped into primary or secondary debris. Primary plastics are in their original or close-to-original form, whereas secondary ones consist of pieces of plastics broken down via environmental degradation. Plastic debris can also be grouped into micro-plastic (<5 mm) and macro-plastic debris (>5 mm). The longevity of most of the classical plastics has led to the development of biodegradable plastics. Some examples of commercially available biodegradable plastics include polybutylene adipate terephthalate, polycaprolactone, polyhydroxyalkanoate (PHA), and polylactic acid (PLA) (Rujnić-Sokele and Pilipović 2017).

Some of the most widely used plastics include polyethylene, polypropylene, polystyrene, polyvinyl chloride, polyurethane, polyethylene terephthalate, polybutylene terephthalate, and nylons. Some examples of synthetic plastics that are biodegradable include aliphatic polyesters (e.g., PLA), polyamides, and polyvinyl alcohol (PVA) (Rujnić-Sokele and Pilipović 2017). Among the vinyl polymers produced industrially, PVA is the only one known to be mineralized by microorganisms (Kawai and Hu 2009).

Several enzyme systems were reported to be involved in PVA degradation by different bacterial strains. Two categories of enzymes stand out, i.e., extracellular and intracellular depolymerases. The microbial extracellular enzymes first break the large molecules of the polymer into smaller ones, such as monomers, dimers, or oligomers, which can diffuse through the microbial cell wall. These molecules are metabolized by β-oxidation and the TCA cycle to produce CO_2 and H_2O under aerobic conditions, and CO_2, H_2O, and CH_4 under (methanogenic) anaerobic conditions (Kawai and Hu 2009).

PHAs and polyhydroxybutyrate are examples of naturally occurring polyesters (or biodegradable plastics). A number of aerobic and anaerobic microorganisms have been found to degrade these polyesters. Some examples from soil include the fungus *Aspergillus fumigatus*, and the bacteria *Pseudomonas lemonignei* and *Variovorax paradoxus*. The ability of the bacteria to metabolize these compounds is facilitated by their PHA depolymerases. Usually, PHA depolymerases have only one substrate-binding domain, but some that had two such domains were thought to increase the substrate specificity or adsorption of enzymes (Numata et al. 2009).

23.6 DEGRADATIVE GENES AND THEIR EXPRESSION IN SOIL MICROBIAL COMMUNITIES

It is a challenge to predict the biodegradative potential of any terrestrial environment as the possibilities of culturing microbial communities to study their pollutant-degrading capabilities are often limited. However, the new molecular genetic technologies (as discussed in Chapter 14) allow researchers to examine the soil microbiome phylogenetic and biochemical properties, including biodegradative potential, by culture-independent metagenomics.

23.6.1 Activity-Based and Sequence-Based Metagenomics

There are two approaches to study the pollutant-degrading catabolic pathways of soil microbiomes via metagenomics (Figure 23.10). In the first approach, activity-based metagenomics, cloning of large DNA fragments (from about 25 to >100 kbp) from the microbiome of a polluted site is followed by the screening of pollutant-degrading activities of the resultant clone library. In the second

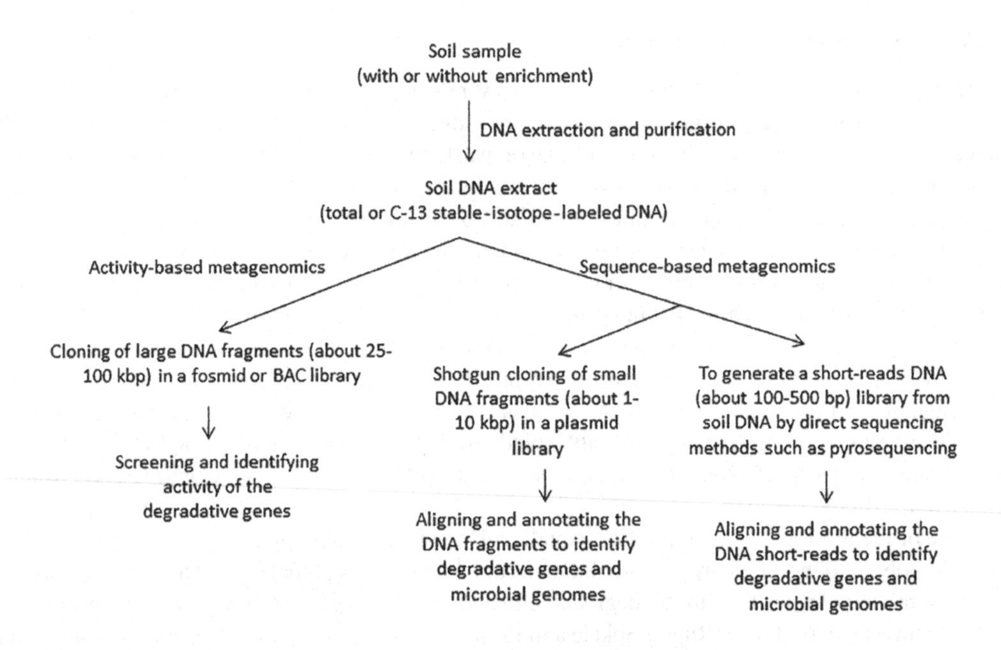

FIGURE 23.10 Metagenomics approaches to study soil microbiomes and their degradative genes.

approach, sequence-based metagenomics is used, in which functional annotation is performed for DNA fragments (100–500 bp for short-reads to 1–10 kbp for shotgun cloning methods) isolated from a polluted soil. Both approaches require DNA from microbiomes of polluted soils.

Using the first approach, Ono et al. (2007) constructed a library (of about 24,000 clones) from a petroleum-contaminated soil, Sagara oil field, Japan. Subsequent screening of the clones for naphthalene dioxygenase activity showed that one 25-kbp clone possessed naphthalene-degrading activity and contained the naphthalene degradation upper pathway operon.

Usually, such screenings are done by chemical analysis of the reaction substrates and products by high-pressure liquid chromatography, gas chromatography, and/or mass spectrometry. However, these approaches are labor-intensive and challenging. One rapid screening method was designed by Uchiyama et al. (2005). They constructed a metagenomics cloning vector that contains a green fluorescent protein biosensor gene (*gfp*) downstream of the cloning site. By determining the *gfp* expression, 33 clones were found to be induced by benzoate and two clones by naphthalene (from more than 150,000 clones). Recently, other biosensor-based methods have also been developed for screening benzamide-degrading and naphthalene-degrading genes from metagenomics clone libraries (Uchiyama and Miyazaki 2010, Yun and Ryu 2005).

Recently, a gene-targeted sequence-based metagenomics approach was used to improve the discovery of pollutant-degrading genes. Thus, using a set of PCR primers to target biphenyl dioxygenase genes in the metagenomics samples extracted from a PCB-contaminated soil, Iwai et al. (2010) created a library of 713 PCB-degrading gene (*bphA1*) sequences by the pyrosequencing method. Out of the 76 *bphA1* clusters in the library, 25 clusters were novel to the existing *bphA1* database.

23.6.2 Stable-Isotope Probing–Metagenomics

Stable-isotope probing (SIP) was extensively described in Chapter 17. Here, we examine its use in the exploration of soil for biodegrading organisms. Amendment of soil with pollutants labeled with a stable isotope (such as ^{13}C) will enhance the metabolism of potential pollutant-degrading populations which will assimilate the ^{13}C into their DNA. Upon extraction, the ^{13}C-labeled DNA that represents the pollutant-degrading population can be separated from the ^{12}C-labeled DNA of the non-degradative population by CsCl ultracentrifugation. Thus, the proportion of degradative genes in the ^{13}C-labeled DNA will be significantly higher than in that without ^{13}C enrichment. Using a ^{13}C-naphthalene enrichment approach, Wang et al. (2012) combined SIP and activity-based metagenomics, enabling them to identify several naphthalene degradation operons and an unculturable *Acidovorax* species as the dominant naphthalene degraders in the samples.

23.6.3 Expression of Degradative Genes in Soil

As outlined in Chapter 15, it remains challenging to measure the *in situ* expression of genes in soil. This also pertains to biodegradative genes. Genetically labeled microorganisms via gene fusion technology, such as *lux* system fusions with the promoter of *nah* and other degradative genes (Close et al. 2012), have been used to determine gene expression in soil. However, in most approaches the effective extraction of mRNA from soil is required. Once purified RNA is obtained, it is first converted to cDNA by reverse transcription. Then, the cDNA from specific degradative genes can be determined by quantitative PCR, microarrays, or parallel sequencing. The latter allows sequencing and quantifying the cDNA simultaneously.

Reverse transcription-quantitative PCR (RT-qPCR) has been successfully applied to determine the expression of a specific degradative gene in soil. Thus, Monard et al. (2013) showed that the expression levels of the (atrazine degradation) *atzD* gene in five agricultural soils were positively correlated to the degradation of atrazine in these soils. However, qPCR primers may not always be able to capture all natural sequence divergence of a degradative gene in soil. Furthermore, this method is also restricted to monitoring only a limited number of genes simultaneously.

Moreover, the microarray approach (see Chapters 12 and 15) is another well-accepted tool that allows the screening of expression of thousands of degradative genes simultaneously. Using the so-called GeoChip, a microarray that contains 6,465 probes of aromatic degradative genes, Leigh et al. (2007) showed that 27 degradative genes associated with the degradation of biphenyls, benzoates, and aromatic hydrocarbons were actively expressed in the root zone of an Austrian pine tree grown in a PCB-contaminated soil. Although this approach can rapidly detect the expression of thousands of degradative genes simultaneously, it is still limited by its (low) sensitivity and the insufficient number of documented degradative genes and their sequence divergence.

Metatranscriptomics has been described in Chapter 15 of this book. Using metatranscriptomics, Page et al. (2015) investigated the rhizo-stimulating effect of *Salix purpurea* (willow) on the microbial degradation of chlorinated and non-chlorinated hydrocarbon contaminants. Thus, a DNA as well as a cDNA sequence library was generated from the rhizosphere of *S. purpurea* grown in a petrochemical-contaminated soil, as well as from the bulk soil. Based on the metatranscriptomics libraries, ten microbial oxygenase genes predicted to be involved in the degradation of hydrocarbon contaminants were chosen to compare their expression between the rhizosphere and the bulk soil. All ten genes, either at the microbial community level or in specific taxonomic orders (such as the Sphingomonadales and Actinomycetales), showed higher expression in the rhizosphere than in the bulk soil settings. This shows that the metatranscriptomics method can be used effectively to determine the expression of a diverse group of oxygenase genes in soil at a high sensitivity level.

23.7 BIOREMEDIATION

According to the United States Environmental Protection Agency (USEPA), bioremediation is a "treatment that uses naturally occurring organisms to break down hazardous substances into less toxic or nontoxic substances." The technology should be cost-effective. One of the main reasons for the popularity of bioremediation techniques for cleaning up spills in soils is that the costs are reasonable compared to other conventional cleanup technologies, such as transport and incineration. Some estimates reason that the cost of bioremediation can be as little as 20% of the conventional methods. Bioremediation also gains public acceptance as an environmentally friendly remediation technology.

Bioremediation processes can be classified broadly into *in situ* and *ex situ* techniques. *In situ* bioremediation is the simplest method that treats contaminated soil in its natural state. Alternatively, *ex situ* treatment involves the excavation of the contaminated soil that is then treated either on-site or off-site. The fate of the contaminants depends on the approaches chosen and other constraints, such as bioavailability, microbial degradation ability, and physicochemical conditions of the treatment approaches. Generally speaking, however, *in situ* methods are favored, if applicable, due to these being more sustainable, cost-effective and environmentally friendly.

Regardless of where the soil is treated, there are three general approaches that can be taken. The first and undoubtedly the simplest method is *natural attenuation*. This involves allowing the extant biome to degrade the pollutants without any intervention. Though this is a valid approach, bioremediation practices often strive to overcome the natural barriers that may be present, in order to promote the efficient removal of contaminants. The other two general approaches (that both enhance the rates of biodegradation) are *biostimulation* and *bioaugmentation*. Biostimulation is the process in which the microbial flora existing in soil is stimulated to degrade the pollutant(s), for example by addition of nutrients and/or small amounts of the contaminant itself to contaminated sites. Bioaugmentation, on the other hand, is the process in which the contaminated soil is inoculated with single microorganisms, or a consortium thereof, capable of degrading the pollutant(s). Bacteria can be either foreign or, if possible, indigenous to the site due to the growth selectivity of the pollutant. Foreign bacteria often have the advantage of having catabolic processes more specific to the pollutant; however, the local bacteria may survive better in the environmental conditions present. Therefore, indigenous bacteria in the soil can be selected, extracted, stimulated, and then reintroduced. In addition to this procedure, advanced technologies exist wherein genetically engineered microorganisms are produced for degradation of target chemicals and then introduced into the contaminated sites. The application of this technology is limited due to restrictions placed by regulatory agencies and is therefore not widely adopted. When used together, bioaugmentation and biostimulation can be effective for degrading pollutants, as summarized in Table 23.2.

TABLE 23.2
Some Common Bioremediation Methods

Method (Type)	Principle and Process	Advantages and Disadvantages
Natural attenuation (*In situ*)	• Degradation of pollutants by indigenous microorganisms without stimulation or treatment.	Adv: Minimal disturbance to contamination sites; minimal cost. Disadv: Slow remediation rate; high unpredictability; requires constant monitoring.
Bioventing (*In situ* biostimulation)	• For unsaturated soils (vadose zone). • Gas (air, oxygen, or methane) is pumped through the contaminated soil to encourage growth and activity of degradative microorganisms. • Volatile pollutants will also be vented out.	Adv: Does not require excavation; shorter remediation time than the natural attenuation; effective for contaminations in deeper subsurface soils. Disadv: Heterogeneity of the soils can affect the effectiveness of the process; aboveground biofilters and/or bioreactors may be needed.
Electroremediation (*In situ* biostimulation)	• Insertion of electrodes into contaminated soil generating an electric field. • Move inoculated bacteria deeper into soil through electroosmosis. • Increase mass transfer of pollutants (increasing bioavailability). • Can also transport pollutants out of the soil.	Adv: Increases efficiency of bioaugmentation and is itself biostimulating. Soil does not have to be removed. Disadv: Soil must have the right pore size and be conductive. Cannot increase mass transfer of insoluble pollutants. Ion concentrations must have ideal ratio.
Biosparing (*In situ* biostimulation)	• Gas is pumped into the saturated zone of the contaminated soil to stimulate biodegradation.	Adv: Similar to bioventing. Disadv: Similar to bioventing.
Biobarrier (*In situ*)	• Permeable biological active barriers are placed in front of contaminant plumes to degrade and stop the spread of the contaminants.	Adv: Prevent spreading of contaminants; better control of the conditions of the biobarriers. Disadv: Limited to point source contamination; limited by the depth of contamination plumes.
Phytoremediation (*In situ* biostimulation)	• Using plants and/or plant-stimulated rhizosphere microorganisms to degrade contaminants in soils. • Pollutants can be broken down by both plant and microorganisms • Plant can uptake pollutants and remove from the soil through roots • Plant supplies O_2 and root exudates to stimulate the growth of microorganisms in the root zone and increases biodegradation.	Adv: Less costly than other bioremediation technologies (except natural attenuation); a more environmentally friendly method. Can treat a variety of pollutants (organic/inorganic). Disadv: Limited by the growth conditions of the sites; more commonly used for heavy metal contaminations than organic contaminations; limited to shallow contaminations.

(*Continued*)

TABLE 23.2 (*Continued*)
Some Common Bioremediation Methods

Method (Type)	Principle and Process	Advantages and Disadvantages
Land farming (*Ex situ* bioaugmentation and/or biostimulation)	• Contaminated soils are excavated, mixed, watered, aerated, and spread in a somewhat controlled environment to degrade the contaminants. Biostimulation and/or bioaugmentation can be used to speed up the bioremediation process.	Adv: Faster remediation because of better control of the soil physicochemical and biological conditions. Disadv: Requires sufficient space to spread the contaminated soils.
Composting/biopiling (*Ex situ* bioaugmentation and/or biostimulation)	• Biodegradation of contaminants by aerobic thermophilic microorganisms. • Contaminated soils are excavated, put in static piles (windrows), and periodically mixed and watered to enhance microbial activity. Aerobic biodegradation can be improved by pumping air or mixing bulking agents into the piles.	Adv: Requires less space and less attention than land farming treatment; still maintains a reasonable rate of remediation. Disadv: Unevenness of microbial activity inside the piles; too much reliance on the thermophilic microorganisms.
Bioreactor (*Ex situ* bioaugmentation and/or biostimulation)	• Contaminated soils are excavated and treated in bioreactors, in which the physicochemical conditions (i.e., nutrients, substrates, water, oxygen, temperature, pH, and other parameters) are optimized for biodegradation. Bioaugmentation can be used to speed up the bioremediation process.	Adv: Optimal remediation rate; full control of the reactor's conditions; more consistent degradation results. Disadv: More costly than other bioremediation technologies; size of the reactors can be a limiting factor.

When selecting a bioremediation method, it is important to consider the specific properties of the contaminants. For example, remediation of PCE or PCB pollution is difficult because it requires both anaerobic and aerobic conditions. This problem can be successfully resolved by utilizing the aerobic and anaerobic environments generated in the root zone of phytoremediation process (Becker and Seagren 2010, Passatore et al. 2014). The effectiveness and versatility of phytoremediation are further discussed in Box 23.2.

BOX 23.2 PHYTOREMEDIATION OF PCBS

There is an increasing interest in using phytoremediation to remove PCBs from contaminated soil because the root zones of plants provide a unique environment that includes both aerobic and anaerobic niches that are essential for the biodegradation of PCBs (Passatore et al. 2014). Anaerobic rhizobacteria dechlorinate the PCBs to congeners containing zero to three chlorine atoms. The lower chlorinated congeners can be completely degraded by aerobic rhizobacteria. The root system stimulates the microbial processes involved in PCB degradation by:

- Generating microzones with both aerobic and anaerobic environments allowing the growth of both types of bacteria.
- Root exudates supplying organic and inorganic nutrients which act as both carbon and energy sources of PCB-degrading bacteria.
- Increasing the growth of PCB microorganisms by introducing PCB like organic compounds (such as terpene) that act as inducer molecules for enzymes involved in PCB biodegradation.
- Releasing biological surfactants that increase the bioavailability of PCBs.

These factors increase the microbial density and diversity as well as the biodegradation of PCB compared to bulk soil samples. A recent study showed that a 2-year phytoremediation treatment of a PCB-contaminated site in San Giuliano Terme Municipality (Pisa, Italy) by three plant species (*Populus alba, Cytisus scoparius, Paulownia tomentosa*) lowered the concentration of PCBs by an average of 70%. The PCB concentrations after remediation decreased below the regulated government mandated requirement (5 mg kg^{-1}). Most notably, the *Poplar alba* plant lowered the concentration by almost 100% showing the importance of plant selection in phytoremediation (Macci et al. 2016). Further studies are required to increase the effectiveness of this approach in field conditions.

In particular situations, bioremediation may be disadvantageous, as it may be too slow in cases where a fast cleanup is required. Moreover, bioremediation comes with inherent difficulties as the outcome is not always predictable. For instance, the contaminants may not be eliminated to zero by bioremediation techniques because strongly sorbed contaminants are not amenable to microbial degradation. Over a long period of time, residuals may thus be slowly released from contaminated sites resulting in both regulatory and public concern about bioremediation techniques.

23.7.1 Environmental Factors Affecting Bioremediation

While the indigenous microbial community determines the biodegradative potential of specific pollutants in a polluted soil, physical and chemical factors determine the bioavailability and distribution of pollutants in soil. Moreover, growth, survival, and activity of soil microorganisms also depend on the environmental conditions in the system. Some of the major factors that determine the properties and biodegradation potential of the soil ecosystem include bioavailability of the pollutants, concentration of nutrients (such as C, N, and P) and specific electron acceptors, soil texture, porosity, moisture content, pH, temperature, redox potential, presence/absence of plants, and the microbial community structure. The effects of these factors on the activities of soil microorganisms and the bioavailability of pollutants are summarized in Table 23.3.

TABLE 23.3
Major Factors Affecting Bioremediation of Organic Pollutants in Soil

Environmental Factor	Effects on Biodegradative Microorganisms and Pollutants
Nutrients	• If the organic pollutants are the primary substrates of the degradative microbial community and present in high concentrations, other nutrients such as N, P, and oxygen are needed to maintain the growth of the degradative population. The optimal C:N:P ratio is roughly about 100:10:1. • If the pollutants are too low in concentration to support growth of the degradative population or are the secondary substrates in a co-metabolism process, a primary organic substrate should be added to stimulate the growth of the degradative community.
Oxygen	• Aerobic microorganisms grow preferably at dissolved $[O_2] > 0.2$ mg L^{-1} and when $[O_2] < 1\%$ by volume, anaerobic metabolisms are preferable. • Generally, aerobic biodegradation is faster than anaerobic biodegradation.
Electron acceptors	• When O_2 is absent, other oxidizing agents can serve as alternative electron acceptors, in the order NO_3^-, Mn^{4+}, Fe^{3+}, SO_4^{2-}, and HCO_3^-. • Highly chlorinated organic compounds such as PCE and PCBs can also be used as electron acceptors in reductive dechlorination or dechlororespiration processes.

(Continued)

TABLE 23.3 (*Continued*)
Major Factors Affecting Bioremediation of Organic Pollutants in Soil

Environmental Factor	Effects on Biodegradative Microorganisms and Pollutants
Redox potential (Eh)	• Aerobic microorganisms are active at Eh > 50 mV, and anaerobic microorganisms are preferable at Eh < 50 mV.
Soil texture and structure	• Soil texture and structure control the porosity of soil and in turn determine the water holding capacity and permeability of the soil. • Clay soils have high water holding capacity, and hence, the permeability (or mobility) of pollutants is low.
Soil moisture content	• Soil moisture or water is essential for the biodegradative reactions of soil microbes, and optimum moisture level is about 30%–90% of the soil's holding capacity. • Soil moisture content also controls the pore space and diffusion of oxygen in the soil, and a water-occupied pore space of 38%–81% is usually optimal for aerobic microorganisms.
Temperature	• Most degradative microorganisms are active under mesophilic conditions (15°C–45°C). However, some hydrocarbons have been shown to be degraded at near freezing temperatures. • Temperature also affects the physical nature of hydrocarbon pollutants and microbial composition at a contaminated site.
Soil pH	• Bacterial growth is optimal between pH 6.5 and 8.0. Fungi have a more important role in biodegradation at lower pH (5–6.5). • pH can also change the charge of some organic pollutants, and this affects their sorption on soil and bioavailability to microbial communities.
Bioavailability of pollutants	• Bioavailability of an organic pollutant depends on (1) its sorption on the soil surfaces and (2) its ability to form nonaqueous-phase liquid which reduces its solubility in water and its biodegradability in soil. • Bioavailability of pollutants can be enhanced by biosurfactants that are secreted naturally by soil microorganisms. Synthetic surfactants can be used to increase the bioavailability of some pollutants.

23.8 CONCLUDING REMARKS

Despite the advances of our understanding on the biodegradation pathways of microorganisms and their capabilities to metabolize a wide range of hazardous organic pollutants, the ever-increasing release of new and novel synthetic chemicals such as pesticides and pharmaceuticals to the environment may challenge the biodegrading capability of microbial communities. Therefore, further research has to address the biodegradability and the reaction mechanisms of these pollutants. Recent development of the metagenomics technology has improved our ability to discover novel biodegradation genes from microbial communities in the environment. With the advancement of the non-cloning DNA sequencing technology, the metagenomics approach may provide a rapid assessment of the biodegradability of a pollutant with an affordable cost. Although many bioremediation methods have been developed to clean up petroleum pollutants in soil with high degree of success, highly chlorinated organic pollutants, such as PCE and PCBs, are always challenging for *in situ* bioremediation, as they require both anaerobic and aerobic conditions for complete degradation. Some studies have reported success with respect to the biodegradation of PCBs by phytoremediation. This approach has provided an opportunity to remediate highly chlorinated organic pollutants in soil efficiently; however, further research is required to develop an effective model that can optimize the selection of plant species and its application in field conditions.

ACKNOWLEDGMENTS

This work is supported by NSERC Discovery Grant. We also thank Devon Prontack and Brian Hicks for conducting a thorough literature review and helping to prepare the manuscript.

REFERENCES

Alexander, M. 1999. *Biodegradation and Bioremediation*, 2nd ed., Academic Press: San Diego, CA.

Becker, J.G. and Seagren, E.A. 2010. Bioremediation of hazardous organics. In: Mitchell, R. and Gu, J.D. (eds.), *Environmental Microbiology*, 2nd ed., Wiley-Blackwell: Hoboken, NJ, 177–212.

Cai, M. and Xun, L. 2002. Organization and regulation of pentachlorophenol-degrading genes in *Sphingobium chlorophenolicum* ATCC 39723. *J Bacteriol* 184, 4672–4680.

Carson, R. 1964. *Silent Spring, Fawcett Crest Book*, Fawcett Crest: New York.

Close, D., Xu, T., Smartt, A., et al. 2012. The evolution of bacterial luciferase gene cassette (*lux*) as a real-time bioreporter. *Sensors* 12, 732–752.

Daughton, C.G. and Hsieh, D.P.H. 1977. Parathion utilization by bacterial symbionts in a chemostat. *Appl Envrion Microbiol* 34, 175–184.

De, A., Bose, R., Kumar, A. and Mozumdar, S. 2014. *Targeted Delivery of Pesticides Using Biodegradable Polymeric Nanoparticles*, Springer: New York, pp. 5–6.

Freedman, D.L. and Gossett, J.M. 1989. Biological reductive dechlorination of tetrachloroethylene and trichloroethylene to ethylene under methanogenic conditions. *Appl Environ Microbiol* 55, 2144–2151.

Gao, J., Ellis, L.B.M. and Wackett, L.P. 2010. The University of Minnesota Biocatalysis/Biodegradation Database: improving public access. *Nucleic Acids Res* 38, D488–D491.

Gibson, J. and Harwood, C.S. 2002. Metabolic diversity in aromatic compound utilization by anaerobic microbes. *Annu Rev Microbiol* 56, 345–369.

Iwai, S., Chai, B., Sul, W.J., Cole, J.R., Hashsham, S.A. and Tiedje, J.M. 2010. Gene-targeted-metagenomics reveals extensive diversity of aromatic dioxygenase genes in the environment. *ISME J* 4, 279–285.

Jenkins, R.R., Guignet, D. and Walsh, P.J. 2014. Prevention, cleaning, and reuse benefits from the federal UST Program. U.S. Environmental Protection Agency, National Centre for Environmental Economics, Washington, D.C. U.S.A. Environmental Economic Working Paper Series #14–05. www.epa.gov/sites/production/files/2015-01/documents/prevention_cleanup_and_reuse_benefits_from_the_federal_ust_program.pdf

Kaseem, M., Hamad, K. and Deri, F. 2012. Thermoplastic starch blends: A review of recent works. *Polym Sci Ser A* 54, 165–176.

Kawai, F. and Hu, X. 2009. Biochemistry of microbial polyvinyl alcohol degradation. *Appl Microbiol Biotechnol* 84, 227–237.

Leigh, M.B., Pellizari, V.H., Uhlik, O., et al. 2007. Biphenyl-utilizing bacteria and their functional genes in a pine root zone contaminated with polychlorinated biphenyls (PCBs). *ISME J* 1, 134–148.

Li, J., Dodgen, L., Ye, Q. and Gan, J. 2013. Degradation kinetics and metabolites of carbamazepine in soil. *Environ Sci Technol* 47, 3678–3684.

Lopez-Echartea, E., Macek, T., Demnerova, K. and Uhlik, O. 2016. Bacterial biotransformation of pentachlorophenol and micropollutants formed during its production process. *Int J Environ Res Public Health* 13, 1146.

Macci, C., Peruzzi, E., Doni, S., Poggio, G. and Masciandaro, G. 2016. The phytoremediation of an organic and inorganic polluted soil: A real scale experience. *Int J Phytorem* 18, 378–386.

Mompelat, S., Le Bot, B. and Thomas, O. 2009. Occurrence and fate of pharmaceutical products and by-products, from resource to drinking water. *Environ Int* 35, 803–814.

Monard, D., Martin-Laurent, F., Lima, O., Devers-Lamrani, M. and Binet, F. 2013. Estimating the biodegradation of pesticide in soils by monitoring pesticide-degrading gene expression. *Biodegradation* 24, 203–213.

Nielsen, T.K., Sorensen, S.R. and Hensen, L.H. 2015. Draft genome sequence of isoproturon-mineralizing *Sphingomonas* sp. SRS2, isolated from an agricultural field in the United Kingdom. *Genome Announc* 3, e00569-15.

Numata, K., Abe, H. and Iwata, T. 2009. Biodegradability of poly(hydroxyalkanoate) materials. *Materials* 2, 1104–1126.

Ono, A., Miyazaki, R., Sota, M., Ohtsubo, Y., Nagata, Y. and Tsuda, M. 2007. Isolation and characterization of naphthalene-catabolic genes and plasmids from oil-contaminated soil by using two cultivation-independent approaches. *Appl Microbiol Biotechnol* 74, 501–510.

Page, A.P., Yergeau, E. and Greer, C.W. 2015. *Salix purpurea* stimulates the expression of specific bacterial xenobiotic degradation genes in a soil contaminated with hydrocarbons. *PLoS One* 10, e0132062.

Passatore, L., Rossetti, S., Juwarkar, A.A. and Massacci, A. 2014. Phytoremediation and bioremediation of polychlorinated biphenyls (PCBs): State of knowledge and research perspectives. *J Hazard Mater* 278, 189–202.

Rujnić-Sokele, M. and Pilipović, A. 2017. Challenges and opportunities of biodegradable plastics: A mini review. *Waste Manage Res* 35, 132–140.

Simon-Delso, N., Amaral-Rogers, V., Belzunces, L.P., et al. 2014. Systemic insecticides (neonicotinoids and fipronil): Trends, uses, mode of action and metabolites. *Environ Sci Pollut Res* 22, 5–34.

Singh, B.K., Walker, A., Morgan, J.A.W. and Wright, D.J. 2004. Biodegradation of chlorpyrifos by *Enterobacter* strain B-14 and its use in bioremediation of contaminated soils. *Appl Environ Microbiol* 70, 4855–4863.

Sittig, S., Kasteel, R., Groeneweg, J., et al. 2014. Dynamics of transformation of the veterinary antibiotic sulfadiazine in two soils. *Chemosphere* 95, 470–477.

Song, W. and Guo, M. 2014. Residual veterinary pharmaceuticals in animal manures and their environmental behaviors in soils. In: He, Z. and Zhang, H. (eds.), *Applied Manure and Nutrient Chemistry for Sustainable Agriculture and Environment*, Springer Science+Business Media Dordrecht: The Netherlands, 23–52.

Symons, Z.C. and Bruce, N.C. 2006. Bacterial pathways for degradation of nitroaromatics. *Nat Prod Rep* 23, 845–850.

Tappe, W., Herbst, M., Hofmann, D., et al. 2013. Degradation of sulfadiazine by *Microbacterium lacus* strain SDZm4 isolated from lysimeters previously manured with slurry from sulfadiazine-medicated pigs. *Appl Environ Microbiol* 79, 2572–2577.

Uchiyama, T., Abe, T., Ikemura, T. and Watanabe, K. 2005. Substrate-induced gene-expression screening of environmental metagenome libraries for isolation of catabolic genes. *Nat Biotechnol* 23, 88–93.

Uchiyama, T. and Miyazaki, K. 2010. Product-induced gene expression, a product-responsive reporter assay used to screen metagenomic libraries for enzyme-encoding genes. *Appl Environ Microbiol* 76, 7029–7035.

Wagner, S. 2016. *Environmental Fate of Imidacloprid, Environmental Monitoring and Pest Management Branch*, Department of Pesticide Regulation: Sacramento, CA. www.cdpr.ca.gov/docs/emon/pubs/fatememo/Imidacloprid_2016.pdf

Wang, Y., Chen, Y., Zhou, Q., et al. 2012. A culture-independent approach to unravel uncultured bacteria and functional genes in a complex microbial community. *PLoS One* 7, e47530.

Whitby, C. 2010. Microbial naphthenic acid degradation. *Adv Appl Microbiol* 70, 93–125.

Whitehorn, P.R., O'Connor, S., Wackers, F.L. and Goulson, D. 2012. Neonicotinoid pesticide reduces bumble bee colony growth and queen production. *Science* 336, 351–352.

Yun, J. and Ryu, S. 2005. Screening for novel enzymes from metagenome and SIGEX, as a way to improve it. *Microb Cell Fact* 4, 8.

24 The Impact of Metal Contamination on Soil Microbial Community Dynamics

David C. Gillan
Mons University

Rob Van Houdt
Belgian Nuclear Research Centre (SCK•CEN)

CONTENTS

24.1 Introduction 404
24.2 Soil Metals, Concentrations, and Bioavailability 404
24.2.1 Soil Metals and Concentrations 404
24.2.1.1 Arsenic 405
24.2.1.2 Cadmium 406
24.2.1.3 Chromium 406
24.2.1.4 Copper 406
24.2.1.5 Lead 406
24.2.1.6 Mercury 407
24.2.1.7 Nickel 407
24.2.1.8 Zinc 407
24.3 Bioavailability 407
24.4 Duration of Exposure 408
24.5 Effects of Metals on Microbial Diversity 408
24.6 Effects of Metals on Soil Microbial Biomass 409
24.7 Effect on Metabolic Activity 410
24.8 Metal Resistance and Adaptation 410
24.8.1 Metal Resistance Systems 410
24.8.1.1 Extracellular Systems 410
24.8.1.2 Outer Membrane Systems 412
24.8.1.3 Periplasmic Systems 413
24.8.1.4 Cytoplasmic Membrane Systems 413
24.8.1.5 Cytoplasmic Resistance Systems 414
24.8.2 Metal Resistance Carriers 415
24.8.3 Genetic Adaptation 415
24.8.4 Interactions between Metals and Antibiotics 416
24.9 Conclusions and Future Prospects 417
References 417

24.1 INTRODUCTION

"Heavy metals" is a widely used term to group metals (and metalloids) that are associated with contamination and potential toxicity. Although it is not a scientifically accurate description, since it implies that the pure metal and all its compounds have the same physicochemical, biological, and toxicological properties, the term is being increasingly used despite the ongoing debate and calls to stop using it. For the purpose of discussing soil contamination, we define metals as the 38 transition elements of the d block in the periodic table (i.e., Sc to Zn in period 4, Y to Cd in period 5, Hf to Hg in period 6, and Rf to Cn in period 7), also including six elements of the p block, namely Pb, Tl, Bi, Sn, Sb, and As. Lanthanides (rare earth elements) and actinides are not included in this definition as they belong to the f block. All these elements may be found in soils, sometimes in concentrations higher than usual, not only in the form of ions (e.g., Cu^{2+}) and oxyanions (e.g., AsO_4^{3-}) but also as mineral particles (e.g., MnOOH), including nanoparticles. Different sources contribute to metals in soils, such as natural (rock weathering, volcanic eruptions, windblown dust particles, sea sprays, and aerosols), agricultural (inorganic fertilizer, pesticide, sewage, wastewater, and fungicide), industrial (mining, refineries), domestic (e-waste, batteries, effluents, biomass burning), and miscellaneous sources. Metal-contaminated soils are of great concern, with more than 5 million metal-contaminated soil sites reported and an estimated worldwide economic cost of more than US $10 billion per year. Since soil microbial communities and linked properties play essential roles in soil functioning, it is of importance to determine the effects of metals on them. Relevant reviews are cited, and the reader may consult these for further knowledge of the topic.

24.2 SOIL METALS, CONCENTRATIONS, AND BIOAVAILABILITY

24.2.1 Soil Metals and Concentrations

Metals are naturally present in soils. According to the fraction considered (sand, silt, and clay), the most abundant metals are usually Fe and Mn (several g kg^{-1}). Pb, Cd, Cu, Zn, Ni, and Cr can be found in the mg kg^{-1} concentration range, while other metals (including the rare earth elements) are found in the μg kg^{-1} concentration range. It is, however, difficult to generalize, as metal concentrations in natural soils may vary by several orders of magnitude (Wuana and Okieimen 2011, Kabata-Pendias 2010, Kabata-Pendias and Stzeke 2015). Many metals in the soil environment may be considered to be contaminants or xenobiotics. A *xenobiotic* is an element or chemical substance found within an environment that is not naturally expected in that environment. It can also cover substances that are present in much higher concentrations than usual. Soils are major sinks for metals released into the environment by anthropogenic activities, essentially since the industrial revolution starting in the 1700s, but also to ancient regional metallurgical activities. Unlike organic contaminants, which may be completely oxidized to CO_2 by microbial action, most metals do not undergo microbial or chemical degradation, and their total concentration in soils persists for a long time after their introduction (Adriano 2001).

Eight metals (As, Cd, Cr, Cu, Pb, Hg, Ni, and Zn) will be highlighted in this section (Table 24.1; Figure 24.1). These metals are among the 20 contaminants of greatest concern (EPA's Superfund program; www.epa.gov/superfund). They are also of concern to Europe, where risk levels of metals are defined differently. Finnish standard values represent a good approximation of the mean values of different national systems in Europe (Toth et al. 2016). In this system, "guideline values" are defined. If these guidelines are exceeded, soils have a contamination level that presents ecological or health risks. Different guideline values are set for industrial and transport areas (higher guideline value) compared with other land uses (lower guideline value). The higher guideline values for the metals listed in Table 24.1 are as follows (in mg kg^{-1}): As, 100; Cd, 20; Cr, 300; Cu, 200; Pb, 750; Hg, 5; Ni, 150; and Zn, 400 (Toth et al. 2016).

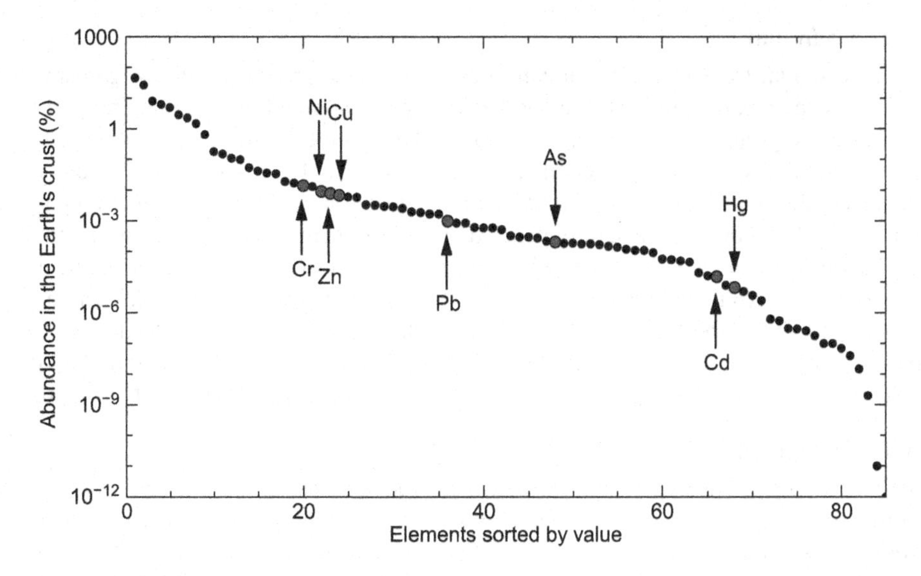

FIGURE 24.1 Abundance of elements in the Earth's crust.

TABLE 24.1
General Characteristics of Metals of Great Concern in Contaminated Soils

Metal	Crust[a]	Soil Level[a]	Soil Max[a]	Main Oxidation State	Essential for Life	Oxyanions	NPL[b]
As	2	0.1–1,000	18,500	+3, +5	no	AsO_4^{3-}	1,182
Cd	0.1	0.5	100	+2	no	/	869
Cr	100–300	5–3,000	43,200	+3, +6	no	CrO_4^{2-}	1,060
Cu	50	14	20,000	+1, +2	yes	/	770
Pb	10–30	10–100	10,000	+2, +4	no	/	1,552
Hg	0.08	0.05	119	0, +1, +2	no	/	1,103
Ni	90	13–40	26,000	+2	yes	/	710
Zn	70	10–300	26,000	+2	yes	/	820

References are given in the text. The soil maximum column gives examples of extreme contaminations, usually near mines or industrial sites.

[a] Level in mg kg^{-1}.

[b] Number of sites on National Priorities List, EPA (statistics established in November 2017).

24.2.1.1 Arsenic

The average natural As level in the Earth's crust is about 2 mg kg^{-1}, and soils may contain between 0.1 and 1,000 mg kg^{-1}. The levels in contaminated soils can be up to 18,100 mg kg^{-1}. As has a variety of oxidation states (–3, 0, +3, +5), but the +5 (arsenate) and +3 (arsenite) states are the most common in nature. Arsenate resembles phosphate: As and P are both elements in group 15 of the periodic table, and both form oxyanions (AsO_4^{3-} and PO_4^{3-}). Arsenate is toxic for the glycolysis pathway because it can replace inorganic phosphate, leading to the loss of adenosine triphosphate (ATP) generation. Arsenic can also appear naturally in the form of organic derivatives, which implies the presence of C–As bounds. A number of microorganisms are able to methylate arsenic, giving rise to mono-, di-, and/or trimethyl derivatives (Paez-Espino et al. 2009).

24.2.1.2 Cadmium

Cd is a nonessential trace element that can interfere with the growth of microorganisms. It is in many respects chemically similar to Zn and Hg, elements classified in the same group of the periodic table, that is, group 12. The average natural Cd level in the Earth's crust is about 0.1 mg kg^{-1}, and uncontaminated soils usually contain about 0.5 mg kg^{-1} (dw) of Cd. Cd levels in soils impacted by human activities may reach 19 mg kg^{-1} (dw), and natural black shales may contain up to 100 mg kg^{-1} of Cd (dw). The oxidation state of soluble Cd in the biosphere is +2. Cd is released into the environment in variable amounts from natural (e.g., volcanoes and fires) and anthropogenic activities. Anthropogenic sources include the use of phosphate fertilizers, fossil fuel combustion, metallurgical works, wastes from the cement industry, sewage sludge, municipal and industrial wastes, mining, and smelting and processing of metal ores. Cd is a frequent soil contaminant in the vicinity of Zn smelters, as it is a minor component in most Zn ores and therefore a by-product of Zn production.

24.2.1.3 Chromium

Although Cr was proposed as an essential element for humans in 1950, the precise mechanism of action is unknown and its status as an essential element is no longer supported. In the microbial world, Cr is not an essential element, but it can be oxidized or reduced by specific microbial enzymes. The average natural Cr level in the Earth's crust ranges from 100 to 300 mg kg^{-1}, and soils may contain between 5 and 3,000 mg kg^{-1}. Levels may reach 43,200 mg kg^{-1} in extremely contaminated areas. Although Cr can occur in the environment in several oxidation states, the most common ones are trivalent Cr^{3+} and hexavalent Cr^{6+}. In contrast to Cr^{3+}, which is relatively insoluble under environmental conditions, Cr^{6+} is soluble and more toxic, as it can pass through cell membranes. The latter exerts its toxicity in the cell by generating reactive oxygen species (ROS) that may damage DNA and proteins (Viti et al. 2014). Cr^{6+} usually forms the chromate oxyanion (CrO_4^{2-}, structurally similar to sulfate) that enters the cell using the sulfate transport pathway and gets reduced to Cr^{3+} by various enzymatic and nonenzymatic processes. Cr^{3+} ions tend to form octahedral complexes and readily forms insoluble oxyhydroxides ($Cr(OH)_3$) above pH ~ 5.5. As a result, biological membranes are nearly impermeable to Cr^{3+}.

24.2.1.4 Copper

Cu is an essential element used in many enzymes. For instance, it is found in superoxide dismutase, as well as in complex IV of the electron transport system used during aerobic respiration in heterotrophic bacteria. Moreover, it is found in plastocyanin, involved in photosynthesis. Cu is chemically similar to Ag and Au, and these elements are grouped together in the periodic table (group 11). The main oxidation states of soluble Cu in the biosphere are +1 in reduced environments and +2 in oxidized environments. Cu is present in the Earth's crust in proportions of about 50 mg kg^{-1} (dw). Worldwide background Cu levels in soils are up to 110 mg kg^{-1} depending on parent material and soil texture, with an average of 14 mg kg^{-1}. Cu-contaminated soils, particularly near ancient Cu smelting sites, may reach 20,000 mg kg^{-1}. Anthropogenic activities such as agriculture (e.g., the use of fungicides such as the Bordeaux mixture), pig farming (dietary supplements), industrialization, mining, traffic, and domestic heating can make a significant contribution to the accumulation of Cu in soil.

24.2.1.5 Lead

The average natural Pb level in the Earth's crust ranges from 10 to 30 mg kg^{-1}, and it is usually found as a mineral combined with other elements, such as sulfur or oxygen. In the biosphere, Pb shows two main oxidation states: +4 and +2, the latter being dominant in soils and oxygenated waters. The typical average Pb concentration in soils ranges from 10 to 100 mg kg^{-1} but it can reach up to 10,000 mg kg^{-1}. However, its biologically available concentration is usually low because of low solubility (especially lead phosphate is insoluble, with a solubility product of 10^{-54}). When bioavailable, Pb is a highly toxic nonessential metal that exerts its toxicity by interacting with nucleic acids and proteins, inhibiting enzyme activity, disrupting membrane functions and oxidative phosphorylation, altering the osmotic balance, and displacing Ca^{2+} and Zn^{2+} in proteins (Martinez-Finley et al. 2012).

24.2.1.6 Mercury

Hg is not an essential element, and the Earth's crust contains only about 0.08 mg kg^{-1} (dw). This element is very toxic for life as Hg^{2+} has a very high affinity for thiol groups and may thus interfere with many proteins. It is classified in group 12 of the periodic table, together with Zn and Cd. Hg exists in three main oxidation states in the biosphere: 0, +1, and +2. The latter is the major oxidation state in aerobic soils and aquatic environments. According to the redox conditions, Hg forms many complexes with other compounds, with $HgCl_4^{2-}$ and $HgOH^+$ being dominant in aerobic conditions, whereas sulfur-related forms (HgS, HgS^{2-} and CH_3HgS^-) prevail in reduced conditions. Hg can also occur in organic compounds such as CH_3Hg^+ (methylmercury) or $Hg(CN)_2$. Hg occurs naturally and is found in the air, water, and soil. It is released into the environment from various natural sources such as volcanic activity, weathering of rocks, and volatilization from the ocean (Kim et al. 2016). However, the main cause of Hg release is human activity, particularly coal fired power stations, residential coal burning for heating and cooking, industrial processes, waste incinerators, and mining activities for gold and other metals. In soils, Hg levels can vary by several orders of magnitude, from 0.05 to 119 mg kg^{-1} (dw) in contaminated areas (Kim et al. 2016).

24.2.1.7 Nickel

Ni is an essential element for some bacteria, archaea, fungi, algae, and higher plants. It is a cofactor of at least nine enzymes involved in diverse cellular processes such as energy metabolism and virulence, including [NiFe] hydrogenase, urease, Ni-SOD, and CO dehydrogenase (Boer et al. 2014). No enzymes or cofactors that include Ni have been identified in higher organisms so far. Ni is classified, together with Pd and Pt, in group 10 of the periodic table, and its abundance in the Earth's crust is about 90 mg kg^{-1}. The total Ni content in uncontaminated soils ranges between 13 and 40 mg kg^{-1} (dw). Some soils are naturally enriched with Ni, and these may contain more than 10,000 mg kg^{-1}. The highest Ni contamination, up to 26,000 mg kg^{-1}, was reported for top soils near a smelter in Canada (Rinklebe and Shaheen 2017), and the largest nickel deposits on the Earth are located at Norilsk, Russia. Ni in soil solution generally occurs in its free ionic form (Ni^{2+}), which is stable over a range of redox conditions. Ni is widely used in industry for electroplating, battery manufacturing, stainless steel, and coin production, and it is mainly introduced into soils through atmospheric deposition, sewage sludge, and industrial compost.

24.2.1.8 Zinc

Zinc is an essential element that occurs naturally in air, water, rocks, and soil. It is essential, as many enzymes use it as a cofactor. It has roles in the metabolism of RNA and DNA, signal transduction, and gene expression (Frassinetti et al. 2006). The oxidation state of soluble zinc in the biosphere is +2. The average natural zinc level in the Earth's crust is about 70 mg kg^{-1} (dw), and soils usually contain 10–300 mg kg^{-1}. Zinc has been concentrated to much higher levels at some locations by natural geological and chemical as well as anthropogenic processes related to the wide use of zinc compounds in industry, agriculture, and medicine. For instance, Zn levels reached 8,247 mg kg^{-1} in the soil of a medieval foundry (Gillan et al. 2017). Despite its essential role as a trace element in various biological processes, excess zinc has significant toxicity and acts as a potent disrupter of biological systems. This duality of zinc properties requires a tight regulation of its intracellular homeostasis.

24.3 BIOAVAILABILITY

Metals in soils are not always able to enter living organisms, and—consequently—biological effects may be limited or completely absent. It is therefore important to determine metal bioavailability to understand any soil environment. *Bioavailability* can be defined as "the maximum amount of a contaminant which is available, or solubilized, in the environment of an organism." As this definition was used in the gastrointestinal environment, it may be adapted to "a bioavailable compound is one that is freely available to cross an organism's cellular membrane from the medium the organism inhabits at a given time" (Semple et al. 2004).

In soils, metals can be (1) dissolved in the pore water, (2) adsorbed to the surface of mineral and organic particles, or (3) embedded deeply in the latter particles (occluded in minerals). Dissolved, next to adsorbed, metals are frequently thought to be the most bioavailable, and these metals may be called "easily exchangeable" (Filgueiras et al. 2002). In contrast, metals that are intimately associated with minerals, such as silicates, are believed to be mobilized slowly, only as a result of weathering (Filgueiras et al. 2002). However, bacteria are able to form biofilms on sediment particles (Gillan and Pernet 2007), and as a result of their activity, metals deeply embedded in particles may also be reached to some extent.

To estimate metal bioavailability in a soil, the simplest approach is to use a diluted HCl treatment (e.g., HCl 1M) for 24h. This treatment will mainly assess dissolved and adsorbed metals (Roosa et al. 2014). After filtration through a 0.45 μm filter, metal concentrations are measured using inductively coupled plasma—atomic emission spectroscopy. Another method is to use biosensors, also called whole-cell bioreporters. Biosensors are further discussed in Chapter 25. Metal biosensors employ genes from metal-responsive operons as sensing elements together with genes encoding reporter enzymes to produce a measurable signal (e.g., bioluminescent, colorimetric, or fluorescent). Many biosensors have been designed for metals. The difficulty is in the development of a quantitative and specific biosensor because the regulatory proteins generally respond to diverse metals. Another difficulty is the presence of soil particles that may interfere with the biosensor.

Once bioavailability has been determined, bioavailable metal concentrations may be compared with various microbial parameters in soil, such as the abundance of selected bacterial species, total biomass, microbial activity, or gene levels as determined by quantitative polymerase chain reaction (PCR). For instance, it was determined that total Co did not correlate well with any microbial parameter, contrary to bioavailable Co that correlated well with mRNA levels coding for a Cd–Zn–Co efflux pump (CzcA) as measured by quantitative PCR in a sediment microbial community (Roosa et al. 2014).

24.4 DURATION OF EXPOSURE

The duration of exposure to a metal is a key parameter to assess metal impact. A recent and acute metal contamination caused by human activities or natural processes may have large effects on the soil community. Many studies have shown that soil microbial communities are affected after a few hours or days of exposure. In contrast, in soils that have been exposed to high metal levels for a long period of time, the established microbial communities have already been adapted and an equilibrium has possibly been reached. For instance, in areas of glaciated terrain in Scandinavia, where sulfide minerals are found close to the surface, a situation of naturally high metal content in the surface soil is observed. In such areas, it is possible that high metal levels in the soil have existed since the last ice age, that is, several thousands of years (Baath et al. 2005). The analysis of soils exposed to metals for long periods of time is important so as to identify possible core bacterial communities, that is, a set of taxa common to many different metal-contaminated soils. For instance, such a set of taxa was proposed for contaminated soils located more than 40km apart in Poland. The core set, as determined by 16S rRNA gene sequencing, comprised members of *Sphingomonas*, Candidatus *Solibacter*, and *Flexibacter* (Golebiewski et al. 2014).

24.5 EFFECTS OF METALS ON MICROBIAL DIVERSITY

Some studies have suggested that metals can significantly reduce the diversity of soil microbial communities, even after prolonged exposure (Abdu et al. 2017). For instance, Gans et al. (2005)

suggested that metal pollution reduces soil microbial diversity more than 99.9% via reanalysis of data generated from bacterial community DNA from pristine versus metal-contaminated soils. Interestingly, the total bacterial biomass remained unchanged, at about 2×10^9 cells per gram of soil, despite metal exposure. The major effect was the elimination of rare taxa: in the pristine soil, taxa with abundance values $<10^5$ cells per gram accounted for 99.9% of the diversity, and genetic diversity from this fraction of the community appeared to have been purged by metal pollution (Gans et al. 2005). However, the reliability of this study was questioned, as it was suggested that the experimental data did not support the conclusion (Volkov et al. 2006, Bunge et al. 2006). Other examples of reduced diversity of soil microbiomes after long periods of metal exposure include 11 years of exposure to Cu and Zn (Singh et al. 2014) and 40 years of exposure to As and Cr (Sheik et al. 2012). Singh et al. (2014) demonstrated—by metagenomics and functional assays—that long-term metal stresses result in a significant loss of diversity. Even at a moderate loss of diversity, some key specialized functions (carried out by specific microbial groups) were compromised. This is a fundamental issue: although diversity is not much affected, the elimination of rare species that carry out specialized, but essential, ecosystem functions may be detrimental to the system. The effect of metals on functionality (i.e., bacterial metabolism) will be further discussed later.

In contrast, some reports concluded that microbial diversity is not affected by metals on the long term, or it recovers from the exposure. For instance, Berg et al. (2012) investigated the impact of copper on bacterial community composition and diversity within a Cu gradient (20–3,537 μg g^{-1}) in soil stemming from industrial contamination more than 85 years ago, via 16S rRNA gene amplicon sequencing. The results did not demonstrate any significant correlation between bioavailable Cu and bacterial OTU richness, indicating that Cu does not reduce bacterial diversity in soil (Berg et al. 2012). However, the composition of the microbial communities was impacted by copper. The relative abundance of members of several phyla or candidate phyla, including Proteobacteria, Bacteroidetes, Verrucomicrobia, Chloroflexi, WS3, and Planctomycetes, decreased with increasing bioavailable Cu, whereas members of the dominant Actinobacteria phylum showed no response and Acidobacteria members showed a marked increase in relative abundance (Berg et al. 2012). The same conclusion was drawn for soil microbial communities exposed for more than 400 years to a cocktail of metals, such as Cu, Zn, and Pb, as the microbial community structure, and thus diversity, was little affected, contrary to the set of metal-resistance genes (Gillan et al. 2017). Clearly, microbial diversity may be affected by metals at the onset of soil contamination, but over the course of time, diversity may recover and metal-resistance genes may spread across members of the microbiome. Such a general adaptation process was also observed in other environments, for instance, in metal-contaminated marine sediments and in river sediments (Gillan et al. 2015).

24.6 EFFECTS OF METALS ON SOIL MICROBIAL BIOMASS

Some studies have suggested that the total microbial biomass in soil can be affected by metal contamination, whereas others reported the opposite. For instance, Knight et al. (1997) determined the microbial biomass after a 3-year period in soils with a range of pHs and amended with Cu, Cd, or Zn at concentrations around the maximum permissible values in agricultural land receiving sewage sludge. There was no reduction in microbial biomass either due to pH or to metal treatment in any of the soils, except for the Cu treatment in which the microbial biomass was reduced at low pH. However, metal bioavailability was not determined, and microbial biomass was determined by the fumigation extraction technique. Other studies similarly observed no microbial biomass changes after exposure to metal contamination (e.g., Baath et al. 2005, Hesse et al. 2018).

On the other hand, there are studies supporting the conclusion that metal contamination is associated with a decreased soil microbial biomass (Abdu et al. 2017). Gough et al. (2008) evaluated the biomass changes in a zinc-exposed soil via direct total microscopic counts as well as measured various metal fractions (pore water extracts, sequential extractions, and total extracts). The results indicated that microbial biomass was inversely correlated with the pore water concentrations of zinc and arsenic. Furthermore, other parameters known to influence biomass in sediments (e.g., organic carbon and nitrogen concentrations, pH, sediment texture, and macrophytes) showed no differences that could explain the observed biomass trends. Hence, the microbial abundance in a contaminated sediment may be strongly controlled by pore water metal concentrations.

24.7 EFFECT ON METABOLIC ACTIVITY

It is known for a long time that metals can influence the metabolism of soil residents (Abdu et al. 2017). The effects of metals on microbial metabolism may have huge consequences at the ecosystem level, as soil microbes exert key ecological functions by contributing to global element cycling, plant nutrition and health, organic matter turnover, the breakdown of xenobiotics, and the formation of soil aggregates. For instance, several studies showed litter accumulation on forest floors near smelters in contrast to the "normal" degradation. More specifically, the decomposition rate of Scots pine needle litter in two metal pollution gradients in Sweden, one near a brass mill and the other around a primary smelter, was strongly influenced by metal pollution, resulting in a decreased rate of mass loss (Berg et al. 1991). The decomposition rate of natural polymers such as cellulose and starch was reduced by metal contamination, possibly due to an effect of metals on extracellular enzymes. In contrast, the decomposition of simple substances like glucose or albumin generally proceeds at a similar rate in polluted and nonpolluted soils, although the short-term decomposition patterns may also be affected (Nordgren et al. 1988, Berg et al. 1991). Metals may inhibit enzymatic reactions by binding to the substrate, combining with the protein-active groups of the enzymes, or reacting with enzyme–substrate complexes (Caldwell 2005, Nannipieri et al. 2012). Affected enzymes include, but are not limited to, urease, alkaline phosphatase, and invertase. Numerous field studies demonstrated the adverse effects of metal contamination on soil microbial activities. For example, reductions (10- to 50-fold) in enzyme activity, including that of N-acetylglucosaminidase, β-glucosidase, endocellulase, and acid and alkaline phosphatases, paralleled the increase in heavy metal concentrations for a grassland ecosystem with a wide range of metal concentrations ranging from 7.2 to 48.1 mmol kg^{-1} (As, Cd, Cr, Cu, Ni, Pb, and Zn). Furthermore, total and fluorescein diacetate (FDA) active fungal biomass, FDA-active bacterial biomass, and substrate-induced respiration were lower in the polluted soils. These results demonstrate that metal contamination of soil has adverse effects on microbial activity involved in organic matter decomposition and nutrient cycling.

24.8 METAL RESISTANCE AND ADAPTATION

Survival of the bacterial soil community in metal-contaminated soils is an interplay between many factors, including intrinsic biochemical and structural properties, physiological and/or genetic adaptations, and environmental modifications of metal speciation. A variety of resistance mechanisms are used by soil bacteria to cope with metal toxicity, including efflux, bioaccumulation, sequestration, and modification (Figure 24.2). Table 24.2 lists 22 metal resistance systems that are known in bacteria. These are localized in the extracellular environment, the outer membrane and periplasm of Gram-negative bacteria, the cytoplasmic membrane, and the cytoplasm (Gillan 2016).

24.8.1 METAL RESISTANCE SYSTEMS

24.8.1.1 Extracellular Systems

Sulfide Production

Sulfide production by bacteria can remove the dissolved metals from solution. For instance, metal speciation (i.e., the different, defined species, forms or phases in which an element occurs) was controlled by the most active zone of sulfate reduction in a metal-contaminated salt marsh. Sulfide production is usually performed by δ-Proteobacteria (sulfate-reducing bacteria) that may be found in anaerobic zones of soil particles.

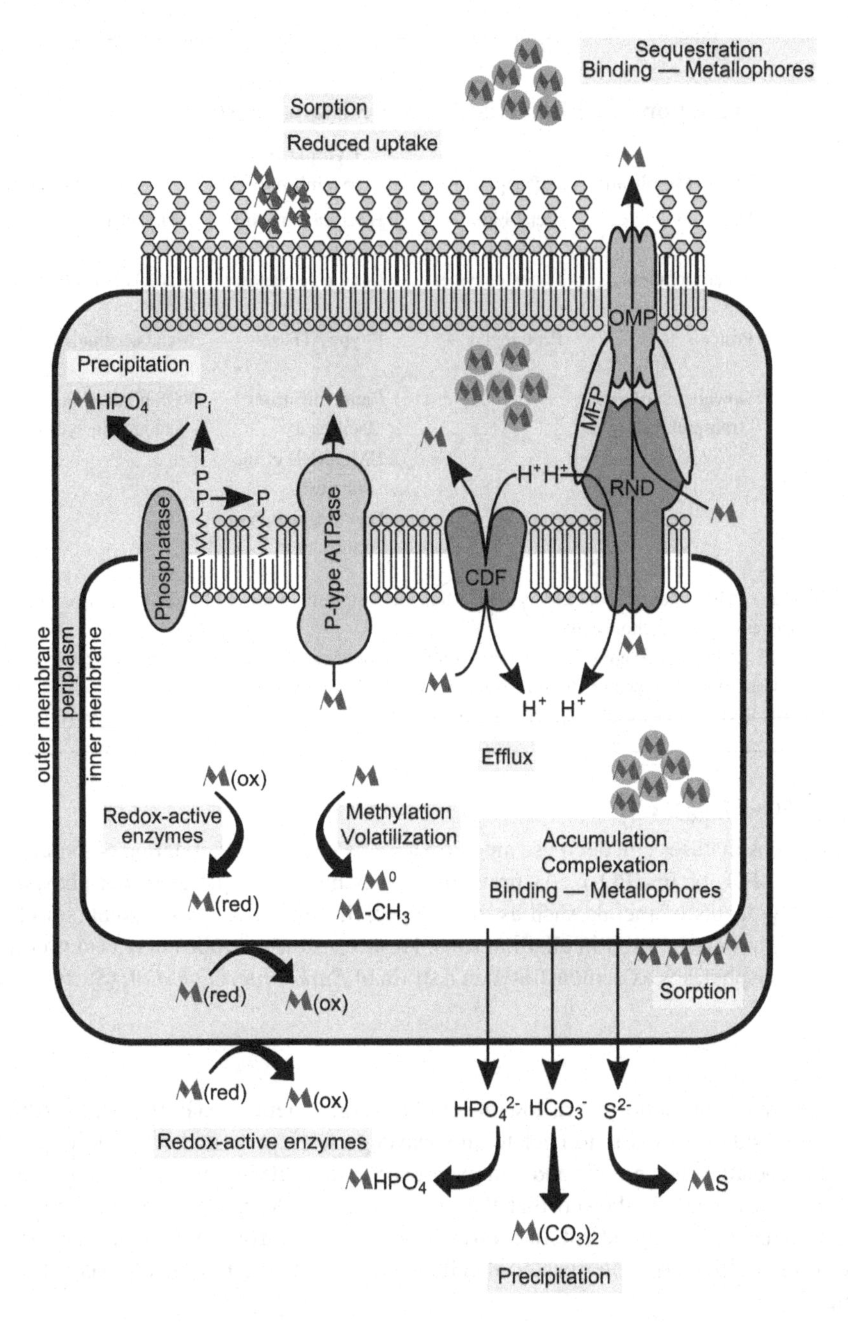

FIGURE 24.2 Overview of metal (M)-resistance mechanisms in Gram-negative bacteria.

Calcite Precipitation

Calcite precipitation induced by bacteria is a widespread phenomenon, particularly in the presence of ureolytic bacteria. This process could protect bacteria, as extracellular calcite is able to trap various toxic metals. This resistance mechanism could be important for microbial communities living in calcium-rich soils at elevated pH. Consumption of 1 mol of urea by ureolytic bacteria will result in 1 mol of ammonia (increasing the pH) and 1 mol of H_2CO_3; the latter can form calcium carbonate in the presence of Ca^{2+}.

TABLE 24.2
Localization and Function of 22 Bacterial Metal Resistance Systems

Extracellular Environment	Outer Membrane	Periplasm	Cytoplasmic Membrane	Cytoplasm
Sulfide production	LPS	Metal-binding proteins	Electron transport chains	Redox changes
Calcite precipitation	Outer membrane vesicles	Redox enzymes	Lipid phosphatases	Methylation–volatilization
Extracellular phosphatases	Proteins	Peptidoglycan[a]	P-type ATPases	Metal accumulation
Extracellular metallophores	Downregulation of transporters		Cation diffusion facilitators	ROS production, DNA reparation, and protein hydrolysis
EPS			HME-RND efflux systems[b]	
			Downregulation of transporters	

[a] The peptidoglycan of Gram-negative bacteria is localized in the periplasm; however, for Gram-positive bacteria, it is part of the cell wall together with teichoic acids.

[b] Although the HME-RND proteins are located in the cytoplasmic membrane, the efflux systems are spread across the outer and inner (cytoplasmic) membranes of Gram-negative bacteria. EPS, extracellular polymeric substances; LPS, lipopolysaccharides; HME-RND, heavy metal–resistance–nodulation–division efflux systems; ROS, reactive oxygen species.

Extracellular Phosphatases

Extracellular phosphatases can decrease metal toxicity by causing precipitation of metal phosphates such as $NH_4UO_2PO_4$ or $Pb_3(PO_4)_2$. As previous studies have demonstrated that phosphatases are inhibited by low levels of metals such as zinc, especially in acidic soils (Kucharski et al. 2011), this process might only be used in alkaline soils. However, soil microbes may also produce metal-resistant acid phosphatases as demonstrated in a strain of *Burkholderia gladioli* (Story and Brigmon 2017).

Metallophores

Metallophores and siderophores are low-molecular-weight metal-chelating agents. Although the classical role of siderophores is to chelate and scavenge ferric iron under Fe-limiting conditions, they can also chelate other metals and can even exert a detoxifying effect on many other metals. Hesse et al. (2018) recently showed that the proportion of siderophore-producing taxa increases along a natural heavy metal gradient. In contrast to siderophore–iron complexes, siderophore–metal complexes do not efficiently enter bacterial cells, thereby reducing free metal concentration in the environment.

Extracellular Polymeric Substances

Extracellular polymeric substances of bacteria are known to complex metals and protect against excessive metal stress. Once complexed, minerals may nucleate and provide additional protection, particularly for Fe/Mn oxyhydroxides.

24.8.1.2 Outer Membrane Systems

Lipopolysaccharides

Lipopolysaccharides are found in the outer membrane of Gram-negative bacteria. These large molecules, consisting of lipid and polysaccharide moieties, have been proposed to be involved in metal resistance by preventing cellular uptake, for example, via nucleation of minerals like $NH_4UO_2PO_4$ as demonstrated in the bacterium *Citrobacter* sp. N14.

Outer Membrane Vesicle

Outer membrane vesicle production can also be used to expel metals complexed with periplasmic or outer membrane proteins. However, little is known about the genes involved in this process related to metal resistance. As many as 150 genes may be involved in vesiculation, with an important role for genes involved in lipopolysaccharide and enterobacterial common antigen structures.

Redox-Active Enzymes

Redox-active enzymes in the outer membrane can reduce metals, leading to mineral precipitation. For instance, in many Gram-negative bacteria, a trans-outer membrane porin-cytochrome protein complex (Pcc) or an MtrABC extracellular electron conduit may be used to transfer electrons from the cytoplasmic membrane to the exterior of the cell. Such a system is used by dissimilatory metal reducers such as *Geobacter* spp. Metals can also be oxidized by specific enzymes in the outer membrane, such as c-type cytochromes in iron-oxidizing bacteria. In *Roseobacter*, animal heme peroxidases are responsible for Mn^{2+} oxidation.

Downregulation of Transporters

Downregulation of transporters is another way to avoid the entry of excessive metals. For instance in *Escherichia coli*, endogenous resistance after exposure to silver resulted from the loss of the OmpF/C porins and derepression of the CusCFBA efflux system elicited by point mutations in *ompR* and *cusS*, respectively. Downregulation of porins was also observed in the presence of Cu for the acidophilic bacterium *Acidithiobacillus ferrooxidans*.

24.8.1.3 Periplasmic Systems

Periplasmic Proteins

Periplasmic proteins such as metallo-chaperones and redox-active enzymes play a role in metal resistance via the binding of free metal ions in the periplasm and reduction, respectively. Examples of Cu metallo-chaperones include CusF from *E. coli*, CueP from *Salmonella enterica* serovar Typhimurium, CopK from *Cupriavidus metallidurans* CH34, and CopM from *Synechocystis* sp. PCC 6803. Examples of redox-active periplasmic enzymes include multicopper oxidases such as CueO and CuiD, and CopA1 from *C. metallidurans* CH34. In addition, metal reduction was also shown for a series of periplasmic soluble cytochromes that formed a redox network. Once reduced, oxidized, or simply complexed, periplasmic minerals may nucleate, as observed for Pd, Fe, Mn, and Pb.

Peptidoglycan

Peptidoglycan is a major structural component of the cell wall of most bacteria and is known to complex metals. In Gram-positive bacteria, peptidoglycan is associated with teichoic acids that participate in metal complexation. Once complexed, mineral phases may form in the periplasm as mentioned earlier.

24.8.1.4 Cytoplasmic Membrane Systems

Electron Transport Chain

Electron transport chain enzymes can directly oxidize or reduce metals, a process that may lead to mineral formation in the periplasm or the cytoplasm as in the case of tellurite.

Lipid Phosphatases

Lipid phosphatases in the cytoplasmic membrane can also sequester metals as phosphate salts in the periplasm. The best example is the undecaprenyl pyrophosphate phosphatase PbrB that acts together with the P-type ATPase PbrA in *C. metallidurans* CH34 to detoxify lead (Hynninen et al. 2009).

P-type ATPases

P-type ATPases transport (metal) cations across membranes and translocate lipids between lipid bilayer leaflets, by using ATP hydrolysis as energy source (Palmgren and Nissen 2011). The most common P-type ATPases in bacteria are the P_{IB}-ATPase group, which detoxifies the cytoplasm by regulating metal ion concentrations and provides metal ions to periplasmic or secreted metalloproteins. Two phylogenetic clusters can be distinguished: one specific to monovalent metal ions (Cu^+) and the other one including transporters of bivalent metal ions (Co^{2+}, Cu^{2+}, Zn^{2+}). Because of the chemical similarities among metal ions, some P-type ATPases can also transport nonphysiological metal ions such as Ag^+, which is transported by Cu^+-ATPases, and Cd^{2+} and Pb^{2+}, that are transported by Zn^{2+}-ATPases.

Cation Diffusion Facilitators

Cation diffusion facilitators (CDFs) are antiporters that use a pH gradient, a membrane potential, or a K^+ gradient to drive the active transport of metal ions through the plasma membrane. They are mainly involved in metal resistance and/or homeostasis (Nies 2003). CDF proteins specifically transport bivalent metal cations including Zn^{2+}, Fe^{2+}, Mn^{2+}, Cd^{2+}, Co^{2+}, Ni^{2+}, and Cu^{2+}. The CDF transporter CzcD from *C. metallidurans* CH34 was one of the first identified members. It is part of a large cluster (*czcNICBADRSEJP*) involved in bacterial resistance to Co^{2+}, Zn^{2+}, and Cd^{2+}. It mediates low-level resistance to Co^{2+}, Zn^{2+}, and Cd^{2+} in the absence of the resistance–nodulation–cell division (RND)-driven system CzcCBA (Mergeay and Van Houdt 2015).

HME-RND-Driven Efflux Systems

HME-RND-driven efflux systems are tripartite efflux pumps that spread across the outer and inner (cytoplasmic) membranes of Gram-negative bacteria. These systems are driven by RND proteins located in the cytoplasmic membrane. The RND superfamily contains nine phylogenetic families involved in different transport processes. Members involved in the efflux of metal ions (HME-RND) and of hydrophobic/amphiphilic compounds (HAE1-RND) have been extensively studied in Gram-negative bacteria. These RND proteins are accompanied by proteins belonging to the outer membrane factor and the membrane fusion protein, and the assembled tripartite transport complexes catalyze the efflux of substrates out of the cell. The archetypal HME-RND efflux system is CzcCBA in *C. metallidurans* CH34, which mediates resistance to Co^{2+}, Zn^{2+}, and Cd^{2+}. Other examples include the CusCBA and SilCBA systems, which are involved in Cu^+ and Ag^+ resistance, and the CnrCBAT system, which is involved in Co^{2+} and Ni^{2+} resistance (Mergeay and Van Houdt 2015).

Downregulation of Transporters

Downregulation of transporters located in the cytoplasmic membrane, preventing the entry of particular metals, can also increase resistance. For instance, Cr^{6+} resistance could be obtained by mutation of the sulfate transporters.

24.8.1.5 Cytoplasmic Resistance Systems

Redox-Active Enzymes

Redox-active enzymes can detoxify metals that entered the cytoplasm, such as As, Sb, Cr, Hg, Se, and Te. For instance, ArsC reduces As^{5+} to As^{3+}, the arsenite oxidase AioA oxidizes As^{3+}, and ArsH confers resistance to organo-arsenicals. For Cr, specialized Cr^{6+} reductases have been found in some bacteria, and several components of the cytoplasm such as NADH, flavoproteins, and other heme proteins also reduce Cr^{6+}. Thioredoxin is a cytoplasmic electron donor, and thioredoxin reductase has been reported to be upregulated under toxic metal stress. Thioredoxin, but also glutaredoxin, is known to reduce ArsC. For Hg, the MerA protein reduces Hg^{2+} to the less toxic Hg^0, which diffuses out of the cell through the cytoplasmic membrane.

Methylation and Volatilization

Methylation and volatilization of As, Sn, and Hg have been reported in both aerobic and anaerobic bacteria. Together with *advection* (i.e., the removal of metals through pore water movements) and phytoextraction, volatilization is one of the few processes able to completely remove some metals from contaminated soils.

Metal Accumulation and Complexation

Metal accumulation and complexation is also used to detoxify the cytoplasm, resulting in the formation of minerals such as magnetite, iron sulfides, or uranium phosphate, and different Cd and Pb complexes. Complexants include, among others, glutathione and bacterial metallothioneins.

ROS Scavenging, DNA Repair, and Protein Hydrolysis

ROS scavenging, DNA repair, and protein hydrolysis are general stress-induced defense systems of bacteria in response to many kinds of stress. Such systems are thus indirect metal resistance systems that repair metal-induced damages.

24.8.2 Metal Resistance Carriers

Metal resistance determinants carried by mobile genetic elements (MGEs) have been extensively described, with the resistance to mercury, nickel, and cobalt carried by R plasmids already described 50 years ago. MGEs are further discussed in Chapter 7. The purpose of this section is to give some examples that highlight their diversity, roles, and implications for metal resistance. Two megaplasmids studied in detail are pMOL28 and pMOL30 from *C. metallidurans* type strain CH34. Plasmid pMOL28 (171 kb) confers resistance to nickel and cobalt, chromate, and mercury. Plasmid pMOL30 (234 kb) confers resistance to mercury, zinc, cadmium and cobalt, lead, silver, and copper. These plasmids are collectors of resistance genes, clustering different mechanisms and including many satellite genes (with unknown functions) (Mergeay and Van Houdt 2015). Numerous other examples of plasmids conferring metal resistance have been found, such as those conferring resistance to mercury, copper, and multiple metals, highlighting their importance in the adaptation to metal pollution in soil (Heuer and Smalla 2012).

Moreover, different transposons have been associated with metal resistance. Examples are the transposons involved in the worldwide dissemination of mercury resistance (*mer* operons) in environmental bacteria (Yurieva et al. 1997). In addition, integrative and conjugative elements (ICEs), such as R391, CMGI-1 from *C. metallidurans* CH34, and members of the Tn*4371* family, also carry metal resistance determinants. The Tn*4371* ICE family members identified in β- and γ-Proteobacteria (including soil isolates) contain accessory genes coding for a wide variety of functions. These include metal resistance such as Czc-like and Ag^+/Cu^+ RND transporters, multicopper oxidase, P-type ATPase, phosphatase, mercury and arsenic reductase, and arsenic efflux (Van Houdt et al. 2013).

24.8.3 Genetic Adaptation

Horizontal gene transfer (HGT; see Chapter 7) has played an important role in the spread of metal resistance among microbial communities in groundwater (Hemme et al. 2016) and in radionuclide- and metal-contaminated subsurface soils (Sobecky and Coombs 2009). Similarly, long-term metal exposure has been shown to increase the incidence of plasmid-borne metal resistance determinants (Lakzian et al. 2007, Gillan et al. 2015), as well as plasmid mobilization (Top et al. 1995). Recently, Klümper et al. (2017) showed, by uncoupling plasmid transfer from selective forces, that Zn^{2+}, Cd^{2+}, Cu^{2+}, Ni^{2+}, and As^{3+} affected the permissiveness of the broad-host-range IncP-type plasmid pKJK5.

One frequently encountered type of mutation results from the displacement of transposable elements from a donor to a target site, potentially leading to important phenotypic changes (Vandecraen et al. 2017). Transposition of insertion sequences (IS), one of the simplest MGEs, can have different outcomes, from gene inactivation to constitutive expression or repression of adjacently located genes by delivering IS-specified (partial) promoter or terminator sequences (Vandecraen et al. 2017). IS-mediated derepression of metal efflux pumps has been observed in *C. metallidurans* AE126, by inactivation of *cnrY* or *cnrX*, coding for a membrane-bound antisigma factor (CnrY) and a sensor protein (CnrX), respectively. This results in the release of the extracytoplasmic function family sigma factor CnrH that directs the transcription of the *cnrCBAT* efflux pump and increased (nonspecific) Zn^{2+} efflux. Furthermore, subinhibitory zinc and cadmium concentrations induced the promoter activity of three involved IS elements. One of them, IS*Rme5*, also carries a complete outward-directed promoter that could drive the expression of the *cnrCBAT* efflux pump. Next to IS transposition, other mutations can confer increased resistance to metals. One example is a mutation in the stimulus-sensing sensor kinase of a two-component regulatory system, which can result in the constitutive and stress-independent transcriptional activation of cognate RND-driven efflux systems, resulting in increased metal resistance.

24.8.4 Interactions between Metals and Antibiotics

Recently, attention has been given to the co-occurrence and co-selection of metal and antibiotic resistance in different environments (Pal et al. 2017, Poole 2017). Co-selection can occur, when resistance genes to both metals and antibiotics are harbored by one cell (co-resistance) or when a single resistance mechanism confers resistance to both metals and antibiotics (cross-resistance) (Figure 24.3) (Pal et al. 2017, Poole 2017). Many studies have shown a positive correlation between the presence of metal- and antibiotic-resistance genes, next to an increased occurrence of antibiotic-resistance genes in metal-contaminated compared to non-contaminated soils and agricultural settings. For instance, 46 randomly selected soils from a Scottish archive, as well as 90 garden soils from Western Australia, showed a link between soil metal (copper, chromium, nickel, lead, arsenic, mercury, and iron) levels and the presence of antibiotic-resistance genes (Knapp et al. 2011, 2016). Similarly, long-term (4–5 years) copper and nickel exposure increased the abundance of antibiotic-resistance genes as well as MGEs, suggesting that copper and nickel exposure might enhance their HGT potential (Hu et al. 2017). Class 1 integrons could be important factors in the co-selection mechanism, as they are frequently associated with gene cassettes in which both metal- and antibiotic-resistance genes are found. Thus, bacteria with class 1 integrons may have a selective advantage compared to the rest of the bacterial community (Gillings 2014). Although, as pointed

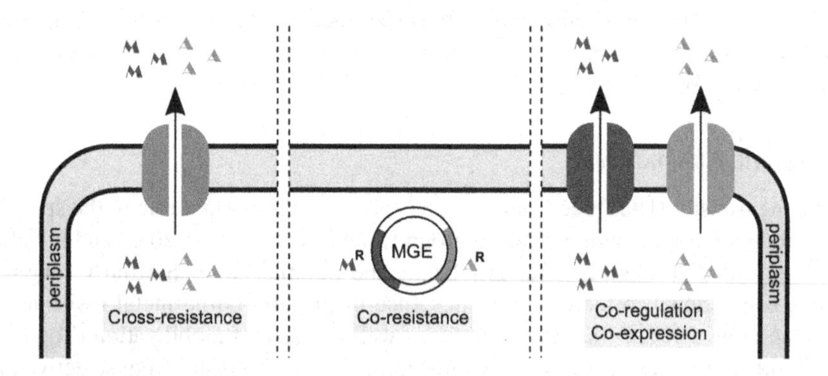

FIGURE 24.3 Co-selection of metal (**M**) and antibiotic (**A**) resistance caused by cross-resistance (one system confers resistance to both), co-resistance [resistances are physically colocated on a mobile genetic element (MGE)], and coregulation/expression by a common regulator. (Adapted from Pal et al. 2017.)

out by Pal et al. (2017), correlation does not imply causality and conclusive data demonstrating a clinical impact of metal use/exposure on antibiotic resistance are still lacking, these observations stimulate further experiments that assess the diversity of links between MGEs and antibiotic and metal resistances.

24.9 CONCLUSIONS AND FUTURE PROSPECTS

From this general overview, it is clear that soil metal contamination is a widespread concern and has severe impact on the residing microbial population and soil functioning. The effects on microbial abundance, diversity, and metabolic and resistance potential are remarkable. The final outcome of metal contamination of soil is modulated by many soil chemical and physical factors and subsequently affects soil fertility and structure. The resilience and survival of certain population members to metal exposure is mediated by the presence and horizontal spread of many metal-resistance genes. Apparently, not all metal-resistance genes are equally important at the community level. However, more detailed studies, including controlled field-based studies, are needed to confirm this. Future prospects should concentrate on the collaboration of individual microbial species during metal exposure and adaptation, the pleiotropic effects of (multiple) metal contamination, the characterization of novel resistance and associated accessory genes, and the association between antibiotic and biocide/metal resistance in microorganisms.

REFERENCES

Abdu, N., Abdullahi, A. A. and A. Abdulkadir. 2017. Heavy metals and soil microbes. *Environ Chem Lett* 15, 65–84.

Adriano, D. C. 2001. *Trace Elements in Terrestrial Environments*, 2 ed., Springer-Verlag: New York.

Baath, E., Diaz-Ravina, M. and L. R. Bakken. 2005. Microbial biomass, community structure and metal tolerance of a naturally Pb-enriched forest soil. *Microb Ecol* 50, 496–505.

Berg, B., Ekbohm, G., Soderstrom, B. and H. Staaf. 1991. Reduction of decomposition rates of scots pine needle litter due to heavy-metal pollution. *Water Air Soil Pollut* 59, 165–177.

Berg, J., Brandt, K. K., Al-Soud, W. A. et al. 2012. Selection for Cu-tolerant bacterial communities with altered composition, but unaltered richness, via long-term Cu exposure. *Appl Environ Microbiol* 78, 7438–7446.

Boer, J. L., Mulrooney, S. B. and R. P. Hausinger. 2014. Nickel-dependent metalloenzymes. *Arch Biochem Biophys* 544, 142–152.

Bunge, J., Epstein, S. S. and D. G. Peterson. 2006. Comment on "Computational improvements reveal great bacterial diversity and high metal toxicity in soil". *Science* 313, 918.

Caldwell, B. A. 2005. Enzyme activities as a component of soil biodiversity: A review. *Pedobiologia* 49, 637–644.

Filgueiras, A. V., Lavilla, I. and C. Bendicho. 2002. Chemical sequential extraction for metal partitioning in environmental solid samples. *J Environ Monit* 4, 823–857.

Frassinetti, S., Bronzetti, G. L., Caltavuturo, L., Cini, M. and C. Della Croce. 2006. The role of zinc in life: A review. *J Environ Pathol Toxicol Oncol* 25, 597–610.

Gans, J., Wolinsky, M. and J. Dunbar. 2005. Computational improvements reveal great bacterial diversity and high metal toxicity in soil. *Science* 309, 1387–1390.

Gillan, D. C. 2016. Metal resistance systems in cultivated bacteria: Are they found in complex communities? *Curr Opin Biotechnol* 38, 123–130.

Gillan, D. C. and P. Pernet. 2007. Adherent bacteria in heavy metal contaminated marine sediments. *Biofouling* 23, 1–13.

Gillan, D. C., Roosa, S., Kunath, B., Billon, G. and R. Wattiez. 2015. The long-term adaptation of bacterial communities in metal-contaminated sediments: A metaproteogenomic study. *Environ Microbiol* 17 (6), 1991–2005.

Gillan, D. C., van Camp, C., Mergeay, M. et al. 2017. Paleomicrobiology to investigate copper resistance in bacteria: Isolation and description of *Cupriavidus necator* B9 in the soil of a medieval foundry. *Environ Microbiol* 19, 770–787.

Gillings, M. R. 2014. Integrons: Past, present, and future. *Microbiol Mol Biol Rev* 78, 257–277.

Golebiewski, M., Deja-Sikora, E., Cichosz, M., Tretyn, A. and B. Wrobel. 2014. 16S rDNA pyrosequencing analysis of bacterial community in heavy metals polluted soils. *Microb Ecol* 67, 635–647.

Gough, H. L., Dahl, A. L., Nolan, M. A., Gaillard, J. F. and D. A. Stahl. 2008. Metal impacts on microbial biomass in the anoxic sediments of a contaminated lake. *J Geophys Res Biogeosci* 113 (G2), G02017.

Hemme, C. L., Green, S. J., Rishishwar, L. et al. 2016. Lateral gene transfer in a heavy metal-contaminated groundwater microbial community. *MBio* 7, e02234–15.

Hesse, E., O'Brien, S., Tromas, N. et al. 2018. Ecological selection of siderophore-producing microbial taxa in response to heavy metal contamination. *Ecol Lett* 21, 117–127.

Heuer, H. and K. Smalla. 2012. Plasmids foster diversification and adaptation of bacterial populations in soil. *FEMS Microbiol Rev* 36, 1083–1104.

Hu, H. W., Wang, J. T., Li, J. et al. 2017. Long-term nickel contamination increases the occurrence of antibiotic resistance genes in agricultural soils. *Environ Sci Technol* 51, 790–800.

Hynninen, A., Touze, T., Pitkanen, L., Mengin-Lecreulx, D. and M. Virta. 2009. An efflux transporter PbrA and a phosphatase PbrB cooperate in a lead-resistance mechanism in bacteria. *Mol Microbiol* 74, 384–394.

Kabata-Pendias, A. 2010. *Trace Elements in Soils and Plants*, 4th ed., CRC Press: Boca Raton, FL.

Kabata-Pendias, A. and B. Stzeke. 2015. *Trace Elements in Abiotic and Biotic Environments*, CRC Press: Boca Raton, FL.

Kim, K. H., Kabir, E. and S. A. Jahan. 2016. A review on the distribution of Hg in the environment and its human health impacts. *J Hazard Mater* 306, 376–385.

Klümper, U., Dechesne, A., Riber, L. et al. 2017. Metal stressors consistently modulate bacterial conjugal plasmid uptake potential in a phylogenetically conserved manner. *ISME J* 11, 152–165.

Knapp, C. W., Callan, A. C., Aitken, B., Shearn, R., Koenders, A. and A. Hinwood. 2016. Relationship between antibiotic resistance genes and metals in residential soil samples from Western Australia. *Environ Sci Pollut Res* 24, 2484–2494.

Knapp, C. W., McCluskey, S. M., Singh, B. K., Campbell, C. D., Hudson, G. and D. W. Graham. 2011. Antibiotic resistance gene abundances correlate with metal and geochemical conditions in archived Scottish soils. *PLoS One* 6, e27300.

Knight, B. P., McGrath, S. P. and A. M. Chaudri. 1997. Biomass carbon measurements and substrate utilization patterns of microbial populations from soils amended with cadmium, copper, or zinc. *Appl Environ Microbiol* 63, 39–43.

Kucharski, J., Wieczorek, K. and J. Wyszkowska. 2011. Changes in the enzymatic activity in sandy loam soil exposed to zinc pressure. *J Elementol* 16, 577–589.

Lakzian, A., Murphy, P. and K. E. Giller. 2007. Transfer and loss of naturally-occurring plasmids among isolates of *Rhizobium leguminosarum* bv. *viciae* in heavy metal contaminated soils. *Soil Biol Biochem* 39, 1066–1077.

Martinez-Finley, E. J., Chakraborty, S., Fretham, S. J. and M. Aschner. 2012. Cellular transport and homeostasis of essential and nonessential metals. *Metallomics* 4, 593–605.

Mergeay, M. and R. Van Houdt (eds.) 2015. *Metal Response in Cupriavidus Metallidurans, Volume I: From Habitats to Genes and Proteins, Springerbriefs in Biometals*, Springer International Publishing: Switzerland.

Nannipieri, P., Giagnoni, L., Renella, G. et al. 2012. Soil enzymology: Classical and molecular approaches. *Biol Fertil Soils* 48, 743–762.

Nies, D. H. 2003. Efflux-mediated heavy metal resistance in prokaryotes. *FEMS Microbiol Rev* 27, 313–339.

Nordgren, A., Baath, E. and B. Soderstrom. 1988. Evaluation of soil respiration characteristics to assess heavy metal effects on soil microorganisms using glutamic acid as a substrate. *Soil Biol Biochem* 20, 949–954.

Paez-Espino, D., Tamames, J., de Lorenzo, V. and D. Canovas. 2009. Microbial responses to environmental arsenic. *Biometals* 22, 117–130.

Pal, C., Asiani, K., Arya, S. et al. 2017. Metal resistance and its association with antibiotic resistance. *Adv Microb Physiol* 70, 261–313.

Palmgren, M. G. and P. Nissen. 2011. P-type ATPases. *Annu Rev Biophys* 40, 243–266.

Poole, K. 2017. At the nexus of antibiotics and metals: The impact of Cu and Zn on antibiotic activity and resistance. *Trends Microbiol* 25, 820–832.

Rinklebe, J. and S. M. Shaheen. 2017. Redox chemistry of nickel in soils and sediments: A review. *Chemosphere* 179, 265–278.

Roosa, S., Wattiez, R., Prygiel, E., Lesven, L., Billon, G. and D. C. Gillan. 2014. Bacterial metal resistance genes and metal bioavailability in contaminated sediments. *Environ Pollut* 189, 143–51.

Semple, K. T., Doick, K. J., Jones, K. C., Burauel, P., Craven, A. and H. Harms. 2004. Defining bioavailability and bioaccessibility of contaminated soil and sediment is complicated. *Environ Sci Technol* 38, 228A–231A.

Sheik, C. S., Mitchell, T. W., Rizvi, F. Z. et al. 2012. Exposure of soil microbial communities to chromium and arsenic alters their diversity and structure. *PLoS One* 7, e40059.

Singh, B. K., Quince, C., Macdonald, C. A. et al. 2014. Loss of microbial diversity in soils is coincident with reductions in some specialized functions. *Environ Microbiol* 16, 2408–2420.

Sobecky, P. A. and J. M. Coombs. 2009. Horizontal gene transfer in metal and radionuclide contaminated soils. In M. B. Gogarten (ed.), *Horizontal Gene Transfer: Genomes in Flux*, Humana Press: New York, 455–472.

Story, S. and R. L. Brigmon. 2017. Influence of triethyl phosphate on phosphatase activity in shooting range soil: Isolation of a zinc-resistant bacterium with an acid phosphatase. *Ecotoxicol Environ Saf* 137, 165–171.

Top, E. M., Derore, H., Collard, J. M. et al. 1995. Retromobilization of heavy-metal resistance genes in unpolluted and heavy-metal polluted soil. *FEMS Microbiol Ecol* 18, 191–203.

Toth, G., Hermann, T., da Silva, M. R. and L. Montanarella. 2016. Heavy metals in agricultural soils of the European Union with implications for food safety. *Environ Int* 88, 299–309.

Van Houdt, R., Toussaint, A., Ryan, M. P., Pembroke, J. T., Mergeay, M. and C. C. Adley. 2013. The Tn*4371* ICE family of bacterial mobile genetic elements. In A. P. Roberts and P. Mullany (eds.), *Bacterial Integrative Mobile Genetic Elements*, Landes Bioscience: Austin, TX, 179–200.

Vandecraen, J., Chandler, M., Aertsen, A. and R. Van Houdt. 2017. The impact of insertion sequences on bacterial genome plasticity and adaptability. *Crit Rev Microbiol* 43, 1–22.

Viti, C., Marchi, E., Decorosi, F. and L. Giovannetti. 2014. Molecular mechanisms of Cr(VI) resistance in bacteria and fungi. *FEMS Microbiol Rev* 38, 633–659.

Volkov, I., Banavar, J. R. and A. Maritan. 2006. Comment on "Computational improvements reveal great bacterial diversity and high metal toxicity in soil". *Science* 313, 918.

Wuana, R. A. and F. E. Okieimen. 2011. Heavy metals in contaminated soils: A review of sources, chemistry, risks and best available strategies for remediation. *ISRN Ecology* 2011, 1–20.

Yurieva, O., Kholodii, G., Minakhin, L. et al. 1997. Intercontinental spread of promiscuous mercury-resistance transposons in environmental bacteria. *Mol Microbiol* 24, 321–329.

25 Management Strategies for Soil Used for Cultivation, Including Modulation of the Soil Microbiome

Alexandre Soares Rosado
Federal University of Rio de Janeiro

Paolo Nannipieri
University of Firenze

Jan Dirk van Elsas
University of Groningen

CONTENTS

25.1 Introduction 421
25.2 Soil Management Strategies—How to Foster Soil Quality 422
25.2.1 Soil Management, Smart Farming, and the Microbiome 422
25.2.2 The Microbiome in Agroecosystems and Its Modulation 423
25.3 What Advanced Management and Monitoring Methods Offer Us 424
25.3.1 Cutting-Edge Plant/Soil Observation Methods and Deep Learning Robotics 424
25.3.2 Biosensors 425
25.3.3 System-Level Soil Microbiome Modulation 425
25.4 Land Use and Its Effects on Soil Microbiomes—How to Apply Novel Tools and Insights to Direct Land-Use Management 426
25.4.1 Land Use as a Modulator of Soil Microbiomes—Organic versus Conventional Agricultural Practices 426
25.4.2 Modulation of Soil Microbiomes by Changes in Management 427
25.5 The Importance of an Integrated Approach That Enables the Monitoring of Soil Microbiome Modulation Approaches 428
25.6 Final Remarks and Outlook 428
References 429

25.1 INTRODUCTION

Agricultural intensification has resulted in food production security for a global population that is expected to increase to more than nine billion people by 2050 (UN DESA 2017). The growing production has been achieved mainly by the conversion of natural areas into intensively managed farmland (Scherr and McNeely 2008). Moreover, there have been significant contributions of chemical fertilizers and high-yield crop varieties to conventional agricultural techniques (Chaparro et al. 2012). Notwithstanding these advances, there is a growing awareness of the adverse environmental impacts of these intensified agricultural practices. Soils constitute limited resources that are subjected to degradation and erosion as a result of overuse. Hence, sustainable land use, in which soil quality (including soil health) is preserved, is an indispensable strategy to ensure the continuity of our food production (Tillman et al. 2011, Blaser et al. 2016).

As we have learned in this book, the soil microbiome is a key element for soil health and quality, as it—by both composition and activity—directs the soil processes relevant for these. Soil microbiome parameters should be included as major factors in the equation to achieve the sustainable agriculture that is essential for food production. However, many forms of microbial life are—unfortunately—still poorly understood and explored across soils and agroecosystems. There is, therefore, an increasing demand for improved soil analysis and preservation methods. Such improved tools that are capable of monitoring soil quality will then alert both competent authorities and the public about the directions taken by production systems and their probable consequences. Recently, Schloter et al. (2018) addressed this issue, pointing to particular functional aspects of the soil microbiome as the major soil facet to focus on.

There are also rapid developments in robotics, imaging, and artificial intelligence that enable a precise, "within-field" type of agriculture, in which farmers may be able to regularly monitor crop growth almost at the individual plant level and place interventions when required. Whereas this development will soon allow plant disease and quality measurements, inclusion of soil quality parameters will be a necessary next step.

In this chapter, we evaluate these developments, including how one can take profit of the emerging new "precision" soil management approaches in order to foster the sustainability of land use.

25.2 SOIL MANAGEMENT STRATEGIES—HOW TO FOSTER SOIL QUALITY

25.2.1 Soil Management, Smart Farming, and the Microbiome

The concepts of soil quality and health have been defined and discussed in Chapters 2 and 21. Both the quality and the health of soil are driven by the collective effects of a suite of parameters, such as soil organic matter (SOM) content, soil pH, the levels of soil aggregation, and those of several key nutrients (see Schloter et al. 2018). Together, these parameters determine the quality and level of the key life support processes of soil.

How can soil management drive the aforementioned parameters to the desired levels? First, SOM is a key factor. High-quality soils typically have high SOM levels, whereas prolonged intensive agricultural practices tend to reduce or erode these. Hence, management strategies that promote the return to the soil of plant residues, where possible minimizing mechanical preparation, have been recommended. Among these approaches, no-tillage, minimum cultivation, winter vegetable cover, the use of intercropping pasture, and the rotation of annual crops and pastures stand out (Hobbs et al. 2008).

Another parameter that determines soil (agricultural) quality is the level of agrochemicals, added as pesticides and herbicides to the soil, as well as the crops used for food or feed. Today, contamination of food and the environment with toxic chemicals that may affect public health across generations is considered to be unacceptable. Sustainable solutions are required, leading to a decreased, or regulated, application of agrochemicals (Carvalho 2017), as exemplified by the recent uproar about the use of the herbicide glyphosate (commercial name: Round-up) in agriculture. Here, the use of biological pest control agents and the principles of organic agriculture may progressively come into play provided high production levels can be maintained (Reganold and Wachter 2016).

To foster the quality of agricultural soils at lower scale, concepts denoted "smart" and/or "precision" farming have been developed. Smart farming advocates the use of advanced interdisciplinary methods to assess and foster soil quality at fine level, in order to improve agricultural production within a field (Wolfert et al. 2017). Its central promises are targeted and site-specific interventions, with on-the-spot highly automated (robots and drones) agents that monitor crops—via advanced imaging techniques—at individual plant level and intervene at this level in case of possible (nutritional or health) problems in the crop. Here, the observational agents (robots and sensors) yield massive data that are provided to machine and deep learning algorithms, so as to provide robust algorithms that direct on-farm management. Unfortunately, this latter aspect of the advanced robotics-driven management, that is, decision-making with respect to interventions, is still in its infancy.

In these highly advanced endeavors, it is important that, next to the observational sustainable soil management approaches, the quality of the agroecosystem microbiome is also taken into consideration (Figure 25.1). The clear objective is to foster the expression of beneficial traits in the soil microbiome that affect the plant (e.g., the microorganisms that improve plant growth, nutrient use efficiency, abiotic stress tolerance, and disease resistance), next to that of the bulk soil (e.g., nutrient cycling, removal of contaminants) (Busby et al. 2017). About 12 years ago, Garbeva et al. (2006) already noticed a clear effect of cropping regime on soil microbiome characteristics, with grass cover concomitantly promoting microbial diversity and suppressiveness of the phytopathogen *Rhizoctonia solani* AG3. Hartman et al. (2018) recently found additional evidence that shows how different cropping practices enable the modulation of the soil and root microbiomes, thus paving the way for the implementation of microbiome management strategies into smart cropping systems. However, different types of land management and cultivation intensities exert significantly different influences on the dominant bacteria and fungi in soil. Hence, in many cases it is uncertain which beneficial traits and cropping regime-sensitive microbes are selected by the management practice.

25.2.2 The Microbiome in Agroecosystems and Its Modulation

As argued in the foregoing, soil microbiomes are essential for the quality of agroecosystems, as they drive the belowground processes that support crop development. Given their sensitivity to soil environmental conditions, any major management step can have severe effects on function. However, we are still confronted with big gaps in our knowledge in this area. The challenge is to discern patterns in soil microbiome properties and relate these to the biotic and abiotic parameters driven by soil

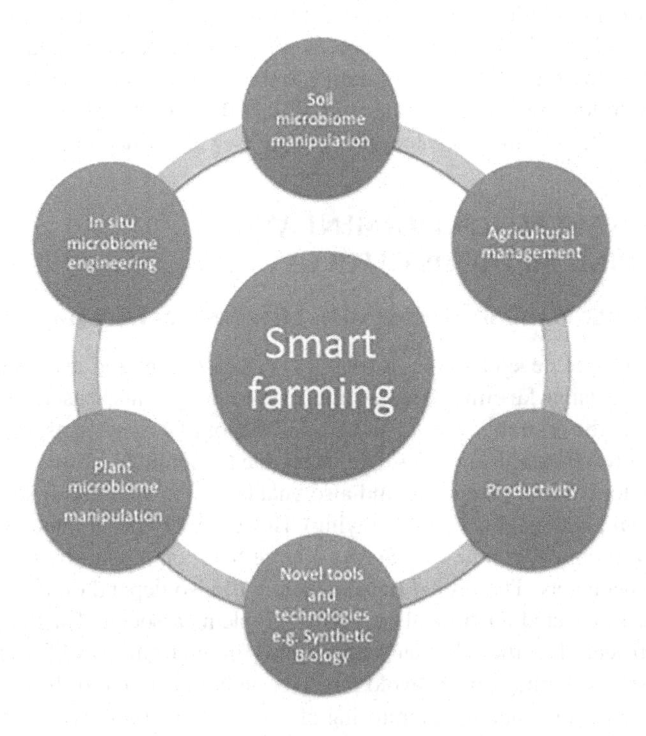

FIGURE 25.1 Some factors key to chasing the smart farming concept.

management, in order to establish connections between these. Another challenge lies in the detection of the *in situ* activities resulting from management, through space and time. Understanding of the match between management step and mechanisms of microbiome diversity, activity, and interaction will eventually allow us to manipulate soil function in the desired manner.

In order to stimulate the development of soil microbiome management practices for agricultural production, Busby et al. (2017) proposed five key agricultural microbiome research priorities:

1. Develop model host plant–microbiome systems for crop and non-crop plants with associated microbial culture collections and reference genomes.
2. Define core microbiomes and metagenomes in model host plant–microbiome systems.
3. Elucidate the rules of synthetic, functionally programmable microbiome assembly.
4. Determine functional mechanisms of host plant–microbiome interactions.
5. Characterize and refine plant genotype-by-environment-by-microbiome-by-management interactions.

The dedicated aim of these research priorities is to make use of beneficial bacteria and fungi, delivering these to soil/plant systems (Turner et al. 2013; see Chapter 22). Alternatively, the soil microbiome might be modulated *in situ* in order to boost plant growth, increase plant resistance to drought, disease, and pests, and reduce farmers' dependency on fertilizers and pesticides.

The extent to which (and how) soil microbiomes can be modulated through different cropping practices is variable and dependent on the system being studied. Several studies have shown that crop rotations, as well as the type of crop, influence soil fertility and microbiome diversity (e.g., Chaparro et al. 2012). As a case in point, Garbeva et al. (2006) showed that the type of crop and cropping regime affect the diversity and composition of microbiomes in soil, with grass having a diversity-promoting effect. However, to persist, the plant effect may need repeated cropping or cover of soil over several years. For example, at plant harvest, bacterial diversity was not affected by *Carex arenaria*, a non-mycorrhizal plant species (chosen to avoid the confounding effect by differential mycorrhizal colonization), cultivated in ten soils, but it depended on soil properties (De Ridder-Duine et al. 2005). Rhizodeposition as well as plant residues added to soil may cause temporary changes in the microbiome, with return to the original state once the plant is harvested. As discussed in the following, also Delmont et al. (2014) showed the importance of soil properties in shaping the composition of the microbial community (see also Chapter 2 of this book).

25.3 WHAT ADVANCED MANAGEMENT AND MONITORING METHODS OFFER US

25.3.1 Cutting-Edge Plant/Soil Observation Methods and Deep Learning Robotics

The aforementioned rapid developments in robotics, visualization, and artificial intelligence hold great promise for precision farming. Clearly, the high-tech systems, based on optical/radiation-based observational static or mobile (aerial and ground) robot-placed systems, enable high-precision monitoring of crop growth, status, and possible stress due to abiotic (e.g., water, nutrient) and biotic (e.g., pathogens) factors. Such systems should also enable on-the-spot and instantaneous intervention, at the individual plant level (e.g., by spraying). However, key developments are still required, as it is not trivial to monitor crop quality across a large field, on a plant-by-plant basis, and make proper on-the-spot decisions. The proper development will also depend on the crop species that is being monitored, the expected nutritional or disease problems associated with it, and the solutions to these that are foreseen. It is thought that such systems, in the future, will foster the sustainability of our cropping systems, as they aim to avoid fertilizer and/or pesticide spilling. Unfortunately, the systems are not yet able to include the monitoring of the *in situ* availability of nutrients or stressing/toxic compounds, nor the microbiome quality, composition, and activity, in the soil.

25.3.2 Biosensors

To enable the linking of microbiome activity assessments (see, e.g., Chapters 15 and 16) to soil management practices, biosensors can be of great help. Biosensors are "devices" that enable researchers to measure the (small-scale) *in situ* levels of particular compounds or conditions. There are several types of biosensors. Biosensors based on whole bacterial cells have been traditionally developed to assess pollutant bioavailability in soil (Poulsen et al. 2007). A whole-cell biosensor is a living cell that produces a measurable response to the compound or condition in question. The response could, for example, be the phenotype of an expressed gene or an electrochemical current. The main advantage of using whole-cell biosensors is their ability to report on the immediate environmental conditions that the cells are exposed to in soil. Therefore, they provide an accurate picture of the conditions, for example, with regard to the bioavailability of compounds, encountered in the microenvironments in soil where soil microorganisms reside. Whole-cell biosensors have been used to monitor compounds in soil such as heavy metals, polycyclic aromatic hydrocarbons, and many natural products (Poulsen et al. 2007). Besides those using whole cells, other types of biosensors have also been developed. These include biosensors that combine subcellular components or macromolecules and intact tissue with a transducing element that produces a signal in response to an environmental change. They do not give any information on the bioavailability of a compound and are not readily applicable to soil; hence, they will not be discussed further.

Biosensors can also provide valuable information about the quality of unplanted and planted soils, for example, information on the level of particular nutrient as well as oxygen in the soil. Therefore, they are useful tools to monitor both soil quality and soil toxicity. For example, Van Overbeek et al. (1995) used a *lacZ*-based root exudate-responsive biosensor, a derivative of *Pseudomonas fluorescens*, to show that the effect of plant roots extended only a few millimeters into the bulk soil. The compound in the root exudates responsible for the induction of the reporter was found to be proline. In a later study, they revealed the use of a starvation-inducible biosensor to show the general lack of available nutrients for bacteria in bulk soil (Van Overbeek et al. 1997). Højberg et al. (1999) constructed an oxygen-responsive biosensor by fusing an anaerobically inducible promoter to the *lacZ* gene and then inserting this construct into the root-colonizing bacterium *P. fluorescens*. The authors were able to show lower oxygen content in bulk soil when the water content increased. A similar low oxygen response by the biosensors was observed when the bulk density of the soil was increased by *compaction*. Biosensors undoubtedly are key tools in cases small-scale measurements of local conditions in the soil are required; potential applications lie in the monitoring of soil nutritional or oxygen status, as illustrated with the aforementioned examples.

25.3.3 System-Level Soil Microbiome Modulation

Given the fact that the individual factors that drive overall function in soil microbiomes are often unclear, "system-level" approaches to modulate the soil microbiome have recently been proposed (Sheth et al. 2016). Such *in situ* microbiome engineering methods allow the manipulation and detailed study of soil microbiomes in their native context, without the need for prior isolation of the microbes. A variety of approaches, based on chemical compounds (prebiotics, varied nutrients, xenobiotics), cells ("probiotics"), and DNA (phages, plasmids, modified DNA), can be applied in order to modulate soil microbiomes *in situ* (Figure 25.2). However, critical knowledge gaps still need to be filled in these approaches (Sheth et al. 2016). In particular, the modulation of gene expression patterns appears as a promising approach, as core microbiomes in soils may be relatively stable and so a large part of soil function is determined by similar metagenomes (Chaparro et al. 2012). However, this is still a challenging task as microbiome function as well as composition may critically depend on soil properties (Sheth et al. 2016, Foo et al. 2017). In experiments in which sterilized soil was inoculated with microbiomes from the same or different soils, the final soil microbiome compositions were those of the receiving soil (Delmont et al. 2014). Modification of key soil

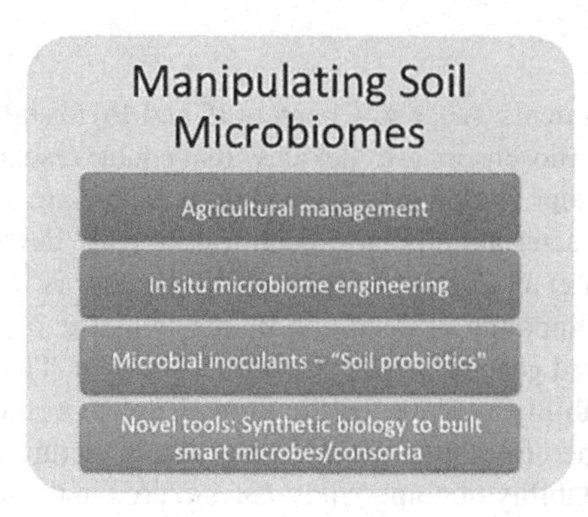

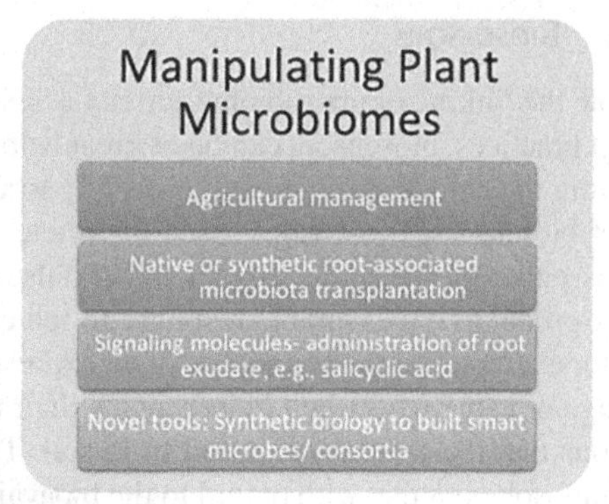

FIGURE 25.2 Strategies for engineering of microbiomes in agroecosystems.

properties, such as SOM or soil pH (affecting the composition of microbial communities), may be a key step to modulate soil microbiomes.

25.4 LAND USE AND ITS EFFECTS ON SOIL MICROBIOMES—HOW TO APPLY NOVEL TOOLS AND INSIGHTS TO DIRECT LAND-USE MANAGEMENT

25.4.1 Land Use as a Modulator of Soil Microbiomes—Organic versus Conventional Agricultural Practices

Soil microbiomes will respond to land-use changes (e.g., deforestation, logging, tillage, revegetation, and changed cropping systems), as such changes generally result in changed soil nutritional, biological, and physicochemical properties (Foo et al. 2017, Hartman et al. 2018, Peralta et al. 2018; see Figure 25.3) The changes are reflected in the biogeochemical cycles of nutrients (resulting, for instance, in changed emissions of greenhouse gases like N_2O), influencing soil productivity and sustainability (Aboim et al. 2008, Chaparro et al. 2012, Navarrete et al. 2015, Hartman et al. 2018). Clear examples of farm management regimes that result in soil microbiome modulations are the "organic" versus "conventional" agricultural practices. The organic management practices aim to achieve optimal productivity, under the premise that the factor "sustainability" is included in the equation. This often implies low input and low tillage practices. In contrast, the conventional management practices require a significant amount of fertilizers, pesticides, and energy to maximize the

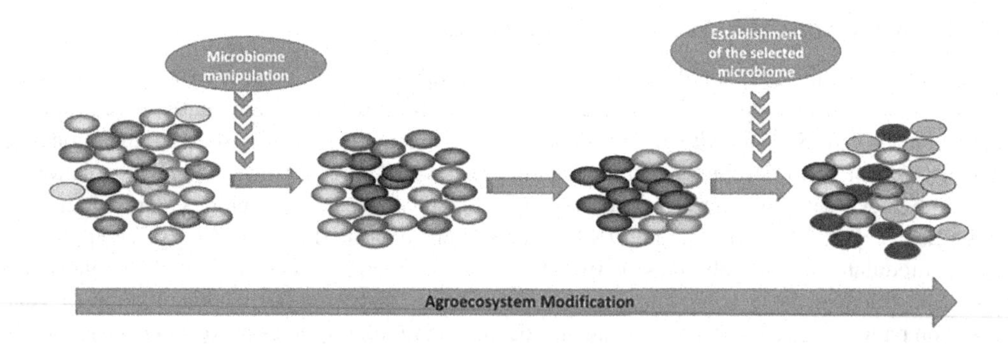

FIGURE 25.3 **(See color insert)** Representation of modifications in the soil microbiome after soil modulation events. The soil management strategy used is an important selection factor for target species and functions, or for the "core microbiome." In addition to being directly stimulated, activity of a particular community member may also generate adequate substrate for growth of other populations. The strategy applied should consider the specific soil microbiome changes, next to soil physicochemical data.

yield of a particular crop or set of crops. The high efficiency of the latter systems goes hand in hand with major costs to the environment and the ecological sustainability of the landscape.

25.4.2 Modulation of Soil Microbiomes by Changes in Management

There are several examples of successful modulation of soil microbiomes to improve agricultural yields or foster sustainable approaches. For instance, Shen et al. (2015) recently applied pyrosequencing of 16S ribosomal RNA and internal transcribed spacer sequences to fields subjected to two consecutive years of biofertilizer application, compared to chemical fertilizer or pig manure applications. The alpha-diversity of the respective (rhizosphere) bacteriomes increased, whereas that of fungomes decreased. Most notable was the decrease of abundance of *Fusarium* sp., the causal pathogen of *Fusarium* wilt disease in banana. Moreover, the planting of cover crops in an apple orchard changed the composition of the respective microbiomes, increasing the relative abundance of genes involved in the degradation of cellulose and hemicellulose (as a result of the increase of plant residues from the cover crops). The apple yield, as well as the soil organic C and total N contents, increased by the presence of the cover crops (Zheng et al. 2018).

Another example comes from agroforestry, where litter decomposition below tree stands is influenced by phosphorus availability, soil water content, temperature, pH, C/N ratio, and lignin and cellulose contents (Rachid et al. 2015). Ndaw et al. (2009) showed that mixing *Acacia mangium* and *Eucalyptus grandis* trees—instead of having mono-stands—affected soil productivity, with litter of the latter trees having a lower quality and being more recalcitrant than that of the former ones. One main factor that causes *A. mangium* to be a good soil conditioner is that it associates symbiotically with nitrogen-fixing *Rhizobium* and *Bradyrhizobium* spp., which results in the production of high (N)-quality litter. Decaying tissue of these trees, when deposited on the soil, promotes nutritional enrichment of the soil, with elevation of N content and acceleration of organic matter cycling. In the experiment by Rachid et al. (2015), the introduction of *A. mangium* in the *E. grandis* plantation incited a significant increase in the fungal richness, in the diversity index, and in the frequency of occurrence of several genera (not found in the mono-stands).

In *Eucalyptus*, 55% of the identifiable sequences, on average, were classified as belonging to the genus *Pisolithus*, and about 17% belonging to the genus *Scleroderma*. These two genera did not occur at a relevant frequency in *Acacia* treatment. In contrast, non-cultivable genera of the family Thelephoraceae accounted for one-third of all sequences identified in the *Acacia* treatment. These organisms represented only 5% of the populations in *Eucalyptus*. Only few genera thus showed a cosmopolitan distribution, remaining constant regardless of the vegetation cover, as, for example, observed with the genus *Tomentella*.

The most abundant fungal genera found in this work can establish ectomycorrhizal associations with the *Eucalyptus* and *Acacia* hosts. It is well known that many ectomycorrhizal associations are specific to the plant species and that these plants may attract a given fungal group via chemical signaling and physiological compatibility. Therefore, it was thought that, due to these associations, each tree species selected specific groups of fungi (Rachid et al. 2015). The inter-stand systems thus shifted to a higher fungal diversity status than the mono-stand one, which modulated soil quality.

The modulation of soil microbiomes also constitutes a promising approach to be further explored for biotechnology purposes. For instance, some pulp and paper companies are currently replicating the mixed plantation model of *Acacia* and *Eucalyptus* to test it for enrichment of the soil for biodegradation purposes. Such systems can be the key to achieving directed soil microbiome management, culminating in a powerful biotechnological tool. Through the system, microbial groups of interest are introduced or selected in the soil microbiomes, with the association of the two tree species—one of commercial interest and the other one of ecological interest—helping to foster the abundance of key microbial groups. Such microbial groups can be further explored with respect to the traits they have, for further ecological or bioprocess-oriented exploration. There is an urgent demand for prolonged research efforts, in order to fully unravel the potential of such new eco-bio-technological approaches.

25.5 THE IMPORTANCE OF AN INTEGRATED APPROACH THAT ENABLES THE MONITORING OF SOIL MICROBIOME MODULATION APPROACHES

According to Narayanasamy et al. (2015), agricultural management approaches that integrate the different "omics" methods should be the future standard, as they enable the large-scale characterization of soil microbiomes. The data obtained in an integrated way allow the "deconvolution" of structure–function relations, identifying the main members and functions. In the foregoing, we have provided arguments for a possible way forward in our quest to promote the environmentally friendly and sustainable agricultural use of soil. The development of smart farming approaches (see Section 25.2.2) is a key facet of our current and future endeavors. In brief, in the light of the in-field differences in soil conditions that many farmers experience in their cropping practice, there is a great need to work with on-the-spot, small-scale assessments of soil quality. Such fast, robotics (lab-on-a-chip)-supported approaches may allow for the almost instantaneous "reading" of soil status and relate this to the plant, with respect to growth and health status. Here, it is necessary to also include an assessment of soil abiotic factors (e.g., soil nutritional status, organic matter and water contents, pH), next to the soil microbiome and activity (e.g., by biosensors) analyses. Inclusion of lab-on-a-chip-gathered microbiome status and activity data into the equation that describes soil status at the within-field (and possibly individual plant) scale represents a true step forward in precision farming and management. With the availability of such tools, the farmer will be able to precisely map their fields, in a spatiotemporally explicit manner, with respect to fertility and sustainability.

However, to what extent are soil microbiome data useful in this context? The growing massive data sets that can be generated in soil microbiome monitoring approaches pose a range of methodological challenges. These lie in the interpretation of the data and their coupling to both the environmental and process metadata. Recently, powerful approaches, such as the generation of connecting conclusions based on machine learning (see also Chapter 19 in this book), have been used successfully. For instance, 30 independent soil bacteriome data sets, comprising 1,998 soil samples from 21 countries, were compared (Ramirez et al. 2018). Quite surprisingly, this meta-analysis indicated that rare taxa may be more important for structuring soil communities than abundant taxa and that these rarer taxa are better predictors of community composition than environmental factors. The latter were idiosyncratic across the data sets. In this way, these new approaches to data analysis can help to propose hypotheses about the factors that shape bacterial biogeography that have been neglected in former studies.

These novel approaches, in which artificial intelligence will play increasingly larger roles, will ultimately allow the integration of information from the soil metagenome, down to the proteome as well as the metabonome, into algorithms that predict the status of soil as the support matrix for crop development and production. At the microbiome side, it will allow the establishment of associations between the genetic potential and the final "collective phenotype" that drives the soil system. In this respect, the scale at which the assessments are performed is crucial to the precision of the monitoring and the subsequent decisions with respect to the farming approach.

25.6 FINAL REMARKS AND OUTLOOK

Our paradigms on the soil quality required to achieve agricultural sustainability (in which soil/plant quality and health are warranted) are evolving. In contrast to what was previously possible, we are now in a position to precisely characterize the soil microbiome and couple such findings to soil metadata that enable a robust assessment of soil status. This will even be possible at rather small scale, within a field or at the individual plant basis. Here, we argue that the soil microbiome is to be considered as a true "functional extension" of the host plant. Also, in a broad sense, the soil microbiome is the engine of soil functioning. The inclusion of data that describe the dynamic role of the microbiome in "smart and precision farming" models that safeguard and predict production is the key challenge here (Turner et al. 2013).

In the ideal smart farming approach, the farmer has precise knowledge with respect to any single site in his field, even at the level of individual plants. There are multiple ways to achieve this aim. For instance, a grid overlay over the field may assist in the monitoring program. The mesh of the grid that is applied will guide the sampling, which may change along development of the crop. It obviously will also depend on the methodological possibilities that are available. Here, both the developments in observation robotics, deep learning, sampling and observation techniques, and the ever-expanding advanced soil microbiome analysis methods, including biosensors, come into play.

Moreover, advancements in *in situ* microbiome modulation methods could eventually point to a way forward. It may be helped by using principles from synthetic biology, in which microbes may be designed that have desired trait combinations for agricultural applications. Interestingly, Toju et al. (2018) propose the use of interdisciplinary research strategies to optimize the functioning of the microbiome in agroecosystems. Bioinformatics tools now allow us to identify members and features of "core microbiomes," which can be used to organize uncontrollable dynamics of resident microbiomes. The integration of other techniques from non-biology disciplines, such as microfluidics, robotics, and machine learning, may provide new ways of harnessing the microbiomes in order to increase resource efficiency and resistance to stress factors by agroecosystems.

We are experiencing a great momentum in developments in the area of smart farming. The large multidisciplinary arsenal of techniques that is now available will enable understanding the functioning of soil microbiomes to promote sustainability in agriculture. A move towards significantly improved intelligent approaches will enable to finally have "smart farms" (or site-adapted agriculture). On the basis of the efforts to better monitor and improve microbiomes of soils and crop plants, microbiome engineering will be a key strategy that is integrated into the strategies in advanced agriculture future decades.

REFERENCES

Aboim, M.C.R., Coutinho, H.L.C., Peixoto, R.S., Barbosa, J.C. and A.S. Rosado. 2008. Soil bacterial community structure and soil quality in a slash-and-burn cultivation system in southeastern Brazil. *Appl Soil Ecol* 38, 100–108.

Blaser, M.J., Cardon, Z.G., Cho, M.K. et al. 2016. Toward a predictive understanding of Earth's microbiomes to address 21st century challenges. *mBio* 7, e00714–e00716.

Busby, P.E., Soman, C., Wagner, M.R. et al. 2017. Research priorities for harnessing plant microbiomes in sustainable agriculture. *PLoS Biol* 15, e2001793.

Carvalho, F.P. 2017. Pesticides, environment, and food safety. *Food Energy Secur* 6, 48–60.

Chaparro, J.M., Sheflin, A.M., Manter, D.K. and J.M. Vivanco. 2012. Manipulating the soil microbiome to increase soil health and plant fertility. *Biol Fertil Soils* 48, 489–499.

Delmont, T.O., Francioli, D., Jacquesson, N. et al. 2014. Microbial community development and unseen diversity recovery in inoculated sterile soil. *Biol Fertil Soils* 50, 1069–1076.

De Ridder-Duine, A.S., Kowalchuk, G.A., Klein Gunnewick, P.J.A., Smant, W., Van Veen, J.A. and W. de Boer. 2005. Rhizosphere bacterial community composition in natural stands of *Carex arenaria* (sand Sedge) is determined by bulk soil community composition. *Soil Biol Biochem* 37, 349–357.

Foo, J.L., Ling, H., Lee, Y.S. and M.W. Chang. 2017. Microbiome engineering: Current applications and its future. *Biotechnol J* 12, 1600099.

Garbeva, P., Postma, J. and J.D. van Elsas. 2006. Effect of above-ground plant species on soil microbial community structure and its impact on suppression of *Rhizoctonia solani* AG3. *Environ Microbiol* 8, 233–246.

Hartman, K., van der Heijden, M.G.A., Wittwer, R.A., Banerjee, S., Walser, J.-C. and K. Schlaeppi. 2018. Cropping practices manipulate abundance patterns of root and soil microbiome members paving the way to smart farming. *Microbiome* 6, 14.

Hobbs, P.R., Sayre, K. and R. Gupta. 2008. The role of conservation agriculture in sustainable agriculture. *Philos Trans R Soc Lond B Biol Sci* 363, 543–555.

Højberg, O., Schnider, U., Winteler, H.V., Sørensen, J. and D. Haas. 1999. Oxygen- sensing reporter strain of *Pseudomonas fluorescens* for monitoring the distribution of low-oxygen habitats in soil. *Appl Environ Microbiol* 65, 4085–4093.

Narayanasamy, S., Muller, E.E.L., Sheik, A.R. and P. Wilmes. 2015. Integrated omics for the identification of key functionalities in biological wastewater treatment microbial communities. *Microb Biotechnol* 8, 363–368.

Navarrete, A.A., Tsai, S.M., Mendes, L.W. et al. 2015. Soil microbiome responses to the short-term effects of Amazonian deforestation. *Mol Ecol* 24, 2433–2448.

Ndaw, S.M., Gama-Rodrigues, A.C., Gama-Rodrigues, E.F., Sales, K.R.N. and A.S. Rosado. 2009. Relationships between bacterial diversity, microbial biomass, and litter quality in soils under different plant covers in northern Rio de Janeiro State, Brazil. *Can J Microbiol* 55, 1089–1095.

Peralta, A.L., Sun, Y., McDaniel, M.D. and J.T. Lennon. 2018. Crop rotational diversity increases disease suppressive capacity of soil microbiomes. *Ecosphere* 9, e02235.

Poulsen, P.H.B., Hansen, L.H. and S.J. Sorensen. 2007. Biosensors to monitor soil health or toxicity. In: *Modern Soil Microbiol II*, J.D. van Elsas, J.K. Jansson, and J.T. Trevors (eds.), CRC Press, New York, NY, 435–454.

Rachid, C.T.C.C., Balieiro, F.C., Fonseca, E.S. et al. 2015. Intercropped silviculture systems, a key to achieving soil fungal community management in Eucalyptus plantations. *PLoS One* 10, e0118515.

Ramirez, K.S., Knight, C.G., de Hollander, M. et al. 2018. Detecting macroecological patterns in bacterial communities across independent studies of global soils. *Nat Microbiol* 3, 189–196.

Reganold, J.P. and J.M. Wachter. 2016. Organic agriculture in the twenty-first century. *Nat Plants* 2, 15221.

Scherr, S.J. and J.A. McNeely. 2008. Biodiversity conservation and agricultural sustainability: Towards a new paradigm of 'ecoagriculture' landscapes. *Philos Trans R Soc Lond B Biol Sci* 363, 477–494.

Schloter, M., Nannipieri, P., Sørensen, S.J. and J.D. van Elsas. 2018. Microbial indicators for soil quality. *Biol Fertil Soils* 54, 1–10.

Shen, Z., Ruan, Y., Chao, X., Zhang, J., Li, R. and Q. Shen. 2015. Rhizosphere microbial community manipulated by 2 years of consecutive biofertilizer application associated with banana *Fusarium* wilt disease suppression. *Biol Fertil Soils* 51, 553–562.

Sheth, R.U., Cabral, V., Chen, S.P. and H.H. Wang. 2016. Manipulating bacterial communities by in situ microbiome engineering. *Trends Genet* 32, 189–200.

Tillman, D., Balzer, C., Hill, J. and B.L. Befort. 2011. Global food demand and the sustainable intensification of agriculture. *Proc Natl Acad Sci USA* 108, 20260–20264.

Toju, H., Peay, K.G., Yamamichi, M. et al. 2018. Core microbiomes for sustainable agroecosystems. *Nat Plants* 4, 247–257.

Turner, T.R., Euan, K.J. and S.P. Philip. 2013. The plant microbiome. *Genome Biol* 20, 209.

Van Overbeek, L.S. and J.D. van Elsas. 1995. Root exudate-induced promoter activity in *Pseudomonas fluorescens* mutants in the wheat rhizosphere. *Appl Environ Microbiol* 61, 890–898.

Van Overbeek, L.S., van Elsas, J.D. and J.A. van Veen. 1997. *Pseudomonas fluorescens* Tn5-B20 mutant RA92 responds to carbon limitation in soil. *FEMS Microbiol Ecol* 24, 57–71.

Wolfert, S., Ge, L., Verdouw, C. and M.J. Bogaardt. 2017. Big data in smart farming - a review. *Agric Syst* 153, 69–80.

United Nations, Department of Economic and Social Affairs, Population Division. 2017. World Population Prospects: The 2017 Revision, Key Findings and Advance Tables. Working Paper No. ESA/P/WP/248.

Zheng, W., Zhao, Z., Gon, Q., Zhai, B. and Z. Li. 2018. Effects of cover crop in an apple orchard on microbial community composition network, and potential genes involved with degradation of crop residues in soil. *Biol Fertil Soils* 54(6), 743–759.

Glossary

ACC deaminase: An enzyme found in some bacteria and fungi that can cleave the compound 1-aminocyclopropane-1-carboxylate (ACC), which is the immediate precursor of ethylene in plants), into ammonia and alpha-ketobutyrate.

Accessory genes: Genes that belong to the so-called accessory gene pool, the pool of genes that are not part of the core gene pool in an organism.

Acetotroph: Microorganism that consumes acetate. For instance, acetotrophic methanogens that produce methane in soil according to the following reaction: $CH_3COOH \rightarrow CH_4 + CO_2$.

Acidophile: Microorganism that grows optimally under acidic conditions (low pH).

Acidophilic: Optimally active at low pH.

Acyl homoserine lactone (AHL): A chemical compound used by some Gram-negative bacteria as a signal of cell density in quorum sensing.

Acyl homoserine lactone (AHL) signal mimic: A compound that mimics the signaling function of acyl homoserine lactones (*see separate definition*) in quorum sensing.

α-Diversity: The biodiversity within a local habitat.

Adaptive radiation: Biological diversification due to divergent natural selection.

Advection: Removal of compounds (e.g., metals) through pore water movement.

Aerenchyma: Channels through which air can diffuse from plant shoots to roots, which are caused by cell wall and protoplast removal due to programmed cell death in plant roots (usually following flooding).

Aerobic: Living or active in the presence of oxygen.

Aerotropism: The growth or movement of organisms towards the air.

Agglomeration: The process of associating one case (or variable) with another one in an ordered way, based on the relative similarity of all individual and previously agglomerated subsets. Different approaches include simple linkage, complete linkage, and unpaired group mean association (UPGMA) algorithms.

AHL: See acyl homoserine lactone.

Alicyclic compound: Compound having a carbocyclic ring structure that may be saturated or unsaturated, but not a benzenoid or aromatic compound.

Algivore: (Obligate) consumer of algae.

Alkaliphile: Microorganism that grows optimally at high pH (alkaline conditions).

Alkaliphilic: Optimally active at high pH.

α Log-series index: Diversity index based on a log-serial species abundance distribution model.

Alternative hypothesis: A statistical hypothesis representing a statement of measured effect or difference.

Amensalism: The process in which one microbial group limits the survival of another one, for example, by the production of toxins.

Amoebae: Protozoan subgroup of polyphyletic origin, divided into naked and testate amoebae that feed and move on solid surfaces by the protrusion of parts of the cell forming pseudopodia.

Amplicon: DNA fragment synthesized using amplification techniques such as the polymerase chain reaction (PCR).

Anaerobic: Living or active in the absence of oxygen.

Anammox: Anaerobic oxidation of ammonium in which ammonium and nitrite are converted directly to nitrogen gas.

Anamorph: The asexual stage of a mitosporic fungus (*see separate definition*).

Antagonist (-ic): Microorganism that reduces the growth or activity of another microorganism.

Antagonism: An interactive strategy that involves the inhibitory action of one species on another.

Antibiosis: An antagonistic relationship between organisms in which one produces compounds, for instance, antibiotics, harmful to the others.

Antibiotic: Compound produced by a microorganism that negatively affects other organisms, being either –cidal (killing) or –static (growth-inhibiting).

Apoptosis (programmed cell death): An ordered process involving specialized cellular machinery that enables multicellular organisms to control cell number and eliminate cells that threaten the organism's survival.

Arbuscular mycorrhiza: The most ancient and widespread form of symbiotic mycorrhizal association between a wide range of plant hosts and fungi from the order Glomeromycota—characterized by the formation of branching structures or arbuscules (*see separate definition*).

Arbuscules: Repeatedly branched hyphal structures formed by glomeromycotan fungi within individual plant root cells—thought to be the main sites of nutrient exchange between arbuscular mycorrhizal fungi and their plant hosts.

Arbutoid mycorrhiza: Symbiotic mycorrhizal associations between fungi that are normally ectomycorrhizal and plants in the genera *Arbutus* and *Arctostaphylos* and the family *Pyrolaceae*, where extensive intracellular fungal proliferation occurs.

Archaea (formerly archaebacteria): One of the three main divisions of life, separate from the Bacteria and the Eukarya.

Archaeome: The entire archaeal community within a habitat.

Area under disease progress curve (AUDPC): An integrated measure of disease progress, which is generally nonlinear, for a variety of reasons such as rate of disease increase, source and amount of initial disease, and final level of disease.

Ascomycota: The largest fungal phylum, containing fungi that have a septate mycelium and produce their sexual spores inside sac-like structures called asci.

Associative interaction: The association of two organisms, such as a bacterium and a plant, that is not obligatory, but that can result in a mutual interaction, such as exchange of nutrients.

Autochthonous: Common, indigenous, inhabitant of an environment (e.g., soil) that tends to remain at relatively constant levels, despite fluctuations in local conditions, including organic matter levels.

Autogamy: A process of meiosis and fertilization that takes place within an organism.

Autoinducer: Signaling compound involved in quorum sensing (*see separate definition*), such as acyl homoserine lactones in Gram-negative bacteria and some peptides in Gram-positive bacteria.

Autophagy: Process of sequestering organelles and long-lived proteins in a double-membrane vesicle inside the cell, where the contents are subsequently delivered to the lysosome for degradation. Is part of everyday normal cell growth and development.

Autoregulation: The self-regulation of gene expression within a microbial (bacterial) population. Often, autoregulation is similar to quorum sensing (*see separate definition*).

Autoregulator: Compound involved in autoregulation.

Autotroph: Primary producer, able to fix carbon by photosynthesis.

Auxiliary metabolic gene (AMG): Metabolic gene inserted into a host genome by a mobile genetic element (e.g., a bacteriophage).

Auxin: Any of a group of plant hormones promoting cell growth. The primary auxin present in most plants is indole-3-acetic acid.

Bacteria: One of the three main divisions of life, separate from the Archaea and the Eukarya.

Bacterial artificial chromosome (BAC): A cloning vector that allows cloning of large DNA inserts and replicates as a bacterial chromosome.

Bacteriocin: Proteinaceous compound produced by bacteria, which is toxic to closely related bacteria.

Bacteriome: The entire bacterial community within a habitat.

Bacteriophage (or phage): Virus that infects a bacterium.

Bacterivore: (Obligate) consumer of bacteria.

Basidiomycota: A large fungal phylum containing fungi that have septate hyphae and bear their sexual spores on the outside of spore-producing bodies called basidia.

Bdelloplast: A specific complex structure that is essential in the life cycle of predatory *Bdellovibrio*, a bacterium that feeds on other bacteria.

β-Diversity: The biodiversity (e.g., species composition) along a gradient of habitats.

Binary data: Measurement scale where a variable may be scored with only one of two values, 0 or 1.

Bioassay: A test used *in vivo* or *in vitro* to determine the biological activity of a substance or of organisms against a standard culture of living cells (test organism).

Bioaugmentation: Addition of microorganisms to a contaminated site to enhance bioremediation or biodegradation.

Bioavailability: A measure of the physicochemical access of a compound to the biological processes of an organism. The less the bioavailability of a toxicant, the less its toxic effect on an organism or its ability to be degraded.

Biobarrier: Placement of bioactive materials that contain an active population of degradative microorganisms, in the path of a contamination source to reduce the spread of contamination.

Biocontrol agent (biological control agent [BCA]): An organism used to eliminate or control populations of pests, weeds, or pathogens or to decrease the damage caused by them.

Biodegradation: The use of organisms or enzymes to degrade pollutant compounds, such as recalcitrant compounds in soil.

Biodiversity (biological diversity): Collective variation at all levels of biological organization, from genetic variations within populations and species, species within communities, to communities that compose an ecosystem. Describes both the information content and how this information is distributed, in biological systems such as microbial assemblages or natural communities of other organisms.

Biofertilization: The use of biological agents (usually microorganisms) to increase the availability and/or uptake of nutrients for plants.

Biofertilizer: A material containing living microorganisms that can provide nutrients or increase the supply of nutrients to a plant.

Biofilm: An assemblage of microbial cells on a surface.

Biogeochemical cycling: The pathways by which compounds move through both biotic ("bio-") and abiotic ("geo-") compartments of an ecosystem.

Bioindicator: Biological parameter that reports on the phytosanitary status, for example, of soil.

BIOLOG: A commercial approach to assess carbon source utilization patterns of single microbial species or microbial communities that uses 96-well microplates containing different carbon sources and a tetrazolium-based redox dye technology. The method can provide a measure of metabolic potential.

Biological control (-agent): (Organism that enables) control of pests, pathogens, or weeds.

Bioluminescence: Emission of visible light by a biochemical process catalyzed by luciferase enzymes or by a living organism possessing luciferase enzymes.

Biomagnification: An increase in concentration of a chemical, often xenobiotic, in tissues of organisms during passage from the lower to the higher levels of a food chain.

Biopesticide: Organism that prevents or reduces the proliferation of plant pathogens.

Biopiling: A bioremediation process where contaminated soil is placed in piles and various treatments, such as aeration, fertilization, and/or bioaugmentation, can be applied to enhance the removal of contaminants.

Bioremediation: A biological process (usually involving microorganisms) that reduces or removes harmful chemicals or their toxicity in an environment.

Biosensor: A living organism that is used to measure the presence and/or the concentration of a certain chemical or physical condition. The response of a microbial biosensor is usually monitored by the expression of reporter genes.

Biosorption: The removal or recovery of free metal ions from a solution by a biosorbing material containing a variety of functional sites.

Biosparging: Injecting gaseous nutrients, such as air, oxygen, or methane, through the saturated zone of a contaminated soil to stimulate biodegradation or bioremediation.

Biostimulation: Addition of nutrient(s) to stimulate microbial metabolism at a contaminated site to enhance bioremediation or biodegradation rates.

Biosurfactant: A surface-active compound produced by microorganisms that can decrease the surface tension of water.

Biotrophic pathogen: A pathogen that feeds on another organism.

Bioventing: Injecting gaseous nutrients, such as air, oxygen, or methane, through the vadose zone of a contaminated soil to enhance biodegradation or bioremediation rates (*see biosparging*).

Bivariate correlation coefficient: A measure of association summarizing the relationship between two variables, for example, Pearson's r or Kendall's τ.

Bootstrapping: A set of methods used to estimate a confidence interval that is commonly applied in phylogenetic analysis to a dendrogram in order to indicate the consistency with which particular cases are associated at different branch points.

Box plot: A graph of a data set where the sample distribution is projected. Box plots typically employ a median line, a box that outlines the interquartile range, and "whiskers" that extend from the upper and lower ends of the box to the 90th and 10th percentile values, respectively.

Broad-host-range plasmid: A plasmid that can be transferred to and replicate in more than one microbial taxon.

Bromodeoxyuridine (BrdU) immunocapture: A method that relies on the incorporation of a thymidine nucleotide analogue, bromodeoxyuridine (BrdU), into growing cells during DNA synthesis. The newly synthesized DNA can then be isolated using immunocapture.

Brown-rot fungi: Fungi that decay wood, degrading most of the cellulose and hemicelluloses, but that are unable to decay lignin, leaving it as a more or less intact brown substance.

Browsing: Feeding by protists moving over biofilms of attached bacteria; depending on the locomotion of the predator, free-swimming, transient, or gliding browsers have been classified.

BTEX: An abbreviation for benzene, toluene, ethylbenzene, and xylene.

Bulk soil: The soil outside the rhizosphere.

Candidate division: Phylogenetic group, which is characterized only by directly obtained sequences with no cultured representatives and therefore has an unconfirmed status.

Canonical correspondence analysis (CCA): Type of constrained ordination technique (also collectively called "direct gradient analysis") that assumes that species have unimodal (i.e., Gaussian; bell-shaped) response curves to environmental gradients.

Capture probe: Probe linked to magnetic beads that specifically binds to single-stranded DNA and is used to "capture" DNA from solutions containing inhibitory substances, for example, humic substances.

Carbon cycle: The biological transformations of different forms of carbon.

Cardinal (measurement scale): Measurement scale where numbers are ordered and differences between numbered levels are equal throughout the length of the scale.

Carrying capacity: The amount of life that can be sustained in a given environment (e.g., soil) as defined by the number of habitable sites and available energy sources; the steady-state abundance of a species over time in an environment.

Catabolic diversity: The range, and evenness, of the catabolic functions that are expressed by microbial assemblages or communities.

Central tendency: A description of the middle subset of data values, typically summarized by the mean or median.

Cell-specific activity (vc): Metabolic rate per microbial cell in terms of mole substrate per time and cell.

Cellulase: An enzyme that cleaves β-1,4 links in cellulose (*see separate definition*).

Cellulose: A complex carbohydrate composed of long unbranched chains of glucose [(1,4)-linked-β-D-glucose] molecules that constitutes the chief part of the cell walls of higher plants.

Cesium chloride (CsCl) density gradient ultracentrifugation: High-speed centrifugation step used in DNA extraction protocols to separate DNA from proteins and cell debris and/or plasmid DNA from chromosomal DNA in a density gradient of CsCl.

Cetyl trimethyl ammonium bromide (CTAB): Detergent that complexes with both polysaccharide and residual protein that is used to remove inhibitory substances from crude DNA extracts.

Chaperone: A protein that facilitates folding of other proteins or keeps them from misfolding in a cell.

Charge-coupled device (CCD): A light-sensitive image sensor that stores and displays data for an image in such a way that each pixel (i.e., picture element) in the image is converted into an electrical charge, the intensity of which is related to light intensity or to a color in the color spectrum.

Chelating agent: A chemical compound that forms more than one coordinate bond with metals in solution.

Chemo(auto)troph: An organism that obtains its energy from the oxidation of chemical compounds, for example, nitrifiers.

Chemolithotroph: An organism that obtains its energy through oxidation of inorganic compounds and fixes carbon dioxide for cell biosynthesis.

Chemotaxis: Movement of an organism towards or away from a chemical compound.

Chimeragenesis: The formation of a DNA sequence chimera, or sequence anomaly.

Chitin: A polysaccharide consisting of a polymer of β-1,4-linked N-acetyl-D-glucose-2-amine that is a major constituent of fungal cell walls and of exoskeletons of insects and other arthropods.

Chitinase: An enzyme that hydrolyzes glycosidic bonds in chitin to form oligomers and dimers that can be taken up by the cell, providing sources of carbon and nitrogen.

Chytridiomycota: The most ancient of the fungal phyla, containing the only "true fungi" that produce motile, flagellate zoospores.

Ciliate: Protozoan subgroup consisting of members that swim and feed using hair-like cilia.

Cirri: Large organelles of heterotrichous ciliates formed from congregated cilia, which occur on the flattened underside of the cell to propel the organism, for example, on the surface of soil particles.

Clade: A branch in a cladogram (*see separate definition*) that describes the evolutionary relationships among organisms based on data from a cladistic analysis.

Cladistics: See phyletics.

Cladogram: A dendrogram derived from a phyletic analysis. Also called a phylogram.

Clamp connection: Lateral connection between two adjacent cells of a dikaryotic fungal hypha, which ensures that each cell contains two genetically dissimilar nuclei.

Classification: The act of defining subsets of items within a larger set based on one or more measurable variables.

Cloning vector: A genetic element, often plasmid based, which allows the insertion of DNA for introduction into a host cell.

Co-metabolism: Transformation of an organic compound by a microorganism that is not capable of using the substrate for energy or growth.

Commensalism: A unidirectional interaction between microorganisms, in which one population benefits, whereas the latter is unaffected, for example, production of vitamins by one population that is used to the sole benefit of the latter.

Community-level physiological profiling (CLPP): A method used to classify microbial communities based on the ability of the community members to use a series of substrates. See *BIOLOG* for an example.

Compensation concentration: Steady-state concentration of a specific gas in soil, at which the rate of production of that gas equals the rate of its consumption.

Compensatory mutation(s): A mutation that restores the fitness loss of another mutation elsewhere in the genome by epistatic interaction.

Competence (for transformation): The capacity to acquire exogenous DNA.

Competition: A negative interaction in which more than one partner requires a limiting (nutrient) factor, resulting in a situation where at least one of the partners is adversely affected (e.g., growth limited).

Competitive: "Involved" or "efficient" in competition.

Competitive PCR: A method to quantify target DNA molecules that relies on the addition of an internal standard, the competitive template, which is co-amplified by PCR using the same primers in competition with the environmental target DNA sequence. Comparison of the relative amount of the two templates is then used to estimate the target copy number in the sample.

Complementary DNA (cDNA): DNA synthesized from an mRNA template, using reverse transcriptase (yielding c (copy) DNA).

Complementation analysis: Analysis of the capacity of a gene to add or restore a particular function when introduced into an organism in which this function is not present or has been deleted, for example, by mutation.

Concentric circle plot: Use of concentric circles to represent one or more aspects of one or more genome sequences.

Conducive soil: A soil that allows the development of plant disease symptoms; the opposite of a suppressive soil (*see separate definition*).

Confocal laser scanning microscopy (CLSM): Type of microscopy based on a confocal laser, allowing an enhancement of resolution and a visualization of field depth.

Congruence: The situation where similar phylogenetic groupings are found based on more than one shared feature.

Conjugation: The transfer of DNA between bacterial cells via direct cell-to-cell contact.

Conjugative transposon: A transposable element capable of mediating its own excision, conjugal transfer, and integration.

Contig: Contiguous DNA sequence (generated from overlapping sequence reads).

Continuous variable: A variable that can assume any fractional value within some specified range, for example, temperature.

Convergent evolution: The development of similar traits in two taxonomically distinct groups with no direct common ancestor.

Cophenetic correlation: A measure of the association between the original resemblance matrix and one derived from a hierarchical dendrogram. Values close to 1.0 indicate that the dendrogram (or portion thereof) is a fair model of the more complex numerical associations described in the resemblance matrix.

Copiotrophic (organism): Organism that requires access to plentiful fermentable organic material for growth.

Core gene: Gene present in all members of a particular taxon.

Core microbiome: Microbiome minimally present across a set of habitats.

Correspondence analysis (CA): A constrained ordination technique that aims to explain the organization of one set of variables based on another set of explanatory variables. The technique has been referred to as a combination of ordination and multiple linear regression.

Cortex: The tissue of a root confined externally by the epidermis and internally by the endodermis.

Cosmid: A hybrid cloning vector containing elements of both plasmids and bacteriophage.

5-Cyano-2,3-ditolyl tetrazolium chloride (CTC): A colorless compound that is reduced by electron transport activity to red fluorescent insoluble CTC formazan. Used as a stain for actively respiring "viable" bacteria, for example, in environmental samples.

Cyclic lipopeptides (CLPs): Biosurfactant molecules (wetting agents) that facilitate root colonization by some microorganisms and may have a role in the biocontrol of plant pathogens.

Cytokinin: Any one of a group of plant hormones chemically related to the purine adenine.

Cytopharynx: A microtubular organelle that transports food vacuoles further into the cell in some ciliates.

Cytopyge, also called cytoproct: The cell anus of protozoa on the cell surface where the cell membrane making up the food vacuole rejoins the outer cell membrane and the food vacuole contents (nondigested food parts such as bacterial cell walls and organelles) are excreted into the surrounding medium.

Cytostome: Invagination in the cell membrane of some protozoa that forms food vacuoles containing collected food particles.

Dechlorination: Removal of chlorine from an organic compound by microorganisms.

Deletion: Loss of one base pair or a segment of DNA leading to a reduction in DNA length and in some cases loss of function.

Demultiplexing: Assignment of sequence reads to samples on the basis of barcodes.

Denaturing gradient gel electrophoresis (DGGE): Microbial community fingerprinting technique that relies on the separation of DNA molecules according to their change in conformation when exposed to an increasing gradient of denaturant (e.g., formamide and urea) during polyacrylamide gel electrophoresis.

Dendrogram: A tree-like graph that projects a summary of multidimensional associations among cases (or variables) in a multivariate analysis. In a rectangular view, horizontal "branches" that are connected by vertical lines are said to "cluster" together, being related in some way.

Descriptive statistics: Summary values that describe some aspect of a data set, for example, central tendency or dispersion of data.

Detritusphere: Area in soil where decaying material occurs.

Diazotroph: A nitrogen-fixing organism (bacterium or Archeon).

Dikaryon: Fungal hypha or mycelium in which each hyphal compartment contains two nuclei representing different mating compatibility groups.

Direct extraction (of DNA): Isolation of community DNA directly from soil without prior extraction of cells.

Direct viable count (DVC): A microscopic assay to enumerate viable cells based on their ability to grow by elongation when exposed to antibiotics inhibiting cellular division.

Disease-suppressive (soil): Soil in which a pathogen causes little or no damage to a host plant, or only transient disease symptoms, although the pathogen may persist in the soil.

Discrete variable: Variable that can assume only a limited set of unit values within a specified range.

Dispersion: The amount of variation in a data set.

Dissolved organic matter: Organic matter occurring in solution in soil water.

Diversity: A general term for describing the number and relative abundance of different categories of entities, for example, living species. May be measured by many indices that must take into account both richness and the relative abundance of categories in a sample. Abundance distributed evenly among categories results in high diversity.

Division cyst: A protozoan cyst that provides shelter during multiplication. It is formed by the mother cell turning into 2-16 offspring.

DNA microarray: An orderly arrangement (size <200 μm) of up to thousands of spots containing nucleic acids with known identities consisting of either cDNA (reverse-transcribed messenger RNA; 500 to ~5,000 bases long), 20-mer to 80-mer oligonucleotides, or peptide nucleic acids immobilized on a solid surface. The arrays are exposed to labeled DNA, and after hybridization, the abundance of hybridizing sequences is determined.

DNA restriction/modification: Bacterial defense mechanism against foreign DNA and/or phage integration that consists of a restriction enzyme that specifically cuts unmethylated DNA, and a methylase that adds a methyl group to the restriction enzyme target, rendering the DNA insensitive to the restriction enzyme.

Dormancy: A cellular resting state characterized by low metabolic activity, which is often a response of microbes to stress conditions such as starvation.

Drilosphere: Area in soil influenced by earthworms.

Dwarf cell: A cell (mostly bacterial) of minute dimensions, often less than a micrometer in diameter.

Ecochips: Functional gene arrays for specific detection and quantification of functional groups (guilds), for example, to monitor the expression of whole sets of genes in response to environmental changes.

ECO plates: Commercially available CLPP assays with a limited number of substrates (*see community-level physiological profiling*).

Ecospecies: A taxonomic species defined by its ecological characteristics that usually contains more than one interbreeding ecotype.

Ecosystem function: Any output resulting from the biota within an ecosystem. Often applied to goods and services delivered by an ecosystem of benefit to humans, for example, the production of biomass for food or the recycling of nutrients.

Ecotype: Subspecific form within a species, resulting from selection within a particular habitat and therefore genetically adapted to that habitat.

Ectendomycorrhiza: Mycorrhizal roots that exhibit some of the structural characteristics of both ectomycorrhizas and endomycorrhizas, in particular a high degree of intracellular fungal penetration.

Ectomycorrhiza: Ectotrophic mycorrhizal associations formed between long-lived, woody perennial plants and fungi predominantly from the *Basidiomycota* and *Ascomycota*.

Ectotrophic: Mycorrhizal association in which the fungus grows on the outside of the root, with no or only very limited intracellular penetration.

Edge (in correlation network): Correlational connection between network nodes.

Efficacy: Power or capacity to produce a desired effect, effectiveness. For example, used when defining how effective a chemical or biocontrol agent is at controlling pests, weeds, or diseases.

Electropherogram: A graphic recording of the separated components of a mixture of molecules produced by electrophoresis.

Elicitor: A surface structure or diffusible compound from a plant pathogen that induces the plant's systemic resistance to the pathogen as an active defense mechanism.

Endobacteria: Intracellular bacteria occurring inside roots or fungal hyphae.

Endocytobiotic: Biotic interaction involving the incorporation of an organism into the cytoplasm of another (eukaryotic) organism by endocytosis.

Endodermis: A single layer of cells enclosing the vascular tissue of the root.

Endophyte: A microorganism that lives within the plant, often without showing any symptoms of its presence.

Endoplant: "Within the plant." Term used to indicate the plant interior.

Endosphere: The area within the plant. Plant interior.

Endospore: Nongrowing cellular form used by some species of Gram-positive bacteria that allows for the genome of the cells to be protected during harsh conditions, such as those often occurring in soil. Once conditions become favorable again, it can convert back into a vegetative cell.

Endosymbiont: Symbiotic organism that lives within tissue or cells of another organism.

Engulfing: Enclosing large amounts of prey (bacteria), or large prey items in a food vacuole, but also gradual digestion of prey items too long to fit into a food vacuole (e.g., filaments).

Epifluorescence microscopy (EFM): A microscopic method used to view fluorescing objects, such as cells stained with a fluorescent dye. Fluorescence is detected using appropriate optical filters to select the correct wavelength of the incoming beam exiting the dye and to eliminate the incident wavelength from the emitted fluorescent light source (i.e., the specimen).

Epinastic curvature: Wilting or downward curvature of plant leaves, usually following flooding.

Epiparasite: Plant living parasitically, via fungal hyphae, off another organism.

Ergosterol: A sterol present in the cell membranes of yeasts and certain fungi.

Ericoid mycorrhiza: Mycorrhizal associations formed in three plant families, the *Ericaceae*, the *Empetraceae*, and *Epacridaceae*, all belonging to the *Ericales*.

Estimator (of community diversity or richness): Mathematical tool to describe diversity or richness in a single parameter.

Euclidian distance: A dissimilarity coefficient based on the average of one or more scores for the two cases being considered.

Eukaryvore (-ic): Consumer (protist) of whole other protists, fungi, and algae.

Evenness: A description of how evenly distributed individuals are in a community; a common measure of diversity.

Exonuclease: Enzyme that degrades DNA from one end.

Expression screening: The screening (analysis) of (meta)genomes for gene expression.

Extremophilic: (Organism) optimally active in extreme environments.

Extrusome: Membrane-bound structure that may be triggered to discharge its content, often toxic.

Factor analysis: An ordination technique similar to principal components analysis, but differing in that the unique variance (i.e., which is attributable to one or more variables) in a data set is separated from the common variance.

Fatty acid methyl ester (FAME): Fatty acid that has been chemically derivatized to a methyl ester prior to analysis by gas chromatography (GC) or GC in combination with mass spectrometry (GC-MS). Used as a signature molecule for characterization of the composition of microbial communities.

Fick's law: A modification of Ohm's law to describe the movement (flux) of gases across a surface. The law states that flux is directly proportional to the gradient across the surface and inversely proportional to the resistance to flow.

Filter feeding: See interception feeding.

Flagellate: Protozoan subgroup of polyphyletic origin of which the members possess one or more long, slender flagellae used for locomotion and catching prey.

Flavonoid: A specific class of plant metabolites, encompassing compounds that affect gene expression in some microorganisms. For instance, several compounds that induce nodulation genes in rhizobia are flavonoids.

Flow cytometry (FCM): A technique used to detect and/or quantify thousands of suspended bacterial cells within seconds as they are suspended in a narrow, rapidly flowing stream of fluid passing a laser beam. Different properties of individual cells can be detected by different detectors, that is, forward scatter (size), side scatter (shape/granularity), and fluorescence properties.

Fluorescence-activated cell sorting (FACS): Use of flow cytometry (*see separate definition*) to physically sort individual cells based on their specific light scattering and/or fluorescent characteristics.

Fluorescence *in situ* hybridization (FISH): Use of fluorescently labeled oligonucleotide probes specific for species or phylogenetic groups, in combination with fluorescence microscopy, to detect and quantify single cells of interest.

Fluorescence *in situ* hybridization – Secondary ion mass spectrometry (FISH-SIMS): A technique to simultaneously identify microorganisms using fluorescently labeled DNA probes and epifluorescent microscopy (FISH, *see separate definition*), and to determine the isotopic composition for functional analysis using secondary ion mass spectrometry (SIMS).

Fluorescent antibody (FA): Covalent modification of an antibody molecule with a fluorescent dye that enables the antibody to be visible under fluorescent light and that can be used to detect specific cells in environmental samples.

Fluorescent dye: Dye that fluoresces under light of specific wavelength, often ultraviolet.

Formulation: The use of inert materials in a product to "pack," for instance, biological or chemical pesticides.

Fosmid: Cloning vector derived from bacteriophage for large DNA inserts (up to approximately 45 Kb) used, for example, for the construction of metagenomic libraries.

Fractal: Object with fractional dimension that possesses self-similarity, composed of several parts, each of which is a small-scale copy of the whole. For example, it can be applied to describe soil texture.

Free-air CO_2 enrichment (FACE): A type of experiment for assessing the impact of increased CO_2 partial pressures on the ecosystem, using the controlled release of CO_2 from air tanks into the free air.

Fruiting body: A multicellular structure in fungi, or sporocarp, on which spore-producing structures (e.g., basidia, asci) are borne as part of the sexual phase of the fungal life cycle. When visible to the naked eye, especially fruiting bodies of a more or less agaricoid morphology, they are often referred to as mushrooms. Also occur in the myxobacteria (*see separate definition for myxospore*).

Functional core microbiome: Core microbiome defined by consistency of function (across samples or habitats).

Functional diversity: The occurrence and distribution of physiological and metabolic traits among members of a community. Can also take into account numbers of different functional groups (guilds) present.

Functional module: Structural component of a mobile genetic element encoding specific transfer and/or stable maintenance function(s) of the element.

Functional potential: Sum of all protein-encoding genes in a single cell.

Functional redundancy: The presence of a functional capacity in different (bacterial) taxa.

Fungicide: Chemical substance used to kill fungi.

Fungistasis: Inhibition of fungal spore germination or growth in a soil.

Fungivore: (Obligate) consumer of fungi.

Fungome: The entire fungal community within a habitat.

GacS/GacA signal transduction pathway: A global regulatory system in some bacteria that regulates gene expression as a response to environmental signals, directing the expression of a range of genes and operons in the cell.

γ-Diversity: Biodiversity over a geographic region comprising many different habitats.

G+C mole percentage: The amount of guanine plus cytosine bases in the DNA as a fraction of the total bases.

General linear model (GLM): A linear model where a variable responds to one or more specified fixed variables, that is, discrete variables that are manipulated in a defined way, such as fertilizer levels. The model may also include consideration of up to one random effect variable, that is, a discrete variable, the values of which are selected at random from a larger population.

Gene targeting: Targeting of specific genes to direct, for example, the metagenomic screening towards desired products.

Genetic hub: Connection point for horizontal gene transfer among organisms in a network or habitat.

Gene transfer agent (GTA): Bacteriophage carrying a fragment(s) of a host bacterial genome. These phage do not have the ability to productively infect bacterial cells, but instead allow the transfer of bacterial genes to recipient genomes.

Generalist: Organism that is able to survive in diverse environments and different geographical locations.

Generalized transduction: The process in which a phage incorporates host DNA instead of phage DNA in a transducing phage particle which then infects a recipient cell, resulting in transfer of host DNA to a recipient bacterium.

General suppression: Suppression of plant disease by soil without a clear link to function.

Genetic drift: Population dynamics of selectively neutral alleles (those defined as having no positive or negative impact on fitness). Drift is observed most strongly in small populations and can, together with natural selection, alter the frequencies of alleles.

Genomic island: Genetic element, with a length between 10 and several hundreds of kilobases, that has been inserted into a chromosome after a lateral transfer event. The heterologous origin is generally evidenced by a G+C content, which is different from that of the remaining bacterial chromosome, and by the presence of mobility genes (i.e., genes involved in transposition, transduction, or conjugative transfer).

Genotoxicity: The amount of damage a genotoxin (agent that damages DNA) can cause to a DNA molecule.

Gibberellin: Any one of a group of plant growth regulatory hormones related in structure to the compound gibberellic acid.

Gibbs free energy (ΔG): The energy change of a reaction that can be used for doing work, for example, producing microbial biomass.

Glomeromycota: A fungal phylum comprising the arbuscular mycorrhizal fungi that are characterized by the formation of dichotomously branched arbuscules.

Gower's index: A general similarity coefficient for use on multivariate data sets containing a mixture of variables measured on some combination of binary, ordinal, and/or cardinal scales.

Gram-negative: A term used to classify bacteria based on their red color after use of Gram stain reagents that stain the cell wall.

Gram-positive: A term used to classify bacteria based on their blue/violet color after use of Gram stain reagents that stain the cell wall.

Great plate count anomaly (GPCA): The observation that there are often 100 to 1,000 times more microorganisms that can be directly counted in environmental samples, such as soil, compared to the number that can currently be cultivated.

Green fluorescent protein (GFP): A protein encoded by the *gfp* marker gene, originally isolated from the jellyfish *Aequorea victoria*, that emits green fluorescence upon excitation with blue light, thus enabling the detection of specific cells, for example, in soil.

Greenhouse gas: A gas, usually of anthropogenic or microbial origin, involved in causing the greenhouse effect (*see separate definition*), for example, CO_2, CH_4, N_2O, and halocarbons.

Green manure: Manure from cover crops plowed under and incorporated into the soil.

Guild: The members of a community that perform the same specific function (functional group).

Gulping: Enclosing whole protists through an oral opening; usually after paralyzing the prey by extrusomes (membrane-bound structures that may be triggered to discharge their content, often toxic).

Habitable pore space: The (pore) space in soil that is favorable to microbial life.

Half-life (of messenger RNA): Time period in which 50% of a given messenger RNA is degraded by RNase activity. This time can vary from seconds to hours depending on the corresponding gene and organism.

Halophilic: Organism that is optimally active at high salt concentrations.

Hartig net: Branched network of hyphae growing between the cortical cells of an ectomycorrhizal root that maximizes contact between the mycorrhizal fungus and its plant host and is thought to be the interface for exchange of nutrients and carbon between the host and the fungus.

Haustoria: Specialized feeding structures of a fungal pathogen that are capable of direct penetration into, and nutrient absorption from, a host plant.

Heat shock: The sudden exposure of cells to a temperature rise, inciting a specific physiological response in which heat shock proteins are involved.

Herbicide: A chemical substance used to kills weeds.

Heterogramic: (Gene transfer) across the Gram barrier.

Heterologous expression system: Recognition of foreign bacterial transcriptional and translational signals for the expression of foreign proteins.

Heterotroph (-ic): An organism that uses organic compounds as an energy source.

Hierarchical cluster analysis: Partitioning of a data set into subsets (clusters). Ideally, the objects (e.g., samples, organisms) in each subset share more common traits or higher similarities within each other than objects belonging to different clusters as determined by a defined distance measure (e.g., Euclidean distance).

High-pressure liquid chromatography (HPLC): Method for chromatographic separation of molecules, such as proteins, through a column under high pressure.

Holobiont: System comprising a (plant) host and its associated organisms.

Homologous recombination: Recombination that promotes the pairing between identical or nearly identical DNA sequences and the subsequent exchange of genetic material between them.

Homology-facilitated illegitimate recombination (HFIR): A process in which short one-sided stretches of homology between donor and recipient genomes leads to strand exchange and recombination events that continue into regions of lower similarity.

Homoplasy: Events such as convergent evolution, parallel trait acquisition, and reversion that can significantly distort perceived evolutionary relationships.

Horizontal gene pool: Pool of genes that are horizontally moved across diverse hosts.

Horizontal gene transfer (HGT): The direct transfer of genes from one organism to another one, that is, not through vertical inheritance.

Hydrogenotroph: Microorganism that consumes hydrogen, for example, hydrogenotrophic methanogens, which produce methane in soil according to the following reaction: $4 H_2 + CO_2 \rightarrow CH_4 + 2 H_2O$.

Hydrophobins: Small, secreted proteins that are highly hydrophobic and appear to play different roles in fungal differentiation.

Hydropriming: Seed priming (soaking of seeds followed by drying back to the original water contents), involving the addition of free water.

Hydroxylation: A chemical or biochemical process in which one or more OH groups are added to a molecule.

Hyperaccumulator (plant): A plant that is able to accumulate large concentrations of certain metals from water or soil.

Hyperparasitism: The parasitism of an organism which itself is a parasite.

Hyperthermophilic: Extremely thermophilic (heat-loving) organism, that is, an organism that is optimally active under extremely warm conditions (often >100°C).

Hyphal growth unit: The unit of growth (often a hypha) of a fungus that grows by hyphal extension.

Hypovirulence: Reduced ability of selected isolates within a population of a fungal pathogen to infect, colonize, kill, and/or multiply on susceptible host tissues.

Hypovirulent: Fungus with reduced pathogenic or infectious potential.

Hysteresis: In soil–water systems, used to explain the different moisture release curves described by a soil during (a) wetting and (b) drying, due to pores often being wider than their pore-neck openings.

Ice nucleation activity: The ability to catalyze ice formation in supercooled water based on the presence of a bacterial-encoded protein that serves as a nucleus for ice crystal formation and that can lead to frost damage in plants.

Illegitimate recombination: Recombination between nonhomologous DNA sequences that are RecA independent.

Immunofluorescence staining: Immunochemical detection technique using fluorescent antibodies (*see separate definition*) that bind to antigenic determinants on target cells. Alternatively, an indirect assay can be performed, in which a fluorescently labeled secondary antibody recognizes the primary antibody and functions as a reporter for the presence of target cells.

Immunomagnetic capture: The use of polyclonal or monoclonal antibodies immobilized on magnetic beads to physically capture specific bacterial cells.

Immunomagnetic separation: Separation of cells based on immunomagnetic capture (*see separate definition*) on magnetic beads.

Indel: A portmanteau of DNA insertions and deletions, which are mutations that may cause frameshifts in the DNA sequence.

Indirect extraction: Extraction of DNA from cells that have been previously separated from the soil matrix.

Indole acetic acid (IAA): A plant growth hormone (auxin type) that is also produced in some microorganisms.

Induced systemic resistance (ISR): A resistance mechanism (typically towards fungal pathogens) in plants initiated by environmental compounds (signals) from other sources than the pathogen. Encompasses a broad range of phenomena elicited by nonpathogenic organisms that allow the plant to restrict the pathogen to certain tissues.

Infectious potential (infectivity potential): The diversity of pathogenic populations present in soil (*see separate definition for inoculum potential*).

Infectious potential unit (IPU): Volume of soil needed to cause disease in 50% of plants that can be measured in a bioassay based on successive dilutions of naturally infested soil in disinfested soil and cultivation of a highly susceptible plant.

Inoculum potential: Growth energy available to a pathogen to enable it to infect a plant host. This term refers to the fact that the presence of the inoculum, although necessary, is not sufficient to explain disease, as it is affected by environmental conditions.

Insecticide: A chemical substance used to kill insects.

Insertion: The addition of one to several base pairs of DNA into the genome.

In silico mining: Analysis of directly sequenced cloned DNA fragments.

Integrative conjugative element (ICE): Element that excises by site-specific recombination into a circular form, transfers by conjugation, and subsequently integrates into the genome of the new host, for example, conjugative transposons, integrative plasmids, and genomic islands.

Integrative plasmid: A plasmid with the capacity to integrate into the host bacterial genome.

Integron: A genetic structure that allows the directed insertion of exogenous DNA, so-called gene cassettes (e.g., antibiotic resistance genes), by site-specific recombination.

Interception feeding: Feeding on suspended bacteria, either passively (filter feeding) or actively by the moving predator (raptorial or direct interception feeding).

Intergenic spacer (IGS): DNA sequence located between genes or operons with no known function. In microorganisms, the ribosomal RNA intergenic spacer region has been used for microbial community fingerprinting and phylogenetics.

Intrinsic marker: A natural DNA sequence or phenotype that serves as a signature for a particular organism or group of organisms.

Intron: A specific DNA sequence that intervenes between coding regions of genes in eukarya, but that is excised after transcription and not included in the resulting messenger RNA.

Inversion: Reversal of the normal order of a particular DNA sequence, at the level of nucleotides or genes.

Ionophore: A molecule that creates an ion channel in a cell wall.

Island biogeography: Ecological concept that describes evolution in islands to be determined by (parallel) events of mutation selection and migration.

Isotope ratio mass spectrometry (IRMS): An analytical technique used to measure the mass-to-charge ratio of ions, specifically designed to determine the isotopic composition of elements within a sample.

Kill-the-Winner (model for bacteriophage–host interactions): Model that assumes that phages will primarily infect and lyse hosts that are ecologically successful and increasing population density.

K_m: A kinetic parameter, giving the substrate concentration at which the reaction rate reaches half of its maximum rate (V_{max}; *see separate definition*); valid for hyperbolic enzyme kinetics (Michaelis–Menten kinetics), but also used for characterizing microbial metabolic reactions in culture and in the environment.

***K*-mer:** DNA fragment of defined size used to create de Bruijn graphs.

Knockout mutant: A mutant organism in which the function of a particular gene has been eliminated, for example, by deletion or site-directed mutagenesis.

Koch's postulates: Set of criteria (formulated by Koch in 1884) that must be fulfilled to establish a causal relationship between pathogen and disease: (i) the microorganism must be detectable in the infected host at all stages of disease, (ii) it must be isolated from the diseased host and grown in pure culture, (iii) if susceptible healthy hosts are infected with the microorganism, disease symptoms must occur, (iv) the microorganism must be re-isolated from the diseased host and be equal to the original microorganism.

K-Strategist: Organism that can metabolize and grow with a low supply of nutrients as well as in relatively crowded situations.

Land farming: A bioremediation process in which contaminated materials are spread on soils that are exposed to conditions favorable for the degradation or detoxification of the contaminants.

Lichen: Symbiosis between a fungus and a photosynthetic partner that can be either a green alga or a cyanobacterium.

Lignin: A group of high-molecular-weight amorphous compounds, comprising polymers of phenylpropanoid compounds, that is a component of many plant cell walls and that is highly resistant to chemical and enzymatic degradation, providing strength to plant tissues.

Lignocellulose: A major chemical component of wood composed of covalently linked molecules of lignin (*see separate definition*) and cellulose (*see separate definition*).

Linear mixed model (LMM): A linear model with more than one random effects variable.

Linear regression: The process by which a single equation is derived to predict the values of one variable by one or more other variables.

Lipopolysaccharide (LPS): A molecule consisting of carbohydrate and lipid in the outer wall layer (membrane) of Gram-negative bacteria.

Locally acquired resistance (LAR): A resistance mechanism (typically towards fungal pathogens) in plants initiated during a limited (local) infection by the pathogen in specific tissue.

Lysogenic conversion: Insertion of a phage genome into the genome of the host.

Lysogenic phage: Phage that has as its strategy to insert into the host genome.

Lysogeny: The presence of an integrated copy of a bacteriophage genome in a bacterial genome (a prophage), which may give rise to cell lysis upon prophage induction.

Maintenance coefficient (m_E): Metabolic rate (*compare cell-specific activity*) needed to maintain microbial biomass under given conditions, in terms of mole substrate per time and unit of microbial biomass.

Maintenance energy: Same as the maintenance coefficient (*see separate definition*), but in terms of metabolic energy consumed per time and energy content of the biomass.

Matrix-assisted laser desorption/ionization and time-of-flight (MALDI-TOF): A mass spectrometric method for analyzing biomolecules such as proteins and nucleic acids. The sample is mixed with a matrix, ionized with a laser beam, and the time the ionized molecules travel through a field-free region is proportional to their mass-to-charge (*m/z*) ratios.

Mantle: A sheath-like structure of fungal hyphae enclosing an ectomycorrhizal root tip.

Marker gene: DNA sequence introduced into an organism conferring a specific genotype or phenotype allowing it to be specifically detected, for example, in environmental samples.

Matric potential: A water potential component, always of negative value, resulting from capillary, imbibitional, and adsorptive forces in soil.

Maximum likelihood (ML): An algorithm that estimates a parameter based on a predetermined model. In phyletics, used to model evolution in multiple ways, whereby the classifier can specify the types and relative frequencies with which different nucleotide or amino acid changes occurred in the past.

Maximum parsimony: A mathematical algorithm that identifies the minimal number of changes required to explain the current variation in a data set.

Mean: The average value of a sample or population.

Median: The middle value of a sample or population.

Melanin: A range of black or brown pigments of amorphous structure.

Membrane potential dye: A colored or fluorescent dye that either accumulates in the cell or is excluded from a cell, depending on the cellular membrane potential and that is used as a viability stain.

Mesophile (-ic): A microorganism able to grow in the temperature range of 20°C–50°C; optimal growth often occurs at about 37°C.

Messenger RNA: Also referred to as mRNA. A form of RNA that carries the genetic code for a particular protein from the DNA to a ribosome and acts as a template, or pattern, for the formation of that protein.

Metabiosis: Successive conversion of a substrate to another one by organisms in synergy or succession.

Metabolome: Set of collective low-molecular-mass metabolites in a given (micro)organism.

Metabonome: The collective metabolites in a sample.

Metabonomics: Study of the collective (low-molecular-mass) metabolites in a system.

Metagenome: The collective genomes of a total community in a given habitat.

Metagenome-assembled genome (MAG): Genome assembled for metagenomics reads.

Metagenomics: Study of the collective genomes (metagenome, *see separate definition*) from a given sample, such as a soil sample.

Metaphenomics: (Functional) outcome of genetic diversity in a metagenome.

Metaproteome: Collective proteins from all organisms present in a biological system.

Metaproteomics: Study of collective proteins from all organisms present in a biological system.

Metatranscriptome: Collective messenger RNA from all organisms present in a biological system.

Metatranscriptomics: Study of collective messenger RNA from all organisms present in a biological system.

Methanogen: Microorganism producing methane as the energy-yielding reaction for growth, generally belonging to anaerobic archaea.

Methanogenesis: The production of the greenhouse gas methane by methanogenic archaea.

Methanotroph: Microorganism that oxidizes methane as an energy-yielding reaction for growth, usually belonging to aerobic bacteria that are common in soil, but also anaerobic archaea that are common in marine sediments.

Methylotrophic (organism): Methanol-consuming organism.

Microarray analysis: Technology that allows simultaneous measurement of genes, transcripts, or proteins for up to tens of thousands of genes.

Microbial-associated molecular pattern (MAMP): Molecule that enables an organism to interact with a host, increasing its fitness.

Microbial diversity: A measure of the variety of microorganisms in a community, based on one of several mathematical formulae that account for both numbers of species and numbers of individuals within species. High diversity results from high numbers of species (*see separate definition for richness*) and an even distribution of numbers within species (*see separate definition for evenness*).

Microbial loop: A term used to describe the turnover of carbon and nutrients in food chains via consumption of organic matter by bacteria that are in turn rapidly consumed by protozoa, excreting nutrients and CO_2. The released nutrients can then be utilized by other microorganisms.

Microbiome: The entire microbial community within a habitat.

Microcolony: A small assemblage of microbial cells, often containing tens to hundreds of cells.

Microcolony assay: A technique, whereby active, viable cells are identified by their ability to perform a limited number of cell divisions when placed on a suitable growth medium.

Microcosm: A laboratory system used to simulate biotic and abiotic components of a particular environment.

Microdiversity: Diversity of phylogenetically closely related (below species level) microbial populations, which are ecophysiologically different and occupy different niches.

Microsite: A microscopic area with physical and ecological characteristics that distinguish it from its immediate surrounding area.

Mineralization: Catabolism of an organic compound into inorganic components, such as water, carbon dioxide, and sometimes salts, by microorganisms.

Mitosporic fungi: Ascomycetous fungi that produce conidia but whose sexual stages are absent, rare, or unknown; previously referred to as the *Deuteromycota* or "Fungi imperfecti."

Mixotroph (-ic): Organism (often a protist) that can switch from autotrophy to heterotrophy (consumer).

Mobile genetic element: A genetic element that has the capacity to move within the same replicon, or between different replicons within the cell, and/or from cell to cell.

Mode: The most frequently occurring score in a distribution.

Molecular marker: DNA sequence that enables specific detection of an organism.

Monophyletic: A group of species that all have a single common ancestor.

Monotropoid mycorrhiza: Symbiotic mycorrhizal association between fungi and achlorophyllous plants in the family *Monotropaceae*.

Most probable number (MPN) method: Technique for enumeration of microorganisms, for example, in soil, on the basis of multiple end-point dilutions and statistical inference of estimated cell numbers. The number of positive observations in different sample dilutions is entered in an MPN table (available as a computer program), to obtain an estimate of the number of microorganisms in the sample.

MRR-like estimators: Species richness estimators based on mark, release, and recapture methods of wildlife ecology, where the total number of an organism cannot be directly measured.

Mucigel: Gelatinous layer that covers the root surface.

Mucilage: A high-molecular-weight mixture of various polysaccharides secreted from the root cap and border cells.

Multi-photon laser scanning microscopy (MP-LSM): Fluorescence laser scanning microscopic approach allowing for deep specimen imaging (up to hundreds of μm) and for generation of 3D images without out-of-focus light and with high contrast. Detection requires simultaneous excitation of the fluorescent dye by two or three low-intensity photons in order to generate fluorescence.

Multivariate: A data set or analysis technique considering measurements taken on more than one variable.

Multivariate analysis: An analysis that attempts to simultaneously examine the behavior of more than one dependent variable. Largely used in community ecology to search for patterns of species' responses to environmental gradients.

Mutation: A change in DNA that can result in loss of one or more functions. Mutations can occur spontaneously or be caused by many factors including environmental ones such as radiation and mutagenic chemicals.

Mutator (phenotype): Bacterial strain with elevated mutation frequency compared to other members of the population.

Mutualism: An interaction between organisms in which the partners receive mutual benefits (*see separate definitions of symbiosis, synergism, and syntrophy as examples*).

Mutualistic: Pertaining to interaction in which the partners receive mutual benefits from the interaction (*see separate definitions of symbiosis, synergism, and syntrophy as examples*).

Mycobiont: The fungal partner in a symbiotic association.

Mycoheterotrophic (plants): Plants that are parasitic on fungi, exploiting them as their principal source of carbon. This carbon may often be provided by a different autotrophic plant connected to the same fungal mycelium.

Mycoparasitism: Parasitism of one fungus by another fungus, whereby the parasite gains some or all of its nutrients while conferring no benefit in return.

Mycorrhiza: Symbiotic relationship between plant roots and certain groups of fungi.

Mycorrhiza/mycorrhizal fungus: Fungus that forms a symbiotic association with plant roots.

Mycorrhizosphere: The soil volume influenced by the combined activities of a root and its associated mycorrhizal hyphae.

Mycosphere: Area in soil influenced by fungi.

Mycosphere competence: Ability to survive and/or grow in the mycosphere.

Mycota: The classical taxonomic kingdom of true fungi, sometimes referred to as the Eumycota.

Mycovirus: Virus that infects fungi.

Myxospore: Complex structure in the life cycle of myxobacteria, consisting of a rounded cell with a thick cell wall that forms from a vegetative myxobacterial cell when nutrients are scarce; a process mediated by contact signaling.

***N*-Acyl homoserine lactone:** Chemical compound used by some Gram-negative bacteria as a signal of cell density (*see also separate definitions for quorum sensing and acyl homoserine lactone*).

Naked amoebae: Amoebae lacking an outer test or shell.

Natural transformation: The physiologically regulated uptake of extracellular (plasmid or chromosomal) DNA into transformation-competent bacteria, that is, bacteria that have reached the state of competence (*see separate definition*).

Necrosis: The death of cells or tissues, usually resulting from acute cellular injury.

Necrotrophic: Strictly, feeding on dead cells and tissues; however, the term is also used to distinguish *necrotrophic* pathogens (those killing their hosts before feeding on the dead tissues) from *biotrophic* pathogens that feed on living host tissues.

Neighbor joining: An algorithm for modeling an evolutionary tree based on the minimum evolution criterion; that is, the topology that gives the least total branch length is preferred at each step of the algorithm. Neighbor joining is a bottom-up clustering method usually used for trees based on DNA or protein sequence data; the algorithm requires knowledge of the distance between each pair of taxa (e.g., species or sequences) in the tree.

Nernst equation: A mathematical description of electrode behavior, relating the total potential (E, mV) developed between sensing and reference electrodes to a range of factors such as electrode type, temperature, pressure, charge, and "activity" of the ion to which the electrode is responding.

Neutralism: Lack of interaction between microbial populations, for example, due to spatial separation, the inability to contact or sense each other, or lack of growth.

Niche: An ecological role of an organism in a community, or a space in an ecosystem that is possible to be inhabited. For example, microhabitats in soil.

Niche construction: Modulation of conditions by an organism, resulting in an established niche.

Niche exclusion: The principle that a niche can only support one type of organism (species, genotype, or ecotype).

Nitric oxide (NO): A nitrogen oxide that plays an important role in the chemistry of the atmosphere and is produced mainly by nitrifying and denitrifying bacteria.

Nitrification: The oxidation of reduced forms of nitrogen to nitrate.

Nitrifier: A microorganism involved in the transformation of reduced forms of nitrogen to nitrate.

Nitrogen cycle: The cyclic steps involved in the transformation of different forms of organic and inorganic nitrogen.

Nitrous oxide (N_2O): A greenhouse gas (*see separate definition*) produced mainly by nitrifying and denitrifying bacteria.

Node (in correlation network): Defined OTU or feature in the network.

Nod factor: A molecule produced in some bacteria, such as rhizobia, serving as a plant-responsive signal during nodule (*see separate definition*) formation.

Nod genes: Nodulation genes associated with the production of nodulation factors (*see separate definition for Nod factor*) in some bacteria, such as rhizobia.

Nodule: A structure formed within plant tissue, for example, in roots of legumes, during invasion by some bacteria, such as rhizobia.

Nominal: Measurement scale where names represent mutually exclusive categories.

(Non-metric) multidimensional scaling (NMDS): Distance-based ordination method that assumes that dissimilarity of samples is monotonically related to ecological distance. An MDS algorithm starts with a matrix of item–item similarities, then assigns a location of each item in a low-dimensional space.

Nonparametric: A statistic or test that does not assume one or more characteristics about the underlying distribution of data.

Null hypothesis: A statistical hypothesis representing a statement of no effect or no difference.

Numerical taxonomy: See taxometrics.

Nycodenz™: A density gradient material that is used to separate bacteria and archaea from soil particles, fungal hyphae, and debris during centrifugation.

Oleophilic (fertilizer): A fertilizer that adheres strongly to oil or oily substances.

Oligotrophic (environment): Environment with low nutrient levels that is not conducive to high population densities and activities of microorganisms.

Oligotrophic (organism): Organism that can metabolize and grow under conditions of scarcity of resources (i.e., under nutrient limitation) with a relatively low growth rate.

Omnivore: Consumer (protist) of eukaryotes and bacteria.

Open reading frame: A potential gene identified on the basis of the DNA sequence beginning with a start codon and ending with a stop codon.

Operational taxonomic unit (OTU): Similarity group based on the numerical analysis of genotypic or phenotypic data; functional group or grouping at different levels of phylogenetic relatedness.

Orchid mycorrhiza: Mycorrhizal fungus associated with orchids.

Ordinal: Measurement scale where numbers are ordered, but differences between levels are not necessarily equal across the measurement scale.

Ordination method: The collective term for multivariate techniques that arrange samples with respect to one or more axes in a diagram on the basis of their data on species composition. Commonly used in macroecology, such techniques have been recently employed in molecular microbial community fingerprinting.

Ordination plot: A graph that projects a summary of multidimensional associations among cases (or variables) in a multivariate analysis. Though it looks like a simple scatter plot, the axes represent composite variables (or cases), and individual cases (or variables) plotted close to one another can be classified as related in some way.

Osmotic potential: The potential energy of a solution relative to a pure water reference. Thus, in soil, osmotic potential is always negative or zero, depending on the amount of solutes present.

Osmotrophy: Absorbing nutrients through the cell membrane; osmotrophs are commonly parasites, endophytes, or saprotrophs.

OTU picking (*de novo*, closed reference, or open reference): Assignment of operational taxonomic unit (OTU) type to sequences in a sample.

Oxalotrophy: The utilization of oxalate as a carbon (and energy) source.

Oxidation: A chemical or biochemical process in which electrons or hydrogen are lost from a molecule.

Pangenome (of a taxon): Total genetic diversity found among members of the taxon.

Paralogous genes: Genes at different chromosomal locations in the same organism that are similar, indicating that they are originated from a common ancestral gene by gene duplication and have since diverged from the parent copy.

Parametric: A statistic or test that specifically assumes one or more characteristics about the underlying distribution of data. Typical assumptions are that data values are normally distributed around the mean.

Parasite: Consumer living at the expense of a host.

Parasitism: An interaction between organisms in which one partner gains an advantage at the expense of the other.

Parsimony analyses: A group of phyletic classification techniques that describes lines of descent based on a calculation of the number of steps required to generate the current spectrum of character states from a single state of a hypothetical progenitor. Such analyses assume generally that models with the least number of evolutionary steps represent the best model of evolutionary descent.

Particulate organic matter: Organic matter occurring as (non-soluble) particles.

Pathogen: Any disease-causing (micro)organism.

Pathogen-assisted molecular pattern (PAMP): Molecule that enhances pathogen resistance.

Pathogen suppressive: Situation in soil in which a pathogen does not establish or persist.

Pearson's correlation coefficient: Also known as *Pearson's product–moment correlation coefficient (r)*. In statistics, a measure of how well a linear equation describes the relation between two variables X and Y measured on the same object or organism. It is used in software packages for gel analysis to calculate the congruence between arrays of values, typically densitometric arrays.

Peptidoglycan: A polymer (also known as murein) that forms a homogenous layer lying outside the plasma membrane in Gram-negative cells and is cross-linked into several layers in Gram-positive cells. It serves a structural role in bacterial cell walls giving these shape and strength, counteracting the osmotic pressure of the cytoplasm. It is made up of three parts: a backbone, composed of alternating N-acetylglucosamine and N-acetylmuramic acid; a set of identical tetrapeptide side chains attached to N-acetylmuramic acid; and a set of identical peptide cross-bridges.

Percoll™: A density gradient material that is used to separate bacteria and archaea from soil particles, fungal hyphae, and debris during centrifugation.

Permutation test: *Also called Monte Carlo test.* A mode of randomization test whose purpose is to determine, for a given data set, whether the value of a statistics (e.g., F, t, r) is greater (or lower) than the value expected by chance (i.e., if the null hypothesis is true).

Pesticide: A chemical substance used to kill pests, such as insects, weeds, microbial pathogens, and others.

pH: Measure of the acidity of a solution. pH is equal to the negative logarithm of the concentration of hydrogen ions in a solution. A pH of 7 is neutral. Values less than 7 are acidic, and values greater than 7 are basic.

Phage (bacteriophage): Virus that infects bacteria or archaea.

Phagocytosis: Ingestion of particles in a food vacuole.

Phagotrophic: Mode of heterotrophic feeding of protists.

Phenetics: Classification of organisms based on expressed traits or phenotypes.

Phenolics: Aromatic compounds that contain a phenolic subgroup. Relating to, derived from, or containing phenol.

Phenotypic marker: A unique phenotype (e.g., presence of a specific protein, antibiotic/metal resistance, substrate utilization, or enzymatic activity) that can be used to specifically detect an organism or a group of organisms.

Pheromone: Any chemical compound or set of compounds produced by a living organism that transmits a message to other members of the same species.

Phospholipid fatty acid (PFLA): A fatty acid present in the membranes of all living cells that can serve as a biomarker. The PFLA composition of a system can provide information about phylogeny of microorganisms and provide profiles of microbial communities.

Photoautotroph (-ic): A microorganism that derives energy from photosynthesis and obtains carbon from CO_2. For example, cyanobacteria.

Photobiont: The photosynthetic partner in a symbiotic association.

Phyletics: Classification of organisms and genes based on current state information and various models of evolutionary descent, the parameters of which are selected by the classifier. Also called phylogenetics and/or cladistics.

Phyllosphere: The habitat associated with the leaf surface.

Phylochip: A DNA microarray based on phylogenetic (e.g., ribosomal RNA) gene probes for high-throughput screening of temporal and spatial variation in microbial diversity and community structure.

Phylogenetic tree: Pairwise map of the evolutionary relatedness of organisms. Normally based on small subunit ribosomal RNA sequences.

Phylogeny: The evolutionary relationship among organisms, inferred from the patterns of lineage branching resulting from the true evolutionary history of the organisms being considered.

Phylo-phenetic species: Monophyletic and genomically coherent cluster of individual organisms that show a high degree of overall similarity in many independent characteristics, and is diagnosable by a discriminative phenotypic property.

Phylotype: Monophyletic cluster of strains or uncultured organisms with a high degree of sequence similarity in genes and/or sequences (e.g., ribosomal RNA) used for phylogenetic analysis.

Phytoalexin: Low-molecular-mass lipophilic antimicrobial compound that accumulates at some sites of pathogen infection.

Phytoextraction: Extraction from soil of compounds (e.g., metals) by plants.

Phytohormones: Plant hormones such as auxin, gibberellin, cytokinin, abscisic acid, and ethylene.

Phytoremediation: The use of plants to remove toxicants (either metals or organics) from soil or water. The plant roots may directly take up some pollutant molecules, such as heavy metals, or the roots can serve as a vehicle for spreading degrading bacteria in the soil.

Phytoremediator: Any compound or microorganism that facilitates or performs phytoremediation.

Phytostimulators: Compounds or microorganisms that promote plant growth.

Picoplankton: Term used in aquatic microbiology to indicate organisms in the water column with sizes ranging from 0.2 to 2.0 μm.

Piggy-back-the-winner (PtW) model: Model that assumes that phage integration into a host genome is the prevalent viral strategy in crowded conditions.

Pinocytosis: Ingestion of nutrients in solution in a food vacuole.

Plant endophyte: *See endophyte.*

Plant growth-promoting bacteria (PGPB): Bacteria that colonize roots and increase plant growth. They may act directly by improving nutrient availability or through hormone production, or indirectly by acting as biocontrol agents.

Plasmid: A self-replicating non-chromosomal genetic element that confers genetic flexibility to a cell.

Plasmodia: Multinucleate networks of protoplasm exhibiting rapid, rhythmical surges of protoplasmic streaming. Formed by plasmodial slime molds (*Myxomycota*).

Point mutation: Mutation in the DNA at a defined spot of limited size, often one to a few nucleotides.

Polar lipid-derived fatty acid-based stable isotope probing (PLFA-SIP): A method to detect microbes carrying out a specific metabolic process, where a stable isotope-labeled compound is used as a tracer, and detected in microbial polar lipid fatty acids.

Polyacrylamide gel electrophoresis: A method used to separate small molecules (e.g., proteins and oligonucleotides) based on their charges and sizes as they pass through a small pore size gel matrix made up of polyacrylamide.

Polymerase chain reaction (PCR): Enzymatic amplification of a defined segment of DNA during a series of temperature cycles using a forward and reverse primer that flank the sequence of interest. DNA is first melted at a high temperature (e.g., 95°C), cooled to allow primers to anneal (e.g., 40–50°C), and heated to 72°C to enable a thermostable DNA polymerase (Taq polymerase) to replicate the DNA segment flanked by the primers. This temperature cycle is repeated approximately 30 times resulting in a billion-fold amplification of the original target DNA.

Polyphasic approach: The use of multiple complementary methods, for example, to study a specific ecosystem or group of organisms.

Polyphyletic: A group of microbial species that descended from more than one common ancestor.

Polyvinyl polypyrrolidone (PVPP): Chemical used in DNA extraction to remove inhibitory substances such as humic acids, by binding to phenolic groups.

Posttranslational modification: Modification of a protein after its translation, for example, by the addition of side groups. It is one of the later steps in protein biosynthesis for many proteins.

Potential activity: The potential of a microorganism to perform a specific activity, for example, upon the addition of a substrate, which is not necessarily exhibited *in situ* at the time of measurement.

Predation: An interaction between organisms in which one partner consumes the other. For example, the predator organism may engulf, attack, or digest the prey organism.

Pressure potential: Pressure from external forces, including gravity. For example, standing water can add additional pressure to soil over that due to the atmosphere, resulting in positive pressure.

Primer(s): In DNA replication, a short single strand of RNA to which DNA polymerases can add new nucleotides. In PCR (*see separate definition*), two short sequences of DNA (or oligonucleotides) that are used to flank the target DNA sequence to be amplified (one on each strand) and that serve as starting points for DNA replication.

Priming (of seeds): A process whereby the hydration level within seeds is controlled so that the metabolic activity necessary for germination can occur, but radicle emergence is prevented.

Principle components analysis (PCA): Classification technique that uses unconstrained ordination to organize cases (or variables) along a series of meaningful gradients, called principal components, derived from a larger set of variables (or cases). The first component represents the dominant gradient, whereas the second, orthogonal (i.e., non-correlated) to the first, explains some of the residual variation as the third axis does and so on.

Programmed cell death: See *Apoptosis.*

Prokaryote: Term used to classify organisms without a cell nucleus (= karyon) that has traditionally been used to distinguish bacteria and archaea from Eukaryotes (Eukarya). *Note that this term is growing out of favor, and instead, Bacteria and Archaea should be substituted when possible.*

Prophage: Phage genome in integrated form in host genome.

Proteome: The collective proteins found in a particular cell type or community under a particular set of environmental conditions.

Protozoa: Term to describe single-celled Eukarya (traditionally termed "protista") that show characteristics of animals, most notably mobility and heterotrophy.

Protozoome: The entire protozoan community within a habitat.

Pseudopodium ("false foot"): A temporary projection of the cytoplasm of an amoeba used to move on a surface or to feed on bacteria. Various typical shapes have traditionally been used as an aid in classical taxonomy.

Psychrophile: A microorganism that grows optimally at low temperatures.

Psychrophilic: Optimally active at low temperatures.

Pulse labeling: The delivery of isotope tracers into an ecosystem for a short duration to monitor the rates and pathways of movement of the tracer molecule through the biota, for example, to determine metabolic routes and interactions between microorganisms.

Q10 relationship: An approximate doubling of enzymatic activity with every 10°C rise in temperature between 0°C and 30/35°C.

Quality filtering: Removal of sequencing errors or primer sequences from DNA sequence reads.

Quantile: An even portion of a distribution. Referred to as quartiles, deciles, and percentiles when a distribution is divided into four, ten, or one hundred quantiles, respectively.

Quantitative PCR: PCR protocol used to estimate the initial number of target sequences in a sample (e.g., competitive PCR, real-time PCR; *see separate definitions*).

Quorum sensing (QS): The regulation of bacterial gene expression via the production and sensing of signaling molecules, in response to fluctuations in cell densities within the population.

Quorum quenching (QQ): The inhibition of a quorum sensing signal, for example, by degradation.

Range: The difference between the highest and the lowest score in a distribution.

Raptorial feeding: See interception feeding.

Rarefaction curves: Mathematical curves representing the cumulative number of operational taxonomic units (OTUs—genotypes, phenotypes, taxa) that are obtained as the sample size increases. The curve asymptote, if reached, provides an estimate of the number of different OTUs (richness) in the total community.

Real-time PCR: A method used to detect product formation in real time during PCR, that is, not based on end-point detection. Formation of double-stranded (ds) DNA is proportional to a fluorescent signal detected directly after each cycle. Fluorescent dyes can be introduced into the assay using a specific probe (Taqman [quantitative] qPCR) or a dye that intercalates into the PCR product (e.g., SYBR Green qPCR).

Recombinase: Enzyme that catalyzes intramolecular DNA recombination.

Recombination: The natural process of breaking and rejoining DNA strands to produce new combinations of genes and, thus, generate genetic variation.

Redox potential: The potential of a reversible oxidation–reduction electrode measured with respect to a reference electrode, corrected to the hydrogen electrode, in a given electrolyte.

Reduction: A chemical or biochemical process where hydrogen or electrons are added to a molecule.

Redundancy: With respect to clones in a metagenomic library (*see separate definition*), it refers to the degree of occurrence of repeats of clones or sequences in the library.

Regulon: Global regulatory system that results in the expression of several genes in a cell simultaneously in response to a change in environmental conditions.

Reporter gene: A gene encoding a quantifiable protein and/or expressing a detectable phenotype, used, for example, to quantify inducible gene expression when cloned behind an inducible reporter.

Reproductive isolation: Prevention of interbreeding between individuals belonging to the same species resulting from different times or patterns of reproduction.

Resemblance matrix: A data matrix containing the similarity (or dissimilarity) coefficients for every pair of cases (or variables) in a multivariate data set.

Residual pores: Soil pores with neck diameters less than 0.3 µm that remain water filled under the influence of matric potential (*see separate definition*).

Resilience (ecological): The ability and rate at which a system returns to the original (steady or cyclic) state following a perturbation.

Resistance (ecological): The capacity of a system to withstand adverse (stress) conditions.

Response regulator (protein): A protein that, following an environmental trigger (sensed by a membrane-bound sensory protein), incites the cellular response to the trigger, often by modulating gene expression.

Resting cyst: Special cell structure in protozoa that is formed to withstand adverse conditions such as drought and heat.

Restriction/modification: *see DNA restriction/modification.*

Retromobilize: Transfer of DNA in the non-canonical direction, that is, from recipient to donor.

Reverse transcriptase: Enzyme involved in reverse transcription.

Reverse transcription: Action of RNA-dependent DNA polymerase—an enzyme that uses an RNA molecule as a template for the synthesis of a complementary DNA (cDNA) strand.

Rhizobia: A group of nitrogen-fixing bacteria that can form a symbiotic association with legume plants. In most cases, nitrogen fixation occurs in specialized structures called nodules (*see separate definition*) on the plant root.

Rhizodeposit: Organic carbon released from roots of a growing plant to soil.

Rhizodeposition: The release of organic substrates from the root, which may amount to 10%–40% of the plant's photosynthate.

Rhizopine: Organic compound serving as a unique carbon source and thus offering a competitive advantage to some rhizobia (*see separate definition*) capable of utilizing the compound.

Rhizoplane: Surface of the root, encompassing epidermis and rhizosheath.

Rhizosheath: Soil adhered by root hairs and mucilage.

Rhizosphere (effect): The narrow soil region influenced by plant roots that is the site of many key interactions between plants and microorganisms.

Rhizosphere competence: The ability of a microorganism to grow and compete in the rhizosphere. In some cases, this may also include growth on the root surface (rhizoplane, *see separate definition*) or within the root as an endophyte (*see separate definition*).

Ribonuclease (RNase): Enzyme that catalyzes the degradation of RNA to nucleotides.

Ribonuclease protection assay: Assay that determines RNA via complement synthesis, annealing, and enzymatic digestion of single-stranded RNA.

Ribosomal RNA (rRNA): A class of RNA found in the ribosomes of all living cells. Ribosomal RNA gene sequences are the current standard for determining the phylogenetic relatedness of organisms.

Ribotype: Cluster of closely related organisms (species to subspecies level) based on ribosomal RNA (rRNA) gene restriction patterns or rRNA gene internally transcribed spacer (ITS) similarities, which can be distinguished from other clusters according to the technique employed.

Richness: The number of categories found within a given sample (e.g., the number of bands in a DGGE profile), or the number of different taxa in a community; also used as a measure of diversity.

Root border cells: Cells detached from the root tips that disperse immediately into suspension after their contact with water.

Root exudate: Material released by plant roots containing a variety of monomeric compounds (sugars, amino acids, fatty acids, carboxylic acids) that are important for bacterial growth in the rhizosphere (*see separate definition*).

Root hair: Long tubular-shaped outgrowth from root epidermal cells.

***r*-Strategist:** Organism that can metabolize and grow fast (has a high growth rate) when nutrient supply is high and that has a survival advantage in uncrowded situations.

Reverse transcription PCR (RT-PCR): Amplification of a DNA strand (cDNA) that is first produced by the action of the enzyme reverse transcriptase on RNA.

Saprotroph: An organism that feeds on dead organic matter.

Scatter plot: A graph of a data set where each data value is plotted as a point measured along one or more axes representing variables.

Scintillation counting: The counting of ionizing radiation or light in a scintillation counter, a machine that detects and quantifies radiation.

Selective enrichment: A method used to allow specific organisms with the capacity to grow under defined conditions to become dominant, for example, by repeated additions of a particular carbon source to soil to select and enrich for bacteria that can use this compound as a carbon source.

Selective media: Nutrient solutions used to culture specific groups of microbes, based on their specific nutrient requirements or properties.

Selective sweep: Process of an allele becoming more prevalent in a population due to positive selection.

Sensor protein: A membrane-bound protein involved in the sensing of extracellular signals or conditions, which relays a signal to another protein in the cell, a response regulator (*see separate definition*).

Shannon–Weaver index (H_s): A diversity index that takes into account richness and evenness of taxons. $H_s = -\Sigma\, p_i \ln p_i$, where $p_i = n_i/N$ = the number of individuals within a species (n_i) divided by the total number of individuals (N) present in the entire sample.

Sheath (of fungi): A mantle-like structure of fungal hyphae enclosing an ectomycorrhizal root tip.

Shotgun sequencing: Sequencing of DNA prepared without prior selection that is randomly fragmented and either cloned and sequenced or sequenced directly; often used for sequencing of whole genomes or metagenomes.

Shuttle vector: Cloning vector designed to allow transfer of DNA from one host to another. Can often replicate in several hosts.

Siderophores: Iron-chelating organic molecules of different structures produced by many microorganisms that sequester or chelate the essential nutrient iron from their surroundings.

Sigma factor: A protein that regulates the correct attachment of the RNA polymerase enzyme to the DNA (promoter region) where transcription is initiated in bacteria.

Signal transduction: A series of cellular responses to an external signal, involving a sensor protein that is activated and transfers a signal to a response regulator (*see separate definitions*). This in turn induces the expression of genes, producing a specific physiological response to the external signal.

Signature lipid biomarker (SLB): A fatty acid or lipid biomarker of a specific microorganism, or group, that can be used as a marker for the detection of that group in the environment.

Simple matching coefficient: A bivariate measure of association that quantifies the proportion of instances when two samples (or variables) score the same value for multiple variables (or samples).

Simpson's index: A diversity index that takes account of the dominance (D) of individual species. D measures the probability that two individuals randomly selected from a sample will belong to the same species (or some category other than species). There are two formulas to calculate D. D, $1 - D$, and $1/D$ are all used as "loosely defined" Simpson's indices.

Single nucleotide polymorphism (SNP): Single nucleotide difference within a genome between members of a species or taxon.

Small subunit (SSU) ribosomal RNA (rRNA): Refers to different forms of RNA, for example, the 5S and 16S ribosomal RNA (for the bacteria and archaea) and 18S ribosomal RNA (for the eukarya), which are present in the small subunit of the ribosome.

Social motility: (Regulation of) movement in groups.

Soft-rot fungi: Fungi growing on wood in damp environments, usually within the lumen of individual woody cells but with little ability to degrade lignin (*see separate definition*).

Soil disinfestation: The (partial) destruction of pathogens and microorganisms in soil by, for example, fumigation or steam under water seal and plastic film. Alternatively, incorporation into the soil of specific crops with toxic breakdown products that restrict the development of (pathogenic) soilborne fungi.

Soil microbiostasis: The propensity of soil to inhibit the establishment and survival of incoming organisms.

Soil receptivity: With respect to soilborne diseases, the capacity of soil to allow saprophytic growth and infectious activity of pathogenic populations present or introduced into soil. A suppressive soil has a low receptivity, whereas a conducive soil has a high receptivity.

SOS response: DNA repair response to DNA damage in bacteria controlled by the RecA and LexA proteins that regulate expression of several proteins involved in DNA repair.

Southern blot/hybridization: Method used to identify the presence of specific DNA sequences, for example, on a chromosome or plasmid, by hybridization of labeled DNA probes to DNA fragments separated during gel electrophoresis and subsequently transferred to a nylon or nitrocellulose membrane for hybridization.

Specialist: Organism that has characteristics that restrict its survival to particular environmental conditions or geographic locations.

Specialized transduction: Process in which phage particles contain flanking host DNA in addition to phage DNA. Integration occurs either via phage site-specific recombination or via homologous recombination into the recipient genome.

Species: A group of closely related organisms that are capable of interbreeding and that are reproductively isolated from other groups of organisms; the basic unit of biological classification. Note: the bacterial and archael species definition is complex due to horizontal gene transfer between "species" and other factors (*see text for a discussion*).

Specific suppression: Specific suppression operates in many cases against a background of general suppression but is more qualitative, owing to specific effects of individual groups of organisms antagonistic (like antibiotic producers) to a pathogen during some stage in its life cycle.

Spermosphere competence: The ability of a microorganism to grow on the seed or in the region of the soil influenced by the seed.

Stable isotope probing (SIP): A method to detect microorganisms carrying out a specific metabolic process, in which a stable isotope (typically ^{13}C)-labeled compound is used as a tracer, and detected in microbial nucleic acids (DNA or RNA) or phospholipid fatty acids.

Standing crop value: The biomass of protozoa present at any given time in a system.

Storage pores: Soil pores larger than 0.3 μm that empty under the influence of matric potential (*see separate definition*).

Stress ethylene: In plants, a large burst of ethylene produced following a biotic or abiotic stress.

Stringent control phase: The first phase of a starvation-induced response in bacteria that includes a decline in the synthesis of macromolecules and an increase in protein degradation.

Substrate-induced respiration (SIR): The measured respiration response of a soil community upon exposure of the soil to an easily degradable substrate such as glucose.

Subtractive suppressive PCR: A PCR-based molecular method that applies subtractive hybridization and PCR to identify differences between genomes of organisms or between transcriptomes (*see separate definition*).

Sucking: Describes a turgor-driven uptake of cell content after piercing preyed cells.

Suppressiveness: Inhibition or restriction of infectious activity (suppressiveness to a disease) or of the development of a pathogen (suppressiveness to a pathogen) in soil due to biotic and/or abiotic factors.

Suppressive soil: A soil in which disease severity or incidence remains low, in spite of the presence of a virulent pathogen, a susceptible host plant, and climatic conditions favorable for disease development.

SYBR Green: Fluorescent dye that binds to double-stranded nucleic acids, for example, used for real-time PCR (*see separate definition*).

Symbiont: One of the partners in a symbiotic interaction (*see separate definition for symbiosis)*.

Symbiosis: A close and usually obligatory association of two organisms of different species living together, not necessarily to their mutual benefit, but often used exclusively for associations in which both partners benefit, which is more correctly termed *mutualism* (*see separate definition*).

Synergism: A facultative metabolic interaction between microorganisms in which two or more microbial populations are not capable of growth and survival in the absence of the other. For example, one population may produce a metabolic compound that the other population can further metabolize, allowing the growth of both populations.

Synteny: The preserved order of genes in genomes of organisms.

Syntrophy (-ic): The phenomenon that one species lives off the metabolic products of another species, leading to nutritional interdependency of that population.

Systemic acquired resistance (SAR): Long-lasting and broad-host-range defense mechanism in plants that can be brought about by previously treating a plant, or parts of it, with microorganisms that induce localized cell death.

Take-all decline (TAD): Spontaneous decrease in the incidence and severity of take-all disease, for example, occurring with monoculture of wheat or barley after a severe disease outbreak. TAD relates to the build-up of 2,4 diacetylphloroglucinol (DAPG)-producing fluorescent pseudomonads.

Taqman probe: A DNA oligonucleotide probe used for real-time PCR (*see separate definition*), which is labeled with a reporter dye on one end and a quencher dye (called TAMRA) on the other. A signal is generated when the reporter and quencher dyes are separated during DNA elongation steps of the PCR.

Taxometrics: Classification of cases (or variables) based on their current state, measured for one or more variables (or cases).

Taxon–area relationship (TAR): A power-law relationship between the number of taxa in an area and the size of that area. Generally, the number of species increases with an increase in sample area.

Telomorph: The sexual stage of a mitosporic fungus.

Testate amoebae: Amoebae partially enclosed in a shell, or test, made up of organic material, agglutinated particles, calcium carbonate, or silica.

Test statistic: A value used to calculate the probability that one or more samples came from a single, much larger population.

Thermophile: A microorganism that grows optimally at high temperatures; an important source of thermostable enzymes.

Thermophilic: Active at high temperatures.

Thin layer chromatography (TLC): Chromatographic technique using thin layers of a matrix on a solid surface for separating molecules according to their chemical properties.

Trans/cis ratio: Ratio between trans- and cis-fatty acids in the cell membrane, used by cells as a response to environmental stress or conditions.

Transcript: Messenger RNA transcribed from a single gene or operon.

Transcriptome: The collective transcripts that are formed in a cell or biological system at a certain time point.

Transduction: The transfer of DNA from a donor to a recipient cell population via a bacteriophage vector.

Transformation: The uptake of extracellular DNA by a bacterial cell (*see also natural transformation*).

Transmission pores: The main conduits for water and nutrient flow in soil.

Transposition: The process in which a transposable element moves from one site to another in the host genome.

Transposon: A mobile genetic element that is able to excise from the genome, or to make an additional copy of itself, and that can insert itself at a different locus in the genome.

Transposon-facilitated sequencing: DNA sequencing using transposon (*see separate definition*)-specific primers to localize transposons directly in the genomic DNA, thus generating a set of overlapping sequence contigs that can be assembled to a long genomic fragment.

Trapping: Feeding strategy (of protists) where agile prey is immediately immobilized upon contact with sit-and-wait predators (often aided by extrusomes) before being ingested.

Trophic level: The position that an organism occupies in a food chain using as a criterion what it uses as a food source and how it is consumed.

Tropism: Growth movement of a sessile plant, animal, or fungus, usually curvature towards (positive) or away (negative) from the source of stimulus.

Two-component (regulatory) system: A cellular regulatory system consisting of a sensor protein and a regulator protein that allows the cell to express particular genes as a response to environmental conditions (*see separate definition for signal transduction*).

Two-photon laser scanning microscopy: (*see separate definition for multi-photon laser scanning microscopy*).

Type IV secretion system (T4SS): A family of macromolecule transport systems enabling transport (often of DNA or proteins) from the cytosol, across the cell envelope, to other cells.

Vampyrellidae: A family of naked amoebae with a spherical body shape that in soil have been associated with feeding on (plant pathogenic) fungi.

Vesicles: Spherical or ovoid swellings in the hyphae of some arbuscular mycorrhizal fungi.

Vesicular arbuscular (mycorrhizae): An older name for arbuscular mycorrhizal fungi (*see separate definition*), derived from the fact that some Glomeromycotan fungi, in addition to forming arbuscules, also form swollen storage structures termed vesicles.

Viable-but-nonculturable (VBNC) state: A state in which a microorganism is alive but unable to grow on medium that normally supports its growth and that is reversible, meaning that the cells can be resuscitated to be able to grow if conditions become favorable.

Virome: The entire viral community within a habitat.

Virus-like particle (VLP): Particle detected by (electron) microscopy that has the appearance of a virion.

V_{max}: A kinetic parameter, giving the maximum rate of a reaction, which cannot be further enhanced by increasing the concentration of the substrate.

White-rot fungi: Fungi that are able to achieve the complete degradation of lignin.

Xenobiotic: Man-made molecule that is foreign to life, for example, a range of pesticides of varying chemical nature and polychlorinated biphenyls (PCBs).

Zygomycota: A diverse group of fungi characterized by sexual reproduction by fusion of gametangia to produce resting sexual spores (zygospores) and asexual reproduction by nonmotile spores.

Zygospore: Thick-walled sexual spore resulting from the fusion of gametangia in zygomycetous fungi.

Zymogenous (microorganism): Fast-growing microorganism that consumes organic substrates.

Index

A

Abiotic soil factors, 347–348
Acacia mangium, 427
Acacia treatment, 427
Acanthamoeba, 126–127, 133
Accessory genes, 110
Acidithiobacillus ferrooxidans, 413
Acidobacteria, 52–53, 57, 301, 302
Acinetobacter, 150
Acinetobacter baylyi, 108, 110
Acinetobacter calcoaceticus, 369
Active microbial communities, 253
Activity-based metagenomics, 394–395
N-acyl-l-homoserine lactones (AHLs), 146–148, 373–374
Adventurous motility, 158
Aerobic oxidation, 332
Agaricomycetes, 69
Agaricomycotina, 70
Agaricus bisporus, 154
Agaricus brunnescens, 154
Agglomerative clustering, 322
Agricultural intensification, 421
Agricultural management regime, 31
Agricultural practices, 355–357
Agrobacterium tumefaciens, 174, 373
Agroecosystems, 423–424
Algorithm, 232, 323, 325–326
Alicyclic hydrocarbons, biodegradation of, 384
Aliphatic hydrocarbons, biodegradation of, 383–384
Alkanes, biodegradation of, 383, 384
Alkenes, biodegradation of, 383
Allium, 356
Allomyces macrogynus, 68
α-diversity metrics, 220, 221, 223
α-Proteobacterium, 183
Alternative hypothesis, 317–318
Alternative solidifying agents, 298
Alveolata, 128–129, 135
AM, *see* Arbuscular mycorrhiza (AM)
AMF, *see* Arbuscular mycorrhizal fungi (AMF)
1-aminocyclopropane-1-carboxylate (ACC) deaminase, 174, 369–370
Ammonia, 185
Ammonia monooxygenase (AMO), 185–186
Ammonia-oxidizing archaea (AOA), 58, 185
Ammonia-oxidizing bacteria (AOB), 58, 185
Ammonium, 185
Ammonium nitrate, 185
Amoebozoa, 126–127
Amphizonella, 127
Anabaena, 185
Anaerobic ammonia oxidation (Anammox), 28, 183–184
Anaerobic soil bacteria, 188
Anaerobiosis, 15
Anaeromyxobacter (δ-Proteobacteria), 187
Analyses of variance (ANOVA), 318–320
Antagonism, 349–350
Antagonistic interactions, 25–26, 151–153
Antibiosis, 350
Antibiotics, 151–152, 159, 370–371
- metals and, 416–417
- production of, 74

AOA, *see* Ammonia-oxidizing archaea (AOA)
AOB, *see* Ammonia-oxidizing bacteria (AOB)
Aphanomyces euteiches, 348, 355
Apicomplexa, 128
Arabidopsis, 170, 171, 173
Arabidopsis thaliana, 138, 171–172, 366, 368
Arbuscular mycorrhiza (AM), 68, 77–78, 81
Arbuscular mycorrhizal fungi (AMF), 137, 150, 183
Arbuscules, 68
Arbutoid mycorrhiza, 80
Archaea, 49–50, 62–63, 188, 189–190, *see also* Bacteria
- adaptations to soil conditions, 58–59
- gene expression mechanisms in, 50
- lifestyle and survival strategies, 54
- methane production, 58
- organic carbon, 51
- soil microbiomes diversity, 216
- virus interaction with, 95

Archaeal diversity, 57–58
Arctic systems, global change, 334–336
Area under the disease progress curve (AUDPC), 346
Armillaria bulbosa, 71
Armillaria mellea, 76
Aromatic hydrocarbon, biodegradation of, 384, 385
Arsenic, 405
Arthrobacter, 353
Arthrobacter chlorophenolicus, 60, 262
Arthrobacter globiformis, 156
Arthrobotrys oligospora, 154
Arthropods, virus interaction with, 95
Artificial intelligence, 428
Ascomycota, 69, 70, 78
Aspergillus flavus, 152
Aspergillus fumigatus, 394
Aspergillus nidulans, 263
Assays determining bacterial viability, 209–210
Associative interactions, 25–26
Atmospheric compositions, use of, 301
Autochthonous microorganisms, 51
Autotrophs, 50
Azolla, 185
Azospirillum, 56, 367–368
Azospirillum (α-Proteobacteria), 184
Azospirillum brasilense, 370
Azotobacter (γ-Proteobacteria), 184, 189

B

Bacillus, 45, 56, 59, 95–96
Bacillus amyloliquefaciens, 370
Bacillus anthracis, 56
Bacillus atrophaeus, 261

Bacillus cereus, 108
Bacillus mycoides, 142, 156
Bacillus subtilis, 39, 45, 59, 110, 114, 145, 148, 150–151, 368
Bacillus thuringiensis, 56, 148, 156
Bacteria, 49–50, 62–63
acidobacteria, 52–53
adaptations to soil conditions, 58–59
cultivation and enrichment approaches, 54–57
cyclohexane mineralization in, 384
fungal interaction with, 82
generalists *vs.* specialists, 51–53
growth strategies, 50–51
inoculants, 61–62
lifestyle and survival strategies, 54
metabolic versatility of, 55
mutation rate in, 106–107
physiological response to nutrient limitation, 59–61
and predatory protozoa, 156–157
recombination *vs.* mutation (r/m) ratio, 107–108
soil microbiomes diversity, 216
uncultured, 296–297
viable but nonculturable state, 61
virus interaction with, 95
Bacterial cells extraction, 203–204
Bacterial communities, at root–soil interface, *see* Root-associated bacterial communities
Bacterial diversity, 57
Bacterial evolution, 106
Bacterial–fungal interactions, 152–156
Bacterial genomes, 105–106
Bacterial interactions, 6
Bacterial isolation, 297–298
Bacterial traits for root colonization, 172–173
Bacterial viability and activity, 208–210
Bacteriocins, 152, 159
Bacteriophages, 91, 372–373
Bacterivory, 135
Baeyer–Villiger monooxygenation reaction, 384
Bar graphs, 313
Basidiomycota, 69, 70, 78
BCAs, *see* Biological control agents (BCAs)
Bdellovibrio, 157–158
Beggiatoa, 189
Beneficial bacteria, 174
Benzene, biodegradation of, 384, 385
Bergey's Manual of Systematics of Archaea and Bacteria, 55
β-diversity metrics, 220, 221, 224
β-glucosidase, 74
β-Proteobacteria, 56
Betula pendula, 79
BHR plasmids, *see* Broad-host-range (BHR) plasmids
Binning process, 219
Bioaugmentation, 119, 396
Bioavailability, 407–408
Biobarrier, 397
Biocontrol agents, 354–355
Biodegradation
of chlorinated hydrocarbons, 385–388
degradative genes in soil, 394–396
hydrocarbon
aliphatic and alicyclic compounds, 383–384
under anaerobic conditions, 384, 385
aromatic compounds, 384, 386
of nitroaromatic compounds, 388
of organic pollutants, 382
of synthetic organic pollutants
pesticides, 389–391
pharmaceuticals, 391–393
plastics, 393–394
Biofertilization, 87
Biofilms, 101
formation of, 143–144
gene expression effect on, 145
plasmid transfer in, 113–114
Biogeochemical cycles, fungi, 84
Bioindicators, 354
Bioinformatics tools, 429
Biological control agents (BCAs), 86
Bioluminescence markers, 207
Bioreactor, 398
Bioremediation, 86–87, 382
definition of, 396
environmental factors and, 399–400
methods, 397–398
in situ and *ex situ* techniques, 396
Biosensors (bioreporter), 207, 408, 425
Biosorption, 86
Biosparing, 397
Biostimulation, 396
Biosystematics, 321
Biotic soil factors, 347–348
Biotrophic pathogen, 76
Bioventing, 397
Bivariate correlation coefficient, 311
Blastocladiomycota, 68
Bootstrapping, 324–325
Border cells, 168
Bottom-up approach, 259
Box plots, 313
Brachypodium distachyon, 171
Bradyrhizobium, 53, 296, 427
Bradyrhizobium japonicum, 59
Brassica, 188
Broad-host-range (BHR) plasmids, 109
Bromus, 334
Brown-rot fungi, 74
Burkholderia, 365
Burkholderia cepacia R34, 388
Burkholderia gladioli, 412
Buttiauxella, 353

C

Cadmium, 406
Caenorhabditis elegans, 371
CAHs, *see* Chlorinated aliphatic hydrocarbons (CAHs)
Calcite precipitation, 411
Calvin–Benson–Bassham cycle, 181
Candidatus Methanoflorens stordalenmirensis, 239
Carbamazepine, biodegradation of, 392
Carbon, 27, 382
energy source, 10–11
loss, 335
Carbon cycle, 180, 275–286
CO_2 fixation, 181
methane cycle, 181–182

organic carbon cycling, 181
Carbon monoxide (CO), 181
Carbon polymers, nutrient deprivation, 60
Carbon use efficiency (CUE), 27, 338
Cardinal scales, 309
Carex arenaria, 29, 424
"Carrying capacity," 28
Catalyzed reporter deposition–fluorescence *in situ* hybridization (CARD-FISH), 168, 169
Cation diffusion facilitators (CDFs), 414
Cell lysis, 197
Cellular interactions, categories of, 148
Cellulose, 74
Cell wall-degrading enzymes, 371–372
Cenarchaeum symbiosum, 58
Central tendency, 310
Centrifugation, 204
Cercomonas, 127
Cercozoa, 128, 130
Cetyl trimethyl ammonium bromide, 198
Chemical cell lysis, 197
Chemotrophs, 50
Chitin, 71, 86
Chitosan, 86
Chlorinated aliphatic hydrocarbons (CAHs), 385–386
Chlorinated aromatic hydrocarbons, 386–388
Chlorinated phenols, 387, 388
2-Chlorobiphenyl, biodegradation of, 387
Chromium, 406
Chromosomal DNA
horizontal gene transfer of, 108–109
recombination rates, 108
Chytridiomycota, 68
Chytrids, 68
^{13}C-labeled DNA, 395
Cladistics, 321
Cladogram, 324
Clamp connections, 69
Classification methods, 324–325
Clavaria genus, 78
Cloning approach, 229
Closed-reference OTU picking, 219
Clostridium, 56
CLSM, *see* Confocal laser scanning microscopy (CLSM)
Cluster analyses approach, 322–323
Clustering, 325
^{13}C-naphthalene enrichment approach, 395
CO_2 fixation, 181
Colicins, 152
Collective phenotype, 428
Collimonas species, 153
Colony-forming units (CFUs), 319
Comammox bacteria, 184
Combined omics, 264
Co-metabolism, 382
Commensal bacteria, 175
Commercialization, 239–240
Community genome arrays, 199
Community richness, 220
Comparative metagenomics, 238
Compensatory mutations, 119
Competitive RT-PCR (cRT-PCR), 251
Complementary DNA (cDNA), 251, 395
Complexation, 415
Composting/biopiling, 398
Concentric circle plots, 313
CONCOCT simulator software, 231, 233
CoNet (Correlation Networks), 222
Confocal laser scanning microscopy (CLSM), 204–205
Conjugation, bacterial, 108
mechanism, 111–112
in soil, 112–113
Conjugative MGEs, 109, 113
Conjugative plasmids, 111–113
Conosa, 126–127
Contiguous sequences, 232
Co-phenetic correlation, 324
Copiotrophic microorganisms, 51
Copper, 406
Coprinus quadrifidus, 154
Correlation analysis, 220–222
Correspondence analysis, 321
Corticata, 128–129
Co-selection of metal, 418
Crenarchaeota, 58
CRISPR-Cas systems, 240
Cryphonectria parasitica, 94
Cultivation-dependent methods, 210–211
Culturability, 298, 300–301
Cupriavidus metallidurans, 260, 261, 265, 266, 413
Cupriavidus strain MBT14, 202
Cutting-edge plant/soil observation methods, 424
Cyanobacteria, 50
Cyberlindnera saturnus, 369
Cytokinins, 368
Cytoplasmic membrane systems, 413–415
Cytoplasmic resistance systems, 414–415

D

Dactylella gephyropaga, 154
DAL method, *see* Double-agar-layer (DAL) method
Database resources, 217
Data normalization, 219–220
Deamination of amino acids, 11
Deconvolution, 428
Deep learning, 326
Deep learning robotics, 424
Defense reaction, stimulation, 350
Degradative genes, in soil, 394–396
Dehalococcoides species, 387
Delft School of Microbiology, 55
Demultiplexing, 218–219
Denaturing gradient gel electrophoresis (DGGE), 272
Dendrograms, 314–316
Denitrification, 186–187
De novo assembly, 232, 247
De novo OTU picking, 219
Descriptive statistics, 308, 310–313
DESeq2 software package, 220
Detection methods, extraction of nucleic acids, *see also* Direct detection methods
cultivation-dependent methods, 210–211
DNA from soil, 197–199
hybridization, 199
polymerase chain reaction (PCR), 199–200
digital droplet, 201–203
quantitative, 200–201

Development of soil microbiology, 31, 32
2,4-diacetylphloroglucinol (2,4-DAPG), 371
Dickeya dadantii, 174
Dickeya solani, 100, 174
Dictyostelium discoideum, 70
Differential exudation, 171
Diffusion feeding, 156
Digital droplet polymerase chain reaction (ddPCR), 201–203
Dikaryon, 69
Diphoda, 125–126
  Corticata, 128–129
  Excavata, 128
Direct detection methods, 203
  bacterial cells extraction, 203–204
  bacterial viability and activity, 208–210
  fluorescence *in situ* hybridization, 206
  IF (fluorescent antibody) methods, 207–208
  marker genes, 207
  microscale/single-cell analysis
    epifluorescence microscopy, 204–205
    flow cytometry, 205–206
  total bacterial (direct) counts, 206
Direct extraction procedure, 197
Direct isolation procedure, 197
Direct mechanism, 365–366
Direct sequencing approaches
  genomes construction, 232–233
  metagenomics data, 232
  microbial population, abundance, 231
  sequence coverage, 231
  statistical significance, 233
  target group of interest, 231
  techniques, 230–231
Discosea, 126
Disease suppression, *see* Soilborne diseases
Disease-suppressive soils, 343
Dispersion, 310
Dissimilatory nitrate reduction to ammonium (DNRA), 183, 186
Dissolved organic matter (DOM), 100
Diversity, 310, 350–351, *see also* Soil microbiomes
DNA, 270
DNA extraction, 197–199
  hybridization, 199
  polymerase chain reaction (PCR), 199–200
    digital droplet, 201–203
    quantitative, 200–201
DNA isolation, 229–230
DNA repair, 415
DNA-SIP, 272–273, 275, 286–289
DOM, *see* Dissolved organic matter (DOM)
Double-agar-layer (DAL) method, 92
Downregulation of transporters, 413, 414
Duration of exposure, 408
DVC procedure, 209

E

Earth System Models (ESMs), 337–338
Ecological resilience, 45–46
Ecological resistance, 45–46
Ecosystem-scale process, 238–239
Ecosystem services for preagricultural soils, 238
Ectendomycorrhizas, 80
Ectomycorrhizas, 78–79, 81, 84
Effector-triggered immunity (ETI), 171
EFM, *see* Epifluorescence microscopy (EFM)
Eigen analysis, 321
18S ribosomal RNA (rRNA) gene sequence, 43
18S rRNA genes, 217
Electron transport chain, 413
Electroremediation, 397
Electrospray mass spectrometry (ESI), 259
Endobacteria, 82
Endoglucanases, 74
Endoparasitic fungi, 154
Endophytes, 80, 168
Endophytic bacteria, 362–363
Energy flow, 28–29
Energy sources, 9–11
Engulfing, 134
Enterobacter (γ-Proteobacteria), 185, 189
Enterobacter agglomerans, 119
Enumeration, cultivation-dependent methods, 210–211
Environmental proteomics
  bottom-up approach, 259
  definitions and developments, 258–259
  high mass-accuracy tandem spectrometers, 259–260
  proteome database, 260
  stable-isotope probing, 260
  top-down approach, 259
  two-dimensional electrophoresis, 259
Enzymatic cell lysis, 197
Eocronartium, 77
Epifluorescence microscopy (EFM), 92, 204–205, 207
Epiparasites, 80
Ergosterol, 71
Ericoid mycorrhizas, 78
Erwinia amylovora, 174
Escherichia coli, 60, 92, 107, 112, 119, 152, 156, 235, 413
  IncP-1-type plasmid transfer in, 113
  O157:H7, 45
ETI, *see* Effector-triggered immunity (ETI)
"E-type" mycorrhizas, 80
Eucalyptus, 427
Eucalyptus grandis, 427
Eukaryotes, 246
Euryarchaeota, 58
Evenness, 313
Excavata, 128
Exocellulases, 253
Exopolysaccharides (EPSs), 173
Exoproteomics, 258
Extracellular phosphatases, 412
Extracellular polymeric substances (EPSs), 203, 412
Extracellular proteins, 25, 258–259
Extracellular systems, 410–412
Extraction efficiency, 203
Extraction protocols, 248–249
Extreme soil environments, 13

F

FACS, *see* Fluorescence-activated cell sorting (FACS)
Factor analysis, 321
Feeding patterns, 156
Fe-oxidizing bacteria, 13

Fick's law, 15
Filter feeding, 156
Firmicutes, 45
Fisculla terrestris, 132
FISH, *see* Fluorescence *in situ* hybridization (FISH)
Flow cytometry, 205–206
Flow of energy, 28–29
Fluorescence-activated cell sorting (FACS), 205–206
Fluorescence *in situ* hybridization (FISH), 206
Fluorescent antibody (FA) technique, 207–208
Fluorescent dyes, 250
Fluorescent marker genes, 207
Fluorescent pseudomonads, 175
Folsomia candida, 83
Foraminifera, 128
Fortuitous metabolism, 382
Fractal system, 5
Frankia, 57, 184–185
Functional core microbiome, 173
Functional extension, 429
Functional gene arrays, 199
Functional potential, 245
Functional redundancy, 44
Functional screening, 236–237
Function-driven microbial communities, 144
Fungal biomass, 179
Fungi, 66
  activity, 71–72
  activity in soil, 31
  in biofertilization, 86
  in biogeochemical and nutrient cycling, 84
  in biological control, 85
  biomass measurement, 71
  in bioremediation, 85–86
  diversity of, 66–70
  features of, 67, 69
  filamentous nature of, 86
  identification of, 72
  interactions with other organisms, 82–83
  metabolism, 72–74
  metabolites, 73
  parasites and pathogens, 76–77
  as saprotrophs, 74–76, 87
  soil microbiomes diversity, 216
  symbioses, 77–81
  virus interaction with, 95
Fungistasis, 350
Fungivores, 135
Fungus-like organisms, 69–70
Fusarium, 31, 351, 427
Fusarium graminearum, 155
Fusarium moniliforme, 368
Fusarium oxysporum, 348–351, 354
Fusarium solani, 347, 370

G

Gaeumannomyces graminis, 76, 347
γ-Proteobacteria, 56
Gene expression, 145
Generalists, 52–53
Generalized linear models, 318
General suppressiveness, 348
Genes encoding proteins, 27
Gene-targeted assembly, 232
Genetic adaptation, 415–416
Genetic drift, 106
Genetic redundancy, 44
Gene transfer agents (GTAs), 114
Genome reconstruction, 238
Genomes, construction of, 232–233
Geobacter, 413
Geosiphon pyrifomis, 68
Gibberella fujikuroi, 368
Gibberellins, 368–369
Gliding browsers, 134
Gliocladium, 350
Global change, 334
  permafrost thaw, 335–336
  vegetation change and microbial communities, 332–334
  wildfire effects, 334–335
Glomeromycota, 68–69, 88
Gluconacetobacter, 56, 365
Gluconacetobacter (α-Proteobacteria), 185
Gluconacetobacter diazotrophicus, 173
Gold-fluorescence *in situ* hybridization (FISH), 62
Gram-negative bacteria, 60, 413
Gram-negative (G–) bacteria., 111–112
Gram-positive (G+) bacteria, 50, 56, 111–112, 146
Gram-positive type IV secretion system, 112
Graphical descriptions, 313–314
Great plate count anomaly (GPCA), 215, 295–296
The great plate count anomaly, 206
Green manures, 356
Growth media
  atmospheric compositions, use of, 301
  development of, 297–300
  micronutrients, addition of, 301
  oxidation-protective agents, 301
  traditional media, dilution of, 301
GTAs, *see* Gene transfer agents (GTAs)
Gulping, 134
Gunnera, 185

H

Harosa, 128–129
Hartig net, 79
Hebeloma crustuliniforme, 79
Herbaspirillum (β-Proteobacteria), 185
Herbaspirillum seropedicae, 173
Heterobasidion annosum, 76
Heterodera schachtii, 351
Heterogeneity, degree of, 247–248
Heterogramic plasmid transfer, 113
Heterokonta, 129
Heterolobosean amoebae, 128
Heterotrophic bacteria, 51
Heterotrophic nitrification, 183
Hexadecyl methylammonium bromide (CTAB), 249
HGT, *see* Horizontal gene transfer (HGT)
Hierarchical clustering, 322
High mass-accuracy tandem spectrometers (MS/MS), 259–260
High-molecular-weight (HMW) proteins, 259
High-throughput meta-omics techniques, 351–353
Histograms, 313

Hitherto-uncultured bacteria, 298
HME-RND-driven efflux systems, 414
Holobiont, plant, 163
Homogenizer, 229
"Homology-facilitated illegitimate recombination," 108
"Horizontal gene pool," 109
Horizontal gene transfer (HGT), 92, 101, 106, 119–120, 415
  of chromosomal DNA, 108–109
  experimental studies of, 116
  identification of, 115
  mechanisms in soil
    conjugation, 111–114
    natural transformation, 110–111
    transduction, 114
  of mobile genetic elements, 109
  population-scale considerations of, 116–119
  selection of, 119
  temporal effects of, 114
Hosts, 235–236
"Hot moments," 29
Hot spots for activity, in soil, 26
HTSSIP, 290
Humicola grisea, 261
Humicola insolens, 261
Hybridization-based detection, 199
Hydrocarbon biodegradation
  aliphatic and alicyclic compounds, 383–384
  under anaerobic conditions, 384, 385
  aromatic compounds, 384, 386
Hydrogen cyanide (HCN), 370–371
Hydrophobins, 73–74
Hydroxamate siderophores, 73
Hydroxylamine oxidoreductase (HAO), 186
Hypholoma fasciculare, 75
Hypothesis testing, 314–315
  null and alternative hypotheses, 317–318
  parametric and nonparametric analyses, 318–320
  sample data and populations, 317
Hypovirulence, 349
Hypovirulent, 94
Hysteresis, 8

I

ICEs, *see* Integrative conjugative elements (ICEs)
IF (fluorescent antibody) methods, 207–208
Illegitimate recombination, 108
Imidacloprid biodegradation, 390–391
IncQ plasmids, 112–113, 119
Incubation, 273, 275, 298
Indirect mechanism, 363
Indole-3-acetic acid (IAA), 367–368
Induced systemic resistance (ISR), 367, 370
Infectious potential unit (IPU), 345–346
Innate immune system, 171
Inoculant bacteria, 61–62
Inoculation, 188
Inoculum, 344
Insertion sequences (IS), 416
In silico analyses, 199
In situ microbial activity, 272
In situ microbiome engineering methods, 425
Integrated approach, 428
Integrated Taxonomic Information System (ITIS), 129
Integrative conjugative elements (ICEs), 109
  conjugative transfer of, 112
Interception feeding, 134
International Committee on Taxonomy of Viruses (ICTV), 97
Intracellular activities in soil microbiomes, 24–25
Intrinsic marker, 196
In vitro detection, 196
IPU, *see* Infectious potential unit (IPU)
Iron sequestration, 365–367
Island biogeography, 142, 148
ISO-11063 Soil quality method, 199
Isotopes, 270, 273
IS*Rme5,* 416
ITS sequence-based analysis, 217

K

Kill-the-winner (KtW) model, 100
k-mers, 232
Kraken, 134
Kruskal–Wallis test, 320
K-strategists, 51, 56

L

Laccaria bicolor, 83, 153
LAM, *see* Liquid aliquot method (LAM)
Land farming, 398
Land-use changes, 426
Lead, 406–407
Leguminous shrubs, 334
Lepista nuda, 154
Lichen, 80
Light energy, 14
Lignin, 339
Linear mixed models, 319
Linear regression, 311
Line graphs, 313
Linkage algorithms, 322
Lipid phosphatases, 413
Lipopolysaccharides, 412–413
Liquid aliquot method (LAM), 131
Live/Dead staining protocol, 210
Lysobacter capsici, 372
Lysogenic conversion, 101
Lysogenic phages, 101
Lytic phages, role of, 100

M

Machine learning approaches, 325–326
Malawimonas, 125
MAMPs, *see* Microbial-associated molecular patterns (MAMPs)
Management, *see* Soil management
Marker genes, 207
Matric potential of water, 8
Matrix-assisted laser desorption–ionization (MALDI), 259
Maximum-likelihood approach, 324
Maximum parsimony approach, 323
Measurement scale, 308–309

Melanin, 74
Meloidogyne incognita, 372
Membrane integrity and potential, 210
Mercury, 407
Messenger RNA (mRNA), 246–247, 249, 253–255
  extraction protocols, 248–249
  metatranscriptomes, 252–253
  microarrays, 252
  mRNA-SIP, 270
  PCR-based approaches, 251–252
  purification, 249–250
  quantification of RNA, 250
  reverse transcription, 250–251
  sampling and storage of samples, 247–248
  techniques for, 254
  translation of, 257
MetaBAT, 233
Metabiosis, 87
Metabolic activity, 410
Metabolic redundancy, 44
Metabonomics, 354
Metagenome-assembled genomes (MAGs), 232, 233
Metagenome libraries, 235–236
Metagenomes, 272
MetagenomeSeq software package, 220
Metagenomics approach, 383, 394–396, *see also* Soil metagenomics
Metal accumulation, 415
Metal contamination
  bioavailability, 407–408
  duration of exposure, 408
  genetic adaptation, 415–416
  metabolic activity, 410
  metals and antibiotics, 416–417
  microbial diversity, 408–409
  resistance carriers, 415
  resistance systems, 412
    cytoplasmic membrane systems, 413–414
    cytoplasmic resistance systems, 414–415
    extracellular systems, 410–412
    outer membrane systems, 412–413
    periplasmic systems, 413
  soil metals and concentrations, 404–405
    arsenic, 405
    cadmium, 406
    chromium, 406
    copper, 406
    lead, 406–407
    mercury, 407
    nickel, 407
    zinc, 407
  soil microbial biomass, 409–410
Metallophores, 412
Metal resistance carriers, 415
Metals and antibiotics, 416–417
Meta-omics techniques, 351–353
"Metaphenomics" concept, 113
MetaProSIP, 288
Metaproteomics, 257–258, 272, 354, *see also* Soil metaproteomics
Metarhizium, 86
Metarhizium robertsii, 86
MetaSim simulator software, 231
Metatranscriptomes, 252–253, 272
Metatranscriptomics approach, 29, 353, 396
Methane, 181
Methane cycle, 181–182
Methanocella, 58
Methanogenesis pathway, 181–182
Methanogens, 58
Methanosarcina, 58
Methylation, 415
2-Methyl-4-chlorophenoxy acetic acid (MCPA) mineralization, 202
Methylocaldum, 287
Methylomicrobium, 287
MGEs, *see* Mobile genetic elements (MGEs)
Microarrays, 199, 252
Microbacterium lacus strain SDZm4, 393
Microbe-associated molecular pattern (MAMP)-triggered immunity (MTI), 171
Microbial antagonism, 349–350
Microbial assemblages, 143
Microbial-associated molecular patterns (MAMPs), 155–156
Microbial biofilms, 18
Microbial cells, 180
Microbial communities, 331–338
Microbial diversity, 408–409
Microbial function, 144
Microbial genomes, 239
Microbial habitat/activity, in soil, 3–4, 18
  energy and nutrient source, 9–11
  soil atmosphere and redox potential, 14–16
  soil light, 14
  soil pH, 16–18
  soil temperature, 11–13
  spatiotemporal aspects, 4–6
  water, 6–9
Microbial interactions, 142, 158–159
  antagonistic interactions, 151–153
  bacterial–fungal interactions, 152–156
  categories of cellular interactions, 148
  at different scales, 142–143
  and environmental conditions, 142
  examples of, 149
  molecular sensing and signaling, 145–146
  mutualistic interactions, 150–151
  predatory interactions, 156–158
  quorum sensing, 146–148
  and spatial structure of soil, 142
  stringent *vs.* relaxed, 158
  suppressiveness
    biotic *vs.* abiotic factors, 347–348
    general *vs.* specific suppressiveness, 348
    mechanisms, 349–350
    organisms identification, 348–349
Microbial loop, 136
Microbial population, abundance, 231
Microbial RNA, 245–246
Microbiome diversity analysis, 220
Microbiomes, 37, *see also* Soil microbiome
Micrococcus species, 60
Microcolonies, 143–144
Microcolony assay, 209
Microflora, 356
Micronutrients, addition of, 301
Microorganisms, 333–334

Microorganisms cycling soil nutrients, 179–180
nitrogen cycle, 182–184
denitrification, 186–187
nitrification, 185–186
nitrogen fixation, 184–185
phosphorus, 187–188
soil carbon cycle, 180
CO_2 fixation, 181
methane cycle, 181–182
organic carbon cycling, 181
sulfur, 188–189
Microscale/single-cell analysis
epifluorescence microscopy, 204–205
flow cytometry, 205–206
Mobile genetic elements (MGEs), 109, 110, 118, 415, 416
Modulation of soil microbiomes, 427
Molecular hydrogen, 301
Molecular markers, 216–217
Molecular recognition mechanisms, 155–156
Monotropid mycorrhiza, 79–80
Moore's law, 337
Most probable number (MPN) method, 131
MTI, *see* Microbe-associated molecular pattern (MAMP)-triggered immunity (MTI)
m-toluate degradation, 86
Mucilage, 166, 168
Mucoromycota, 68
Multivariate techniques, 320–321
Mutational process, 106–108
Mutator phenotypes, 107
Mutualistic interactions, 150–151
and niche construction, 155
Mycobacterium, 92
Mycoheterotrophs, 79–80
Mycorrhizal fungi, 70, 77–80
functional effects of, 81
interaction with bacteria, 82
Mycoviruses, 94
Myxobacteria, 158
Myxococcus xanthus, 158
Myxomycetes, 127

N

NaOH, 263
National Center for Biotechnology Information (NCBI), 263
Natural attenuation, 396, 397
Natural phenomena
descriptive statistics, 310–313
graphical descriptions, 313–314
variables and measurement scales, 308–309
Natural selection, 106
Necrotrophic pathogens, 76
Nematicidal activity (NA), 372
Nematodes, virus interaction with, 95
Nematophagous fungus, 83
^{15}N-enriched fertilizer, 26
Neonicotinoid pesticides, 390
Nernst equation, 15–16
Niche construction, mutualistic interactions and, 142
Nickel, 407
nifH, 184
nirK, 187
nirS, 187
Nitrification, 18, 144, 150, 185–186
Nitrite, 183
Nitrite-oxidizing bacteria (NOB), 185
Nitroaromatic compounds, biodegradation of, 388
Nitrobacter (α-Proteobacteria), 185
Nitrogen, 28, 33, 182
Nitrogenase, 363–364
Nitrogen cycle, 182–184, 276, 286–287
denitrification, 186–187
nitrification, 185–186
nitrogen fixation, 184–185
Nitrogen fixation, 184–185, 364
Nitrogen loss, 335
Nitrolancea hollandica (Chloroflexi), 185
Nitrosococcus, 28
Nitrosomonadaceae, 302
Nitrosomonas, 28, 186, 286
Nitrosopumilus maritimus, 58
Nitrososphaera, 286
Nitrososphaera viennensis, 58, 186
Nitrosospira, 28, 186, 286
Nitrospira, 28, 184
Nitrospirae, 185
Nodules, 184
Nominal scales, 311
Nonoverlapping clustering, 322
Nonparametric analysis, 318–320
Non-vegetated soil, 164–165
nosZ, 187
Novel cultivation approaches, 303
Novel isolation approaches, 302
Nucleic acids for detection methods
DNA from soil, 197–199
hybridization, 199
polymerase chain reaction, 199–200
digital droplet, 201–203
quantitative, 200–201
NUE, *see* N-use efficiency (NUE)
Null hypothesis, 317–318
Numerical taxonomy, 321
N-use efficiency (NUE), 29
Nutrient cycles, fungi in, 84
Nutrient limitation, physiological response to, 59–61
Nutrient source, 9–11
Nycodenz technique, 204

O

Obazoa, 127
Oligonucleotide microarrays, 199
Oligotrophic microorganisms, 51
Omics techniques, 296
Oomycota, 69
Open-reference OTU picking, 219
Operational taxonomic units (OTUs), 43, 44, 179, 190, 218–219
data normalization, 219–220
microbiome diversity analysis, 220
Opimoda, 125–126
Amoebozoa, 126–127
Obazoa, 127
"Opisthokonta," 127
Orchid mycorrhiza, 79–80

Ordinal scales, 309
Ordination plots, 314
Ordination techniques, 321–322
Organic C, 27, 28
Organic carbon cycling, 181
Organic matter, 9–11
Organic *vs.* conventional agricultural practices, 426–427
(micro)Organisms, introduction of, 30
Osmotic potential of water, 8
Osmotrophy, 135
OTUs, *see* Operational taxonomic units (OTUs)
Outer membrane systems, 412–413
Outer membrane vesicle, 413
Oxalic acid production, 73
Oxidation-protective agents, 301
Oxidative stress-protective agents, 298

P

Paenibacillus, 56
PAHs, *see* Polyaromatic hydrocarbons (PAHs)
Paraburkholderia, 154
Paraburkholderia phytofirmans, 367
Paraburkholderia rhizoxinica, 155
Paraburkholderia terrae, 59, 153–155
Paraburkholderia terrae BS001, 102
Paracoccus denitrificans, 183
Paramecium caudatum, 156
Parametric analysis, 318–320
Parasites, fungal, 76–77
Parenchelys terricola, 127
Parsimony analyses, 323
Particulate organic matter (POM), 100
Pathogen-associated molecular patterns, 155
Pathogenic bacteria, 174–175
Pathogen suppressive, 347
Pattern-recognition receptors (PRRs), 171
Paxillus involutus, 79, 83
PCBs, *see* Polychlorinated biphenyls (PCBs)
PCE, *see* Perchloroethene (PCE)
PCP, *see* Pentachlorophenols (PCP)
PCR, *see* Polymerase chain reaction (PCR)
Pectobacterium carotovorum, 174, 263
Pelotomaculum schinkii, 144
Penicillium, 31
Pentachlorophenols (PCP), 387, 388
Peptidoglycan, 413
Peptidomics, 259
Perchloroethene (PCE), 385–386
Periplasmic proteins, 413
Periplasmic systems, 413
Permafrost thaw, 335–336
Pesticides, 382
    biodegradation of, 389–391
Petroleum hydrocarbons, 383
PGPB, *see* Plant growth-promoting bacteria (PGPB)
PHA, *see* Polyhydroxyalkanoate (PHA)
Phages, 91–92
    abundance and diversity of, 95–100
    in biofilm modulation, 101
    controllers of host population densities, 100
    as genetic parasites, 101
    in horizontal gene transfer, 101
Phagocytosis, 133
Phanerochaete velutina, 71
Pharmaceuticals, 382
    biodegradation of, 391–393
Phenazine, 371
Phenetics, 321
Phenol–chloroform extraction/purification, 198
Phenol solution, 263
Phenylobacterium, 276
Phlebiopsis gigantea, 86
pH of soil, 16–18, 30–31
Phosphatase activity, 11
Phosphate-solubilizing bacteria (PSB), 150, 365
Phospholipid fatty acids-stable isotope probing (PLFA-SIP), 272
Phospholipids, 72
Phosphorus, 187–188
Photoautotrophic soil microorganisms, 14
Photorhabdus, 369
Photorhabdus temperata, 369
Photosynthetic cyanobacteria, 184
Phototrophs, 50
Phyletics, 321
Phyllosphere, 135, 136
Phylogenetic analysis, 43–44, 323–324
Phylogenetics, 321
Physarum polycephalum, 133
Physical disruption cell lysis, 197
Physicochemical conditions, in soil, 30–31
Physiological stratification, 253
Phytohormones/phytostimulators, 367–369
Phytophthora (Oomycota), 76
Phytoremediation, 397
    of polychlorinated biphenyls, 398–399
Phytosanitary soil quality, 355
Piggy-back-the-winner (PtW) model, 96
Pinus sylvestris, 80
Pisolithus, 427
Plantae, 128
Plant defense pathways, 370
Plant diseases, *see* Soilborne diseases
Plant growth-promoting bacteria (PGPB), 174
Plant growth promotion, 369
Plant immune system, 171–172
"Plant microbiome," 138
Plant nutrient uptake, protists in, 137
Plant–pathogen interaction, 343
Plants
    holobiont, 163
    and microorganisms, 188
    pathogens, 87
    protists and, 136
    as sessile organisms, 164
    tolerance, 369–370
    virus interaction with, 95
Plasmid pMOL30, 415
Plasmodiophora brassicae, 70
Plastics, biodegradation of, 393–394
Platycladus orientalis, 368
Pleistocene soils, 336
Pleurotus ostreatus, 154
PLFA-SIP, 287–288
Plant growth-promoting bacteria (PGPB)
    ACC deaminase synthesis, 369–370
    antibiotics and hydrogen cyanide, 370–371

Plant growth-promoting bacteria (PGPB) (*cont.*)
  applications
    market considerations, 376
    outcome, 375–376
    sustainability of production, 377
  bacteriophages, 372–373
  cell wall-degrading enzymes, 371–372
  direct *vs.* indirect mechanisms, 363
  endophytic bacteria, 362–363
  iron sequestration, 365–367
  nitrogen fixation, 363
  phosphate solubilization, 365
  phytohormones/phytostimulators, 367–369
  plant defense pathways, 370
  quorum sensing and quenching, 373–374
  rhizospheric bacteria, 362
  symbiotic bacteria, 362
  volatile compounds, 372
Point mutational process, 107
Polyaromatic hydrocarbons (PAHs), 383, 384
Polychlorinated biphenyls (PCBs), 387
  phytoremediation of, 398–399
Polyhydroxyalkanoate (PHA), 393
Polyhydroxybutyrate, 394
Polymerase chain reaction (PCR), 29, 181, 350
  DNA extraction, 199–200
    digital droplet, 201–203
    quantitative, 200–201
  mRNA from soils, 251–252
  real-time, 309
Polyols, 73
Polyphasic approach, 354
Polythetic agglomerative hierarchical clustering, 322
Polyvinyl alcohol (PVA) degradation, 393–394
Polyvinyl polypyrrolidone (PVPP), 198, 249
POM, *see* Particulate organic matter (POM)
Potassium (K), 187
Potato rhizosphere, colonization, 205
Predatory interactions, 156–158
Press disturbances, 332, 333
Pressure potential of water, 8
Priming effect, 26
Principal components analyses, 321–322
Principle of Soil Microbiology, 22
PrnD, 351
PromA-type plasmid, 112–113
Protein distribution, 264–265
Protein hydrolysis, 415
Protein-SIP, 273, 288–289
Proteobacteria, 55–57
Proteogenomics, 264
Proteome database, 260
Proteomics, *see* Environmental proteomics
Protists, 125, 138
  autotrophic, 135
  classification of, 129
  feeding modes, 133–135
  identification of, 130
  molecular environmental sampling of, 129–130
  "plant microbiome," 138
  in plant nutrient uptake, 137
  and plants, 136
  quantification methods, 131
  sexuality in, 131–133
  in soil food web, 135–137
  trophic levels in, 134
Protozoa
  predators of bacteria, 156–157
  virus interaction with, 95
PRRs, *see* Pattern-recognition receptors (PRRs)
PSB, *see* Phosphate-solubilizing bacteria (PSB)
Pseudobactin, 367
Pseudomonadaceae, 352
Pseudomonas, 154, 352, 353, 355, 366, 369, 371
Pseudomonas aeruginosa, 107, 382
Pseudomonas chlororaphis, 371
Pseudomonas fluorescens, 101, 110, 112, 207, 324–325, 366, 367, 371, 425
Pseudomonas fluorescens F113, 152
Pseudomonas lemonignei, 394
Pseudomonas protegens, 209
Pseudomonas putida, 150, 261, 262
Pseudomonas putida F1, 386
Pseudomonas putida KT2440, 101
Pseudomonas strain, 207
Pseudomonas stutzeri, 382
Pseudomonas syringae, 351
Pseudostellaria heterophylla, 148
Pteris vittata, 366
PtW model, *see* Piggy-back-the-winner (PtW) model
P-type ATPases, 414
Pucciniomycotina, 70
Pulse disturbances, 332, 334
Purification protocols, 204
PVA degradation, *see* Polyvinyl alcohol (PVA) degradation
Pyrinomonas, 301
Pythium, 345

Q

qCO_2 concept, 27
QIIME package, 218
qPCR, *see* quantitative polymerase chain reaction (qPCR)
Q10 relationship, 12
QS system, *see* Quorum sensing (QS) system
Quality filtering, 218
Quantiles, 313
quantitative polymerase chain reaction (qPCR), 131, 200–203
quantitative real-time RT-PCR (qRT-PCR), 251
Quenching, 373
Quorum quenching (QQ), 363
Quorum sensing (QS), 6, 61, 373–374
  signals, 205
  system, 146–148, 159

R

R2A, 300
Radiolaria, 128
Ralstonia metallidurans, 53
Ralstonia solanacearum, 152, 174, 373
Ralstonins, 152
Raman-SIP, 288
Range, 311
Raptorial feeding, 156
Reactive oxygen species (ROS), 353
"Recalcitrance," of soil, 30

Receptivity, 343–346
Recombination *vs.* mutation (r/m) ratio, 107–108
Redox-active enzymes, 413, 414
Redox potential, 15–16
Relaxed interaction, 158
Relic DNA, 45
Representative soil samples, 39
Residual pores, 4
Resistance–nodulation–cell division (RND), 414
Reverse transcription, 250–251
Reverse transcription-PCR (RT-PCR) approach, 250, 251
Reverse transcription-quantitative PCR (RT-qPCR), 395
Rhizaria, 128
Rhizobia, 364
Rhizobium, 156, 427
Rhizobium tumefaciens, 56
Rhizoctonia, 356
Rhizoctonia solani, 348, 351–353, 356, 423
Rhizodeposition, 424
Rhizophlyctis rosea, 68
Rhizoplane, 167–168
  competence, 172
Rhizopus microsporus, 155
Rhizosheath, 166–167
Rhizosphere, 6, 29, 135, 136, 164, 166
Rhizospheric bacteria, 362
Rhodoplanes, 286
RiboGreen, 250
Ribonuclease (RNase) protection assays, 250
Ribosomal Database Project, 296
Ribosomal RNA (rRNA), 43, 246
Ribulose-1,5-bisphosphate carboxylase/oxygenase (RubisCO), 181
Rice root-associated microbiomes, 170
r/m ratio, *see* Recombination *vs.* mutation (r/m) ratio
RNA, 269
  cDNA synthesis, 251
  detection, 197
  microbial, 245–246
  quantification of, 250
RNA-SIP, 272–273,, 287–289
Robotics-based heterologous expression screening, 229
Root-associated bacterial communities, 163–164, 175
  bacterial traits for root colonization, 172–173
  beneficial bacteria, 174
  commensal bacteria, 175
  core microbiome, 173–174
  illustration of, 165
  lateral patterns, 170–171
  non-vegetated soil, 164–165
  pathogenic bacteria, 174–175
  plant genotype, 170
  plant immune system, 171–172
  radial patterns, 170
  rhizosphere, 166–168
  root endosphere, 168–169
  root exudates, 169, 171
  separation techniques, 169
  three-step enrichment model, 170
Root border cells, 168
Root colonization by bacteria, *see* Root-associated bacterial communities
Root core microbiome, 173–174
Root endophyte, 168
  competence, 172–173
Root endosphere, 168–169
Root exudates, 169, 171
Roseobacter, 413
ROS scavenging, 415
Rothamsted soil, 33
r-strategists, 51, 56
RT-qPCR, *see* Reverse transcription-quantitative PCR (RT-qPCR)

S

S-adenosyl-methionine, 370
Salix purpurea, 396
Sample data and populations, 317
Sampling, 39–40, 317
Saprotrophic fungi, 74–76, 82, 84, 87
Saprotrophs, 135
Scatter plots, 313
Schiermonnikoog soil chronosequence case study, 223–224
Schizosaccharomyces pombe, 70
Scleroderma, 427
Sclerotinia sclerotiorum, 371
Sclerotium cepivorum, 356
Scots pine, ectomycorrhizosphere of, 87
SDS–PAGE, 263
Seed inoculation, 368
Selective enrichment, 55
Sensor kinases, 146
Sequence alignments, 313–314
Sequence-based metagenomics, 394–395
Sequence-based screening, 236
Sequence coverage, 231
Sequential agglomerative hierarchical, 322
Serendipita indica, 150–151
Serpula lacrymans, 82
Serratia marcescens, 148
Seven grand questions of Waksmann, 23–24
“Short patch double illegitimate recombination,” 108
Shuttle vectors, 235
Siderophores, 145, 366–367, 412
Silent Spring, 382
Single-nucleotide polymorphisms (SNPs), 118
SIP, *see* Stable isotope probing (SIP)
16S ribosomal RNA (rRNA) gene sequence, 43, 49, 57, 204, 216, 296, 297
Small subunit (SSU) rRNA, 43, 44
Smart farming approach, 422–423, 428, 429
SNPs, *see* Single-nucleotide polymorphisms (SNPs)
Sodium dodecyl sulfate (SDS), 259
Soft-rot fungi, 74
Software packages, 217
Soil amendments, 30
Soil atmosphere, 14–16
Soil Biology and Biochemistry, 22
Soilborne diseases
  agricultural practices, 355–357
  high-throughput meta-omics techniques, 351–353
  isolation and biocontrol agents, 354–355
  metaproteomics and metabonomics, 354
  microbial interactions
    biotic *vs.* abiotic factors, 347–348
    general *vs.* specific suppressiveness, 348

Soilborne diseases (*cont.*)
mechanisms, 349–350
organisms identification, 348–349
phytosanitary soil quality, 355
soil infectious potential, 344–346
soil receptivity, 344–346
studies and diversity, 350–351
Soil carbon cycle, 180, 337–338
CO_2 fixation, 181
methane cycle, 181–182
organic carbon cycling, 181
Soil chronosequence, *see* Schiermonnikoog soil chronosequence case study
Soil community metagenome, 228
Soil contamination, 357
Soil environment, 344
Soil fertility, SOM transformations and, 26
Soil food web, protists in, 135–137
Soil functionality, 257
nitrogen in, 28, 33
organic C in, 27
Soil hot spots, plasmid transfer in, 113–114
Soil infectious potential, 344–346
Soil inoculum potential, 344
Soil management, 421–422
biosensors, 425
cutting-edge plant/soil observation methods, 424
deep learning robotics, 424
microbiome, 423–424
integrated approach, 428
to land-use changes, 426
modulation of, 427–428
system-level approach, 425–426
smart farming, 422–423
Soil metabolic activity, 257
Soil metagenomics, 227–228
algorithms, 232
calcium-dependent antibiotics, 234
commercialization, 239–240
direct sequencing approaches
genomes construction, 232–233
metagenomics data, 232
microbial population, abundance, 231
sequence coverage, 231
statistical significance, 233
target group of interest, 231
techniques, 230–231
ecosystem-scale process, regulation, 238–239
functional screening, 236–237
genome reconstruction, 238
metagenome libraries, vectors, and hosts, 235–236
microbial ecology and evolution, 237–238
preagricultural soils, ecosystem services for, 238
principles of, 228–230
to scientific aims, 240
sequence-based screening, 236
soil microbiome DNA isolation, 229–230
wheat straw hydrolysis, 239–240
Soil metals and concentrations, 404–405
arsenic, 405
cadmium, 406
chromium, 406
copper, 406
lead, 406–407
mercury, 407
nickel, 407
zinc, 407
Soil metaproteomics, 257–258
direct/indirect methods, 262
model soil and laboratory systems, 260–261
protein distribution, 264–265
state of the art and technical challenges, 261–264
Soil microbial biomass, 409–410
Soil microbiomes, 37–39, 215, 422–424
data analysis, 217–218
correlation analysis, 220–222
data normalization, 219–220
microbiome diversity analysis, 220
operational taxonomic units, raw data to, 218–219
determinants of, 41–42
diversity
archaea, 216
bacteria, 216
fungi, 216–217
Schiermonnikoog soil chronosequence case study, 223–224
DNA isolation, 229–230
ecological resistance and resilience, 45–46
ideal (beneficial) status of, 46
integrated approach, 428
intracellular activities in, 24–25
to land-use changes, 426
modulation of, 29–30, 427
operational taxonomic unit, 43
phylogenetic signal, use of, 44
sampling of, 39–40
structure, 41, 42
and soil functioning, 44–45
system-level approach, 425–426
taxonomic and functional descriptions of, 46
tree of life, 43–44
ubiquity of, 40–41
Soil microbiostasis, 30
Soil microorganisms
classification of, 13
habitat (*see* Microbial habitat, in soil)
photoautotrophic, 14
Soil moisture, 8, 11
Soil organic matter (SOM), 26, 137, 143, 261, 331, 422
Soil pH, 16–18, 30–31
Soil–plant system, interconversion of energy, 14
Soil pores, 4
Soil proteomes, 263–264
Soil quality, 27
Soil receptivity, 343–345
Soil respiration, 181
Soil temperature, 11–13
Soil virome, 37
Soil water, 6–9, 31
Solar light, 14
Solid media, 297
Solid-phase purification, 198
SOM, *see* Soil organic matter (SOM)
Sorangium cellulosum, 59
S-oxidizing bacteria, 13
Spatial scale, 5–6
Specialists, 53
Specific suppressiveness, 348

Sphingobium chlorophenolicum, 387
Sphingomonas, 154, 276
Sphingopyxis strain MD2, 202
Stable isotope choice, 273
Stable isotope probing (SIP), 72, 260, 353, 395
applications, 270–272, 274–275, 277–286
defined, 269–270
high-throughput sequencing data, 288–289
processing, 271
standards, 289
techniques, 274–275
types
carbon cycling studies, 275–276
incubation technique choice, 273, 275
nitrogen cycling studies, 286
PLFA-SIP, 287–288
protein-SIP, 288
Raman-SIP, 288
RNA-SIP, 287
stable isotope choice, 273
target molecule choice, 272–275
uses, 270
Standard deviation, 311
Staphylococcus aureus, 234
Statistical Analyses of Metagenomic Profiles (STAMP), 233
Stenotrophomonas, 353
Stenotrophomonas maltophilia, 112
Stimulation, 350
Storage pores, 4
Stramenopiles, 129
Streptomyces, 236, 372
Streptomyces acidiscabies, 174
Streptomyces genus, 56
Streptomyces ipomoeae, 174
Streptomyces plasmids, 112
Streptomyces scabies, 174, 371
Streptomyces turgidiscabies, 174
Stress tolerance, 59–60
Stringent-type interaction, 158
Stylopage hadra, 154
Substrate utilization assays, 72
Sucking, 134
Suillus bovinus, 80, 82, 83
Sulfadiazine, biotransformation of, 392, 393
Sulfatase activity, 11
Sulfide, 189
Sulfide production, 410
Sulfolobus solfataricus, 236
Sulfur, 188–189
Supervised machine learning approach, 325
Suppressiveness, 343–344
agricultural practices, 355–357
high-throughput meta-omics techniques, 351–353
isolation and biocontrol agents, 354–355
metaproteomics and metabonomics, 354
microbial interactions
biotic *vs.* abiotic factors, 347–348
general *vs.* specific suppressiveness, 348
mechanisms, 349–350
organisms identification, 348–349
phytosanitary soil quality, 355
soil infectious potential, 344–346
soil receptivity, 344–346
studies and diversity, 350–351
types, 346–347
Surface-reactive particles, in soil, 25
SYBR Green-based detection systems, 200, 205
Symbiotic bacteria, 362
Symbiotic mycorrhizal fungi, 66
Synthetic compounds, 382
Synthetic organic pollutants, biodegradation of
pesticides, 389–391
pharmaceuticals, 391–393
plastics, 393–394
Syntrophy, 144
System-level approach, 425–426
SYTO-9 and propidium iodide (PI), 210
SYTOX, 210

T

Take-all decline soils (TAD), 347, 350
Taqman probe approach, 200, 201
Target group of interest, 231
Target molecule choice, 272–275
Taxometrics, 321
TCE, *see* Trichloroethene (TCE)
TEM method, *see* Transmission electron microscopy (TEM) method
Temperature, in soil, 11–13
Temporal scale, 5–6
Terminal restriction fragment length polymorphisms (T-RFLP), 272, 351
tfdA gene, 201, 202
Thaumarchaeota, 58
Thermophilic species, 68
Thiobacillus, 189
Thiobacillus acidophilus, 13
Thiobacillus ferrooxidans, 13
Thiobacillus thiooxidans, 13
Thioredoxin, 414
Tn*4371* ICE family, 415
Tomentella, 427
Top-down approach, 259
Total bacterial (direct) counts, 206
Traditional media, dilution of, 301
Transcript analyses
challenges, 246–247
microbial RNA, 245–246
mRNA from soils
extraction protocols, 248–249
metatranscriptomes, 252–253
microarrays, 252
PCR-based approaches, 251–252
purification, 249–250
quantification of RNA, 250
reverse transcription, 250–251
sampling and storage of samples, 247–248
Transduction, bacterial, 108, 114
Transfer *(tra)* operons, 111–112
Transformation, bacterial, 108
Transient browsers, 134
Transmission electron microscopy (TEM), 24, 92
Transmission pores, 4
Trapping, 135
Tree of life, 43–44
Trichloroethene (TCE), 382, 385–386

Trichoderma, 86, 350
Trichoderma asperellum, 366
Trifolium repens, 78
T4SS, *see* Type IV secretion systems (T4SS)
Tubulinea, 126
28S rRNA genes, 217
Two-dimensional (2D) electrophoresis mapping, 258–261
Type IV secretion systems (T4SS), 111

U

Ultra-cytochemical activity tests, 24
Ultraviolet (UV) sensitivity, 375
Ulva australis, 174
Uncultured bacteria, 296–297
United States Environmental Protection Agency (USEPA), 396
UNITE website, 217
Unsupervised machine learning approach, 325
Urease activity, 11
Ustilaginomycotina, 70

V

Variables, 38–39
Variance, 311
Variosea, 127, 132
Variovorax paradoxus, 394
VBNC state, *see* Viable but nonculturable (VBNC) state
Vectors, 235
Vegetation change and microbial communities, 332–334
Verrucomicrobia, 301
Verrucomicrobia, 57
Verticillium dahliae, 351
Viable bacterial cells, 209
Viable but nonculturable (VBNC), 61, 208, 261
Vibrio parahaemolyticus, 108
Virome metagenomics approach, 94
Virus, 91–92
  approaches to soil virome studies, 101–102
  classification, 97
  detection of, 92
  features of, 95
  and phages (*see* Phages)
  virus–host interaction
    soil habitat structure and, 92–93
    soil microbiome diversity, 93–95
    spatiotemporal aspects of, 102–103
Virus-like particles (VLPs), 94, 96
Vitamins-xylan-gellan gum (VXG), 301
Volatile compounds, 372
Volatile organic compounds (VOCs), 369, 372
Volatilization, 415

W

Warming, 332
Water, 31
  and biological activity, 6–7
  potential, 8–9
Water–oil emulsion droplet technology, 201–202
Water-saturated soils, 15
Wheat straw hydrolysis, 239–240
White-rot fungi, 74, 86
Wildfire effects, 334–335

X

Xanthomonas axonopodis, 174
Xanthomonas campestris, 174
Xanthomonas oryzae, 174
Xenobiotic compounds in soil, 55, 61
Xenobiotics, 119, 404
Xylella fastidiosa, 174

Y

Yedoma, 336

Z

Zinc, 407
Zoopagomycota, 68
Zygomycetes, 68
Zygomycota, 70
Zygospores, 68
Zymogenous microorganisms, 6, 51

Printed in the United States
by Baker & Taylor Publisher Services